HANDBOOK OF THE LOGISTIC DISTRIBUTION

STATISTICS: Textbooks and Monographs

A Series Edited by

Vol. 1: The Generalized Jackknife Statistic, *H. L. Gray and W. R. Schucany*
Vol. 2: Multivariate Analysis, *Anant M. Kshirsagar*
Vol. 3: Statistics and Society, *Walter T. Federer*
Vol. 4: Multivariate Analysis: A Selected and Abstracted Bibliography, 1957-1972, *Kocherlakota Subrahmaniam and Kathleen Subrahmaniam* (out of print)
Vol. 5: Design of Experiments: A Realistic Approach, *Virgil L. Anderson and Robert A. McLean*
Vol. 6: Statistical and Mathematical Aspects of Pollution Problems, *John W. Pratt*
Vol. 7: Introduction to Probability and Statistics (in two parts), Part I: Probability; Part II: Statistics, *Narayan C. Giri*
Vol. 8: Statistical Theory of the Analysis of Experimental Designs, *J. Ogawa*
Vol. 9: Statistical Techniques in Simulation (in two parts), *Jack P. C. Kleijnen*
Vol. 10: Data Quality Control and Editing, *Joseph I. Naus* (out of print)
Vol. 11: Cost of Living Index Numbers: Practice, Precision, and Theory, *Kali S. Banerjee*
Vol. 12: Weighing Designs: For Chemistry, Medicine, Economics, Operations Research, Statistics, *Kali S. Banerjee*
Vol. 13: The Search for Oil: Some Statistical Methods and Techniques, *edited by D. B. Owen*
Vol. 14: Sample Size Choice: Charts for Experiments with Linear Models, *Robert E. Odeh and Martin Fox*
Vol. 15: Statistical Methods for Engineers and Scientists, *Robert M. Bethea, Benjamin S. Duran, and Thomas L. Boullion*
Vol. 16: Statistical Quality Control Methods, *Irving W. Burr*
Vol. 17: On the History of Statistics and Probability, *edited by D. B. Owen*
Vol. 18: Econometrics, *Peter Schmidt*
Vol. 19: Sufficient Statistics: Selected Contributions, *Vasant S. Huzurbazar (edited by Anant M. Kshirsagar)*
Vol. 20: Handbook of Statistical Distributions, *Jagdish K. Patel, C. H. Kapadia, and D. B. Owen*
Vol. 21: Case Studies in Sample Design, *A. C. Rosander*
Vol. 22: Pocket Book of Statistical Tables, *compiled by R. E. Odeh, D. B. Owen, Z. W. Birnbaum, and L. Fisher*

Vol. 23: The Information in Contingency Tables, *D. V. Gokhale and Solomon Kullback*
Vol. 24: Statistical Analysis of Reliability and Life-Testing Models: Theory and Methods, *Lee J. Bain*
Vol. 25: Elementary Statistical Quality Control, *Irving W. Burr*
Vol. 26: An Introduction to Probability and Statistics Using BASIC, *Richard A. Groeneveld*
Vol. 27: Basic Applied Statistics, *B. L. Raktoe and J. J. Hubert*
Vol. 28: A Primer in Probability, *Kathleen Subrahmaniam*
Vol. 29: Random Processes: A First Look, *R. Syski*
Vol. 30: Regression Methods: A Tool for Data Analysis, *Rudolf J. Freund and Paul D. Minton*
Vol. 31: Randomization Tests, *Eugene S. Edgington*
Vol. 32: Tables for Normal Tolerance Limits, Sampling Plans, and Screening, *Robert E. Odeh and D. B. Owen*
Vol. 33: Statistical Computing, *William J. Kennedy, Jr. and James E. Gentle*
Vol. 34: Regression Analysis and Its Application: A Data-Oriented Approach, *Richard F. Gunst and Robert L. Mason*
Vol. 35: Scientific Strategies to Save Your Life, *I. D. J. Bross*
Vol. 36: Statistics in the Pharmaceutical Industry, *edited by C. Ralph Buncher and Jia-Yeong Tsay*
Vol. 37: Sampling from a Finite Population, *J. Hajek*
Vol. 38: Statistical Modeling Techniques, *S. S. Shapiro*
Vol. 39: Statistical Theory and Inference in Research, *T. A. Bancroft and C.-P. Han*
Vol. 40: Handbook of the Normal Distribution, *Jagdish K. Patel and Campbell B. Read*
Vol. 41: Recent Advances in Regression Methods, *Hrishikesh D. Vinod and Aman Ullah*
Vol. 42: Acceptance Sampling in Quality Control, *Edward G. Schilling*
Vol. 43: The Randomized Clinical Trial and Therapeutic Decisions, *edited by Niels Tygstrup, John M. Lachin, and Erik Juhl*
Vol. 44: Regression Analysis of Survival Data in Cancer Chemotherapy, *Walter H. Carter, Jr., Galen L. Wampler, and Donald M. Stablein*
Vol. 45: A Course in Linear Models, *Anant M. Kshirsagar*
Vol. 46: Clinical Trials: Issues and Approaches, *edited by Stanley H. Shapiro and Thomas H. Louis*
Vol. 47: Statistical Analysis of DNA Sequence Data, *edited by B. S. Weir*
Vol. 48: Nonlinear Regression Modeling: A Unified Practical Approach, *David A. Ratkowsky*
Vol. 49: Attribute Sampling Plans, Tables of Tests and Confidence Limits for Proportions, *Robert E. Odeh and D. B. Owen*
Vol. 50: Experimental Design, Statistical Models, and Genetic Statistics, *edited by Klaus Hinkelmann*
Vol. 51: Statistical Methods for Cancer Studies, *edited by Richard G. Cornell*
Vol. 52: Practical Statistical Sampling for Auditors, *Arthur J. Wilburn*
Vol. 53: Statistical Signal Processing, *edited by Edward J. Wegman and James G. Smith*
Vol. 54: Self-Organizing Methods in Modeling: GMDH Type Algorithms, *edited by Stanley J. Farlow*
Vol. 55: Applied Factorial and Fractional Designs, *Robert A. McLean and Virgil L. Anderson*
Vol. 56: Design of Experiments: Ranking and Selection, *edited by Thomas J. Santner and Ajit C. Tamhane*
Vol. 57: Statistical Methods for Engineers and Scientists. Second Edition, Revised and Expanded, *Robert M. Bethea, Benjamin S. Duran, and Thomas L. Boullion*
Vol. 58: Ensemble Modeling: Inference from Small-Scale Properties to Large-Scale Systems, *Alan E. Gelfand and Crayton C. Walker*

Vol. 59: Computer Modeling for Business and Industry, *Bruce L. Bowerman and Richard T. O'Connell*
Vol. 60: Bayesian Analysis of Linear Models, *Lyle D. Broemeling*
Vol. 61: Methodological Issues for Health Care Surveys, *Brenda Cox and Steven Cohen*
Vol. 62: Applied Regression Analysis and Experimental Design, *Richard J. Brook and Gregory C. Arnold*
Vol. 63: Statpal: A Statistical Package for Microcomputers – PC-DOS Version for the IBM PC and Compatibles, *Bruce J. Chalmer and David G. Whitmore*
Vol. 64: Statpal: A Statistical Package for Microcomputers – Apple Version for the II, II+, and IIe, *David G. Whitmore and Bruce J. Chalmer*
Vol. 65: Nonparametric Statistical Inference, Second Edition, Revised and Expanded, *Jean Dickinson Gibbons*
Vol. 66: Design and Analysis of Experiments, *Roger G. Petersen*
Vol. 67: Statistical Methods for Pharmaceutical Research Planning, *Sten W. Bergman and John C. Gittins*
Vol. 68: Goodness-of-Fit Techniques, *edited by Ralph B. D'Agostino and Michael A. Stephens*
Vol. 69: Statistical Methods in Discrimination Litigation, *edited by D. H. Kaye and Mikel Aickin*
Vol. 70: Truncated and Censored Samples from Normal Populations, *Helmut Schneider*
Vol. 71: Robust Inference, *M. L. Tiku, W. Y. Tan, and N. Balakrishnan*
Vol. 72: Statistical Image Processing and Graphics, *edited by Edward J. Wegman and Douglas J. DePriest*
Vol. 73: Assignment Methods in Combinatorial Data Analysis, *Lawrence J. Hubert*
Vol. 74: Econometrics and Structural Change, *Lyle D. Broemeling and Hiroki Tsurumi*
Vol. 75: Multivariate Interpretation of Clinical Laboratory Data, *Adelin Albert and Eugene K. Harris*
Vol. 76: Statistical Tools for Simulation Practitioners, *Jack P. C. Kleijnen*
Vol. 77: Randomization Tests, Second Edition, *Eugene S. Edgington*
Vol. 78: A Folio of Distributions: A Collection of Theoretical Quantile-Quantile Plots, *Edward B. Fowlkes*
Vol. 79: Applied Categorical Data Analysis, *Daniel H. Freeman, Jr.*
Vol. 80: Seemingly Unrelated Regression Equations Models : Estimation and Inference, *Virendra K. Srivastava and David E. A. Giles*
Vol. 81: Response Surfaces: Designs and Analyses, *Andre I. Khuri and John A. Cornell*
Vol. 82: Nonlinear Parameter Estimation: An Integrated System in BASIC, *John C. Nash and Mary Walker-Smith*
Vol. 83: Cancer Modeling, *edited by James R. Thompson and Barry W. Brown*
Vol. 84: Mixture Models: Inference and Applications to Clustering, *Geoffrey J. McLachlan and Kaye E. Basford*
Vol. 85: Randomized Response: Theory and Techniques, *Arijit Chaudhuri and Rahul Mukerjee*
Vol. 86: Biopharmaceutical Statistics for Drug Development, *edited by Karl E. Peace*
Vol. 87: Parts per Million Values for Estimating Quality Levels, *Robert E. Odeh and D. B. Owen*
Vol. 88: Lognormal Distributions: Theory and Applications, *edited by Edwin L. Crow and Kunio Shimizu*
Vol. 89: Properties of Estimators for the Gamma Distribution, *K. O. Bowman and L. R. Shenton*
Vol. 90: Spline Smoothing and Nonparametric Regression, *Randall L. Eubank*
Vol. 91: Linear Least Squares Computations, *R. W. Farebrother*
Vol. 92: Exploring Statistics, *Damaraju Raghavarao*

Vol. 93: Applied Time Series Analysis for Business and Economic Forecasting, *Sufi M. Nazem*

Vol. 94: Bayesian Analysis of Time Series and Dynamic Models, *edited by James C. Spall*

Vol. 95: The Inverse Gaussian Distribution: Theory, Methodology, and Applications, *Raj S. Chhikara and J. Leroy Folks*

Vol. 96: Parameter Estimation in Reliability and Life Span Models, *A. Clifford Cohen and Betty Jones Whitten*

Vol. 97: Pooled Cross-Sectional and Time Series Data Analysis, *Terry E. Dielman*

Vol. 98: Random Processes: A First Look, Second Edition, Revised and Expanded, *R. Syski*

Vol. 99: Generalized Poisson Distributions: Properties and Applications, *P.C. Consul*

Vol. 100: Nonlinear L_p-Norm Estimation, *René Gonin and Arthur H. Money*

Vol. 101: Model Discrimination for Nonlinear Regression Models, *Dale S. Borowiak*

Vol. 102: Applied Regression Analysis in Econometrics, *Howard E. Doran*

Vol. 103: Continued Fractions in Statistical Applications, *K.O. Bowman and L.R. Shenton*

Vol. 104: Statistical Methodology in the Pharmaceutical Sciences, *Donald A. Berry*

Vol. 105: Experimental Design in Biotechnology, *Perry D. Haaland*

Vol. 106: Statistical Issues in Drug Research and Development, *edited by Karl E. Peace*

Vol. 107: Handbook of Nonlinear Regression Models, *David A. Ratkowsky*

Vol. 108: Robust Regression: Analysis and Applications, *edited by Kenneth D. Lawrence and Jeffrey L. Arthur*

Vol. 109: Statistical Design and Analysis of Industrial Experiments, *edited by Subir Ghosh*

Vol. 110: *U*-Statistics: Theory and Practice, *A. J. Lee*

Vol. 111: A Primer in Probability, Second Edition, Revised and Expanded, *Kathleen Subrahmaniam*

Vol. 112: Data Quality Control: Theory and Pragmatics, *edited by Gunar E. Liepins and V. R. R. Uppuluri*

Vol. 113: Engineering Quality by Design: Interpreting the Taguchi Approach, *Thomas B. Barker*

Vol. 114: Survivorship Analysis for Clinical Studies, *Eugene K. Harris and Adelin Albert*

Vol. 115: Statistical Analysis of Reliability and Life-Testing Models, Second Edition, *Lee J. Bain and Max Engelhardt*

Vol. 116: Stochastic Models of Carcinogenesis, *Wai-Yuan Tan*

Vol. 117: Statistics and Society: Data Collection and Interpretation, Second Edition, Revised and Expanded, *Walter T. Federer*

Vol. 118: Handbook of Sequential Analysis, *B. K. Ghosh and P. K. Sen*

Vol. 119: Truncated and Censored Samples: Theory and Applications, *A. Clifford Cohen*

Vol. 120: Survey Sampling Principles, *E. K. Foreman*

Vol. 121: Applied Engineering Statistics, *Robert M. Bethea and R. Russell Rinehart*

Vol. 122: Sample Size Choice: Charts for Experiments with Linear Models, Second Edition, *Robert E. Odeh and Martin Fox*

Vol. 123: Handbook of the Logistic Distribution, *edited by N. Balakrishnan*

ADDITIONAL VOLUMES IN PREPARATION

HANDBOOK OF THE LOGISTIC DISTRIBUTION

edited by

N. BALAKRISHNAN

McMaster University
Hamilton, Ontario
Canada

CRC Press
Taylor & Francis Group
Boca Raton London New York

CRC Press is an imprint of the
Taylor & Francis Group, an **informa** business

First published 1992 by Marcel Dekker, Inc.

Published 2019 by CRC Press
Taylor & Francis Group
6000 Broken Sound Parkway NW, Suite 300
Boca Raton, FL 33487-2742

First issued in paperback 2019

ISBN 13: 978-0-367-45045-8 (pbk)
ISBN 13: 978-0-8247-8587-1 (hbk)

Visit the Taylor & Francis Web site at
http://www.taylorandfrancis.com

and the CRC Press Web site at
http://www.crcpress.com

Library of Congress Cataloging-in-Publication Data

Handbook of the logistic distribution/edited by N. Balakrishnan.
p. cm. -- (Statistics, textbooks and monographs; v. 123)
Includes bibliographical references (p.) and indexes.
ISBN 0-8247-8587-8 (acid-free paper)
1. Logistic distribution. I. Balakrishnan, N.
II. Series.
QA273.6. H35 1991
519.2'4--dc20 **91-35297**
CIP

To my sister, Malathi,
and my brother-in-law, Subramanian

Preface

Over the last 150 years or so, considerable research has been carried out regarding the theory, methodology, and applications of the logistic distribution. We have made a sincere effort in this volume to consolidate most of these contributions and simultaneously present some new developments in this area. It is shown that the logistic distribution, in addition to possessing mathematically tractable features, arises as a natural model in numerous situations.

In this volume, we have stressed both theoretical developments and practical applications. A detailed description of various univariate and multivariate generalizations of the distribution has been included. Several numerical illustrative examples have also been incorporated which should be of particular benefit to graduate students and practitioners. For helpful referencing on this topic, an up-to-date, comprehensive, and collective bibliography has been presented at the end of this volume.

This volume has been prepared in order to highlight various developments on the logistic distribution and to illustrate the utility of these results. We hope that this elaborate and exhaustive treatment will be of great value to researchers, practicing statisticians, actuarial scientists, and graduate students. We also hope that this volume will renew interest in this distribution and stimulate some further research and applications.

In reporting on various advances on this distribution, I have tried to the best of my knowledge to give due credit and to prioritize various related problems and issues. I sincerely apologize for any oversights and will welcome corrections or any information regarding omission of relevant references.

N. Balakrishnan

Acknowledgments

First, I would like to thank Mr. Maurits Dekker for extending me an invitation to prepare this volume. Next, I would like to express my deep appreciation and sincere thanks to Ms. Maria Allegra (Associate Acquisitions Editor, Marcel Dekker, New York) for her immense patience, kind support, and efficient management throughout the preparation of the volume. All the authors who participated in this volume showed great enthusiasm when approached and made timely contributions, and I would like to commend them for their support and effort. My special thanks go to Mrs. Edna Pathmanathan for her excellent typing of a good part of this volume and to Mr. Chan Ping Shing for his assistance in graphical and computational works. I would also like to thank the Natural Sciences and Engineering Research Council of Canada for funding all my research work associated with this volume.

Shanti S. Gupta would like to acknowledge the financial support of the NSF and the Office of Naval Research through their respective grants DMS-8702620 and N00014-88-K-0170 at Purdue University. A. K. Md. E. Saleh, K. M. Hassanein, and M. Masoom Ali would like to acknowledge that their research was supported by grants GR-5 and NSERC grant No. A3088 of the former and Ball State University academic year grant of the latter. S. Panchapakesan would like to acknowledge the financial support of the NSF through grant DMS-8702620 at Purdue University. Colleen D. Cutler acknowledges the support of the Natural Sciences and Engineering Research Council of Canada. B. K. Shah would like to express his sincere thanks to N. Balakrishnan for providing computer assistance in determining the percentage points presented in Table 18.3.2. B. Krishna Kumar would like to thank the Council of Scientific and Industrial Research of India for providing financial assistance to carry out his research.

Furthermore, I (along with all the authors) would like to offer thanks to the following individuals/bodies for kindly giving permission to reproduce previously published tables and charts:

Editor of *Australian Journal of Statistics*
Editor of *Journal of Information & Optimization Sciences*
Editor of *Statische Hefte*
Prof. Gunnar Kulldorff, University of Umea, Sweden
Marcel Dekker, Inc., New York
The American Statistical Association
The Biometrika Trustees
The Royal Statistical Society
The Technometrics Management Committee

Finally, I would like to offer my sincere thanks to Ms. Deirdre Griese (Production Editor, Marcel Dekker, Inc.) for her patient, dedicated, and sincere efforts to get the volume into its present form.

N. Balakrishnan

Contents

Preface *v*
Acknowledgments *vii*
Contributors *xiii*

Chapter 1 Introduction and Historical Remarks 1
N. Balakrishnan

Chapter 2 Logistic Order Statistics and Their Properties 17
Shanti S. Gupta and N. Balakrishnan

Chapter 3 Maximum Likelihood Estimation Based on Complete and Type II Censored Samples 49
N. Balakrishnan

Chapter 4 Linear Estimation Based on Complete and Censored Samples 79
N. Balakrishnan, Smiley W. Cheng, A. K. Md. Ehsanes Saleh, Khatab M. Hassanein, and M. Masoom Ali

Chapter 5 Reliability Estimation Based on MLEs for Complete and Censored Samples 123

L. J. Bain, N. Balakrishnan, James A. Eastman, Max Engelhardt, and Charles E. Antle

Chapter 6 Ranking and Selection Procedures 145

S. Panchapakesan

Chapter 7 Characterizations 169

Janos Galambos

Chapter 8 Translated Families of Distributions 189

Norman L. Johnson and Pandu R. Tadikamalla

Chapter 9 Univariate Generalized Distributions 209

Daniel Zelterman and N. Balakrishnan

Chapter 10 Some Related Distributions 223

E. Olusegun George and Meenakshi Devidas

Chapter 11 Multivariate Logistic Distributions 237

Barry C. Arnold

Chapter 12 Outlier and Robustness of Estimators 263

N. Balakrishnan

Chapter 13 Goodness-of-Fit Tests 291

Ralph B. D'Agostino and Joseph M. Massaro

Chapter 14 Tolerance Limits and Sampling Plans Based on Censored Samples 373

N. Balakrishnan and Karen Y. Fung

Chapter 15 Logistic Stochastic Growth Models and Applications 397

Wai-Yuan Tan

Chapter 16 Logistic Growth Models and Related Problems 427

Dieter Rasch

Chapter 17 Applications in Health and Social Sciences 449

Chris P. Tsokos and Peter S. DiCroce

Chapter 18 Some Other Applications 495

Kenneth J. Koehler, James T. Symanowski, Colleen D. Cutler, B. K. Shah, John F. Monahan, Leonard A. Stefanski, P. R. Parthasarathy, and B. Krishna Kumar

Bibliography *553*

Author Index *587*

Subject Index *597*

Contributors

M. Masoom Ali Department of Mathematical Sciences, Ball State University, Muncie, Indiana

Charles E. Antle Department of Statistics, The Pennsylvania State University, University Park, Pennsylvania

Barry C. Arnold Department of Statistics, University of California, Riverside, California

Lee J. Bain Department of Mathematics and Statistics, University of Missouri-Rolla, Rolla, Missouri

N. Balakrishnan Department of Mathematics and Statistics, McMaster University, Hamilton, Ontario, Canada

Smiley W. Cheng Department of Statistics, University of Manitoba, Winnipeg, Manitoba, Canada

Colleen D. Cutler Department of Statistics and Actuarial Science, University of Waterloo, Waterloo, Ontario, Canada

Ralph B. D'Agostino Department of Mathematics, Boston University, Boston, Massachusetts

Meenakshi Devidas Department of Mathematical Sciences, Memphis State University, Memphis, Tennessee

Peter S. DiCroce Department of Mathematics, University of South Florida, Tampa, Florida

James A. Eastman* Department of Mathematics and Statistics, University of Missouri-Rolla, Rolla, Missouri

Max Engelhardt Department of Mathematics and Statistics, University of Missouri-Rolla, Rolla, Missouri

Karen Y. Fung Department of Mathematics and Statistics, University of Windsor, Windsor, Ontario, Canada

Janos Galambos Department of Mathematics, Temple University, Philadelphia, Pennsylvania

E. Olusegun George Department of Mathematical Sciences, Memphis State University, Memphis, Tennessee

Shanti S. Gupta Department of Statistics, Purdue University, West Lafayette, Indiana

Khatab M. Hassanein Department of Biometry, Kansas University Medical Center, Kansas City, Kansas

Norman L. Johnson Department of Statistics, University of North Carolina at Chapel Hill, Chapel Hill, North Carolina

Kenneth J. Koehler Department of Statistics, Iowa State University, Ames, Iowa

B. Krishna Kumar Department of Mathematics, Indian Institute of Technology, Madras, India

---

**Current affiliation*: Department of Operations and Planning, New York City Transit Authority, New York, New York

Joseph M. Massaro Department of Mathematics, Boston University, Boston, Massachusetts

John F. Monahan Department of Statistics, North Carolina State University, Raleigh, North Carolina

S. Panchapakesan Department of Mathematics, Southern Illinois University, Carbondale, Illinois

P. R. Parthasarathy Department of Mathematics, Indian Institute of Technology, Madras-600 036, India

Dieter Rasch Institute of Biometry, Rostock, Germany

A. K. Md. Ehsanes Saleh Department of Mathematics and Statistics, Carleton University, Ottawa, Ontario, Canada

B. K. Shah Pearl River, New York

Leonard A. Stefanski Department of Statistics, North Carolina State University, Raleigh, North Carolina

James T. Symanowski Department of Statistical and Mathematical Services, Lilly Research Laboratories, Greenfield, Indiana

Pandu R. Tadikamalla Joseph M. Katz Graduate School of Business, University of Pittsburgh, Pittsburgh, Pennsylvania

Wai-Yuan Tan Department of Mathematical Sciences, Memphis State University, Memphis, Tennessee

Chris P. Tsokos Department of Mathematics, University of South Florida, Tampa, Florida

Daniel Zelterman School of Public Health, Division of Biostatistics, University of Minnesota, Minneapolis, Minnesota

1
Introduction and Historical Remarks

N. Balakrishnan
McMaster University, Hamilton, Ontario, Canada

1.1 HISTORICAL REMARKS

The logistic growth function was first proposed as a tool for use in demographic studies by Verhulst (1838, 1845) and was given its present name by Reed and Berkson (1929), two of the function's early enthusiasts. Like Verhulst (1838, 1845), some other authors also used the logistic function for estimating the growth of human population; for example, see Pearl and Reed (1920), Pearl et al. (1940), and Schultz (1930), and more recently Oliver (1982). The function was also applied as a growth model in biology by Pearl and Reed (1924) and to model the yeast cell growth, in particular, by Schultz (1930). The yeast cell growth data of Schultz was also utilized by Oliver (1964). The logistic growth function was used by Oliver (1969) to model the spread of an innovation. Some applications of the logistic function in bioassay problems were given by Pearl (1940), Emmens (1941), Wilson and Worcester (1943), Berkson (1944, 1951, 1953), and Finney (1947, 1952). While Plackett (1959) used the logistic function in the analysis of survival data, Fisk (1961) used it in studying the income distributions. Perks (1932) proposed a very general class of distributions which includes the logistic and used it in the graduation of mortality statistics. The logistic growth function was used by Schultz (1930) and Oliver (1964) to model agricultural production data. Two more interesting applications of the logistic model, among many others, were due to Dyke and Patterson (1952) and Grizzle (1961) in the area of public health.

1.2 INTRODUCTION

The probability density function (pdf) of a logistic $L(\mu, \sigma^2)$ variable X is simply given by

$$f(x; \mu, \sigma) = \frac{\pi}{\sigma\sqrt{3}} \frac{e^{-\pi(x-\mu)/\sigma\sqrt{3}}}{\{1 + e^{-\pi(x-\mu)/\sigma\sqrt{3}}\}^2}, \quad \begin{array}{l} -\infty < x < \infty, \\ -\infty < \mu < \infty, \sigma > 0, \end{array} \tag{1.1}$$

and the corresponding cumulative distribution function (cdf) is

$$F(x; \mu, \sigma) = \frac{1}{1 + e^{-\pi(x-\mu)/\sigma\sqrt{3}}}, \quad \begin{array}{l} -\infty < x < \infty, \\ -\infty < \mu < \infty, \sigma > 0. \end{array} \tag{1.2}$$

It is easily verified that (1.2) defines a proper cumulative distribution function with

$$\lim_{x \to -\infty} F(x; \mu, \sigma) = 0 \quad \text{and} \quad \lim_{x \to \infty} F(x; \mu, \sigma) = 1.$$

It will be shown in the next section that μ and σ in (1.1) are respectively the mean and standard deviation of the logistic variable X.

By using the fact that $\tanh x = (e^x - e^{-x})/(e^x + e^{-x})$ and hence

$$\frac{1}{2}\left\{1 + \tanh\left(\frac{x}{2}\right)\right\} = \frac{1}{1 + e^{-x}},$$

we can write the cdf of the logistic $L(\mu, \sigma^2)$ population given in Eq. (1.2) equivalently as

$$F(x; \mu, \sigma) = \frac{1}{2}\left\{1 + \tanh\left(\frac{\pi(x - \mu)}{2\sigma\sqrt{3}}\right)\right\}, \qquad -\infty < x < \infty;$$

the pdf in Eq. (1.1) can then be equivalently written as

$$f(x; \mu, \sigma) = \frac{\pi}{4\sigma\sqrt{3}} \operatorname{sech}^2\left(\frac{\pi(x - \mu)}{2\sigma\sqrt{3}}\right), \qquad -\infty < x < \infty.$$

It is for this reason that the logistic distribution is sometimes called as the sech-squared distribution.

The probability density function of a logistic $L(0, 1)$ variable $Y = (X - \mu)/\sigma$ is given by

$$f(y) = \frac{\pi}{\sqrt{3}} \frac{e^{-\pi y/\sqrt{3}}}{(1 + e^{-\pi y/\sqrt{3}})^2}, \qquad -\infty < y < \infty, \tag{1.3}$$

and the corresponding cumulative distribution function is

$$F(y) = \frac{1}{1 + e^{-\pi y/\sqrt{3}}}, \qquad -\infty < y < \infty. \tag{1.4}$$

This is the "standardized form" of the logistic distribution, as it will be shown in the next section that this random variable Y has mean 0 and variance 1.

Sometimes it is simpler to work with the random variable $Z = \pi(X - \mu)/\sigma\sqrt{3} = \pi Y/\sqrt{3}$, which has as its density function

$$f^*(z) = \frac{e^{-z}}{(1 + e^{-z})^2}, \qquad -\infty < z < \infty, \tag{1.5}$$

and its distribution function is

$$F^*(z) = \frac{1}{1 + e^{-z}}, \qquad -\infty < z < \infty. \tag{1.6}$$

Equations (1.5) and (1.6) are also treated as standard forms for the logistic distribution for the sake of mathematical ease.

The probability density function and the cumulative distribution function of the logistic $L(0, 1)$ variable Y given in Eqs. (1.3) and (1.4) have been graphed in Figs. 1.1 and 1.2, respectively. From Eqs. (1.3) and (1.4), we derive the hazard function of the logistic $L(0, 1)$ variable Y as

$$h(y) = \frac{f(y)}{1 - F(y)} = \frac{\pi}{\sqrt{3}(1 + e^{-\pi y \sqrt{3}})}. \tag{1.7}$$

Its graph is shown in Fig. 1.3.

From Fig. 1.1, it is clear that the logistic density function in (1.3) is symmetric about zero and is more peaked in the center than the normal density function; see Chew (1968). Further, it is easy to observe that the hazard function in (1.7) is simply proportional to the cumulative distribution function in (1.4). This particular characterizing property of the logistic distribution makes it useful as a growth curve model as shall be seen in later chapters.

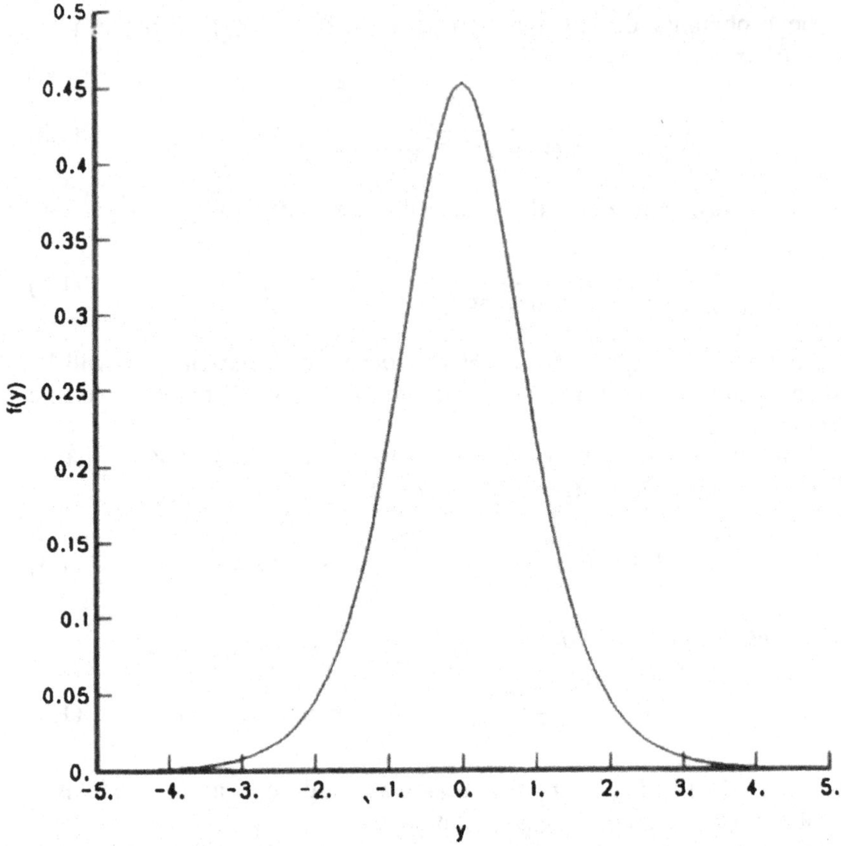

Fig. 1.1 The density function of the logistic $L(0, 1)$ variable.

1.3 BASIC CHARACTERISTICS AND PROPERTIES

For mathematical ease, we shall study the basic characteristics and properties of the logistic population through the random variable Z, whose pdf and cdf are given in Eqs. (1.5) and (1.6). The moment generating function of Z is

$$\begin{aligned} M_Z(t) &= E(e^{tZ}) = \int_{-\infty}^{\infty} \frac{e^{-(1-t)z}}{(1 + e^{-z})^2}\, dz \\ &= \int_0^1 u^t(1 - u)^t\, du \qquad \left(\text{with } u = \frac{1}{1 + e^{-z}}\right) \\ &= B(1 + t, 1 - t) \\ &= \Gamma(1 + t)\Gamma(1 - t). \end{aligned} \tag{1.8}$$

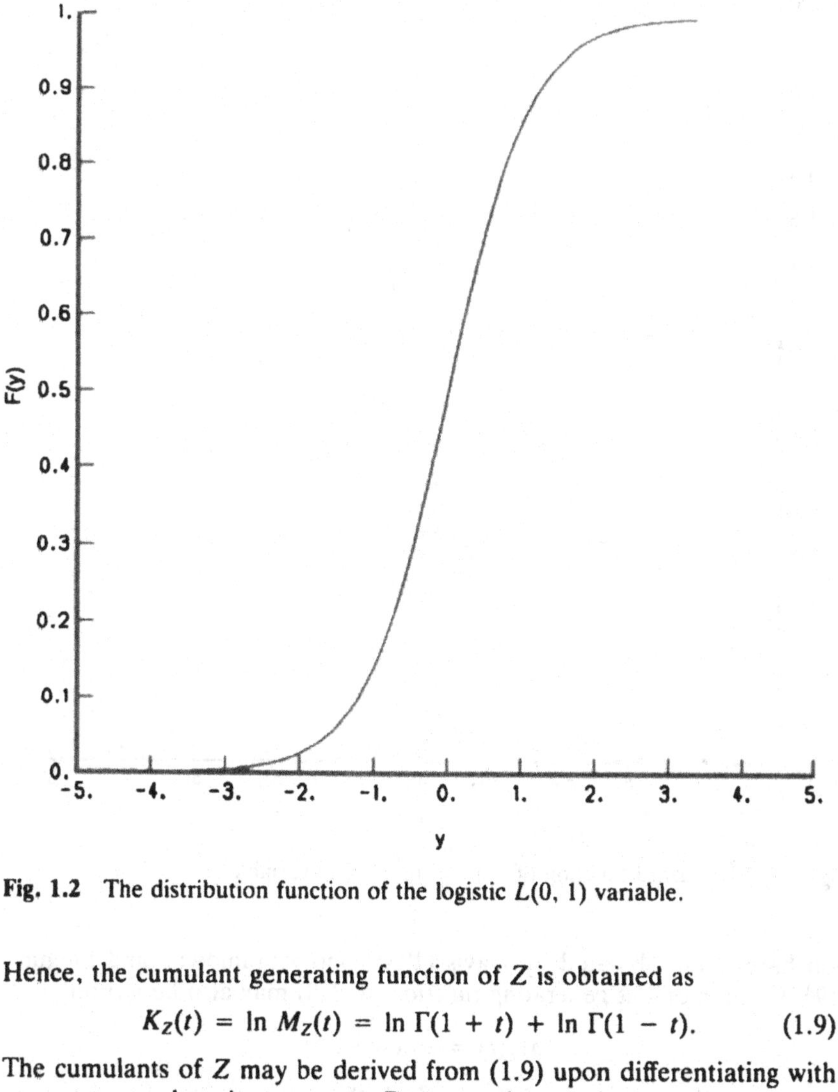

Fig. 1.2 The distribution function of the logistic $L(0, 1)$ variable.

Hence, the cumulant generating function of Z is obtained as

$$K_Z(t) = \ln M_Z(t) = \ln \Gamma(1 + t) + \ln \Gamma(1 - t). \tag{1.9}$$

The cumulants of Z may be derived from (1.9) upon differentiating with respect to t and setting t to zero. For example, we obtain

$$E(Z) = \Gamma'(1) - \Gamma'(1) = 0$$

and

$$\text{Var}(Z) = 2\{\Gamma''(1) - (\Gamma'(1))^2\},$$

where $\Gamma'(\cdot)$ and $\Gamma''(\cdot)$ are first and second derivatives of the gamma function. Similarly, higher-order cumulants can be expressed in terms of higher-order derivatives of the gamma function. These derivatives of gamma func-

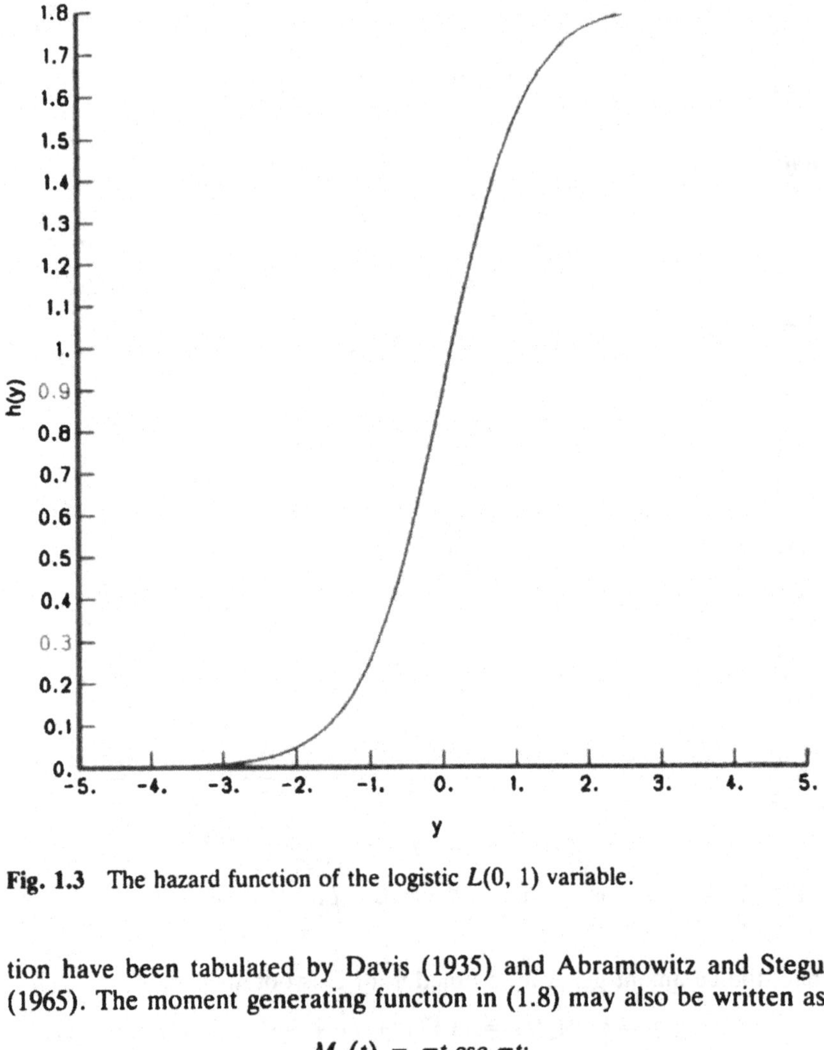

Fig. 1.3 The hazard function of the logistic $L(0, 1)$ variable.

tion have been tabulated by Davis (1935) and Abramowitz and Stegun (1965). The moment generating function in (1.8) may also be written as

$$M_Z(t) = \pi t \csc \pi t;$$

the characteristic function of Z, $E(e^{itZ})$, is then given by $\pi t \operatorname{csch} \pi t$.

The moments of Z may also be determined from the density in (1.5) by direct integration. First of all, the odd-order moments of Z are all zero since the distribution of Z is symmetric about zero. Let us now consider the even-order moment of Z given by

$$E(Z^{2r}) = \int_{-\infty}^{\infty} z^{2r} \frac{e^{-z}}{(1 + e^{-z})^2} dz$$

$$= 2\int_0^\infty z^{2r} \frac{e^{-z}}{(1+e^{-z})^2}\,dz \quad \text{(since the integrand is an even function in } z)$$

$$= 2\int_0^\infty z^{2r}\left\{\sum_{j=1}^\infty (-1)^{j-1} j e^{-jy}\right\} dy$$

$$= 2\Gamma(2r+1)\sum_{j=1}^\infty \frac{(-1)^{j-1}}{j^{2r}}$$

$$= 2\Gamma(2r+1)\left(1 - \frac{1}{2^{2r-1}}\right)\mathfrak{h}(2r) \tag{1.10}$$

for $r = 1, 2, \ldots$, where $\mathfrak{h}(s) = \sum_{j=1}^\infty j^{-s}$ is the Riemann zeta function. From Eq. (1.10), we obtain in particular

$$\text{Var}(Z) = E(Z^2) = 2 \times 2\left(1 - \frac{1}{2}\right)\mathfrak{h}(2) = 2\left(\frac{\pi^2}{6}\right) = \frac{\pi^2}{3}$$

and

$$E(Z^4) = 2 \times 24\left(1 - \frac{1}{8}\right)\mathfrak{h}(4) = 42\left(\frac{\pi^4}{90}\right) = \frac{7\pi^4}{15}.$$

As a result, we have the coefficients of skewness and kurtosis of the logistic population as

$$\sqrt{\beta_1} = 0 \quad \text{and} \quad \beta_2 = \frac{63}{15} = 4.2.$$

That the value of β_2 for the logistic distribution is 4.2 (which is considerably larger than the β_2 value of 3 for the normal distribution) meaning that the logistic distribution has "longer tails" than the normal distribution. The heavier tails of the logistic distribution, even though they have a considerable effect on the fourth central moment and hence the coefficient of kurtosis, have only a very small effect on the distribution function.

Since $Z = \pi(X - \mu)/\sigma\sqrt{3}$, we immediately see that the logistic $L(\mu, \sigma^2)$ variable X whose pdf and cdf are given in Eqs. (1.1) and (1.2), has its mean and variance

$$E(X) = \mu + \frac{\sigma\sqrt{3}}{\pi} E(Z) = \mu$$

and

$$\text{Var}(X) = \frac{3\sigma^2}{\pi^2}\text{Var}(Z) = \sigma^2.$$

By comparing the cdf of the standard normal variable with the cdf of the standard logistic variable in Eq. (1.4), Johnson and Kotz (1970, p. 6) have shown that the two are very close and that the maximum value of the difference is about 0.0228, which occurs when $y = 0.7$. Johnson and Kotz have also pointed out that the maximum difference can be reduced to a value less than 0.01 by changing the scale of y in the cdf of the standard normal variable to $16y/15$. A graphical comparison of the two cdf's has been presented by Johnson and Kotz (1970, p. 6). By using the fact that $\beta_2 = 4.2$ for the logistic distribution, Mudholkar and George (1978) observed that the logistic distribution very closely approximates Student's t-distribution with 9 degrees of freedom. A similar reasoning has been given by George and Ojo (1980) and George et al. (1986) to use generalized logistic distributions to closely approximate Student's t-distribution with ν degrees of freedom; more details are presented in Chapter 9.

1.4 NOTATION

In spite of the fact that more than 30 authors have participated in the preparation of this volume, notation used throughout has been made as consistent as possible. This should enable the reader to move from one chapter to another rather easily without having to first resolve differences in notation.

For example, the notation $X \rightarrow L(\mu, \sigma^2)$ is used to imply that the random variable X is distributed as logistic with mean μ and variance σ^2 and with pdf and cdf

$$f(x; \mu, \sigma) = \frac{\pi}{\sigma\sqrt{3}} \frac{e^{-\pi(x-\mu)/\sigma\sqrt{3}}}{\{1 + e^{-\pi(x-\mu)/\sigma\sqrt{3}}\}^2}, \qquad -\infty < x < \infty,$$

and

$$F(x; \mu, \sigma) = \frac{1}{1 + e^{-\pi(x-\mu)/\sigma\sqrt{3}}}, \qquad -\infty < x < \infty.$$

Similarly, the notation $Y \rightarrow L(0, 1)$ is used to imply that the random variable Y is distributed as logistic with mean 0 and variance 1 and with pdf and cdf

$$f(y) = \frac{\pi}{\sqrt{3}} \frac{e^{-\pi y/\sqrt{3}}}{(1 + e^{-\pi y/\sqrt{3}})^2}, \qquad -\infty < y < \infty,$$

and

$$F(y) = \frac{1}{1 + e^{-\pi y/\sqrt{3}}}, \qquad -\infty < y < \infty.$$

The notation $Z \to L(0, \pi^2/3)$ is used to indicate that the random variable Z is distributed as logistic with mean 0 and variance $\pi^2/3$ and with pdf and cdf

$$f^*(z) = \frac{e^{-z}}{(1 + e^{-z})^2}, \qquad -\infty < z < \infty,$$

and

$$F^*(z) = \frac{1}{1 + e^{-z}}, \qquad -\infty < z < \infty.$$

The samples from these three logistic populations are respectively denoted by $(X_1, X_2, \ldots, X_n)$, $(Y_1, Y_2, \ldots, Y_n)$, and $(Z_1, Z_2, \ldots, Z_n)$, and the corresponding order statistics by $(X_{1:n}, X_{2:n}, \ldots, X_{n:n})$, $(Y_{1:n}, Y_{2:n}, \ldots, Y_{n:n})$, and $(Z_{1:n}, Z_{2:n}, \ldots, Z_{n:n})$, respectively. The moments of the order statistics $Y_{i:n}$, viz., $E(Y_{i:n})$, $E(Y_{i:n}Y_{j:n})$, $\mathrm{Var}(Y_{i:n})$, and $\mathrm{Cov}(Y_{i:n}, Y_{j:n})$, are denoted by $\alpha_{i:n}^{(k)}$, $\alpha_{i,j:n}$, $\beta_{i,i:n}$ and $\beta_{i,j:n}$. Similarly, the moments of the order statistics $Z_{i:n}$, viz., $E(Z_{i:n}^k)$, $E(Z_{i:n}Z_{j:n})$, $\mathrm{Var}(Z_{i:n})$, and $\mathrm{Cov}(Z_{i:n}, Z_{j:n})$, are denoted by $\alpha_{i:n}^{*(k)}$, $\alpha_{i,j:n}^*$, $\beta_{i,i:n}^*$ and $\beta_{i,j:n}^*$.

As a general rule, ^ is used to denote the MLE (maximum likelihood estimator) of a parameter, ‾ to denote an approximate MLE, * to denote the BLUE (best linear unbiased estimator), and ** to denote an approximate BLUE.

Some exceptions to these notations are made when they do not result in any confusion.

1.5 SCOPE OF THE VOLUME

Several theoretical, methodological, and applied issues relating to the logistic distribution or the logistic model are addressed in great detail in this volume and explained, when possible, with many numerical examples.

In Chapter 2 a detailed discussion of order statistics from the logistic and truncated logistic distributions and some of their properties is presented. While the percentage points and modes of order statistics are presented in Section 2.2, exact and explicit expressions for the single and product moments of order statistics are derived in Section 2.3 in terms of the gamma function and its successive derivatives. Some simple recurrence relations satisfied by the single and the product moments of order statistics are presented in Section 2.4. The distributions of some systematic statistics such as the sample range and the rth quasi-range are discussed in Section 2.5. The results given in Section 2.4 are extended in Section 2.6 by establishing several recurrence relations satisfied by the single and the product moments of order statistics from a general doubly truncated logistic distribution. In Section 2.7 a description of tables available in this context is provided. Finally, a method of using the moments of logistic order statistics in order to approximate the moments of order statistics from an arbitrary continuous population is described in Section 2.8.

In Chapter 3 the maximum likelihood estimation of the mean and standard deviation of the logistic population based on complete and Type II censored samples is discussed. In Section 3.2, in addition to studying the bias and mean square error of the MLEs for small sample sizes over various choices of censoring through Monte Carlo simulations, formulas for asymptotic variances and covariance of the MLEs are derived and are also computed for various proportions of censoring. In Section 3.3 the approximate maximum likelihood estimators of the mean and standard deviation of the logistic population are derived based on doubly Type II censored samples. These estimators, in addition to being simple and explicit estimators, are shown to be jointly as efficient as the MLEs even in case of small sample sizes. The bias and mean square error of these estimators are examined for small sample sizes through Monte Carlo simulations. Also, formulas for asymptotic variances and covariance of the AMLEs are derived and computed for various proportions of censoring. Some examples are presented to illustrate the two methods of estimation.

In Chapter 4 the problem of linear estimation is discussed in great length. In Section 4.2 the best linear unbiased estimation of the mean and standard deviation of the logistic population based on complete and doubly Type II censored samples is described. In Section 4.3 the best linear unbiased estimation and the asymptotic best linear unbiased estimation based on k optimally selected order statistics from the sample are described. Similar work for the estimation of the quantiles of the logistic population is presented in Section 4.4. Some linear estimators with polynomial coefficients are derived in Section 4.5. Finally, in Section 4.6 some simplified linear estimators are presented for the estimation of the mean and standard deviation of the logistic population. These estimators, which are in terms of quasi-midrange and quasi-range, are shown to be highly efficient in case of small sample sizes.

In Chapter 5, methods of constructing confidence intervals for μ and σ (the mean and standard deviation, respectively), lower and upper tolerance limits for the distribution, and an estimate of the reliability function are described based on Type II right-censored samples. The distributions of some pivotal quantities that are necessary for these developments are determined by Monte Carlo simulations, and some asymptotic approximations for these are also presented. Finally, an example involving lifetimes of certain incandescent lamps given by Davis (1952) is used to illustrate the methods of inference discussed in this chapter.

In Chapter 6, ranking and selection procedures relating to the logistic distribution are discussed. To start with, the two basic formulations in ranking and selection methodology, viz., the subset formulation and the indifference zone formulation, are discussed in Section 6.2. Next, in Sections 6.3–6.5 procedures for selecting the "best population" (one with the largest mean) from several logistic populations with a common known variance are described under the subset, indifference zone, and the restricted subset formulations, respectively. In Section 6.6 a two-stage procedure using the indifference zone formulation is described in which the first stage involves a subset approach to eliminate inferior population. In Section 6.7, a selection procedure for selecting the population with the smallest q-quantile is presented, which will be particularly useful in the context of logistic distribution, as it is used to model quantal response in experiments involving quantitative treatment factors. In Section 6.8, selection procedures for a family of distributions which are partially ordered with respect to the logistic distribution are presented. Finally, in Section 6.9 some concluding remarks are made, including some possible further research relating to this area.

Characterizations of distributions are very important, as some of them can be used to distinguish that distribution from others theoretically while some other characterization results can be utilized in model building. Characterizations of both these sorts for the logistic distribution are discussed in Chapter 7. The characterization of the logistic distribution as a model for population growth is discussed in Section 7.2. In Section 7.3, some characterizations based on properties of order statistics are presented. Some characterization results based on samples with random sample size are discussed in Section 7.4, and some miscellaneous characterizations of the logistic distribution are presented in Section 7.5.

Johnson (1949) defined the well-known Johnson systems of S distributions by ascribing the standard normal distribution to logarithmic (S_L system), logit (S_B system), and arcsinh (S_U system) transformations. In Chapter 8, similar systems of L distributions are developed by ascribing the standard logistic distribution to logarithmic (L_L system), logit (L_B system), and arcsinh (L_U system) transformations. In Sections 8.2–8.4, the L_L, L_B, and L_U systems of distributions, respectively, are discussed, and their properties and characteristics analyzed. After a general discussion on the fitting of these systems of distributions in Section 8.5, the fitting of the L_L, L_B, and L_U systems are individually discussed in Sections 8.6–8.8, respectively. Finally, a brief discussion on the choice among these three families of distributions is presented in Section 8.9.

In Chapter 9, four types of univariate generalized logistic distributions are presented, and their properties and characteristics are also discussed.

In Section 9.2, some characterization results are presented for these generalized logistic distributions. For the remainder of this chapter, discussions are restricted to the Type I generalized logistic distribution. The maximum likelihood estimation, the approximate maximum likelihood estimation, and the best linear unbiased estimation of the parameters of the distribution are discussed in Section 9.3. Finally, a discussion on the order statistics and their moments from this Type I generalized logistic distribution is provided in Section 9.4.

In Chapter 10, relationships of the logistic distribution to some other distributions are discussed. In Section 10.2, some relationships of the logistic distribution with the exponential, double exponential, and extreme value distributions are presented. In Section 10.3, relationship of the logistic distribution with the geometric distribution is explained. Finally, in Section 10.4 some relationships between order statistics from the logistic and half-logistic distributions are presented.

In Chapter 11, multivariate forms of the logistic distribution are presented, and some properties are discussed. In Section 11.2, the classical multivariate logistic distribution as defined by Gumbel (1961) and Malik and Abraham (1973) is presented, and some of its characteristics are discussed. In Section 11.3, a mixture representation for the multivariate logistic vector is provided. In Section 11.4, a multivariate logistic random vector is constructed as differences of independent multivariate extreme random vector. In Section 11.5, a multivariate geometric distribution is used in conjunction with geometric minimization and maximization to produce a multivariate distribution with logistic marginals. Frailty and Archimedean models of the multivariate logistic distribution are discussed in Section 11.6. While the Farlie-Gumbel-Morgenstern type multivariate logistic distribution is presented in Section 11.7, a more flexible model is discussed in Section 11.8. In Section 11.9, the limiting distribution of coordinatewise extremes of a sample from the flexible model in Section 11.8 is derived. The simulational problems associated with the various multivariate forms are addressed in Section 11.10. The estimation of all the parameters in the multivariate forms is discussed in Section 11.11. The conditional distributions are derived in Section 11.12 from the multivariate forms, and some of their properties are observed. Finally, some miscellaneous notes are made in Section 11.13.

In Chapter 12, the effects of a single outlier in a sample on various estimators of the mean and standard deviation of the logistic distribution are studied. First, the density function and the joint density function of order statistics from a single outlier model are presented in Section 12.2. In Section 12.3, the single location-outlier and scale-outlier logistic models are explained and the evaluation of the means, variances, and covariances of order statistics under these models is described. The functional behavior of order statistics from the single location-outlier and the scale-outlier models is discussed in Section 12.4. Finally, in Section 12.5 the effects of a single outlier on various estimators of the mean and standard deviation of the logistic population are examined through their bias and mean square error.

In Chapter 13, various goodness-of-fit techniques for the logistic distribution are discussed in great detail and illustrated with many numerical examples. Section 13.2 starts the discussion of the graphical informal analysis techniques with the definition of the empirical distribution function of a random sample and a discussion of the uses of its graph for the goodness-of-fit evaluation. The logistic probability plot is developed in Section 13.3. Sections 13.4–13.7 deal with various examples, uses of the informal techniques, and also methods for treating censored data. In Section 13.8, a discussion on the formal goodness-of-fit tests is presented with chi-squared tests being covered in Section 13.9, empirical distribution function procedures in Section 13.10, procedures based on regression and correlation statistics in Section 13.12. Finally, some recommendations are made in Section 13.13 with regard to the use of these goodness-of-fit techniques.

In Chapter 14, one-sided and two-sided tolerance limits and sampling plans are developed for the logistic distribution based on complete and Type II censored samples by using best linear unbiased estimators of the mean and standard deviation of the logistic distribution based on these samples. In Section 14.2, the construction of one-sided tolerance limits using the BLUEs is described, and the necessary tables are presented. A large-sample normal approximation for the one-sided tolerance limits is presented in Section 14.3 and is shown to provide very good approximation to the tolerance limits for large sample sizes. In Section 14.4, the construction of two-sided tolerance limits using the BLUEs is explained, and the necessary tables are also included. In Section 14.5, it is shown that acceptance sampling plans for the logistic distribution may be set up by using the appropriate tolerance limits. Two examples are presented in Section 14.6 which illustrate the construction of tolerance limits by using the best linear unbiased estimators of the mean and standard deviation computed from the given complete or Type II censored sample.

In Chapter 15, a discussion on the logistic stochastic growth models and their applications is presented. In Section 15.2, a description of the stochastic logistic growth models is given and the transition probabilities, absolute probabilities, and probability distributions of first absorption times associated with this model are derived. In Section 15.3, the determination of moments and cumulants is described. In Section 15.4, a diffusion approximation is discussed to study the stationary distribution, the absorption probability distribution and the moments of first absorption times.

The logistic growth function is one of the oldest growth functions in the literature and is used to describe both population growth as well as organismic growth. In Chapter 16, several inference procedures are discussed for the parameters involved in the logistic growth model as originally proposed by Verhulst (1838). In Section 16.2, the point estimation of parameters is described and illustrated with a few numerical examples. In Section 16.3, the interval estimation of parameters and some testing procedures are discussed. In addition to some numerical examples, the section also includes a discussion on the robustness of the procedures under departures from the assumed normal distribution for the error. Section 16.4 deals with some experimental design questions associated with this problem and presents *D*-optimum experimental designs for the logistic function.

In Chapter 17, various applications of the logistic model in health and social sciences are illustrated in detail. In Section 17.2, the point estimation of the parameters of the logistic function is described. In Section 17.3, the point estimation of the parameters of the logistic growth function is explained. Section 17.4 deals with the estimation of the parameters and discrimination procedures of the logistic regression model. After an overview of goodness-of-fit techniques for the logistic function in Section 17.5, some specific goodness-of-fit methods for the logistic regression and for the logistic distribution are discussed in Sections 17.6 and 17.7, respectively. Finally, in Section 17.8 some applications of the logistic function to population growth, bioassay, medical diagnosis, and public health are explained and illustrated with some practical examples.

Finally, in Chapter 18 some miscellaneous applications of the logistic distribution and the logistic function are described. In Section 18.1, a bivariate logistic regression model is presented which is suitable for the analysis of bivariate ordered categorical data with potentially nonzero correlation between the bivariate responses. In Section 18.2, it is shown that a generalized logistic distribution (see Chapter 9 for details) arises (asymptotically) in a natural way from statistics based on the *k*th nearest neighbor distance through which the advantage of using higher-order nearest neighbors is displayed. Section 18.3 deals with the analysis of bioavailability data when successive samples arise from logistic population. In Section 18.4, a convenient method of computing the logistic-normal integral is described which makes use of a normal scale mixture representation of the logistic distribution. Finally, in Section 18.5 the role of some logistic growth models in ecology is reviewed.

REFERENCES

Abramowitz, M. and Stegun, I. A. (eds.) (1965). *Handbook of Mathematical Functions with Formulas, Graphs, and Mathematical Tables*. Dover, New York.

Berkson, J. (1944). Application of the logistic function to bioassay. *J. Amer. Statist. Assoc.*, *37*, 357–365.

Berkson, J. (1951). Why I prefer logits to probits. *Biometrics*, *7*, 327–339.

Berkson, J. (1953). A statistically precise and relatively simple method of estimating the bio-assay with quantal response, based on the logistic function. *J. Amer. Statist. Assoc.*, *48*, 565–599.

Chew, V. (1968). Some useful alternatives to the normal distribution. *Amer. Statist.*, *22*, 22–24.

Davis, D. J. (1952). An analysis of some failure data. *J. Amer. Statist. Assoc.*, *47*, 113–150.

Davis, H. T. (1935). *Tables of the Higher Mathematical Functions*, Vols. 1, 2. Principia Press, Bloomington.

Dyke, G. V. and Patterson, H. D. (1952). Analysis of factorial arrangements when the data are proportions. *Biometrics*, *8*, 1–12.

Emmens, C. W. (1941). The dose-response relation for certain principles of the pituitary gland, and of the serum and urine of pregnancy. *J. of Endocrinology*, *2*, 194–225.

Finney, D. J. (1947). The principles of biologial assay. *J. Roy. Statist. Soc., Ser. B*, *9*, 46–91.

Finney, D. J. (1952). *Statistical Methods in Biological Assay*. Hafner, New York.

Fisk, P. R. (1961). The graduation of income distributions. *Econometrica*, *29*, 171–185.

George, E. O., El-Saidi, M., and Singh, K. (1986). A generalized logistic approximation of the Student *t* distribution. *Commun. Statist.—Simul. Comput.*, *15*, 1199–1208.

George, E. O. and Ojo, M. O. (1980). On a generalization of the logistic distribution. *Ann. Inst. Statist. Math.*, *32*, 161–169.

Grizzle, J. E. (1961). A new method of testing hypotheses and estimating parameters for the logistic model. *Biometrics*, *17*, 372–385.

Gumbel, E. J. (1961). Bivariate logistic distributions. *J. Amer. Statist. Assoc.*, *56*, 335–349.

Johnson, N. L. (1949). Systems of frequency curves generated by methods of translation. *Biometrika*, *36*, 149–176.

Johnson, N. L. and Kotz, S. (1970). *Distributions in Statistics: Continuous Univariate Distributions*, Vol. 2. Wiley, New York.

Malik, H. J. and Abraham, B. (1973). Multivariate logistic distributions. *Ann. Statist.*, *1*, 588–590.

Mudholkar, G. S. and George, E. O. (1978). A remark on the shape of the logistic distribution. *Biometrika*, *65*, 667–668.

Oliver, F. R. (1964). Methods of estimating the logistic growth function. *Appl. Statist.*, *13*, 57–66.

Oliver, F. R. (1969). Another generalization of the logistic growth function. *Econometrica*, *37*, 144–147.

Oliver, F. R. (1982). Notes on the logistic curve for human populations. *J. Roy. Statist. Soc.*, *Ser. A*, *145*, 359–363.

Pearl, R. (1940). *Medical Biometry and Statistics*. Saunders, Philadelphia.

Pearl, R. and Reed, L. J. (1920). On the rate of growth of the population of the United States since 1790 and its mathematical representation. *Proc. Nat. Acad. Sci.*, *6*, 275–288.

Pearl, R. and Reed, L. J. (1924). *Studies in Human Biology*. Williams and Wilkins, Baltimore.

Pearl, R., Reed, L. J., and Kish, J. F. (1940). The logistic curve and the census count of 1940. *Science*, *92*, 486–488.

Perks, W. F. (1932). On some experiments in the graduation of mortality statistics. *J. Inst. Actuaries*, *58*, 12–57.

Plackett, R. L. (1959). The analysis of life test data. *Technometrics*, *1*, 9–19.

Reed, L. J. and Berkson, J. (1929). The application of the logistic function to experimental data. *J. Phys. Chem.*, *33*, 760–779.

Schultz, H. (1930). The standard error of a forecast from a curve. *J. Amer. Statist. Assoc.*, *25*, 139–185.

Verhulst, P. J. (1838). Notice sur la lois que la population suit dans sons accroissement. *Corr. Math. Phys.*, *10*, 113–121.

Verhulst, P. J. (1845). Recherches mathématiques sur la loi d'accroissement de la population. *Acad. Bruxelles*, *18*, 1–38.

Wilson, E. B. and Worcester, J. (1943). The determination of L.D.50 and its sampling error in bio-assay. *Proc. National Acad. Sci.*, *29*, 79–85.

2

Logistic Order Statistics and Their Properties

Shanti S. Gupta
Purdue University, West Lafayette, Indiana

N. Balakrishnan
McMaster University, Hamilton, Ontario, Canada

2.1 INTRODUCTION

Let $Z_1, Z_2, \ldots, Z_n$ be a random sample of size n from the logistic $L(0, \pi^2/3)$ population with pdf $f^*(z)$ and cdf $F^*(z)$ as given in Eqs. (1.5) and (1.6), respectively. Let $Z_{1:n} \le Z_{2:n} \le \cdots \le Z_{n:n}$ denote the order statistics obtained by arranging the above sample in increasing order of magnitude. Then the density function of $Z_{i:n}$ $(1 \le i \le n)$ is given by

$$f^*_{i:n}(z_i) = \frac{n!}{(i-1)!(n-i)!} \times \{F^*(z_i)\}^{i-1}\{1 - F^*(z_i)\}^{n-i}f^*(z_i), \qquad -\infty < z_i < \infty, \quad (2.1.1)$$

and the joint density function of $Z_{i:n}$ and $Z_{j:n}$ $(1 \le i < j \le n)$ is given by

$$f^*_{i,j:n}(z_i, z_j) = \frac{n!}{(i-1)!(j-i-1)!(n-j)!} \times \{F^*(z_i)\}^{i-1}\{F^*(z_j) - F^*(z_i)\}^{j-i-1}\{1 - F^*(z_j)\}^{n-j} \times f^*(z_i)f^*(z_j), \qquad -\infty < z_i < z_j < \infty. \quad (2.1.2)$$

Let us now denote the single moments $E(Z^k_{i:n})$ by $\alpha^{*(k)}_{i:n}$ for $1 \le i \le n$, $k \ge 1$, and the product moments $E(Z_{i:n}Z_{j:n})$ by $\alpha^*_{i,j:n}$ for $1 \le i < j \le n$. For convenience, let us also denote $E(Z_{i:n})$ by $\alpha^*_{i:n}$ and $E(Z^2_{i:n})$ by $\alpha^*_{i,i:n}$ for $1 \leq i \leq n$. Further, let us denote $\mathrm{Cov}(Z_{i:n}, Z_{j:n})$ by $\beta^*_{i,j:n}$.

Order statistics $Z_{i:n}$ and their moments have been studied in great detail by several authors, including Birnbaum and Dudman (1963), Gupta and Shah (1965), Tarter and Clark (1965), Shah (1966, 1970), Gupta et al. (1967), Malik (1980), George and Rousseau (1987), and Balakrishnan and Malik (1991). Birnbaum and Dudman (1963) derived explicit expression for the cumulants of order statistics and tabulated the means and standard deviations for sample sizes up to 10 and for some large sample sizes as well. They then summarized these quantities in graphs to facilitate interpolation to other sample sizes. Gupta and Shah (1965) derived exact expressions for the moments of order statistics in terms of Bernoulli and Stirling numbers of the first kind and used them to tabulate the first four moments for sample sizes up to 10. They also expressed the cumulants in terms of polygamma functions, as was originally pointed out by Plackett (1958). Plackett (1958) used these explicit expressions of the moments of logistic order statistics to develop a method of approximating the moments of order statistics from an arbitrary continuous distribution. Distribution of the sample range has been studied by Gupta and Shah (1965), who also provided a short table of its percentage points for $n = 2$ and 3. By generalizing this result, Malik (1980) derived the exact formula for the cumulative distribution function of the rth quasi-range, viz., $Z_{n-r:n} - Z_{r+1:n}$ for $r = 0, 1, \ldots, [(n-1)/2]$. In an independent study, Tarter and Clark (1965) reproduced some of the results of Gupta and Shah (1965) and then studied the distribution of the sample median in detail. George and Rousseau (1987) recently examined the distribution of the sample midrange, viz., $(Z_{1:n} + Z_{n:n})/2$, and established several relationships in distribution between the midrange and sample median of the logistic and Laplace random variables. A series expression for the covariance of two order statistics has been provided by both Gupta and Shah (1965) and Tarter and Clark (1965). Shah (1966) tabulated the covariances for sample sizes up to 10 and Gupta, Qureishi, and Shah (1967) extended this table for sample sizes up to 25. However, by means of some recurrence formulas Kjelsberg (1962) had already derived exact numerical results for the covariances from samples of size 5 or less.

By using the fact that $f^*(z)$ and $F^*(z)$ given in Eqs. (1.5) and (1.6) satisfy the relation

$$f^*(z) = F^*(z)\{1 - F^*(z)\}, \tag{2.1.3}$$

Shah (1966, 1970) established several recurrence relations satisfied by the single and the product moments of order statistics. Recently, Balakrishnan and Malik (1991) prepared tables of means, variances, and covariances for sample sizes up to 50 by applying these relations in a simple and systematic recursive way. Some of the results in the references cited have also been summarized in a review article by Malik (1985).

The truncated logistic distribution plays a role in a variety of applications, as has been mentioned by Kjelsberg (1962). Order statistics and their moments from a general truncated logistic distribution have been studied by Tarter (1966). He derived exact and explicit expressions for the means, variances, and covariances of order statistics in terms of a finite series involving logarithms and dilogarithms of the constants of truncation. By following the lines of Shah (1966, 1970), Balakrishnan and Joshi (1983a,b) established several recurrence relations satisfied by the single and the product moments of order statistics from a symmetrically truncated logistic distribution. Then, Balakrishnan and Kocherlakota (1986) generalized these results to the doubly truncated logistic distribution and displayed that these recurrence relations could be used systematically in order to evaluate the means, variances, and covariances of all order statistics for all sample sizes.

In this chapter we present a detailed discussion of order statistics from the logistic distribution and some of their properties. In Section 2.2 we give the percentage points and modes of order statistics. In Section 2.3 we derive exact and explicit expressions for the single and the product moments of order statistics. These work in terms of the gamma function and its successive derivatives. In Section 2.4 we present some recurrence relations satisfied by the single and the product moments of order statistics which would enable one to compute the means, variances, and covariances in a simple recursive way. The distribution function of the sample range, as derived by Gupta and Shah (1965), is presented in Section 2.5, and the distribution of the rth quasi-range derived by Malik (1980) is also given for the sake of completeness. In Section 2.6 we give some relations between moments for the case of the doubly truncated logistic distribution that are due to Balakrishnan and Kocherlakota (1986). In Section 2.7 we present details of tables that are available in this context. Finally, in Section 2.8 we describe Plackett's (1958) method of approximating the moments of order statistics from an arbitrary continuous distribution by using the moments of logistic order statistics.

2.2 PERCENTAGE POINTS AND MODES

The distribution function of $Z_{1:n}$ is given by

$$F^*_{1:n}(z_1) = 1 - \{1 - F^*(z_1)\}^n. \tag{2.2.1}$$

From (2.2.1) we obtain the 100α percentage point of $Z_{1:n}$ to be

$$z_{1:n(\alpha)} = \ln\{1 - (1 - \alpha)^{1/n}\} - \ln(1 - \alpha)^{1/n}, \qquad 0 < \alpha < 1. \qquad (2.2.2)$$

Next, the distribution function of $Z_{n:n}$ is given by

$$F^*_{n:n}(z_n) = \{F^*(z_n)\}^n. \qquad (2.2.3)$$

From (2.2.3) we obtain the 100α percentage point of $Z_{n:n}$ as

$$z_{n:n(\alpha)} = \ln(\alpha^{1/n}) - \ln(1 - \alpha^{1/n}), \qquad 0 < \alpha < 1. \qquad (2.2.4)$$

Similarly, the distribution function of $Z_{i:n}$ $(2 \le i \le n - 1)$ is given by

$$\begin{aligned} F^*_{i:n}(z_i) &= I_{F^*(z_i)}(i, n - i + 1) \\ &= \frac{1}{B(i, n - i + 1)} \int_0^{F^*(z_i)} u^{i-1}(1 - u)^{n-i}\, du. \qquad (2.2.5) \end{aligned}$$

Now, let $B_\alpha(i, n - i + 1)$ denote the 100α percentage point of the $B(i, n - i + 1)$ distribution. From (2.2.5) we then have the 100α percentage point of $Z_{i:n}$ $(2 \le i \le n - 1)$:

$$\begin{aligned} z_{i:n(\alpha)} &= \ln B_\alpha(i, n - i + 1) \\ &\quad - \ln\{1 - B_\alpha(i, n - i + 1)\}, \qquad 0 < \alpha < 1. \qquad (2.2.6) \end{aligned}$$

By using the symmetric relation satisfied by the incomplete beta functions, we observe from Eqs. (2.2.2), (2.2.4), and (2.2.6) that the 100α percentage point of $Z_{i:n}$ is simply the negative of 100(1 − α) percentage point of $Z_{n-i+1:n}$. While the percentage points of $Z_{1:n}$ and $Z_{n:n}$ may be obtained easily from (2.2.2) and (2.2.4), respectively, the percentage points of $Z_{i:n}$ for $2 \le i \le n - 1$ may be obtained from (2.2.6) either by using the extensive tables of incomplete beta function prepared by Karl Pearson (1934) and Pearson and Hartley (1970) or by using the algorithm given by Cran, Martin, and Thomas (1977). Gupta and Shah (1965) have tabulated some percentage points of all order statistics for sample sizes up to 10 and some selected order statistics for sample sizes up to 25.

Next, by differentiating the density function of $Z_{i:n}$ in (2.1.1) with respect to z_i and using the relation in (2.1.3), we get

$$\begin{aligned} \frac{df^*_{i:n}(z_i)}{dz_i} &= \frac{n!}{(i - 1)!(n - i)!} \{F^*(z_i)\}^{i-1}\{1 - F^*(z_i)\}^{n-i} \\ &\quad \times [i - (n + 1)F^*(z_i)]. \qquad (2.2.7) \end{aligned}$$

Upon equating (2.2.7) to zero and solving for z_i, we obtain the mode of $Z_{i:n}$ $(1 \leq i \leq n)$:

$$m^*_{i:n} = \ln\left(\frac{i}{n - i + 1}\right) = \ln\left(\frac{p_i}{q_i}\right), \tag{2.2.8}$$

where $p_i = 1 - q_i = i/(n + 1)$. Due to the symmetry of the logistic distribution, we observe once again that the mode of $Z_{i:n}$ is simply the negative of the mode of $Z_{n-i+1:n}$.

2.3 MOMENTS AND CUMULANTS

From (2.1.1) we obtain the moment generating function of $Z_{i:n}$ $(1 \leq i \leq n)$:

$$\begin{aligned} M^*_{i:n}(t) &= E\{e^{tZ_{i:n}}\} = \frac{1}{B(i, n - i + 1)} \int_{-\infty}^{\infty} \frac{e^{-(n-i+1)z+tz}}{(1 + e^{-z})^{n+1}}\, dz \\ &= \frac{B(i + t, n - i + 1 - t)}{B(i, n - i + 1)} = \frac{\Gamma(i + t)}{\Gamma(i)} \frac{\Gamma(n - i + 1 - t)}{\Gamma(n - i + 1)}, \end{aligned} \tag{2.3.1}$$

where $B(\cdot, \cdot)$ and $\Gamma(\cdot)$ are the usual complete beta and gamma functions, respectively. An alternative expression of $M^*_{i:n}(t)$ involving Bernoulli numbers and Stirling numbers of the first kind has been given by Gupta and Shah (1965). From the expression of the moment generating function in (2.3.1), we obtain the following:

$$\alpha^*_{i:n} = E(Z_{i:n}) = \psi(i) - \psi(n - i + 1), \tag{2.3.2}$$

$$\begin{aligned} \alpha^{*(2)}_{i:n} = E(Z^2_{i:n}) &= \psi'(i) + \psi'(n - i + 1) \\ &\quad + \{\psi(i) - \psi(n - i + 1)\}^2, \end{aligned} \tag{2.3.3}$$

and

$$\beta^*_{i,i:n} = \text{Var}(Z_{i:n}) = \psi'(i) + \psi'(n - i + 1), \tag{2.3.4}$$

where

$$\psi(z) = \frac{d}{dz} \ln \Gamma(z) = \frac{\Gamma'(z)}{\Gamma(z)}$$

and

$$\psi'(z) = \frac{d^2}{dz^2} \ln \Gamma(z) = \frac{\Gamma''(z)}{\Gamma(z)} - \psi^2(z)$$

are the digamma and trigamma functions, respectively. Thus, from Eqs. (2.3.2) to (2.3.4), one may compute the means and variances of order statistics either by using the extensive tables of digamma and trigamma functions prepared by Davis (1935) and Abramowitz and Stegun (1965) or by using the algorithms given by Bernardo (1976) and Schneider (1978). Gupta and Shah (1965) have given exact expressions for the first four moments of order statistics for sample sizes up to 10, and the values of mean and variance have been tabulated recently by Balakrishnan and Malik (1991) for sample sizes up to 50. We may note here that the moment generating function in (2.3.1) may be used to obtain higher-order single moments also by involving polygamma functions.

From (2.3.1) we obtain the cumulant generating function of $Z_{i:n}$ ($1 \le i \le n$):

$$\begin{aligned} K^*_{i:n}(t) &= \ln M^*_{i:n}(t) \\ &= \ln \Gamma(i + t) + \ln \Gamma(n - i + 1 - t) \\ &\quad - \ln \Gamma(i) - \ln \Gamma(n - i + 1). \end{aligned} \tag{2.3.5}$$

From (2.3.5) we obtain the kth cumulant of $Z_{i:n}$ ($1 \le i \le n$):

$$\begin{aligned} \kappa^{*(k)}_{i:n} &= \frac{d^k}{dt^k} \ln \Gamma(i + t)\Big|_{t=0} + \frac{d^k}{dt^k} \ln \Gamma(n - i + 1 - t)\Big|_{t=0} \\ &= \psi^{(k-1)}(i) + (-1)^k \psi^{(k-1)}(n - i + 1), \end{aligned} \tag{2.3.6}$$

where

$$\psi^{(k-1)}(z) = \frac{d^k}{dz^k} \ln \Gamma(z) \qquad \text{for } k = 1, 2, \ldots,$$

and

$$\psi^{(0)}(z) \equiv \psi(z) = \frac{\Gamma'(z)}{\Gamma(z)}.$$

It is clear from (2.3.6) that for $k = 1, 2, \ldots,$

$$\kappa^{*(2k-1)}_{i:n} = -\kappa^{*(2k-1)}_{n-i+1:n} \tag{2.3.7}$$

and

$$\kappa^{*(2k)}_{i:n} = \kappa^{*(2k)}_{n-i+1:n}. \tag{2.3.8}$$

These may also be observed simply by using the symmetry of the logistic distribution. By applying the series expansions for $\psi(z)$ and $\psi^{(k-1)}(z)$ given by

$$\psi(z) = \sum_{l=1}^{\infty} \left\{ \frac{1}{l} - \frac{1}{l + z - 1} \right\}$$

and

$$\psi^{(k-1)}(z) = (-1)^k (k-1)! \sum_{l=1}^{\infty} \frac{1}{(l + z - 1)^k}, \qquad k \geq 2,$$

we obtain from (2.3.6) that for $n - i + 1 > i$,

$$\kappa^*_{i:n} \equiv \kappa^{*(1)}_{i:n} = -\left\{ \frac{1}{i} + \frac{1}{i+1} + \cdots + \frac{1}{n-i} \right\} \tag{2.3.9}$$

and

$$\kappa^{*(k)}_{i:n} = (-1)^k (k-1)! \times \left\{ \sum_{l=1}^{\infty} \frac{1}{(l + i - 1)^k} + (-1)^k \sum_{l=1}^{\infty} \frac{1}{(l + n - i)^k} \right\}. \tag{2.3.10}$$

The above formulas for the first four cumulants were originally given by Plackett (1958). From (2.3.9) we get

$$\kappa^*_{1:n} = -\left\{ 1 + \frac{1}{2} + \frac{1}{3} + \cdots + \frac{1}{n-1} \right\},$$

which was also given by Gumbel (1958). The exact and explicit expression for the cumulants of logistic order statistics given in Eqs. (2.3.9) and (2.3.10) will later be used in Section 2.8 for developing some series approximations for the moments of order statistics from an arbitrary continuous distribution.

From the joint density of $Z_{i:n}$ and $Z_{j:n}$ $(1 \leq i < j \leq n)$ in (2.1.2), we have the joint moment generating function of $Z_{i:n}$ and $Z_{j:n}$:

$$\begin{aligned} M^*_{i,j:n}(t_1, t_2) &= E\{e^{t_1 Z_{i:n} + t_2 Z_{j:n}}\} \\ &= \frac{n!}{(i-1)!(j-i-1)!(n-j)!} \int_{-\infty}^{\infty} \int_{-\infty}^{z_j} e^{t_1 z_i + t_2 z_j} \\ &\quad \times \{F^*(z_i)\}^{i-1} \{F^*(z_j) - F^*(z_i)\}^{j-i-1} \{1 - F^*(z_j)\}^{n-j} \\ &\quad \times f^*(z_i) f^*(z_j) \, dz_i \, dz_j. \end{aligned} \tag{2.3.11}$$

Making the transformations

$$u = F^*(z_i) = \frac{1}{1 + e^{-z_i}} \qquad \text{and} \qquad v = F^*(z_j) = \frac{1}{1 + e^{-z_j}}$$

and thence, noting that

$$e^{z_i} = \frac{u}{1-u} \quad \text{and} \quad e^{z_j} = \frac{v}{1-v},$$

we can rewrite Eq. (2.3.11) as

$$M^*_{i,j:n}(t_1, t_2) = \frac{n!}{(i-1)!(j-i-1)!(n-j)!} \int_0^1 \int_0^v \frac{u^{t_1}}{(1-u)^{t_1}} \times \frac{v^{t_2}}{(1-v)^{t_2}} u^{i-1}(v-u)^{j-i-1}(1-v)^{n-j}\, du\, dv.$$

By expanding $(1-u)^{-t_1}$ as an infinite series in powers of u, we obtain

$$M^*_{i,j:n}(t_1, t_2) = \frac{n!}{(i-1)!(j-i-1)!(n-j)!} \times \sum_{l=0}^{\infty} \frac{(t_1+l-1)^{(l)}}{l!} \int_0^1 \int_0^v u^{t_1+i-1+l}(v-u)^{j-i-1} v^{t_2} \times (1-v)^{n-j-t_2}\, du\, dv, \tag{2.3.12}$$

where

$$(t_1+l-1)^{(l)} = 1 \qquad \text{if } l = 0,$$
$$= t_1(t_1+1)\cdots(t_1+l-1) \qquad \text{if } l \geq 1.$$

By noting that

$$\int_0^v u^{t_1+i-1+l}(v-u)^{j-i-1}\, du = v^{j+t_1+l-1} B(t_1+i+l, j-i),$$

we may rewrite Eq. (2.3.12) as

$$M^*_{i,j:n}(t_1, t_2) = \frac{n!}{(i-1)!(j-i-1)!(n-j)!} \times \sum_{l=0}^{\infty} \frac{(t_1+l-1)^{(l)}}{l!} B(t_1+i+l, j-i) \times \int_0^1 v^{j+t_1+t_2+l-1}(1-v)^{n-j-t_2}\, dv$$

$$= \frac{n!}{(i-1)!(j-i-1)!(n-j)!} \times \sum_{l=0}^{\infty} \frac{(t_1+l-1)^{(l)}}{l!} B(t_1+i+l, j-i) \times B(j+t_1+t_2+l, n-j-t_2+1)$$

$$= \frac{\Gamma(n+1)}{\Gamma(i)\Gamma(n-j+1)} \sum_{l=0}^{\infty} \frac{(t_1+l-1)^{(l)}}{l!} \frac{\Gamma(t_1+i+l)}{\Gamma(t_1+j+l)} \times \frac{\Gamma(t_1+t_2+j+l)\Gamma(n-j+1-t_2)}{\Gamma(n+1+t_1+l)}. \tag{2.3.13}$$

From the above expression for the joint moment generating function of $Z_{i:n}$ and $Z_{j:n}$, one can obtain the product moments as follows:

$$\alpha_{i,j:n}^{*(k_1,k_2)} = E(Z_{i:n}^{k_1} Z_{j:n}^{k_2}) = \frac{\partial^{k_1+k_2}}{\partial t_1^{k_1}\,\partial t_2^{k_2}} M_{i,j:n}^{*}(t_1, t_2)\big|_{t_1=t_2=0}. \tag{2.3.14}$$

The case $k_1 = k_2 = 1$ is of particular importance, and we get

$$\alpha_{i,j:n}^{*} = \psi'(j) + \{\psi(i) - \psi(n+1)\}\{\psi(j) - \psi(n-j+1)\} + \sum_{l=1}^{\infty} \frac{1}{l} \frac{(i+l-1)^{(l)}}{(n+l)^{(l)}} \{\psi(j+l) - \psi(n-j+1)\}. \tag{2.3.15}$$

Shah (1966) tabulated the covariances of order statistics for sample sizes up to 10 while Gupta et al. (1967) extended up to 25. Recently, Balakrishnan and Malik (1991) provided tables of means, variances, and covariances for sample sizes of 50 and less. Balakrishnan and Leung (1988a,b) derived series expressions similar to the ones given in Eqs. (2.3.2)–(2.3.4) and (2.3.15) for the single and the product moments of order statistics from a generalized logistic distribution and provided tables of means, variances, and covariances for sample sizes up to 15.

2.4 RECURRENCE RELATIONS FOR MOMENTS

We present some recurrence relations satisfied by the single and the product moments that were established by Shah (1966, 1970) and show that one may evaluate these moments for all order statistics from all sample sizes in a simple and systematic recursive manner.

Relation 2.4.1 For $n \geq 1$ and $k = 0, 1, 2, \ldots,$

$$\alpha_{1:n+1}^{*(k+1)} = \alpha_{1:n}^{*(k+1)} - \frac{k+1}{n} \alpha_{1:n}^{*(k)} \tag{2.4.1}$$

with $\alpha_{1:n}^{*(0)} \equiv 1$ for $n = 1, 2, \ldots.$

Proof. From (2.1.1) we have for $n \geq 1$ and $k \geq 0$,

$$\alpha_{1:n}^{*(k)} = n \int_{-\infty}^{\infty} z^k \{1 - F^*(z)\}^{n-1} f^*(z)\, dz. \tag{2.4.2}$$

By using the relation (2.1.3) in Eq. (2.4.2), we get

$$\alpha_{1:n}^{*(k)} = n \int_{-\infty}^{\infty} z^k F^*(z) \{1 - F^*(z)\}^{n}\, dz,$$

which, upon integrating by parts, yields

$$\begin{aligned}\alpha_{1:n}^{*(k)} &= \frac{n}{k+1}\left[n \int_{-\infty}^{\infty} z^{k+1} F^*(z)\{1 - F^*(z)\}^{n-1} f^*(z)\, dz \right.\\ &\qquad \left. - \int_{-\infty}^{\infty} z^{k+1}\{1 - F^*(z)\}^{n} f^*(z)\, dz \right]\\ &= \frac{n}{k+1}\{\alpha_{1:n}^{*(k+1)} - \alpha_{1:n+1}^{*(k+1)}\}.\end{aligned}$$

The recurrence relation in (2.4.1) is obtained by rewriting the above equation.

Relation 2.4.2 For $1 \leq i \leq n$ and $k = 0, 1, 2, \ldots,$

$$\alpha_{i+1:n+1}^{*(k+1)} = \alpha_{i:n+1}^{*(k+1)} + \frac{(k+1)(n+1)}{i(n-i+1)}\alpha_{i:n}^{*(k)}. \tag{2.4.3}$$

Proof. From (2.1.1) we have for $1 \leq i \leq n$ and $k \geq 0$,

$$\alpha_{i:n}^{*(k)} = \frac{n!}{(i-1)!(n-i)!}\int_{-\infty}^{\infty} z^k \{F^*(z)\}^{i-1}\{1 - F^*(z)\}^{n-i} f^*(z)\, dz. \tag{2.4.4}$$

By using the relation (2.1.3) in Eq. (2.4.4), we get

$$\alpha_{i:n}^{*(k)} = \frac{n!}{(i-1)!(n-i)!}\int_{-\infty}^{\infty} z^k \{F^*(z)\}^{i}\{1 - F^*(z)\}^{n-i+1}\, dz,$$

which, upon integrating by parts, yields

$$\begin{aligned}\alpha_{i:n}^{*(k)} &= \frac{n!}{(i-1)!(n-i)!(k+1)}\\ &\quad \times \left[(n-i+1)\int_{-\infty}^{\infty} z^{k+1}\{F^*(z)\}^{i}\{1 - F^*(z)\}^{n-i} f^*(z)\, dz \right.\end{aligned}$$

$$- i \int_{-\infty}^{\infty} z^{k+1}\{F^*(z)\}^{i-1}\{1 - F^*(z)\}^{n-i+1} f^*(z)\, dz\Big]$$

$$= \frac{i(n - i + 1)}{(k + 1)(n + 1)} \{\alpha_{i+1:n+1}^{*(k+1)} - \alpha_{i:n+1}^{*(k+1)}\}.$$

The recurrence relation in (2.4.3) follows by rewriting the above equation.

With the values of $\alpha_{1:1}^{*(j)}$ ($j = 1, 2, \ldots, k$) known, one may be able to use Relations 2.4.1 and 2.4.2 in a simple recursive way to compute the first k single moments of all order statistics from all sample sizes. Thus, for example, by starting with the values of $\alpha_{1:1}^{*(1)} = 0$ and $\alpha_{1:1}^{*(2)} = \pi^2/3$, one may employ Relations 2.4.1 and 2.4.2 to evaluate the first two single moments and, thence, the variances of all order statistics from all sample sizes in a simple and systematic recursive process. These computations may be checked by using the identities (David, 1981, p. 39; Arnold and Balakrishnan, 1989, p. 6)

$$\sum_{i=1}^{n} \alpha_{i:n}^{*(j)} = n\alpha_{1:1}^{*(j)}, \qquad j = 1, 2, \ldots; \tag{2.4.5}$$

see also Balakrishnan and Malik (1986).

Relation 2.4.3 For $1 \leq i \leq n - 1$,

$$\alpha_{i,i+1:n+1}^{*} = \alpha_{i:n+1}^{*(2)} + \frac{n + 1}{n - i + 1}\left\{\alpha_{i,i+1:n}^{*} - \alpha_{i:n}^{*(2)} - \frac{1}{n - i}\alpha_{i:n}^{*}\right\}. \tag{2.4.6}$$

Proof. For $1 \leq i \leq n - 1$, we may write from (2.1.2) that

$$\alpha_{i:n}^{*} = E(Z_{i:n}Z_{i+1:n}^{0})$$

$$= \frac{n!}{(i - 1)!(n - i - 1)!} \int_{-\infty}^{\infty} z_1\{F^*(z_1)\}^{i-1} f^*(z_1)K(z_1)\, dz_1, \tag{2.4.7}$$

where

$$K(z_1) = \int_{z_1}^{\infty} \{1 - F^*(z_2)\}^{n-i-1} f^*(z_2)\, dz_2. \tag{2.4.8}$$

By using the relation (2.1.3) in Eq. (2.4.8) and integrating by parts, we get

$$K(z_1) = -z_1\{1 - F^*(z_1)\}^{n-i} + z_1\{1 - F^*(z_1)\}^{n-i+1}$$

$$+ (n - i) \int_{z_1}^{\infty} z_2\{1 - F^*(z_2)\}^{n-i-1} f^*(z_2)\, dz_2$$

$$- (n - i + 1) \int_{z_1}^{\infty} z_2\{1 - F^*(z_2)\}^{n-i} f^*(z_2)\, dz_2.$$

Upon substituting the above expression of $K(z_1)$ in (2.4.7) and simplifying the resulting equation, we get

$$\alpha^*_{i:n} = -(n-i)\alpha^{*(2)}_{i:n} + \frac{(n-i)(n-i+1)}{(n+1)}\alpha^{*(2)}_{i:n+1} + (n-i)\alpha^*_{i,i+1:n} - \frac{(n-i)(n-i+1)}{(n+1)}\alpha^*_{i,i+1:n+1}.$$

The recurrence relation in (2.4.6) follows by rewriting the above equation.

Relation 2.4.4 For $1 \le i \le n-1$,

$$\alpha^*_{i+1,i+2:n+1} = \alpha^{*(2)}_{i+2:n+1} + \frac{n+1}{i+1}\left\{\frac{1}{i}\alpha^*_{i+1:n} + \alpha^*_{i,i+1:n} - \alpha^{*(2)}_{i+1:n}\right\}. \quad (2.4.9)$$

Proof. For $1 \le i \le n-1$, we may write from (2.1.2) that

$$\begin{aligned}\alpha^*_{i+1:n} &= E(Z^0_{i:n}Z_{i+1:n}) \\ &= \frac{n!}{(i-1)!(n-i-1)!}\int_{-\infty}^{\infty} z_2 \\ &\quad \times \{1-F^*(z_2)\}^{n-i-1}f^*(z_2)K(z_2)\,dz_2, \end{aligned} \quad (2.4.10)$$

where

$$K(z_2) = \int_{-\infty}^{z_2} \{F^*(z_1)\}^{i-1}f^*(z_1)\,dz_1. \quad (2.4.11)$$

By using the relation (2.1.3) in Eq. (2.4.11) and integrating by parts, we get

$$K(z_2) = z_2\{F^*(z_2)\}^i - z_2\{F^*(z_2)\}^{i+1} - i\int_{-\infty}^{z_2} z_1\{F^*(z_1)\}^{i-1}f^*(z_1)\,dz_1 + (i+1)\int_{-\infty}^{z_2} z_1\{F^*(z_1)\}^i f^*(z_1)\,dz_1.$$

Upon substituting the above expression of $K(z_2)$ in (2.4.10) and simplifying the resulting equation, we get

$$\alpha^*_{i+1:n} = i\alpha^{*(2)}_{i+1:n} - \frac{i(i+1)}{(n+1)}\alpha^{*(2)}_{i+2:n+1} - i\alpha^*_{i,i+1:n} + \frac{i(i+1)}{(n+1)}\alpha^*_{i+1,i+2:n+1}.$$

The recurrence relation in (2.4.9) is obtained by rewriting the above equation.

In particular, by setting $i = n - 1$ in Relation 2.4.4 we get for $n \geq 2$,

$$\alpha^*_{n,n+1:n+1} = \alpha^{*(2)}_{n+1:n+1} + \frac{n+1}{n} \times \left\{ \frac{1}{n-1} \alpha^*_{n:n} + \alpha^*_{n-1,n:n} - \alpha^{*(2)}_{n:n} \right\}. \qquad (2.4.12)$$

The recurrence relations in (2.4.6) and (2.4.12) are sufficient for the evaluation of all the product moments of order statistics from all sample sizes. By starting with the result that $\alpha^*_{1,2:2} = \alpha^{*2}_{1:1} = 0$ (Govindarajulu, 1963; Joshi, 1971), the recurrence relations in (2.4.6) and (2.4.12) will enable one to compute all the immediate upper-diagonal product moments $\alpha^*_{i,i+1:n}$ $(1 \leq i \leq n-1)$ for all sample sizes in a simple recursive way. All the remaining product moments, viz., $\alpha^*_{i,j:n}$ for $1 \leq i < j \leq n$ and $j - i \geq 2$, may be determined systematically by employing the well-known recurrence relation (David, 1981, p. 48; Arnold and Balakrishnan, 1989, p. 10)

$$(i-1)\alpha^*_{i,j:n} + (j-i)\alpha^*_{i-1,j:n} + (n-j+1)\alpha^*_{i-1,j-1:n} = n\alpha^*_{i-1,j-1:n-1}, \qquad (2.4.13)$$

which is true for any arbitrary distribution. These computations may then be checked by using the identity (David, 1981, p. 39; Arnold and Balakrishnan, 1989, p. 10)

$$\sum_{i=1}^{n-1} \sum_{j=i+1}^{n} \alpha^*_{i,j:n} = \binom{n}{2} \alpha^{*2}_{1:1}; \qquad (2.4.14)$$

see also Balakrishnan and Malik (1986).

By proceeding on similar lines, we may also establish the following recurrence relations.

Relation 2.4.5 For $1 \leq i < j \leq n$ and $j - i \geq 2$,

$$\alpha^*_{i,j:n+1} = \alpha^*_{i,j-1:n+1} + \frac{n+1}{n-j+2} \times \left\{ \alpha^*_{i,j:n} - \alpha^*_{i,j-1:n} - \frac{1}{n-j+1} \alpha^*_{i:n} \right\}. \qquad (2.4.15)$$

Relation 2.4.6 For $1 \leq i < j \leq n$ and $j - i \geq 2$,

$$\alpha^*_{i+1,j+1:n+1} = \alpha^*_{i+2,j+1:n+1} + \frac{n+1}{i+1} \left\{ \frac{1}{i} \alpha^*_{j:n} + \alpha^*_{i,j:n} - \alpha^*_{i+1,j:n} \right\}. \qquad (2.4.16)$$

One may employ Relations 2.4.5 and 2.4.6 to determine all the product moments other than the immediate upper-diagonal product moments, viz., $\alpha^*_{i,j:n}$ for $1 \le i < j \le n$ and $j - i \ge 2$, instead of the recurrence relation in (2.4.13).

2.5 DISTRIBUTIONS OF SOME SYSTEMATIC STATISTICS

In this section we first present the distribution of the sample range as derived by Gupta and Shah (1965). Then, we give the expression of the distribution of the rth quasi-range derived by Malik (1980).

Let us denote the sample range $Z_{n:n} - Z_{1:n}$ by W_n. The cumulative distribution function of W_n can be written down as (David, 1981, p. 12)

$$\Pr(W_n \le w) = n \int_{-\infty}^{\infty} \{F^*(z + w) - F^*(z)\}^{n-1} \times f^*(z)\, dz, \qquad 0 \le w < \infty. \tag{2.5.1}$$

Expanding $\{F^*(z + w) - F^*(z)\}^{n-1}$ binomially, we get from (2.5.1) that

$$\begin{aligned}\Pr(W_n \le w) &= n \sum_{k=0}^{n-1} (-1)^k \binom{n-1}{k} \times \int_{-\infty}^{\infty} \{F^*(z + w)\}^{n-1-k} \{F^*(z)\}^k f^*(z)\, dz \\ &= n \sum_{k=0}^{n-1} (-1)^k \binom{n-1}{k} \times \int_{-\infty}^{\infty} \frac{e^{-z}}{(1 + e^{-w}e^{-z})^{n-1-k}(1 + e^{-z})^{k+2}}\, dz. \end{aligned} \tag{2.5.2}$$

By substituting $u = 1/(1 + e^{-w}e^{-z})$ in the integral in (2.5.2), we get

$$\Pr(W_n \le w) = n \sum_{k=0}^{n-1} (-1)^k \binom{n-1}{k} e^{-(k+1)w} A_{k,n}(w), \qquad 0 \le w < \infty, \tag{2.5.3}$$

where, with $a = e^{-w} - 1$,

$$\begin{aligned} A_{k,n}(w) &= \int_0^1 u^{n-1}(1 + au)^{-k-2}\, du \\ &= \frac{1}{(-a)^n} \left[(-1)^{k+1} \binom{n-1}{k+1} \ln(1 + a) + \sum_{\substack{l=0 \\ l \ne k+1}}^{n-1} \right.\end{aligned}$$

$$\times (-1)^l \binom{n-1}{l} \frac{1}{(l-k-1)} \{(1+a)^{l-k-1} - 1\}\Bigg]$$

$$= \frac{1}{(1-e^{-w})^n} \Bigg[(-1)^k \binom{n-1}{k+1} w + \sum_{\substack{l=0 \\ l \neq k+1}}^{n-1}$$

$$\times (-1)^l \binom{n-1}{l} \frac{1}{(l-k-1)} \{e^{-(l-k-1)w} - 1\}\Bigg]. \tag{2.5.4}$$

In (2.5.4), $\binom{n-1}{k+1}$ should be set to zero if $k > n - 2$. By substituting (2.5.4) into (2.5.3), we derive the cumulative distribution function of the sample range W_n:

$$\Pr(W_n \le w) = \frac{n}{(1-e^{-w})^n} \sum_{k=0}^{n-1} (-1)^k \binom{n-1}{k}$$

$$\times \Bigg[(-1)^k \binom{n-1}{k+1} we^{-(k+1)w} + \sum_{\substack{l=0 \\ l \neq k+1}}^{n-1} (-1)^l \binom{n-1}{l}$$

$$\times \frac{1}{(l-k-1)} \{e^{-lw} - e^{-(k+1)w}\}\Bigg], \qquad 0 \le w < \infty. \tag{2.5.5}$$

In particular, for $n = 2$ and 3, we obtain from (2.5.5) that

$$\Pr(W_2 \le w) = \frac{1 - e^{-2w} - 2we^{-w}}{(1-e^{-w})^2} \tag{2.5.6}$$

and

$$\Pr(W_3 \le w) = \frac{1 + 9e^{-w} - 9e^{-2w} - e^{-3w} - 6we^{-w}(1+e^{-w})}{(1-e^{-w})^3}. \tag{2.5.7}$$

Gupta and Shah (1965) tabulated the probability integrals of the range for $n = 2$ and 3 from (2.5.6) and (2.5.7), respectively.

By differentiating the distribution function of the sample range in (2.5.5) with respect to w, we derive the density function of the sample range W_n as

$$f_{W_n}(w) = -\frac{n^2 e^{-w}}{(1-e^{-w})^{n+1}} \sum_{k=0}^{n-1} (-1)^k \binom{n-1}{k}$$

$$\times \Bigg[(-1)^k \binom{n-1}{k+1} we^{-(k+1)w} + \sum_{\substack{l=0 \\ l \neq k+1}}^{n-1} (-1)^l \binom{n-1}{l}$$

$$\times \frac{1}{(l-k-1)} \{e^{-lw} - e^{-(k+1)w}\}\Bigg]$$

$$+ \frac{n}{(1-e^{-w})^n} \sum_{k=0}^{n-1} (-1)^k \binom{n-1}{k}$$

$$\times \left[(-1)^k \binom{n-1}{k+1} e^{-(k+1)w}\{1-(k+1)w\}\right.$$

$$- \sum_{\substack{l=0 \\ l \neq k+1}}^{n-1} (-1)^l \binom{n-1}{l} \frac{1}{(l-k-1)}$$

$$\times \{le^{-lw} - (k+1)e^{-(k+1)w}\}\Bigg], \qquad 0 \le w < \infty, \qquad (2.5.8)$$

where, as before, $\binom{n-1}{k+1}$ should be set to zero if $k > n-2$.

Proceeding exactly on similar lines, Malik (1980) derived the cumulative distribution function of the rth quasi-range $W_{n,r} = Z_{n-r:n} - Z_{r+1:n}$ ($r = 0, 1, \ldots, [(n-1)/2]$) to be

$$\Pr(W_{n,r} \le w) = \sum_{k=0}^{r} \frac{\prod_{i=0}^{2r-k}(n-i)}{r!(r-k)!} \left[\sum_{j=0}^{n-2r+k-1} (-1)^j \binom{n-2r+k-1}{j}\right.$$

$$\times e^{-(r+j+1)w} \left\{\sum_{l=0}^{r-k} \frac{(-1)^{l+1}\binom{r-k}{l}}{(1-e^{-w})^{n-r+k+l}}\right.$$

$$\times \left((-1)^{r+j} \binom{n-r+k+l-1}{r+j+1} w\right.$$

$$+ \sum_{\substack{m=0 \\ m \neq r+j+1}}^{n-r+k+l-1} (-1)^m \binom{n-r+k+l-1}{m}$$

$$\times \left.\left.\left.\frac{e^{-(m-r-j-1)w} - 1}{m-r-j-1}\right)\right\}\right], \qquad 0 \le w < \infty, \qquad (2.5.9)$$

where

$$\binom{n-r+k+l-1}{r+j+1}$$

should be set to zero if $j > n - 2r + k + l - 2$. The distribution function of the sample range in (2.5.8) may be derived as a special case from (2.5.9) by setting $r = 0$.

2.6 RESULTS FOR TRUNCATED DISTRIBUTIONS

In this section we start with order statistics from a doubly truncated logistic distribution and present the results of Balakrishnan and Joshi (1983a) and Balakrishnan and Kocherlakota (1986). In addition to generalizing the relations given in Section 2.4, these results will enable one to evaluate the single and the product moments of all order statistics from all sample sizes in a simple recursive manner.

Let $Z^*_{1:n} \le Z^*_{2:n} \le \cdots \le Z^*_{n:n}$ be order statistics from a random sample of size n from a doubly truncated logistic distribution with probability density function

$$f^{**}(z) = \begin{cases} \dfrac{1}{P-Q}\dfrac{e^{-z}}{(1+e^{-z})^2}, & Q_1 \le z \le P_1, \\ 0 & \text{otherwise}, \end{cases} \tag{2.6.1}$$

and cumulative distribution function

$$F^{**}(z) = \frac{1}{P-Q}\left\{\frac{1}{(1+e^{-z})} - Q\right\}, \qquad Q_1 \le z \le P_1, \tag{2.6.2}$$

where Q and $1 - P$ are the proportions of truncation on the left and the right of the standard logistic density function in (1.5). With this notation,

$$Q_1 = \ln\left(\frac{Q}{1-Q}\right), \qquad P_1 = \ln\left(\frac{P}{1-P}\right), \tag{2.6.3}$$

and let

$$Q_2 = \frac{Q(1-Q)}{P-Q}, \qquad P_2 = \frac{P(1-P)}{P-Q}. \tag{2.6.4}$$

From (2.6.1) and (2.6.2) we observe the relations

$$f^{**}(z) = (1 - 2Q)F^{**}(z) - (P - Q)\{F^{**}(z)\}^2 + Q_2, \tag{2.6.5}$$

$$f^{**}(z) = (2P - 1)\{1 - F^{**}(z)\} - (P - Q) \times \{1 - F^{**}(z)\}^2 + P_2, \tag{2.6.6}$$

$$f^{**}(z) = (2P - 1)F^{**}(z)\{1 - F^{**}(z)\} + (P + Q - 1) \times \{1 - F^{**}(z)\}^2 + P_2. \tag{2.6.7}$$

Let us now denote the single moments $E(Z^{*k}_{i:n})$ by $\alpha^{**(k)}_{i:n}$ ($1 \le i \le n$, $k \ge 1$) and the product moments $E(Z^*_{i:n}Z^*_{j:n})$ by $\alpha^{**}_{i,j:n}$ ($1 \le i < j \le n$). For convenience, let us also use $\alpha^{**}_{i:n}$ for $\alpha^{**(1)}_{i:n}$ and $\alpha^{**}_{i,i:n}$ for $\alpha^{**(2)}_{i:n}$. Then, these moments satisfy the following recurrence relations.

Relation 2.6.1 For $k = 0, 1, 2, \ldots,$

$$\alpha_{1:2}^{**(k+1)} = Q_1^{k+1} + \frac{1}{P-Q}[P_2\{P_1^{k+1} - Q_1^{k+1}\} + (2P-1) \times \{\alpha_{1:1}^{**(k+1)} - Q_1^{k+1}\} - (k+1)\alpha_{1:1}^{**(k)}]. \quad (2.6.8)$$

Proof. For $k \geq 0$, let us consider

$$\alpha_{1:1}^{**(k)} = \int_{Q_1}^{P_1} z^k f^{**}(z)\, dz.$$

Upon using (2.6.6) in the above equation and then integrating by parts, we get

$$\alpha_{1:1}^{**(k)} = \frac{1}{k+1}[(2P-1)\{\alpha_{1:1}^{**(k+1)} - Q_1^{k+1}\} - (P-Q) \times \{\alpha_{1:2}^{**(k+1)} - Q_1^{k+1}\} + P_2\{P_1^{k+1} - Q_1^{k+1}\}]. \quad (2.6.9)$$

The recurrence relation in (2.6.8) follows by rewriting Eq. (2.6.9).

Relation 2.6.2 For $k = 0, 1, 2, \ldots,$

$$\alpha_{2:2}^{**(k+1)} = P_1^{k+1} - \frac{1}{P-Q}[Q_2\{P_1^{k+1} - Q_1^{k+1}\} + (1-2Q)\{P_1^{k+1} - \alpha_{1:1}^{**(k+1)}\} - (k+1)\alpha_{1:1}^{**(k)}]. \quad (2.6.10)$$

Proof. For $k \geq 0$, let us consider

$$\alpha_{1:1}^{**(k)} = \int_{Q_1}^{P_1} z^k f^{**}(z)\, dz.$$

Upon using (2.6.5) in the above equation and then integrating by parts, we get

$$\alpha_{1:1}^{**(k)} = \frac{1}{k+1}[Q_2\{P_1^{k+1} - Q_1^{k+1}\} + (1-2Q) \times \{P_1^{k+1} - \alpha_{1:1}^{**(k+1)}\} - (P-Q)\{P_1^{k+1} - \alpha_{2:2}^{**(k+1)}\}]. \quad (2.6.11)$$

The recurrence relation in (2.6.10) is obtained by rewriting Eq. (2.6.11).

Relation 2.6.3 For $n \geq 2$ and $k = 0, 1, 2, \ldots,$

$$\alpha_{1:n+1}^{**(k+1)} = Q_1^{k+1} + \frac{1}{P-Q}\left[P_2\{\alpha_{1:n-1}^{**(k+1)} - Q_1^{k+1}\} + (2P-1)\{\alpha_{1:n}^{**(k+1)} - Q_1^{k+1}\} - \frac{k+1}{n}\alpha_{1:1}^{**(k)}\right]. \quad (2.6.12)$$

Proof. For $n \geq 2$ and $k \geq 0$, let us consider

$$\alpha_{1:n}^{**(k)} = n \int_{Q_1}^{P_1} z^k \{1 - F^{**}(z)\}^{n-1} f^{**}(z)\, dz.$$

Upon using (2.6.6) in the above equation and then integrating by parts, we get

$$\alpha_{1:n}^{**(k)} = \frac{n}{k+1} [P_2\{\alpha_{1:n-1}^{**(k+1)} - Q_1^{k+1}\} + (2P - 1) \times \{\alpha_{1:n}^{**(k+1)} - Q_1^{k+1}\} - (P - Q)\{\alpha_{1:n+1}^{**(k+1)} - Q_1^{k+1}\}]. \quad (2.6.13)$$

The recurrence relation in (2.6.12) follows by rewriting Eq. (2.6.13).

Relation 2.6.4 For $k = 0, 1, 2, \ldots,$

$$\alpha_{2:3}^{**(k+1)} = \alpha_{1:3}^{**(k+1)} + \frac{3}{P - Q} \left[P_2\{P_1^{k+1} - \alpha_{1:1}^{**(k+1)}\} + \frac{2P - 1}{2} \{\alpha_{2:2}^{**(k+1)} - \alpha_{1:2}^{**(k+1)}\} - \frac{k+1}{2} \alpha_{2:2}^{**(k)} \right]. \quad (2.6.14)$$

Proof. For $k \geq 0$, let us consider

$$\alpha_{2:2}^{**(k)} = 2 \int_{Q_1}^{P_1} z^k F^{**}(z) f^{**}(z)\, dz.$$

Upon using (2.6.6) in the above equation and then integrating by parts, we get

$$\alpha_{2:2}^{**(k)} = \frac{1}{k+1} \left[(2P - 1)\{\alpha_{2:2}^{**(k+1)} - \alpha_{1:2}^{**(k+1)}\} - \frac{2}{3}(P - Q) \times \{\alpha_{2:3}^{**(k+1)} - \alpha_{1:3}^{**(k+1)}\} + 2P_2\{P_1^{k+1} - \alpha_{1:1}^{**(k+1)}\} \right]. \quad (2.6.15)$$

The recurrence relation in (2.6.14) is obtained by rewriting Eq. (2.6.15).

Relation 2.6.5 For $n \geq 3$ and $k = 0, 1, 2, \ldots,$

$$\alpha_{2:n+1}^{**(k+1)} = \alpha_{1:n+1}^{**(k+1)} + \frac{n+1}{P - Q} \left[\frac{P_2}{n-1} \{\alpha_{2:n-1}^{**(k+1)} - \alpha_{1:n-1}^{**(k+1)}\} + \frac{2P - 1}{n} \{\alpha_{2:n}^{**(k+1)} - \alpha_{1:n}^{**(k+1)}\} - \frac{k+1}{n(n-1)} \alpha_{2:n}^{**(k)} \right]. \quad (2.6.16)$$

Proof. For $n \geq 3$ and $k \geq 0$, let us consider

$$\alpha_{2:n}^{**(k)} = n(n-1) \int_{Q_1}^{P_1} z^k F^{**}(z)\{1 - F^{**}(z)\}^{n-2} f^{**}(z)\, dz.$$

Upon using (2.6.6) in the above equation and then integrating by parts, we get

$$\alpha_{2:n}^{**(k)} = \frac{n-1}{k+1}\Bigg[(2P-1)\{\alpha_{2:n}^{**(k+1)} - \alpha_{1:n}^{**(k+1)}\} - (P-Q)\frac{n}{n+1}\{\alpha_{2:n+1}^{**(k+1)} - \alpha_{1:n+1}^{**(k+1)}\} + P_2\{\alpha_{2:n}^{**(k+1)} - \alpha_{1:n}^{**(k+1)}\}\Bigg]. \tag{2.6.17}$$

The recurrence relation in (2.6.16) follows by rewriting Eq. (2.6.17).

Relation 2.6.6 For $2 \leq i \leq n-1$ and $k = 0, 1, 2, \ldots,$

$$\alpha_{i+1:n+1}^{**(k+1)} = \frac{n+1}{i(2P-1)}\Bigg[\frac{k+1}{n-i+1}\alpha_{i:n}^{**(k)} - \frac{nP_2}{n-i+1}\{\alpha_{i:n-1}^{**(k+1)} - \alpha_{i-1:n-1}^{**(k+1)}\} - \frac{1}{n+1}\{(n+1)(P+Q-1) - i(2P-1)\} \times \alpha_{i:n+1}^{**(k+1)} + (P+Q-1)\alpha_{i-1:n}^{**(k+1)}\Bigg]. \tag{2.6.18}$$

Proof. For $2 \leq i \leq n-1$ and $k \geq 0$, let us consider

$$\alpha_{i:n}^{**(k)} = \frac{n!}{(i-1)!(n-i)!}\int_{Q_1}^{P_1} z^k\{F^{**}(z)\}^{i-1}\{1 - F^{**}(z)\}^{n-i} f^{**}(z)\, dz.$$

Upon using (2.6.7) in the above equation and then integrating by parts, we get

$$\alpha_{i:n}^{**(k)} = \frac{1}{k+1}\Bigg[\frac{i(n-i+1)}{n+1}(2P-1)\{\alpha_{i+1:n+1}^{**(k+1)} - \alpha_{i:n+1}^{**(k+1)}\} - \frac{(n-i+1)(n-i+2)}{n+1}(P+Q-1)\{\alpha_{i:n+1}^{**(k+1)} - \alpha_{i-1:n+1}^{**(k+1)}\} + nP_2\{\alpha_{i:n-1}^{**(k+1)} - \alpha_{i-1:n-1}^{**(k+1)}\}\Bigg]. \tag{2.6.19}$$

If we now use the well-known relation (David, 1981, p. 46; Arnold and Balakrishnan, 1989, p. 6)

$$(n - i + 2)\{\alpha_{i:n+1}^{**(k+1)} - \alpha_{i-1:n+1}^{**(k+1)}\} = (n + 1)\{\alpha_{i:n+1}^{**(k+1)} - \alpha_{i-1:n}^{**(k+1)}\}$$

in (2.6.19) and simplify the resulting equation, we derive the recurrence relation in (2.6.18).

Relation 2.6.7 For $n \geq 2$ and $k = 0, 1, 2, \ldots,$

$$\begin{aligned}\alpha_{n+1:n+1}^{**(k+1)} = \frac{n+1}{n(2P-1)} \Bigg[& (k+1)\alpha_{n:n}^{**(k)} - nP_2\{P_1^{k+1} - \alpha_{n-1:n-1}^{**(k+1)}\} \\ & - \frac{1}{n+1}\{(n+1)(P+Q-1) - n(2P-1)\} \\ & \times \alpha_{n:n+1}^{**(k+1)} + (P+Q-1)\alpha_{n-1:n}^{**(k+1)}\Bigg]. \end{aligned} \quad (2.6.20)$$

Proof. For $n \geq 2$ and $k \geq 0$, let us consider

$$\alpha_{n:n}^{**(k)} = n \int_{Q_1}^{P_1} z^k \{F^{**}(z)\}^{n-1} f^{**}(z)\, dz.$$

Upon using (2.6.7) in the above equation and then integrating by parts, we get

$$\begin{aligned}\alpha_{n:n}^{**(k)} = \frac{1}{k+1}\Bigg[& \frac{n}{n+1}(2P-1)\{\alpha_{n+1:n+1}^{**(k+1)} - \alpha_{n:n+1}^{**(k+1)}\} \\ & + \frac{2}{n+1}(P+Q-1)\{\alpha_{n:n+1}^{**(k+1)} - \alpha_{n-1:n+1}^{**(k+1)}\} \\ & + nP_2\{P_1^{k+1} - \alpha_{n-1:n-1}^{**(k+1)}\}\Bigg]. \end{aligned} \quad (2.6.21)$$

If we now use the well-known relation (David, 1981, p. 46; Arnold and Balakrishnan, 1989, p. 6)

$$2\{\alpha_{n:n+1}^{**(k+1)} - \alpha_{n-1:n+1}^{**(k+1)}\} = (n + 1)\{\alpha_{n:n+1}^{**(k+1)} - \alpha_{n-1:n}^{**(k+1)}\}$$

in (2.6.21) and simplify the resulting equation, we derive the recurrence relation in (2.6.20).

By starting with the values of $\alpha_{1:1}^{**(j)}$ ($j = 1, 2, \ldots, k$), one can employ Relations 2.6.1–2.6.7 in a simple and systematic recursive way to compute the first k single moments of all order statistics from all sample sizes for any choice of Q and P.

We may also establish the following recurrence relations satisfied by the product moments of order statistics.

Relation 2.6.8 We have

$$\alpha_{1,2:3}^{**} = \alpha_{1:3}^{**(2)} + \frac{3}{2(P - Q)}\,[2P_2\{P_1\alpha_{1:1}^{**(1)} - \alpha_{1:1}^{**(2)}\}$$
$$+ (2P - 1)\{\alpha_{1,2:2}^{**} - \alpha_{1:2}^{**(2)}\} - \alpha_{1:2}^{**(1)}]. \tag{2.6.22}$$

Proof. Let us start with

$$\alpha_{1:2}^{**(1)} = E(Z_{1:2}^{**}Z_{2:2}^{**0}) = 2\int_{Q_1}^{P_1} z_1 f^{**}(z_1)K(z_1)\,dz_1, \tag{2.6.23}$$

where

$$K(z_1) = \int_{z_1}^{P_1} f^{**}(z_2)\,dz_2. \tag{2.6.24}$$

Upon using (2.6.6) in (2.6.24) and integrating by parts, we get

$$K(z_1) = (2P - 1)\left\{\int_{z_1}^{P_1} z_2 f^{**}(z_2)\,dz_2 - z_1\{1 - F^{**}(z_1)\}\right\}$$
$$- (P - Q)\left\{2\int_{z_1}^{P_1} z_2\{1 - F^{**}(z_2)\}f^{**}(z_2)\,dz_2\right.$$
$$\left.- z_1\{1 - F^{**}(z_1)\}^2\right\} + P_2(P_1 - z_1).$$

Upon substituting the above expression of $K(z_1)$ in (2.6.23) and simplifying the resulting equation, we derive the recurrence relation in (2.6.22).

Relation 2.6.9 We have

$$\alpha_{2,3:3}^{**} = \alpha_{3:3}^{**(2)} + \frac{3}{2(P - Q)}$$
$$\times\,[\alpha_{2:2}^{**(1)} - 2Q_2\{\alpha_{1:1}^{**(2)} - Q_1\alpha_{1:1}^{**(1)}\}$$
$$- (1 - 2Q)\{\alpha_{2:2}^{**(2)} - \alpha_{1,2:2}^{**}\}]. \tag{2.6.25}$$

Proof. Let us start with

$$\alpha_{2:2}^{**(1)} = E(Z_{1:2}^{**0}Z_{2:2}^{**}) = 2\int_{Q_1}^{P_1} z_2 f^{**}(z_2)K(z_2)\,dz_2, \tag{2.6.26}$$

where

$$K(z_2) = \int_{Q_1}^{z_2} f^{**}(z_1)\, dz_1. \qquad (2.6.27)$$

Upon using (2.6.5) in (2.6.27) and integrating by parts, we get

$$\begin{aligned} K(z_2) = (1 - 2Q)&\left\{ z_2 F^{**}(z_2) - \int_{Q_1}^{z_2} z_1 f^{**}(z_1)\, dz_1 \right\} \\ &- (P - Q)\left[z_2\{F^{**}(z_2)\}^2 - 2\int_{Q_1}^{z_2} z_1 F^{**}(z_1) f^{**}(z_1)\, dz_1 \right] \\ &+ Q_2(z_2 - Q_1). \end{aligned}$$

Upon substituting the above expression of $K(z_2)$ in (2.6.26) and simplifying the resulting equation, we derive the recurrence relation in (2.6.25).

Relation 2.6.10 For $1 \le i \le n - 2$,

$$\begin{aligned} \alpha_{i,i+1:n+1}^{**} = \alpha_{i:n+1}^{**(2)} &+ \frac{(n + 1)}{(n - i + 1)(P - Q)} \\ &\times \left[\frac{n}{n - i} P_2\{\alpha_{i,i+1:n-1}^{**} - \alpha_{i:n-1}^{**(2)}\} + (2P - 1) \right. \\ &\left. \times \{\alpha_{i,i+1:n}^{**} - \alpha_{i:n}^{**(2)}\} - \frac{1}{n - i}\alpha_{i:n}^{**(1)} \right]. \end{aligned} \qquad (2.6.28)$$

Proof. For $1 \le i \le n - 2$, let us consider

$$\begin{aligned} \alpha_{i:n}^{**(1)} &= E(Z_{i:n}^{*} Z_{i+1:n}^{*0}) \\ &= \frac{n!}{(i - 1)!(n - i - 1)!} \\ &\quad \times \int_{Q_1}^{P_1} z_1\{F^{**}(z_1)\}^{i-1} f^{**}(z_1) K(z_1)\, dz_1, \end{aligned} \qquad (2.6.29)$$

where

$$K(z_1) = \int_{z_1}^{P_1} \{1 - F^{**}(z_2)\}^{n-i-1} f^{**}(z_2)\, dz_2. \qquad (2.6.30)$$

Upon using (2.6.6) in (2.6.30) and integrating by parts, we get

$$K(z_1) = (2P - 1)\Big[(n - i)\int_{z_1}^{P_1} z_2\{1 - F^{**}(z_2)\}^{n-i-1}f^{**}(z_2)\, dz_2$$

$$- z_1\{1 - F^{**}(z_1)\}^{n-i}\Big] - (P - Q)\Big[(n - i + 1)\int_{z_1}^{P_1} z_2$$

$$\times \{1 - F^{**}(z_2)\}^{n-i}f^{**}(z_2)\, dz_2 - z_1\{1 - F^{**}(z_1)\}^{n-i+1}\Big]$$

$$+ P_2\Big[(n - i - 1)\int_{z_1}^{P_1} z_2\{1 - F^{**}(z_2)\}^{n-i-2}f^{**}(z_2)\, dz_2$$

$$- z_1\{1 - F^{**}(z_1)\}^{n-i-1}\Big].$$

Upon substituting the above expression of $K(z_1)$ in (2.6.29) and simplifying the resulting equation, we derive the recurrence relation in (2.6.28).

Relation 2.6.11 For $n \geq 2$,

$$\alpha^{**}_{n-1,n:n+1} = \alpha^{**(2)}_{n-1:n+1} + \frac{(n + 1)}{2(P - Q)}\Big[nP_2\{P_1\alpha^{**(1)}_{n-1:n-1} - \alpha^{**(2)}_{n-1:n-1}\}$$

$$+ (2P - 1)\{\alpha^{**}_{n-1,n:n} - \alpha^{**(2)}_{n-1:n}\} - \alpha^{**(1)}_{n-1:n}\Big]. \tag{2.6.31}$$

Proof. For $n \geq 2$, let us consider

$$\alpha^{**(1)}_{n-1:n} = E(Z^{*}_{n-1:n}Z^{*0}_{n:n})$$

$$= n(n - 1)\int_{Q_1}^{P_1} z_1\{F^{**}(z_1)\}^{n-2}f^{**}(z_1)K(z_1)\, dz_1, \tag{2.6.32}$$

where

$$K(z_1) = \int_{z_1}^{P_1} f^{**}(z_2)\, dz_2. \tag{2.6.33}$$

Upon using (2.6.6) in (2.6.33) and integrating by parts, we get

$$K(z_1) = (2P - 1)\Big[\int_{z_1}^{P_1} z_2 f^{**}(z_2)\, dz_2 - z_1\{1 - F^{**}(z_1)\}\Big]$$

$$- (P - Q)\Big[2\int_{z_1}^{P_1} z_2\{1 - F^{**}(z_2)\}f^{**}(z_2)\, dz_2$$

$$- z_1\{1 - F^{**}(z_1)\}^2\Big] + P_2(P_1 - z_1).$$

Upon substituting the above expression of $K(z_1)$ in (2.6.32) and simplifying the resulting equation, we derive the recurrence relation in (2.6.31).

Relation 2.6.12 For $n \geq 2$,

$$\alpha^{**}_{2,3:n+1} = \alpha^{**(2)}_{3:n+1} + \frac{n+1}{2(P-Q)}\left[\alpha^{**(1)}_{2:n} - nQ_2 \times \{\alpha^{**(2)}_{1:n-1} - Q_1\alpha^{**(1)}_{1:n-1}\} - (1-2Q)\{\alpha^{**(2)}_{2:n} - \alpha^{**}_{1,2:n}\}\right]. \quad (2.6.34)$$

Proof. For $n \geq 2$, let us start with

$$\alpha^{**(1)}_{2:2} = E(Z^{*(1)}_{1:2}Z^{*}_{2:2}) = 2\int_{Q_1}^{P_1} z_2 f^{**}(z_2)K(z_2)\,dz_2, \quad (2.6.35)$$

where

$$K(z_2) = \int_{Q_1}^{z_2} f^{**}(z_1)\,dz_1. \quad (2.6.36)$$

Upon using (2.6.5) in (2.6.36) and integrating by parts, we get

$$K(z_2) = Q_2(z_2 - Q_1) + (1-2Q)\left\{z_2F^{**}(z_2) - \int_{Q_1}^{z_2} z_1 f^{*}(z_1)\,dz_1\right\}$$
$$- (P-Q)\left[z_2\{F^{**}(z_2)\}^2 - 2\int_{Q_1}^{z_2} z_1F^{**}(z_1)f^{**}(z_1)\,dz_1\right].$$

Upon substituting the above expression of $K(z_2)$ in (2.6.35) and simplifying the resulting equation, we derive the recurrence relation in (2.6.34).

Relation 2.6.13 For $2 \leq i \leq n-1$,

$$\alpha^{**}_{i+1,i+2:n+1} = \alpha^{**(2)}_{i+2:n+1} + \frac{(n+1)}{(i+1)(P-Q)}$$
$$\times\left[\frac{1}{i}\alpha^{**(1)}_{i+1:n} - \frac{nQ_2}{i}\{\alpha^{**(2)}_{i:n-1} - \alpha^{**}_{i-1,i:n-1}\}\right.$$
$$\left. - (1-2Q)\{\alpha^{**(2)}_{i+1:n} - \alpha^{**}_{i,i+1:n}\}\right]. \quad (2.6.37)$$

Proof. For $2 \le i \le n-1$, let us consider

$$\alpha_{i+1:n}^{**(1)} = E(Z_{i:n}^{*0} Z_{i+1:n}^{*})$$
$$= \frac{n!}{(i-1)!(n-i-1)!} \int_{Q_1}^{P_1} z_2$$
$$\times \{1 - F^{**}(z_2)\}^{n-i-1} f^{**}(z_2) K(z_2)\, dz_2, \qquad (2.6.38)$$

where

$$K(z_2) = \int_{Q_1}^{z_2} \{F^{**}(z_1)\}^{i-1} f^{**}(z_1)\, dz_1. \qquad (2.6.39)$$

Upon using (2.6.5) in (2.6.39) and integrating by parts, we get

$$K(z_2) = Q_2\left[z_2\{F^{**}(z_2)\}^{i-1} - (i-1)\int_{Q_1}^{z_2} z_1\{F^{**}(z_1)\}^{i-2} f^{**}(z_1)\, dz_1\right]$$
$$+ (1-2Q)\left[z_2\{F^{**}(z_2)\}^{i} - i\int_{Q_1}^{z_2} z_1\{F^{**}(z_1)\}^{i-1} f^{**}(z_1)\, dz_1\right]$$
$$- (P-Q)\left[z_2\{F^{**}(z_2)\}^{i+1} - (i+1)\int_{Q_1}^{z_2} z_1\right.$$
$$\left.\times \{F^{**}(z_1)\}^{i} f^{**}(z_1)\, dz_1\right].$$

Upon substituting the above expression of $K(z_2)$ in (2.6.38) and simplifying the resulting equation, we derive the recurrence relation in (2.6.37).

In particular, by setting $i = n-1$ in (2.6.37), we obtain the recurrence relation

$$\alpha_{n,n+1:n+1}^{**} = \alpha_{n+1:n+1}^{**(2)} + \frac{n+1}{n(P-Q)}$$
$$\times \left[\frac{1}{n-1}\alpha_{n:n}^{**(1)} - \frac{nQ_2}{n-1}\{\alpha_{n-1:n-1}^{**(2)} - \alpha_{n-2,n-1:n-1}^{**}\}\right.$$
$$\left. - (1-2Q)\{\alpha_{n:n}^{**(2)} - \alpha_{n-1,n:n}^{**}\}\right], \qquad n \ge 3. \qquad (2.6.40)$$

By starting with the result that $\alpha_{1,2:2}^{**} = \alpha_{1:1}^{**2}$ (Govindarajulu, 1963; Joshi, 1971), one may employ Relations 2.6.8–2.6.13 to compute all the immediate upper-diagonal product moments, viz., $\alpha_{i,i+1:n}^{**}$ $(1 \le i \le n-1)$, in a simple recursive way for all sample sizes. As mentioned in Section 2.4, this is sufficient for the evaluation of all the product moments, as the remaining product moments, viz., $\alpha_{i,j:n}^{**}$ for $1 \le i < j \le n$ and $j - i \ge 2$, may be determined by using the recurrence relation in (2.4.13). However, for the sake of completeness we present here some more recurrence relations satisfied by the general product moments. These results may be established by following exactly the same steps as used in proving Relations 2.6.8–2.6.13.

Relation 2.6.14 For $1 \le i \le n - 2$,

$$\alpha_{i,n:n+1}^{**} = \alpha_{i,n-1:n+1}^{**} + \frac{n+1}{2(P-Q)} [nP_2\{P_1\alpha_{i:n-1}^{**(1)} - \alpha_{i,n-1:n-1}^{**}\} + (2P-1)\{\alpha_{i,n:n}^{**} - \alpha_{i,n-1:n}^{**}\} - \alpha_{i:n}^{**(1)}]. \tag{2.6.41}$$

Relation 2.6.15 For $1 \le i < j \le n - 1$ and $j - i \ge 2$,

$$\alpha_{i,j:n+1}^{**} = \alpha_{i,j-1:n+1}^{**} + \frac{n+1}{(n-j+2)(P-Q)} \times \left[\frac{nP_2}{n-j+1}\{\alpha_{i,j:n-1}^{**} - \alpha_{i,j-1:n-1}^{**}\} + (2P-1) \times \{\alpha_{i,j:n}^{**} - \alpha_{i,j-1:n}^{**}\} - \frac{1}{n-j+1}\alpha_{i:n}^{**(1)}\right]. \tag{2.6.42}$$

Relation 2.6.16 For $3 \le j \le n$,

$$\alpha_{2,j+1:n+1}^{**} = \alpha_{3,j+1:n+1}^{**} + \frac{n+1}{2(P-Q)} \times [\alpha_{j:n}^{**(1)} - nQ_2\{\alpha_{1,j-1:n-1}^{**} - Q_1\alpha_{j-1:n-1}^{**(1)}\} - (1-2Q)\{\alpha_{2,j:n}^{**} - \alpha_{1,j:n}^{**}\}]. \tag{2.6.43}$$

Relation 2.6.17 For $2 \le i < j \le n$ and $j - i \ge 2$,

$$\alpha_{i+1,j+1:n+1}^{**} = \alpha_{i+2,j+1:n+1}^{**} + \frac{n+1}{(i+1)(P-Q)} \times \left[\frac{1}{i}\alpha_{j:n}^{**(1)} - \frac{nQ_2}{i}\{\alpha_{i,j-1:n-1}^{**} - \alpha_{i-1,j-1:n-1}^{**}\} - (1-2Q)\{\alpha_{i+1,j:n}^{**} - \alpha_{i,j:n}^{**}\}\right]. \tag{2.6.44}$$

2.7 DETAILS OF AVAILABLE TABLES

We list below the tables that are currently available on order statistics and their moments.

a. Table of 100α% points (for $\alpha = 0.50, 0.75, 0.90, 0.95, 0.975, 0.990$) for all order statistics for sample sizes up to 10 and for extreme and central order statistics for sample sizes from 11 to 25 has been given by Gupta and Shah (1965).
b. Table of probability integrals of the sample range W_n for $n = 2$ and 3 evaluated at $w = 0.20(0.20)1.00(0.50)4.00$ has been given by Gupta and Shah (1965).
c. Table of means and standard deviations of order statistics for sample sizes up to 10 has been given by Birnbaum and Dudman (1963). Table of covariances of order statistics for sample sizes up to 10 has been given by Shah (1966). These two tables have been extended by Gupta et al. (1967) for sample sizes up to 25. Recently, Balakrishnan and Malik (1991) have prepared tables of means, variances, and covariances for sample sizes 50 and less in which the values are reported to 10 decimal places.
d. By using the results presented in Section 2.6, Balakrishnan and Joshi (1983b) have given tables of means, variances, and covariances for the symmetrically truncated logistic distribution (with $Q = 1 - P = 0.01, 0.05(0.05)0.20$) for sample sizes up to 10.
e. Balakrishnan (1985) has handled the half logistic distribution (case when $Q = 1/2$ and $P = 1$) and has presented tables of means, variances, and covariances for sample sizes up to 15. He has also given tables of 100α% points (for $\alpha = 0.01, 0.05, 0.10(0.10)0.90, 0.95, 0.99$) for extreme order statistics for sample sizes up to 15. In addition, he has presented a table of modes of all order statistics for sample sizes up to 15.

2.8 PLACKETT'S APPROXIMATION

David and Johnson (1954) and Clark and Williams (1958) have developed some series approximations for moments of order statistics from an arbitrary continuous distribution. These have been developed by applying the probability integral transformation and then using the known moments of order statistics from the uniform distribution. Plackett (1958), instead, has used the logit transformation which transforms an order statistic $T_{i:n}$ from an arbitrary continuous distribution into the order statistic $Z_{i:n}$ from the logistic $L(0, \pi^2/3)$ distribution to develop some series approximations for the moments of $T_{i:n}$ in terms of the moments of logistic order statistics $Z_{i:n}$.

We have already seen in Section 2.3 that the moments and the cumulants of the logistic order statistics $Z_{i:n}$ are all available in explicit form. Now, by realizing that the logit transformation

$$Z = \ln\left\{\frac{F_T(t)}{1 - F_T(t)}\right\} \tag{2.8.1}$$

transforms the order statistic $T_{i:n}$ from an arbitrary continuous distribution with cdf $F_T(t)$ into the logistic order statistic $Z_{i:n}$ and, therefore, expanding $T_{i:n}$ in a Taylor series about the point $E(Z_{i:n}) = \kappa_{i:n}^{*(1)}$, we derive

$$\begin{aligned} T_{i:n} &= t^{(0)} + t^{(1)}(Z_{i:n} - \kappa_{i:n}^{*(1)}) + \frac{1}{2}t^{(2)} \\ &\times (Z_{i:n} - \kappa_{i:n}^{*(1)})^2 + \frac{1}{6}t^{(3)}(Z_{i:n} - \kappa_{i:n}^{*(1)})^3 \\ &+ \frac{1}{24}t^{(4)}(Z_{i:n} - \kappa_{i:n}^{*(1)})^4 + \cdots, \end{aligned} \tag{2.8.2}$$

where, for $n - i + 1 > i$,

$$\kappa_{i:n}^{*(1)} = -\left\{\frac{1}{i} + \frac{1}{i+1} + \cdots + \frac{1}{n-i}\right\}$$

as derived in Section 2.3, and $t^{(j)}$ is the value of the jth derivative of t with respect to Z at $Z = \kappa_{i:n}^{*(1)}$. Now, by taking expectation on both sides of (2.8.2) and upon using the exact and explicit expressions for the cumulants of logistic order statistics derived in Section 2.3, we obtain the series approximation

$$\begin{aligned} E(T_{i:n}) &\simeq t^{(0)} + \frac{1}{2}t^{(2)}\kappa_{i:n}^{*(2)} + \frac{1}{6}t^{(3)}\kappa_{i:n}^{*(3)} \\ &+ \frac{1}{24}t^{(4)}\{\kappa_{i:n}^{*(4)} + 3(\kappa_{i:n}^{*(2)})^2\}. \end{aligned} \tag{2.8.3}$$

The derivatives appearing as coefficients in the approximation in (2.8.3) are easy to obtain as in the case of approximations due to David and Johnson (1954) and Clark and Williams (1958). For example, for the standard normal distribution with probability density function $\phi(t)$ and cumulative distribution function $\Phi(t)$, we have

$$t^{(0)} = \Phi^{-1}\left\{\exp\left[\frac{\kappa_{i:n}^{*(1)}}{1 + \kappa_{i:n}^{*(1)}}\right]\right\},$$

$$t^{(1)} = \frac{\Phi(1 - \Phi)}{\phi},$$
$$t^{(2)} = t^{(1)}\{tt^{(1)} - (2\Phi - 1)\},$$
$$t^{(3)} = (t^{(1)})^3 + 2tt^{(1)}t^{(2)} + t^{(2)}(1 - 2\Phi) - 2t^{(1)}\Phi(1 - \Phi),$$

and

$$t^{(4)} = 5(t^{(1)})^2 t^{(2)} + t^{(3)}\{2tt^{(1)} - (2\Phi - 1)\}$$
$$+ 2t^{(2)}\{tt^{(2)} - 2\Phi(1 - \Phi)\} + 2t^{(1)}(2\Phi - 1)\Phi(1 - \Phi).$$

The above given derivatives are all bounded. As pointed out by Blom (1958), suppose we include the first $j - 1$ terms in the series expansion for $E(T_{i:n})$ obtained from (2.8.2), then the absolute value of the remainder after $j - 1$ terms is at most

$$\frac{1}{j!} \max|t^{(j)}| \, E|Z_{i:n} - \kappa_{i:n}^{*(1)}|^j. \tag{2.8.4}$$

Since $E|Z_{i:n} - \kappa_{i:n}^{*(1)}|^{2j}$ is known and

$$\{E|Z_{i:n} - \kappa_{i:n}^{*(1)}|^{2j-1}\}^{1/(2j-1)} \le \{E|Z_{i:n} - \kappa_{i:n}^{*(1)}|^{2j}\}^{1/2j},$$

we will be able to present bounds to $E(T_{i:n})$ for all values of j.

REFERENCES

Abramowitz, M. and Stegun, I. A. (eds.) (1965). *Handbook of Mathematical Functions with Formulas, Graphs, and Mathematical Tables*. Dover, New York.

Arnold, B. C. and Balakrishnan, N. (1989). *Relations, Bounds and Approximations for Order Statistics*, Lecture Notes in Statistics No. 53, Springer-Verlag, New York.

Balakrishnan, N. (1985). Order statistics from the half logistic distribution. *J. Statist. Comput. Simul.*, *20*, 287–309.

Balakrishnan, N. and Joshi, P. C. (1983a). Single and product moments of order statistics from symmetrically truncated logistic distribution. *Demonstratio Math.*, *16*, 833–841.

Balakrishnan, N. and Joshi, P. C. (1983b). Means, variances and covariances of order statistics from symmetrically truncated logistic distribution. *J. Statist. Res.*, *17*, 51–61.

Balakrishnan, N. and Kocherlakota, S. (1986). On the moments of order statistics from the doubly truncated logistic distribution. *J. Statist. Plann. Inf.*, *13*, 117–129.

Balakrishnan, N. and Leung, M. Y. (1988a). Order statistics from the Type I generalized logistic distribution. *Commun. Statist.—Simul. Comput.*, *17*(1), 25–50.

Balakrishnan, N. and Leung, M. Y. (1988b). Means, variances and covariances of order statistics, BLUE's for the Type I generalized logistic distribution, and some applications. *Commun. Statist.—Simul. Comput.*, *17*(1), 51–84.

Balakrishnan, N. and Malik, H. J. (1986). A note on moments of order statistics. *Amer. Statist.*, *40*, 147–148.

Balakrishnan, N. and Malik, H. J. (1991). Means, variances, and covariances of logistic order statistics for sample sizes up to fifty. To appear in *Selected Tables in Mathematical Statistics*.

Bernardo, J. M. (1976). Psi (digamma) function, Algorithm AS 103. *Appl. Statist.*, *25*, 315–317.

Birnbaum, A. and Dudman, J. (1963). Logistic order statistics. *Ann. Math. Statist.*, *34*, 658–663.

Blom, G. (1958). *Statistical Estimates and Transformed Beta-Variables*. Almqvist and Wiksell, Uppsala, Sweden.

Clark, C. E. and Williams, G. T. (1958). Distribution of members of an ordered sample. *Ann. Math. Statist.*, *29*, 862–870.

Cran, G. W., Martin, K. J., and Thomas, G. E. (1977). A remark on algorithms, Algorithm AS 109. *Appl. Statist.*, *26*, 111–114.

David, F. N. and Johnson, N. L. (1954). Statistical treatment of censored data. I: Fundamental formulae. *Biometrika*, *41*, 228–240.

David, H. A. (1981). *Order Statistics*. Sec. ed., Wiley, New York.

Davis, H. T. (1935). *Tables of the Higher Mathematical Functions*, Vols. 1, 2. Principia Press, Bloomington.

George, E. O. and Rousseau, C. C. (1987). On the logistic midrange. *Ann. Inst. Statist. Math.*, *39*, 627–635.

Govindarajulu, Z. (1963). On moments of order statistics and quasi-ranges from normal populations. *Ann. Math. Statist.*, *34*, 633–651.

Gumbel, E. J. (1944). Ranges and midranges. *Ann. Math. Statist.*, *15*, 414–422.

Gumbel, E. J. (1958). *Statistics of Extremes*. Columbia University Press, New York.

Gupta, S. S., Qureishi, A. S., and Shah, B. K. (1967). Best linear unbiased estimators of the parameters of the logistic distribution using order statistics. *Technometrics*, *9*, 43–56.

Gupta, S. S. and Shah, B. K. (1965). Exact moments and percentage points of the order statistics and the distribution of the range from the logistic distribution. *Ann. Math. Statist.*, *36*, 907–920.

Joshi, P. C. (1971). Recurrence relations for the mixed moments of order statistics. *Ann. Math. Statist.*, *42*, 1096–1098.

Kjelsberg, M. O. (1962). *Estimation of the parameters of the logistic distribution under truncation and censoring*, Doctoral Thesis, University of Minnesota.

Malik, H. J. (1980). Exact formula for the cumulative distribution function of the quasi-range from the logistic distribution. *Commun. Statist.—Theor. Meth.*, *A9*(14), 1527–1534.

Malik, H. J. (1985). Logistic distribution, in *Encyclopedia of Statistics*, S. Kotz and N. L. Johnson, (eds.), Wiley, New York.

Pearson, E. S. and Hartley, H. O. (1970). *Bometrika Tables for Statisticians*, Vol. I, 3rd ed., Cambridge University Press, England.

Pearson, K. (1934). *Tables of the Incomplete B-Function*. Cambridge University Press, England.

Plackett, R. L. (1958). Linear estimation from censored data. *Ann. Math. Statist.*, *29*, 131–142.

Schneider, B. E. (1978). Trigamma function, Algorithm AS 121. *Appl. Statist.*, *27*, 97–99.

Shah, B. K. (1966). On the bivariate moments of order statistics from a logistic distribution. *Ann. Math. Statist.*, *37*, 1002–1010.

Shah, B. K. (1970). Note on moments of a logistic order statistics. *Ann. Math. Statist.*, *41*, 2151–2152.

Tarter, M. E. (1966). Exact moments and product moments of the order statistics from the truncated logistic distribution. *J. Amer. Statist. Assoc.*, *61*, 514–525.

Tarter, M. E. and Clark, V. A. (1965). Properties of the median and other order statistics of logistic variates, *Ann. Math. Statist.*, *36*, 1779–1786. Correction *8*, 935.

3

Maximum Likelihood Estimation Based on Complete and Type II Censored Samples

N. Balakrishnan
McMaster University, Hamilton, Ontario, Canada

3.1 INTRODUCTION

In the last chapter order statistics from the logistic distribution and their properties have been discussed in great detail. In this chapter we consider a doubly Type II censored sample from the logistic $L(\mu, \sigma^2)$ distribution and explain in Section 2 the maximum likelihood estimation of μ and σ as developed by Harter and Moore (1967). The bias and mean square error of these estimators, determined by Monte Carlo simulations, are presented for sample sizes 10 and 20 and over various choices of censoring. The expressions of the asymptotic variances and covariance of the estimators derived by Harter and Moore (1967) are given, and the asymptotic values of n Var$(\hat{\mu})/\sigma^2$, n Var$(\hat{\sigma})/\sigma^2$, n Cov$(\hat{\mu}, \hat{\sigma})/\sigma^2$, n Var$(\hat{\mu}|\sigma)/\sigma^2$, and n Var$(\hat{\sigma}|\mu)/\sigma^2$, determined from these formulae are also presented for various proportions of censoring. In Section 3 we describe the approximate maximum likelihood estimators of μ and σ derived by Balakrishnan and Cohen (1990) based on doubly Type II censored samples. We present the values of the bias and the variance of these estimators, determined by Monte Carlo simulations, for sample sizes 10 and 20 and over various choices of censoring. The expressions of the asymptotic variances and covariance of the estimators are obtained; these expressions are then used to compute the asymptotic values of n Var$(\tilde{\mu})/\sigma^2$, n Var$(\tilde{\sigma})/\sigma^2$, n Cov$(\tilde{\mu}$,

$\tilde{\sigma})/\sigma^2$, n Var$(\tilde{\mu}|\sigma)/\sigma^2$, and n Var$(\tilde{\sigma}|\mu)/\sigma^2$, for various proportions of censoring. By comparing these values with the corresponding values of the maximum likelihood estimators given in Section 2, we show that the approximate maximum likelihood estimators $\tilde{\mu}$ and $\tilde{\sigma}$ are jointly as efficient as the maximum likelihood estimators $\hat{\mu}$ and $\hat{\sigma}$. Examples to illustrate the two methods of estimation are also presented.

3.2 MAXIMUM LIKELIHOOD ESTIMATORS OF μ and σ

Let

$$X_{r+1:n} \le X_{r+2:n} \le \cdots \le X_{n-s:n} \tag{3.2.1}$$

be a doubly Type II censored sample from the logistic $L(\mu, \sigma^2)$ population with pdf $f(x; \mu, \sigma)$ and cdf $F(x; \mu, \sigma)$ as given in Eqs. (1.1) and (1.2), respectively; here, the smallest r and the largest s observations in a sample of size n have been censored. Harter and Moore (1967) have studied the maximum likelihood estimators, $\hat{\mu}$ and $\hat{\sigma}$, of μ and σ based on the doubly Type II censored sample in (3.2.1). Their results are described in this section.

Let $Z = \pi(X - \mu)/\sqrt{3}\sigma$ be the logistic $L(0, \pi^2/3)$ variable with pdf $f^*(z)$ and cdf $F^*(z)$ as given in Eqs. (1.5) and (1.6), respectively. Then

$$Z_{r+1:n} \le Z_{r+2:n} \le \cdots \le Z_{n-s:n} \tag{3.2.2}$$

is the Type II censored sample corresponding to the censored sample in (3.2.1). The likelihood function, based on the censored sample in (3.2.1), is given by

$$L = \frac{n!}{r!s!}\{F(X_{r+1:n}; \mu, \sigma)\}^r\{1 - F(X_{n-s:n}; \mu, \sigma)\}^s \times \prod_{i=r+1}^{n-s} f(X_{i:n}; \mu, \sigma). \tag{3.2.3}$$

Upon using Eqs. (1.1) and (1.2) and the relation

$$F^*(z)\{1 - F^*(z)\} = f^*(z) \tag{3.2.4}$$

in (3.2.3), we may write the log-likelihood function as

$$\ln L = \ln n! - \ln r! - \ln s! + (n - r - s)\ln\left(\frac{\pi}{\sqrt{3}\sigma}\right) + r\ln F^*(Z_{r+1:n}) + s\ln\{1 - F^*(Z_{n-s:n})\} - \sum_{i=r+1}^{n-s} Z_{i:n} + 2\sum_{i=r+1}^{n-s} \ln F^*(Z_{i:n}). \tag{3.2.5}$$

From (3.2.5) we obtain the maximum likelihood equations for μ and σ

$$\frac{\partial \ln L}{\partial \mu} = \frac{\pi}{\sqrt{3}\sigma}\left[-r\frac{f^*(Z_{r+1:n})}{F^*(Z_{r+1:n})} + s\frac{f^*(Z_{n-s:n})}{1 - F^*(Z_{n-s:n})} + (n - r - s) - 2\sum_{i=r+1}^{n-s}\frac{f^*(Z_{i:n})}{F^*(Z_{i:n})}\right] = 0 \qquad (3.2.6)$$

and

$$\frac{\partial \ln L}{\partial \sigma} = \frac{1}{\sigma}\left[-(n - r - s) - rZ_{r+1:n}\frac{f^*(Z_{r+1:n})}{F^*(Z_{r+1:n})} + sZ_{n-s:n}\frac{f^*(Z_{n-s:n})}{1 - F^*(Z_{n-s:n})} + \sum_{i=r+1}^{n-s} Z_{i:n} - 2\sum_{i=r+1}^{n-s} Z_{i:n}\frac{f^*(Z_{i:n})}{F^*(Z_{i:n})}\right] = 0. \qquad (3.2.7)$$

From (3.2.5) we also obtain the second partial derivatives of ln L with respect to the parameters μ and σ as

$$\frac{\partial^2 \ln L}{\partial \mu^2} = \frac{\pi^2}{3\sigma^2}\left[r\frac{f^{*\prime}(Z_{r+1:n})}{F^*(Z_{r+1:n})} - r\left\{\frac{f^*(Z_{r+1:n})}{F^*(Z_{r+1:n})}\right\}^2 - s\frac{f^{*\prime}(Z_{n-s:n})}{1 - F^*(Z_{n-s:n})} - s\left\{\frac{f^*(Z_{n-s:n})}{1 - F^*(Z_{n-s:n})}\right\}^2 + 2\sum_{i=r+1}^{n-s}\frac{f^{*\prime}(Z_{i:n})}{F^*(Z_{i:n})} - 2\sum_{i=r+1}^{n-s}\left\{\frac{f^*(Z_{i:n})}{F^*(Z_{i:n})}\right\}^2\right], \qquad (3.2.8)$$

$$\frac{\partial^2 \ln L}{\partial \mu\, \partial \sigma} = \frac{\pi}{\sqrt{3}\sigma^2}\left[rZ_{r+1:n}\frac{f^{*\prime}(Z_{r+1:n})}{F^*(Z_{r+1:n})} + r\frac{f^*(Z_{r+1:n})}{F^*(Z_{r+1:n})} - rZ_{r+1:n}\left\{\frac{f^*(Z_{r+1:n})}{F^*(Z_{r+1:n})}\right\}^2 - sZ_{n-s:n}\frac{f^{*\prime}(Z_{n-s:n})}{1 - F^*(Z_{n-s:n})} - s\frac{f^*(Z_{n-s:n})}{1 - F^*(Z_{n-s:n})} - sZ_{n-s:n}\left\{\frac{f^*(Z_{n-s:n})}{1 - F^*(Z_{n-s:n})}\right\}^2 - (n - r - s) + 2\sum_{i=r+1}^{n-s} Z_{i:n}\frac{f^{*\prime}(Z_{i:n})}{F^*(Z_{i:n})} + 2\sum_{i=r+1}^{n-s}\frac{f^*(Z_{i:n})}{F^*(Z_{i:n})} - 2\sum_{i=r+1}^{n-s} Z_{i:n}\left\{\frac{f^*(Z_{i:n})}{F^*(Z_{i:n})}\right\}^2\right], \qquad (3.2.9)$$

and

$$\frac{\partial^2 \ln L}{\partial \sigma^2} = \frac{1}{\sigma^2}\Bigg[(n - r - s) + rZ_{r+1:n}^2 \times \frac{f^{*\prime}(Z_{r+1:n})}{F^*(Z_{r+1:n})} + 2rZ_{r+1:n}\frac{f^*(Z_{r+1:n})}{F^*(Z_{r+1:n})} - r\left\{Z_{r+1:n}\frac{f^*(Z_{r+1:n})}{F^*(Z_{r+1:n})}\right\}^2 - sZ_{n-s:n}^2\frac{f^{*\prime}(Z_{n-s:n})}{1 - F^*(Z_{n-s:n})} - 2sZ_{n-s:n}\frac{f^*(Z_{n-s:n})}{1 - F^*(Z_{n-s:n})} - s\left\{Z_{n-s:n}\frac{f^*(Z_{n-s:n})}{1 - F^*(Z_{n-s:n})}\right\}^2 - 2\sum_{i=r+1}^{n-s} Z_{i:n} + 2\sum_{i=r+1}^{n-s} Z_{i:n}^2\frac{f^{*\prime}(Z_{i:n})}{F^*(Z_{i:n})} + 4\sum_{i=r+1}^{n-s} Z_{i:n} \times \frac{f^*(Z_{i:n})}{F^*(Z_{i:n})} - 2\sum_{i=r+1}^{n-s}\left\{Z_{i:n}\frac{f^*(Z_{i:n})}{F^*(Z_{i:n})}\right\}^2\Bigg], \tag{3.2.10}$$

where $f^{*\prime}(z) = (d/dz)f^*(z)$.

Table 3.2.1 Simulated Values of $\text{Bias}(\hat{\mu})/\sigma$, $\text{Bias}(\hat{\sigma})/\sigma$, $\text{Var}(\hat{\mu})/\sigma^2$, $\text{Var}(\hat{\sigma})/\sigma^2$, $\text{Cov}(\hat{\mu}, \hat{\sigma})/\sigma^2$, $\text{Var}(\hat{\mu}|\sigma)/\sigma^2$, and $\text{Var}(\hat{\sigma}|\mu)/\sigma^2$ When $n = 10$

p_r	p_s	$\text{Bias}(\hat{\mu})/\sigma$	$\text{Bias}(\hat{\sigma})/\sigma$	$\text{Var}(\hat{\mu})/\sigma^2$	$\text{Var}(\hat{\sigma})/\sigma^2$	$\text{Cov}(\hat{\mu},\hat{\sigma})/\sigma^2$	$\text{Var}(\hat{\mu}\mid\sigma)/\sigma^2$	$\text{Var}(\hat{\sigma}\mid\mu)/\sigma^2$
0.0	0.0	0.00	− 0.06	0.092	0.071	− 0.001	0.092	0.073
0.0	0.1	− 0.00	− 0.07	0.093	0.079	− 0.000	0.092	0.082
0.0	0.2	− 0.01	− 0.08	0.094	0.085	0.002	0.094	0.090
0.0	0.3	− 0.02	− 0.10	0.097	0.095	0.008	0.096	0.101
0.0	0.4	− 0.05	− 0.13	0.105	0.106	0.017	0.102	0.113
0.0	0.5	− 0.08	− 0.16	0.113	0.121	0.028	0.107	0.126
0.0	0.6	− 0.14	− 0.22	0.147	0.146	0.056	0.126	0.142
0.0	0.7	− 0.26	− 0.31	0.231	0.184	0.110	0.163	0.153
0.0	0.8	− 0.54	− 0.48	0.394	0.224	0.194	0.228	0.167
0.1	0.1	− 0.00	− 0.08	0.093	0.090	0.000	0.092	0.094
0.1	0.2	− 0.01	− 0.10	0.094	0.099	0.003	0.094	0.105
0.1	0.3	− 0.02	− 0.12	0.097	0.113	0.009	0.096	0.120
0.1	0.4	− 0.05	− 0.16	0.105	0.126	0.022	0.102	0.135
0.1	0.5	− 0.09	− 0.21	0.115	0.153	0.034	0.107	0.157
0.1	0.6	− 0.17	− 0.30	0.153	0.187	0.071	0.126	0.179
0.1	0.7	− 0.35	− 0.48	0.242	0.221	0.132	0.163	0.196
0.2	0.2	0.00	− 0.12	0.096	0.111	− 0.001	0.096	0.120
0.2	0.3	− 0.01	− 0.16	0.098	0.130	0.005	0.098	0.142
0.2	0.4	− 0.04	− 0.22	0.106	0.149	0.017	0.104	0.168
0.2	0.5	− 0.10	− 0.30	0.114	0.189	0.035	0.108	0.203
0.2	0.6	− 0.22	− 0.48	0.149	0.215	0.068	0.128	0.241
0.3	0.3	0.00	− 0.21	0.099	0.150	0.001	0.099	0.170
0.3	0.4	− 0.03	− 0.31	0.106	0.173	0.016	0.105	0.210
0.3	0.5	− 0.10	− 0.48	0.114	0.204	0.028	0.110	0.262
0.4	0.4	0.00	− 0.50	0.108	0.210	0.010	0.108	0.299

The maximum likelihood equations in (3.2.6) and (3.2.7) do not admit explicit solutions. But, the maximum likelihood estimates of μ and σ may be obtained by solving these two equations iteratively. By generating one thousand pseudorandom samples of size $n = 10$ and $n = 20$ from the logistic $L(0, 1)$ population and then employing the rule of false position (iterative linear interpolation) to solve the likelihood equations in (3.2.6) and (3.2.7), Harter and Moore (1967) determined the bias, variances, covariance, and conditional variances of the maximum likelihood estimates $\hat{\mu}$ and $\hat{\sigma}$. They determined these values for various proportions of censoring $p_r = r/n$ and $p_s = s/n$. These tables have also been reproduced by Harter (1970). The values of Bias($\hat{\mu}$)/σ, Bias($\hat{\sigma}$)/σ, Var($\hat{\mu}$)/σ^2, Var($\hat{\sigma}$)/σ^2, Cov($\hat{\mu}$, $\hat{\sigma}$)/σ^2, Var($\hat{\mu}|\sigma$)/σ^2, and Var($\hat{\sigma}|\mu$)/σ^2, have been taken from their tables for $n = 10$ and $n = 20$ for various choices of censoring and presented in Tables 3.2.1 and 3.2.2, respectively.

Table 3.2.2 Simulated Values of Bias($\hat{\mu}$)/σ, Bias($\hat{\sigma}$)/σ, Var($\hat{\mu}$)/σ^2, Var($\hat{\sigma}$)/σ^2, Cov($\hat{\mu}$, $\hat{\sigma}$)/σ^2, Var($\hat{\mu}|\sigma$)/σ^2, and Var($\hat{\sigma}|\mu$)/σ^2 When $n = 20$

p_r	p_s	Bias($\hat{\mu}$)/σ	Bias($\hat{\sigma}$)/σ	Var($\hat{\mu}$)/σ^2	Var($\hat{\sigma}$)/σ^2	Cov($\hat{\mu},\hat{\sigma}$)/σ^2	Var($\hat{\mu}\|\sigma$)/σ^2	Var($\hat{\sigma}\|\mu$)/σ^2
0.0	0.0	0.00	− 0.03	0.047	0.035	− 0.001	0.047	0.035
0.0	0.1	0.00	− 0.04	0.047	0.038	− 0.001	0.047	0.038
0.0	0.2	− 0.00	− 0.04	0.048	0.043	0.001	0.047	0.044
0.0	0.3	− 0.01	− 0.05	0.048	0.050	0.004	0.048	0.052
0.0	0.4	− 0.02	− 0.06	0.052	0.061	0.009	0.051	0.061
0.0	0.5	− 0.03	− 0.07	0.060	0.077	0.020	0.055	0.074
0.0	0.6	− 0.06	− 0.10	0.083	0.097	0.042	0.065	0.082
0.0	0.7	− 0.12	− 0.14	0.125	0.131	0.078	0.078	0.088
0.0	0.8	− 0.27	− 0.24	0.243	0.183	0.157	0.111	0.091
0.0	0.9	− 0.72	− 0.47	0.740	0.323	0.411	0.214	0.100
0.1	0.1	0.00	− 0.04	0.047	0.041	− 0.001	0.047	0.042
0.1	0.2	− 0.00	− 0.05	0.048	0.047	0.000	0.047	0.049
0.1	0.3	− 0.01	− 0.06	0.048	0.057	0.003	0.048	0.059
0.1	0.4	− 0.02	− 0.07	0.052	0.070	0.009	0.051	0.071
0.1	0.5	− 0.03	− 0.09	0.060	0.091	0.021	0.055	0.087
0.1	0.6	− 0.08	− 0.14	0.085	0.119	0.047	0.065	0.100
0.1	0.7	− 0.16	− 0.22	0.134	0.172	0.099	0.078	0.110
0.1	0.8	− 0.47	− 0.50	0.264	0.232	0.188	0.111	0.114
0.2	0.2	0.00	− 0.06	0.048	0.055	− 0.001	0.048	0.056
0.2	0.3	− 0.00	− 0.08	0.049	0.067	− 0.001	0.049	0.069
0.2	0.4	− 0.02	− 0.10	0.052	0.087	0.007	0.051	0.087
0.2	0.5	− 0.04	− 0.13	0.060	0.115	0.022	0.056	0.111
0.2	0.6	− 0.10	− 0.22	0.086	0.157	0.057	0.066	0.133
0.2	0.7	− 0.28	− 0.47	0.143	0.229	0.120	0.079	0.152
0.3	0.3	0.00	− 0.11	0.049	0.082	− 0.002	0.049	0.085
0.3	0.4	− 0.01	− 0.15	0.052	0.113	0.006	0.052	0.114
0.3	0.5	− 0.04	− 0.23	0.060	0.162	0.025	0.056	0.160
0.3	0.6	− 0.16	− 0.49	0.090	0.218	0.072	0.067	0.218
0.4	0.4	0.01	− 0.25	0.053	0.165	− 0.003	0.053	0.173
0.4	0.5	− 0.04	− 0.50	0.060	0.243	0.019	0.058	0.292

By noting that as $n \to \infty$ (with p_r and p_s fixed), $Z_{r+1:n} \to \delta_1$ and $Z_{n-s:n} \to \delta_2$ where $\delta_1 = F^{*-1}(p_r)$ and $\delta_2 = F^{*-1}(1 - p_s)$, and expressing the limits (as $n \to \infty$) of the expected values of the sums in Eqs. (3.2.8)–(3.2.10) as integrals between the limits δ_1 and δ_2, and then performing the necessary integrations, Harter and Moore (1967) and Harter (1970) have shown that

$$
\begin{aligned}
V_1 &= \lim_{n\to\infty} E\left(-\frac{\sigma^2}{n}\frac{\partial^2 \ln L}{\partial \mu^2}\right) \\
&= \frac{\pi^2}{3}\left\{\left[\frac{2}{3(1+e^t)^3} - \frac{1}{(1+e^t)^2}\right]_{\delta_1}^{\delta_2}\right. \\
&\quad \left. - f^{*\prime}(\delta_1) + \frac{1}{p_r} f^{*2}(\delta_1) + f^{*\prime}(\delta_2) + \frac{1}{p_s} f^{*2}(\delta_2)\right\}, \qquad (3.2.11)
\end{aligned}
$$

$$
\begin{aligned}
V_2 &= \lim_{n\to\infty} E\left(-\frac{\sigma^2}{n}\frac{\partial^2 \ln L}{\partial \mu\, \partial \sigma}\right) \\
&= \frac{\pi}{\sqrt{3}}\left\{(1 - p_r - p_s) - \frac{1}{3}\left[\frac{t}{1+e^t} + \ln(1+e^{-t})\right]_0^{\delta_2}\right. \\
&\quad + \frac{1}{3}\left[\frac{t}{1+e^{-t}} - \ln(1+e^t)\right]_{\delta_1}^{0} - \frac{1}{3}\left[\frac{2te^{-2t}}{(1+e^{-t})^3} - \frac{te^{-t}}{(1+e^{-t})^2}\right. \\
&\quad \left. + \frac{2e^{-t}}{(1+e^{-t})^2} + \frac{3}{1+e^{-t}}\right]_{\delta_1}^{\delta_2} - f^*(\delta_1) - \delta_1 f^{*\prime}(\delta_1) + \frac{\delta_1}{p_r} f^{*2}(\delta_1) \\
&\quad \left. + f^*(\delta_2) + \delta_2 f^{*\prime}(\delta_2) + \frac{\delta_2}{p_s} f^{*2}(\delta_2)\right\}, \qquad (3.2.12)
\end{aligned}
$$

and

$$
\begin{aligned}
V_3 &= \lim_{n\to\infty} E\left(-\frac{\sigma^2}{n}\frac{\partial^2 \ln L}{\partial \sigma^2}\right) \\
&= -(1 - p_r - p_s) - \frac{1}{3}\left[\frac{t^2}{1+e^t} + 2t\ln(1+e^{-t})\right]_0^{\delta_2} \\
&\quad + \frac{1}{3}\left[\frac{t^2}{1+e^{-t}} - 2t\ln(1+e^t)\right]_{\delta_1}^{0} + \frac{2}{3}(I_1 - I_2) \\
&\quad - \frac{1}{3}\left[\frac{2t^2e^{-2t}}{(1+e^{-t})^3} - \frac{t^2e^{-t}}{(1+e^{-t})^2}\right. \\
&\quad \left. + \frac{4te^{-t}}{(1+e^{-t})^2} - \frac{4}{1+e^{-t}}\right]_{\delta_1}^{\delta_2}
\end{aligned}
$$

$$- \delta_1^2 f^{*\prime}(\delta_1) - 2\delta_1 f^*(\delta_1) + \frac{\delta_1^2}{p_r} f^{*2}(\delta_1) + \delta_2^2 f^{*\prime}(\delta_2)$$

$$+ 2\delta_2 f^*(\delta_2) + \frac{\delta_2^2}{p_s} f^{*2}(\delta_2), \tag{3.2.13}$$

where I_1 and I_2 are definite integrals which, upon expanding the integrands in infinite series and then integrating term by term, become

$$I_1 = \begin{cases} \left[e^t - \frac{1}{2^2} e^{2t} + \frac{1}{3^2} e^{3t} - \frac{1}{4^2} e^{4t} + \cdots \right]_{\delta_1}^{0} & \text{if } \delta_1 < 0, \\ 0 & \text{if } \delta_1 = 0, \\ -\left[\frac{t^2}{2} - e^{-t} + \frac{1}{2^2} e^{-2t} - \frac{1}{3^2} e^{-3t} + \frac{1}{4^2} e^{-4t} - \cdots \right]_0^{\delta_1} & \text{if } \delta_1 > 0, \end{cases}$$

and

$$I_2 = \begin{cases} \left[e^{-t} - \frac{1}{2^2} e^{-2t} + \frac{1}{3^2} e^{-3t} - \frac{1}{4^2} e^{-4t} + \cdots \right]_0^{\delta_2} & \text{if } \delta_2 > 0, \\ 0 & \text{if } \delta_2 = 0, \\ -\left[\frac{t^2}{2} - e^{t} + \frac{1}{2^2} e^{2t} - \frac{1}{3^2} e^{3t} + \frac{1}{4^2} e^{4t} - \cdots \right]_{\delta_2}^{0} & \text{if } \delta_2 < 0. \end{cases}$$

Since the asymptotic variance-covariance matrix of the maximum likelihood estimators $\hat{\mu}$ and $\hat{\sigma}$ is given by

$$\begin{pmatrix} E\left(-\frac{\partial^2 \ln L}{\partial \mu^2}\right) & E\left(-\frac{\partial^2 \ln L}{\partial \mu\, \partial \sigma}\right) \\ & E\left(-\frac{\partial^2 \ln L}{\partial \sigma^2}\right) \end{pmatrix}^{-1}$$

it is possible for us to write the asymptotic variances and covariance of $\hat{\mu}$ and $\hat{\sigma}$ as

$$A\ \mathrm{Var}(\hat{\mu}) = \frac{\sigma^2}{n} \frac{V_3}{V_1 V_3 - V_2^2}, \tag{3.2.14}$$

$$A\ \mathrm{Var}(\hat{\sigma}) = \frac{\sigma^2}{n} \frac{V_1}{V_1 V_3 - V_2^2}, \tag{3.2.15}$$

and

$$A\ \mathrm{Cov}(\hat{\mu}, \hat{\sigma}) = -\frac{\sigma^2}{n} \frac{V_2}{V_1 V_3 - V_2^2}. \tag{3.2.16}$$

Then, by making use of Eqs. (3.2.11)–(3.2.13), respectively, Harter and Moore (1967) computed the values of $nA\ \mathrm{Var}(\hat{\mu})/\sigma^2$, $nA\ \mathrm{Var}(\hat{\sigma})/\sigma^2$, and $nA\ \mathrm{Cov}(\hat{\mu}, \hat{\sigma})/\sigma^2$ and the values of $nA\ \mathrm{Var}(\hat{\mu}|\sigma)/\sigma^2$ and $nA\ \mathrm{Var}(\hat{\sigma}|\mu)/\sigma^2$ for various choices of the proportions of censoring p_r and p_s. These values are presented in Table 3.2.3. The values are given for $p_r = 0.0(0.1)0.4$ and $p_s = p_r(0.1)0.9 - p_r$. Note that the values of the asymptotic variances and covariance for the case when $p_s < p_r$ are also available indirectly from Table 3.2.3, as interchanging p_r and p_s leaves the variances and the absolute value of the covariance unchanged, but changes the sign of the covariance. Upon comparing the values of the variances and the covariance of the estimators $\hat{\mu}$ and $\hat{\sigma}$ computed from Table 3.2.3 for $n = 20$ to the corresponding simulated values presented in Table 3.2.2, we observe that the asymptotic expressions of the variances and covariance given in Eqs. (3.2.14)–(3.2.16) provide close approximations to the actual variances and

Table 3.2.3 Values of $nA\ \mathrm{Var}(\hat{\mu})/\sigma^2$, $nA\ \mathrm{Var}(\hat{\sigma})/\sigma^2$, $nA\ \mathrm{Cov}(\hat{\mu}, \hat{\sigma})/\sigma^2$, $nA\ \mathrm{Var}(\hat{\mu}|\sigma)/\sigma^2$, and $nA\ \mathrm{Var}(\hat{\sigma}|\mu)/\sigma^2$

p_r	p_s	$n\ A\mathrm{Var}(\hat{\mu})/\sigma^2$	$n\ A\mathrm{Var}(\hat{\sigma})/\sigma^2$	$n\ A\mathrm{Cov}(\hat{\mu},\hat{\sigma})/\sigma^2$	$n\ A\mathrm{Var}(\hat{\mu}\vert\sigma)/\sigma^2$	$n\ A\mathrm{Var}(\hat{\sigma}\vert\mu)/\sigma^2$
0.0	0.0	0.911891	0.699322	0.000000	0.911891	0.699322
0.0	0.1	0.912876	0.769054	0.007452	0.912803	0.768994
0.0	0.2	0.920737	0.876663	0.036169	0.919245	0.875242
0.0	0.3	0.946701	1.029899	0.098945	0.937195	1.019557
0.0	0.4	1.012343	1.246821	0.217955	0.974242	1.199896
0.0	0.5	1.163950	1.562093	0.436172	1.042161	1.398644
0.0	0.6	1.513872	2.046448	0.847222	1.163126	1.572310
0.0	0.7	2.386396	2.866066	1.691620	1.387961	1.666944
0.0	0.8	5.025345	4.520589	3.777594	1.868628	1.680940
0.0	0.9	18.049503	9.509748	11.817224	3.364910	1.772871
0.1	0.1	0.913718	0.854083	0.000000	0.913718	0.854083
0.1	0.2	0.921155	0.988235	0.031160	0.920172	0.987181
0.1	0.3	0.946972	1.185761	0.102228	0.938159	1.174725
0.1	0.4	1.015380	1.479166	0.243534	0.975284	1.420756
0.1	0.5	1.182939	1.936695	0.519938	1.043353	1.708167
0.1	0.6	1.604612	2.720513	1.094087	1.164611	1.974522
0.1	0.7	2.825894	4.322156	2.491149	1.390077	2.126099
0.1	0.8	8.034268	9.220459	7.537549	1.872465	2.148919
0.2	0.2	0.926718	1.169419	0.000000	0.926718	1.169419
0.2	0.3	0.949261	1.448721	0.078894	0.944964	1.442164
0.2	0.4	1.015735	1.893218	0.250308	0.982641	1.831535
0.2	0.5	1.198442	2.664977	0.625186	1.051777	2.338839
0.2	0.6	1.743912	4.257226	1.556113	1.175117	2.868688
0.2	0.7	4.020561	9.157141	4.893916	1.405070	3.200158
0.3	0.3	0.963944	1.880831	0.000000	0.963944	1.880831
0.3	0.4	1.018918	2.643180	0.203958	1.003180	2.602354
0.3	0.5	1.209178	4.229832	0.752396	1.075343	3.761663
0.3	0.6	2.055657	9.132028	2.787791	1.204611	5.351351
0.4	0.4	1.045746	4.221988	0.000000	1.045746	4.221988
0.4	0.5	1.214726	9.122277	0.907717	1.124403	8.443975

covariance of $\hat{\mu}$ and $\hat{\sigma}$ even for a sample of size 20 as long as censoring in the sample is not extremely heavy. The results presented in Tables 3.2.1–3.2.3 agree closely with those for the normal distribution given by Harter and Moore (1966). As rightly pointed out by Harter (1970), this close agreement between the results for normal and logistic populations is not surprising since with the proper choice of parameters the supremum of the absolute difference between the distribution functions of these two populations is less than 0.01, as has been shown by Hailey (1952); see also Chapter 1.

Example 3.2.1 Sarhan and Greenberg (1962) have given data resulting from an experiment in which students were learning to measure strontium-90 concentrations in samples of milk. The test substance was supposed to contain 9.22 micromicrocuries per liter. The measurements, each involving readings and calculations, were made, but, since the measurement error was known to be relatively larger at the extremes, especially the upper one, a decision was made to censor the two smallest and the three largest observations, leaving the following censored sample:

$$8.2,\ 8.4,\ 9.1,\ 9.8,\ 9.9$$

Sarhan and Greenberg (1962) assumed a normal distribution for the above data and proceeded to calculate the best linear unbiased estimators of μ and σ from the doubly censored sample, while Gupta, Qureishi, and Shah (1967) assumed a logistic distribution for the data and calculated the best linear unbiased estimators of μ and σ.

By taking $n = 10$, $r = 2$, and $s = 3$, using the above censored sample in the maximum likelihood equations (3.2.6) and (3.2.7), and then solving the two equations for μ and σ, we get the maximum likelihood estimates of μ and σ (see also Harter and Moore, 1967): $\hat{\mu} = 9.2718$ and $\hat{\sigma} = 1.5678$. By using Table 3.2.1, we obtain unbiased estimates of σ and μ:

$$\hat{\sigma}_U = \frac{\hat{\sigma}}{0.84} = 1.8664 \qquad \text{and} \qquad \hat{\mu}_U = \hat{\mu} + 0.01\hat{\sigma}_U = 9.2905.$$

We also obtain the standard errors of the above estimates as

$$\text{S.E.}(\hat{\sigma}_U) = \frac{\hat{\sigma}_U\sqrt{0.130}}{0.84} = 0.8011$$

and

$$\begin{aligned}\text{S.E.}(\hat{\mu}_U) &= \hat{\sigma}_U\left\{0.098 + 0.130\left(\frac{0.01}{0.84}\right)^2 + 2(0.005)\left(\frac{0.01}{0.84}\right)\right\}^{1/2} \\ &= 0.5847.\end{aligned}$$

These may be compared to the best linear unbiased estimates of μ and σ (see Gupta et al., 1967 and Chapter 4) given by

$$\mu^* = 9.3031 \qquad \text{and} \qquad \sigma^* = 1.8765$$

and their standard errors

$$\text{S.E.}(\mu^*) = \sigma^*\sqrt{0.09921} = 0.5911$$

and

$$\text{S.E.}(\sigma^*) = \sigma^*\sqrt{0.18338} = 0.8036.$$

Refer to Chapter 4 for details on the best linear unbiased estimates of μ and σ.

Example 3.2.2 Davis (1952) has given lifetimes in hours of 417 forty-watt incandescent lamps taken from 42 weekly forced-life test samples. As an example, let us consider the first 20 observations from this data. Suppose 50% censoring had occurred in this data and that only the following first 10 observations were available:

785, 855, 905, 918, 919, 920, 929, 936, 948, 950

By taking $n = 10$, $r = 0$, and $s = 10$, using the above censored sample in the maximum likelihood equations (3.2.6) and (3.2.7), and then solving the two equations for μ and σ, we obtain the maximum likelihood estimates of μ and σ (also see Chapter 5):

$$\hat{\mu} = 948 \qquad \text{and} \qquad \hat{\sigma} = 59.$$

By using Table 3.2.2, we obtain unbiased estimates of σ and μ:

$$\hat{\sigma}_U = \frac{\hat{\sigma}}{0.93} = 63.44 \qquad \text{and} \qquad \hat{\mu}_U = \hat{\mu} + 0.03\hat{\sigma}_U = 949.90.$$

We also obtain the standard errors of the above estimates as

$$\text{S.E.}(\hat{\sigma}_U) = \frac{\hat{\sigma}_U\sqrt{0.077}}{0.93} = 18.93$$

and

$$\begin{aligned}\text{S.E.}(\hat{\mu}_U) &= \hat{\sigma}_U\left\{0.060 + 0.077\left(\frac{0.03}{0.93}\right)^2 + 2(0.020)\left(\frac{0.03}{0.93}\right)\right\}^{1/2} \\ &= 15.72.\end{aligned}$$

Instead, if we use the asymptotic values of the variances and the covariance of the MLEs presented in Table 3.2.3, we obtain the (approximate) standard errors of the estimates to be

$$\text{S.E.}(\hat{\sigma}_U) = \frac{\hat{\sigma}_U\sqrt{0.078104}}{0.93} = 19.06$$

and

$$\text{S.E.}(\hat{\mu}_U) = \hat{\sigma}_U\left\{0.0581975 + 0.078104\left(\frac{0.03}{0.93}\right)^2 + 2(0.0218086)\left(\frac{0.03}{0.93}\right)\right\}^{1/2}$$
$$= 15.50.$$

These standard errors are very close to the values obtained by using the finite sample results (based on simulations, of course) in Table 3.2.2.

The above computed estimates may be compared to the best linear unbiased estimates of μ and σ (see Chapter 4 for details) given by $\mu^* = 956$ and $\sigma^* = 66$ and their standard errors

$$\text{S.E.}(\mu^*) = \sigma^*\sqrt{0.06168} = 16.391$$

and

$$\text{S.E.}(\sigma^*) = \sigma^*\sqrt{0.08679} = 19.444.$$

3.3 APPROXIMATE MAXIMUM LIKELIHOOD ESTIMATORS OF μ AND σ

In the last section we discussed the maximum likelihood estimation of the parameters μ and σ based on complete and Type II censored samples. By noting that the maximum likelihood equations do not admit explicit solutions, we studied the bias, variances, and covariance of the maximum likelihood estimators for small sample sizes through Monte Carlo simulations and for large sample sizes through asymptotic arguments. In this section, we describe the approximate maximum likelihood estimators of μ and σ proposed by Balakrishnan and Cohen (1990). These estimators, derived by using some linear approximations in the maximum likelihood equations (3.2.6) and (3.2.7), are explicit estimators and have been displayed by Balakrishnan and Cohen (1990) to be almost as efficient as the maximum likelihood estimators and the best linear unbiased estimators. Here, we also derive some asymptotic expressions for the variances and covariance of these estimators and use these formulas to establish that these estimators are asymptotically fully efficient as compared to the maximum likelihood estimators.

In order to describe the method, let us consider the maximum likelihood equations (3.2.6) and (3.2.7). By using the relation between $f^*(z)$ and $F^*(z)$ given in (3.2.4), we can rewrite the two maximum likelihood equations as follows:

$$\frac{\partial \ln L}{\partial \mu} = -\frac{\pi}{\sqrt{3}\sigma}\left[r\{1 - F^*(Z_{r+1:n})\} - sF^*(Z_{n-s:n}) + \sum_{i=r+1}^{n-s}\{1 - 2F^*(Z_{i:n})\}\right] = 0 \tag{3.3.1}$$

and

$$\frac{\partial \ln L}{\partial \sigma} = -\frac{1}{\sigma}\left[A + rZ_{r+1:n}\{1 - F^*(Z_{r+1:n})\} - sZ_{n-s:n}F^*(Z_{n-s:n}) + \sum_{i=r+1}^{n-s} Z_{i:n}\{1 - 2F^*(Z_{i:n})\}\right] = 0, \tag{3.3.2}$$

where $A = n - r - s$ is the number of observations left in the Type II censored sample. Equations (3.3.1) and (3.3.2), not surprisingly, do not admit explicit solutions. But, by expanding the function $F^*(Z_{i:n})$ in a Taylor series around the point $F^{*-1}(p_i) = \ln(p_i/q_i)$ and then approximating it by

$$F^*(Z_{i:n}) \simeq \gamma_i + \delta_i Z_{i:n}, \tag{3.3.3}$$

where

$$p_i = \frac{i}{n+1}, \quad q_i = 1 - p_i,$$

$$\gamma_i = p_i - p_i q_i \ln\left(\frac{p_i}{q_i}\right), \quad \text{and} \quad \delta_i = p_i q_i,$$

we may approximate the maximum likelihood equations (3.3.1) and (3.3.2) as follows:

$$\frac{\partial \ln L}{\partial \mu} \simeq \frac{\partial \ln L^*}{\partial \mu} = -\frac{\pi}{\sqrt{3}\sigma}\left[r(1 - \gamma_{r+1}) - s\gamma_{n-s} + A - 2\sum_{i=r+1}^{n-s}\gamma_i - r\delta_{r+1}Z_{r+1:n} - s\delta_{n-s}Z_{n-s:n} - 2\sum_{i=r+1}^{n-s}\delta_i Z_{i:n}\right] = 0 \tag{3.3.4}$$

and

$$\begin{aligned}\frac{\partial \ln L}{\partial \sigma} &\simeq \frac{\partial \ln L^*}{\partial \sigma}\\ &= -\frac{1}{\sigma}\Bigg[A + r(1-\gamma_{r+1})Z_{r+1:n} - s\gamma_{n-s}Z_{n-s:n}\\ &\quad - \sum_{i=r+1}^{n-s}(2\gamma_i - 1)Z_{i:n} - r\delta_{r+1}Z_{r+1:n}^2\\ &\quad - s\delta_{n-s}Z_{n-s:n}^2 - 2\sum_{i=r+1}^{n-s}\delta_i Z_{i:n}^2\Bigg]\\ &= 0. \end{aligned} \tag{3.3.5}$$

Upon solving Eqs. (3.3.4) and (3.3.5), we derive the approximate maximum likelihood estimators of μ and σ:

$$\tilde{\mu} = B - \frac{\sqrt{3}}{\pi}C\tilde{\sigma} \tag{3.3.6}$$

and

$$\tilde{\sigma} = \frac{\pi}{\sqrt{3}}\left\{\frac{D + (D^2 + 4AE)^{1/2}}{2A}\right\}, \tag{3.3.7}$$

where

$$A = n - r - s,$$

$$m = r\delta_{r+1} + s\delta_{n-s} + 2\sum_{i=r+1}^{n-s}\delta_i, \tag{3.3.8}$$

$$B = \frac{1}{m}\left\{r\delta_{r+1}X_{r+1:n} + s\delta_{n-s}X_{n-s:n} + 2\sum_{i=r+1}^{n-s}\delta_i X_{i:n}\right\}, \tag{3.3.9}$$

$$C = \frac{1}{m}\left\{n - s - r\gamma_{r+1} - s\gamma_{n-s} - 2\sum_{i=r+1}^{n-s}\gamma_i\right\}, \tag{3.3.10}$$

$$\begin{aligned} D &= \sum_{i=r+1}^{n-s}(2\gamma_i - 1)(X_{i:n} - B)\\ &\quad - r(1-\gamma_{r+1})(X_{r+1:n} - B) + s\gamma_{n-s}(X_{n-s:n} - B)\\ &= \sum_{i=r+1}^{n-s}(2\gamma_i - 1)X_{i:n} - r(1-\gamma_{r+1})X_{r+1:n}\\ &\quad + s\gamma_{n-s}X_{n-s:n} + mBC, \end{aligned} \tag{3.3.11}$$

and

$$
\begin{aligned}
E &= r\delta_{r+1}(X_{r+1:n} - B)^2 + s\delta_{n-s}(X_{n-s:n} - B)^2 \\
&\quad + 2\sum_{i=r+1}^{n-s} \delta_i(X_{i:n} - B)^2 \\
&= r\delta_{r+1}X_{r+1:n}^2 + s\delta_{n-s}X_{n-s:n}^2 \\
&\quad + 2\sum_{i=r+1}^{n-s} \delta_i X_{i:n}^2 - mB^2.
\end{aligned} \tag{3.3.12}
$$

Upon solving (3.3.5) we obtain a quadratic equation in σ which has two roots; however, one of them drops out since $\delta_i > 0$ ($i = r + 1, \ldots, n - s$) and hence, $E > 0$ from Eq. (3.3.12).

Furthermore, we observe that

$$\gamma_i = p_i - p_i q_i \ln\left(\frac{p_i}{q_i}\right) = 1 - \gamma_{n-i+1}$$

and, hence, in the case of symmetric Type II censoring, that is, when $r = s$, we have

$$\sum_{i=r+1}^{n-s} \gamma_i = \sum_{i=r+1}^{n-r} \gamma_i = \frac{n - 2r}{2}.$$

In this case, we get from Eq. (3.3.10) that

$$
\begin{aligned}
C &= \frac{1}{m}\left\{n - s - r\gamma_{r+1} - s\gamma_{n-s} - 2\sum_{i=r+1}^{n-s} \gamma_i\right\} \\
&= \frac{1}{m}\left\{n - r - r\gamma_{r+1} - r(1 - \gamma_{r+1}) - (n - 2r)\right\} \\
&= 0.
\end{aligned}
$$

As a result, for the case of symmetric Type II censoring the estimator $\tilde{\mu}$ in (3.3.6) simply becomes

$$\tilde{\mu} = B,$$

which is easily observed to be an unbiased estimator of μ.

In order to examine the performance of the approximate maximum likelihood estimators $\tilde{\mu}$ and $\tilde{\sigma}$ given in Eqs. (3.3.6) and (3.3.7), respectively, we simulated 10,000 pseudorandom samples of size $n = 10$ and $n = 20$ from the logistic $L(0, 1)$ population and determined the values of Bias($\tilde{\mu}$)/σ, Bias($\tilde{\sigma}$)/σ, Var($\tilde{\mu}$)/σ^2, Var($\tilde{\sigma}$)/σ^2, Cov($\tilde{\mu}$, $\tilde{\sigma}$)/σ^2, Var($\tilde{\mu}|\sigma$)/σ^2, and Var($\tilde{\sigma}|\mu$)/σ^2. These values are presented in Tables 3.3.1 and 3.3.2 for $n =$

10 and $n = 20$, respectively, for $r = 0(1)(n/2) - 1$ and $s = r(1)n - 2 - r$. Upon comparing these values with the corresponding entries of the maximum likelihood estimators given in Tables 3.2.1 and 3.2.2, we simply observe that the approximate maximum likelihood estimators presented here are almost as efficient as the maximum likelihood estimators even for a sample of size as small as 10.

The conditional bias of $\tilde{\mu}$ can be determined exactly from Eq. (3.3.6) since

$$
\begin{aligned}
E(\tilde{\mu}|\sigma) &= E(B) - \frac{\sqrt{3}}{\pi} C\sigma \\
&= \mu + \frac{\sigma\sqrt{3}}{\pi m} \\
&\quad \times \left\{ r\delta_{r+1}\alpha^*_{r+1:n} + s\delta_{n-s}\alpha^*_{n-s:n} + 2 \sum_{i=r+1}^{n-s} \delta_i \alpha^*_{i:n} \right\} - \frac{\sqrt{3}}{\pi} C\sigma,
\end{aligned}
$$

Table 3.3.1 Simulated Values of Bias($\tilde{\mu}$)/σ, Bias($\tilde{\sigma}$)/σ, Var($\tilde{\mu}$)/σ^2, Var($\tilde{\sigma}$)/σ^2, Cov($\tilde{\mu}$, $\tilde{\sigma}$)/σ^2, Var($\tilde{\mu}|\sigma$)/σ^2, and Var($\tilde{\sigma}|\mu$)/σ^2 When $n = 10$

r	s	Bias($\tilde{\mu}$)/σ	Bias($\tilde{\sigma}$)/σ	Var($\tilde{\mu}$)/σ^2	Var($\tilde{\sigma}$)/σ^2	Cov($\tilde{\mu}$,$\tilde{\sigma}$)/σ^2	Var($\tilde{\mu}|\sigma$)/σ^2	Var($\tilde{\sigma}|\mu$)/σ^2
0	0	−0.004366	− 0.022507	0.094101	0.073791	0.000109	0.094101	0.073791
0	1	−0.013923	− 0.042927	0.094085	0.079432	0.000425	0.094083	0.079430
0	2	−0.024348	− 0.059064	0.095006	0.089451	0.003260	0.094887	0.089339
0	3	−0.039362	− 0.079437	0.097350	0.102653	0.008610	0.096628	0.101892
0	4	−0.060309	− 0.103908	0.104937	0.122487	0.020430	0.101529	0.118509
0	5	−0.092256	− 0.135989	0.121922	0.147909	0.040537	0.110812	0.134432
0	6	−0.150350	− 0.187646	0.157762	0.183630	0.075250	0.126925	0.147737
0	7	−0.256864	− 0.269111	0.234788	0.231672	0.134500	0.156703	0.154623
0	8	−0.510277	− 0.434479	0.413680	0.285834	0.231859	0.225603	0.155882
1	1	−0.003963	− 0.070418	0.094061	0.084596	0.000404	0.094059	0.084594
1	2	−0.015375	− 0.092659	0.095070	0.095068	0.003233	0.094960	0.094958
1	3	−0.032732	− 0.121875	0.097486	0.109019	0.008616	0.096805	0.108258
1	4	−0.058623	− 0.159450	0.105160	0.131022	0.021130	0.101752	0.126776
1	5	−0.101085	− 0.213047	0.122790	0.161316	0.043630	0.110990	0.145813
1	6	−0.183992	− 0.305996	0.160364	0.200672	0.081635	0.127155	0.159115
1	7	−0.361159	− 0.481900	0.235267	0.234650	0.135461	0.157067	0.156656
2	2	−0.004283	− 0.121007	0.096102	0.109468	−0.000115	0.096102	0.109468
2	3	−0.022731	− 0.161448	0.098259	0.127362	0.005522	0.098020	0.127052
2	4	−0.052555	− 0.217564	0.105502	0.156533	0.019497	0.103074	0.152930
2	5	−0.106502	− 0.306247	0.122965	0.192950	0.044547	0.112680	0.176812
2	6	−0.228449	− 0.487227	0.159532	0.227491	0.082990	0.129256	0.184319
3	3	−0.005431	− 0.216738	0.100581	0.155210	−0.001829	0.100560	0.155176
3	4	−0.038848	− 0.307910	0.106919	0.190910	0.012816	0.106059	0.189374
3	5	−0.111000	− 0.481250	0.122749	0.226304	0.038942	0.116048	0.213950
4	4	−0.004819	− 0.489191	0.111389	0.223225	−0.001651	0.111377	0.223200

Table 3.3.2 Simulated Values of $\text{Bias}(\bar{\mu})/\sigma$, $\text{Bias}(\bar{\sigma})/\sigma$, $\text{Var}(\bar{\mu})/\sigma^2$, $\text{Var}(\bar{\sigma})/\sigma^2$, $\text{Cov}(\bar{\mu}, \bar{\sigma})/\sigma^2$, $\text{Var}(\bar{\mu}|\sigma)/\sigma^2$, and $\text{Var}(\bar{\sigma}|\mu)/\sigma^2$ When $n = 20$

r	s	$\text{Bias}(\bar{\mu})/\sigma$	$\text{Bias}(\bar{\sigma})/\sigma$	$\text{Var}(\bar{\mu})/\sigma^2$	$\text{Var}(\bar{\sigma})/\sigma^2$	$\text{Cov}(\bar{\mu},\bar{\sigma})/\sigma^2$	$\text{Var}(\bar{\mu}\|\sigma)/\sigma^2$	$\text{Var}(\bar{\sigma}\|\mu)/\sigma^2$
0	0	−0.001748	− 0.006819	0.046395	0.037204	−0.000688	0.046382	0.037194
0	1	−0.004180	− 0.013050	0.046444	0.038493	−0.000498	0.046438	0.038487
0	2	−0.006676	− 0.018112	0.046531	0.040265	−0.000175	0.046531	0.040264
0	3	−0.008989	− 0.021689	0.046700	0.042623	0.000404	0.046697	0.042619
0	4	−0.011572	− 0.025316	0.047011	0.045809	0.001370	0.046970	0.045769
0	5	−0.015329	− 0.030997	0.047741	0.048733	0.002782	0.047583	0.048571
0	6	−0.019109	− 0.035909	0.048433	0.053289	0.004558	0.048043	0.052860
0	7	−0.023110	− 0.040490	0.049993	0.058033	0.007196	0.049101	0.056997
0	8	−0.028686	− 0.046793	0.051996	0.062777	0.010204	0.050338	0.060775
0	9	−0.035453	− 0.053928	0.055421	0.069387	0.014883	0.052228	0.065390
0	10	−0.042804	− 0.060733	0.059640	0.077270	0.020566	0.054166	0.070178
0	11	−0.055809	− 0.073072	0.066901	0.086437	0.028620	0.057425	0.074193
0	12	−0.072555	− 0.087681	0.077297	0.098167	0.039481	0.061419	0.078002
0	13	−0.094501	− 0.105402	0.094284	0.113853	0.055581	0.067151	0.081088
0	14	−0.126063	− 0.129015	0.119535	0.133448	0.077511	0.074514	0.083187
0	15	−0.174122	− 0.162204	0.163464	0.158299	0.110075	0.086922	0.084175
0	16	−0.251413	− 0.211011	0.234599	0.192872	0.158810	0.103836	0.085368
0	17	−0.396682	− 0.294432	0.368983	0.232077	0.230723	0.139607	0.087808
0	18	−0.726500	− 0.462868	0.622258	0.271139	0.332138	0.215397	0.093856
1	1	0.002036	− 0.019889	0.046126	0.037565	−0.000028	0.046126	0.037565
1	2	−0.000467	− 0.025507	0.046208	0.039523	0.000319	0.046206	0.039521
1	3	−0.003098	− 0.030458	0.046334	0.041984	0.000832	0.046317	0.041969
1	4	−0.006053	− 0.035429	0.046717	0.045181	0.001914	0.046636	0.045103
1	5	−0.009433	− 0.040645	0.047143	0.048858	0.003144	0.046941	0.048648
1	6	−0.013539	− 0.046534	0.047970	0.053606	0.005064	0.047491	0.053072
1	7	−0.018203	− 0.052621	0.049366	0.058225	0.007548	0.048387	0.057071
1	8	−0.024455	− 0.060417	0.051239	0.064160	0.010816	0.049415	0.061876
1	9	−0.031735	− 0.068762	0.054634	0.071120	0.015598	0.051214	0.066667
1	10	−0.042551	− 0.080809	0.059621	0.079970	0.022125	0.053499	0.071759
1	11	−0.057327	− 0.096271	0.066872	0.091512	0.031128	0.056284	0.077022
1	12	−0.074843	− 0.112975	0.077924	0.106002	0.043609	0.059983	0.081596
1	13	−0.101400	− 0.136941	0.094491	0.123719	0.060497	0.064909	0.084986
1	14	−0.138204	− 0.167732	0.122443	0.146437	0.085375	0.072668	0.086908
1	15	−0.201884	− 0.218069	0.170254	0.178780	0.124228	0.083932	0.088135
1	16	−0.321573	− 0.306750	0.246858	0.216011	0.177194	0.101505	0.088821
1	17	−0.578577	− 0.484503	0.373930	0.241745	0.239479	0.136696	0.088374

(*continued*)

so that

$$\frac{1}{\sigma}\,\text{Bias}(\bar{\mu}|\sigma) = \frac{\sqrt{3}}{\pi}\left[\frac{1}{m}\left\{r\delta_{r+1}\alpha^*_{r+1:n} + s\delta_{n-s}\alpha^*_{n-s:n} + 2\sum_{i=r+1}^{n-s}\delta_i\alpha^*_{i:n}\right\} - C\right]. \quad (3.3.13)$$

But it is difficult to evaluate the conditional bias of $\bar{\sigma}$ exactly. However, the conditional bias of $\bar{\sigma}$ may be approximated by

$$\text{Bias}(\bar{\sigma}|\mu) \simeq E\left(\frac{\partial \ln L^*}{\partial\sigma}\right)\bigg/ E\left(-\frac{\partial^2 \ln L^*}{\partial\sigma^2}\right); \quad (3.3.14)$$

Table 3.3.2 (*continued*)

r	s	Bias($\tilde{\mu}$)/σ	Bias($\tilde{\sigma}$)/σ	Var($\tilde{\mu}$)/σ^2	Var($\tilde{\sigma}$)/σ^2	Cov($\tilde{\mu},\tilde{\sigma}$)/$\sigma^2$	Var($\tilde{\mu}\mid\sigma$)/σ^2	Var($\tilde{\sigma}\mid\mu$)/σ^2
2	2	0.000213	− 0.028509	0.046807	0.043663	0.000303	0.046804	0.043661
2	3	−0.002411	− 0.034008	0.047017	0.046346	0.000964	0.046997	0.046326
2	4	−0.005461	− 0.039777	0.047287	0.050094	0.001937	0.047212	0.050015
2	5	−0.009034	− 0.045996	0.048025	0.054800	0.003753	0.047768	0.054506
2	6	−0.013859	− 0.054126	0.048879	0.059480	0.005680	0.048337	0.058820
2	7	−0.019095	− 0.061875	0.050048	0.064911	0.008119	0.049033	0.063594
2	8	−0.026309	− 0.072117	0.052261	0.071792	0.011940	0.050275	0.069064
2	9	−0.035038	− 0.083473	0.055154	0.080053	0.016744	0.051652	0.074970
2	10	−0.047217	− 0.098608	0.060532	0.091657	0.024517	0.053974	0.081727
2	11	−0.063153	− 0.117115	0.068583	0.105232	0.034833	0.057053	0.087541
2	12	−0.085322	− 0.141178	0.080389	0.124438	0.049698	0.060541	0.093714
2	13	−0.118899	− 0.175654	0.100274	0.147434	0.070834	0.066242	0.097396
2	14	−0.170948	− 0.225968	0.131520	0.176614	0.100735	0.074064	0.099458
2	15	−0.265010	− 0.312497	0.184187	0.215157	0.145598	0.085659	0.100063
2	16	−0.467367	− 0.490092	0.265963	0.247505	0.199666	0.104889	0.097610
3	3	0.000680	− 0.041589	0.048294	0.048798	−0.000402	0.048291	0.048795
3	4	−0.002701	− 0.049125	0.048615	0.052367	0.000613	0.048608	0.052360
3	5	−0.006227	− 0.055902	0.049082	0.057347	0.002070	0.049007	0.057259
3	6	−0.011043	− 0.064759	0.049950	0.062404	0.004089	0.049682	0.062070
3	7	−0.016984	− 0.074936	0.051308	0.069673	0.007140	0.050576	0.068679
3	8	−0.025274	− 0.088411	0.053526	0.078038	0.011352	0.051875	0.075630
3	9	−0.035811	− 0.104287	0.056737	0.088131	0.016933	0.053483	0.083077
3	10	−0.049495	− 0.123500	0.062282	0.101421	0.025392	0.055925	0.091069
3	11	−0.066386	− 0.145332	0.070666	0.120123	0.037733	0.058813	0.099974
3	12	−0.094967	− 0.181113	0.083780	0.142312	0.054582	0.062846	0.106752
3	13	−0.136638	− 0.230202	0.105460	0.171573	0.079586	0.068542	0.111513
3	14	−0.214671	− 0.318606	0.138343	0.204831	0.112736	0.076295	0.112963
3	15	−0.372068	− 0.489956	0.189945	0.242242	0.157765	0.087198	0.111205
4	4	−0.000365	− 0.056469	0.046644	0.055992	0.000179	0.046643	0.055992
4	5	−0.004434	− 0.066046	0.047231	0.061435	0.001889	0.047173	0.061359
4	6	−0.009178	− 0.076012	0.048302	0.069102	0.004680	0.047985	0.068648
4	7	−0.015172	− 0.087622	0.049552	0.077538	0.007838	0.048760	0.076298
4	8	−0.023321	− 0.102503	0.051536	0.087103	0.012084	0.049860	0.084270
4	9	−0.034123	− 0.120843	0.054808	0.099301	0.018278	0.051443	0.093206
4	10	−0.049594	− 0.145851	0.060300	0.116188	0.027764	0.053665	0.103405
4	11	−0.073570	− 0.182725	0.068861	0.139287	0.041635	0.056416	0.114113
4	12	−0.109447	− 0.234888	0.082376	0.167949	0.061101	0.060146	0.122628
4	13	−0.169536	− 0.318586	0.104773	0.201417	0.088610	0.065790	0.126476
4	14	−0.300741	− 0.495951	0.136928	0.226458	0.119538	0.073829	0.122102

(*continued*)

see, for example, Kendall and Stuart (1961).

Furthermore, we obtain from Eqs. (3.3.4) and (3.3.5) that

$$E\left(-\frac{\partial^2 \ln L}{\partial \mu^2}\right) \simeq E\left(-\frac{\partial^2 \ln L^*}{\partial \mu^2}\right) = \frac{m\pi^2}{3\sigma^2}, \tag{3.3.15}$$

$$E\left(-\frac{\partial^2 \ln L}{\partial \mu\, \partial \sigma}\right) \simeq E\left(-\frac{\partial^2 \ln L^*}{\partial \mu\, \partial \sigma}\right) = \frac{m\pi}{\sqrt{3}\sigma^2} W_1, \tag{3.3.16}$$

Table 3.3.2 (*continued*)

r	s	Bias($\bar{\mu}$)/σ	Bias($\bar{\sigma}$)/σ	Var($\bar{\mu}$)/σ^2	Var($\bar{\sigma}$)/σ^2	Cov($\bar{\mu},\bar{\sigma}$)/σ^2	Var($\bar{\mu}\mid\sigma$)/σ^2	Var($\bar{\sigma}\mid\mu$)/σ^2
5	5	0.000978	− 0.075110	0.049482	0.068376	−0.000401	0.049480	0.068373
5	6	−0.004081	− 0.087907	0.050018	0.076530	0.001594	0.049985	0.076480
5	7	−0.010742	− 0.103403	0.051264	0.086432	0.004975	0.050977	0.085950
5	8	−0.019462	− 0.122174	0.053250	0.099168	0.009896	0.052262	0.097328
5	9	−0.032665	− 0.149241	0.056547	0.115073	0.017008	0.054034	0.109958
5	10	−0.051363	− 0.185186	0.061120	0.137167	0.026844	0.055866	0.125377
5	11	−0.080392	− 0.238213	0.070242	0.166051	0.042917	0.059149	0.139829
5	12	−0.132426	− 0.329699	0.085897	0.201986	0.066817	0.063794	0.150010
5	13	−0.231758	− 0.498191	0.110133	0.239566	0.098358	0.069751	0.151724
6	6	−0.002255	− 0.107142	0.049994	0.088033	−0.000863	0.049986	0.088018
6	7	−0.009219	− 0.127308	0.051041	0.101765	0.002767	0.050965	0.101615
6	8	−0.018938	− 0.153292	0.053015	0.116589	0.008097	0.052453	0.115353
6	9	−0.033183	− 0.189076	0.055899	0.137243	0.015625	0.054120	0.132875
6	10	−0.054963	− 0.240586	0.060850	0.167685	0.027643	0.056293	0.155127
6	11	−0.095296	− 0.332513	0.069622	0.203283	0.045280	0.059536	0.173834
6	12	−0.173768	− 0.505075	0.084929	0.227122	0.067547	0.064841	0.173400
7	7	−0.002983	− 0.159980	0.051152	0.112610	0.000146	0.051152	0.112610
7	8	−0.013189	− 0.195814	0.052746	0.131632	0.005531	0.052513	0.131052
7	9	−0.029652	− 0.250211	0.055745	0.157911	0.014394	0.054433	0.154194
7	10	−0.056773	− 0.334777	0.060984	0.192650	0.028095	0.056887	0.179707
7	11	−0.112898	− 0.503862	0.069322	0.226676	0.046237	0.059891	0.195837
8	8	0.000658	− 0.240224	0.053117	0.166007	0.000063	0.053117	0.166007
8	9	−0.016813	− 0.324643	0.055457	0.202158	0.009318	0.055027	0.200592
8	10	−0.052690	− 0.490222	0.059688	0.239128	0.022947	0.057486	0.230306
9	9	0.000120	− 0.497751	0.057626	0.229983	0.001608	0.057615	0.229938

and

$$E\left(-\frac{\partial^2 \ln L}{\partial\sigma^2}\right) \simeq E\left(-\frac{\partial^2 \ln L^*}{\partial\sigma^2}\right) = \frac{m}{\sigma^2} W_2, \tag{3.3.17}$$

where

$$W_1 = \frac{2}{m}\left\{r\delta_{r+1}\alpha^*_{r+1:n} + s\delta_{n-s}\alpha^*_{n-s:n} + 2\sum_{i=r+1}^{n-s} \delta_i\alpha^*_{i:n}\right\} - C \tag{3.3.18}$$

and

$$\begin{aligned} W_2 = \frac{1}{m}\Bigg[&3\left\{r\delta_{r+1}\alpha^{*(2)}_{r+1:n} + s\delta_{n-s}\alpha^{*(2)}_{n-s:n} + 2\sum_{i=r+1}^{n-s} \delta_i\alpha^{*(2)}_{i:n}\right\} \\ &+ 2\left\{\sum_{i=r+1}^{n-s} (2\gamma_i - 1)\alpha^*_{i:n} + s\gamma_{n-s}\alpha^*_{n-s:n}\right. \\ &\left. - r(1 - \gamma_{r+1})\alpha^*_{r+1:n}\right\} - A\Bigg]. \end{aligned} \tag{3.3.19}$$

From Eqs. (3.3.15) to (3.3.19), we obtain approximate expressions for the variances and the covariance of the estimators $\tilde{\mu}$ and $\tilde{\sigma}$ as

$$\text{Var}(\tilde{\mu}) = \frac{3\sigma^2}{m\pi^2}\left\{\frac{W_2}{W_2 - W_1^2}\right\}, \tag{3.3.20}$$

$$\text{Var}(\tilde{\sigma}) = \frac{\sigma^2}{m}\left\{\frac{1}{W_2 - W_1^2}\right\}, \tag{3.3.21}$$

and

$$\text{Cov}(\tilde{\mu}, \tilde{\sigma}) = -\frac{\sqrt{3}\sigma^2}{m\pi}\left\{\frac{W_1}{W_2 - W_1^2}\right\}. \tag{3.3.22}$$

In the case of symmetric censoring, that is, $r = s$, we have already noted that $C = 0$; in this case, we further observe from Eq. (3.3.18) that $W_1 = 0$ since we have $\delta_i = \delta_{n-i+1}$ and $\alpha^*_{i:n} = -\alpha^*_{n-i+1:n}$. Therefore, when the censoring in the sample is symmetric, we see from (3.3.22) that the estimators $\tilde{\mu}$ and $\tilde{\sigma}$ are uncorrelated.

By making use of all these expressions, we computed the values of (1) {exact conditional bias of $\tilde{\mu}$}/σ, (2) {approximate conditional bias of $\tilde{\sigma}$}/σ, (3) {approximate variance of $\tilde{\mu}$}/σ^2, (4) {approximate variance of $\tilde{\sigma}$}/σ^2, (5) {approximate covariance of $\tilde{\mu}$ and $\tilde{\sigma}$}/σ^2, (6) {approximate conditional variance of $\tilde{\mu}$}/σ^2, and (7) {approximate conditional variance of $\tilde{\sigma}$}/σ^2 for sample sizes $n = 10$ and $n = 20$. These values are presented in Tables 3.3.3 and 3.3.4, respectively, for $0 \le r \le s \le 4$. Upon comparing these values with the corresponding values of the maximum likelihood estimators in Tables 3.2.1 and 3.2.2 and those of the best linear unbiased estimators given by Gupta et al. (1967) and Balakrishnan (1990) (see also Chapter 4), we find the estimators $\tilde{\mu}$ and $\tilde{\sigma}$ to be remarkably efficient. Further, by comparing the values in Tables 3.3.3 and 3.3.4 with the values of the modified maximum likelihood estimators given by Tiku (1968), we find the estimators $\tilde{\mu}$ and $\tilde{\sigma}$ to be slightly more efficient than his estimators despite the fact that the two sets of estimators are of similar form.

Furthermore, for the case of symmetric censoring (i.e., $r = s$), we have presented in Table 3.3.5 the values of {approximate variance of $\tilde{\mu}$}/σ^2 determined from Eq. (3.3.20), {exact variance of $\tilde{\mu}$}/σ^2 determined directly from Eq. (3.3.6), {variance of the best linear unbiased estimator of μ}/σ^2 taken from the tables prepared by Gupta et al. (1967) and Balakrishnan (1990), and {mean square error of the maximum likelihood estimator of μ}/σ^2 taken from Tables 3.2.1 and 3.2.2. In Table 3.3.5, we have given the values of the above quantities for $n = 10$ and $n = 20$ and $r = s = 1(1)4$. It is quite apparent from this that $\tilde{\mu}$ is an unbiased estimator which

Table 3.3.3 Values of (1) {exact cond. bias of $\tilde{\mu}\}/\sigma$, (2) {appr. cond. bias of $\tilde{\sigma}\}/\sigma$, (3) {appr. variance of $\tilde{\mu}\}/\sigma^2$, (4) {appr. variance of $\tilde{\sigma}\}/\sigma^2$, (5) {appr. covariance of $\tilde{\mu}$ and $\tilde{\sigma}\}/\sigma^2$, (6) {appr. cond. variance of $\tilde{\mu}\}/\sigma^2$, and (7) {appr. cond. variance of $\tilde{\sigma}\}/\sigma^2$, When $n = 10$

r	s	(1)	(2)	(3)	(4)	(5)	(6)	(7)
0	0	0.0000	0.0623	0.08359	0.04864	0.00000	0.08359	0.04864
0	1	−0.0084	0.0560	0.08406	0.05578	0.00227	0.08397	0.05572
0	2	−0.0164	0.0542	0.08579	0.06434	0.00647	0.08514	0.06385
0	3	−0.0247	0.0550	0.09003	0.07522	0.01362	0.08756	0.07316
0	4	−0.0340	0.0580	0.09917	0.08946	0.02541	0.09195	0.08295
1	1	0.0000	0.0475	0.08436	0.06522	0.00000	0.08436	0.06522
1	2	−0.0079	0.0439	0.08581	0.07688	0.00458	0.08554	0.07664
1	3	−0.0160	0.0430	0.08978	0.09229	0.01285	0.08799	0.09045
1	4	−0.0249	0.0448	0.09898	0.11343	0.02730	0.09241	0.10590
2	2	0.0000	0.0388	0.08674	0.09290	0.00000	0.08674	0.09290
2	3	−0.0080	0.0366	0.08999	0.11492	0.00909	0.08927	0.11400
2	4	−0.0166	0.0374	0.09853	0.14670	0.02626	0.09383	0.13970
3	3	0.0000	0.0330	0.09195	0.14753	0.00000	0.09195	0.14753
3	4	−0.0085	0.0330	0.09872	0.19747	0.01954	0.09679	0.19360
4	4	0.0000	0.0331	0.10217	0.28152	0.00000	0.10217	0.28152

Table 3.3.4 Values of (1) {exact cond. bias of $\tilde{\mu}\}/\sigma$, (2) {appr. cond. bias of $\tilde{\sigma}\}/\sigma$, (3) {appr. variance of $\tilde{\mu}\}/\sigma^2$, (4) {appr. variance of $\tilde{\sigma}\}/\sigma^2$, (5) {appr. covariance of $\tilde{\mu}$ and $\tilde{\sigma}\}/\sigma^2$, (6) {appr. cond. variance of $\tilde{\mu}\}/\sigma^2$, and (7) {appr. cond. variance of $\tilde{\sigma}\}/\sigma^2$, When $n = 20$

r	s	(1)	(2)	(3)	(4)	(5)	(6)	(7)
0	0	0.0000	0.0353	0.04352	0.02848	0.00000	0.04352	0.02848
0	1	−0.0023	0.0323	0.04355	0.03024	0.00032	0.04355	0.03024
0	2	−0.0044	0.0309	0.04366	0.03217	0.00085	0.04364	0.03215
0	3	−0.0065	0.0301	0.04388	0.03438	0.00162	0.04380	0.03432
0	4	−0.0086	0.0298	0.04429	0.03692	0.00270	0.04409	0.03676
1	1	0.0000	0.0289	0.04358	0.03222	0.00000	0.04358	0.03222
1	2	−0.0021	0.0271	0.04367	0.03442	0.00054	0.04366	0.03441
1	3	−0.0042	0.0260	0.04388	0.03695	0.00134	0.04383	0.03691
1	4	−0.0063	0.0254	0.04428	0.03988	0.00249	0.04412	0.03974
2	2	0.0000	0.0251	0.04375	0.03693	0.00000	0.04375	0.03693
2	3	−0.0021	0.0237	0.04394	0.03983	0.00083	0.04392	0.03981
2	4	−0.0041	0.0228	0.04431	0.04322	0.00202	0.04422	0.04313
3	3	0.0000	0.0222	0.04409	0.04319	0.00000	0.04409	0.04319
3	4	−0.0020	0.0211	0.04442	0.04715	0.00122	0.04439	0.04712
4	4	0.0000	0.0197	0.04468	0.05183	0.00000	0.04468	0.05183

Table 3.3.5 Comparison of Some Estimators of μ for $n = 10$ and 20 When $r = s$

	n = 10				n = 20			
r =	1	2	3	4	1	2	3	4
Appr. $\text{Var}(\bar{\mu})/\sigma^2$	0.084	0.087	0.092	0.102	0.044	0.044	0.044	0.045
Exact $\text{Var}(\bar{\mu})/\sigma^2$	0.094	0.096	0.101	0.112	0.046	0.046	0.047	0.047
Var(BLUE of μ)/σ^2	0.094	0.096	0.101	0.112	0.046	0.046	0.047	0.047
MSE(MLE of μ)/σ^2	0.093	0.096	0.099	0.108	-	0.047	-	0.048

is just as efficient as the best linear unbiased estimator and almost as efficient as the maximum likelihood estimator.

Next, we shall derive some asymptotic expressions for $n\ \text{Var}(\bar{\mu})/\sigma^2$, $n\ \text{Var}(\bar{\sigma})/\sigma^2$, and $n\ \text{Cov}(\bar{\mu}, \bar{\sigma})/\sigma^2$. For this purpose, let us set

$$V_1^* = \frac{m}{n}, \qquad V_2^* = \frac{m}{n} W_1, \qquad \text{and} \qquad V_3^* = \frac{m}{n} W_2.$$

Then as $n \to \infty$, with $p_r = r/n$ and $p_s = s/n$ fixed, we obtain the following:

$$\begin{aligned}
\lim_{n\to\infty} V_1^* &= \lim_{n\to\infty} \frac{1}{n}\left\{ r\delta_{r+1} + s\delta_{n-s} + 2\sum_{i=r+1}^{n-s} \delta_i \right\} \\
&= p_r\delta_r + p_s\delta_{n-s} + 2\left\{ \lim_{n\to\infty} \sum_{i=r+1}^{n-s} \delta_i \right\} \\
&= p_r\delta_r + p_s\delta_{n-s} + 2\int_{p_r}^{1-p_s} u(1-u)\, du \\
&= p_r\delta_r + p_s\delta_{n-s} + 2\left[\frac{1}{2}\{(1-p_s)^2 - p_r^2\} \right. \\
&\quad \left. - \frac{1}{3}\{(1-p_s)^3 - p_r^3\} \right],
\end{aligned} \tag{3.3.23}$$

since $\delta_i = p_i q_i$ and, hence $\delta(u) = u(1-u)$. By using a similar integral representation for the sum, we derive

$$\begin{aligned}
\lim_{n\to\infty} \frac{1}{n} \sum_{i=r+1}^{n-s} \delta_i \alpha_{i:n}^* &= \lim_{n\to\infty} \frac{1}{n} \sum_{i=r+1}^{n-s} \delta_i F^{*-1}(p_i) \\
&= \lim_{n\to\infty} \frac{1}{n} \sum_{i=r+1}^{n-s} \delta_i(\ln p_i - \ln(1-p_i))
\end{aligned}$$

$$= \int_{p_r}^{1-p_s} u(1-u)(\ln u - \ln(1-u))\,du$$

$$= H(p_r, p_s) \quad \text{(say)} \tag{3.3.24}$$

and

$$\lim_{n\to\infty} \frac{1}{n} \sum_{i=r+1}^{n-s} \gamma_i = \int_{p_r}^{1-p_s} [u - u(1-u)(\ln u - \ln(1-u))]\,du$$

$$= \frac{1}{2}((1-p_s)^2 - p_r^2) - H(p_r, p_s) \tag{3.3.25}$$

since $\gamma_i = p_i - p_i q_i \ln(p_i/q_i)$ and, hence, $\gamma(u) = u - u(1-u) \ln(u/(1-u))$; here,

$$\begin{aligned} H(p_r, p_s) = {} & \tfrac{1}{4}\{p_r^2(1 - 2\ln p_r) - p_{n-s}^2(1 - 2\ln p_{n-s}) \\ & - p_s^2(1 - 2\ln p_s) + p_{n-r}^2(1 - 2\ln p_{n-r})\} \\ & + \tfrac{1}{8}\{p_{n-s}^3(1 - 3\ln p_{n-s}) - p_r^3(1 - 3\ln p_r) \\ & - p_{n-r}^3(1 - 3\ln p_{n-r}) + p_s^3(1 - 3\ln p_s)\}. \end{aligned} \tag{3.3.26}$$

By making use of (3.3.24)–(3.3.26), we obtain

$$\begin{aligned} \lim_{n\to\infty} V_2^* &= \lim_{n\to\infty} \left[\frac{2}{n}\left\{r\delta_{r+1}\alpha_{r+1:n}^* + s\delta_{n-s}\alpha_{n-s:n}^* \right.\right. \\ &\qquad \left.\left. + 2\sum_{i=r+1}^{n-s} \delta_i \alpha_{i:n}^*\right\} - \left(\frac{m}{n}\right) C\right] \\ &= \lim_{n\to\infty} \left[\frac{2}{n}\left\{r\delta_{r+1}F^{*-1}(p_{r+1}) + s\delta_{n-s}F^{*-1}(p_{n-s}) \right.\right. \\ &\qquad \left. + 2\sum_{i=r+1}^{n-s} \delta_i F^{*-1}(p_i) \right\} \\ &\qquad \left. - \frac{1}{n}\left\{n - s - r\gamma_{r+1} - s\gamma_{n-s} - 2\sum_{i=r+1}^{n-s} \gamma_i\right\}\right] \\ &= 2\left\{p_r\delta_r \ln\left(\frac{p_r}{1-p_r}\right) + p_s\delta_{n-s}\ln\left(\frac{p_{n-s}}{1-p_{n-s}}\right)\right\} \\ &\qquad - \{1 - p_s + p_r^2 - p_{n-s}^2 - p_r\gamma_r - p_s\gamma_{n-s}\} \\ &\qquad + 2H(p_r, p_s), \end{aligned} \tag{3.3.27}$$

where $H(p_r, p_s)$ is as given in (3.3.26). Proceeding similarly, we derive

$$\begin{aligned}\lim_{n\to\infty}\frac{1}{n}\sum_{i=r+1}^{n-s}\delta_i\alpha_{i:n}^{*(2)} &= \lim_{n\to\infty}\sum_{i=r+1}^{n-s}\delta_i\{F^{*-1}(p_i)\}^2\\ &= \int_{p_r}^{1-p_s} u(1-u)\{\ln u - \ln(1-u)\}^2\,du\\ &= \int_{\ln(p_r/(1-p_r))}^{\ln((1-p_s)/p_s)} t^2\left\{\frac{e^{-t}}{(1+e^{-t})^2}\right\}^2 dt\\ &= K(p_r, p_s) \quad \text{(say)},\end{aligned} \tag{3.3.28}$$

$$\begin{aligned}\lim_{n\to\infty}\frac{1}{n}\sum_{i=r+1}^{n-s}\gamma_i\alpha_{i:n}^{*} &= \lim_{n\to\infty}\frac{1}{n}\sum_{i=r+1}^{n-s}\gamma_i F^{*-1}(p_i)\\ &= \int_{p_r}^{1-p_s}\left\{u - u(1-u)\ln\left(\frac{u}{1-u}\right)\right\}\ln\left(\frac{u}{1-u}\right)du\\ &= \frac{1}{4}\{p_r^2(1-2\ln p_r) - p_{n-s}^2(1-2\ln p_{n-s})\\ &\quad - p_{n-r}^2(1-2\ln p_{n-r}) + p_s^2(1-2\ln p_s)\\ &\quad - p_s(1-\ln p_s) + p_{n-r}(1-\ln p_{n-r})\}\\ &\quad - K(p_r, p_s),\end{aligned} \tag{3.3.29}$$

and

$$\begin{aligned}\lim_{n\to\infty}\frac{1}{n}\sum_{i=r+1}^{n-s}\alpha_{i:n}^{*} &= \lim_{n\to\infty}\frac{1}{n}\sum_{i=r+1}^{n-s}F^{*-1}(p_i)\\ &= \int_{p_r}^{1-p_s}(\ln u - \ln(1-u))\,du\\ &= (1-p_s)\ln(1-p_s) - p_r\ln p_r + p_s\ln p_s\\ &\quad - (1-p_r)\ln(1-p_r).\end{aligned} \tag{3.3.30}$$

By making use of (3.3.28)–(3.3.30) and using a numerical integration method to evaluate $K(p_r, p_s)$ from (3.3.28), we obtain

$$\begin{aligned}\lim_{n\to\infty} V_3^* &= 2K(p_r, p_s) + 1 - p_r - p_s + p_r^2(1-p_r)\\ &\quad\times\left\{\ln\left(\frac{p_r}{1-p_r}\right)\right\}^2 + p_s^2(1-p_s)\left\{\ln\left(\frac{1-p_s}{p_s}\right)\right\}^2.\end{aligned} \tag{3.3.31}$$

By using the limiting expressions of V_1^*, V_2^*, and V_3^* derived in Eqs. (3.3.23), (3.3.27), and (3.3.31), respectively, we may compute the asymp-

Table 3.3.6 Asymptotic Values of $nA\ \mathrm{Var}(\tilde{\mu})/\sigma^2$, $nA\ \mathrm{Var}(\tilde{\sigma})/\sigma^2$, $nA\ \mathrm{Cov}(\tilde{\mu}, \tilde{\sigma})/\sigma^2$, $nA\ \mathrm{Var}(\tilde{\mu}|\sigma)/\sigma^2$, and $nA\ \mathrm{Var}(\tilde{\sigma}|\mu)/\sigma^2$

p_r	p_s	H	$n\ \mathrm{AVar}(\tilde{\mu})/\sigma^2$	$n\ \mathrm{AVar}(\tilde{\sigma})/\sigma^2$	$n\ \mathrm{ACov}(\tilde{\mu},\tilde{\sigma})/\sigma^2$	$n\ \mathrm{AVar}(\tilde{\mu}\|\sigma)/\sigma^2$	$n\ \mathrm{AVar}(\tilde{\sigma}\|\mu)/\sigma^2$
0.0	0.01	0.21367	0.911892	0.704504	0.000060	0.911892	0.704504
0.0	0.025	0.20954	0.911905	0.713132	0.000393	0.911905	0.713132
0.0	0.05	0.20013	0.912008	0.729480	0.001674	0.912005	0.729477
0.0	0.10	0.17848	0.912876	0.769054	0.007452	0.912803	0.768993
0.0	0.15	0.15785	0.915397	0.817869	0.018493	0.914979	0.817496
0.0	0.20	0.14052	0.920737	0.876663	0.036169	0.919245	0.875242
0.0	0.25	0.12718	0.930455	0.946721	0.062223	0.926365	0.942560
0.0	0.30	0.11779	0.946701	1.029899	0.098945	0.937195	1.019557
0.0	0.35	0.11191	0.972522	1.128759	0.149430	0.952739	1.105798
0.0	0.40	0.10881	1.012342	1.246821	0.217955	0.974242	1.199896
0.0	0.45	0.10766	1.072767	1.388966	0.310585	1.003318	1.299046
0.0	0.50	0.10749	1.163950	1.562093	0.436172	1.042161	1.398644
0.0	0.55	0.10732	1.302048	1.776225	0.608065	1.093886	1.492254
0.0	0.60	0.10617	1.513873	2.046448	0.847222	1.163126	1.572310
0.0	0.65	0.10307	1.846239	2.396517	1.188196	1.257130	1.631821
0.0	0.70	0.09718	2.386395	2.866066	1.691620	1.387961	1.666944
0.0	0.75	0.08780	3.311405	3.526483	2.472894	1.577324	1.679772
0.0	0.80	0.07446	5.025344	4.520589	3.777594	1.868628	1.680940
0.0	0.85	0.05713	8.623871	6.181689	6.221066	2.363175	1.693951
0.0	0.90	0.03650	18.049454	9.509726	11.817191	3.364909	1.772871
0.0	0.95	0.01485	58.727207	19.504873	31.949337	6.393601	2.123486
0.0	0.975	0.00544	174.762787	39.502609	80.069229	12.467654	2.818134
0.0	0.99	0.00130	670.757507	99.504875	252.365997	30.702404	4.554610
0.01	0.01	0.21237	0.911892	0.709763	0.000000	0.911892	0.709763
0.01	0.025	0.20824	0.911906	0.718522	0.000335	0.911906	0.718522
0.01	0.05	0.19882	0.912009	0.735120	0.001624	0.912006	0.735117

(continued)

totic variances and covariance of $\tilde{\mu}$ and $\tilde{\sigma}$ as

$$A\ \mathrm{Var}(\tilde{\mu}) = \frac{\sigma^2}{n} \frac{V_3^*}{V_1^* V_3^* - V_2^{*2}}, \tag{3.3.32}$$

$$A\ \mathrm{Var}(\tilde{\sigma}) = \frac{\sigma^2}{n} \frac{V_1^*}{V_1^* V_3^* - V_2^{*2}}, \tag{3.3.33}$$

and

$$A\ \mathrm{Cov}(\tilde{\mu}, \tilde{\sigma}) = -\frac{\sigma^2}{n} \frac{V_2^*}{V_1^* V_3^* - V_2^{*2}}. \tag{3.3.34}$$

From Eqs. (3.3.32) to (3.3.34), we have computed the values of $nA\ \mathrm{Var}(\tilde{\mu})/\sigma^2$, $nA\ \mathrm{Var}(\tilde{\sigma})/\sigma^2$, and $nA\ \mathrm{Cov}(\tilde{\mu}, \tilde{\sigma})/\sigma^2$ and the values of $nA\ \mathrm{Var}(\tilde{\mu}|\sigma)/\sigma^2$ and $nA\ \mathrm{Var}(\tilde{\sigma}|\mu)/\sigma^2$ for various choices of the proportions of censoring p_r and p_s. These values are presented in Table 3.3.6. If we interchange p_r and p_s in Table 3.3.6, the variances and the absolute value of the covariance remains unchanged while the sign of the covariance

Table 3.3.6 (*continued*)

p_r	p_s	H	$n\,AVar(\tilde{\mu})/\sigma^2$	$n\,AVar(\tilde{\sigma})/\sigma^2$	$n\,ACov(\tilde{\mu},\tilde{\sigma})/\sigma^2$	$n\,AVar(\tilde{\mu}\mid\sigma)/\sigma^2$	$n\,AVar(\tilde{\sigma}\mid\mu)/\sigma^2$
0.01	0.10	0.17717	0.912876	0.775325	0.007446	0.912804	0.775264
0.01	0.15	0.15655	0.915398	0.824963	0.018582	0.914980	0.824586
0.01	0.20	0.13922	0.920745	0.884816	0.036429	0.919245	0.883375
0.01	0.25	0.12588	0.930486	0.956232	0.062765	0.926366	0.951998
0.01	0.30	0.11649	0.946788	1.041158	0.099936	0.937196	1.030609
0.01	0.35	0.11060	0.972733	1.142288	0.151120	0.952740	1.118811
0.01	0.40	0.10751	1.012808	1.263337	0.220727	0.974243	1.215233
0.01	0.45	0.10635	1.073735	1.409475	0.315039	1.003319	1.317041
0.01	0.50	0.10618	1.165889	1.588055	0.443267	1.042162	1.419527
0.01	0.55	0.10602	1.305861	1.809831	0.619385	1.093887	1.516050
0.01	0.60	0.10486	1.521344	2.091128	0.865492	1.163127	1.598749
0.01	0.65	0.10177	1.861052	2.457927	1.218356	1.257131	1.660318
0.01	0.70	0.09588	2.416646	2.954188	1.743250	1.387963	1.696692
0.01	0.75	0.08649	3.376591	3.660570	2.566385	1.577327	1.709984
0.01	0.80	0.07315	5.179116	4.742758	3.962426	1.868633	1.711194
0.01	0.85	0.05582	9.048034	6.603369	6.643985	2.363183	1.724680
0.01	0.90	0.03520	19.633663	10.540915	13.095321	3.364922	1.806558
0.01	0.95	0.01355	71.700829	24.357613	39.883911	6.393668	2.172004
0.01	0.975	0.00413	282.961945	65.912292	133.524872	12.467790	2.904209
0.025	0.025	0.20411	0.911919	0.727499	0.000000	0.911919	0.727499
0.025	0.05	0.19469	0.912021	0.744518	0.001298	0.912019	0.744516
0.025	0.10	0.17304	0.912883	0.785782	0.007181	0.912818	0.785725
0.025	0.15	0.15242	0.915400	0.836803	0.018458	0.914993	0.836431
0.025	0.20	0.13509	0.920747	0.898435	0.036568	0.919259	0.896982
0.025	0.25	0.12175	0.930508	0.972134	0.063349	0.926380	0.967821
0.025	0.30	0.11236	0.946879	1.060003	0.101239	0.937210	1.049179
0.025	0.35	0.10647	0.972995	1.164962	0.153555	0.952755	1.140729
0.025	0.40	0.10338	1.013446	1.291058	0.224931	0.974258	1.241135
0.025	0.45	0.10222	1.075146	1.443965	0.322013	1.003335	1.347520
0.025	0.50	0.10205	1.168836	1.631824	0.454622	1.042179	1.454997
0.025	0.55	0.10189	1.311833	1.866675	0.637807	1.093906	1.556577
0.025	0.60	0.10073	1.533320	2.167048	0.895644	1.163149	1.643884
0.025	0.65	0.09764	1.885275	2.562948	1.268793	1.257157	1.709049
0.025	0.70	0.09175	2.467095	3.106304	1.830851	1.387994	1.747615
0.025	0.75	0.08236	3.487729	3.895353	2.727918	1.577367	1.761719
0.025	0.80	0.06902	5.449134	5.140956	4.290328	1.868689	1.763004
0.025	0.85	0.05169	9.829425	7.392321	7.429145	2.363273	1.777324
0.025	0.90	0.03107	22.864864	12.668049	15.716995	3.365103	1.864401
0.025	0.95	0.00942	111.099358	39.200096	64.065964	6.394326	2.256162

(*continued*)

changes. This is due to the symmetry of the logistic distribution, and this will enable us to get the values for the case when $p_s < p_r$, also from Table 3.3.6. Upon comparing the values in Table 3.3.6 to the corresponding entries in Table 3.2.3, we observe that the approximate maximum likelihood estimators $\tilde{\mu}$ and $\tilde{\sigma}$ are asymptotically fully efficient as compared to the maximum likelihood estimators $\hat{\mu}$ and $\hat{\sigma}$ with efficiency being measured either by the trace efficiency or the determinant efficiency. Moreover, we note that the asymptotic values in Table 3.3.6 provide very close approximations to the simulated values of the variances and covariance of the estimators even when $n = 20$ as long as the amount of censoring is not extremely heavy.

Table 3.3.6 *(continued)*

p_r	p_s	H	$n\,AVar(\hat{\mu})/\sigma^2$	$n\,AVar(\hat{\sigma})/\sigma^2$	$n\,ACov(\hat{\mu},\hat{\sigma})/\sigma^2$	$n\,AVar(\hat{\mu}\mid\sigma)/\sigma^2$	$n\,AVar(\hat{\sigma}\mid\mu)/\sigma^2$
0.05	0.05	0.18528	0.912119	0.762349	0.000000	0.912119	0.762349
0.05	0.10	0.16362	0.912962	0.805652	0.005957	0.912918	0.805613
0.05	0.15	0.14300	0.915448	0.859338	0.017454	0.915093	0.859005
0.05	0.20	0.12567	0.920763	0.924402	0.036005	0.919360	0.922994
0.05	0.25	0.11233	0.930513	1.002515	0.063561	0.926483	0.998174
0.05	0.30	0.10294	0.946944	1.096088	0.102733	0.937315	1.084943
0.05	0.35	0.09705	0.973289	1.208494	0.157110	0.952864	1.183133
0.05	0.40	0.09396	1.014322	1.344450	0.231757	0.974372	1.291497
0.05	0.45	0.09280	1.077315	1.510661	0.334030	1.003456	1.407093
0.05	0.50	0.09264	1.173704	1.716902	0.474964	1.042310	1.524698
0.05	0.55	0.09247	1.322206	1.977922	0.671770	1.094050	1.636617
0.05	0.60	0.09132	1.554958	2.316994	0.952598	1.163311	1.733414
0.05	0.65	0.08822	1.930574	2.773036	1.366340	1.257346	1.806026
0.05	0.70	0.08233	2.564750	3.416306	2.004837	1.388225	1.849148
0.05	0.75	0.07295	3.711804	4.387697	3.060058	1.577665	1.864947
0.05	0.80	0.05961	6.025696	6.016925	5.000988	1.869107	1.866386
0.05	0.85	0.04228	11.674195	9.296332	9.303290	2.363942	1.882441
0.05	0.90	0.02165	32.647728	19.205879	23.714396	3.366456	1.980405
0.10	0.10	0.14197	0.913718	0.854083	0.000000	0.913718	0.854083
0.10	0.15	0.12135	0.916050	0.914486	0.011807	0.915898	0.914334
0.10	0.20	0.10402	0.921155	0.988235	0.031160	0.920172	0.987181
0.10	0.25	0.09068	0.930680	1.077575	0.060287	0.927307	1.073669
0.10	0.30	0.08129	0.946972	1.185761	0.102228	0.938159	1.174725
0.10	0.35	0.07540	0.973473	1.317418	0.161250	0.953736	1.290707
0.10	0.40	0.07231	1.015380	1.479166	0.243534	0.975284	1.420756
0.10	0.45	0.07115	1.080813	1.680702	0.358314	1.004423	1.561913
0.10	0.50	0.07099	1.182939	1.936695	0.519938	1.043353	1.708167
0.10	0.55	0.07082	1.344122	2.270346	0.751759	1.095200	1.849893
0.10	0.60	0.06966	1.604611	2.720513	1.094087	1.164611	1.974522
0.10	0.65	0.06657	2.042515	3.357436	1.622052	1.258865	2.069292
0.10	0.70	0.06068	2.825894	4.322157	2.491149	1.390077	2.126100
0.10	0.75	0.05129	4.375113	5.944987	4.076343	1.580058	2.147012
0.10	0.80	0.03795	8.034269	9.220462	7.537552	1.872464	2.148918
0.10	0.85	0.02062	20.892750	19.137205	18.827820	2.369312	2.170227

(continued)

We consider the two examples presented in the last section and illustrate the method of estimation discussed in this section.

Example 3.3.1 Let us consider the censored sample given in Example 3.2.1:

$$8.2,\ 8.4,\ 9.1,\ 9.8,\ 9.9$$

In this case, we have $n = 10$, $r = 2$, and $s = 3$ and the following:

i	p_i	q_i	γ_i	δ_i
3	3/11	8/11	0.467272	0.198347
4	4/11	7/11	0.493134	0.231405
5	5/11	6/11	0.499749	0.247934
6	6/11	5/11	0.500251	0.247934
7	7/11	4/11	0.506866	0.231405

Table 3.3.6 (*continued*)

p_r	p_s	H	n AVar($\check{\mu}$)/σ^2	n AVar($\check{\sigma}$)/σ^2	n ACov($\check{\mu},\check{\sigma}$)/$\sigma^2$	n AVar($\check{\mu}\mid\sigma$)/σ^2	n AVar($\check{\sigma}\mid\mu$)/σ^2
0.15	0.15	0.10073	0.918088	0.983730	0.000000	0.918088	0.983730
0.15	0.20	0.08340	0.922753	1.068996	0.019888	0.922383	1.068567
0.15	0.25	0.07006	0.931720	1.173372	0.050431	0.929552	1.170642
0.15	0.30	0.06067	0.947424	1.301376	0.095218	0.940457	1.291807
0.15	0.35	0.05478	0.973524	1.459561	0.159422	0.956111	1.433454
0.15	0.40	0.05169	1.015702	1.657573	0.250755	0.977768	1.595666
0.15	0.45	0.05053	1.083121	1.910060	0.381164	1.007058	1.775923
0.15	0.50	0.05036	1.191239	2.240226	0.570026	1.046196	1.967459
0.15	0.55	0.05020	1.367638	2.686976	0.850657	1.098333	2.157876
0.15	0.60	0.04904	1.665282	3.320688	1.284836	1.168155	2.329381
0.15	0.65	0.04595	2.196647	4.282699	1.999625	1.263006	2.462424
0.15	0.70	0.04006	3.238673	5.904032	3.299144	1.395128	2.543289
0.15	0.75	0.03067	5.660599	9.180878	6.115800	1.586586	2.573271
0.15	0.80	0.01733	13.956242	19.106421	15.188890	1.881641	2.576010
0.20	0.20	0.06606	0.926718	1.169419	0.000000	0.926718	1.169419
0.20	0.25	0.05273	0.934722	1.293845	0.031499	0.933955	1.292783
0.20	0.30	0.04334	0.949261	1.448721	0.078894	0.944964	1.442164
0.20	0.35	0.03745	0.974196	1.643637	0.148551	0.960770	1.620985
0.20	0.40	0.03435	1.015735	1.893218	0.250308	0.982641	1.831534
0.20	0.45	0.03320	1.084309	2.220676	0.400086	1.012228	2.073053
0.20	0.50	0.03303	1.198442	2.664977	0.625186	1.051777	2.338839
0.20	0.55	0.03287	1.393481	3.296625	0.976070	1.104485	2.612933
0.20	0.60	0.03171	1.743912	4.257226	1.556113	1.175117	2.868687
0.20	0.65	0.02862	2.431432	5.878307	2.611609	1.271149	3.073170
0.20	0.70	0.02273	4.020561	9.157140	4.893916	1.405070	3.200157
0.20	0.75	0.01334	9.401025	19.089218	12.203518	1.599456	3.247770
0.25	0.25	0.03939	0.941306	1.445244	0.000000	0.941306	1.445244
0.25	0.30	0.03000	0.953970	1.637054	0.049215	0.952491	1.634516
0.25	0.35	0.02411	0.976723	1.883845	0.124078	0.968551	1.868082
0.25	0.40	0.02102	1.016299	2.208822	0.237408	0.990781	2.153363
0.25	0.45	0.01986	1.084633	2.650985	0.411145	1.020868	2.495135
0.25	0.50	0.01969	1.204445	3.280939	0.685767	1.061109	2.890489
0.25	0.55	0.01953	1.423414	4.240480	1.144008	1.114781	3.321033
0.25	0.60	0.01837	1.857270	5.861501	1.982446	1.186778	3.745443
0.25	0.65	0.01528	2.863539	9.141939	3.799037	1.284805	4.101782
0.25	0.70	0.00939	6.262877	19.078705	9.610514	1.421775	4.331175

(*continued*)

Further, we have $A = 5$, $m = 3.404959$, $C = -0.114447$, $B = 9.168932$, $D = 2.228374$, and $E = 1.836713$, and by substituting these values in (3.3.6) and (3.3.7), we obtain the approximate maximum likelihood estimates of μ and σ to be

$$\check{\mu} = 9.2683 \qquad \text{and} \qquad \check{\sigma} = 1.5755.$$

By using Table 3.3.1, we obtain unbiased estimates of σ and μ to be

$$\check{\sigma}_U = \frac{\check{\sigma}}{0.838552} = 1.8788$$

and

$$\check{\mu}_U = \check{\mu} + 0.022731\check{\sigma}_U = 9.3110$$

Table 3.3.6 (*continued*)

p_r	p_s	H	$n\,AVar(\dot{\mu})/\sigma^2$	$n\,AVar(\dot{\sigma})/\sigma^2$	$n\,ACov(\dot{\mu},\dot{\sigma})/\sigma^2$	$n\,AVar(\dot{\mu}\|\sigma)/\sigma^2$	$n\,AVar(\dot{\sigma}\|\mu)/\sigma^2$
0.30	0.30	0.02061	0.963944	1.880831	0.000000	0.963944	1.880831
0.30	0.35	0.01472	0.983208	2.203206	0.078712	0.980396	2.196905
0.30	0.40	0.01163	1.018918	2.643180	0.203958	1.003180	2.602354
0.30	0.45	0.01047	1.084707	3.271413	0.407143	1.034036	3.118592
0.30	0.50	0.01031	1.209178	4.229832	0.752396	1.075343	3.761663
0.30	0.55	0.01014	1.461638	5.850575	1.391885	1.130501	4.525116
0.30	0.60	0.00898	2.055658	9.132030	2.787792	1.204611	5.351349
0.30	0.65	0.00589	4.075935	19.071875	7.268630	1.305732	6.109703
0.35	0.35	0.00883	0.997419	2.640668	0.000000	0.997419	2.640668
0.35	0.40	0.00574	1.026591	3.266899	0.135018	1.021011	3.249141
0.35	0.45	0.00458	1.085973	4.223896	0.373243	1.052991	4.095613
0.35	0.50	0.00442	1.212600	5.843957	0.825975	1.095858	5.281335
0.35	0.55	0.00425	1.520635	9.125758	1.831162	1.153197	6.920658
0.35	0.60	0.00309	2.589497	19.067511	5.090612	1.230414	9.060036
0.40	0.40	0.00265	1.045746	4.221987	0.000000	1.045746	4.221987
0.40	0.45	0.00149	1.092159	5.840827	0.273846	1.079320	5.772164
0.40	0.50	0.00132	1.214726	9.122276	0.907717	1.124403	8.443974
0.40	0.55	0.00116	1.662023	19.064877	3.016163	1.184851	13.591284
0.45	0.45	0.00033	1.115121	9.121157	0.000000	1.115121	9.121157
0.45	0.50	0.00017	1.215687	19.063631	0.999232	1.163311	18.242315

We also obtain the standard errors of the above estimates to be

$$\text{S.E.}(\tilde{\sigma}_U) = \frac{\tilde{\sigma}_U\sqrt{0.127362}}{0.838552} = 0.7996$$

and

$$\text{S.E.}(\tilde{\mu}_U) = \tilde{\sigma}_U\left\{0.098259 + 0.127362\left(\frac{0.022731}{0.838552}\right)^2 + 2(0.005522)\left(\frac{0.022731}{0.838552}\right)\right\}^{1/2} = 0.5901.$$

We observe that the estimates of μ and σ and their standard errors are very close to those calculated in Example 3.2.1 based on the maximum likelihood estimation method.

Example 3.3.2 Let us consider the Type II censored sample given in Example 3.2.2:

785, 855, 905, 918, 919, 920, 929, 936, 948, 950

In this case, we have $n = 20$, $r = 0$, and $s = 10$ and the following:

i	p_i	q_i	γ_i	δ_i
1	1/21	20/21	0.183479	0.045351
2	2/21	19/21	0.289227	0.086168
3	3/21	18/21	0.362256	0.122449
4	4/21	17/21	0.413584	0.154195
5	5/21	16/21	0.449098	0.181406
6	6/21	15/21	0.472713	0.204082
7	7/21	14/21	0.487366	0.222222
8	8/21	13/21	0.495449	0.235828
9	9/21	12/21	0.499024	0.244898
10	10/21	11/21	0.499964	0.249433

Further, we have $A = 10$, $m = 5.986394$, $B = 934.525771$, $C = -0.551912$, $D = 219.005212$, and $E = 4409.36436$, and by substituting these values in (3.3.6) and (3.3.7), we obtain the approximate maximum likelihood estimates of μ and σ to be

$$\tilde{\mu} = 953.63982 \quad \text{and} \quad \tilde{\sigma} = 62.81626.$$

By using Table 3.3.2, we obtain unbiased estimates of σ and μ:

$$\tilde{\sigma}_U = \frac{\tilde{\sigma}}{0.939267} = 66.8780$$

and

$$\tilde{\mu}_U = \tilde{\mu} + 0.042804\tilde{\sigma}_U = 956.5025.$$

We also obtain the standard errors of the above estimates as

$$\text{S.E.}(\tilde{\sigma}_U) = \frac{\tilde{\sigma}_U\sqrt{0.077270}}{0.939267} = 19.792$$

and

$$\text{S.E.}(\tilde{\mu}_U) = \tilde{\sigma}_U\left\{0.059640 + 0.077270\left(\frac{0.042804}{0.939267}\right)^2 + 2(0.020566)\left(\frac{0.042804}{0.939267}\right)\right\}^{1/2}$$

$$= 16.609$$

REFERENCES

Balakrishnan, N. (1990). Best linear unbiased estimates of the location and scale parameters of logistic distribution for complete and censored samples of sizes 2(1)25(5)40, submitted for publication.

Balakrishnan, N. and Cohen, A. C. (1990). *Order Statistics and Inference: Estimation Methods*. Academic Press, Boston.

Davis, D. J. (1952). An analysis of some failure data. *J. Amer. Statist. Assoc.*, *47*, 113–150.

Gupta, S. S., Qureishi, A. S., and Shah, B. K. (1967). Best linear unbiased estimators of the parameters of the logistic distribution using order statistics. *Technometrics*, *9*, 43–56.

Hailey, D. C. (1952). Estimation of the dosage mortality relationship when the cost is subject to error. *Technical Report No. 15*, Applied Mathematics Laboratory, Stanford University.

Harter, H. L. (1970). *Order Statistics and Their Uses in Testing and Estimation*, Vol. 2. U.S. Government Printing Office, Washington, D.C.

Harter, H. L. and Moore, A. H. (1966). Iterative maximum-likelihood estimation of the parameters of normal populations from singly and doubly censored samples. *Biometrika*, *53*, 205–213.

Harter, H. L. and Moore, A. H. (1967). Maximum-likelihood estimation, from censored samples, of the parameters of a logistic distribution. *J. Amer. Statist. Assoc.*, *62*, 675–684.

Kendall, M. G. and Stuart, A. (1961). *The Advanced Theory of Statistics*, Vol. 2. Charles Griffin, London.

Sarhan, A. E. and Greenberg, B. G. (eds.) (1962). *Contributions to Order Statistics*. Wiley, New York.

Tiku, M. L. (1968). Estimating the parameters of normal and logistic distributions from censored samples. *Austral. J. Statist.*, *10*, 64–74.

4

Linear Estimation Based on Complete and Censored Samples

4.1 INTRODUCTION

N. Balakrishnan
McMaster University, Hamilton, Ontario, Canada

In the last chapter we discussed the maximum likelihood and an approximate maximum likelihood estimation of the location and scale parameters based on complete and Type II censored samples. In this chapter, we discuss in detail the linear estimation of the location and scale parameters based on complete and censored samples. In Section 4.2 we describe the best linear unbiased estimation based on complete and doubly Type II censored samples. In Section 4.3 the best linear unbiased estimation and the asymptotically best linear unbiased estimation based on optimally selected order statistics are discussed. The best linear unbiased estimation and the asymptotically best linear unbiased estimation of the quantiles of the logistic $L(\mu, \sigma^2)$ population based on optimally selected order statistics are described in Section 4.4. In Section 4.5 we derive linear estimators with polynomial coefficients for the location and scale parameters by using the method of Downton (1966). Finally, in Section 4.6 we present some simplified linear estimators for the location and scale parameters proposed by Raghunandanan and Srinivasan (1970) which are in terms of quasi-midrange and quasi-range, respectively. These estimators, which are simple in form and also highly efficient, are similar to those suggested by Dixon (1957) and Raghunandanan and Srinivasan (1971) for the normal and the double exponential distributions, respectively.

4.2 BEST LINEAR UNBIASED ESTIMATION BASED ON CENSORED SAMPLES

N. Balakrishnan
McMaster University, Hamilton, Ontario, Canada

Let

$$X_{r+1:n} \le X_{r+2:n} \le \cdots \le X_{n-s:n} \tag{4.2.1}$$

be the doubly Type II censored sample available from the logistic $L(\mu, \sigma^2)$ population with pdf and cdf as given in Eqs. (1.1) and (1.2), respectively. By using $\mathbf{1}$ to denote a column vector of $(n - r - s)$ 1's, we may write

$$\xi(\mathbf{X}) = \mu\mathbf{1} + \sigma\boldsymbol{\alpha} \qquad \text{and} \qquad \mathcal{V}(\mathbf{X}) = \sigma^2\boldsymbol{\beta}. \tag{4.2.2}$$

Then, by considering the generalized variance

$$(\mathbf{X} - \mu\mathbf{1} - \sigma\boldsymbol{\alpha})^T\boldsymbol{\beta}^{-1}(\mathbf{X} - \mu\mathbf{1} - \sigma\boldsymbol{\alpha}) \tag{4.2.3}$$

and minimizing it with respect to μ and σ, we obtain the equations

$$\mu\mathbf{1}^T\boldsymbol{\beta}^{-1}\mathbf{1} + \sigma\boldsymbol{\alpha}^T\boldsymbol{\beta}^{-1}\mathbf{1} = \mathbf{1}^T\boldsymbol{\beta}^{-1}\mathbf{X} \tag{4.2.4}$$

and

$$\mu\boldsymbol{\alpha}^T\boldsymbol{\beta}^{-1}\mathbf{1} + \sigma\boldsymbol{\alpha}^T\boldsymbol{\beta}^{-1}\boldsymbol{\alpha} = \boldsymbol{\alpha}^T\boldsymbol{\beta}^{-1}\mathbf{X}. \tag{4.2.5}$$

Upon solving Eqs. (4.2.4) and (4.2.5), we derive the best linear unbiased estimators of μ and σ (Lloyd, 1952; David, 1981; Balakrishnan and Cohen, 1990) to be

$$\begin{aligned}\mu^* &= \left\{\frac{\boldsymbol{\alpha}^T\boldsymbol{\beta}^{-1}\boldsymbol{\alpha}\mathbf{1}^T\boldsymbol{\beta}^{-1} - \boldsymbol{\alpha}^T\boldsymbol{\beta}^{-1}\mathbf{1}\boldsymbol{\alpha}^T\boldsymbol{\beta}^{-1}}{(\boldsymbol{\alpha}^T\boldsymbol{\beta}^{-1}\boldsymbol{\alpha})(\mathbf{1}^T\boldsymbol{\beta}^{-1}\mathbf{1}) - (\boldsymbol{\alpha}^T\boldsymbol{\beta}^{-1}\mathbf{1})^2}\right\}\mathbf{X} \\ &= -\boldsymbol{\alpha}^T\boldsymbol{\Delta}\mathbf{X} \\ &= \sum_{i=r+1}^{n-s} a_i X_{i:n} \end{aligned} \tag{4.2.6}$$

and

$$\begin{aligned}\sigma^* &= \left\{\frac{\mathbf{1}^T\boldsymbol{\beta}^{-1}\mathbf{1}\boldsymbol{\alpha}^T\boldsymbol{\beta}^{-1} - \mathbf{1}^T\boldsymbol{\beta}^{-1}\boldsymbol{\alpha}\mathbf{1}^T\boldsymbol{\beta}^{-1}}{(\boldsymbol{\alpha}^T\boldsymbol{\beta}^{-1}\boldsymbol{\alpha})(\mathbf{1}^T\boldsymbol{\beta}^{-1}\mathbf{1}) - (\boldsymbol{\alpha}^T\boldsymbol{\beta}^{-1}\mathbf{1})^2}\right\}\mathbf{X} \\ &= \mathbf{1}^T\boldsymbol{\Delta}\mathbf{X} \\ &= \sum_{i=r+1}^{n-s} b_i X_{i:n}, \end{aligned} \tag{4.2.7}$$

where $\boldsymbol{\Delta}$ is a skew-symmetric matrix of order $n - r - s$ given by

$$\boldsymbol{\Delta} = \left\{\frac{\boldsymbol{\beta}^{-1}(\mathbf{1}\boldsymbol{\alpha}^T - \boldsymbol{\alpha}\mathbf{1}^T)\boldsymbol{\beta}^{-1}}{(\boldsymbol{\alpha}^T\boldsymbol{\beta}^{-1}\boldsymbol{\alpha})(\mathbf{1}^T\boldsymbol{\beta}^{-1}\mathbf{1}) - (\boldsymbol{\alpha}^T\boldsymbol{\beta}^{-1}\mathbf{1})^2}\right\}. \tag{4.2.8}$$

Furthermore, from Eqs. (4.2.6) and (4.2.7) we obtain the variances and covariance of the estimators to be

$$\mathrm{Var}(\mu^*) = \sigma^2\left\{\frac{\boldsymbol{\alpha}^T\boldsymbol{\beta}^{-1}\boldsymbol{\alpha}}{(\boldsymbol{\alpha}^T\boldsymbol{\beta}^{-1}\boldsymbol{\alpha})(\mathbf{1}^T\boldsymbol{\beta}^{-1}\mathbf{1}) - (\boldsymbol{\alpha}^T\boldsymbol{\beta}^{-1}\mathbf{1})^2}\right\}, \tag{4.2.9}$$

$$\mathrm{Var}(\sigma^*) = \sigma^2\left\{\frac{\mathbf{1}^T\boldsymbol{\beta}^{-1}\mathbf{1}}{(\boldsymbol{\alpha}^T\boldsymbol{\beta}^{-1}\boldsymbol{\alpha})(\mathbf{1}^T\boldsymbol{\beta}^{-1}\mathbf{1}) - (\boldsymbol{\alpha}^T\boldsymbol{\beta}^{-1}\mathbf{1})^2}\right\}, \tag{4.2.10}$$

and

$$\mathrm{Cov}(\mu^*, \sigma^*) = -\sigma^2\left\{\frac{\boldsymbol{\alpha}^T\boldsymbol{\beta}^{-1}\mathbf{1}}{(\boldsymbol{\alpha}^T\boldsymbol{\beta}^{-1}\boldsymbol{\alpha})(\mathbf{1}^T\boldsymbol{\beta}^{-1}\mathbf{1}) - (\boldsymbol{\alpha}^T\boldsymbol{\beta}^{-1}\mathbf{1})^2}\right\}. \tag{4.2.11}$$

For the case when the censoring in the sample is symmetric (that is, $r = s$), some simplification in formulas (4.2.6)–(4.2.11) is possible since the logistic distribution is symmetric. In this case, since $\alpha_{i:n} = -\alpha_{n-i+1:n}$ and $\beta_{i,j:n} = \beta_{n-j+1,n-i+1:n}$, we can show that

$$\boldsymbol{\alpha}^T\boldsymbol{\beta}^{-1}\mathbf{1} = 0. \tag{4.2.12}$$

Hence, we note from (4.2.11) that the estimators μ^* and σ^* are uncorrelated when the sample in (4.2.1) is symmetrically Type II censored. Also, the estimators simplify to

$$\mu^* = \frac{\mathbf{1}^T\boldsymbol{\beta}^{-1}}{\mathbf{1}^T\boldsymbol{\beta}^{-1}\mathbf{1}}\mathbf{X} \tag{4.2.13}$$

and

$$\sigma^* = \frac{\boldsymbol{\alpha}^T\boldsymbol{\beta}^{-1}}{\boldsymbol{\alpha}^T\boldsymbol{\beta}^{-1}\boldsymbol{\alpha}}\mathbf{X}, \tag{4.2.14}$$

and their variances simplify to

$$\mathrm{Var}(\mu^*) = \frac{\sigma^2}{\mathbf{1}^T\boldsymbol{\beta}^{-1}\mathbf{1}} \tag{4.2.15}$$

and

$$\mathrm{Var}(\sigma^*) = \frac{\sigma^2}{\boldsymbol{\alpha}^T\boldsymbol{\beta}^{-1}\boldsymbol{\alpha}}. \tag{4.2.16}$$

The coefficients a_i and b_i of the estimators μ^* and σ^* in Eqs. (4.2.6) and (4.2.7), respectively, and the values of $\mathrm{Var}(\mu^*)/\sigma^2$, $\mathrm{Var}(\sigma^*)/\sigma^2$, and

$\text{Cov}(\mu^*, \sigma^*)/\sigma^2$ computed from Eqs. (4.2.9)–(4.2.11), have been tabulated by Gupta, Qureishi, and Shah (1967) for sample sizes $n = 2(1)5(5)25$ and some selected choices of r and s. Recently, Balakrishnan (1991) has prepared more exhaustive tables covering sample sizes $n = 2(1)25(5)40$ and all possible choices of r and s. We illustrate the application of these tables in the following two examples, which have already been considered in Chapter 3.

Example 4.2.1 Sarhan and Greenberg (1962, p. 212) have given the measurements resulting from an experiment to measure the strontium-90 concentrations in samples of milk. The test substance was supposed to contain 9.22 micromicrocuries per liter. Ten measurements were considered, but owing to relatively larger measurement error known to exist at the extremes, the two smallest and the three largest observations were censored. The remaining five observations, in order of magnitude, are given below:

$$8.2,\ 8.4,\ 9.1,\ 9.8,\ 9.9.$$

In this case, we have $n = 10$, $r = 2$, and $s = 3$. By using Tables 1 and 2 of Balakrishnan (1991), we obtain the best linear unbiased estimates of μ and σ to be

$$\begin{aligned}\mu^* &= 0.15237(8.2) + 0.13241(8.4) + 0.15377(9.1)\\ &\quad + 0.16153(9.8) + 0.39992(9.9)\\ &= 9.3032\end{aligned}$$

and

$$\begin{aligned}\sigma^* &= -0.93064(8.2) - 0.17440(8.4) - 0.04984(9.1)\\ &\quad + 0.07908(9.8) + 1.07581(9.9)\\ &= 1.8758.\end{aligned}$$

From Table 3 of Balakrishnan (1991), we obtain estimates of the variances and covariance of the above estimates to be

$$\begin{aligned}\widehat{\text{Var}}(\mu^*) &= (1.8758)^2(0.09921) = 0.3491,\\ \widehat{\text{Var}}(\sigma^*) &= (1.8758)^2(0.18338) = 0.6452,\end{aligned}$$

and

$$\widehat{\text{Cov}}(\mu^*, \sigma^*) = (1.8758)^2(0.01161) = 0.0409.$$

We, therefore, obtain the standard errors of the estimates to be

$$\text{S.E.}(\mu^*) = (0.3491)^{1/2} = 0.5908$$

and

$$\text{S.E.}(\sigma^*) = (0.6452)^{1/2} = 0.8033.$$

Example 4.2.2 Davis (1952) has given lifetimes in hours of 417 forty-watt lamps taken from 42 weekly forced-life test samples. As in Example 3.2.2, let us consider the first 20 observations from this data. Suppose 50% censoring had occurred in this data and that only the following first 10 observations were available:

785, 855, 905, 918, 919, 920, 929, 936, 948, 950.

By taking $n = 20$, $r = 0$, and $s = 10$ and using Tables 1 and 2 of Balakrishnan (1991), we obtain the best linear unbiased estimates of μ and σ to be

$$\begin{aligned}\mu^* &= -0.04336(785) - 0.02968(855) - 0.01300(905)\\ &\quad + 0.00454(918) + 0.02200(919) + 0.03882(920)\\ &\quad + 0.05460(929) + 0.069021(936) + 0.08176(948) + 0.81530(950)\\ &= 956.28\end{aligned}$$

and

$$\begin{aligned}\sigma^* &= -0.16913(785) - 0.17466(855) - 0.16574(905) - 0.14860(918)\\ &\quad - 0.12598(919) - 0.09961(920) - 0.07070(929)\\ &\quad - 0.04027(936) - 0.00916(948) + 1.00385(950)\\ &= 65.67.\end{aligned}$$

From Table 3 of Balakrishnan (1991), we obtain estimates of the variances and covariance of the above estimates to be

$$\widehat{\text{Var}}(\mu^*) = (65.67)^2(0.06168) = 265.998,$$
$$\widehat{\text{Var}}(\sigma^*) = (65.67)^2(0.08679) = 374.286,$$

and

$$\widehat{\text{Cov}}(\mu^*, \sigma^*) = (65.67)^2(0.02632) = 113.506.$$

We, therefore, obtain the standard errors of the estimates to be

$$\text{S.E.}(\mu^*) = (265.998)^{1/2} = 16.309$$

and

$$\text{S.E.}(\sigma^*) = (374.286)^{1/2} = 19.346.$$

REFERENCES

Balakrishnan, N. (1991). Best linear unbiased estimates of the location and scale parameters of logistic distribution for complete and censored samples of sizes 2(1)25(5)40. Submitted for publication.

Balakrishnan, N. and Cohen, A.C. (1990). *Order Statistics and Inference: Estimation Methods*. Academic Press, Boston.

David, H. A. (1981). *Order Statistics*, 2nd ed. Wiley, New York.

Davis, D. J. (1952). An analysis of some failure data. *J. Amer. Statist. Assoc.*, *47*, 113–150.

Dixon, W. J. (1957). Estimates of the mean and standard deviation of a normal population. *Ann. Math. Statist.*, *28*, 806–809.

Downton, F. (1966). Linear estimates with polynomial coefficients. *Biometrika*, *53*, 129–141.

Gupta, S. S., Qureishi, A. S., and Shah, B. K. (1967). Best linear unbiased estimators of the parameters of the logistic distribution using order statistics. *Technometrics*, *9*, 43–56.

Lloyd, E. H. (1952). Least-squares estimation of location and scale parameters using order statistics. *Biometrika*, *39*, 88–95.

Raghunandanan, K. and Srinivasan, R. (1970). Simplified estimation of parameters in a logistic distribution. *Biometrika*, *57*, 677–678.

Raghunandanan, K. and Srinivasan, R. (1971). Simplified estimation of parameters in a double exponential distribution. *Technometrics*, *13*, 689–691.

Sarhan, A. E. and Greenberg, B. G. (eds.) (1962). *Contributions to Order Statistics*. Wiley, New York.

4.3 LINEAR ESTIMATION BASED ON SELECTED ORDER STATISTICS

Smiley W. Cheng
University of Manitoba, Winnipeg, Manitoba, Canada

4.3.1 Introduction

Ever since Lloyd (1952) and Ogawa (1951) introduced the basic theory of the best linear unbiased estimator (BLUE) and the asymptotically best linear unbiased estimator (ABLUE), many statisticians have devoted their research work in these areas. For the logistic distribution, they are Kulldorff (1964), Gupta and Gnanadesikan (1966), Gupta, Qureishi and Shah (1967), Chan (1969), Hassanein (1969, 1974), Chan, Chan, and Mead (1971, 1973), Chan and Cheng (1972, 1974), Hassanein and Sebaugh (1973), and Cheng (1975).

This section mainly deals with the problem of the best linear unbiased estimators of the location and/or scale parameter based on a few order statistics selected from a complete sample and a censored sample in which some of the observations are missing.

Developments on these problems in general have been discussed in great detail in the recent monograph by Balakrishnan and Cohen (1990).

We will divide the section into three parts: finite sample case—the BLUEs; asymptotic case—the ABLUEs; other linear estimators. However, we first define some notation.

Notations

$$\boldsymbol{\lambda} = (\lambda_1, \lambda_2, \ldots, \lambda_k);$$

$$\lambda_0 = 0 < \lambda_1 < \lambda_2 < \cdots < \lambda_k < 1 = \lambda_{k+1};$$

$$\boldsymbol{\lambda}^0 = \left(\frac{1}{k+1}, \frac{2}{k+1}, \ldots, \frac{k}{k+1}\right);$$

$$\mathbf{I} = [\alpha_{11}, \alpha_{12}] \cup [\alpha_{21}, \alpha_{22}] \cup \cdots \cup [\alpha_{J1}, \alpha_{J2}],$$

$$0 \le \alpha_{11} < \alpha_{12} < \alpha_{21} < \alpha_{22} < \cdots < \alpha_{J1} < \alpha_{J2} \le 1;$$

$[a]$: Gauss's symbol = the integral part of a.

$$\mathbf{n}_k = (n_1, n_2, \ldots, n_k);\ 0 < n_1 < n_2 < \cdots < n_k \le n;$$

$$\mathbf{Y}_k^T = (Y_{n_1:n}, Y_{n_2:n}, \ldots, Y_{n_k:n});$$

$$\mathbf{X}_k^T = (X_{n_1:n}, X_{n_2:n}, \ldots, X_{n_k:n});$$

$$\boldsymbol{\alpha}_k^T = (\alpha_{n_1:n}, \alpha_{n_2:n}, \ldots, \alpha_{n_k:n});$$

$\mathbf{A}_k = E\{(\mathbf{Y}_k - \boldsymbol{\alpha}_k)(\mathbf{Y}_k - \boldsymbol{\alpha}_k)^T\}$, where E is the expectation operator;

$$\mathbf{1}^T = (1, 1, \ldots, 1).$$

BLUE

Lloyd (1952) first used the generalized least-squares method to derive the best linear unbiased estimators (BLUE) of the parameters based on k

order statistics selected from a given sample below:

$$X_{r+1:n} < X_{r+2:n} < \cdots < X_{n-s:n}. \tag{4.3.1}$$

1. The BLUE μ^* when σ is unknown:

$$\mu^* = \frac{\mathbf{1}^T A_k^{-1}\mathbf{X}_k}{\mathbf{1}^T\mathbf{A}_k^{-1}\mathbf{1}} - \frac{\mathbf{1}^T\mathbf{A}_k^{-1}\boldsymbol{\alpha}_k}{\mathbf{1}^T\mathbf{A}_k^{-1}\mathbf{1}}\sigma,$$

$$\mathrm{Var}(\mu^*) = \frac{\sigma^2}{\mathbf{1}^T\mathbf{A}_k^{-1}\mathbf{1}},$$

$$\mathrm{R.E.}(\mu^*) = \frac{\mathrm{Var[BLUE\ based\ on\ (4.3.1)]}}{\mathrm{Var}(\mu^*)}.$$

2. The BLUE σ^* when μ is unknown:

$$\sigma^* = \frac{\boldsymbol{\alpha}_k^T\mathbf{A}_k^{-1}\mathbf{X}_k}{\boldsymbol{\alpha}_k^T\mathbf{A}_k^{-1}\boldsymbol{\alpha}_k} - \frac{\boldsymbol{\alpha}_k^T\mathbf{A}_k^{-1}\mathbf{1}}{\boldsymbol{\alpha}_k^T\mathbf{A}_k^{-1}\boldsymbol{\alpha}_k}\mu,$$

$$\mathrm{Var}(\sigma^*) = \frac{\sigma^2}{\boldsymbol{\alpha}_k^T\mathbf{A}_k^{-1}\boldsymbol{\alpha}_k},$$

$$\mathrm{R.E.}(\sigma^*) = \frac{\mathrm{Var[BLUE\ based\ on\ (4.3.1)]}}{\mathrm{Var}(\sigma^*)}.$$

3. The BLUE(μ^*, σ^*) when both μ and σ are unknown:

$$\mu^* = \frac{1}{\Delta}\{(\boldsymbol{\alpha}_k^T\mathbf{A}_k^{-1}\boldsymbol{\alpha}_k)(\mathbf{1}^T\mathbf{A}_k^{-1}) - (\boldsymbol{\alpha}_k^T\mathbf{A}_k^{-1}\mathbf{1})(\boldsymbol{\alpha}_k^T\mathbf{A}_k^{-1})\}\mathbf{X}_k,$$

$$\sigma^* = \frac{1}{\Delta}\{(\mathbf{1}^T\mathbf{A}_k^{-1}\mathbf{1})(\boldsymbol{\alpha}_k^T\mathbf{A}_k^{-1}) - (\mathbf{1}^T\mathbf{A}_k^{-1}\boldsymbol{\alpha}_k)(\mathbf{1}^T\mathbf{A}_k^{-1})\}\mathbf{X}_k,$$

where

$$\Delta = (\mathbf{1}^T\mathbf{A}_k^{-1}\mathbf{1})(\boldsymbol{\alpha}_k^T\mathbf{A}_k^{-1}\boldsymbol{\alpha}_k) - (\mathbf{1}^T\mathbf{A}_k^{-1}\boldsymbol{\alpha}_k)^2$$

$$= \frac{\sigma^4}{\mathrm{G.Var}(\mu^*, \sigma^*)},$$

$$\mathrm{Var}(\mu^*) = \frac{\boldsymbol{\alpha}_k^T\mathbf{A}_k^{-1}\boldsymbol{\alpha}_k}{\Delta}\sigma^2,$$

$$\mathrm{Var}(\sigma^*) = \frac{\mathbf{1}^T\mathbf{A}_k^{-1}\mathbf{1}}{\Delta}\sigma^2,$$

$$\mathrm{Cov}(\mu^*, \sigma^*) = -\frac{\mathbf{1}^T\mathbf{A}_k^{-1}\boldsymbol{\alpha}_k}{\Delta}\sigma^2,$$

$$\text{R.E.}(\mu^*) = \frac{\text{Var[BLUE based on (4.3.1)]}}{\text{Var}(\mu^*)},$$

$$\text{R.E.}(\sigma^*) = \frac{\text{Var[BLUE based on (4.3.1)]}}{\text{Var}(\sigma^*)},$$

$$\text{G.R.E.}(\mu^*, \sigma^*) = \frac{\text{G.Var[BLUE based on (4.3.1)]}}{\text{G.Var}(\mu^*, \sigma^*)}.$$

From the formulas above, we can see that these estimators, the corresponding variances, and the relative efficiencies can easily be obtained if the first two moments of order statistics are available. As for the logistic distribution, these can be found in Birnbaum and Dudman (1963), Gupta, Qureishi, and Shah (1967), Shah (1966), and Balakrishnan and Malik (1991).

A FORTRAN computer program was written by Chan and Cheng (1982) to determine the best ranks of the order statistics that will provide the BLUEs and the associated measurements.

Tables 4.3.1–4.3.3 are given to show a few cases of the BLUEs. Extensive results for sample sizes $n = 1(1)20$ and number of selected order statistics $k = 1(1)4$ were done in Chan, Chan, and Mead (1973).

ABLUE

Ogawa (1951) applied the Gauss-Markov theorem and the results by Mosteller (1946) to get the ABLUE of the parameters based on k order statistics chosen from the sample below:

$$X_{[n\alpha+1]:n} < X_{[n\alpha+2]:n} < \cdots < X_{[n\beta+1]:n}, \qquad 0 \le \alpha < \beta \le 1. \quad (4.3.2)$$

Table 4.3.1 Optimum Ranks n_i, the Corresponding Coefficients a_i, the Variance, and the Relative Efficiency of the BLUE μ^* Based on Four Order Statistics Chosen from a Type II Censored Sample of Size 20

r	s	n_1	n_2	n_3	n_4	a_1	a_2	a_3	a_4	Var (μ^*)	R.E. (μ^*)
0	0	4	8	12	16	.1690	.2933	.3061	.2317	.0479	.9635
0	6	4	8	11	14	.1713	.2598	.2369	.3320	.0486	.9819
2	5	4	8	11	15	.1698	.2575	.2744	.2983	.0484	.9794
2	10	3	6	8	10	.1055	.1735	.1634	.5575	.0541	.9952
4	10	5	7	8	10	.2003	.1152	.1233	.5612	.0544	.9984
5	8	6	8	10	12	.2490	.1548	.1648	.4313	.0510	.9977
10	6	11	12	13	14	.5294	.0873	.0837	.2995	.0558	1.0000
13	0	14	15	17	19	.7017	.1253	.1205	.0525	.0663	.9977
14	1	15	16	17	18	.7616	.0806	.0996	.0582	.0740	.9990

Table 4.3.2 Optimum Ranks n_i, the Corresponding Coefficients b_i, the Variance, and the Relative Efficiency of the BLUE σ^* Based on Three Order Statistics Chosen from a Type II Censored Sample of Size 15

r	s	n_1	n_2	n_3	b_1	b_2	b_3	Var (σ^*)	R.E. (σ^*)
0	0	1	3	14	−.1476	−.3098	.3841	.0597	.8304
0	5	1	3	10	−.2099	−.5004	.6027	.0850	.9302
1	1	2	4	14	−.2988	−.2195	.4137	.0633	.9018
2	2	3	4	13	−.4173	−.1943	.5703	.0717	.9610
5	2	6	11	13	−.8390	.3365	.6683	.1029	.9907
7	3	9	11	12	−1.1118	.3701	1.2648	.1763	.9955
8	0	9	13	15	−.3923	.6140	.2874	.1162	.9299
10	1	11	13	14	−.1002	.3895	.5844	.1437	.9912
11	1	12	13	14	.0574	.2912	.5866	.1441	1.0000

He also gave these estimators:

1. The ABLUE μ^* when σ is unknown:

$$\mu^* = \sum_{i=1}^{k} a_i X_{n_i:n} - \frac{K_3}{K_1}\sigma,$$

$$\text{A.Var}(\mu^*) = \frac{\sigma^2}{nK_1},$$

Table 4.3.3 Optimum Ranks n_i, the Corresponding Coefficients c_i and d_i, the Variances, the Covariance, the Relative Efficiencies, and the Generalized Relative Efficiency of the BLUE (μ^*, σ^*) Based on Two Order Statistics Chosen from a Type II Censored Sample of Sizes 10 and 12

n	r	s	n_1	n_2	c_1	d_1	V(μ^*)	V(σ^*)	Cov(μ^*,s^*)	R.E.(μ^*)	R.E.(σ^*)	G.R.E.(μ^*,s^*)
10	0	0	2	9	.5000	−.5279	.1379	.1047	.0000	.6778	.7337	.4973
10	0	4	2	6	.1043	−.9457	.1133	.1961	.0304	.9586	.7624	.7156
10	3	3	4	7	.5000	−1.4706	.1035	.2546	.0000	.9767	.9991	.9768
12	0	0	3	10	.5000	−.6824	.0969	.1026	.0000	.8002	.6139	.4912
12	0	6	2	6	−.0946	−1.0292	.1105	.2084	.0570	.9709	.7496	.7211
12	1	3	2	9	.3144	−.6447	.0978	.1090	−.0018	.8171	.9238	.7503
12	4	0	5	11	.7911	−.7438	.0931	.1310	−.0077	.8988	.7947	.6992
12	7	1	8	11	1.3590	−1.2778	.1508	.2849	−.1166	.9991	.9525	.9378
12	10	0	11	12	2.7682	−1.6626	1.3195	.9192	−1.0034	1.0000	1.0000	1.0000

Note: $c_2 = 1 - c_1$, $d_2 = -d_1$.

where

$$n_i = [n\lambda_i] + 1, \lambda = F^*(z),$$

$$a_i = \frac{\lambda_i(1 - \lambda_i)(\lambda_{i+1} - \lambda_{i-1})}{K_1},$$

$$K_3 = \sum_{i=1}^{k+1} (1 - \lambda_i - \lambda_{i-1})(f_i^* z_i - f_{i-1}^* z_{i-1}),$$

$$K_1 = \sum_{i=1}^{k+1} (1 - \lambda_i - \lambda_{i-1})^2(\lambda_i - \lambda_{i-1}),$$

$$f_i^* = \lambda_i(1 - \lambda_i), z_i = \ln \frac{\lambda_i}{1 - \lambda_i}.$$

2. The ABLUE σ^* when μ is unknown:

$$\sigma^* = \sum_{i=1}^{k} b_i X_{n_i:n} - \frac{K_3}{K_2} \mu,$$

$$\text{A.Var}(\sigma^*) = \frac{\sigma^2}{nK_2},$$

where

$$b_i = \left[\frac{f_i^* z_i - f_{i-1}^* z_{i-1}}{\lambda_i - \lambda_{i-1}} - \frac{f_{i+1}^* z_{i+1} - f_i^* z_i}{\lambda_{i+1} - \lambda_i} \right] \frac{\lambda_i(1 - \lambda_i)}{K_2},$$

$$K_2 = \sum_{i=1}^{k+1} \frac{(f_i^* z_i - f_{i-1}^* z_{i-1})^2}{\lambda_i - \lambda_{i-1}},$$

3. The ABLUE(μ^*, σ^*) when both μ and σ are unknown:

$$\mu^* = \sum_{i=1}^{k} c_i X_{n_i,n},$$

$$\sigma^* = \sum_{i=1}^{k} d_i X_{n_i:n},$$

$$\text{A.Var}(\mu^*) = \frac{\sigma^2 K_2}{n\Delta},$$

$$\text{A.Var}(\sigma^*) = \frac{\sigma^2 K_1}{n\Delta},$$

$$\text{A.Cov}(\mu^*, \sigma^*) = -\frac{\sigma^2 K_3}{n\Delta},$$

where

$$\Delta = K_1K_2 - K_3^2,$$

$$c_i = \frac{f_i^*}{\Delta}\left\{\left[\frac{f_i^* - f_{i-1}^*}{\lambda_i - \lambda_{i-1}} - \frac{f_{i+1}^* - f_i^*}{\lambda_{i+1} - \lambda_i}\right] K_2 - \left[\frac{f_i^* z_i - f_{i-1}^* z_{i-1}}{\lambda_i - \lambda_{i-1}} - \frac{f_{i+1}^* z_{i+1} - f_i^* z_i}{\lambda_{i+1} - \lambda_i}\right] K_3\right\},$$

$$d_i = \frac{f_i^*}{\Delta}\left\{\left[\frac{f_i^* z_i - f_{i-1}^* z_{i-1}}{\lambda_i - \lambda_{i-1}} - \frac{f_{i+1}^* z_{i+1} - f_i^* z_i}{\lambda_{i+1} - \lambda_i}\right] K_1 - \left[\frac{f_i^* - f_{i-1}^*}{\lambda_i - \lambda_{i-1}} - \frac{f_{i+1}^* - f_i^*}{\lambda_{i+1} - \lambda_i}\right] K_3\right\}.$$

To obtain the ABLUE, we simply try to find the optimum spacing $\boldsymbol{\lambda}$ that is obtained by minimizing A.Var(μ^*), A.Var(σ^*), and G.A.Var(μ^*, σ^*), in turn, by maximizing K_1, K_2, and Δ, respectively.

Only case 1 had been solved (Chan and Cheng, 1972; Chan, 1969; Gupta and Gnanadesikan, 1966). The results are in the tables below.

Cases 2 and 3 were never completely solved, although many people published results in these areas. However, either special conditions were added, or no analytic solution is provided to explain how to maximize K_2 and Δ. Even then, the results were restricted to complete samples (i.e., with $\alpha = 0$, $\beta = 1$ in (4.3.2)) only. We summarize some of these results in Tables 4.3.4–4.3.6.

Cheng (1975) provided a unified approach in finding the optimum spacing for the ABLUE, including the logistic mean μ. Chan and Cheng (1974) developed an algorithm for the ABLUE μ^* based on multiply censored data—i.e., the ABLUE μ^* uses k order statistics selected from a large multiply censored sample

$$X_{m_1:n} < X_{m_1+1:n} < \cdots < X_{m_1+l_1:n} < X_{m_2:n} < X_{m_2+1:n} < \cdots < X_{m_2+l_2:n} < \cdots < X_{m_j+l_j}, \quad (4.3.3)$$

where $m_i = [n\alpha_{i1}] + 1, m_i + l_i = [n\alpha_{i2}] + 1$. Their algorithm is as follows.

The sample (4.3.3) can be represented by $\mathbf{I}$. If $\boldsymbol{\lambda}^0 \in \mathbf{I}$, then $\boldsymbol{\lambda}^c = \boldsymbol{\lambda}^0$. If $\boldsymbol{\lambda}^0 \notin \mathbf{I}$, an optimum spacing $\boldsymbol{\lambda}^c$ which maximizes K_1 defined on $\mathbf{I}$ must be one of the following

$$\sum_{j=1}^{2J} \binom{k}{j}\binom{2J}{j}$$

Table 4.3.4 Optimum Spacings $\boldsymbol{\lambda}$ and the Corresponding Coefficients a_i of the ABLUE μ^* for Known σ

Sample	Proportions of censoring					
	α	$1-\beta$	λ_i	a_i	ARE(μ^*)	K_3
Complete	0	0	$\frac{i}{k+1}$	$\frac{2i}{k^2}\left(1-\frac{i}{k+1}\right)$	$\frac{k(k+2)}{(k+1)^2}$	$\frac{2}{k+1}\sum_{i=1}^{k} g_i$
Right-	0	$\beta \ge \frac{k}{k+1}$	$\frac{i}{k+1}$	$\frac{2i}{k^2}\left(1-\frac{i}{k+1}\right)$	$\frac{k(k+2)}{(k+1)^2}$	$\frac{2}{k+1}\sum_{i=1}^{k} g_i$
Censored	0	$\beta < \frac{k}{k+1}$	$\frac{i}{k}\beta$	$\frac{2i}{k^2}\left(1-\frac{i}{k+1}\right)$	$3\left(\frac{k^2-1}{3k^2}\beta^3-\beta^2+\beta\right)$	$\frac{2\beta}{k}\sum_{i=1}^{k-1} g_i+\left(1-\frac{k-1}{k}\beta\right)g_k$
Left-	$\alpha \le \frac{1}{k+1}$	0	$\frac{i}{k+1}$	$\frac{2i}{k^2}\left(1-\frac{i}{k+1}\right)$	$\frac{k(k+2)}{(k+1)^2}$	$\frac{2}{k+1}\sum_{i=1}^{k} g_i$
Censored	$\alpha > \frac{1}{k+1}$	0	$\frac{i-1}{k}(1-\alpha)+\alpha$	$\frac{2\{(i-1)(1-\alpha)+k\alpha\}\{k-i+i\alpha\}}{k^2(k+1)(1-\alpha)^2}$	$(1-\alpha^3)-\frac{1}{k^2}(1-\alpha)^3$	$\frac{2(1-\alpha)}{k}\sum_{i=2}^{k} g_i+\left(\frac{1}{k}+\frac{k-1}{k}\alpha\right)g_1$
	$\alpha \le \frac{1}{k+1}$	$\beta \ge \frac{k}{k+1}$	$\frac{i}{k+1}$	$\frac{2i}{k^2}\left(1-\frac{i}{k+1}\right)$	$\frac{k(k+2)}{(k+1)^2}$	$\frac{2}{k+1}\sum_{i=1}^{k} g_i$
Doubly	$\alpha \le \frac{\beta}{k}$	$\beta < \frac{k}{k+1}$	$\frac{i}{k}\beta$	$\frac{2i}{k^2}\left(1-\frac{i}{k+1}\right)$	$3\left(\frac{k^2-1}{3k^2}\beta^3-\beta^2+\beta\right)$	$\frac{2\beta}{k}\sum_{i=1}^{k-1} g_i+\left(1-\frac{k-1}{k}\beta\right)g_k$
Censored	$\alpha > \frac{1}{k+1}$	$\beta \ge \frac{k-1-\alpha}{k}$	$\frac{i-1}{k}(1-\alpha)+\alpha$	$\frac{2\{(i-1)(1-\alpha)+k\alpha\}\{k-i+i\alpha\}}{k^2(k+1)(1-\alpha)^2}$	$(1-\alpha^3)-\frac{1}{k^2}(1-\alpha)^3$	$\frac{2(1-\alpha)}{k}\sum_{i=2}^{k} g_i+\left(\frac{1}{k}+\frac{k-1}{k}\alpha\right)g_1$
	$\alpha > \frac{\beta}{k}$	$\beta < \frac{k}{k+1}$	$\frac{(k-i)\alpha+(i-1)\beta}{k-1}$	$\frac{\{2(k-i)\alpha+(i-1)\beta\}\{2(i-1)\alpha+(k-i)\beta\}}{k^2(k+1)(\beta-\alpha)^2}$	$\frac{k(k-2)(\beta^3-\alpha^3)+3\alpha\beta(\beta-\alpha)}{(k-1)^2}-3\beta^2+3\beta$	$\frac{2(\beta-\alpha)}{k-1}\sum_{i=2}^{k-1} g_i+\left(\frac{k-2}{k-1}\alpha+\frac{1}{k-1}\beta\right)\cdot g_1+\left(1-\frac{1}{k-1}\alpha-\frac{\kappa-2}{k-1}\beta\right)g_k$
	$\alpha > \frac{1}{k+1}$	$\beta < \frac{k-1-\alpha}{k}$	$\frac{(k-i)\alpha+(i-1)\beta}{k-1}$	$\frac{\{2(i-i)\alpha+(i-1)\beta\}\{2(i-1)\alpha+(k-i)\beta\}}{k^2(k+1)(\beta-\alpha)^2}$	$\frac{k(k-2)(\beta^3-\alpha^3)+3\alpha\beta(\beta-\alpha)}{(k-1)^2}-3\beta^2+3\beta$	$\frac{2(\beta-\alpha)}{k-1}\sum_{i=2}^{k-1} g_i+\left(\frac{k-2}{k-1}\alpha+\frac{1}{k-1}\beta\right)\cdot g_1+\left(1-\frac{1}{k-1}\alpha-\frac{\kappa-2}{k-1}\beta\right)g_k$

$g_i = \lambda_i(1-\lambda_i)[\ln\lambda_i - \ln(1-\lambda_i)]$, $i = 1, 2, \ldots, k$.

Table 4.3.5 Optimum Spacing $\boldsymbol{\lambda}$, the Corresponding Coefficients b_i, the Asymptotic Variance, the Asymptotic Relative Efficiency, and K_3/K_2 of the ABLUE σ^* for Known μ

k	1	2	3	4	5	6	7	8	9	10
λ_1	0.0832	0.1029	0.0376	0.0404	0.0190	0.0197	0.0108	0.0110	0.0066	0.0068
b_1	−0.4168	−0.2309	−0.0746	−0.0724	−0.0321	−0.0321	−0.0170	−0.0171	−0.0101	−0.0101
λ_2		0.8971	0.1576	0.1690	0.0821	0.0850	0.0480	0.0490	0.0303	0.0308
b_2		0.2309	−0.1778	−0.1699	−0.0843	−0.0842	−0.0466	−0.0468	−0.0283	−0.0285
λ_3			0.8914	0.8310	0.2048	0.2124	0.1197	0.1223	0.0767	0.0778
b_3			0.2186	0.1699	−0.1346	−0.1331	−0.0816	−0.0818	−0.0518	−0.0521
λ_4				0.9596	0.8259	0.7876	0.2379	0.2434	0.1504	0.1526
b_4				0.0724	0.1680	0.1331	−0.1091	−0.1087	−0.0757	−0.0759
λ_5					0.9584	0.9150	0.7834	0.7556	0.2627	0.2669
b_5					0.0721	0.0842	0.1326	0.1087	−0.0915	−0.0914
λ_6						0.9803	0.9134	0.8777	0.7532	0.7331
b_6						0.0321	0.0843	0.0818	0.1085	0.0914
λ_7							0.9799	0.9510	0.8762	0.8474
b_7							0.0323	0.0468	0.0820	0.0759
λ_8								0.9890	0.9503	0.9222
b_8								0.0171	0.0470	0.0521
λ_9									0.9888	0.9692
b_9									0.0172	0.0285
λ_{10}										0.9932
b_{10}										0.0101
K_3/K_2	−0.4168	0.0000	−0.0338	0.0000	−0.0109	0.0000	−0.0050	0.0000	−0.0027	0.0000
$\frac{n}{\sigma^2}\text{AVar}(\sigma^*)$	2.2767	1.0226	0.9187	0.8284	0.7991	0.7709	0.7580	0.7453	0.7384	0.7315
A.R.E.σ^*	0.3071	0.6838	0.7612	0.8442	0.8751	0.9071	0.9225	0.9383	0.9471	0.9561

possible spacing ($\binom{k}{j} = 0$ if $j > k$), each of which can be obtained by repeating the following five steps:

Step 1. For every $j = 1, 2, \ldots, 2J$, set each of the $\binom{k}{j}$ subsets $\boldsymbol{\lambda}_j = \{\lambda_{i_1}, \lambda_{i_2}, \ldots, \lambda_{i_j}\}$ of $\boldsymbol{\lambda}$ equal each of the $\binom{2J}{j}$ subsets $\mathbf{I}_j = \{\alpha_{r_1s_1}, \alpha_{r_2s_2}, \ldots, \alpha_{r_js_j}\}$; i.e., $\lambda_{i_1} = \alpha_{r_1s_1}, \lambda_{i_2} = \alpha_{r_2s_2}, \ldots, \lambda_{i_j} = \alpha_{r_js_j}$.

Step 2. If $\lambda_{i_1} = \alpha_{r_1s_1} = 0$ or $\lambda_{i_j} = \alpha_{r_js_j} = 1$, $\boldsymbol{\lambda}_j$ is disregarded.

Step 3. If $\lambda_{i_1} = \alpha_{r_1s_1} > 0$, then

$$\lambda_t = \frac{t\alpha_{r_1s_1}}{i_1}, \qquad t = 1, 2, \ldots, (i_1 - 1).$$

Table 4.3.6 Optimum Spacing $\boldsymbol{\lambda}$, the Corresponding Coefficients c_i and d_i, and the Asymptotic Variances of the ABLUE(μ^*, σ^*)

k	2	3	4	5	6	7	8	9	10
λ_1	.1873	.1213	.0752	.0517	.0363	.0265	.0199	.0153	.0120
λ_2		.5000	.3001	.2086	.1490	.1112	.0849	.0663	.0527
λ_3				.5000	.3554	.2657	.2042	.1609	.1292
λ_4						.5000	.3863	.3050	.2455
λ_5								.5000	.4060
c_1	.5000	.1801	.0686	.0326	.0163	.0089	.0051	.0031	.0019
c_2		.6398	.4314	.2355	.1271	.0734	.0442	.0277	.0179
c_3				.4638	.3566	.2357	.1551	.0989	.0663
c_4						.3640	.2996	.2207	.1569
c_5								.2992	.2570
d_1	.3407	.2526	.1424	.0929	.0619	.0434	.0318	.0239	.0184
d_2		.0000	.1686	.1724	.1370	.1054	.0809	.0629	.0496
d_3				.0000	.0977	.1213	.1551	.0974	.0816
d_4						.0000	.0639	.0889	.0915
d_5								.0000	.0451
$\frac{n}{\sigma^2} A\ \mathrm{Var}(\mu^*)$	4.0414	3.3790	3.2877	3.1830	3.1405	3.1067	3.0858	3.0699	3.0584
$\frac{n}{\sigma^2} A\ \mathrm{Var}(\sigma^*)$	1.1736	1.0312	.8876	.8361	.7988	.7768	.7609	.7497	.7413

Note: For $i > (k+1)/2$, $\lambda_i = 1 - \lambda_{k-i+1}$, $c_i = c_{k-i+1}$, $d_i = -d_{k-i+1}$.

Step 4. If $\alpha_{rs} < \alpha_{r's'}$ are two consecutive elements of α_j and $\lambda_i < \lambda_{i'}$ are two elements of $\boldsymbol{\lambda}_j$ such that $\lambda_i = \alpha_{rs}$, $\lambda_{i'} = \alpha_{r's'}$, then

$$\lambda_{i+t} = \frac{(i' - i - t)\alpha_{rs} + t\alpha_{r's'}}{i' - i},$$
$$t = 1, 2, \ldots, (i' - i - 1).$$

Step 5. If $\lambda_{i_j} = \alpha_{r_j s_j} < 1$, then

$$\lambda_{i_j+t} = \frac{(k - i_j + 1 - t)\alpha_{r_j s_j} + t}{k - i_j + 1},$$
$$t = 1, 2, \ldots, (k - i_j).$$

Verification of the Algorithm

For any positive integers ν, h and constants $0 \le \alpha_1 < \alpha_2 \le 1$, let

$$K_1(\alpha_1, \alpha_2: \nu) \equiv \sum_{i=h}^{h+\nu} \frac{(f_i - f_{i-1})^2}{\lambda_i - \lambda_{i-1}}$$
$$= \sum_{i=h}^{h+\nu} (\lambda_i - \lambda_{i-1})(1 - \lambda_i - \lambda_{i-1})^2$$

for the logistic distribution, where

$$\alpha_1 \equiv \lambda_{h-1} < \lambda_h < \cdots < \lambda_{h+\nu-1} < \lambda_{h+\nu} \equiv \alpha_2.$$

[In particular, $K_1'(0, 1; k)$ with $h = 1$ is the K_1 for the complete sample case.] It can be shown that (cf. Chan, 1969 and Chan and Cheng, 1972):

(a) If $K_1'(\alpha_1, \alpha_2; \nu)$ is considered to be defined on the closure of $\alpha_1 < \lambda_h < \lambda_{h+1} < \cdots < \lambda_{h+\nu-1} < \alpha_2$, we let $\lambda_h = \alpha_1$, $\lambda_{h+\nu-1} = \alpha_2$ or $\lambda_i = \lambda_{i+1}$ for some $i = h, h+1, \ldots, h+\nu-1$, while the remaining λ_i are unchanged and let it be denoted by $K_1'^{-}$, then $K_1' - K_1'^{-} > 0$.

(b) Being continuous on the preceding closure implies that $K_1'(\alpha_1, \alpha_2; \nu)$ attains its maximum at some point of the closure. By (a) it must attain its maximum at some interior point. Hence, this point must be a solution of $\partial K_1'/\partial\lambda_i = 0$, $i = h, h+1, \ldots, h+\nu-1$. Each $\partial K_1'/\partial\lambda_i$ can be expressed as the product of a strictly nonzero term and

$$\begin{aligned} M(\lambda_{i-1}, \lambda_i, \lambda_{i+1}) &\equiv \frac{f_{i+1} - f_i}{\lambda_{i+1} - \lambda_i} + \frac{f_i - f_{i-1}}{\lambda_i - \lambda_{i-1}} - 2\frac{df_i}{d\lambda_i} \\ &\equiv -(\lambda_{i+1} - \lambda_i) + (\lambda_i - \lambda_{i-1}) \end{aligned} \tag{4.3.4}$$

for the logistic distribution. So this point is the solution of

$$M(\lambda_{i-1}, \lambda_i, \lambda_{i+1}) = 0, \qquad i = h, h+1, \ldots, h+\nu-1, \tag{4.3.5}$$

which has a unique solution. In particular, for the complete sample, the unique solution of (4.3.5) is $\boldsymbol{\lambda}^0$.

Now consider the K_1 defined on $\mathbf{I}$. By (a), the optimum spacing $\boldsymbol{\lambda}^0$ must satisfy the inequalities $0 < \lambda_1^c < \lambda_2^c < \cdots < \lambda_k^c < 1$. If $\boldsymbol{\lambda}^c \neq \boldsymbol{\lambda}^0$, it must be a boundary point of $\mathbf{I}$; i.e., $\lambda_i^c = \alpha_{rs}$ for some i, r, and s, because otherwise $\boldsymbol{\lambda}^c$ is an interior point of $\mathbf{I}$, which implies that it is a solution of $M(\lambda_{i-1}, \lambda_i, \lambda_{i+1}) = 0$, $i = 1, 2, \ldots, k$, $\lambda_0 = 0$, $\lambda_{k+1} = 1$ with $\boldsymbol{\lambda}$ belonging to the interior of $\mathbf{I}$; this contradicts the uniqueness of $\boldsymbol{\lambda}^0$. Therefore, $\boldsymbol{\lambda}^c$ is one of the

$$\sum_{j=1}^{2J} \binom{k}{j}\binom{2J}{j}$$

spacings described in Step 1.

Suppose exactly j of the $k\lambda_i^c$ equal j of the $2J\alpha_{rs}$; i.e.,

$$\lambda_{i1}^c = \alpha_{r_1s_1}, \qquad \lambda_{i2}^c = \alpha_{r_2s_2}, \ldots, \lambda_{ij}^c = \alpha_{r_js_j}.$$

It follows from (a) (by letting $\nu = k$, $\alpha_1 = 0$, $\alpha_2 = 1$) that we must have $\alpha_{r_1s_1} > 0$ and $\alpha_{r_js_j} < 1$. So Step 2 is verified. The remaining λ_i^c can be obtained by maximizing

$$K_1 = K_1'(0, \alpha_{r_1s_1}; i_1 - 1) + K_1'(\alpha_{r_1s_1}, \alpha_{r_2s_2}; i_2 - i_1 - 1) + \cdots + K_1'(\alpha_{r_js_j}, 1; k - i_j). \tag{4.3.6}$$

It follows from (a) and (b) that in (4.3.6) each of the right-hand terms $K_1'(\alpha_{rs}, \alpha_{r's'}; i' - i - 1)$ attains its maximum at the unique solution of

$$M(\lambda_{i+t-1}, \lambda_{i+t}, \lambda_{i+t+1}) = 0, t = 1, 2, \ldots, i' - i - 1;$$

Other Estimators

Jung (1955) felt finding the coefficients of the BLUE was difficult and, hence, proposed a linear estimator defined by a continuous weight function. This approximate BLUEs are asymptotically unbiased and asymptotically best. Gupta and Gnanadesikan (1966) modified it to make it unbiased:

$$\sigma^{**} = \frac{9\pi C_n}{n(n+1)^2(3+\pi^2)\sqrt{3}} \sum_{i=1}^{n} \left[-(n+1)^2 + 2i(n+1) + 2i(n+1-i)\log\left(\frac{i}{n+1-i}\right) \right] X_{i:n},$$

where

$$C_n = \frac{-n}{(n+1)^2} \frac{\sqrt{3}}{\pi} \sum_{i=1}^{n-1} \frac{1}{i}.$$

Because the calculations of the BLUEs are time-consuming and mostly involve numerical integrations for small n, say ≤ 10, and are impractical for moderate n, Blom (1958) proposed the nearly best linear estimates which do not necessarily have the minimum variance among all linear unbiased estimators:

$$\mu^{**} = \sum_{i=1}^{n} \frac{6i(n+1-i)}{n(n+1)(n+2)} X_{i:n},$$

and the approximate variance is

$$\text{Var}(\mu^{**}) \doteq \frac{9\sigma^2(n+1)^2}{\pi^2 n(n+2)^2}.$$

Blom's estimator of σ is

$$\sigma^{**} = \sum_{i=1}^{n} \frac{\pi i(n+1-i)(c_i - c_{i-1})}{\sqrt{3}d(n+1)^2} X_{i:n},$$

$$c_i = \frac{i(n+1-i)}{(n+1)^2} t_i - \frac{(i+1)(n-i)}{(n+1)^2} t_{i+1},$$

$$t_i = -\frac{\sqrt{3}}{\pi} \sum_{j=1}^{n-1} \frac{1}{j} = -t_{n-i+1} \qquad (n - i > i - 1),$$

$$d = \sum_{i=0}^{n} c_i^2.$$

Gupta and Gnanadesikan (1966) did some comparisons and found that Blom's estimate of σ has a higher variance than the corresponding modified estimator of Jung. Its relative efficiency when compared with the BLUE is at least 96% for $5 \le n \le 25$. For n between 5 and 25, the relative efficiency of Blom's estimator of μ and Jung's estimate, modified, of σ is never less than 99.1% and increases with n. For $n > 25$, these estimators are essentially as efficient as the BLUEs and are simpler to compute.

REFERENCES

Balakrishnan, N. and Cohen, A. C. (1990). *Order Statistics and Inference: Estimation Methods*. Academic Press, Boston.

Balakrishnan, N. and Malik, H. J. (1991). Means, variances and covariances of logistic order statistics for sample sizes up to fifty, *Selected Tables in Mathematical Statistics* (to appear).

Birnbaum, A. and Dudman, J. (1963). Logistic order statistics. *Ann. Math. Statist.*, *34*, 658–663.

Blom, G. (1958). *Statistical Estimates and Transformed Beta Variables*. Wiley, New York.

Chan, L. K. (1969). Linear quantile estimates of the location and scale parameters of the logistic distribution. *Statistische Hefte*, *10*, 277–282.

Chan, L. K., Chan, N. N., and Mead, E. R. (1971). Best linear unbiased estimates of the parameters of the logistic distribution based on selected order statistics. *J. Amer. Statist. Assoc.*, *66*, 889–892.

Chan, L. K., Chan, N. N., and Mead, E. R. (1973). Tables for the best linear unbiased estimate based on selected order statistics from the normal, logistic, Cauchy and double exponential distribution. *Math. Comp.*, *27*, 445–446.

Chan, L. K. and Cheng, S. W. (1972). Optimum spacing for the asymptotically best linear estimate of the location parameter of the logistic distribution when samples are complete or censored. *Statistische Hefte*, *13*, 41–57.

Chan, L. K. and Cheng, S. W. (1974). An algorithm for determining the asymptotically best linear estimate of the mean from multiply censored logistic data. *J. Amer. Statist. Assoc.*, *69*, 1027–1030.

Chan, L. K. and Cheng, S. W. (1982). The best linear unbiased estimates of parameters using order statistics. *Soochow J. Math. Natural Sci.*, *8*, 1–13.

Cheng, S. W. (1975). A unified approach to choosing optimum quantiles for the ABLE's. *J. Amer. Statist. Assoc.*, *70*, 155–159.

Gupta, S. S. and Gnanadesikan, M. (1966). Estimation of the parameters of the logistic distribution. *Biometrika*, *53*, 565–570.

Gupta, S. S., Qureishi, A. S., and Shah, B. K. (1967). Best linear unbiased estimators of the parameters of the logistic distribution using order statistics. *Technometrics*, *9*, 43–56.

Hassanein, K. M. (1969). Estimation of the parameters of the logistic distribution by sample quantiles. *Biometrika*, *56*, 684–687.

Hassanein, K. M. (1974). Linear estimation of the parameters of the logistic distribution by selected order statistics for very large samples. *Statistische Hefte*, *15*, [illegible]–70.

Hassanein, K. M. and Sebaugh, J. L. (1973). Estimation of the parameters of the logistic distribution from grouped samples. *Skand. Aktuarietidskr*, *56*, 1–10.

Jung, J. (1955). On linear estimates defined by a continuous weight function. *Ark. Mat.*, *3*, 199–209

Kulldorff, G. (1964). Optimum spacing of sample quantiles from a logistic distribution and best linear unbiased estimates of its parameters. *Technical Report*, University of Lund, Lund, Sweden.

Lloyd, E. H. (1952). Least-squares estimation of location and scale parameters using order statistics. *Biometrika*, *39*, 88–95.

Mosteller, F. (1946). On some useful "inefficient" statistics. *Ann. Math. Statist.*, *17*, 377–407.

Ogawa, J. (1951). Contributions to the theory of systematic statistics. I. *Osaka Math. J.*, *3*, 175–213.

Shah, B. K. (1966). On the bivariate moments of order statistics from a logistic distribution. *Ann. Math. Statist.*, *37*, 1002–1011.

4.4 ESTIMATION OF QUANTILES USING SELECTED ORDER STATISTICS

A. K. Md. Ehsanes Saleh
Carleton University, Ottawa, Ontario, Canada

Khatab M. Hassanein
Kansas University Medical Center, Kansas City, Kansas

M. Masoom Ali
Ball State University, Muncie, Indiana

The estimation of parameters based on censored samples and selected order statistics has been discussed in the last two sections. In this section, we discuss the problem of the large sample estimation, i.e., the ABLUE of quantiles based on selected order statistics. The general problem has received much attention in recent years; see for example, Ali et al. (1983), Ali, Umbach, and Saleh (1982), Ali, Umbach, and Hassanein (1981a,b), Koutrovelis (1981), Kubat and Epstein (1980), Saleh (1981), Saleh, Ali, and Umbach (1981, 1983, 1985), and Umbach, Ali, and Hassanein (1981a,b). Recently, Ali and Umbach (1988, 1989) have obtained the BLUEs of quantiles of symmetrically truncated logistic distribution using selected order statistics. The problem has also been discussed in detail in the recent monograph by Balakrishnan and Cohen (1990), and most of the references given above have been cited by the authors.

4.4.1 ABLUE of Quantiles

Let $Y_{1:n}, Y_{2:n}, \ldots, Y_{n:n}$ be the order statistics corresponding to a random sample of size n from (1.1). The quantile function of (1.1) is defined by

$$Q(\xi) = \mu + \left(\frac{\sigma}{\sqrt{3}}\right) \ln\left(\frac{\xi}{1-\xi}\right), \qquad 0 < \xi < 1. \tag{4.4.1}$$

We obtain the ABLUE of $Q(\xi)$ using $k < n$ optimally selected order statistics. We also construct a $100(1 - \alpha)\%$ confidence interval for $Q(\xi)$.

The k-tuple $(\lambda_1, \lambda_2, \ldots, \lambda_k)$, whose elements satisfy the relation $0 < \lambda_1 < \lambda_2 < \cdots < \lambda_k < 1$, is called a spacing for the k order statistics $Y_{n_1:n}, Y_{n_2:n}, \ldots, Y_{n_k:n}$, where $n_i = [n\lambda_i] + 1$, $i = 1, 2, \ldots, k$, and $[\cdot]$ is the greatest integer function. For a given spacing $(\lambda_1, \lambda_2, \ldots, \lambda_k)$, the corresponding order statistics $Y_{n_1:n}, Y_{n_2:n}, \ldots, Y_{n_k:n}$ under (1.1) have a k-variate normal distribution (see Mosteller, 1946) as $n \to \infty$ with mean vector

$(\mu + (\sigma/\sqrt{3}) \ln\{\lambda_1/(1 - \lambda_1)\}, \mu + (\sigma/\sqrt{3}) \ln\{\lambda_2/(1 - \lambda_2)\}, \ldots, \mu + (\sigma/\sqrt{3}) \ln\{\lambda_k/(1 - \lambda_k)\})$, and $k \times k$ covariance matrix.

$$\frac{3}{\pi^2}\frac{\sigma^2}{n}\left(\frac{1}{\lambda_j(1 - \lambda_i)}\right), \qquad i \le j = 1, 2, \ldots, k. \tag{4.4.2}$$

The ABLUE of $Q(\xi)$ with fixed $\lambda_1, \lambda_2, \ldots, \lambda_k$, using the generalized least squares principle (Sarhan and Greenberg, 1962) is given by

$$\hat{Q}(\xi) = \sum_{i=1}^{k} c_i Y_{n_i:n} = \hat{\mu} + \hat{\sigma}\left(\frac{\sqrt{3}}{\pi}\right) \ln\left(\frac{\xi}{1 - \xi}\right),$$

$$\hat{\mu} = \sum_{i=1}^{k} s_{1i} Y_{n_i:n}, \qquad \hat{\sigma} = \sum_{i=1}^{k} s_{2i} Y_{n_i:n}, \tag{4.4.3}$$

where c_i, s_{1i}, and s_{2i} are defined through the following quantities. For $i = 1, 2, \ldots, k$ let

$$t_i = \left(\frac{\pi}{\sqrt{3}}\right) \frac{\lambda_i}{(1 - \lambda_i)}, \qquad q_i = \lambda_i(1 - \lambda_i) \ln\left(\frac{\lambda_i}{1 - \lambda_i}\right). \tag{4.4.4}$$

Then for $i = 1, 2, \ldots, k$,

$$a_i = t_i\left\{\frac{t_i - t_{i-1}}{\lambda_i - \lambda_{i-1}} - \frac{t_{i+1} - t_i}{\lambda_{i+1} - \lambda_i}\right\}, \tag{4.4.5}$$

$$b_i = t_i\left\{\frac{q_i - q_{i-1}}{\lambda_i - \lambda_{i-1}} - \frac{q_{i+1} - q_i}{\lambda_{i+1} - \lambda_i}\right\}, \tag{4.4.6}$$

$$s_{1i} = \frac{K_2 a_i - K_3 b_i}{\Delta}, \qquad s_{2i} = \frac{K_1 b_i - K_3 a_i}{\Delta}, \tag{4.4.7}$$

$$c_i = s_{1i} + \left(\frac{\sqrt{3}}{\pi}\right) \ln\left(\frac{\xi}{1 - \xi}\right) s_{2i}, \qquad \sum_{i=1}^{k} c_i = 1, \tag{4.4.8}$$

and

$$K_1 = \sum_{i=1}^{k} \frac{(t_i - t_{i-1})^2}{\lambda_i - \lambda_{i-1}}, \tag{4.4.9}$$

$$K_2 = \sum_{i=1}^{k} \frac{(q_i - q_{i-1})^2}{\lambda_i - \lambda_{i-1}}, \tag{4.4.10}$$

$$K_3 = \sum_{i=1}^{k} \frac{(t_i - t_{i-1})(q_i - q_{i-1})}{\lambda_i - \lambda_{i-1}}, \tag{4.4.11}$$

$$\Delta = K_1 K_2 - K_3^2. \tag{4.4.12}$$

The asymptotic variance of $\hat{Q}(\xi)$ is given by

$$\text{Var } \hat{Q}(\xi) = \frac{\sigma^2}{n\Delta}\left[K_2 + \frac{3}{\pi^2}\left\{\ln^2\left(\frac{\xi}{1-\xi}\right)\right\} K_1 - \frac{2\sqrt{3}}{\pi}\left\{\ln\left(\frac{\xi}{1-\xi}\right)\right\} K_3\right]. \tag{4.4.13}$$

In order to choose the spacing $\lambda_1, \lambda_2, \ldots, \lambda_k$ optimally, we minimize (4.4.13); i.e., we minimize

$$\frac{1}{\Delta}\left[K_2 + \frac{3}{\pi^2}\left\{\ln^2\left(\frac{\xi}{1-\xi}\right)\right\} K_1 - \frac{2\sqrt{3}}{\pi}\left\{\ln\left(\frac{\xi}{1-\xi}\right)\right\} K_3\right] \tag{4.4.14}$$

with respect to $(u_1, u_2, \ldots, u_k)$, where $u_i = (\sqrt{3}/\pi)\ln(\lambda_i/(1-\lambda_i))$, $i = 1, 2, \ldots, k$.

Confidence Interval for $Q(\xi)$

We note that the joint asymptotic distribution of $\{\sqrt{n}(\hat{\mu} - \mu), \sqrt{n}(\hat{\sigma} - \sigma)\}$ is a bivariate normal with mean (0, 0) and covariance matrix

$$\frac{\sigma^2}{\Delta}\begin{pmatrix} K_2 & -K_3 \\ -K_3 & K_1 \end{pmatrix}. \tag{4.4.15}$$

Thus, the random variable

$$\sqrt{n}\left[\hat{\mu} + \hat{\sigma}\left(\frac{\sqrt{3}}{\pi}\right)\ln\left(\frac{\xi}{1-\xi}\right) - \mu - \sigma\left(\frac{\sqrt{3}}{\pi}\right)\ln\left(\frac{\xi}{1-\xi}\right)\right] = \sqrt{n}(\hat{Q}(\xi) - Q(\xi)) \tag{4.4.16}$$

is asymptotically normally distributed with zero mean and variance

$$\frac{\sigma^2}{\Delta}\left[K_2 + \frac{3}{\pi^2}\left\{\ln^2\left(\frac{\xi}{1-\xi}\right)\right\} K_1 - \frac{2\sqrt{3}}{\pi}\left\{\ln\left(\frac{\xi}{1-\xi}\right)\right\} K_3\right]. \tag{4.4.17}$$

Further, the quantity $n\Omega_0/\sigma^2$, where

$$\Omega_0 = \sum_{i=1}^{k+1} \frac{(t_i Y_{n_i:n} - t_{i-1} Y_{n_{i-1}:n})^2}{\lambda_i - \lambda_{i-1}} - (K_1\hat{\mu}^2 + 2K_3\hat{\mu}\hat{\sigma} + K_2\hat{\sigma}^2) \tag{4.4.18}$$

has asymptotically a chi-square distribution with $k - 2$ degrees of freedom, and $n\Omega_0/(k - 2)$ is an unbiased estimate of σ^2 which is independent of (4.4.16). Hence, the random variable

$$t = \frac{\hat{Q}(\xi) - Q(\xi)}{\sqrt{\Omega_0}} \times \left[\frac{(k-2)\Delta}{K_2 + \frac{3}{\pi^2}\left\{\ln^2\left(\frac{\xi}{1-\xi}\right)\right\} K_1 - \frac{2\sqrt{3}}{\pi}\left\{\ln\left(\frac{\xi}{1-\xi}\right)\right\} K_2}\right]^{1/2} \tag{4.4.19}$$

is asymptotically ($n \to \infty$) a Student t-variable with $k - 2$ degrees of freedom. Hence, a $100(1 - \alpha)\%$ confidence interval for $Q(\xi)$ based on a fixed spacing $\lambda_1, \lambda_2, \ldots, \lambda_k$ is defined by the interval

$$\hat{Q}(\xi) \pm t_{\alpha/2}^{(k-2)} \times \left[\frac{n\Omega_0\left\{K_2 + \frac{3}{\pi^2}\left(\ln^2\left(\frac{\xi}{1-\xi}\right)\right) K_1 - \frac{2\sqrt{3}}{\pi}\left\{\ln\left(\frac{\xi}{1-\xi}\right)\right\} K_3\right.}{(k-2)\Delta}\right]^{1/2} \tag{4.4.20}$$

The expectation of the square of the length of this interval is given by

$$t_{\alpha/2}^{2(k-2)} \frac{\sigma^2}{\Delta}\left[K_2 + \frac{3}{\pi^2}\left\{\ln\left(\frac{\xi}{1-\xi}\right)\right\} K_1 - \frac{2\sqrt{3}}{\pi}\left\{\ln\left(\frac{\xi}{1-\xi}\right)\right\} K_3\right]. \tag{4.4.21}$$

This interval (4.4.20) is smallest among all choices of spacings when (4.4.21), or equivalently (4.4.14), is minimum. Hence, the same spacings are optimum for both ABLUE and confidence interval for $Q(\xi)$.

Optimal Spacings

To determine the optimal spacings for the ABLUE of $Q(\xi)$ and the confidence interval, (4.4.14) is to be minimized. But this minimization problem is mathematically intractable. However, for large k, there is an approximate solution to this problem.

If the spacings are chosen as the $(k + 1)$-tiles of some continuous distribution $h(u)$, then for sufficiently large k and under appropriate restriction on $h(u)$ (see Sarndal, 1962 and Chernoff, 1971),

$$\frac{1}{\Delta}\left[K_2 + K_1 \ln\left(\frac{\xi}{1-\xi}\right) - 2K_3 \ln\left(\frac{\xi}{1-\xi}\right)\right] \simeq \frac{9}{\pi^2}\left\{1 + \frac{3}{3+\pi^2}\ln^2\left(\frac{\xi}{1-\xi}\right)\right]$$
$$- \frac{1}{12k}\int_0^1 \frac{1}{h^2(u)}\left\{\frac{9}{\pi^2}\ln\left(\frac{\xi}{1-\xi}\right)\left[\frac{1-2u}{u(1-u)} - 2\ln\left(\frac{u}{1-u}\right)\right]\right\}^2 du. \tag{4.4.22}$$

Only the last term of (4.4.22) depends on $h(u)$ and may be minimized to obtain an optimal density $h^*(u)$:

$$h^*(u) = \frac{\left[\frac{2}{\sqrt{3}} + \frac{3}{3+\pi^2}\ln\left(\frac{\xi}{1-\xi}\right)\left\{\frac{1-2u}{u(1-u)} - 2\ln\left(\frac{u}{1-u}\right)\right\}\right]^{2/3}}{\int_0^1\left[\frac{2}{\sqrt{3}} + \frac{\sqrt{3}}{3+\pi^2}\ln\left(\frac{\xi}{1-\xi}\right)\left\{\frac{1-2u}{u(1-u)} - 2\ln\left(\frac{u}{1-u}\right)\right\}\right]^{2/3} du}. \tag{4.4.23}$$

The nearly optimal spacings are given by

$$\lambda_i^* = H^{*-1}\left(\frac{i}{k+1}\right), \qquad i = 1, 2, \ldots, k, \tag{4.4.24}$$

where $H^*(u)$ is the cdf corresponding to $h^*(u)$.

Spacings obtained this way by numerical integration of the expression

$$\frac{i}{k+1} = \int_0^{\lambda_i^*} h^*(u)\, du \tag{4.4.25}$$

are then used to iteratively minimize (4.4.14).

Table 4.4.1 provides optimum spacings and the coefficients of ABLUEs for $Q(\xi)$ for various values of ξ and $k = 2(1)10$. The spacings for $\xi > 0.5$ are not provided since they are symmetric. Table 4.4.2 provides the values of K_1, K_2, K_3, etc., which are useful for confidence intervals.

Asymptotic Relative Efficiency

The asymptotic relative efficiency (ARE) of $\hat{Q}(\xi)$ compared to the maximum likelihood estimates $\overline{Q}(\xi) = \overline{\mu} + \overline{\sigma}(\sqrt{3}/\pi)\ln(\xi/(1-\xi))$, where

Table 4.4.1 Asymptotically Optimum Spacings λ_i, Coefficients c_i for the Estimator $\hat{Q}(\xi)$ for the Logistic Distribution

		ξ										
k		.01	.05	.10	.15	.20	.25	.30	.35	.40	.45	.50
2	λ_1	0.149	0.176	0.200	0.225	0.251	0.117	0.152	0.191	0.235	0.282	0.333
	c_1	1.648	1.311	1.176	1.105	1.059	0.363	0.389	0.416	0.444	0.472	0.500
	λ_2	0.934	0.950	0.961	0.971	0.979	0.361	0.428	0.492	0.553	0.612	0.667
	c_2	−0.648	−0.311	−0.176	−0.105	−0.059	0.637	0.611	0.584	0.556	0.528	0.500
3	λ_1	0.160	0.070	0.085	0.100	0.118	0.139	0.089	0.120	0.157	0.200	0.250
	c_1	1.673	0.430	0.400	0.389	0.388	0.394	0.192	0.216	0.242	0.210	0.300
	λ_2	0.906	0.256	0.293	0.329	0.366	0.407	0.263	0.319	0.378	0.439	0.500
	c_2	−0.486	0.890	0.783	0.721	0.677	0.641	0.380	0.389	0.395	0.399	0.400
	λ_3	0.979	0.940	0.953	0.963	0.972	0.981	0.495	0.566	0.633	0.694	0.750
	c_3	−0.187	−0.320	−0.183	−0.110	−0.065	−0.035	0.428	0.395	0.363	0.331	0.300
4	λ_1	0.062	0.076	0.043	0.053	0.065	0.080	0.100	0.081	0.112	0.152	0.200
	c_1	0.542	0.445	0.182	0.181	0.184	0.193	0.205	0.126	0.148	0.173	0.200
	λ_2	0.236	0.274	0.158	0.184	0.213	0.246	0.285	0.225	0.279	0.338	0.400
	c_2	1.157	0.898	0.426	0.407	0.398	0.393	0.392	0.256	0.272	0.287	0.300
	λ_3	0.887	0.912	0.345	0.387	0.430	0.477	0.527	0.403	0.469	0.535	0.600
	c_3	−0.501	−0.248	0.580	0.527	0.486	0.452	0.421	0.325	0.319	0.311	0.300
	λ_4	0.975	0.981	0.949	0.959	0.969	0.978	0.986	0.612	0.682	0.744	0.800
	c_4	−0.198	−0.095	−0.188	−0.115	−0.068	−0.038	−0.018	0.293	0.261	0.229	0.200
	λ_1	0.064	0.038	0.047	0.057	0.039	0.050	0.065	0.086	0.084	0.120	0.167
	c_1	0.551	0.201	0.189	0.187	0.101	0.108	0.118	0.131	0.097	0.118	0.143
	λ_2	0.243	0.145	0.169	0.194	0.137	0.163	0.196	0.235	0.216	0.272	0.333
	c_2	1.163	0.482	0.438	0.416	0.241	0.245	0.252	0.262	0.193	0.211	0.229

(*continued*)

Table 4.4.1 (*continued*)

k		ξ										
		.01	.05	.10	.15	.20	.25	.30	.35	.40	.45	.50
5	λ_3	0.866	0.325	0.365	0.403	0.280	0.320	0.365	0.417	0.366	0.432	0.500
	c_3	−0.398	0.670	0.580	0.526	0.354	0.343	0.335	0.328	0.252	0.256	0.257
	λ_4	0.949	0.904	0.923	0.938	0.471	0.521	0.574	0.629	0.532	0.601	0.667
	c_4	−0.231	−0.255	0.151	−0.094	0.375	0.344	0.315	0.287	0.259	0.245	0.228
	λ_5	0.989	0.979	0.984	0.987	0.967	0.976	0.984	0.991	0.714	0.778	0.833
	c_5	−0.085	−0.098	−0.066	−0.035	−0.071	−0.040	−0.020	−0.008	0.199	0.170	0.143
	λ_1	0.032	0.021	0.027	0.033	0.041	0.033	0.044	0.061	0.064	0.098	0.143
	c_1	0.245	0.107	0.102	0.101	0.104	0.065	0.073	0.084	0.067	0.085	0.107
	λ_2	0.125	0.087	0.104	0.122	0.143	0.114	0.141	0.176	0.172	0.225	0.286
	c_2	0.007	0.276	0.256	0.248	0.247	0.160	0.169	0.181	0.141	0.160	0.179
	λ_3	0.291	0.197	0.227	0.258	0.291	0.230	0.270	0.317	0.296	0.360	0.429
	c_3	0.879	0.445	0.399	0.374	0.357	0.246	0.248	0.250	0.197	0.206	0.214
6	λ_4	0.855	0.359	0.403	0.444	0.486	0.373	0.423	0.477	0.431	0.501	0.572
	c_4	−0.406	0.531	0.455	0.408	0.372	0.294	0.282	0.271	0.223	0.220	0.214
	λ_5	0.945	0.900	0.919	0.934	0.949	0.551	0.606	0.662	0.577	0.649	0.714
	c_5	−0.238	−0.259	−0.154	−0.096	−0.059	0.275	0.249	0.223	0.213	0.197	0.179
	λ_6	0.988	0.978	0.983	0.987	0.990	0.974	0.982	0.990	0.738	0.802	0.857
	c_6	−0.087	−0.100	−0.058	−0.035	−0.021	−0.040	−0.021	−0.009	0.159	0.132	0.107
	λ_1	0.018	0.023	0.028	0.021	0.027	0.034	0.031	0.045	0.050	0.082	0.125
	c_1	0.129	0.109	0.103	0.062	0.063	0.067	0.048	0.057	0.048	0.064	0.083
	λ_2	0.073	0.090	0.107	0.082	0.098	0.118	0.105	0.135	0.139	0.190	0.250
	c_2	0.343	0.281	0.260	0.157	0.159	0.164	0.118	0.129	0.106	0.125	0.143
	λ_3	0.171	0.203	0.233	0.177	0.205	0.236	0.206	0.249	0.244	0.307	0.375
	c_3	0.569	0.450	0.403	0.257	0.252	0.249	0.183	0.190	0.155	0.167	0.179

7	λ_4	0.322	0.370	0.412	0.305	0.342	0.383	0.327	0.378	0.358	0.428	0.500
	c_4	0.702	0.530	0.454	0.328	0.310	0.296	0.229	0.226	0.186	0.189	0.191
	λ_5	0.849	0.878	0.901	0.472	0.517	0.563	0.465	0.522	0.481	0.554	0.625
	c_5	−0.411	−0.208	−0.125	0.331	0.299	0.272	0.241	0.227	0.195	0.188	0.178
	λ_6	0.943	0.955	0.964	0.932	0.946	0.960	0.629	0.687	0.612	0.683	0.750
	c_6	−0.243	−0.119	−0.070	−0.098	−0.061	−0.036	0.203	0.180	0.179	0.161	0.143
	λ_7	0.987	0.990	0.993	0.986	0.989	0.992	0.982	0.989	0.755	0.820	0.875
	c_7	−0.089	−0.043	−0.025	−0.037	−0.022	−0.012	−0.022	−0.009	0.131	0.106	0.083
8	λ_1	0.018	0.014	0.018	0.022	0.018	0.024	0.032	0.034	0.052	0.069	0.111
	c_1	0.131	0.064	0.061	0.062	0.041	0.044	0.049	0.040	0.050	0.049	0.067
	λ_2	0.075	0.058	0.070	0.084	0.070	0.086	0.108	0.107	0.143	0.164	0.222
	c_2	0.347	0.174	0.163	0.160	0.107	0.112	0.120	0.095	0.108	0.099	0.117
	λ_3	0.175	0.135	0.157	0.181	0.150	0.177	0.210	0.200	0.249	0.266	0.333
	c_3	0.573	0.299	0.273	0.260	0.180	0.182	0.185	0.147	0.157	0.137	0.150
	λ_4	0.328	0.244	0.278	0.311	0.254	0.290	0.332	0.308	0.364	0.372	0.445
	c_4	0.073	0.403	0.356	0.330	0.239	0.234	0.230	0.184	0.187	0.161	0.167
	λ_5	0.831	0.395	0.439	0.481	0.381	0.424	0.472	0.426	0.487	0.482	0.556
	c_5	−0.339	0.436	0.370	0.329	0.269	0.254	0.240	0.201	0.195	0.170	0.167
	λ_6	0.919	0.875	0.897	0.917	0.539	0.587	0.638	0.557	0.619	0.595	0.667
	c_6	−0.238	−0.211	−0.127	−0.081	0.249	0.224	0.201	0.193	0.178	0.162	0.150
	λ_7	0.969	0.954	0.963	0.970	0.944	0.958	0.971	0.705	0.764	0.711	0.778
	c_7	−0.131	−0.121	−0.071	−0.045	−0.062	−0.037	−0.019	0.150	0.128	0.135	0.116
	λ_8	0.993	0.990	0.992	0.993	0.989	0.992	0.994	0.988	0.994	0.834	0.889
	c_8	−0.046	−0.044	−0.025	−0.015	−0.023	−0.013	−0.006	−0.010	−0.003	0.087	0.066
	λ_1	0.011	0.014	0.012	0.015	0.019	0.017	0.024	0.026	0.042	0.059	0.100
	c_1	0.077	0.066	0.039	0.040	0.041	0.030	0.034	0.029	0.037	0.039	0.055
	λ_2	0.047	0.059	0.049	0.059	0.071	0.065	0.083	0.086	0.119	0.142	0.200
	c_2	0.212	0.176	0.108	0.107	0.109	0.079	0.086	0.072	0.084	0.080	0.097
	λ_3	0.113	0.137	0.112	0.131	0.153	0.136	0.166	0.164	0.210	0.233	0.300

(continued)

Table 4.4.1 (*continued*)

k		ξ										
		.01	.05	.10	.15	.20	.25	.30	.35	.40	.45	.50
	c_3	0.373	0.302	0.189	0.183	0.181	0.134	0.140	0.114	0.125	0.114	0.127
	λ_4	0.211	0.248	0.199	0.227	0.258	0.227	0.265	0.255	0.309	0.328	0.400
	c_4	0.518	0.405	0.265	0.250	0.241	0.182	0.183	0.150	0.156	0.137	0.146
9	λ_5	0.352	0.401	0.312	0.348	0.386	0.333	0.378	0.356	0.415	0.425	0.500
	c_5	0.581	0.435	0.315	0.289	0.270	0.214	0.208	0.172	0.172	0.150	0.152
	λ_6	0.826	0.859	0.460	0.503	0.546	0.457	0.506	0.465	0.528	0.525	0.600
	c_6	−0.341	−0.175	0.311	0.275	0.247	0.221	0.207	0.179	0.171	0.152	0.145
	λ_7	0.917	0.934	0.895	0.914	0.932	0.605	0.656	0.585	0.647	0.628	0.700
	c_7	−0.241	−0.120	−0.129	−0.083	−0.052	0.189	0.169	0.167	0.151	0.140	0.127
	λ_8	0.968	0.975	0.962	0.969	0.976	0.957	0.969	0.720	0.778	0.734	0.800
	c_8	−0.132	−0.065	−0.072	−0.045	−0.028	−0.038	−0.020	0.127	0.107	0.115	0.097
	λ_9	0.993	0.995	0.992	0.994	0.995	0.991	0.994	0.988	0.994	0.846	0.900
	c_9	−0.047	−0.022	−0.026	−0.016	−0.009	−0.013	−0.007	−0.010	−0.003	0.073	0.054
	λ_1	0.007	0.009	0.012	0.010	0.013	0.013	0.018	0.020	0.034	0.051	0.091

	c_1	0.049	0.042	0.040	0.027	0.028	0.022	0.025	0.022	0.028	0.031	0.046
	λ_2	0.032	0.040	0.049	0.043	0.053	0.049	0.065	0.070	0.100	0.125	0.182
	c_2	0.139	0.116	0.109	0.075	0.077	0.059	0.064	0.055	0.066	0.066	0.082
	λ_3	0.078	0.096	0.114	0.098	0.116	0.107	0.133	0.137	0.179	0.207	0.273
	c_3	0.252	0.207	0.191	0.132	0.133	0.101	0.108	0.091	0.101	0.095	0.109
	λ_4	0.146	0.175	0.202	0.172	0.198	0.181	0.216	0.215	0.266	0.292	0.364
	c_4	0.370	0.295	0.266	0.189	0.185	0.143	0.146	0.122	0.130	0.118	0.127
	λ_5	0.240	0.280	0.316	0.265	0.299	0.269	0.310	0.302	0.359	0.379	0.455
10	c_5	0.464	0.360	0.315	0.233	0.222	0.175	0.174	0.146	0.149	0.133	0.137
	λ_6	0.300	0.421	0.466	0.378	0.418	0.369	0.416	0.396	0.457	0.469	0.546
	c_6	0.493	0.367	0.310	0.254	0.236	0.194	0.187	0.159	0.156	0.138	0.137
	λ_7	0.823	0.856	0.882	0.520	0.564	0.483	0.534	0.497	0.561	0.561	0.636
	c_7	−0.343	−0.177	−0.108	0.235	0.210	0.194	0.180	0.160	0.151	0.135	0.127
	λ_8	0.915	0.933	0.945	0.912	0.930	0.620	0.671	0.607	0.670	0.655	0.727
	c_8	−0.243	−0.122	−0.072	−0.083	−0.053	0.163	0.144	0.145	0.131	0.122	0.109
	λ_9	0.968	0.975	0.980	0.969	0.975	0.956	0.968	0.731	0.790	0.752	0.818
	c_9	−0.134	−0.065	−0.038	−0.046	−0.028	−0.038	−0.021	0.110	0.091	0.099	0.082
	λ_{10}	0.993	0.995	0.996	0.994	0.995	0.991	0.994	0.987	0.994	0.855	0.909
	c_{10}	−0.047	−0.023	−0.013	−0.016	−0.010	−0.013	−0.007	−0.010	−0.003	0.063	0.045

Table 4.4.2 Variances, K_1, K_2, and K_3 for the Estimator $\hat{Q}(\xi)$ for the Logistic Distribution

		ξ										
k		.01	.05	.10	.15	.20	.25	.30	.35	.40	.45	.50
	K_1	0.561	0.582	0.605	0.629	0.657	0.793	0.864	0.915	0.949	0.969	0.975
2	K_2	0.921	0.849	0.768	0.679	0.576	0.424	0.414	0.414	0.419	0.425	0.427
	K_3	−0.122	−0.175	−0.213	−0.240	−0.257	−0.348	−0.282	−0.211	−0.140	−0.069	0.000
	var{$\hat{Q}$}	7.783	3.914	2.719	2.153	1.815	1.364	1.219	1.126	1.068	1.036	1.026
	K_1	0.637	0.738	0.771	0.802	0.834	0.866	0.935	0.980	1.008	1.023	0.028
3	K_2	1.017	1.001	0.943	0.881	0.812	0.737	0.564	0.604	0.642	0.669	0.679
	K_3	−0.066	0.189	−0.212	−0.223	−0.226	−0.220	−0.244	−0.174	−0.111	−0.054	0.000
	var{$\hat{Q}$}	7.413	3.317	2.289	1.808	1.523	1.337	1.155	1.067	1.012	0.982	0.972
	K_1	0.787	0.803	0.847	0.879	0.910	0.940	0.967	1.013	1.037	1.049	1.053
4	K_2	1.163	1.109	1.019	0.976	0.931	0.884	0.835	0.722	0.781	0.822	0.836
	K_3	−0.090	−0.128	−0.197	−0.200	−0.195	−0.182	−0.162	−0.149	−0.091	−0.043	0.000
	var{$\hat{Q}$}	6.349	3.214	2.165	1.709	1.439	1.264	1.145	1.042	0.989	0.959	0.950
	K_1	0.825	0.876	0.896	0.915	0.951	0.979	1.003	1.023	1.053	1.063	1.066
5	K_2	1.207	1.177	1.132	1.085	0.999	0.971	0.943	0.914	0.874	0.924	0.942
	K_3	−0.061	−0.122	−0.138	−0.147	−0.172	−0.156	−0.134	−0.108	−0.078	−0.036	0.000

	var{$\hat{Q}$}	6.241	3.042	2.122	1.689	1.403	1.232	1.117	1.039	0.976	0.947	0.938
6	K_1	0.893	0.917	0.937	0.956	0.975	1.002	1.024	1.042	1.062	1.072	1.074
	K_2	1.265	1.212	1.178	1.143	1.105	1.027	1.013	1.000	0.940	0.997	1.018
	K_3	−0.065	−0.114	−0.125	−0.129	−0.129	−0.138	−0.116	−0.090	−0.068	−0.031	0.000
	var{$\hat{Q}$}	5.926	2.965	2.068	1.645	1.393	1.215	1.102	1.025	0.968	0.940	0.931
7	K_1	0.931	0.945	0.959	0.982	0.999	1.016	1.037	1.053	1.069	1.077	1.080
	K_2	1.293	1.261	1.227	1.178	1.151	1.122	1.062	1.059	0.988	1.051	1.074
	K_3	−0.065	−0.086	−0.098	−0.116	−0.113	−0.105	−0.102	−0.078	−0.061	−0.027	0.000
	var{$\hat{Q}$}	5.786	2.933	2.054	1.621	1.373	1.211	1.093	1.017	0.964	0.935	0.926
8	K_1	0.951	0.970	0.984	0.997	1.015	1.031	1.044	1.061	1.071	1.081	1.083
	K_2	1.317	1.283	1.256	1.228	1.182	1.162	1.141	1.103	1.102	1.093	1.117
	K_3	−0.049	−0.080	−0.089	−0.093	−0.102	−0.093	−0.081	−0.069	−0.046	−0.024	0.000
	var{$\hat{Q}$}	5.737	2.892	2.024	1.615	1.362	1.201	1.091	1.012	0.963	0.932	0.923
9	K_1	0.975	0.985	1.000	1.013	1.025	1.041	1.053	1.067	1.076	1.084	1.086
	K_2	1.334	1.309	1.275	1.252	1.229	1.190	1.176	1.135	1.141	1.126	1.152
	K_3	−0.048	−0.064	−0.082	−0.084	−0.083	−0.084	−0.071	−0.062	−0.041	−0.021	0.000
	var{$\hat{Q}$}	5.662	2.877	2.007	1.601	1.358	1.194	1.086	1.008	0.960	0.930	0.921
10	K_1	0.990	1.001	1.011	1.024	1.036	1.048	1.060	1.071	1.079	1.086	1.088
	K_2	1.344	1.323	1.301	1.269	1.251	1.211	1.202	1.160	1.172	1.152	1.179
	K_3	−0.047	−0.060	−0.067	−0.077	−0.074	−0.076	−0.064	−0.056	−0.036	−0.019	−0.000
	var{$\hat{Q}$}	5.617	2.852	2.000	1.591	1.351	1.190	1.082	1.006	0.957	0.928	0.920

$\overline{\mu}$ and $\overline{\sigma}$ are maximum likelihood estimates of μ and σ, is given by

$$\mathrm{ARE}(\hat{Q}(\xi):\overline{Q}(\xi))$$

$$= \frac{9}{n\pi^2} \frac{\left[1 + \left\{\frac{\pi^2}{3+\pi^2}\right\} \ln^2\left(\frac{\xi}{1-\xi}\right)\right] \Delta}{K_2 + \frac{3}{\pi^2}\left(\ln^2\left(\frac{\xi}{1-\xi}\right)\right) K_1 - \frac{2\sqrt{3}}{\pi} \ln\left(\frac{\xi}{1-\xi}\right) K_3}.$$

Table 4.4.3 provides the ARE values.

Numerical Example

The following data set is used to illustrate the estimation of $Q(\xi)$ for $\xi = 0.5$ and $\xi = 0.9$ with $k = 4$. The data are generated by taking a sample of size 50 from the logistic distribution with mean 10 and standard deviation 0.5. They are given below in ascending order.

8.1267	9.3890	9.8916	10.1820	10.7783
8.1367	9.3920	9.9638	10.2249	10.8070
8.6839	9.5018	9.9735	10.4094	11.1418
8.9145	9.5396	9.9814	10.4290	11.2253
9.0335	9.5988	9.9847	10.4963	11.2537
9.0343	9.6438	9.9979	10.5626	11.3740
9.1099	9.6579	10.0889	10.5690	11.6733
9.1860	9.7688	10.1255	10.6354	11.7996
9.2383	9.8679	10.1537	10.6682	12.3043
9.3745	9.8722	10.1809	10.6994	12.6922

From Table 4.4.1, for $k = 4$, the optimal spacings for $\xi = 0.5$ are $\lambda_1 = 0.2$, $\lambda_2 = 0.4$, $\lambda_3 = 0.6$, and $\lambda_4 = 0.8$, and the coefficients are $c_1 = 0.2$, $c_2 = 0.3$, $c_3 = 0.3$, and $c_4 = 0.2$; $n_i = [n\lambda_i] + 1$ gives $n_1 = 11$, $n_2 = 21$, $n_3 = 31$, and $n_4 = 41$. Hence,

$$\begin{aligned}\hat{Q}(0.5) &= (0.2)(9.3890) + (0.3)(9.8916) \\ &\quad + (0.3)(10.1820) + (0.2)(10.7783) \\ &= 10.05554\end{aligned}$$

From Table 4.4.2, $\mathrm{Var}(\hat{Q}(0.5)) = 4.621\sigma^2/50 = 0.0924\sigma^2$, and from Table 4.4.3, ARE $= 0.972$.

To estimate $Q(0.9)$, we use the optimum spacings and coefficients for $\xi = 0.10$ from Table 4.4.1. The spacings are $\lambda_1 = 0.043$, $\lambda_2 = 0.158$,

Table 4.4.3 Asymptotic Relative Efficiency (ARE) for the Estimator $\hat{Q}(\xi)$ for the Logistic Distribution

k	.01	.05	.10	.15	.20	.25	.30	.35	.40	.45	.50
2	0.694	0.704	0.713	0.721	0.728	0.857	0.873	0.882	0.886	0.888	0.889
3	0.729	0.831	0.847	0.858	0.867	0.874	0.922	0.931	0.935	0.937	0.938
4	0.851	0.857	0.895	0.908	0.917	0.924	0.929	0.954	0.958	0.960	0.960
5	0.865	0.906	0.913	0.919	0.941	0.948	0.953	0.956	0.970	0.972	0.972
6	0.911	0.929	0.937	0.943	0.948	0.961	0.966	0.969	0.978	0.979	0.980
7	0.933	0.939	0.944	0.957	0.961	0.965	0.974	0.977	0.983	0.984	0.984
8	0.941	0.953	0.957	0.961	0.970	0.973	0.975	0.982	0.983	0.987	0.988
9	0.954	0.958	0.966	0.969	0.972	0.978	0.981	0.985	0.987	0.990	0.990
10	0.961	0.966	0.969	0.975	0.978	0.982	0.984	0.988	0.989	0.992	0.992

$\lambda_3 = 0.345$, and $\lambda_4 = 0.949$, which yields $n_1 = 3$, $n_2 = 8$, $n_3 = 18$, and $n_4 = 48$. Using symmetric ranks, we replace $Y_{i:n}$ by $Y_{n-i+1:n}$ and we get the following order statistics and the coefficients: $Y_{48:50} = 11.7996$, $c_1 = 0.182$, $Y_{43:50} = 11.418$, $c_2 = 0.426$, $Y_{33:50} = 10.494$, $c_3 = 0.580$, $Y_{3:50} = 8.6839$, and $c_4 = -0.188$. Then

$$\hat{Q}(0.9) = (0.182)(11.7996) + (0.426)(11.1418) + (0.580)(10.4094) - (0.188)(8.6839) = 11.2988.$$

REFERENCES

Ali, M. Masoom and Umbach, D. (1988). Tables of BLUE's for quantiles of symmetrically truncated logistic distribution. *Technical Report No. 81*, Dept. of Mathematical Sciences, Ball State University.

Ali, M. Masoom and Umbach, D. (1989). Estimation of quantiles of symmetrically truncated logistic distribution using a few optimally selected order statistics. *J. Info. Opt. Sci.*, *10*(2), 303–307.

Ali, M. Masoom, Umbach, D., and Hassanein, K. M. (1981a). Estimation of quantiles of exponential and double exponential distributions based on two order statistics. *Commun. Statist.—Theor. Meth.*, *10*(19), 1921–1932.

Ali, M. Masoom, Umbach, D., and Hassanein, K. M. (1981b). Small sample quantile estimation of Pareto populations using two order statistics. *Aligarh J. Statist.*, *1*(2), 139–164.

Ali, M. Masoom, Umbach, D., and Saleh, A. K. Md. E. (1982). Small sample quantile estimation of the exponential distribution using optimal spacings. *Sankhyā B*, *44*(2), 135–142.

Ali, M. Masoom, Umbach, D., Saleh, A. K. Md. E., and Hassanein, K. M. (1983). Estimating quantiles using optimally selected order statistics. *Commun. Statist.—Theor. Meth.*, *12*(19), 2261–2271.

Balakrishnan, N. and Cohen, A. C. (1990). *Order Statistics and Inference: Estimation Methods*. Academic Press, Boston.

Chernoff, H. (1971). A note on optimal spacings for systematic statistics, *Technical Report No. 70*, Dept. of Statistics, Stanford University.

Koutrovelis, I. A. (1981). Large sample quantile estimation in Pareto laws. *Commun. Statist.—Theor. Meth.*, *10*(2), 189–201.

Kubat, P. and Epstein, B. (1980). Estimation of quantiles of location-scale distribution based on few ordered observations. *Technometrics*, *12*, 345–361.

Mosteller, F. (1946). On some useful "inefficient" statistics. *Ann. Math. Statist.*, *17*, 377–408.

Saleh, A. K. Md. E. (1981). Estimating quantiles of exponential distribution, in *Statistics and Related Topics*, M. Csorgo, D. Dawson, J. N. K. Rao, and A. K. Md. E. Saleh, (eds.), 279–283, North-Holland, Amsterdam.

Saleh, A. K. Md. E., Ali, M. Masoom, and Umbach, D. (1985). Large sample estimation of Pareto quantiles using selected order statistics. *Metrika*, *32*, 49–56.

Saleh, A. K. Md. E., Ali, M. Masoom, and Umbach, D. (1983). Estimating the quantile function of location-scale family of distributions based on a few selected order statistics. *J. Statist. Plann. Infer.*, *8*, 75–87.

Saleh, A. K. Md. E., Ali, M. Masoom, and Umbach, D. (1985). Large sample estimation of Pareto quantiles using selected order statistics. *Metrika*, *32*, 49–56.

Sarhan, A. E. and Greenberg, B. G. (eds.) (1962). *Contributions to Order Statistics*. Wiley, New York.

Sarndal, C. E. (1962). *Information from Censored Samples*. Almqvist and Wiksell, Stockholm, Sweden.

Umbach, D., Ali, M. Masoom, and Hassanein, K. M. (1981a). Small sample estimation of exponential quantiles with two order statistics. *Aligarh J. Statist.*, *1*(2), 113–120.

4.5 LINEAR ESTIMATION WITH POLYNOMIAL COEFFICIENTS

N. Balakrishnan
McMaster University, Hamilton, Ontario, Canada

Downton (1966a) proposed linear estimators for the location and scale parameters in which the general structure of the coefficients had been chosen for mathematical tractability, both in the determination of the coefficients and also in the computation of the standard errors of the estimates so derived. By considering the normal and the extreme value distributions, Downton (1966a,b) showed that these estimators are highly efficient. In this section, we derive these estimators for the location and scale parameters of the logistic distribution and discuss their efficiency.

Based on a complete ordered sample

$$X_{1:n} \le X_{2:n} \le \cdots \le X_{n:n} \tag{4.5.1}$$

from the logistic $L(\mu, \sigma^2)$ population, let us consider linear estimators of the form

$$\mu_* = \sum_{k=0}^{p} (k+1)\theta_k \sum_{i=1}^{n} \frac{(i-1)^{(k)} X_{i:n}}{n^{(k+1)}} \tag{4.5.2}$$

and

$$\sigma_* = \sum_{k=0}^{p} (k+1)\phi_k \sum_{i=1}^{n} \frac{(i-1)^{(k)} X_{i:n}}{n^{(k+1)}}, \tag{4.5.3}$$

where $m^{(r)}$, with m and r integers, denotes the rth factorial power of m; that is,

$$m^{(r)} = m(m-1)\cdots(m-r+1). \tag{4.5.4}$$

Now by making use of the identities

$$\sum_{i=1}^{n} (i-1)^{(k)} = \frac{n^{(k+1)}}{k+1} \tag{4.5.5}$$

and

$$\sum_{i=1}^{n} (i-1)^{(k)} \alpha_{i:n} = \frac{n^{(k+1)} \alpha_{k+1:k+1}}{k+1} \tag{4.5.6}$$

(see Downton, 1966a; Arnold and Balakrishnan, 1989), we get from (4.5.2) and (4.5.3) that

$$E(\mu_*) = \mu \sum_{k=0}^{p} \theta_k + \sigma \sum_{k=0}^{p} \theta_k \alpha_{k+1:k+1} \tag{4.5.7}$$

and

$$E(\sigma_*) = \mu \sum_{k=0}^{p} \theta_k + \sigma \sum_{k=0}^{p} \theta_k \alpha_{k+1:k+1}. \tag{4.5.8}$$

By letting

$$\boldsymbol{\theta} = (\theta_0 \quad \theta_1 \quad \cdots \quad \theta_p)^T,$$

$$\boldsymbol{\phi} = (\phi_0 \quad \phi_1 \quad \cdots \quad \phi_p)^T,$$

$$\boldsymbol{\alpha} = (\alpha_{1:1} \quad \alpha_{2.2} \quad \cdots \quad \alpha_{p+1:p+1})^T$$

and

$$\mathbf{1} = (1 \quad 1 \quad \cdots \quad 1)^T_{(p+1)\times 1},$$

the estimators μ_* and σ_* in (4.5.2) and (4.5.3), respectively, are seen to be unbiased if

$$\left[\begin{array}{c}\boldsymbol{\theta}^T \\ \hline \boldsymbol{\phi}^T\end{array}\right] [\mathbf{1} \mid \boldsymbol{\alpha}] = \left[\begin{array}{c|c} 1 & 0 \\ \hline 0 & 1\end{array}\right]. \tag{4.5.9}$$

If we now let ψ_k be the random variable defined by

$$\psi_k = (k + 1) \sum_{i=1}^{n} \frac{(i - 1)^{(k)} Y_{i:n}}{n^{(k+1)}} \tag{4.5.10}$$

and let $\boldsymbol{\Omega}$ be the symmetric matrix with elements

$$\Omega_{ij} = \Omega_{ji} = \mathrm{Cov}(\psi_i, \psi_j), \qquad i \le j, \tag{4.5.11}$$

the variance-covariance matrix of the estimators μ_* and σ_* is given by

$$\mathrm{Var}\left[\begin{array}{c}\mu_* \\ \hline \sigma_*\end{array}\right] = \sigma^2 \left[\begin{array}{c}\boldsymbol{\theta}^T \\ \hline \boldsymbol{\phi}^T\end{array}\right] \boldsymbol{\Omega} [\boldsymbol{\theta} \mid \boldsymbol{\phi}]. \tag{4.5.12}$$

The best unbiased estimators (in the least-squares sense) of μ and σ having the form (4.5.2) and (4.5.3), obtained by choosing $\boldsymbol{\theta}$ and $\boldsymbol{\phi}$ such that

$$\left[\begin{array}{c}\boldsymbol{\theta}^T \\ \hline \boldsymbol{\phi}^T\end{array}\right] \boldsymbol{\Omega} [\boldsymbol{\theta} \mid \boldsymbol{\phi}] + 2 \left[\begin{array}{c}\boldsymbol{\theta}^T \\ \hline \boldsymbol{\phi}^T\end{array}\right] [\mathbf{1} \mid \boldsymbol{\alpha}] \left[\begin{array}{c|c}\lambda_1 & \lambda_3 \\ \hline \lambda_2 & \lambda_4\end{array}\right] \tag{4.5.13}$$

is a minimum, are given by

$$[\boldsymbol{\theta} \mid \boldsymbol{\phi}] = \boldsymbol{\Omega}^{-1} [\mathbf{1} \mid \boldsymbol{\alpha}] \left[\left[\begin{array}{c}\mathbf{1}^T \\ \hline \boldsymbol{\alpha}^T\end{array}\right] \boldsymbol{\Omega}^{-1} [\mathbf{1} \mid \boldsymbol{\alpha}] \right]^{-1}. \tag{4.5.14}$$

Thus the computation of the estimators μ_* and σ_* requires the evaluation of the elements Ω_{ij} of $\boldsymbol{\Omega}$ followed by the inversion of two matrices, one of order $p + 1$ and the other of order 2. It may also be noted that in this process the variances and covariance of the estimators μ_* and σ_* are also computed, since using (4.5.12) and (4.5.14) we have

$$\begin{bmatrix}\mathrm{Var}(\mu_*) & \mathrm{Cov}(\mu_*, \sigma_*) \\ & \mathrm{Var}(\sigma_*)\end{bmatrix} = \sigma^2 \left[\left[\begin{array}{c}\mathbf{1}^T \\ \boldsymbol{\alpha}^T\end{array}\right] \boldsymbol{\Omega}^{-1} [\mathbf{1} \mid \boldsymbol{\alpha}] \right]^{-1}. \tag{4.5.15}$$

By making use of the identity in (4.5.6), we obtain from (4.5.10) that

$$E(\psi_k) = \alpha_{k+1:k+1}$$

and, hence, for $i \le j$ we have

$$\Omega_{ij} = \Omega_{ji}$$
$$= (i+1)(j+1)\sum_{r=1}^{n}\sum_{s=1}^{n}\frac{(r-1)^{(i)}(s-1)^{(j)}\alpha_{r,s:n}}{n^{(i+1)}n^{(j+1)}} - \alpha_{i+1:i+1}\alpha_{j+1:j+1}.$$

Now by making use of the identities

$$(a+b)^{(m)} = \sum_{r=0}^{m}\binom{m}{r} a^{(r)}b^{(m-r)}$$

and

$$(a-b)^{(m)} = \sum_{r=0}^{m}(-1)^r\binom{m}{r}(a-r)^{(m-r)}b^{(r)},$$

and the relations (Downton, 1966a; Arnold and Balakrishnan, 1989)

$$\sum_{i=1}^{n}(i-1)^{(k)}(n-i)^{(l)}\alpha_{i,i:n} = k!l!\binom{n}{k+l+1}\alpha_{k+1,k+1:k+l+1}$$

and

$$\sum\sum_{i<j}(i-1)^{(k)}(n-j)^{(l)}\alpha_{i,j:n} = k!l!\binom{n}{k+l+2}\alpha_{k+1,k+2:k+l+2},$$

we may write (see Downton, 1966a; Balakrishnan and Cohen, 1990), for $i \le j$,

$$\Omega_{ij} = \Omega_{ji} = (i+1)(j+1)\sum_{s=0}^{i}\frac{b_{ij}^{(s)}(n-j-1)^{(s)}}{n^{(i+1)}}, \qquad (4.5.16)$$

where

$$\begin{aligned} b_{ij}^{(s)} &= i!j!\frac{\alpha_{s+j+1,s+j+1:s+j+1}}{(i-s)!(s+j-i)!s!(s+j+1)} \\ &\quad + i!j!(j+1)^{(i+1-s)}\sum_{r=0}^{s-1}(-1)^r \\ &\quad \times \frac{\alpha_{j+1,j+2:j+2+r}}{(i+1-s)!(j+2+r)!(s-1-r)!} \\ &\quad - i!j!(i+1)^{(i+1-s)}\sum_{r=0}^{i}(-1)^r \\ &\quad \times \frac{\alpha_{s+j-i,s+j-i+1:s+j-i+1+r}}{(i+1-s)!(i-r)!(s+j-i+1+r)!} \\ &\quad + i!j!\frac{\alpha_{i+1:i+1}\{\alpha_{s+j-i:s+j-i} - \alpha_{j+1:j+1}\}}{(i+1-s)!(s+j-i)!s!} \end{aligned} \qquad (4.5.17)$$

Table 4.5.1 Coefficients $b_{ij}^{(s)}$ for Computing Elements of Ω for the Logistic Distribution

	s=0	s=1	s=2	s=3
i=0, j=0	1.000000000	—	—	—
i=0, j=1	0.500000000	—	—	—
i=0, j=2	0.348018224	—	—	—
i=0, j=3	0.272027336	—	—	—
i=1, j=1	0.348018224	0.303963551	—	—
i=1, j=2	0.500000000	0.227972662	—	—
i=1, j=3	0.598018224	0.185755503	—	—
i=2, j=2	0.413363707	0.759908875	0.177312070	—
i=2, j=3	1.000000000	0.937220948	0.147760060	—
i=3, j=3	0.879407749	2.482369003	1.177858760	0.124962794

Table 4.5.2 Values of Ω_{ij} for the Logistic Distribution

n = 10

i \ j	0	1	2	3
0	0.100000000	0.100000000	0.104405467	0.108810934
1	—	0.123543405	0.139720575	0.152226777
2	—	—	0.164747909	0.184268791
3	—	—	—	0.209856888

n = 15

i \ j	0	1	2	3
0	0.066666667	0.066666667	0.069603644	0.072540622
1	—	0.081896083	0.092447769	0.100622047
2	—	—	0.108585044	0.121156206
3	—	—	—	0.137466713

n = 20

i \ j	0	1	2	3
0	0.050000000	0.050000000	0.052202733	0.054405467
1	—	0.061256443	0.069087398	0.075160132
2	—	—	0.081000917	0.090277104
3	—	—	—	0.102257778

Table 4.5.3 Coefficients, Variances, and Relative Efficiencies of the Estimators μ_* and σ_*

n = 10

p		k=0	k=1	k=2	k=3	(Var.)/σ^2	Rel. Eff.
1	$\underline{\theta}^T$	1.00000	0.00000	—	—	0.10000	93.45%
	$\underline{\phi}^T$	−1.81380	1.81380	—	—	0.07745	99.15%
2	$\underline{\theta}^T$	0.25692	2.22924	−1.48616	—	0.09345	100.00%
	$\underline{\phi}^T$	−1.81380	1.81380	0.00000	—	0.07745	99.15%
3	$\underline{\theta}^T$	0.25675	2.22946	−1.48604	−0.00016	0.09345	100.00%
	$\underline{\phi}^T$	−1.57375	0.13294	2.88187	−1.44106	0.07680	99.99%

n = 15

p		k=0	k=1	k=2	k=3	(Var.)/σ^2	Rel. Eff.
1	$\underline{\theta}^T$	1.00000	0.00000	—	—	0.06667	92.73%
	$\underline{\phi}^T$	−1.81380	1.81380	—	—	0.05010	98.94%
2	$\underline{\theta}^T$	0.17418	2.47746	−1.65164	—	0.06182	100.00%
	$\underline{\phi}^T$	−1.81380	1.81380	0.00000	—	0.05010	98.94%
3	$\underline{\theta}^T$	0.17408	2.47738	−1.65104	−0.00043	0.06182	100.00%
	$\underline{\phi}^T$	−1.51331	−0.29016	3.60723	−1.80376	0.04958	99.98%

n = 20

p		k=0	k=1	k=2	k=3	(Var.)/σ^2	Rel. Eff.
1	$\underline{\theta}^T$	1.00000	0.00000	—	—	0.05000	92.34%
	$\underline{\phi}^T$	−1.81380	1.81380	—	—	0.03703	98.84%
2	$\underline{\theta}^T$	0.13171	2.60487	−1.73658	—	0.04618	99.98%
	$\underline{\phi}^T$	−1.81380	1.81380	0.00000	—	0.03703	98.84%
3	$\underline{\theta}^T$	0.13189	2.60496	−1.73747	0.00062	0.04617	100.00%
	$\underline{\phi}^T$	−1.47986	−0.52309	4.00554	−2.00259	0.03660	100.00%

As pointed out by Balakrishnan and Cohen (1990), the first term on the RHS of (4.5.17), as reported by Downton (1966a), contains an error. The coefficients depend only upon diagonal and immediate upper-diagonal product moments of relatively small variance matrices of order statistics and upon the expected values of the largest order statistics. The amount of computation required to obtain estimates of the form (4.5.2) and (4.5.3) is considerably less than what is required to obtain the best linear unbiased estimators given in Section 4.2.

By using the means, variances, and covariances of logistic order statistics tabulated by Balakrishnan and Malik (1991), we computed the values of the coefficients $b_{ij}^{(s)}$, which are needed to evaluate the variance-covariance matrix $\mathbf{\Omega}$. These values are presented in Table 4.5.1 for $0 \le i \le j \le 3$.

By making use of the values of $b_{ij}^{(s)}$ given in Table 4.5.1, we computed the values of Ω_{ij} $(=\Omega_{ji})$ for $0 \le i \le j \le 3$ and for sample size $n = 10$, 15, and 20. These values are presented in Table 4.5.2. These values were used in Eqs. (4.5.14) and (4.5.15) in order to compute the coefficients θ and ϕ and the variances and covariance of the estimators μ_* and σ_* for $n = 10$, 15, and 20. These values are presented in Table 4.5.3. Due to the symmetry of the logistic distribution, we see that the covariance of the estimators μ_* and σ_* is zero. Also given in the last column of Table 4.5.3, are the efficiencies, relative to the best linear unbiased estimators, of the estimators μ_* and σ_*; the necessary values of the variances of the best linear unbiased estimators of μ and σ were taken from the tables of Gupta, Qureishi, and Shah (1967) and Balakrishnan (1991).

From Table 4.5.3, we observe that even a two-term (linear) estimator of μ and σ is highly efficient as compared to the best linear unbiased estimators. The three-term (quadratic) estimator of μ is found to be as efficient as the BLUE of μ. However, due to the symmetry of the logistic distribution there is no advantage in using a three-term (quadratic) estimator of σ rather than a two-term (linear) one. If, on the other hand, a four-term (cubic) estimator of σ is used the estimator σ_* so obtained is almost identical to the BLUE of σ. Thus, we see that the linear unbiased estimators with polynomial coefficients discussed in this section require considerably less computation than the best linear unbiased estimators and are almost as efficient as the BLUEs of μ and σ; but, the method does not yield to generalizations to handle Type II censored samples.

REFERENCES

Arnold, B. C. and Balakrishnan, N. (1989). *Relations, Bounds and Approximations for Order Statistics*. Lecture Notes in Statistics No. 53, Springer-Verlag, New York.

Balakrishnan, N. (1991). Best linear unbiased estimates of the location and scale parameters of logistic distribution for complete and censored samples of sizes 2(1)25(5)40, submitted for publication.

Balakrishnan, N. and Cohen, A. C. (1990). *Order Statistics and Inference: Estimation Methods*. Academic Press, Boston.

Balakrishnan, N. and Malik, H. J. (1991). Means, variances and covariances of logistic order statistics for sample sizes up to fifty, *Selected Tables in Mathematical Statistics* (to appear).

Downton, F. (1966a). Linear estimates with polynomial coefficients. *Biometrika*, *53*, 129–141.

Downton, F. (1966b). Linear estimates of parameters in the extreme value distribution. *Technometrics*, *8*, 3–17.

Gupta, S. S., Qureishi, A. S., and Shah, B. K. (1967). Best linear unbiased estimators of the parameters of the logistic distribution using order statistics. *Technometrics*, *9*, 43–56.

4.6 SIMPLIFIED LINEAR ESTIMATORS

N. Balakrishnan
McMaster University, Hamilton, Ontario, Canada

For estimating the mean and standard deviation of a normal population, Dixon (1957) proposed some simple linear estimators which are highly efficient. These estimators are in terms of quasi-midrange and quasi-range, respectively. Similar work has been carried out by Raghunandanan and Srinivasan (1970, 1971) for the logistic and double exponential distributions.

Let V_i and W_i denote the ith quasi-midrange and the ith quasi-range of the sample, respectively; that is,

$$V_i = \tfrac{1}{2}(X_{i:n} + X_{n-i+1:n}) \tag{4.6.1}$$

and

$$W_i = X_{n-i+1:n} - X_{i:n}. \tag{4.6.2}$$

For $i = 1$, W_1 in (4.6.2) is simply the sample range. Dixon (1957) proposed V_i with the smallest variance as an estimator for the mean of the normal distribution. Next, he proposed an estimator for the standard deviation of the normal distribution to be the one with the smallest variance among linear estimators of the form

$$C \sum_{i=1}^{[n/2]} \delta_i W_i, \tag{4.6.3}$$

where each δ_i can take values 0 or 1, and C is a constant determined by making use of the expected values of normal order statistics so as to make the estimator in (4.6.3) unbiased for σ. Raghunandanan and Srinivasan (1970) carried out a similar work for the logistic distribution. They have given an estimator for μ of the form in (4.6.1). These estimators are presented in Table 4.6.1 for n up to 20. The variance of this estimator and its efficiency relative to the BLUE based on the complete sample are also presented in the last two columns of this table. We note that this simple estimator of μ is quite efficient and has efficiency close to 90% in almost all cases. Raghunandanan and Srinivasan (1970) have also proposed an estimator for σ of the form in (4.6.3). These estimators are given in Table 4.6.2 for n up to 20. The variance of this estimator and its efficiency relative to the BLUE based on the complete sample are also given in the last two columns of this table. From this table, we observe that this estimator of σ

Table 4.6.1 Estimator in (4.6.1) for the Mean of the Logistic Population and Values of Variance and Efficiency Relative to the BLUE

n	i	(Variance)/σ^2	Rel. Eff.
2	1	0.5000	100.0%
3	2	0.3921	83.1%
4	2	0.2599	92.7%
5	2	0.2071	92.2%
6	3	0.1795	88.1%
7	3	0.1482	91.0%
8	3	0.1294	90.8%
9	4	0.1171	88.9%
10	4	0.1035	90.3%
11	4	0.0940	90.5%
12	5	0.0870	89.2%
13	5	0.0795	89.9%
14	5	0.0738	89.9%
15	6	0.0693	89.2%
16	6	0.0645	89.8%
17	6	0.0607	89.7%
18	7	0.0576	89.2%
19	7	0.0542	89.7%
20	7	0.0516	89.5%

Table 4.6.2 Estimator in (4.6.3) for the Standard Deviation of the Logistic Distribution and Values of Variance and Efficiency Relative to the BLUE

n	Estimator	(Variance)/σ^2	Rel. Eff.
2	$0.9069\ W_1$	0.6449	100.0%
3	$0.6046\ W_1$	0.3333	100.0%
4	$0.3887\ (W_1+W_2)$	0.2324	97.0%
5	$0.3109\ (W_1+W_2)$	0.1719	99.1%
6	$0.2694\ (W_1+W_2)$	0.1376	99.5%
7	$0.2101\ (W_1+W_2+W_3)$	0.1159	98.8%
8	$0.1879\ (W_1+W_2+W_3)$	0.0989	99.5%
9	$0.1724\ (W_1+W_2+W_3)$	0.0865	99.7%
10	$0.1608\ (W_1+W_2+W_3)$	0.0773	99.3%
11	$0.1346\ (W_1+W_2+W_3+W_4)$	0.0694	99.7%
12	$0.1266\ (W_1+W_2+W_3+W_4)$	0.0631	99.8%
13	$0.1154\ (W_1+W_2+W_3+W_4+W_6)$	0.0580	99.6%
14	$0.1049\ (W_1+W_2+W_3+W_4+W_5)$	0.0535	99.7%
15	$0.1215\ (W_1+W_2+W_4+W_5)$	0.0506	98.0%
16	$0.0922\ (W_1+W_2+W_3+W_4+W_5+W_7)$	0.0464	99.8%
17	$0.08808\ (W_1+W_2+W_3+W_4+W_5+W_7)$	0.0434	99.9%
18	$0.084572\ (W_1+W_2+W_3+W_4+W_5+W_7)$	0.0409	99.9%
19	$0.076906\ (W_1+W_2+W_3+W_4+W_5+W_6+W_8)$	0.0386	99.9%
20	$0.074073\ (W_1+W_2+W_3+W_4+W_5+W_6+W_8)$	0.0366	100.0%

is remarkably efficient (never less than 97%). Similar estimators have been given by Raghunandanan and Srinivasan (1970) for the case when the sample is symmetrically Type II censored.

REFERENCES

Dixon, W. J. (1957). Estimates of the mean and standard deviation of a normal population. *Ann. Math. Statist.*, *28*, 806–809.

Raghunandanan, K. and Srinivasan, R. (1970). Simplified estimation of parameters in a logistic distribution. *Biometrika*, *57*, 677–678.

Raghunandanan, K. and Srinivasan, R. (1971). Simplified estimation of parameters in a double exponential distribution. *Technometrics*, *13*, 689–691.

5

Reliability Estimation Based on MLEs for Complete and Censored Samples

Lee J. Bain, James A. Eastman* and Max Engelhardt
University of Missouri-Rolla, Rolla, Missouri

N. Balakrishnan
McMaster University, Hamilton, Ontario, Canada

Charles E. Antle
The Pennsylvania State University, University Park, Pennsylvania

5.1 INTRODUCTION

The maximum likelihood estimation of the location and scale parameters (μ and σ) based on complete and Type II censored samples has been discussed in detail in Chapter 3. The asymptotic variance-covariance matrix of the MLEs, $\hat{\mu}$ and $\hat{\sigma}$, has also been presented there. Based on these MLEs, Antle, Klimko, and Harkness (1970) have developed interval estimation procedures for μ and σ for the complete sample case; reference may also be made to Bain (1978), Bain, Eastman, and Engelhardt (1973), Eastman (1972), Engelhardt, Bain, and Smith (1974), Plackett (1959), and Schafer and Sheffield (1973) for inference procedures and some applications to life-testing. In this chapter, by considering Type II right-censored samples we present methods of constructing confidence intervals for μ and σ, lower and upper tolerance limits for the distribution, and an estimate of the reliability function. Although the MLEs of μ and σ are not available explicitly for either complete or censored samples, $(\hat{\mu} - \mu)/\hat{\sigma}$ and $\hat{\sigma}/\sigma$ are pivotal quantities whose distributions are independent of both μ and σ, as pointed out by Antle et al. (1970); similarly, $(\hat{\mu} - \mu)/\hat{\sigma} + k\sigma/\hat{\sigma}$ is also seen to be a pivotal quantity. The distributions of the above variables are determined in this chapter by Monte Carlo Techniques and then are

**Current affiliation*: New York City Transit Authority, New York, New York

used to determine confidence intervals for μ and σ, tolerance limits for the distribution, and confidence limits for reliability. Finally, we consider an example involving lifetimes of certain incandescent lamps given by Davis (1952) and illustrate the methods of inference developed in this chapter.

5.2 MLEs OF μ AND σ

Let $X_{1:n} \le X_{2:n} \le \cdots \le X_{n-s:n}$ be a Type II right-censored sample from the logistic $L(\mu, \sigma^2)$ population with pdf $f(x; \mu, \sigma)$ and cdf $F(x; \mu, \sigma)$ as given in Eqs. (1.1) and (1.2), respectively. Then the maximum likelihood equations for μ and σ based on the above right-censored sample are presented in Chapter 3 (for the special case when $r = 0$); the asymptotic variance-covariance matrix of the MLEs, viz., $\hat{\mu}$ and $\hat{\sigma}$, has also been derived there.

By using simulated samples from the standard logistic $L(0, 1)$ population and denoting the MLEs in this case by $\hat{\mu}_0$ and $\hat{\sigma}_0$, we determined the means and variances of $\hat{\mu}_0$ and $\hat{\sigma}_0$ for sample size $n = 10, 20, 40$, and 80, the proportion of right-censoring $s/n = 0.0, 0.1, 0.3, 0.5$, and 0.7. These values are presented in Table 5.2.1. By realizing that $E(\hat{\mu}_0) = E(\hat{\mu} - \mu)/\sigma$ and $E(\hat{\sigma}_0) = E(\hat{\sigma})/\sigma$, we may use Table 5.2.1 to obtain unbiased estimators

Table 5.2.1 Simulated Values of $E(\hat{\mu}_0)$, $E(\hat{\sigma}_0)$, n Var$(\hat{\mu}_0)$, and n Var$(\hat{\sigma}_0)$, and the Values of Asymptotic Variances

	$E(\hat{\mu}_0)$					$E(\hat{\sigma}_0)$				
n s/n	0.0	0.1	0.3	0.5	0.7	0.0	0.1	0.3	0.5	0.7
10	.000	.000	−.021	−.078	−.256	.943	.934	.899	.839	.702
20	.000	−.001	−.010	−.039	−.120	.971	.966	.951	.920	.858
40	.000	−.001	−.006	−.020	−.060	.986	.983	.975	.960	.929
80	.000	.000	−.001	−.007	−.026	.992	.991	.988	.981	.967
	$n\text{Var}(\hat{\mu}_0)$					$n\text{Var}(\hat{\sigma}_0)$				
10	.93	.93	.98	1.22	2.36	.67	.73	.95	1.35	2.10
20	.94	.94	.99	1.21	2.43	.68	.76	1.00	1.46	2.50
40	.91	.91	.95	1.18	2.38	.70	.77	1.03	1.52	2.68
80	.91	.91	.94	1.18	2.37	.70	.76	1.03	1.56	2.76
∞	.91	.91	.95	1.16	2.39	.70	.77	1.03	1.56	2.87

of μ and σ as $\hat{\mu} - [\hat{\sigma}/E(\hat{\sigma}_0)]E(\hat{\mu}_0)$ and $\hat{\sigma}/E(\hat{\sigma}_0)$, respectively. Also given in Table 5.2.1 are the values of the asymptotic variances of $\hat{\mu}$ and $\hat{\sigma}$ computed from the formulas presented in Chapter 3. From this table, we observe that both $\hat{\mu}$ and $\hat{\sigma}$ become almost unbiased for large sample sizes even for a large proportion of right-censoring. This is to be expected as the MLEs are asymptotically unbiased estimators. However, the estimators $\hat{\mu}$ and $\hat{\sigma}$ need to be corrected for bias in case of small samples. We observe further from Table 5.2.1 that the asymptotic variance of $\hat{\mu}$ provides a very close approximation to Var($\hat{\mu}$) even for a sample of size as small as 10 and with the proportion of right-censoring as large as 0.7. But, the asymptotic variance of $\hat{\sigma}$ seems to provide a good approximation to Var($\hat{\sigma}$) only for sample size as large as 80 when the proportion of right-censoring is more than 0.5; otherwise, it seems to give reasonable approximation for sample size 20 and more.

5.3 CONFIDENCE INTERVALS FOR μ AND σ

The pivotal quantity $\sqrt{n}(\hat{\mu} - \mu)/\hat{\sigma}$ is known to be asymptotically normally distributed with mean 0 and variance V_{11} whose formula is given in Chapter 3. One may then easily determine the asymptotic upper $1 - \gamma$ percentage point of the distribution of $\sqrt{n}(\hat{\mu} - \mu)/\hat{\sigma}$. But for finite sample size, these percentage points cannot be derived explicitly. Hence, we resorted to Monte Carlo simulations in order to determine the upper $1 - \gamma$ percentage point, m_γ, where

$$\Pr\left\{\frac{\sqrt{n}(\hat{\mu} - \mu)}{\hat{\sigma}} \leq m_\gamma\right\} = \gamma. \tag{5.3.1}$$

For sample size $n = 10, 20, 40$, and 80 and the proportion of right-censoring $s/n = 0.0, 0.1, 0.3, 0.5$, and 0.7, we have simulated the values of m_γ for $\gamma = 0.01, 0.025, 0.05, 0.10, 0.25, 0.50, 0.75, 0.90, 0.95, 0.975$, and 0.99. These values are presented in Table 5.3.1. Also given in this table are the values of the corresponding asymptotic upper $1 - \gamma$ percentage points. From Table 5.3.1, we first of all observe that for the complete sample case (when $s/n = 0$) the distribution of $\sqrt{n}(\hat{\mu} - \mu)/\hat{\sigma}$ is symmetric around zero and this is expected due to the symmetry of the logistic distribution. We next observe that the distribution of $\sqrt{n}(\hat{\mu} - \mu)/\hat{\sigma}$ becomes more and more left-skewed as the proportion of right-censoring increases. Finally, we note from Table 5.3.1 that the asymptotic values provide reasonable approximation to the percentage points only for sample size as large as 80 even for the proportion of right-censoring as small as 0.10. The simulated values reported in this table are based on 8000 Monte Carlo runs when $n = 10, 20$, and 40, and 4000 runs when $n = 80$.

By using the values of m_γ presented in Table 5.3.1, one may easily obtain

Table 5.3.1 Simulated Values and Asymptotic Values of m_γ such that $\Pr\{\sqrt{n}(\hat{\mu} - \mu)/\hat{\sigma} \le m_\gamma\} = \gamma$

		γ										
s/n	n	.01	.025	.05	.10	.25	.50	.75	.90	.95	.975	.99
	10	−2.78	−2.24	−1.81	−1.37	−0.70	0.00	0.70	1.37	1.81	2.24	2.78
	20	−2.51	−2.08	−1.70	−1.30	−0.68	0.00	0.68	1.30	1.70	2.08	2.51
0.0	40	−2.39	−1.96	−1.64	−1.25	−0.65	0.00	0.65	1.25	1.64	1.96	2.39
	80	−2.28	−1.93	−1.60	−1.23	−0.65	0.00	0.65	1.23	1.60	1.93	2.28
	∞	−2.22	−1.87	−1.57	−1.22	−0.64	0.00	0.64	1.22	1.57	1.87	2.22
	10	−2.89	−2.35	−1.90	−1.40	−0.70	0.00	0.72	1.42	1.84	2.28	2.83
	20	−2.61	−2.14	−1.74	−1.32	−0.68	0.01	0.69	1.31	1.71	2.06	2.48
0.1	40	−2.43	−2.00	−1.66	−1.26	−0.65	−0.01	0.65	1.26	1.63	1.95	2.39
	80	−2.30	−1.96	−1.60	−1.23	−0.64	0.00	0.67	1.24	1.62	1.90	2.31
	∞	−2.22	−1.87	−1.57	−1.22	−0.64	0.00	0.64	1.22	1.57	1.87	2.22
	10	−3.98	−3.04	−2.40	−1.70	−0.85	−0.08	0.68	1.40	1.86	2.33	2.84
	20	−3.08	−2.47	−2.00	−1.48	−0.77	−0.04	0.66	1.29	1.69	2.05	2.48
0.3	40	−2.64	−2.19	−1.80	−1.37	−0.70	−0.04	0.63	1.24	1.63	1.93	2.38
	80	−2.38	−2.04	−1.67	−1.31	−0.68	−0.01	0.66	1.24	1.60	1.90	2.33
	∞	−2.26	−1.91	−1.60	−1.25	−0.66	0.00	0.66	1.25	1.60	1.91	2.26
	10	−7.84	−5.71	−4.08	−2.92	−1.39	−0.28	0.57	1.30	1.77	2.28	2.83
	20	−4.63	−3.68	−2.88	−2.10	−1.07	−0.17	0.59	1.26	1.66	2.03	2.42
0.5	40	−3.60	−2.90	−2.38	−1.81	−0.93	−0.12	0.61	1.24	1.61	1.96	2.37
	80	−3.16	−2.55	−2.06	−1.61	−0.84	−0.04	0.66	1.26	1.66	1.97	2.30
	∞	−2.51	−2.11	−1.77	−1.38	−0.73	0.00	0.73	1.38	1.77	2.11	2.51
	10			−16.99	−9.78	−4.43	−1.30	0.25	1.14	1.63	2.12	2.64
	20	−12.55	−9.04	−6.76	−4.70	−2.34	−0.69	0.51	1.34	1.75	2.08	2.43
0.7	40	−7.40	−5.58	−4.56	−3.33	−1.73	−0.44	0.63	1.43	1.85	2.19	2.57
	80	−5.60	−4.43	−3.67	−2.78	−1.45	−0.21	0.78	1.56	1.96	2.36	2.73
	∞	−3.59	−3.03	−2.54	−1.98	−1.04	0.00	1.04	1.98	2.54	3.03	3.59

confidence intervals for μ. For example, by using the values of $m_{0.025}$ and $m_{0.975}$ from the table we have

$$\Pr\left\{ m_{0.025} \le \frac{\sqrt{n}(\hat{\mu} - \mu)}{\hat{\sigma}} \le m_{0.975} \right\} = 0.95, \tag{5.3.2}$$

which yields the 95% confidence interval for μ to be

$$\left[\hat{\mu} - m_{0.975} \frac{\hat{\sigma}}{\sqrt{n}}, \ \hat{\mu} + m_{0.025} \frac{\hat{\sigma}}{\sqrt{n}} \right]. \tag{5.3.3}$$

Moreover, we have $m_\gamma \to \sqrt{V_{11}} z_\gamma$ as $n \to \infty$, where z_γ denotes the upper $1 - \gamma$ percentage point of the standard normal distribution. These asymptotic values of m_γ are also presented in Table 5.3.1, and they may be used to construct asymptotic confidence intervals for μ. The asymptotic confidence interval will provide a reasonable approximation for the finite sample case only for sample size as large as 80 even when the proportion of right-censoring is as small as 0.10.

Similarly, the pivotal quantity $\sqrt{n}(\hat{\sigma}/\sigma - 1)$ is asymptotically normally distributed with mean 0 and variance V_{22} whose formula is given in Chapter 3. By using this result, one may easily determine the asymptotic upper $1 - \gamma$ percentage point of the distribution of $\sqrt{n}(\hat{\sigma}/\sigma - 1)$. For finite sample size, however, these percentage points can not be determined explicitly. We, therefore, simulated the value of the upper $1 - \gamma$ percentage point, s_γ, where

$$\Pr\left\{\sqrt{n}\left(\frac{\hat{\sigma}}{\sigma} - 1\right) \le s_\gamma\right\} = \gamma. \tag{5.3.4}$$

For sample size $n = 10, 20, 40$ and 80 and the proportion of right-censoring $s/n = 0.0, 0.1, 0.3, 0.5$, and 0.7, we have simulated the values of s_γ for $\gamma = 0.01, 0.025, 0.05, 0.10, 0.25, 0.50, 0.75, 0.90, 0.95, 0.975$, and 0.99. These values are presented in Table 5.3.2. The corresponding asymptotic upper $1 - \gamma$ percentage points are also given in the table. From this table, we first observe that the distribution of $\sqrt{n}(\hat{\sigma}/\sigma - 1)$ is right-skewed for small samples and that it becomes more right-skewed as the proportion of right-censoring increases. However, the distribution is seen to become nearly symmetrical for large sample sizes (as large as 80) as one would expect it to be due to its asymptotic normality. We once again note from Table 5.3.2 that the asymptotic values provide reasonable approximation to the percentage points only for samples of size as large as 80 even when the proportion of right-censoring is as small as 0.10.

One can easily construct confidence intervals for σ by using the values of s_γ presented in Table 5.3.2. For example, by using the values of $s_{0.025}$ and $s_{0.975}$ from the table we have

$$\Pr\left\{s_{0.025} \le \sqrt{n}\left(\frac{\hat{\sigma}}{\sigma} - 1\right) \le s_{0.975}\right\} = 0.95 \tag{5.3.5}$$

which yields the 95% confidence interval for σ to be

$$\left(\frac{\hat{\sigma}}{1 + s_{0.975}/\sqrt{n}}, \frac{\hat{\sigma}}{1 + s_{0.025}/\sqrt{n}}\right). \tag{5.3.6}$$

Furthermore, by using the result that $s_\gamma \to \sqrt{V_{22}}z_\gamma$ as $n \to \infty$ where z_γ is the upper $1 - \gamma$ percentage point of the standard normal distribution, we determined the asymptotic values of s_γ and have presented them in Table

Table 5.3.2 Simulated Values and Asymptotic Values of s_γ such that $\Pr\{\sqrt{n}(\hat{\sigma}/\sigma - 1) \leq s_\gamma\} = \gamma$

		γ										
s/n	n	.01	.025	.05	.10	.25	.50	.75	.90	.95	.975	.99
	10	− 1.81	−1.59	− 1.40	−1.18	− 0.76	− 0.24	0.33	0.89	1.28	1.61	2.02
	20	− 1.86	−1.62	− 1.40	−1.13	− 0.71	− 0.18	0.39	0.95	1.31	1.62	2.01
0.0	40	− 1.91	−1.63	− 1.41	−1.12	− 0.67	− 0.12	0.45	1.00	1.34	1.66	2.05
	80	− 1.93	−1.66	− 1.42	−1.13	− 0.67	− 0.08	0.48	1.03	1.35	1.65	1.94
	∞	− 1.95	−1.64	− 1.38	−1.07	− 0.56	0.00	0.56	1.07	1.38	1.64	1.95
	10	− 1.91	−1.69	− 1.49	−1.26	− 0.81	− 0.26	0.32	0.91	1.29	1.69	2.11
	20	− 1.96	−1.70	− 1.47	−1.22	− 0.76	− 0.20	0.39	0.99	1.37	1.74	2.18
0.1	40	− 1.97	−1.71	− 1.49	−1.20	− 0.71	− 0.14	0.45	1.04	1.38	1.74	2.11
	80	− 2.04	−1.68	− 1.46	−1.17	− 0.69	− 0.09	0.50	1.06	1.39	1.65	1.97
	∞	− 2.04	−1.72	− 1.44	−1.12	− 0.59	0.00	0.59	1.12	1.44	1.72	2.04
	10	− 2.16	−1.96	− 1.75	−1.50	− 1.03	− 0.41	0.30	0.97	1.43	1.81	2.36
	20	− 2.25	−1.98	− 1.75	−1.47	− 0.92	− 0.29	0.41	1.08	1.56	1.93	2.43
0.3	40	− 2.28	−2.03	− 1.75	−1.43	− 0.87	− 0.20	0.50	1.16	1.59	1.98	2.42
	80	− 2.28	−1.97	− 1.72	−1.39	− 0.82	− 0.14	0.55	1.24	1.65	1.98	2.28
	∞	− 2.36	−1.99	− 1.67	−1.30	− 0.69	0.00	0.69	1.30	1.67	1.99	2.36
	10	− 2.53	−2.34	− 2.14	−1.87	− 1.36	− 0.65	0.19	1.04	1.62	2.04	2.79
	20	− 2.70	−2.40	− 2.15	−1.84	− 1.22	− 0.46	0.38	1.27	1.78	2.26	2.99
0.5	40	− 2.77	−2.46	− 2.18	−1.81	− 1.12	− 0.32	0.54	1.36	1.88	2.35	2.95
	80	− 2.81	−2.45	− 2.09	−1.71	− 1.03	− 0.24	0.64	1.48	1.98	2.43	3.01
	∞	− 2.91	−2.45	− 2.06	−1.60	− 0.84	0.00	0.84	1.60	2.06	2.45	2.91
	10	− 2.98	−2.87	− 2.73	−2.52	− 2.00	− 1.23	− 0.18	1.00	1.78	2.58	3.70
	20	− 3.41	−3.16	− 2.89	−2.51	− 1.78	− 0.83	0.29	1.47	2.25	3.04	3.78
0.7	40	− 3.62	−3.27	− 2.88	−2.42	− 1.62	− 0.60	0.57	1.72	2.46	3.12	3.97
	80	− 3.68	−3.24	− 2.88	−2.34	− 1.48	− 0.39	0.75	1.89	2.65	3.22	3.89
	∞	− 3.94	−3.32	− 2.78	−2.17	− 1.14	0.00	1.14	2.17	2.78	3.32	3.94

5.3.2. These values may be made use of in a similar way to construct asymptotic confidence intervals for σ. As in the case of the interval estimation of μ, one has to be warned here that the asymptotic confidence interval for σ will provide a reasonable approximation for the finite sample case only for samples of size as large as 80 even when the proportion of right-censoring is as small as 0.10.

5.4 TOLERANCE LIMITS BASED ON MLEs

For the logistic distribution, Hall (1975) has discussed the determination of one-sided tolerance limits based on the BLUEs of μ and σ (see Chapter

4 for details) for complete and right-censored samples. Hall's results have been extended by Balakrishnan and Fung (1991) to cover larger sample sizes and also the case when samples are doubly Type II censored; these results are presented in detail in Chapter 14.

The tolerance limit $L(\mathbf{X})$ is said to be a lower γ tolerance limit for proportion β if

$$\Pr\{1 - F(L(\mathbf{X}); \mu, \sigma) \geq \beta\} = \gamma, \tag{5.4.1}$$

or

$$\Pr\{F(L(\mathbf{X}); \mu, \sigma) \leq 1 - \beta\} = \gamma. \tag{5.4.2}$$

For a general location-scale family of distributions, Dumonceaux (1969) has shown that tolerance limits can be constructed based on the pivotal quantity

$$T(\mathbf{X}) = \frac{\hat{\mu} - \mu}{\hat{\sigma}} - \frac{\sigma}{\hat{\sigma}} F^{-1}(1 - \beta), \tag{5.4.3}$$

where F denotes the cumulative distribution function $F(\mu, \sigma)$ when $\mu = 0$ and $\sigma = 1$, and F^{-1} denotes its inverse; also see Bain (1978). Thoman, Bain, and Antle (1970) have, for example, given similar results for the Weibull distribution.

For the logistic distribution, in particular, we have

$$F^{-1}(1 - \beta) = -\frac{\sqrt{3}}{\pi} \ln\left(\frac{\beta}{1 - \beta}\right)$$

so that $T(\mathbf{X})$ in (5.4.3) becomes

$$T(\mathbf{X}) = \frac{\hat{\mu} - \mu}{\hat{\sigma}} + \frac{\sqrt{3}}{\pi} \frac{\sigma}{\hat{\sigma}} \ln\left(\frac{\beta}{1 - \beta}\right). \tag{5.4.4}$$

The lower γ tolerance limit for proportion β of the population is given by

$$L(\mathbf{X}) = \hat{\mu} - t_\gamma \hat{\sigma}, \tag{5.4.5}$$

where t_γ is such that

$$\Pr\{T(\mathbf{X}) \leq t_\gamma\} = \gamma. \tag{5.4.6}$$

In other words, t_γ is simply the upper $1 - \gamma$ percentage point of $T(\mathbf{X})$ in (5.4.4) when the sample comes from a logistic $L(\mu, \sigma^2)$ population, or equivalently, the upper $1 - \gamma$ percentage point of the variable

$$\frac{\hat{\mu}}{\hat{\sigma}} + \frac{\sqrt{3}}{\pi\hat{\sigma}} \ln\left(\frac{\beta}{1 - \beta}\right)$$

when the sample is from a logistic $L(0, 1)$ population. Also, by the symmetry of the logistic distribution, we have the upper γ tolerance limit for proportion β of the population to be

$$U(\mathbf{X}) = \hat{\mu} + t_\gamma \hat{\sigma}, \tag{5.4.7}$$

where t_γ is as given in (5.4.6). We simulated the values of t_γ for $n = 10$, 20, 40, and 80, $s/n = 0.0, 0.3, 0.5$, and 0.7, $\beta = 0.500(.025)0.975$, and $\gamma = 0.75, 0.85, 0.90, 0.95$, and 0.99. These values are presented in Table 5.4.1. The simulated values reported in this table are based on 2,000 Monte Carlo runs when $n = 10$, 20, and 40, and 1000 runs when $n = 80$. Furthermore, the values of t_γ (based on 10,000 Monte Carlo simulations) for the more commonly used values of β and γ, viz., $\beta = 0.80, 0.90$, and 0.95, and $\gamma = 0.90, 0.95, 0.98, 0.99$, and 0.995, are presented in Table 5.4.2 for $n = 10$, 20, and 40 for the case when the complete sample is available. Although the values reported in Table 5.4.1 are based on relatively small simulation size, we note that they agree quite well with the values given in Table 5.4.2. The values of the tolerance factor t_γ presented in Table 5.4.1 should, therefore, be sufficiently accurate for applications to practical problems.

Asymptotically, we have the distribution of

$$\frac{\sqrt{n}}{\sqrt{c}}\left\{T(\mathbf{X}) - \frac{\sqrt{3}}{\pi}\ln\left(\frac{\beta}{1-\beta}\right)\right\} \tag{5.4.8}$$

to be standard normal, where

$$\begin{aligned} c &= n\,\mathrm{Var}(T(\mathbf{X})) \\ &= C_{11} + \{F^{-1}(1-\beta)\}^2 C_{22} + 2F^{-1}(1-\beta)C_{12} \\ &= C_{11} + \left\{\frac{\sqrt{3}}{\pi}\ln\left(\frac{\beta}{1-\beta}\right)\right\}^2 C_{22} - \frac{2\sqrt{3}}{\pi}\ln\left(\frac{\beta}{1-\beta}\right) C_{12}; \end{aligned} \tag{5.4.9}$$

here, $C_{11} = n\,\mathrm{Var}(\hat{\mu})/\sigma^2$, $C_{22} = n\,\mathrm{Var}(\hat{\sigma})/\sigma^2$, and $C_{12} = n\,\mathrm{Cov}(\hat{\mu}, \hat{\sigma})/\sigma^2$, and the asymptotic expressions of these quantities are presented in Chapter 3. The asymptotic values of c have been computed and presented in Table 5.4.3 for the proportion of right-censoring $s/n = 0.0, 0.1, 0.3, 0.5$, and 0.7, and $\beta = 0.500(0.025)0.975$. By using these values of c, one may obtain the asymptotic value of the tolerance factor t_γ to be

$$t_\gamma = \sqrt{\frac{c}{n}}\, z_\gamma + \frac{\sqrt{3}}{\pi}\ln\left(\frac{\beta}{1-\beta}\right), \tag{5.4.10}$$

where, as before, z_γ denotes the upper $1 - \gamma$ percentage point of the standard normal distribution. For example, when $n = 80$ and $\beta = 0.90$,

Table 5.4.1 Tolerance Factors t_γ such that $L(\mathbf{X}) = \hat{\mu} - t_\gamma\hat{\sigma}$, $U(\mathbf{X}) = \hat{\mu} + t_\gamma\hat{\sigma}$, $\gamma = .75$

n	10				20				40				80			
s/n	.0	.3	.5	.7	.0	.3	.5	.7	.0	.3	.5	.7	.0	.3	.5	.7
β																
.500	.22	.20	.16	.06	.15	.15	.13	.12	.096	.093	.089	.092	.077	.072	.076	.088
.525	.28	.26	.22	.12	.21	.21	.19	.17	.153	.149	.146	.149	.133	.129	.131	.142
.550	.34	.33	.29	.19	.27	.27	.25	.23	.209	.205	.203	.205	.188	.187	.187	.195
.575	.41	.40	.35	.26	.33	.33	.31	.29	.266	.265	.260	.262	.245	.244	.243	.249
.600	.47	.47	.43	.33	.39	.39	.37	.35	.325	.325	.319	.319	.304	.303	.300	.303
.625	.54	.54	.52	.41	.45	.46	.44	.41	.384	.384	.380	.377	.364	.363	.359	.363
.650	.61	.62	.59	.49	.52	.52	.51	.48	.446	.447	.443	.440	.427	.425	.420	.424
.675	.68	.69	.67	.59	.59	.59	.58	.55	.513	.516	.509	.506	.490	.490	.485	.484
.700	.75	.77	.76	.68	.65	.66	.65	.63	.581	.585	.578	.575	.557	.557	.553	.550
.725	.84	.86	.85	.79	.73	.74	.73	.71	.654	.656	.651	.647	.627	.628	.625	.623
.750	.92	.95	.95	.90	.81	.82	.82	.80	.732	.735	.731	.722	.703	.704	.702	.699
.775	1.01	1.05	1.06	1.02	.89	.91	.92	.89	.812	.817	.819	.804	.784	.783	.786	.781
.800	1.10	1.16	1.19	1.15	.99	1.01	1.02	1.00	.900	.910	.915	.901	.869	.871	.876	.872
.825	1.21	1.28	1.32	1.34	1.09	1.12	1.14	1.11	.997	1.009	1.018	1.005	.961	.970	.971	.972
.850	1.34	1.42	1.48	1.57	1.20	1.24	1.26	1.26	1.109	1.121	1.139	1.129	1.070	1.077	1.082	1.079
.875	1.48	1.59	1.66	1.82	1.34	1.38	1.41	1.43	1.236	1.252	1.274	1.272	1.193	1.200	1.211	1.210
.900	1.66	1.79	1.89	2.14	1.50	1.56	1.59	1.64	1.385	1.407	1.438	1.447	1.338	1.349	1.365	1.358
.925	1.88	2.04	2.18	2.61	1.70	1.77	1.82	1.90	1.571	1.603	1.642	1.667	1.519	1.534	1.551	1.551
.950	2.19	2.39	2.60	3.18	1.97	2.06	2.15	2.28	1.835	1.872	1.926	1.980	1.770	1.786	1.816	1.830
.975	2.71	2.96	3.28	4.25	2.44	2.54	2.69	2.92	2.265	2.321	2.404	2.507	2.190	2.206	2.254	2.286

(continued)

Table 5.4.1 *(continued)*

$\gamma = .85$

n	10				20				40				80			
s/n	.0	.3	.5	.7	.0	.3	.5	.7	.0	.3	.5	.7	.0	.3	.5	.7
β																
.500	.33	.32	.30	.21	.24	.23	.21	.21	.153	.151	.147	.176	.114	.115	.113	.139
.525	.39	.38	.36	.28	.30	.29	.28	.27	.210	.209	.201	.228	.171	.170	.167	.189
.550	.46	.45	.43	.36	.36	.35	.34	.33	.267	.266	.259	.282	.227	.227	.225	.243
.575	.53	.52	.50	.44	.42	.41	.40	.39	.327	.325	.318	.336	.286	.285	.283	.297
.600	.60	.60	.58	.51	.48	.48	.47	.45	.387	.386	.380	.390	.345	.344	.343	.353
.625	.66	.67	.65	.58	.55	.55	.53	.52	.449	.449	.443	.449	.405	.405	.401	.411
.650	.73	.75	.74	.67	.61	.61	.60	.58	.514	.515	.507	.509	.467	.468	.468	.472
.675	.81	.84	.83	.76	.68	.69	.68	.66	.579	.580	.577	.575	.530	.534	.531	.535
.700	.90	.93	.94	.87	.75	.77	.76	.74	.646	.652	.648	.648	.597	.600	.600	.599
.725	.98	1.03	1.06	1.01	.83	.85	.85	.83	.719	.726	.729	.717	.669	.672	.670	.671
.750	1.08	1.14	1.18	1.16	.91	.94	.94	.91	.796	.808	.812	.800	.742	.747	.746	.746
.775	1.18	1.26	1.30	1.34	1.00	1.03	1.04	1.02	.879	.896	.900	.891	.822	.827	.828	.826
.800	1.29	1.38	1.44	1.51	1.10	1.13	1.15	1.15	.970	.993	.999	.992	.911	.917	.918	.917
.825	1.42	1.51	1.59	1.72	1.21	1.24	1.28	1.29	1.069	1.096	1.107	1.108	1.008	1.015	1.026	1.021
.850	1.55	1.66	1.78	2.01	1.33	1.38	1.42	1.45	1.181	1.215	1.228	1.235	1.116	1.125	1.146	1.137
.875	1.72	1.84	2.02	2.36	1.47	1.53	1.60	1.66	1.311	1.345	1.373	1.386	1.243	1.254	1.274	1.275
.900	1.91	2.07	2.29	2.79	1.64	1.70	1.80	1.90	1.463	1.512	1.553	1.573	1.394	1.406	1.431	1.435
.925	2.15	2.35	2.65	3.38	1.85	1.93	2.06	2.22	1.658	1.716	1.765	1.811	1.582	1.599	1.631	1.648
.950	2.48	2.74	3.11	4.24	2.15	2.27	2.43	2.68	1.936	2.003	2.080	2.144	1.842	1.866	1.912	1.936
.975	3.05	3.39	3.91	5.69	2.65	2.82	3.05	3.43	2.398	2.484	2.595	2.725	2.275	2.317	2.373	2.432

$\gamma = .90$

n	10				20				40				80			
s/n	.0	.3	.5	.7	.0	.3	.5	.7	.0	.3	.5	.7	.0	.3	.5	.7
β																
.500	.44	.43	.39	.33	.30	.30	.28	.29	.190	.190	.189	.229	.146	.140	.140	.177
.525	.50	.50	.46	.40	.35	.35	.34	.35	.251	.249	.248	.280	.202	.198	.195	.228
.550	.56	.58	.54	.46	.41	.41	.40	.40	.314	.308	.304	.328	.258	.257	.250	.280
.575	.63	.65	.62	.53	.48	.48	.46	.46	.371	.369	.363	.380	.314	.313	.309	.334
.600	.70	.72	.70	.61	.54	.54	.53	.52	.430	.430	.427	.434	.371	.373	.367	.389
.625	.77	.80	.79	.71	.60	.61	.61	.59	.493	.495	.488	.492	.432	.434	.430	.443
.650	.85	.89	.89	.82	.67	.68	.68	.66	.556	.562	.555	.553	.495	.496	.494	.503
.675	.93	.98	.98	.93	.75	.76	.76	.74	.620	.633	.630	.618	.561	.560	.559	.566
.700	1.01	1.07	1.09	1.05	.82	.84	.84	.82	.690	.707	.705	.691	.626	.628	.627	.635
.725	1.10	1.16	1.21	1.22	.91	.93	.93	.91	.762	.783	.784	.772	.699	.702	.699	.703
.750	1.19	1.28	1.35	1.42	.99	1.02	1.03	1.01	.842	.864	.871	.859	.772	.777	.776	.776
.775	1.30	1.40	1.50	1.60	1.08	1.13	1.15	1.13	.929	.951	.963	.953	.850	.858	.865	.861
.800	1.41	1.53	1.68	1.82	1.18	1.24	1.27	1.27	1.024	1.050	1.065	1.058	.938	.950	.956	.950
.825	1.54	1.67	1.87	2.11	1.29	1.36	1.40	1.43	1.126	1.160	1.179	1.172	1.033	1.049	1.056	1.052
.850	1.69	1.86	2.07	2.46	1.42	1.50	1.55	1.61	1.242	1.290	1.308	1.309	1.144	1.164	1.173	1.171
.875	1.86	2.05	2.29	2.86	1.56	1.66	1.73	1.83	1.378	1.434	1.461	1.469	1.272	1.292	1.313	1.316
.900	2.07	2.31	2.60	3.43	1.73	1.85	1.95	2.09	1.542	1.609	1.642	1.669	1.426	1.450	1.479	1.487
.925	2.33	2.61	2.98	4.18	1.95	2.09	2.23	2.45	1.744	1.827	1.872	1.927	1.617	1.653	1.687	1.698
.950	2.70	3.02	3.53	5.29	2.27	2.42	2.60	2.96	2.023	2.133	2.195	2.288	1.887	1.926	1.974	2.017
.975	3.32	3.74	4.46	7.13	2.78	2.99	3.25	3.85	2.493	2.640	2.740	2.909	2.333	2.385	2.459	2.546

(continued)

Table 5.4.1 (*continued*)

$\gamma = .95$

n	10				20				40				80			
s/n	.0	.3	.5	.7	.0	.3	.5	.7	.0	.3	.5	.7	.0	.3	.5	.7
β																
.500	.58	.58	.54	.51	.37	.37	.36	.39	.262	.260	.252	.292	.183	.183	.184	.228
.525	.64	.64	.62	.58	.43	.43	.43	.44	.321	.321	.313	.341	.239	.239	.235	.277
.550	.71	.72	.71	.65	.49	.50	.49	.50	.378	.377	.370	.391	.297	.297	.290	.329
.575	.79	.80	.80	.75	.56	.57	.56	.56	.438	.441	.430	.447	.354	.355	.349	.378
.600	.86	.90	.89	.84	.63	.64	.63	.62	.499	.498	.490	.505	.412	.413	.409	.433
.625	.94	1.00	1.01	.96	.70	.72	.71	.69	.561	.562	.554	.562	.473	.472	.469	.487
.650	1.03	1.10	1.11	1.10	.77	.79	.79	.76	.628	.628	.623	.622	.536	.537	.528	.544
.675	1.11	1.20	1.24	1.27	.85	.88	.88	.85	.693	.699	.697	.696	.602	.602	.596	.608
.700	1.20	1.32	1.38	1.42	.92	.96	.98	.95	.766	.775	.777	.766	.671	.667	.666	.675
.725	1.30	1.43	1.52	1.62	1.00	1.05	1.08	1.07	.842	.863	.859	.851	.737	.739	.737	.742
.750	1.41	1.55	1.69	1.88	1.10	1.15	1.20	1.20	.925	.954	.949	.940	.815	.814	.816	.816
.775	1.54	1.69	1.85	2.20	1.20	1.26	1.32	1.35	1.014	1.050	1.046	1.042	.899	.901	.907	.903
.800	1.68	1.86	2.05	2.59	1.30	1.39	1.46	1.53	1.108	1.154	1.157	1.161	.991	1.001	1.004	1.002
.825	1.82	2.03	2.23	3.03	1.43	1.51	1.62	1.70	1.217	1.266	1.278	1.303	1.091	1.101	1.112	1.106
.850	1.96	2.21	2.45		1.56	1.65	1.79	1.92	1.345	1.395	1.424	1.447	1.202	1.219	1.233	1.235
.875	2.14	2.43	2.76		1.71	1.83	2.01	2.19	1.490	1.544	1.582	1.637	1.335	1.356	1.375	1.385
.900	2.37	2.71	3.13		1.90	2.04	2.26	2.53	1.665	1.726	1.778	1.862	1.494	1.517	1.547	1.582
.925	2.64	3.06	3.60		2.15	2.34	2.58	2.93	1.880	1.952	2.014	2.143	1.691	1.722	1.769	1.813
.950	3.03	3.56	4.23		2.48	2.73	3.00	3.57	2.186	2.263	2.360	2.561	1.965	2.005	2.066	2.140
.975	3.70	4.36	5.42		3.02	3.38	3.71	4.65	2.689	2.786	2.954	3.224	2.417	2.477	2.562	2.717

$\gamma = .99$

n	10				20				40				80			
s/n	.0	.3	.5	.7	.0	.3	.5	.7	.0	.3	.5	.7	.0	.3	.5	.7
β																
.500	.89	.91	.95	.84	.56	.55	.54	.53	.362	.369	.375	.398	.240	.239	.246	.328
.525	.96	1.01	1.04	.99	.62	.64	.61	.60	.428	.432	.432	.455	.285	.294	.296	.375
.550	1.04	1.09	1.15	1.11	.69	.71	.69	.68	.497	.491	.490	.507	.343	.343	.352	.425
.575	1.12	1.20	1.24	1.29	.77	.79	.78	.77	.556	.558	.550	.559	.406	.405	.411	.470
.600	1.21	1.30	1.37	1.50	.85	.88	.87	.87	.624	.621	.620	.622	.470	.469	.472	.514
.625	1.30	1.41	1.52	1.83	.92	.97	.97	.97	.694	.692	.693	.699	.529	.535	.536	.566
.650	1.39	1.51	1.67	2.04	1.01	1.07	1.07	1.07	.764	.772	.778	.779	.592	.598	.599	.621
.675	1.49	1.62	1.91	2.25	1.09	1.14	1.17	1.18	.835	.857	.860	.856	.658	.676	.672	.683
.700	1.60	1.76	2.13		1.20	1.23	1.29	1.29	.916	.946	.939	.935	.738	.752	.750	.755
.725	1.70	1.96	2.37		1.31	1.36	1.42	1.45	1.014	1.036	1.025	1.016	.819	.834	.825	.827
.750	1.86	2.12	2.58		1.41	1.50	1.55	1.62	1.104	1.122	1.121	1.127	.899	.916	.913	.915
.775	2.00	2.30			1.52	1.63	1.70	1.82	1.198	1.210	1.231	1.266	.984	1.004	1.006	1.010
.800	2.18	2.48			1.66	1.81	1.87	2.12	1.300	1.320	1.359	1.420	1.085	1.099	1.106	1.111
.825	2.34	2.69			1.80	1.95	2.06	2.41	1.416	1.447	1.487	1.609	1.191	1.208	1.218	1.226
.850	2.56	2.99			1.96	2.15	2.30	2.68	1.542	1.592	1.643	1.809	1.308	1.328	1.342	1.371
.875	2.81	3.29			2.13	2.36	2.53		1.692	1.774	1.838	2.031	1.435	1.476	1.495	1.546
.900	3.08	3.68			2.35	2.60	2.87		1.875	1.977	2.081	2.310	1.581	1.644	1.688	1.779
.925	3.46	4.19			2.60	2.93	3.28		2.109	2.247	2.358	2.644	1.788	1.867	1.923	2.035
.950	3.98	4.80			2.97	3.38	3.89		2.437	2.620	2.758	3.150	2.069	2.190	2.255	2.401
.975	4.84	5.84			3.63	4.19	4.90		2.979	3.209	3.432	4.063	2.530	2.714	2.819	3.058

Table 5.4.2 Tolerance Factors t_γ Such That $L(\mathbf{X}) = \hat{\mu} - t_\gamma\hat{\sigma}$, $U(\mathbf{X}) = \hat{\mu} + t_\gamma\hat{\sigma}$, When $s/n = 0$

β	.8			.9			.95		
n	10	20	40	10	20	40	10	20	40
γ									
.9	1.459	1.192	1.044	2.111	1.753	1.557	2.744	2.287	2.045
.95	1.691	1.318	1.119	2.429	1.919	1.663	3.145	2.495	2.167
.98	2.018	1.462	1.207	2.851	2.111	1.773	3.654	2.732	2.310
.99	2.252	1.560	1.272	3.138	2.258	1.854	4.019	2.920	2.404
.995	2.495	1.690	1.333	3.520	2.404	1.931	4.480	3.087	2.504

we computed the asymptotic values of the tolerance factor t_γ from Eq. (5.4.10) for the proportion of right-censoring $s/n = 0.0, 0.3, 0.5$, and 0.7, and $\gamma = 0.75, 0.85, 0.90, 0.95$, and 0.99. These values are presented in Table 5.4.4. By comparing the values in Table 5.4.4 with the corresponding entries in Table 5.4.1, we observe that the asymptotic formula in (5.4.10) gives a fairly good approximation for t_γ for a sample of size 80 whenever the proportion of right-censoring does not exceed 0.5. In the case when the censoring on the right is so heavy (exceeding 0.5), one needs to have a much larger sample size in order for the asymptotic formula in (5.4.10) to yield a satisfactory approximation for the tolerance factor t_γ. If one does not have such a large sample, then one has to resort to a simulation method (as done in constructing Table 5.4.1) for determining the necessary values of t_γ.

5.5 RELIABILITY ESTIMATION BASED ON MLEs

In life-testing problems the probability that a certain item survives at least a specified time t is denoted by the reliability function

$$R(t) = \Pr(X \geq t) = 1 - F(t; \mu, \sigma). \tag{5.5.1}$$

It is recognized that a relationship exists between the tolerance limits and confidence limits for the reliability function in (5.5.1). While discussing the case of the normal distribution, for example, Lloyd and Lipow (1962, p. 204) have stated that a lower confidence limit for the reliability function $R(t)$ can be obtained by setting $t = L(\mathbf{X})$ and then determining for a specified γ what value of β would make the statement in (5.4.6) to be true.

Table 5.4.3 Values of c Such That $t_\gamma = \sqrt{c}z_\gamma/\sqrt{n} + (\sqrt{3}/\pi)\ln[\beta/(1-\beta)]$

	s/n				
β	.0	.1	.3	.5	.7
.500	0.912	0.913	0.947	1.164	2.386
.525	0.914	0.914	0.939	1.121	2.208
.550	0.920	0.921	0.937	1.087	2.047
.575	0.931	0.932	0.942	1.062	1.902
.600	0.947	0.948	0.954	1.047	1.773
.625	0.967	0.970	0.973	1.042	1.661
.650	0.993	0.997	0.999	1.048	1.566
.675	1.025	1.032	1.034	1.066	1.488
.700	1.064	1.074	1.079	1.097	1.431
.725	1.112	1.125	1.135	1.144	1.397
.750	1.168	1.186	1.205	1.209	1.389
.775	1.237	1.260	1.291	1.295	1.412
.800	1.320	1.351	1.397	1.410	1.475
.825	1.423	1.462	1.530	1.560	1.589
.850	1.551	1.602	1.699	1.758	1.772
.875	1.717	1.782	1.920	2.026	2.056
.900	1.938	2.023	2.218	2.400	2.494
.925	2.254	2.368	2.648	2.953	3.199
.950	2.755	2.915	3.340	3.864	4.447
.975	3.765	4.020	4.749	5.775	7.246

Table 5.4.4 Asymptotic Values of $t_\gamma = (\sqrt{c/n})z_\gamma + (\sqrt{3}/\pi)\ln(\beta/(1-\beta))$ when $n = 80$ and $\beta = 0.90$

	γ				
s/n	.75	.85	.90	.95	.99
0.0	1.316	1.373	1.411	1.467	1.573
0.3	1.324	1.384	1.425	1.485	1.599
0.5	1.328	1.391	1.433	1.496	1.614
0.7	1.330	1.394	1.438	1.502	1.622

The value of β thus determined would then be the lower confidence limit for $R(t)$. This is a rather general result, although it is not completely obvious how far it can be extended. For example, let us consider a joint γ tolerance region for proportion β for the joint exponential distribution

$$f(x_1, \ldots, x_k) = \frac{1}{\prod_{i=1}^k \theta_i} \exp\left\{-\sum_{i=1}^k \frac{x_i}{\theta_i}\right\}.$$

In this case, Bain (1967) has shown that an exact tolerance region is obtained by letting $l_i = 2c_i n\bar{x}_i$, where the c_i's satisfy $\Pr\{\Sigma\, c_i\chi_i^2(2n) \le -\ln\beta\} = \gamma$; note that the factor 2 was inadvertantly left out by Bain (1967). For common $c_i = c$, the approximation for a linear combination of chi-square variables yields $c = (-\ln\beta)/\chi_\gamma^2(2nk)$. Now a system of a series of two independent components each exponentially distributed would have a reliability at time τ_0 of $R(\tau_0) = e^{-\tau_0\Sigma 1/\theta_i}$ (Grubbs, 1971), and this corresponds to the probability that each component fails after time τ_0. Hence, it seems that a lower γ confidence limit for the reliability can be obtained by setting $l_i = \tau_0$ or $c_i = \tau_0/2n\bar{x}_i$ and then solving for β that satisfies $\Pr\{\Sigma\, c_i\chi_i^2(2n) \le -\ln\beta\} = \gamma$. It is not clear how this lower limit for the reliability could have been obtained by some other more direct approach. This is the same limit that has been derived by Grubbs (1971) as the fiducial limit.

In general, it is not quite clear how a confidence limit for reliability obtained by an indirect method as described above will correspond to a confidence limit obtained by a more direct approach as mentioned earlier. It may, therefore, be worth verifying that at least for location-scale parameter family of distributions, if the tolerance limits are based on the MLEs as given in Section 5.4, then the associated confidence limits for the reliability will indeed correspond to the usual confidence limit based on $\hat{R}(t)$.

To this end, let us denote $\hat{R}_1$ for the observed value of the MLE $\hat{R}(t)$ for a given sample. Then, by setting

$$L(\mathbf{X}) = \hat{\mu} - t_\gamma \hat{\sigma} = t,$$

we get

$$t_\gamma = -\left(\frac{t - \hat{\mu}}{\hat{\sigma}}\right) = -F^{-1}(1 - \hat{R}_1). \tag{5.5.2}$$

Thence, we choose β to satisfy (see Eq. (5.4.6))

$$\Pr\left\{\frac{\hat{\mu} - \mu}{\hat{\sigma}} - \frac{\sigma}{\hat{\sigma}} F^{-1}(1 - \beta) \leq -F^{-1}(1 - \hat{R}_1)\right\} = \gamma. \tag{5.5.3}$$

Now if $R(t) = \beta$, then $F^{-1}(1 - \beta) = (t - \mu)/\sigma$, so that (5.5.3) becomes

$$\Pr\left\{\frac{\hat{\mu} - \mu}{\hat{\sigma}} - \frac{t - \mu}{\sigma}\frac{\sigma}{\hat{\sigma}} \leq -F^{-1}(1 - \hat{R}_1)\right\} = \gamma$$

$$\Leftrightarrow \Pr\left\{\frac{t - \hat{\mu}}{\hat{\sigma}} \geq F^{-1}(1 - \hat{R}_1)\right\} = \gamma$$

$$\Leftrightarrow \Pr\left\{F\left(\frac{t - \hat{\mu}}{\hat{\sigma}}\right) \geq 1 - \hat{R}_1\right\} = \gamma$$

$$\Leftrightarrow \Pr\left\{1 - F\left(\frac{t - \hat{\mu}}{\hat{\sigma}}\right) \leq \hat{R}_1\right\} = \gamma$$

$$\Leftrightarrow \Pr\{1 - \hat{F}(t; \mu, \sigma) \equiv \hat{R}(t) \leq \hat{R}_1\} = \gamma. \tag{5.5.4}$$

We thus have $R(t) = \beta$ to be the value of $R(t)$, which makes (5.5.4) to be true, and hence β is the lower confidence limit for $R(t)$ by the general method of setting a confidence interval (see Mood, 1950). Such a value exists and is independent of the unknown parameters μ and σ, as Dumonceaux (1969) has shown that the distribution of $\hat{R}(t)$ depends only on the unknown parameter $R(t)$ for location-scale parameter family of distributions.

Table 5.4.1 can, therefore, be used to determine confidence limits for reliability based on $\hat{R}(t)$. For observed values of $\hat{\mu}$ and $\hat{\sigma}$, set

$$t_\gamma = -\left(\frac{t - \hat{\mu}}{\hat{\sigma}}\right) = \frac{\sqrt{3}}{\pi}\ln\left(\frac{\hat{R}}{1 - \hat{R}}\right); \tag{5.5.5}$$

then the value of β such that $\Pr\{T(\mathbf{X}) \leq t_\gamma\} = \gamma$ is the lower limit for the reliability.

We simulated the mean and the variance of the point estimator $\hat{R}(t)$ for sample size $n = 10$, 20, 40, and 80, the proportion of right-censoring $s/n = 0.0$, 0.3, 0.5, and 0.7, and the choices of $R(t) = 0.5$, 0.7, 0.9, and 0.95. These values are presented in Table 5.5.1. We observe from this table that the estimator $\hat{R}(t)$ develops serious bias for small sample sizes and large proportion of right-censoring when we are estimating the reliability at a point near the center of the distribution. The estimator $\hat{R}(t)$ is seen to have negligible bias when the proportion of censoring is small (up to 30%) irrespective of what the sample size is or at what point we are estimating the reliability.

It should be mentioned here that Howlader and Weiss (1989) have worked out Bayesian estimators of the reliability function $R(t)$ in (5.5.1) by using the methods of Lindley (1980) and Tierney and Kadane (1986). They have used both squared-error and log-odds squared-error loss functions in their development. Through Monte Carlo simulations, Howlader and Weiss (1989) have also compared the two Bayesian estimators by inspecting their bias and mean square error for sample size $n = 10$, 20, 30, and 50, and over various choices of t.

Table 5.5.1 Mean and Variance of the Point Estimator $\hat{R}$

		E($\hat{R}$)				V($\hat{R}$)			
	R	.5	.7	.9	.95	.5	.7	.9	.95
s/n	n								
	10	.502	.707	.902	.948	.0208	.0160	.0049	.0021
	20	.501	.704	.901	.949	.0103	.0083	.0026	.0011
.0	40	.500	.702	.900	.949	.0049	.0040	.0013	.0006
	80	.501	.702	.901	.950	.0024	.0020	.0007	.0003
	10	.485	.705	.906	.951	.0259	.0193	.0054	.0023
	20	.493	.703	.903	.951	.0118	.0090	.0029	.0013
.3	40	.496	.702	.902	.950	.0053	.0042	.0015	.0007
	80	.499	.702	.901	.950	.0025	.0020	.0008	.0004
	10	.441	.683	.909	.954	.0374	.0294	.0062	.0025
	20	.471	.695	.905	.952	.0163	.0112	.0032	.0014
.5	40	.485	.698	.903	.951	.0072	.0048	.0016	.0007
	80	.494	.700	.902	.951	.0033	.0022	.0008	.0004
	10	.324	.572	.892	.954	.0614	.0753	.0192	.0054
	20	.407	.647	.903	.954	.0340	.0266	.0047	.0016
.7	40	.452	.678	.903	.953	.0158	.0087	.0019	.0008
	80	.478	.691	.902	.951	.0071	.0034	.0009	.0005

5.6 ILLUSTRATIVE EXAMPLE

Davis (1952) has given lifetimes in hours of 417 forty-watt incandescent lamps taken from 42 weekly forced-life test samples. Davis has indicated that the normal distribution provides a good model for this data, although the true distribution appears to be somewhat more peaked in the center and flatter in the flanks than a normal distribution. As Chew (1968) has pointed out, this is indeed how the logistic distribution compares with the normal, and so the logistic distribution should be a good model for this data. The first 20 observations have been taken from this data and have been ordered in increasing order and presented below:

785, 855, 905, 918, 919, 920, 929, 936, 948, 950, 972,
1035, 1045, 1067, 1092, 1126, 1156, 1162, 1170, 1196

For the purpose of illustration, let us now suppose that 50% censoring had occurred on the right so that only the first 10 observations are available. Then, the MLEs of μ and σ are obtained to be

$$\hat{\mu} = 948 \quad \text{and} \quad \hat{\sigma} = 59.$$

From Table 5.2.1, we calculate an unbiased estimate of σ to be

$$\frac{\hat{\sigma}}{0.920} = \frac{59}{0.920} = 64$$

and an unbiased estimate of μ to be

$$\hat{\mu} - \frac{\hat{\sigma}}{0.920}(-0.039) = 948 + 0.039(64) = 950.5.$$

From Table 5.3.1, we obtain a 95% confidence interval for μ to be

$$\left[948 - \frac{59}{\sqrt{20}}2.03,\ 948 + \frac{59}{\sqrt{20}}3.68\right] = [921.2, 996.5].$$

From Table 5.3.2, we similarly obtain a 95% confidence interval for σ to be

$$\left[\frac{59}{1 + 2.26/\sqrt{20}}, \frac{59}{1 - 2.40/\sqrt{20}}\right] = [39.2, 127.3].$$

A 95% lower tolerance limit for proportion 0.90 of the population is obtained from Table 5.4.1 to be

$$L(\mathbf{X}) = \hat{\mu} - 2.26\hat{\sigma} = 948 - 2.26(59) = 814.66 \text{ hours.}$$

A point estimate of the reliability of the lamps at $t = 800$ hours is given by

$$\hat{R}(800) = 1 - \frac{1}{1 + \exp\left\{-\frac{\pi}{\sqrt{3}}\left(\frac{800 - \hat{\mu}}{\hat{\sigma}}\right)\right\}}$$

$$= 1 - \frac{1}{1 + \exp\left\{-\frac{\pi}{\sqrt{3}}\left(\frac{800 - 948}{59}\right)\right\}}$$

$$= 0.99$$

To determine a 95% lower confidence limit for the reliability at $t = 800$, we set

$$t_{0.95} = -\left(\frac{800 - \hat{\mu}}{\hat{\sigma}}\right) = -\left(\frac{800 - 948}{59}\right) = 2.51;$$

then from Table 5.4.1, we get

$$\beta \doteq 0.925$$

to be the lower 95% confidence limit for the reliability at time $t = 800$ hours.

By using the entire sample of 20 observations and the posterior modes $\tilde{\mu} = 1000$ and $\tilde{\sigma} = 121$, Howlader and Weiss (1989) have provided a Bayesian estimate of the reliability at 800 hours to be 0.9524. They have also used Lindley's method to obtain estimates 0.9428 and 0.9361, and Tierney and Kadane's method to obtain estimates 0.9689 and 0.9610, under squared-error and log-odds loss functions, respectively.

REFERENCES

Antle, C., Klimko, L., and Harkness, W. (1970). Confidence intervals for the parameters of the logistic distribution. *Biometrika*, *57*, 397–402.

Bain, L. J. (1967). Joint tolerance regions for the exponential distribution. *IEEE Trans. Reliab.*, *R-16*, 111–113.

Bain, L. J. (1978). *Statistical Analysis of Reliability and Life-Testing Models—Theory and Practice*. Marcel Dekker, New York.

Bain, L. J., Eastman, J., and Engelhardt, M. E. (1973). A study of life-testing models and statistical analysis for the logistic distribution. *Aerospace Research Laboratories Report ARL 73-0009*, Air Force Systems Command, USAF, Wright Patterson AFB, Ohio.

Balakrishnan, N. and Fung, K. (1991). *Tolerance Limits and Sampling Plans Based on Censored Samples*, Chapter 14.

Chew, V. (1968). Some useful alternatives to the normal distribution. *Amer. Statist.*, *22*, 22–24.

Davis, D. J. (1952). An analysis of some failure data. *J. Amer. Statist. Assoc.*, *47*, 113–150.

Dumonceaux, R. H. (1969). *Statistical inferences for location and scale parameter distributions*. Doctoral Thesis, University of Missouri—Rolla, Rolla, Missouri.

Eastman, J. A. (1972). *Statistical studies of various time-to-fail distributions*. Doctoral Thesis, University of Missouri—Rolla, Rolla, Missouri.

Engelhardt, M. E., Bain, L. J., and Smith, R. M. (1974). A further study of life-testing models and simple estimates for the logistic distribution. *Aerospace Research Laboratories Report ARL 74-0008*, Air Force Systems Command, USAF, Wright Patterson AFB, Ohio.

Grubbs, R. E. (1971). Approximate fiducial bounds for the reliability of a series system for which each component has an exponential time-to-fail distribution. *Technometrics*, *13*, 865–871.

Hall, I. J. (1975). One-sided tolerance limits for a logistic distribution based on censored samples. *Biometrics*, *31*, 873–879.

Howlader, H. A. and Weiss, G. (1989). Bayes estimators of the reliability of the logistic distribution. *Commun. Statist.—Theor. Meth.*, *18*(1), 245–259.

Lindley, D. V. (1980). Approximate Bayesian methods (with discussants). *Trabajos de Estadistica y de Investigacion Operativa*, *31*, 232–245.

Lloyd, D. K. and Lipow, M. (1962). *Reliability: Management, Methods, and Mathematics*. Prentice-Hall, Englewood Cliffs, NJ.

Mood, A. M. (1950). *Introduction to the Theory of Statistics*. McGraw-Hill, New York.

Plackett, R. L. (1959). The analysis of life-testing data. *Technometrics*, *1*, 9–19.

Schafer, R. E. and Sheffield, T. S. (1973). Inferences on the parameters of the logistic distribution. *Biometrika*, *29*, 449–455.

Thoman, D. R., Bain, L. J., and Antle, C. E. (1970). Maximum likelihood estimation, exact confidence intervals for reliability, and tolerance limits in the Weibull distribution. *Technometrics*, *12*, 363–372.

Tierney, L. and Kadane, J. (1986). Accurate approximations for posterior moments and marginals. *J. Amer. Statist. Assoc.*, *81*, 82–86.

6
Ranking and Selection Procedures

S. Panchapakesan
Southern Illinois University, Carbondale, Illinois

6.1 INTRODUCTION

Problems of statistical inference that are now known as ranking and selection problems first came under systematic investigation by statistical researchers in the early 1950s. The classical techniques for testing homogeneity hypotheses were found inadequate to serve, in many practical situations, the experimenter's real purpose, which is often to rank several competing populations (treatments, systems, etc.) or to select the best among them. The attempts to formulate the decision problem to answer such realistic goals set the stage for the development of the ranking and selection theory.

During the last 40 years, the ranking and selection literature has steadily grown with developments dealing with various aspects of the theory and applications. An important part of these developments is the study of ranking and selection problems for specific parametric families of distributions including, of course, logistic distributions. Until recently there has not been much done in the case of logistic distributions. In this chapter, our main objective is twofold, namely, (1) to review available results for logistic distributions, and (2) to provide a *selective overview* of ranking and selection methodology in order to serve as an introduction to the general reader and also to indicate the potential for further investigations in the logistic case.

In Section 6.2, we discuss two basic formulations of ranking and selection problems: the subset formulation and the indifference zone formulation. We also mention some modifications and types of procedures relevant to subsequent discussions. Sections 6.3–6.6 deal with procedures for selecting the population with the largest mean from several logistic populations with a common known variance. Of these, all but Section 6.6 discuss single-stage procedures using the basic subset (Section 6.3), the indifference zone (Section 6.4), and the restricted subset (Section 6.5) formulations. Section 6.6 is concerned with a two-stage procedure using the indifference zone formulation in which the first stage involves a subset approach to eliminate inferior populations. Logistic distribution has been used to model quantal response in experiments involving quantitative treatment factors; a selection problem that arises in this context is discussed in Section 6.7. The next section describes procedures for selecting the population having the largest quantile of a given order from distributions that belong to a restricted family defined by tail-ordering with respect to a known distribution G, including special results for logistic G. Finally, we conclude with a brief discussion on future directions for investigations relating to logistic distributions.

6.2 RANKING AND SELECTION FORMULATIONS

Ranking and selection problems have generally been studied by using either the *indifference zone* approach of Bechhofer (1954) or the so-called *subset selection* approach due mainly to Gupta (1956). In the former approach the number of populations to be selected is predetermined, while in the latter it is random. Suppose there are k (≥ 2) populations $\pi_1, \pi_2, \ldots, \pi_k$, where π_i is characterized by the distribution function F_{θ_i} and θ_i is a real-valued parameter taking a value in the set Θ, $i = 1, 2, \ldots, k$. The θ_i are assumed to be unknown. Let $\theta_{[1]} \leq \theta_{[2]} \leq \cdots \leq \theta_{[k]}$ denote the ordered θ_i and $\pi_{(i)}$ denote the population associated with $\theta_{[i]}$, $i = 1, 2, \ldots, k$. The populations are ranked according to their θ-values. To be specific, we define $\pi_{(j)}$ to be better than $\pi_{(i)}$ if $i < j$, that is, if $\theta_{[i]} \leq \theta_{[j]}$. It is assumed that there is no prior information regarding the true pairing of the ordered and unordered θ_i.

Let us consider the basic problem of selecting the *best* population, namely, the one associated with the largest θ_i. In the indifference zone approach, the goal is to select *one* of the k populations and claim it to be the best.

Let $\Omega = \{\underline{\theta}|\underline{\theta} = (\theta_1, \ldots, \theta_k), \theta_i \in \Theta, i = 1, \ldots, k\}$ be the parameter space and $\Omega_\delta = \{\underline{\theta}|\delta(\theta_{[k]}, \theta_{[k-1]}) \geq \delta^* > 0\}$, where $\delta(\theta_{[k]}, \theta_{[k-1]})$ is an appropriate measure of the separation between the best population $\pi_{(k)}$ and the next best $\pi_{(k-1)}$. A *correct selection* (CS) occurs whenever the selected population is the best population. Let $P(\text{CS}|R)$ denote the probability of a correct selection (PCS) using the rule R. For a rule R to be valid, it is required that

$$P(\text{CS}|R) \geq P^* \qquad \text{whenever } \boldsymbol{\theta} \in \Omega_\delta\,. \tag{6.2.1}$$

The constants δ^* and $P^*(1/k < P^* < 1)$ are specified in advance by the experimenter. The statistical problem is to define a selection rule which typically has three parts: sampling rule, stopping rule for sampling, and decision rule. For a rule based on a single sample of fixed size n from each population, the design aspect of the experiment is to determine the minimum sample size n for which (6.2.1) is satisfied. The region Ω_δ of the parameter space Ω is called the *preference zone*. The complement of Ω_δ is the *indifference zone*, so called because there is no requirement on the PCS when θ lies in it.

In the subset selection approach for selecting the best population, the goal is to select a nonempty subset of the k populations so that the best population is included in the selected subset with a minimum guaranteed probability $P^*(1/k < P^* < 1)$. In other words, any valid rule R should satisfy the condition

$$P(\text{CS}|R) \geq P^* \qquad \text{for any } \boldsymbol{\theta} \in \Omega. \tag{6.2.2}$$

Here, selection of any subset that includes the best population results in a correct selection. In case of a tie for the best population, it is assumed that one of the contenders is tagged as the best. There is no indifference zone in the above formulation. The size of the selected subset S, denoted by $|S|$, is not specified in advance, but is determined by the data themselves. The expected subset size $E(|S|)$ and the expected number of non-best populations (which is equal to $E(|S|)$ − PCS) are natural performance characteristics of a valid rule.

The probability requirements (6.2.1) and (6.2.2) are known as the P^*-condition. An important step in obtaining constant(s) associated with a proposed rule R so that the P^*-condition is satisfied is to evaluate the infimum of the PCS over Ω or Ω_δ depending on the approach. The configuration of $\underline{\theta}$ for which this infimum is attained is called the *least favorable configuration* (LFC).

There are several variations and generalizations of the basic goal in both indifference zone and subset selection formulations. These are discussed in detail in Gupta and Panchapakesan (1979). One such generalization to be discussed in Section 6.5 is the *restricted subset selection* formulation of Santner (1975). The goal in this formulation is to select a nonempty subset of the k populations that contains the best but whose size does not exceed a specified number m ($1 \leq m \leq k$). It is required that the PCS be at least P^* whenever $\theta \in \Omega_{\delta^*}$. This formulation combines the features of the indifference zone and subset selection formulations discussed earlier.

Besides being a goal in itself, selecting a subset containing the best can serve as a first-stage screening in a two-stage procedure designed to select one population as the best. Tamhane and Bechhofer (1977, 1979) have employed this technique for selecting the population with the largest mean from k normal populations with unknown means and a common known variance σ^2, using the indifference zone approach. We will discuss (Section 6.6) a similar procedure for logistic populations. When the common variance σ^2 is unknown in the above normal case, a two-stage procedure is necessary in order to meet the P^*-condition.

Families of distributions can be defined through partial ordering relation with respect to a known distribution. Such families have been called *restricted families* of distributions. Partial ordering such as convex ordering, star-shape ordering, and tail ordering have been considered in the literature. These families are of great importance in reliability theory. Well-known families such as increasing failure rate (IFR) and increasing failure rate average (IFRA) distributions are examples of such families. Selection procedures for such restricted families have been considered by a few authors. A review of these and other procedures applicable to reliability models is given by Gupta and Panchapakesan (1988). We will discuss (Section 6.8) a procedure for selection from a family of distributions which are tail-ordered with respect to a logistic distribution.

As we have pointed out earlier, our objective here is to provide a selective overview of ranking and selection procedures. Several aspects of the theory and related problems have been dealt with by Bechhofer, Kiefer, and Sobel (1968), Büringer, Martin, and Schriever (1980), Gibbons, Olkin, and Sobel (1977), Gupta and Huang (1981), and Gupta and Panchapakesan (1979). The last authors have given a comprehensive survey of developments in the theory with an extensive bibliography. A categorical bibliography is provided by Dudewicz and Koo (1982). Recently, Gupta and Panchapakesan (1985) have given a review and assessment of subset selection procedures.

In the succeeding sections, we discuss specific selection procedures relating to logistic distributions.

6.3 SINGLE-STAGE (UNRESTRICTED) SUBSET SELECTION OF THE POPULATION WITH THE LARGEST MEAN

Let $\pi_1, \ldots, \pi_k$ be k (≥ 2) independent logistic populations $L(\mu_i, \sigma^2)$, $i = 1, 2, \ldots, k$, where the means μ_i are unknown and the common variance σ^2 is assumed to be known. The distribution function associated with π_i ($1 \leq i \leq k$) is

$$F(x; \mu_i, \sigma) = \frac{1}{1 + \exp\{-\pi(x - \mu_i)/\sigma\sqrt{3}\}}, \qquad -\infty < x < \infty. \quad (6.3.1)$$

We assume, without loss of generality, that $\sigma = 1$. The population with the largest μ_i is the best. Let $X_{i,m:n}$ denote the median of a random sample of odd size $n = 2m - 1$ from π_i, $i = 1, \ldots, k$. Lorenzen and McDonald (1981) proposed the rule

$$R_1\text{: Select } \pi_i \text{ if and only if } X_{i,m:n} \geq \max_{1 \leq j \leq k} X_{j,m:n} - d_1, \quad (6.3.2)$$

where $d_1 = d_1(k, m, P^*)$ is the smallest positive constant for which the P^*-condition is satisfied.

For convenience, let $\mu_{[1]} \leq \cdots \leq \mu_{[k]}$ denote the ordered μ_i (we continue with the notation of Section 6.2 for the ranked parameters). Let $X_{(i),m:n}$ denote the sample median from the population associated with $\mu_{[i]}$, $i = 1, \ldots, k$. Then the PCS for the rule R_1 is given by

$$P(\text{CS}|R_1) = \Pr\{X_{(k),m:n} \geq X_{(j),m:n} - d_1, j = 1, \ldots, k - 1\}. \quad (6.3.3)$$

Letting $f_{m:n}(y_m)$ and $F_{m:n}(y_m)$ denote the pdf and the cdf, respectively, of the median of a random sample of size $n = 2m - 1$ from the standard logistic distribution $L(0, 1)$, we can write (6.3.3) as

$$P(\text{CS}|R_1) = \int_{-\infty}^{\infty} \prod_{j=1}^{k-1} F(y_m + \mu_{[k]} - \mu_{[j]} + d_1) f_{m:n}(y_m)\, dy_m. \quad (6.3.4)$$

It is now easy to see that the infimum of the PCS over the parameter space Ω is attained when all the μ_i are equal. Thus the constant $d_1 = d_1(k, m, P^*)$ is given by

$$\int_{-\infty}^{\infty} F_{m:n}^{k-1}(y_m + d_1) f_{m:n}(y_m)\, dy_m = P^*. \quad (6.3.5)$$

Lorenzen and McDonald (1981) have tabulated the d_1-values for $k = 2(1)10$, $m = 1(1)10$, and $P^* = 0.75, 0.90, 0.95, 0.99$.

Alternatively, one can define the following procedure R_2 based on the sample means $\overline{X}_i$, $i = 1, \ldots, k$.

$$R_2\text{: Select } \pi_i \text{ if and only if } \overline{X}_i \geq \max_{1 \leq j \leq k} \overline{X}_j - d_2, \tag{6.3.6}$$

where $d_2 = d_2(k, n, P^*)$ is the smallest positive constant for which the P^*-condition is satisfied. Let $g_n(u)$ and $G_n(u)$ denote the pdf and the cdf, respectively, of $U = \sqrt{n}(\overline{X} - \mu)/\sigma$, where $\overline{X}$ is the mean of a random sample of size n from $L(\mu, \sigma^2)$. Assuming still $\sigma = 1$, the PCS for the rule R_2 is given by

$$P(\text{CS}|R_2) = \int_{-\infty}^{\infty} \prod_{j=1}^{k-1} G_n(u + \sqrt{n}d_2 + \sqrt{n}(\mu_{[k]} - \mu_{[j]}))g_n(u)\, du. \tag{6.3.7}$$

The infimum of the PCS over Ω is attained when the μ_i are equal and the constant $d_2 = d_2(k, n, P^*)$ is given by

$$\int_{-\infty}^{\infty} G_n^{k-1}(u + \sqrt{n}d_2)g_n(u)\, du = P^*. \tag{6.3.8}$$

Lorenzen and McDonald (1981) have considered this rule R_2 in order to study the efficiency of R_1 relative to R_2; however, they considered only asymptotic ($n \to \infty$) case using convergence to normality. Recently, Han (1987) has studied the rule R_2 and has provided tables of values of $h = \sqrt{n}d_2$ for $k = 2(1)10$, $n = 1(1)10$, and $P^* = 0.75, 0.90, 0.95, 0.975, 0.99$; he used the Edgeworth series expansions to the order $O(n^{-3})$ for $G_n(u)$ and $g_n(u)$, the Gauss-Hermite quadrature with 60 nodes for evaluation of the integral, and a modified regula falsi algorithm for solving nonlinear equations.

For comparing the rules R_1 and R_2, Han (1987) considered (1) $E(|S|)$, the expected subset size, (2) $E(|S|)$ − PCS, which is the expected number of non-best populations selected, and (3) $E(T)$, where T is the sum of the ranks of the selected populations. His tables include also the comparison of the expected proportion of populations included in the selected subset, but this is only $E(|S|)/k$. In order to compare R_1 and R_2 in terms of the above performance characteristics, Han (1987) considered two customary types of configurations of the means: (1) the slippage configuration, $\mu_{[1]} = \cdots = \mu_{[k-1]} = \mu_{[k]} - \delta$, $\delta > 0$, and (2) the equally spaced configuration, $\mu_{[i]} = \mu + (i - 1)\delta$, $i = 1, \ldots, k$; $\delta > 0$. His tables of values of the performance characteristics range over $k = 2(1)5, 10$; $n = 3$; $\delta\sqrt{n} = 0.5(0.5)3.0, 4, 5$ (for both configurations); and $P^* = 0.90$. His tables for the ratio of the corresponding performance characteristics of R_1 and R_2 cover $k = 4$; $n = 3, 5$; $\delta\sqrt{n} = 1.5, 3.0$; and $P^* = 0.90, 0.95$. The tables indicate as one would expect that the means procedure R_2 performs better than the medians procedure R_1, the advantage increasing with n.

Lorenzen and McDonald (1981) studied the efficiency of R_1 relative to R_2 for large samples under a slippage configuration described earlier. Let n_1 and n_2 denote the asymptotic sample sizes required by R_1 and R_2, respectively, to satisfy the P^*-condition and to make $E(|S|) - \text{PCS} = \epsilon > 0$. The asymptotic efficiency of R_2 relative to R_1 is defined by

$$\text{ARE}(R_2, R_1; \delta) = \lim_{\epsilon \downarrow 0}\left(\frac{n_2}{n_1}\right). \tag{6.3.9}$$

Lorenzen and McDonald (1981) have shown that $\text{ARE}(R_2, R_1; \delta) = \pi^2/12 \doteq 0.822$. Thus, under a slippage configuration, asymptotically the means procedure requires about 82% of the sample size required by the median procedure to achieve the same expected number of non-best populations in the selected subset. However, the situation can dramatically change in favor of the medians procedure, as Lorenzen and McDonald have shown, when sampling is contaminated in the sense that π_i is logistic with mean θ_i and variance $\alpha + (1 - \alpha)\nu^2$. The savings gained by the medians procedure becomes immense as $\nu \to \infty$.

Lorenzen and McDonald (1981) have also compared the medians procedure R_1 with a rank-sum procedure R_3. Let T_i denote the sum of the ranks of the observations from π_i in the pooled sample obtained from samples of size n from each population (the smallest observation is assigned rank 1 and the largest rank kn). The rule R_3 is defined as follows:

$$R_3\text{: Select } \pi_i \text{ if and only if } T_i \geq \max_{1\leq j\leq k} T_j - d_3, \tag{6.3.10}$$

where d_3 is the smallest positive integer so that the P^*-condition is satisfied. This rule R_3 has been studied by Gupta and McDonald (1970) for location and scale parameter families. Lorenzen and McDonald (1981) have studied the asymptotic efficiency of R_3 relative to R_1 under slippage and equal spacing configurations, using Monte Carlo simulations when $k > 2$. Based on their study, the rank-sum procedure outperforms the median procedure when the means are roughly in a slippage configuration, while the reverse is true when the means are equally spaced.

Properties of Rules R_1 and R_2

Rules R_1 and R_2 possess properties considered desirable in a subset selection rule. We discuss these properties below.

1. *Unbiasedness*. A rule R is said to be *unbiased* if for all $\mu \in \Omega$ and $j < k$,

$$P(\pi_{(k)} \text{ is selected by } R) \geq P(\pi_{(j)} \text{ is selected by } R).$$

2. *Monotonicity*. A rule R is said to be *monotone* if for $\mu \in \Omega$ and $i < j$,

$$P(\pi_{(j)} \text{ is selected by } R) \geq P(\pi_{(i)} \text{ is selected by } R).$$

   Obviously, monotonicity implies unbiasedness.
3. *Strong Monotonicity*. A rule R is said to be *strongly monotone* in $\pi_{(i)}$ if $P(\pi_{(i)}$ is selected by $R)$ is increasing in $\mu_{[i]}$ when all other components of μ are fixed and is decreasing in $\mu_{[j]}$ $(j \neq i)$ when all other components of μ are fixed.
4. *Consistency*. A rule $R(n)$ (based on common sample size n) is said to be *consistent* with respect to $\Omega' \subset \Omega$ if $\inf_{\Omega'} P(\text{CS}|R(n)) \rightarrow 1$ as $n \rightarrow \infty$.

Finally, for the rules R_1 and R_2, the supremum of $E(|S|)$ over Ω is attained when the μ_i are equal. This follows (see Gupta, 1965 and Gupta and Panchapakesan, 1972) from the fact that μ_i is a location parameter in the distributions of $X_{i,m:n}$ and $\overline{X}_i$ $(1 \leq i \leq k)$, and that these distributions have the monotone likelihood ratio (MLR) property. Consequently, this supremum is equal to kP^*.

6.4 SINGLE-STAGE INDIFFERENCE ZONE SELECTION OF THE POPULATION WITH THE LARGEST MEAN

We have k logistic distributions as described in Section 6.3. Under the indifference zone formulation we select one of the k populations so that the PCS $\geq P^*$ whenever $\mu \in \Omega_\delta = \{\underline{\mu}: \mu_{[k]} - \mu_{[k-1]} \geq \delta^* > 0\}$. We consider two procedures R_1' and R_2' which are, respectively, the counterparts of R_1 and R_2 discussed in the previous section. The sampling schemes are same as earlier.

First, we define R_1' based on medians as follows.

$$R_1': \text{Select the population that yields the largest } X_{i,m:n}. \qquad (6.4.1)$$

It is easy to see that

$$P(\mathrm{CS}|R_1') = \int_{-\infty}^{\infty} \prod_{j=1}^{k-1} F_{m:n}(y_m + \mu_{[k]} - \mu_{[j]}) f_{m:n}(y_m)\, dy_m. \quad (6.4.2)$$

Obviously, the infimum of $P(\mathrm{CS}|R_1')$ over Ω_{δ^*} is attained when $\mu_{[1]} = \cdots = \mu_{[k-1]} = \mu_{[k]} - \delta^*$. Thus we need to determine the minimum odd sample size n for which

$$\int_{-\infty}^{\infty} F_{m:n}^{k-1}(y_m + \delta^*) f_{m:n}(y_m)\, dy_m \geq P^*. \quad (6.4.3)$$

A table of n values satisfying (6.4.3) for selected values of k, δ^*, and P^* is not available.

Alternatively, we can define a rule based on sample means, namely,

R_2': Select the population that yields the largest $\overline{X}_i$. (6.4.4)

In this case,

$$P(\mathrm{CS}|R_2') = \int_{-\infty}^{\infty} \prod_{j=1}^{k-1} G_n(u + \sqrt{n}(\mu_{[k]} - \mu_{[j]})) g_n(u)\, du, \quad (6.4.5)$$

and the infimum of the PCS in (6.4.5) is attained when $\mu_{[1]} = \cdots = \mu_{[k-1]} = \mu_{[k]} - \delta^*$. Thus the minimum sample size n required to meet the P^*-condition is given by

$$\int_{-\infty}^{\infty} G_n^{k-1}(u + \sqrt{n}\delta^*) g_n(u)\, du \geq P^*. \quad (6.4.6)$$

Han (1987) has studied the rule R_2' and has tabulated, using the Edgeworth series expansions to the order of $O(n^{-3})$ for $G_n(u)$ and $g_n(u)$, the values of $\hat{n}$ for which (6.4.6) is satisfied with equality of both sides. The required values of n are obtained by rounding up $\hat{n}$ to the next higher integer. His table ranges over $k = 2(1)5, 10, 15$; $\delta^* = 0.1, 0.5, 1.0, 2.0, 4.0$; and $P^* = 0.75, 0.90, 0.95, 0.99$.

One should naturally compare the minimum sample sizes required by R_1' and R_2', and also compare the performance of the two procedures for various typical parametric configurations. This has not yet been done.

Recently, van der Laan (1989) has considered the rule R_2' with $n = 1$, in which case G_n in (6.4.6) is the standardized logistic cdf. He has tabulated the values of δ^* for $k = 2(1)10, 25, 50$ and $P^* = 0.65, 0.75, 0.90, 0.95, 0.99, 0.999$ and compared these with the corresponding values in the normal case available in Bechhofer (1954). However, for $n = 1$, Eqs. (6.4.6) and

(6.3.5) are the same. So the values of δ^* for $k = 2(1)10$, and $P^* = 0.75$, 0.90, 0.95, 0.99 are readily available in Lorenzen and McDonald (1981).

6.5 SINGLE-STAGE RESTRICTED SUBSET SELECTION OF THE POPULATION WITH THE LARGEST MEAN

In Section 6.2, we referred to the restricted subset selection formulation of Santner (1975). The goal here is to select a subset of the k populations whose size does not exceed m ($1 \le m \le k - 1$) and which includes the population with largest mean μ_i. As we mentioned earlier, in this formulation, we introduce a preference zone $\Omega_\delta = \{\mu | \mu_{[k]} - \mu_{[k-1]} \ge \delta^* > 0\}$. We want to define a rule R for which

$$P(\mathrm{CS}|R) \ge P^* \text{ whenever } \mu \in \Omega_\delta . \tag{6.5.1}$$

We note that for $m = 1$, this becomes the indifference zone formulation of Bechhofer (1954). If we allow $m = k$ (unrestricted) and $\delta^* = 0$, we will get the usual subset selection formulation.

Now, for the problem at hand, Han (1987) investigated the following rule R_4 based on the sample means $\overline{X}_i$.

$$R_4\text{: Select } \pi_i \text{ if and only if } \overline{X}_i \ge \max\left\{\overline{X}_{k-m+1:k}, \overline{X}_{k:k} - \frac{d_4\sigma}{\sqrt{n}}\right\}, \tag{6.5.2}$$

where the minimum required sample size n and $d_4 = d_4(k, n, P^*)$ are to be determined so that (6.5.1) is satisfied. We leave out the expression for $P(\mathrm{CS}|R_4)$ for any $\mu \in \Omega$ in order to avoid further notations. It can be shown without much difficulty (see Han, 1987) that the infimum of $P(\mathrm{CS}|R_4)$ over Ω_δ is attained when $\mu \in \Omega_\delta^0 = \{\mu | \mu_{[1]} = \cdots = \mu_{[k-1]} = \mu_{[k]} - \delta^*\}$. Also, $P(\mathrm{CS}|R_4)$ is constant over Ω_δ^0. Assuming that the common $\sigma = 1$,

$$\inf_{\mu\in\Omega_\delta} P(\mathrm{CS}|R_4) = \sum_{i=k-m}^{k} \binom{k-1}{i} \int_{-\infty}^{\infty} G_n^i(t + \delta\sqrt{n})$$
$$\times \{G_n(t + d_4 + \delta^*\sqrt{n}) - G_n(t + \delta^*\sqrt{n})\}^{k-i-1} g_n(t)\, dt. \tag{6.5.3}$$

The minimum sample size n required to satisfy the P^*-condition in (6.5.1) has been tabulated by Han (1987) for $P^* = 0.90$; $k = 5$ with $m = 2(1)4$; $k = 10$ with $m = 2(1)5$; $d_4 = 0.4, 0.7, 1.3, 1.6$ and $\delta = 0.5, 1.0, 2.0$ (the range is for δ/σ if σ is not unity). Han (1987) has further tables regarding some performance characteristics which are not discussed here.

The rule R_4 is strongly monotone in $\pi_{(i)}$ and so is monotone and unbiased. Han (1987) has shown that R_4 is consistent with respect to Ω_{δ^*}. This implies that, for given P^*, k, m, and δ^*, one can meet the P^*-condition by choosing n sufficiently large. Also, not surprisingly, it turns out that, for given δ^*, k, m, and n, the P^*-condition can be met only for $P^* \leq P_1^*(\delta^*, k, m, n) < 1$. The supremum of $E(|S|)$ over Ω is attained when the μ_i are equal. Since we have two constants n and d_4 defining the rule R_4, these can be chosen such that $\inf_{\Omega_{\delta^*}} P(CS|R_4) \geq P^*$ and $\sup_\Omega E(|S||R_4) = 1 + \epsilon$ for a specified ϵ. Han (1987) has some tables for choosing n and d_4 for $\epsilon = 0.01$ and selected values of k, m, P^*, and δ/σ.

6.6 AN ELIMINATION-TYPE TWO-STAGE SELECTION PROCEDURE FOR THE POPULATION WITH THE LARGEST MEAN

As before, we assume that the k logistic populations are $L(\mu_i, \sigma^2)$, $i = 1, \ldots, k$, where the common variance σ^2 is known. We want to select the population with the largest μ_i using the indifference zone formulation. As we mentioned in Section 6.2, we can use a two-stage procedure which screens out bad populations (those with small values of μ_i) by means of a subset selection rule. Such procedures for normal populations with common known variance were studied initially by Cohen (1959) and Alam (1970) whose results were mostly for the case of $k = 2$ populations. Tamhane and Bechhofer (1977, 1979) have studied the problem in depth for $k \geq 2$. Recently, Gupta and Han (1991) have investigated a similar procedure for logistic populations. Procedure R_5 involves the design constants (n_1, n_2, h), where h is a positive constant, and n_1 and n_2 are the sample sizes in the two stages. These constants, depending on k, δ^*, and P^*, are chosen so that the P^*-condition is satisfied and they possess a certain minimax property.

Procedure R_5

At Stage 1, take n_1 independent observations from each of the k populations and compute the sample means $\overline{X}_i^{(1)}$, $i = 1, \ldots, k$. Determine the subset I of $\{1, \ldots, k\}$ where

$$I = \left\{ i \,\middle|\, \overline{X}_i^{(1)} \geq \max_{1 \leq j \leq k} \overline{X}_j^{(1)} - \frac{h\sigma}{\sqrt{n_1}} \right\}. \tag{6.6.1}$$

Let $\Pi_I = \{\pi_i | \, i \in I\}$. If Π_I consists of only one population, stop sampling and select as best the population that yielded the largest $\overline{X}_i^{(1)}$. If Π_I consists of more than one population, proceed to Stage 2.

At Stage 2, take n_2 additional observations from each population in Π_I

and compute the cumulative sample means $\overline{X}_i$ based on $n_1 + n_2$ observations. Select as the best the population that yielded the largest $\overline{X}_i$.

There are an infinite number of choices of (n_1, n_2, h) for which the P^*-condition (6.5.1) is satisfied. Let $|I|$ denote the cardinality of the set I in (6.6.1) and

$$S = \begin{cases} 0 & \text{if } |I| = 1, \\ |I| & \text{if } |I| > 1. \end{cases} \tag{6.6.2}$$

Then the total sample size required is given by

$$T = kn_1 + Sn_2. \tag{6.6.3}$$

Gupta and Han (1991) have adopted an unrestricted minimax criterion to make a choice of (n_1, n_2, h). In other words, for given k, δ^*, and P^*, in order to choose (n_1, n_2, h)

$$\text{minimize} \sup_{\mu\in\Omega} E(T|R_5) \text{ subject to } \inf_{\mu\in\Omega_{\delta^*}} P(\text{CS}|R_5) \geq P^*. \tag{6.6.4}$$

Gupta and Han (1991) have shown that the supremum of $E(T|R_5)$ over Ω is attained when the μ_i are equal. The LFC for the PCS is a slippage configuration; this follows from the result of Bhandari and Chaudhuri (1990) for two-stage selection for the largest population mean when the sample mean has the MLR property. However, the exact evaluation of the PCS under the LFC for rule R_5 is complicated. Gupta and Han (1991) have considered a lower bound to $P(\text{CS}|R_5)$ for which the infimum over Ω_{δ^*} is easily obtained. Using this, a conservative solution can be obtained for the minimization problem in (6.6.4). In other words, a conservative solution for (n_1, n_2, h) is obtained by minimizing

$$kn_1 + n_2 \int_{-\infty}^{\infty} [G_{n_1}^{k-1}(x + h) - G_{n_1}^{k-1}(x - h)] g_{n_1}(x)\, dx \tag{6.6.5}$$

subject to

$$\int_{-\infty}^{\infty} G_{n_1}^{k-1}\left(x + \frac{\delta^*\sqrt{n_1}}{\sigma} + h\right) g_{n_1}(x)\, dx$$
$$\times \int_{-\infty}^{\infty} G_{n_1+n_2}^{k-1}\left(x + \frac{\delta^*\sqrt{n_1 + n_2}}{\sigma}\right) g_{n_1+n_2}(x)\, dx \geq P^*, \tag{6.6.6}$$

where $G_n(x)$ and $g_n(x)$ are so defined earlier in the case of rule R_2 in Section 6.3. Let $(\hat{n}_1, \hat{n}_2, \hat{h})$ be the solution to the minimization of (6.6.5) subject to (6.6.6) when n_1 and n_2 are allowed to be continuous. Then one can use the approximate design constants

$$n_1 = [\hat{n}_1 + 1], \qquad n_2 = [\hat{n}_2 + 1], \qquad h = \hat{h},$$

where $[m]$ denotes the largest integer $\leq m$. Gupta and Han (1991) have tabulated $(\hat{n}_1, \hat{n}_2, \hat{h})$ and $E(T|R_5)$ for $k = 2(1)5, 10, 15$; $P^* = 0.75, 0.90, 0.95, 0.99$ and $\delta^*/\sigma = 0.1, 0.5, 1.0, 2.0, 4.0$; here, the tabulated values of $E(T|R_5)$ are the values of the expression in (6.6.5).

The performance of the two-stage procedure R_5 can be compared with that of the single-stage procedure R_2' defined by (6.4.4). Let $\hat{n}$ be the solution of (6.4.6) with equality. Gupta and Han (1991) considered $E(T|R_5)/k\hat{n}$ as the measure of relative efficiency (RE). If RE < 1, then R_5 is better than R_2'. They have tabulated the values of RE for $k = 2(1)5$, 10, 15 and $P^* = 0.75, 0.90, 0.95, 0.99$ in the cases of the slippage configuration $(\mu, \ldots, \mu, \mu + \delta)$ and the equally spaced configuration $(\mu, \mu + \delta, \ldots, \mu + (k-1)\delta)$ when $\delta/\sigma = 0.1, 0.5, 1.0, 2.0, 4.0$. All the tabulated values of RE are equal to 1 in a very few cases and less than 1 otherwise, thus showing R_5 to be more efficient. The effectiveness of R_5 increases as k increases.

Rule R_5 employs the usual subset selection procedure for eliminating bad populations at the first stage. The size of the selected subset I defined in (6.6.1) can be k. One can use a restricted subset selection procedure of the type discussed in Section 6.5 in order to restrict the size of I. Such a two-stage procedure has been investigated by Han (1987), but it will not be discussed here.

6.7 SELECTION OF THE LOGISTIC QUANTAL RESPONSE WITH THE SMALLEST *q*-QUANTILE (ED100*q*)

Consider an experiment in which the treatment factor is quantitative. Each experimental unit is administered a certain "dose" of the treatment to which the unit either responds (a success) or does not respond (a failure). Such experiments are well known in biological applications as quantal response assays or sensitivity experiments. The probability of response is some unknown function of the dose level x and, denoted by $p(x)$, is called the *quantal response curve*. It is reasonable, in many applications, to assume that $p(x)$ is nondecreasing in x, right-continuous with $p(-\infty) = 0$ and $p(\infty) = 1$. The smallest dose level that induces a response with probability q $(0 < q < 1)$ is called the *100q% effective dose* (ED100q) and is denoted here by $\mu^{(q)}$.

In a selection problem, we are comparing several different quantitative treatments with unknown associated quantal response curves in order to select the best (to be suitably defined) curve. In a nonparametric setup, this problem involves subtle difficulties. These have been discussed by Tamhane (1986), who has given an excellent survey of the literature on quantal response curves. He has discussed problems of estimation and multiple comparisons with a view toward the application to the selection problem. Toward this end, he has also considered two parametric models of which the logistic model is one. We discuss below his formulation of and solution to the problem of selecting the best quantal response curve.

Let $\pi_1, \ldots, \pi_k$ be k populations where π_i has the associated quantal response curve $p_i(x)$ given by

$$p_i(x) = \frac{1}{1 + \exp\{-(\alpha_i + \beta x)\}}, \qquad i = 1, \ldots, k, \tag{6.7.1}$$

and the common value of β is assumed to be *known*. The ED100q of π_i is given by

$$\mu_i^{(q)} = \frac{1}{\beta}\left[\log\left(\frac{q}{1 - q}\right) - \alpha_i\right]. \tag{6.7.2}$$

The quantity $\log\{q/(1 - q)\}$ is referred to as the *logit transform* of q. The goal is to select the population associated with the smallest $\mu_i^{(q)}$. This goal is meaningful when we have several drugs available for a certain ailment, and we want to select the drug that induces a specified success rate q at the lowest dose level. Since all populations have a common β, the problem is equivalent to selecting the population associated with the largest α_i. Consistent with our earlier notations, the ordered α_i are denoted by $\alpha_{[1]} \le \alpha_{[2]} \le \cdots \le \alpha_{[k]}$. Tamhane (1986) has adopted the indifference zone approach, taking $\Omega_{\delta^*} = \{\alpha = (\alpha_1, \ldots, \alpha_k) | \alpha_{[k]} - \alpha_{[k-1]} \ge \beta\delta^*\}$. The P^*-condition is to be satisfied whenever $\underset{\sim}{\alpha} \in \Omega_\delta$.

Tamhane (1986) proposed a single-stage procedure based on n independent observations from each population, those from π_i being taken at equispaced dose levels $x_{i1}, x_{i2}, \ldots, x_{im}$ with $x_{i,j+1} - x_{ij} = d_i$ $(1 \le i \le k$, $1 \le j \le m - 1)$. Let r_{ij} denote the number of successes for population π_i at dose level x_{ij} and let $\hat{p}_{ij} = r_{ij}/n$. The maximum likelihood estimator (MLE) $\hat{\alpha}_i$ of α_i $(1 \le i \le k)$ is obtained by solving $\Sigma_{j=1}^m p_{ij} = \Sigma_{j=1}^m \hat{p}_{ij}$, where $p_{ij} = p_i(x_{ij})$ given by (6.7.1). The rule proposed by Tamhane (1986) is

$$R_6\text{: Select the population that yields the largest } \hat{\alpha}_i. \tag{6.7.3}$$

An exact solution for the minimum sample size n required in order to meet the P^*-condition is not available. Tamhane (1986) has obtained a large sample $(m \to \infty, n \to \infty)$ solution using the fact that, when $d_1 = \cdots = d_k = d$, $\hat{\alpha}_i$ has asymptotically a normal distribution with mean α_i and variance $\beta d/n$. Based on the known results of Bechhofer (1954) for the problem of selecting the largest normal mean when the populations have a common known variance, the solution is given by

$$n = \left\langle \left(\frac{c}{\delta^*}\right) \frac{d}{\beta} \right\rangle, \tag{6.7.4}$$

where $\langle a \rangle$ denotes the smallest integer $\geq a$ and $c = c(k, P^*)$ is given by

$$\int_{-\infty}^{\infty} \Phi^{k-1}(x + c)\, d\Phi(x) = P^*,$$

and Φ is the standard normal cdf. The values of c (or a known multiple of it) have been tabulated by Bechhofer (1954), Gupta (1963), Milton (1963), and Gupta, Nagel, and Panchapakesan (1973).

Tamhane (1986) has also considered a weighted least squares estimator α_i^* of α_i to be used in rule R_6 in the place of $\hat{\alpha}_i$. However, in this case also the exact solution is not available. Asymptotically ($m \to \infty$ $n \to \infty$), α_i^* has the same distribution as $\hat{\alpha}_i$ and so a large sample solution is again given by (6.7.4).

Under the logistic model (6.7.1), the quantal response curves $p_i(x)$ for different α_i's do not intersect. In this case, one can take observations from all populations at the same dose level x_0. The populations can then be considered to be Bernoulli with success probabilities $p_i(x_0)$, $i = 1, \ldots, k$. The selection problem then reduces to the selection of the Bernoulli population associated with the largest success probability. Many procedures are available in the literature for this classical Bernoulli selection problem. A review of these procedures and a complete bibliography have been given by Bechhofer and Kulkarni (1982), who have themselves made significant contributions to this problem. However, the main obstacle in using any of these procedures to our problem at hand is, as pointed out by Tamhane (1986), the specific choice of x_0. With a hapless choice of x_0, it may turn out that $\mu_{[2]}^{(q)} - \mu_{[1]}^{(q)} < \delta^*$; i.e., $\alpha \notin \Omega_{\delta^*}$, thus making it impossible to satisfy the P^*-condition.

If we adopt the subset selection approach, then the problem of the specific choice of x_0 discussed above will not arise. In this case one can use the procedures available for the classical Bernoulli subset selection. For details of such procedures, the reader is referred to Gupta and Panchapakesan (1979, Chapter 13).

6.8 SELECTION FROM A FAMILY OF DISTRIBUTIONS PARTIALLY ORDERED WITH RESPECT TO A LOGISTIC DISTRIBUTION

As we mentioned in Section 6.2, selection procedures for families of distributions which are partially ordered with respect to a known distribution have been considered in the literature. These procedures are of importance in reliability contexts, and a review of these is provided by Gupta and Panchapakesan (1988).

Barlow and Gupta (1969) considered among other things the selection of the population with the largest median (assumed to be stochastically larger than the other populations) from a set of continuous distributions F_i, $i = 1, \ldots, k$, which have *lighter tails* than a specified continuous distribution G with $G(0) = 1/2$. The F_i and G are assumed to have the real line as their support. The definition of F_i having a lighter tail than G used by Barlow and Gupta (1969) implies that F_i centered at its median Δ_i is *tail-ordered* with respect to G; in other words, $G^{-1}F_i(x + \Delta_i) - x$ is nondecreasing in x. The procedure of Barlow and Gupta (1969) has been shown by Gupta and Panchapakesan (1974) to work for this wider class defined by tail ordering. In fact, Gupta and Panchapakesan have also shown a generalized version of this by considering tail-ordering of F_i and G when both are centered at their respective α-quantiles. Formally stated, for $0 < \alpha < 1$, F_i is said to be *α-quantile tail-ordered* with respect to G ($F_i <_{t_\alpha} G$) if $G^{-1}F_i(x + \xi_{i\alpha}) - x - \eta_\alpha$ is nondecreasing in x on the support of F_i, where $\xi_{i\alpha}$ and η_α are the (unique) α-quantiles of F_i and G, respectively. It can be shown (Gupta and Panchapakesan, 1974, Lemma 3.1) that

$$P(a + \eta_\alpha \le Y \le b + \eta_\alpha) \le P(a + \xi_{i\alpha} \le X_i \le b + \xi_{i\alpha}) \quad (6.8.1)$$

for every $a < 0 < b$, where X_i and Y have distributions F_i and G, respectively.

Now, for the discussion of the selection problem, let $\pi_1, \ldots, \pi_k$ be k populations with associated absolutely continuous distributions F_i having unique α-quantile $\xi_{\alpha i}$, $i = 1, \ldots, k$, for a specified $0 < \alpha < 1$. Let G be a specified absolutely continuous distribution G with α-quantile η_α. We assume that the F_i are α-quantile tail-ordered with respect to G. It is also assumed that there is one population among the F_i that is stochastically larger than the remaining ones; consequently, this particular F_i, denoted by $F_{(k)}$, will have the largest α-quantile and is the best population. Our goal is to select the best population using the subset approach.

Let $X_{i1}, \ldots, X_{in}$ be n independent observations from π_i and $X_{i,j:n}$ denote the jth order statistic of the sample from π_i, $i = 1, \ldots, k$, where $j \le (n + 1)\alpha < j + 1$. Then, for selecting the population with the largest α-quantile, Gupta and Panchapakesan (1974) proposed the following rule.

$$R_7\text{: Select } \pi_i \text{ if and only if } X_{i,j:n} \ge \max_{1\le r\le k} x_{r,j:n} - D, \quad (6.8.2)$$

where $D = D(k, P^*, n, j)$ is the smallest positive integer for which the P^*-condition is satisfied no matter what the k-tuple $\{F_1, \ldots, F_k\}$ is.

Since $F_{(k)}$ is stochastically larger than any other F_i,

$$\inf_{\Omega} P(\text{CS}|R_7) = \int_{-\infty}^{\infty} F_{(k),j:n}^{k-1}(t + D) f_{(k),j:n}(t)\, dt, \qquad (6.8.3)$$

where $f_{(k),j:n}$ is the density associated with $F_{(k),j:n}$. Now, since the partial ordering $<_{t_\alpha}$ is preserved by the order statistics, we have $F_{(k),j:n} <_{t_\alpha} G_{j:n}$, where $G_{j:n}$ is the distribution of the jth order statistic in a random sample of size n from G. As a consequence of this,

$$\int_{-\infty}^{\infty} F_{(k),j:n}^{k-1}(t + D) f_{(k),j:n}(t)\, dt \geq \int_{-\infty}^{\infty} G_{j:n}^{k-1}(t + D) g_{j:n}(t)\, dt, \qquad (6.8.4)$$

where $g_{j:n}$ is the density associated with $G_{j:n}$. Thus the constant $D = D(k, P^*, n, j)$ satisfying the P^*-condition is given by

$$\int_{-\infty}^{\infty} G_{j:n}^{k-1}(t + D) g_{j:n}(t)\, dt = P^*. \qquad (6.8.5)$$

The values of D have been tabulated by Gupta and Panchapakesan (1974) for $k = 2(1)10$, $n = 5(2)15$, $j = 1(1)n$, and $P^* = 0.75, 0.90, 0.95, 0.99$ when G is chosen to be the logistic distribution $F^*(z) = [1 + \exp(-z)]^{-1}$, $-\infty < z < \infty$, which is $L(0, \pi^2/3)$. In the particular case of the median (i.e., $\alpha = 1/2$ and $j = (n + 1)/2$), Eq. (6.8.5) is the same as (6.3.5) for rule R_1 of Lorenzen and McDonald (1981) in Section 6.3, except that we now have $F_{j:n}^*$ in the place of $F_{j:n}$ (which is $L(0, 1)$). Thus the constants d_1 and D of rules R_1 and R_7 are related by $D_1 = d_1\pi/\sqrt{3}$.

The infimum of $P(\text{CS}|R_7)$ in (6.8.3) can be evaluated asymptotically as $n \to \infty$ with $j/n \to \alpha$ using the asymptotic normality of the sample quantile (under the assumptions that the densities $f_{(k)}$ and g are differentiable in the neighborhood of their respective α-quantiles and that the densities do not vanish at their α-quantiles). Then, corresponding to (6.8.5), we get

$$\int_{-\infty}^{\infty} \Phi^{k-1}\left[x + Dg(\eta_\alpha)\left\{\frac{n}{\alpha(1-\alpha)}\right\}^{1/2}\right] \phi(x)\, dx = P^* \qquad (6.8.6)$$

where Φ and ϕ denote the standard normal cdf and density.

When G is taken to be $L(0, \pi^2/3)$, $g(\eta_\alpha) = \alpha(1 - \alpha)$, and an approximate value of D is given by

$$\int_{-\infty}^{\infty} \Phi^{k-1}[x + D\{n\alpha(1 - \alpha)\}^{1/2}]\phi(x)\, dx = P^*. \qquad (6.8.7)$$

The D-value satisfying (6.8.7) can be obtained from the tables of Gupta, Nagel, and Panchapakesan (1973); in other words, $D = H[2/n\alpha(1 - \alpha)]^{1/2}$, where H is the table value corresponding to $\rho = 0.5$.

It is relevant to note that the left-hand side of (6.8.5) can be written as $P[Y_k \geq \max_{1\leq r\leq k} Y_r - D]$, where the Y_i are i.i.d. having the distribution function $G_{j:n}$. Further, since $D > 0$, it can be written as $P[\max_{1\leq r\leq k-1} (Y_r - Y_k) \leq D]$. Thus D given by (6.8.5) is the $100P^*\%$ point of the distribution of the maximum of the correlated differences $Y_i - Y_k$, $i = 1, \ldots, k - 1$. A similar probability of interest is $P[\max_{1\leq r\leq k} W_r/W_k \leq a]$, $a \geq 1$, where the W_i are i.i.d. having the distribution of the jth order statistic in a random sample of size n from a distribution G of a continuous *nonnegative* random variable. Such a probability concerning the maximum of correlated ratios arises in the problem of selecting in terms of α-quantiles from k distributions F_i of nonnegative random variables which are *star-shaped* with respect to G. This has been investigated by Barlow and Gupta (1969). When G is exponential, the F_i become IFRA distributions, and the values of $c = a^{-1}$ have been tabulated by Barlow, Gupta, and Panchapakesan (1969). When G is the half-normal, the F_i belong to a subclass of IFRA distributions because the half-normal is IFRA (actually it is IFR, which implies IFRA) and the star-ordering is transitive. In this case, the values of $c = a^{-1}$ have been tabulated by Gupta and Panchapakesan (1975). Another application of immediate interest is to consider the case where G is the half-logistic distribution; i.e., $G(x) = \{2/(1 + \exp(-x))\} - 1$, $x \geq 0$. Since G is IFR, it is also IFRA. Thus our F_i's belong to yet another subclass of IFRA distributions. In this case, tables of a are not available.

6.9 CONCLUDING REMARKS

In our selective overview of the ranking and selection theory in Section 6.2, we have confined ourselves to the basic formulations and some modifications that were relevant to the review of the procedures that have been investigated in the case of logistic populations. Our comments in the present section are meant to indicate the scope for further investigations.

For the problem of selecting the population with the largest mean μ_i using single-stage samples, we have assumed the variances to be *equal and known* and considered *equal sample sizes*. Procedures based on unequal sample sizes are not trivial extensions. Further, the case of unknown variances (known to be equal or not) are important in practice. Also, as mentioned earlier, when the variances are unknown (even if they are known to be equal), a single-stage procedure that guarantees the P^*-condition does not exist under the indifference zone formulation. These problems involve questions relating to distribution theory, determination of the LFC, and computations for implementation.

There are many aspects of the ranking problem that have been studied in the literature such as selecting good populations (which are "close" to the best), selecting populations that are better than a standard or a control, estimation of the actual PCS, estimation after selection (such as estimating the mean of the selected population), and confidence interval for the difference between the selected mean and the largest mean. For selecting the populations better than a standard or a control, we may have prior knowledge of the ordering of the experimental populations even though we may not know the values of the parameter of interest. In this case, the procedures have to exhibit an *isotonic* behavior. Some of the above aspects are of current interest relative to selection problems in general. For discussions of these developments and related references, see Gupta and Panchapakesan (1979, 1985). A few additional recent references that might be of interest are Gupta and Liang (1987, 1991), Gupta and Sohn (1991), Gupta and Panchapakesan (1991), and Liang and Panchapakesan (1991).

Gupta and Sohn (1990) have considered subset selection from Tukey's (symmetric) lambda distributions in terms of their location parameters, assuming that they all have *known* common scale and shape parameters. The lambda family of distributions was suggested by Tukey (1960, 1962) as a wide class of symmetric distributions. Later, Ramberg and Schmeiser (1972, 1974) generalized this family to include both symmetric and asymmetric distributions. The practical usefulness of this family is highlighted by the fact that it has a simple form for the inverse of the cdf, and it can be used to approximate a wide class of densities ranging from the uniform to very heavy tailed ones, of course, *including the logistic* (see also Joiner and Rosenblatt, 1971). Gupta and Sohn (1990) investigated a subset selection rule based on sample medians similar to the rule R_1 in Section 6.3. Based on approximation to the logistic distribution by proper choices of the scale and shape parameters, they have calculated the constants for rule R_1 of Lorenzen and McDonald (1981) when $k = 2, 5, 7$; $m = 2, 5, 7, 9$; and $P^* = 0.90, 0.95$. In 13 out of the 24 cases, the d-values, corrected to three decimal places, agree. In the remaining cases, the approximate value using the lambda distribution is one unit more in the third place. A similar comparison can be made in the case of rule R_2 of Han (1987), described in Section 6.3. Sohn (1985) has verified computationally that the approximation to the cdf of the sample mean of a logistic distribution by using the lambda distribution is as good as that of Goel (1975), who derived the cdf as a series by the method of characteristic functions.

Finally, the use of lambda distribution in approximating a wide class of densities has significant implications in developing versatile software packages for selection and ranking problems. Further, this aspect can be useful in other inference problems related to the logistic distribution.

REFERENCES

Alam, K. (1970). A two-sample procedure for selecting the population with the largest mean from k normal populations. *Ann. Inst. Statist. Math.*, *22*, 127–136.

Barlow, R. E. and Gupta, S. S. (1969). Selection procedures for restricted families of distributions. *Ann. Math. Statist.*, *40*, 905–917.

Barlow, R. E. and Gupta, S. S., and Panchapakesan, S. (1969). On the distribution of the maximum and minimum of ratios of order statistics. *Ann. Math. Statist.*, *40*, 918–934.

Bechhofer, R. E. (1954). A single-sample multiple decision procedure for ranking means of normal populations with known variances. *Ann. Math. Statist.*, *25*, 16–39.

Bechhofer, R. E., Kiefer, J., and Sobel, M. (1968). *Sequential Identification and Ranking Procedures (with Special Reference to Koopman-Darmois Populations)*. University of Chicago Press, Chicago.

Bechhofer, R. E. and Kulkarni, R. (1982). Closed adaptive sequential procedures for selecting the best of $k \geq 2$ Bernoulli populations, in *Statistical Decision Theory and Related Topics—III*, S. S. Gupta and J. O. Berger (eds.), *1*, 61–108, Academic Press, New York.

Bhandari, S. K. and Chaudhuri, A. R. (1990). On two conjectures about two-stage selection problem. *Sankhyā Ser. B*.

Büringer, H., Martin, H., and Schriever, K.-H. (1980). *Nonparametric Sequential Procedures*. Birkhauser, Boston.

Cohen, D. S. (1959). *A Two-Sample Decision Procedure for Ranking Means of Normal Populations with a Common Known Variance*. M.S. Thesis, Dept. of Operations Research, Cornell University, Ithaca, New York.

Dudewicz, E. J. and Koo, J. O. (1982). *The Complete Categorized Guide to Statistical Selection and Ranking Procedures*, Series in Mathematical and Management Sciences, Vol. 6, American Sciences Press, Columbus, Ohio.

Gibbons, J. D., Olkin, I., and Sobel, M. (1977). *Selecting and Ordering Populations: A New Statistical Methodology*. Wiley, New York.

Goel, P. K. (1975). On the distribution of standardized mean of samples from the logistic population. *Sankhyā Ser. B*, *37*, 165–172.

Gupta, S. S. (1956). On A Decision Rule for a Problem in Ranking Means, *Mimeograph Series No. 150*, Institute of Statistics, University of North Carolina, Chapel Hill, North Carolina.

Gupta, S. S. (1963). Probability integrals of the multivariate normal and multivariate t. *Ann. Math. Statist.*, *34*, 792–828.

Gupta, S. S. (1965). On some multiple decision (selection and ranking) rules. *Technometrics*, *7*, 225–245.

Gupta, S. S. and Han, S. (1991). An elimination type two-stage procedure for selecting the population with the largest mean from k logistic populations. *Amer. J. Math. Management Sci.* (to appear).

Gupta, S. S. and Huang, D.-Y. (1981). *Multiple Decision Theory: Recent Developments*, Lecture Notes in Statistics, Vol. 6, Springer-Verlag, New York.

Gupta, S. S. and Liang, T. (1987). On some Bayes and empirical Bayes selection procedures, in *Probability and Bayesian Statistics*, R. Viertl (ed.), Plenum, New York.

Gupta, S. S. and Liang, T. (1991). On a lower confidence bound for the probability of a correct selection: analytical and simulation studies, *Proceedings of the First International Conference on Statistical Computing* held in Turkey, March 30–April 2, 1987, American Sciences Press, Syracuse, New York.

Gupta, S. S. and McDonald, G. C. (1970). On some classes of selection procedures based on ranks, in *Nonparametric Techniques in Statistical Inference*, M. L. Puri (ed.), 491–514, Cambridge University Press, London.

Gupta, S. S., Nagel, K., and Panchapakesan, S. (1973). On the order statistics from equally correlated normal random variables. *Biometrika*, *60*, 403–413.

Gupta, S. S. and Panchapakesan, S. (1972). On a class of subset selection procedures. *Ann. Math. Statist.*, *43*, 814–822.

Gupta, S. S. and Panchapakesan, S. (1974). Inference for restricted families: (a) multiple decision procedures; (b) order statistics inequalities, in *Reliability and Biometry: Statistical Analysis of Lifelength*, F. Proschan and R. J. Serfling (eds.), 503–596, SIAM, Philadelphia.

Gupta, S. S. and Panchapakesan, S. (1975). On a quantile selection procedure and associated distribution of ratios of order statistics from a restricted family of probability distributions, in *Reliability and Fault Tree Analysis: Theoretical and Applied Aspects of System Reliability and Safety Assessment*, R. E. Barlow, J. B. Fussell, and N. D. Singpurwalla (eds.), 557–576, SIAM, Philadelphia.

Gupta, S. S. and Panchapakesan, S. (1979). *Multiple Decision Procedures: Theory and Methodology of Selecting and Ranking Populations*. Wiley, New York.

Gupta, S. S. and Panchapakesan, S. (1985). Subset selection procedures: review and assessment. *Amer. J. Math. Management Sci.*, *5*, 235–311.

Gupta, S. S. and Panchapakesan, S. (1988). Selection and ranking procedures in reliability models, in *Handbook of Statistics 7: Quality Control and Reliability*, P. R. Krishnaiah and C. R. Rao (eds.), 131–156, North-Holland, Amsterdam.

Gupta, S. S. and Panchapakesan, S. (1991). On sequential ranking and selection procedures, in *Handbook of Sequential Methods*, B. K. Ghosh and P. K. Sen (eds.), 363–380, Marcel Dekker, New York.

Gupta, S. S. and Sohn, J. K. (1991). Selection and ranking procedures for Tukey's generalized lambda distributions, in *Frontiers of Modern Statistical Inference Procedures—II*, E. J. Dudewicz and E. Boffinger (eds.), American Sciences Press, Syracuse.

Han, S. (1987). *Contributions to Selection and Ranking Theory with Special Reference to Logistic Populations*. Ph.D. Thesis (also *Technical Report No. 87-38*), Dept. of Statistics, Purdue University, West Lafayette, Indiana.

Joiner, B. L. and Rosenblatt, J. R. (1971). Some properties of the range in samples from Tukey's symmetric lambda distributions. *J. Amer. Statist. Assoc.*, *66*, 394–399.

Liang, T. and Panchapakesan, S. (1991). Isotonic selection with respect to a control: a Bayesian approach, in *Frontiers of Modern Statistical Inference Procedures—II*, E. J. Dudewicz and E. Boffinger (eds.), American Sciences Press, Syracuse.

Lorenzen, T. J. and McDonald, G. C. (1981). Selecting logistic populations using the sample medians. *Comm. Statist. A—Theory Methods*, *10*, 101–124.

Milton, R. C. (1963). Tables of Equally Correlated Multivariate Normal Probability Integral. *Tech. Report No. 27*, Dept. of Statistics, University of Minnesota, Minneapolis, Minnesota.

Ramberg, J. S. and Schmeiser, B. W. (1972). An approximate method for generating symmetric random variables. *Comm. ACM*, *15*, 987–990.

Ramberg, J. S. and Schmeiser, B. W. (1974). An approximate method for generating asymmetric random variables. *Comm. ACM*, *17*, 78–82.

Santner, T. J. (1975). A restricted subset selection approach to ranking and selection problems. *Ann. Statist.*, *3*, 334–349.

Sohn, J. K. (1985). *Multiple Decision Procedures for Tukey's Generalized Lambda Distributions*. Ph.D. Thesis (also *Technical Report No. 85-20*), Dept. of Statistics, Purdue University, West Lafayette, Indiana.

Tamhane, A. C. (1986). A survey of literature on quantal response curves with a view toward application to the problem of selecting the curve with the smallest q-quantile (ED100q). *Comm. Statist. A—Theory Methods*, *15*, 2679–2718.

Tamhane, A. C. and Bechhofer, R. E. (1977). A two-stage minimax procedure with screening for selecting the largest normal mean. *Comm. Statist. A—Theory Methods*, *6*, 1003–1033.

Tamhane, A. C. and Bechhofer, R. E. (1979). A two-stage minimax procedure with screening for selecting the largest mean (II): an improved PCS lower bound and associated tables. *Comm. Statist. A—Theory Methods*, *8*, 337–358.

Tukey, J. W. (1960). The Practical Relationship Between the Common Transformations of Percentages or Fractions and of Amounts. *Tech. Report No. 36*, Statistical Research Group, Princeton.

Tukey, J. W. (1962). The future of data analysis. *Ann. Math. Statist.*, *33*, 1–67.

Van Der Laan, P. (1989). Selection from logistic populations. *Statist. Neerlandica*, *43*, 169–174.

7
Characterizations

Janos Galambos
Temple University, Philadelphia, Pennsylvania

7.1 INTRODUCTION

In order to fully understand a particular distribution, one tries to identify those properties which are characteristic to that distribution. Some of these properties are used only to theoretically distinguish a distribution from the others, while some characteristic properties can be utilized in model building. We shall deal with both types of characterizations.

By monotonic transformations, a characteristic property of one distribution can be carried into a characteristic property of another distribution. Hence, at first, it appears that it suffices to study the characterizations of one particular distribution which can then be restated for the distribution of interest, the logistic distribution in our case. Such an attempt, however, quickly breaks down. While a property can be interesting for one distribution, it may lose its appeal after a transformation. Let us see an important example.

First recall that, if U is a random variable with continuous distribution function $G(x)$, then

$$V = G(U) \qquad \text{and} \qquad W = -\frac{1}{a}\log V, \tag{7.1.1}$$

where $a > 0$ is a constant, are uniform on (0, 1) and exponentially distributed, respectively. That is,

$$P(W \le x) = 1 - e^{-ax}, \qquad x \ge 0. \tag{7.1.2}$$

Now, the simplest characterizing property of the exponential distribution (7.1.2) is the lack of memory (see Sections 1.3 and 1.5 in Galambos and Kotz, 1978 for more details); that is,

$$P(W > x + y | W > x) = P(W > y), \qquad x, y \ge 0, \tag{7.1.3}$$

which, for U, becomes

$$P(U \le G^{-1}(e^{-x-y}) | U \le G^{-1}(e^{-x})) = P(U \le G^{-1}(e^{-y})), \tag{7.1.4}$$

where $x, y \ge 0$ are arbitrary, and $G^{-1}(t) = x$ is the inverse function of $t = G(x)$.

Now, (7.1.4) is a strange-looking characterization of the normal distribution, say (the reader is invited to read the meaning of (7.1.4) for G normal), but it is meaningful for several distributions besides the exponential one. For example, if $G(x) = x$ is uniform on (0, 1), then (7.1.4) becomes *U is uniform on* (0, 1) *if, and only if, for all u and v with* $0 \le u$, $v \le 1$,

$$P(U \le uv | U \le u) = P(U \le v). \tag{7.1.5}$$

Here, we changed the variables x and y of (7.1.4) to $u = e^{-x}$ and $v = e^{-y}$. For the logistic distribution $t = F^*(z)$ of (1.6), the inverse function

$$z = F^{*(-1)}(t) = \log \frac{t}{1-t}, \tag{7.1.6}$$

and thus (7.1.4) takes the form

$$P(e^{-U} + 1 \ge e^{x+y} | e^{-U} + 1 \ge e^{x}) = P(e^{-U} + 1 \ge e^{y}), \qquad x, y \ge 0.$$

Hence, with the transformation

$$T = 1 + e^{-U}, \tag{7.1.7}$$

and with $u = e^x$ and $v = e^y$, we have the following characterization of the logistic distribution from the lack-of-memory property.

Theorem 7.1.1 Let U be a random variable and define T by (7.1.7). Then the distribution function of U is the logistic distribution $F^*(z)$ if, and only if, for all $u \ge 1$ and $v \ge 1$,

$$P(T \ge uv | T \ge u) = P(T \ge v).$$

Even though Theorem 7.1.1 is an artificial characterization of $F^*(z)$, one can argue that its simplicity makes it useful. The variable T of (7.1.7), of course, is just the reciprocal of V of (7.1.1) when $G(x) = F^*(x)$. Thus, Theorem 7.1.1 is an indirect characterization of $F^*(x)$ via the Pareto distribution

$$P(T \le x) = P\left(\frac{1}{V} \le x\right) = P\left(V \ge \frac{1}{x}\right) = 1 - \frac{1}{x}, \qquad x \ge 1.$$

In the next example, the starting point is again a basic characteristic property of the exponential distribution. By the transformations (7.1.1) we shall get a simple, but interesting, characterization of the logistic distribution, whose interpretation, however, will have nothing to do with the initial characterizing property. That is, let U be a random variable such that, with T of (7.1.7), $\log T$ has finite expectation. Assume that the expected residual lifetime of $\log T$,

$$R(x) = E[\log T - x | \log T > x] = c, \tag{7.1.8}$$

where $c > 0$ is a finite constant. Then an easy calculation yields (see Section 1.5 of Galambos and Kotz, 1978 for this and related properties) that $\log T$ is exponentially distributed, from which we get that U is a logistic variable.

This is a characterization theorem, because, for U with distribution function $F^*(x)$, the transformed variable $\log T$ becomes W of (7.1.1), i.e., exponential, for which (7.1.8) holds. Now, if $G(x)$ is the distribution function of U, then

$$P(\log T > x) = P(U < -\log(e^x - 1)) = G(-\log(e^x - 1)),$$

and thus

$$R(x) = \frac{\int_x^{+\infty} P(\log T > t)\, dt}{P(\log T > x)} = \frac{\int_x^{+\infty} G(-\log(e^t - 1))\, dt}{G(-\log(e^x - 1))}.$$

Hence, with the substitution $y = -\log(e^t - 1)$, (7.1.8) with $c = 1$ (this value is the appropriate one due to the definition of T in (7.1.7)) becomes

$$\int_{-\infty}^{z} \frac{G(y)}{1 + e^y}\, dy = G(z), \tag{7.1.9}$$

where $z = -\log(e^x - 1)$, and thus z goes through the whole real line as x goes through the positive numbers. Note that we neglected equation signs in calculating distribution functions. Therefore, (7.1.8) and (7.1.9) are equivalent only if we assume a priori that $G(x)$ is continuous. We thus have that, for continuous population distribution $G(x)$, $G(x) = F^*(x)$ if, and only if, the expected residual lifetime of $\log T$ is unity, or, equivalently, if, and only if, (7.1.9) holds.

When (7.1.9) alone is used in characterizing $F^*(x)$, we do not see any relation to expected residual lifetime, which is the point of this particular characterization statement. Indeed, one can find directly, i.e., without any reference to transformation to exponentiality, that the only solution of (7.1.9) is $F^*(x)$. First, we note that the left-hand side defines a continuous function, so $G(z)$ must be continuous, but then the left-hand side is necessarily differentiable, so $G(z)$ is differentiable. By differentiating, we get from (7.1.9) that

$$\frac{1}{1 + e^{z}} = \frac{G'(z)}{G(z)} \tag{7.1.10}$$

whenever $G(z) > 0$. It thus follows that $G(z) = F^*(z)$ (write the left-hand side as $e^{-z}/(1 + e^{-z})$, and conclude from the actual solution of (7.1.10) that $G(z) = 0$ is impossible since $G(z)$ is continuous).

In addition to the two characterizations of the logistic distribution, obtained in the preceding paragraphs by transformations to exponentiality, one could produce a large number of characterization theorems by transformations to other distributions. However, the result at (7.1.9) clearly shows that a direct approach is better than an artificial reference to a property of a transformation. We shall therefore look at the logistic family of distributions on its own, although the reader may want to go through the results developed in Galambos and Kotz (1978), Azlarov and Volodin (1982), and Arnold (1983).

We conclude this introduction by a convention for this chapter. As in this introduction, the special notations of distributions (distribution functions and densities) defined in (1.1)–(1.6) will be preserved for the logistic distributions. In particular, Z will always be a random variable with the standard form (1.6) of the logistic distribution function $F^*(z)$. A general random variable will be denoted by U and its distribution function by $G(x)$. Observations on U will be denoted by $U_1, U_2, \ldots, U_n$, and their order statistics by $U_{1:n} \leq U_{2:n} \leq \cdots \leq U_{r:n} \leq \cdots \leq U_{n:n}$.

7.2 THE LOGISTIC DISTRIBUTION AS A MODEL FOR POPULATION GROWTH

In order to justify the application of the logistic distribution as a "population growth distribution," one can refer to the following characterization.

Theorem 7.2.1 Assume that the distribution function $G(x)$ has a density $g(x)$ and satisfies the differential equation

$$g(x) = cG(x)[1 - G(x)], \tag{7.2.1}$$

where $c > 0$ is a constant. Then $G(x)$ is the logistic distribution function $F(x, \mu, \sigma)$.

The proof is simple. By standard methods of solving differential equations one has that (7.2.1) has a unique solution, and it indeed is $F(x, \mu, \sigma)$.

N. L. Johnson and S. Kotz (1970, p. 2) describe an application of (7.2.1) in connection with a chemical reaction. As in their case, one rarely faces (7.2.1) directly, but rather a quantity y is known from other considerations (such as a general law of chemistry) to satisfy the "growth rate"

$$\frac{dy}{dt} = c_1(y_0 - y) + c_2y(y_0 - y), \tag{7.2.2}$$

where c_1 and c_2 are constants and y_0 is an initial value of y. By the nature of the constants involved, the right-hand side becomes of the form

$$c_3(y + c_4)[c_5 - (y + c_4)],$$

where c_3, c_4, and c_5 are further constants. Finally, the quantity $y + c_4$ is assumed to represent its own distribution, and then one gets

$$\frac{dG(x)}{dx} = c_4G(x)[c_5 - G(x)],$$

which is similar to (7.2.1). By solving this, one immediately gets that $c_5 = 1$ by $G(+\infty) = 1$. Note that (7.2.2) is usually an accepted general law from a branch of science, but the arbitrary step of assuming that y and its own distribution have the same laws makes the application of Theorem 7.2.1 a crude approximation at best.

The following interpretation of Theorem 7.2.1 has potential applications in model building (for growth curves). Let U be a random variable with distribution function $G(x)$. Define the binary random variables

$$B_U(x) = \begin{cases} 1 & \text{if } U \le x, \\ 0 & \text{if } U > x. \end{cases}$$

Then $E[B_U(x)] = G(x)$ and $V[B_U(x)] = G(x)[1 - G(x)]$. Thus, Theorem 7.2.1 states that $G(x) = G(x, \mu, \sigma)$ if, and only if, the rate of change of the expectation of $B_U(x)$ is proportional to its variance (for every x). Even more striking a property is obtained if we apply Theorem 7.2.1 to the sum

$$S_n(x) = B_{U_1}(x) + B_{U_2}(x) + \cdots + B_{U_n}(x),$$

where $U_1, U_2, \ldots, U_n$ are independent copies of U. Theorem 7.2.1 then states that the logistic, and only the logistic, distributions have the property that the ratio of the rate of change of $E(S_n(x))$ to $V(S_n(x))$ does not depend on either x or n. That is, if the mentioned ratio becomes stable in n, whatever x, then $G(x) = F(x, \mu, \sigma)$, which property can be utilized in empirical justifications of choosing $F(x, \mu, \sigma)$ as the underlying distribution for a model.

7.3 CHARACTERIZATIONS BY PROPERTIES OF ORDER STATISTICS

In this section we describe a general theory of characterizations which is applicable to any underlying population distribution. The first systematic presentation of this theory appeared in Galambos (1975) and Galambos and Kotz (1978), and a recent survey is due to Huang (1989).

Let $U_{1:n} \le U_{2:n} \le \cdots \le U_{n:n}$ be the order statistics of a sample $U_1, U_2, \ldots, U_n$ of size n on a random variable U with distribution function $G(x)$. Set

$$G_{k:n}(x) = P(U_{k:n} \le x) \quad \text{and} \quad E_{k:n} = E(U_{k:n}), \qquad 1 \le k \le n,$$

assuming that the latter is finite for every k. After establishing some formulas concerning $G_{k:n}(x)$, we shall show that $E_{k:n}$ is finite for every k and n if $E(U) = E_{1:1}$ is finite.

First, note that the event $\{U_{k:n} \le x\}$ means that at least k of the events $\{U_j \le x\}$, $1 \le j \le n$, occur; the form of the binomial distribution thus yields

$$G_{k:n}(x) = \sum_{j=k}^{n} \binom{n}{j} G^j(x)[1 - G(x)]^{n-j}. \tag{7.3.1}$$

The right-hand side can be expressed as an integral (integrate by parts below); i.e.,

$$G_{k:n}(x) = k\binom{n}{k} \int_0^{G(x)} y^{k-1}(1 - y)^{n-k}\, dy, \tag{7.3.2}$$

and thus

$$E_{k:n} = \int_{-\infty}^{+\infty} x\, dG_{k:n}(x) = k\binom{n}{k} \int_0^1 G^{-1}(y) y^{k-1}(1 - y)^{n-k}\, dy, \tag{7.3.3}$$

where $G^{-1}(y) = \inf\{x: G(x) \ge y\}$, and the last equality is valid whenever the integral involved is finite.

Now, if we assume that $E(U) = E_{1:1}$ is finite, then (7.3.3) applies with $n = k = 1$. In particular,

$$\int_0^1 |G^{-1}(y)|\, dy < +\infty.$$

But then, for any k and n, $E_{k:n}$ is finite as well, since we increase on the extreme right-hand side of (7.3.3) if we replace $G^{-1}(y)$ by $|G^{-1}(y)|$ and both y and $1 - y$ by 1.

Another immediate consequence of (7.3.3) is the recursive formula (see Arnold and Balakrishnan, 1989)

$$(n - k)E_{k:n} + kE_{k+1:n} = nE_{k:n-1}, \qquad 0 < k < n,\ n \geq 2. \qquad (7.3.4)$$

Indeed, in view of the simple relations

$$t\binom{n}{t} = n\binom{n-1}{t-1} \quad \text{and} \quad \binom{n}{t} = \binom{n}{n-t},$$

we have that

$$(n - k)\binom{n}{k}(1 - y) + (k + 1)\binom{n}{k+1} y = n\binom{n-1}{k}$$

is an identity in y, where $0 < k < n$ and $n \geq 2$, which is needed for (7.3.4) when the expectations occurring in it are replaced by the integral formula of (7.3.3).

We now record two characterization theorems. The first one is due to Huang (1974), while the second is a combination of results by Kadane (1974), Galambos (1975), and Hoeffding (1953).

Theorem 7.3.1 The distribution function $G_{k:n}(x)$ uniquely determines the population distribution $G(x)$.

Proof. Appeal to (7.3.2), from which the theorem is immediate.

Theorem 7.3.2 Let $k(n)$ be a sequence of integers with $1 \leq k(n) \leq n$, $n \geq 1$. Then the sequence $E_{k(n):n}$ of expectations uniquely determines the triangular array $E_{r:n}$, $1 \leq r \leq n$, $n \geq 1$, which in turn uniquely determines the population distribution $G(x)$. In other words, $E_{k(n):n}$, $n \geq 1$, uniquely determines $G(x)$. ($E_{1:1}$ is assumed to be finite.)

Proof. The first part of the theorem follows from the recursive formula (7.3.4). Indeed, for $n = 1$ we must have $k(1) = 1$. Next, out of $E_{1:2}$ and $E_{2:2}$, one is $E_{k(2):2}$, wile the other is determined by (7.3.4) because $E_{1:1}$ is known. Now, if the row $E_{k:n-1}$ has been filled in for all k, then $E_{r:n}$ can be determined by (7.3.4) for all r by moving one step at a time to the left and then to the right from the given value $E_{k(n):n}$.

Turning to the second part of the theorem, we assume that the values $E_{r:n}$ are given for all r and n with $1 \le r \le n$. Let $G_1(x)$ and $G_2(x)$ be two population distribution functions for which the expectations of the order statistics $U_{r:n}$ equal the given values $E_{r:n}$ for all r and n. Then, by (7.3.3),

$$\int_0^1 G_1^{-1}(y)y^{r-1}(1-y)^{n-r}\,dy = \int_0^1 G_2^{-1}(y)y^{r-1}(1-y)^{n-r}\,dy$$

for all $1 \le r \le n$. Fix the value of r. Substitute $u = 1 - y$ in the integrals above, and set $h_i(u) = G_i^{-1}(1-u)(1-u)^{r-1}$, $i = 1$ or 2. We then have

$$\int_0^1 h_1(u)u^k\,du = \int_0^1 h_2(u)u^k\,du, \qquad k = 0, 1, 2, \ldots, \qquad (7.3.5)$$

where we replaced $n - r$ by k. Next, we define $h_i^+(u) = \max(0, h_i(u))$ and $h_i^-(u) = -\min(0, h_i(u))$, $i = 1$ or 2. Set

$$s_1(u) = h_1^+(u) + h_2^-(u) \qquad \text{and} \qquad s_2(u) = h_2^+(u) + h_1^-(u).$$

Then $s_i(u) \ge 0$, and by (7.3.5),

$$\int_0^1 s_1(u)u^k\,du = \int_0^1 s_2(u)u^k\,du, \qquad k = 0, 1, 2, \ldots. \qquad (7.3.6)$$

The case $k = 0$ yields that, with a suitable constant $c > 0$, $cs_1(u)$ and $cs_2(u)$ are density functions. Hence, (7.3.6) can be interpreted as saying that all moments of two random variables, bounded by 0 and 1, are equal. It is well known (see Galambos, 1988, p. 97) that such a property implies that the two distributions involved must be equal; i.e., $s_1(u) = s_2(u)$, which implies that $h_1(u) = h_2(u)$, in both cases, for almost all u in (0, 1). By the definition of $h_i(u)$ we thus have that $G_1(x) = G_2(x)$, and the proof is complete.

Corollary 7.3.1 If $E_{m+1:2m+1} = 0$ and $E_{m:2m} = -1/m$ for all $m \ge 0$, then the population distribution $G(x)$ is the logistic distribution $F^*(x)$.

The corollary is a direct consequence of Theorem 7.3.2. We define the sequence $k(n)$ as $k(2m+1) = m+1$ and $k(2m) = m$. Upon computing the expectations $E_{k(n):n}$ for the logistic distribution $F^*(x)$, we find that these are the given values in the corollary. Consequently, by Theorem 7.3.2, no other distribution would lead to these same expectations of the order statistics.

It is clear from the proof of Theorem 7.3.2 that the characterization of the population distribution by the expectations of the order statistics is equivalent to the moment problem for bounded random variables. Therefore, just as in the case of the moment problem, it is not essential to have, directly or indirectly, the whole triangular array $E_{r:n}$ of expectations for characterizing $G(x)$. We do not discuss this question in detail. For a variety of results, see Arnold and Meeden (1975), Huang (1975, 1989), Hwang (1978), and Lin (1984).

Theorem 7.3.2 can be extended to higher moments. First, note that a formula similar to (7.3.3) applies to $E(U_{k:n}^m)$. Indeed, by replacing $G^{-1}(y)$ by $(G^{-1}(y))^m$ in (7.3.3) we get $E(U_{k:n}^m)$. Therefore, by imposing restrictions among the moments $E(U_{k:n}^m)$, one will arrive at equations like (7.3.5), where $h_1(u)$ involves an unknown inverse distribution function, while $h_2(u)$ is an expression in terms of the special inverse function (7.1.6). One can then argue as at (7.3.5), concluding that the two inverse functions are identical, providing a characterization of the logistic distribution $F^*(x)$, or more generally $F(x, \mu, \sigma)$. The reader can fill in the details of the just described argument, and obtain the following recent result of Lin (1988).

Theorem 7.3.3 Let U be a random variable with distribution function $G(x)$. Assume that $P(U = 0) = 0$ and that $G^{-1}(u)$ is absolutely continuous on (0, 1). Assume that, for some $m \geq 1$, $E(U^m)$ is finite. Then $G(x) = F(x, \mu, \sigma)$ if, and only if, for all $n \geq k$, where $k \geq 2$ is a fixed integer,

$$E(U_{k:n}^m) = E(U_{k-1:n}^m) + \frac{m\sigma n}{(k-1)(n-k+1)} E(U_{k-1:n-1}^{m-1}).$$

Remarks. 1. When $m = 1$, the theorem reduces to characterizing the logistic family by the spacings $E(U_{k:n} - U_{k-1:n}) = \sigma n/(k-1)(n-k+1)$. Note that the value of μ is not involved, confirming earlier results (see Govindarajulu et al., 1975; Madreimov and Petunin, 1983) that the spacings above, with fixed k and all $n \geq k$, characterize $G(x)$ up to a location parameter.

2. Since this theorem also is related to the moment problem, just as in the case of Theorem 7.3.2, the condition "all $n \geq k$" can be relaxed, and it is done in the paper of Lin.

7.4 SAMPLES WITH RANDOM SAMPLE SIZE

We continue to use the notations of the preceding section. However, here n, the sample size, itself is an integer valued random variable, and, unless otherwise stated, n is assumed to be independently distributed of the U_j.

We put

$$p_k = P(n = k), \qquad k \geq 1, \tag{7.4.1}$$

allowing that, for some k, $p_k = 0$. In some cases, we shall have sequences of samples, in which case we write $n(N)$ and $p_k(N)$, $k \geq 1$, where N goes through the integers 1, 2, ... For simplicity of writing, we also put

$$z = \frac{x - \mu}{\sigma} \qquad \text{and} \qquad F^*(z) = F(x, \mu, \sigma).$$

The following theorem is a variant of a result of Kakosyan et al. (1984, p. 82), although not equivalent to their result. The proof is completely new.

Theorem 7.4.1 Assume that, with some $0 < p < 1$, $p_k = p(1 - p)^{k-1}$, $k \geq 1$, in (7.4.1). Then, for $G(x) = F^*(z)$ (recall the relation of x and z), $U_{n:n} + \log p$ has the same distribution $F^*(z)$ as the population. Conversely, assume that, for two values p_1^* and p_2^* of p such that $(\log p_1^*)/\log p_2^*$ is irrational, $U_{n:n} + \log p$ is distributed as the population. Then $G(x) = F^*(z)$.

Proof. By conditioning on $n = k$, the total probability rule yields

$$P(U_{n:n} + \log p \leq z) = \sum_{k=1}^{+\infty} G^k(z - \log p)p(1 - p)^{k-1}$$

$$= \frac{pG(z - \log p)}{1 - (1 - p)G(z - \log p)}.$$

Hence, if $G(z) = F^*(z)$, we indeed have that the distribution of $U_{n:n} + \log p$ is $F^*(z)$ for every p. Next, we assume that the just computed distribution is $G(z)$ for two values of p. That is, for two values of p, p_1^* and p_2^*, say,

$$G(z) = \frac{pG(z - \log p)}{1 - (1 - p)G(z - \log p)}. \tag{7.4.2}$$

Comparing the two sides, we immediately have that $G(z) > 0$ for all z. Therefore, we can rewrite (7.4.2) as

$$\frac{1}{G(z)} - 1 = \frac{1}{p}\left[\frac{1}{G(z - \log p)} - 1\right]. \tag{7.4.3}$$

Changing z to $z - \log p$, and applying (7.4.3) to the expression in brackets, we find that (7.4.3) applies with p^2 in place of p. Upon repeating this argument, we obtain that (7.4.3), and thus (7.4.2) as well, applies when p^t replaces p with $t \geq 1$ an arbitrary integer. Hence, we have from (7.4.2) that, for $p = p_1^*$ and p_2^*, and for every $t \geq 1$,

$$G(z) = \frac{G(z - t \log p)}{G(z - t \log p) + p^{-t}[1 - G(z - t \log p)]}. \tag{7.4.4}$$

Since $\log p < 0$, by letting $t \to +\infty$, we find that

$$\lim p^{-t}[1 - G(z - t \log p)] = e(z) \qquad (t \to +\infty) \tag{7.4.5}$$

exists and

$$G(z) = \frac{1}{1 + e(z)}. \tag{7.4.6}$$

Now, let $s = -m \log p_i^*$ for some fixed m and for $i = 1$ or 2. Then writing $z + s - t \log p = z - (t + m) \log p$ and $t = (t + m) - m$ in (7.4.5), where $p = p_i^*$, we get

$$e(z + s) = e(z)p^m = e(z)e^{-s}, \tag{7.4.7}$$

where z is arbitrary and $s = -m \log p_i^*$, $i = 1$ or 2, and m is an arbitrary integer. We can repeat this argument and get (7.4.7) when s is a linear combination of $\log p_1^*$ and $\log p_2^*$. However, we know from elementary mathematics that, for $(\log p_1^*)/\log p_2^*$ irrational, the set of such linear combinations is dense on the whole real line, and thus, by the evident monotonicity of $e(z)$ (see (7.4.6)), (7.4.7) applies for all z and s. That is, $e(z) = e^{-z}$, which, in view of (7.4.6), concludes the proof.

Kakosyan et al. (1984) prove Theorem 7.4.1 when its assumption is made for a single p but at the expense of assuming that the tail of $G(z)$ is exponentially decreasing. They also treat the case when p itself is random, which, however, results in a random normalization (by $\log p$). There is a danger in random normalizations, however, in particular in limit theorems (see Galambos, 1976), in that the distribution of a randomly normalized random variable may be that of the normalization rather than that of the original statistic. Earlier, Baringhaus (1980) proved a theorem related to Theorem 7.4.1. He assumed that $G(z)$ is a symmetric distribution, and that n has a power series distribution. Then, from the assumption that U_1 and $U_{n:n} - A$, with A a constant, have the same distribution ($G(z)$), he concluded that n must be geometric and $G(z)$ logistic. Baringhaus's theorem has recently been extended by Voorn (1987), who relaxes the assumption on the distribution of n. However, both Baringhaus and Voorn use infinitely many $n = n(N)$, and, thus, there is no direct comparison between Theorem 7.4.1 and their result. Yet, it is significant that one does not have to assume a priori that n is a geometric variable. Below we quote one of the results of Voorn (1987) and give only an outline of his argument.

Theorem 7.4.2 Let $G(x)$ be a nondegenerate symmetric distribution function, and let $n(N)$, $N \geq 1$, be a sequence of sample sizes with distributions $p_k(N)$, defined in (7.4.1), such that, as $N \to +\infty$, $p_1(N) \to 1$. Furthermore, we assume that, for at least one $k \geq 2$, $p_k(N) > 0$ (k may depend on N). Then, with some constants $A(N)$ and $B(N) > 0$, $(U_{n(N):n(N)} - A(N))/B(N)$ and U_1 are identically distributed, i.e., have the common distribution function $G(x)$, if, and only if, each $\{p_k(n)\}$ is geometric and $G(x)$ is logistic.

Remarks. 1. The property that $(U_{n:n} - A)/B$ and U_1 are identically distributed will be referred to as $G(x)$ is max-stable with respect to the distribution $\{p_k\}$.

2. Note that the distributions

$$F_m(x) = (1 + x^{-m})^{-1}, \qquad x > 0,$$

or their extensions by a location and scale parameters, also are max-stable with respect to the geometric distributions. These are not symmetric, and, in regard to Theorem 7.4.1, note that the normalizing constants $A = 0$ and $B = p^{-1/m}$. $F_m(x)$ can be obtained from the logistic distribution $F^*(x)$ by a logarithmic transformation (the name log-logistic is, in fact, being used for $F_m(x)$).

3. The extended logistic distribution functions (see Balakrishnan and Leung, 1988 and Chapter 9 for details)

$$F_t(x) = \left\{1 + \exp\left[-\frac{(x - \mu)}{\sigma}\right]\right\}^{-1/t}, \qquad t \geq 2 \text{ integer},$$

are max-stable with respect to the sample size distributions $p_{1+jt} = c_{j,t}(1 - p^t)p^{jt}$, $0 < p < 1$, $j \geq 0$ integer, and $p_k = 0$ otherwise. Note that $F_t(x)$ is not symmetric and the distributions $\{p_k\}$ are not geometric.

Outline of the Proof of Theorem 7.4.2. We follow Voorn's proof, but leave out the technical details.

By max-stability, after suppressing the parameter N,

$$G\left(\frac{x - A}{B}\right) = G(x)[p_1 + p_2G(x) + \cdots],$$

for which we get that $0 < G((x - A)/B) < 1$ if, and only if, $0 < G(x) < 1$. Hence, if $\alpha = \inf\{x: G(x) > 0\}$ and $\omega = \sup\{x: G(x) < 1\}$ were finite, we would have that $\alpha = (\alpha - A)/B$ and $\omega = (\omega - A)/B$, implying that $A = 0$ and $B = 1$. This, however, contradicts the equation of max-stability above for nondegenerate $G(x)$. Therefore, since $G(x)$ is symmetric, $\omega = -\alpha = +\infty$; i.e., $0 < G(x) < 1$ for all x. Furthermore, $B = 1$, because otherwise, for $x = A/(1 - B)$, we would have $x = (x - A)/B$, implying, via the equation of max-stability, that $G(x) = 0$. One can then prove, and its details are omitted here, that $G(x)$ is differentiable and that there are constants $c_j \geq 0$ with $c_2 + c_3 + \cdots < +\infty$ such that

$$g(x) = \frac{dG(x)}{dx} = c_2[G(x) - G^2(x)] + c_3[G(x) - G^3(x)] + \cdots$$

for almost all x. Choose $x_1 < x_2$ with $G(x_1) = 1 - G(x_2)$ and such that the expression above applies to $g(x_i)$, $i = 1, 2$. By symmetry, $g(x_1) = g(x_2)$. On the other hand, since $g(x) > 0$ for (almost) all x, $G(x_1) < G(x_2)$, and thus, for $j \geq 3$,

$$\begin{aligned} G(x_1) - G^j(x_1) &= G(x_1)[1 - G(x_1)] \\ &\quad \times [1 + G(x_1) + G^2(x_1) + \cdots + G^{j-2}(x_1)] < G(x_1) \\ &\quad \times [1 - G(x_1)][1 + G(x_2) + \cdots + G^{j-2}(x_2)] = G(x_2) - G^j(x_2), \end{aligned}$$

the last equation being due to $1 - G(x_1) = G(x_2)$; i.e.,

$$G(x_1)[1 - G(x_1)] = G(x_1)G(x_2) = [1 - G(x_2)]G(x_2).$$

Consequently, $c_j = 0$ for all $j \geq 3$. That is,

$$g(x) = c_2G(x)[1 - G(x)],$$

which is our basic equation (7.2.1) for characterizing $F(x, \mu, \sigma)$. That is, $G(x)$ is logistic. Now, we know that $U_{n:n} - A$, for some A (we have seen that $B = 1$), has the same logistic distribution $G(x)$ as U_1; that is,

$$G(x) = \sum_{k=1}^{+\infty} G^k(x + A)p_k,$$

but, as was established earlier, $U_{n:n} + \log p$ also has the same distribution as U_1 if $G(x) = F^*(x)$ (we can always bring a logistic distribution into this form by a change of variable) and if n is geometric. That is,

$$G(x) = \sum_{k=1}^{+\infty} G^k(x - \log p)p(1 - p)^{k-1}.$$

Comparing the two expansions, we choose p so that $A = -\log p$. Then we have two Taylor expansions of $G(x)$ in the variable $t = G(x + A)$, and since Taylor's expansion is unique, the distribution p_k, $k \geq 1$, of n is geometric. This completes the proof.

The logistic distribution appears among the limiting distributions in the following theorem of Barndorff-Nielsen (1964) and Mogyoródi (1967). Note that the random sample size n does not have to be independent of the sample elements U_j.

Theorem 7.4.3 Assume that, for fixed sample size k, there are constants a_k and $b_k > 0$ such that, as $k \to +\infty$,

$$\lim P(U_{k:k} \leq a_k + b_k x) = e^{-e^{-x}}, \qquad -\infty < x < +\infty.$$

Let the random sample sizes $n(N)$, which are not necessarily independent of the U_j, be such that, as $N \to +\infty$,

$$\lim P(n(N) \leq Nz) = A(z), \qquad z > 0,$$

where $A(z)$ is a proper distribution function. Then, as $N \to +\infty$,

$$\lim P(U_{n(N):n(N)} \leq a_N + b_N x) = \int_0^{+\infty} e^{-ze^{-x}}\, dA(z). \qquad (7.4.8)$$

The limiting distribution on the right-hand side is logistic if, and only if, $A(z) = 1 - e^{-az}$, $a > 0$.

We do not prove the whole theorem, only the characterization part contained in the last sentence, which has not yet been emphasized in the literature in this generality. For the main part of the theorem, see, in addition to the quoted papers, Chapter 6 of Galambos (1987).

Now, a straight calculation yields that, if $A(z)$ is exponential, then the right-hand side of (7.4.8) equals

$$H(x) = \int_0^{+\infty} e^{-ze^{-x}} a e^{-az}\, dz = \frac{a}{a + e^{-x}} = \frac{1}{1 + e^{-x-\log a}}.$$

Conversely, if the right-hand side of (7.4.8) is logistic, then, by the computation above, one candidate for $A(z)$ is the exponential distribution, $A_1(z)$, say. If another distribution function $A_2(z)$ would result in the same logistic limiting distribution in (7.4.8), we would then have

$$\int_0^{+\infty} e^{-ze^{-x}}\, dA_1(z) = \int_0^{+\infty} e^{-ze^{-x}}\, dA_2(z).$$

However, we can recognize that the integrals above are the Laplace transforms of $A_1(z)$ and $A_2(z)$, respectively, at the variable $t = e^{-x}$. By the uniqueness of Laplace transforms (see Galambos, 1988, p. 126), we must have $A_1(z) = A_2(z)$, i.e., exponential.

In the special case, when the distribution of $n(N)$ for each N is geometric and its parameter $p = p(N) \to 0$ as $N \to +\infty$, then, with $[y]$ signifying the integer part of y,

$$P\left(n(N) > \frac{z}{p(N)}\right) = (1 - p(N))^{[z/p(N)]}$$

$$= \exp\left\{\frac{z \log(1 - p(N))}{p(N)} + o(1)\right\} = e^{-z} + o(1),$$

so the limit in (7.4.8) is logistic whenever the asymptotic distribution of $(U_{k:k} - a_k)/b_k$ is $\exp(-e^{-x})$. Gnedenko and Gnedenko (1982) obtain that this last condition is also necessary for (7.4.8) to be logistic if $n(N)$ is geometric.

Implicit in the preceding computations is the fact that the logistic distribution is a recognizable mixture of the extreme value distribution $H_3(x) = \exp(-e^{-x})$ and the exponential distribution, a fact first pointed out by Dubey (1969). This same fact is behind the appearance of the logistic distribution in extreme value theory for exchangeable variables, a theory not yet well developed but significant for Bayesian statisticians (see Chapter 3 in Galambos, 1987, in particular, Example 3.6.1 on p. 190, which can be extended into a characterization theorem similar to Theorem 7.4.3, in which one has to replace "random sample" by "exchangeable" or "Bayesian observations," and "random sample size" by "random parameter value").

7.5 MISCELLANEOUS CHARACTERIZATIONS

In this section, we present some unrelated characterizations of the logistic distribution, which results do not fit into the previous sections either.

We start with a result of Galambos and Kotz (1978).

Theorem 7.5.1 Let the distribution function $G(x)$ of the random variable U be continuous and symmetric about the origin. Then $G(x)$ is logistic if, and only if, for all $x > 0$,

$$P(-x \le U | U \le x) = 1 - e^{-ax} \quad \text{with some } a > 0. \tag{7.5.1}$$

Proof. If $G(x) = F(x, 0, 1/a)$, then $G(-x) = 1 - G(x)$, and the conditional probability in (7.5.1) equals

$$\frac{F(x, 0, 1/a) - F(-x, 0, 1/a)}{F(x, 0, 1/a)} = \frac{2F(x, 0, 1/a) - 1}{F(x, 0, 1/a)} = 1 - e^{-ax},$$

so the theorem is true for the symmetric logistic distributions. Conversely, let $G(-x) = 1 - G(x)$, $x > 0$, and assume that (7.5.1) holds. Then

$$1 - e^{-ax} = \frac{G(x) - G(-x)}{G(x)} = \frac{2G(x) - 1}{G(x)}, \qquad x > 0,$$

which yields $G(x) = F(x, 0, 1/a)$ for $x > 0$. Since $G(x)$ is assumed to be symmetric about zero, the equation extends to all x, which completes the proof.

Galambos and Kotz (1978) also show that, among the symmetric distributions, only the logistic can satisfy the following extended form of the lack of memory: for all $x, y \geq 0$,

$$\frac{1 - G(x + y)}{[1 - G(x)][1 - G(y)]} = \frac{G(x + y)}{G(x)G(y)}.$$

As a matter of fact, by symmetry, one gets that, for all $x > 0$,

$$1 - H(x) = P(U \leq -x | U \leq x) = \frac{1 - G(x)}{G(x)},$$

and thus the assumption above yields

$$1 - H(x + y) = [1 - H(x)][1 - H(y)], \qquad \text{for all } x, y \geq 0.$$

This is the ordinary lack-of-memory equation, which characterizes the exponential distribution. That is, (7.5.1) holds, and Theorem 7.5.1 yields the claim.

Next, we turn to a result of George and Mudholkar (1982).

Theorem 7.5.2 Let $\varphi(t)$ be the characteristic function of the random variable U with distribution function $G(x)$. Assume that the absolute value of $t\varphi(t)$ is integrable on the whole real line. Then $G(x)$ is the convolution of the distribution function $G_{1:2}(x)$ of $U_{1:2}$ and the unit exponential distribution $E(x) = 1 - e^{-x}, x \geq 0$, if, and only if, $G(x) = F^*(x)$.

Proof. First, note that the characteristic function $\varphi^*(t)$ of $F^*(x)$ can be expressed by the gamma function as

$$\varphi^*(t) = \int_{-\infty}^{+\infty} e^{itx} e^{-x}(1 + e^{-x})^{-2}\, dx = \Gamma(1 - it)\Gamma(1 + it),$$

and the characteristic function $\varphi_{1:2}(t)$ of $U_{1:2}$ has the form $\varphi_{1:2}(t) = (1 - it)\varphi^*(t)$. Since the characteristic function of $E(x)$ is $1/(1 - it)$ and, for convolutions, characteristic functions multiply, we have the claimed property for $G(x) = F^*(x)$.

Conversely, let $G(x)$ be unknown, and assume that the claimed convolution equation holds; that is,

$$\varphi_{1:2}(t) = (1 - it)\varphi(t).$$

By assumption, the right-hand side is absolutely integrable. Consequently, the density of both $U_{1:2}$ and U exist (see Galambos, 1988, p. 101); furthermore, even the density of U is differentiable. Thus, by the inversion formula for densities (loc. cit.),

$$\frac{1}{2\pi}\int_{-\infty}^{+\infty} (1 - it)\varphi(t)e^{-itx}\, dt = g_{1:2}(x) = 2(1 - G(x))G'(x),$$

while, by splitting the left-hand side into two integrals, the same inversion formula yields

$$\frac{1}{2\pi}\int_{-\infty}^{+\infty}(1 - it)\varphi(t)e^{-itx}\,dt = G'(x) + G''(x).$$

Thus, by equating the right-hand sides and integrating, we get

$$c - (1 - G(x))^2 = G(x) + G'(x), \qquad c \text{ constant}.$$

By letting $x \to -\infty$, we get that $c = 1$. Hence, we once more arrived at the basic differential equation (7.2.1)

$$G(x)(1 - G(x)) = G'(x),$$

whose only solution (among distribution functions) is $G(x) = F^*(x)$. The proof is completed.

George and Mudholkar (1982) establish a similar result when the minimum $U_{1:2}$ is replaced by the maximum $U_{2:2}$, and these same authors, George and Mudholkar (1981), extend these results to larger sample sizes.

Finally, we quote a recent result of Johnson and Kotz (1989), which is a characterization theorem of a general nature. It can be specified to characterize the logistic distribution as well as (any) other distribution.

Theorem 7.5.3 Let U and V be two random variables with the same support S, and let Z be a random variable taking the nonnegative integers and such that, for all $t \in S$,

$$P(Z = k|U = t) = P(Z = k|V = t) = u(t)h^k(t), \qquad \text{all } k \geq 0,$$

with some positive functions $u(t)$ and $h(t)$, where $h(t)$ is strictly monotonic. Then U and V are identically distributed.

Proof. By the formula

$$P(Z = k) = E[P(Z = k|U)] = E[P(Z = k|V)],$$

we have

$$\int_{-\infty}^{+\infty} u(t)h^k(t)\,dG_U(t) = \int_{-\infty}^{+\infty} u(t)h^k(t)\,dG_V(t). \qquad (7.5.2)$$

The case $k = 0$ yields that, with a constant $c > 0$, and with $dG_U^*(t) = cu(t)\,dG_U(t)$ and $dG_V^*(t) = cu(t)\,dG_V(t)$, (7.5.2) can be interpreted as all moments of $h(U^*)$ and $h(V^*)$ coincide, where the distribution functions of U^* and V^* are $G_U^*(\cdot)$ and $G_V^*(\cdot)$, respectively. Since $0 < h(t) < 1$ (recall its definition), the stated moments uniquely determine the distributions of

$h(U^*)$ and $h(V^*)$. By the monotonicity of h, we thus have that $G_U^*(\cdot) = G_V^*(\cdot)$, from which $G_U(t) = G_V(t)$ now follows. This completes the proof.

The way in which such a theorem can be applied is that one computes the conditional distributions $P(Z = k|U = t)$ for some Z and U logistic. Now, if these probabilities are of the form $u(t)h^k(t)$ for all $t \in S$ and all $k \geq 0$, and if this same distribution obtains for another random variable V, then V is logistic. The reader is invited to make up such examples.

REFERENCES

Arnold, B. C. (1983). *Pareto Distributions*. Int. Co-op. Publ. House, Baltimore.

Arnold, B. C. and Balakrishnan, N. (1989). *Relations, Bounds and Approximations for Order Statistics*, Lecture Notes in Statistics, Vol. 53, Springer-Verlag, New York.

Arnold, B. C. and Meeden, G. (1975). Characterization of distributions by sets of order statistics. *Ann. Statist.*, *3*, 754–758.

Azlarov, T. A. and Volodin, N. A. (1982). *Characterization Problems Associated with the Exponential Distribution* (in Russian), Fan, Tashkent. English transl., 1986, Springer-Verlag, Berlin.

Balakrishnan, N. and Leung, M. Y. (1988). Order statistics from the Type I generalized logistic distribution. *Commun. Statist.—Simul. Comput.*, *17*(1), 25–50.

Baringhaus, L. (1980). Eine simultane Charakterisierung der geometrischen Verteilung und der logistischen Verteilung. *Metrika*, *27*, 237–242.

Barndorff-Nielsen, O. (1964). On the limit distribution of the maximum of a random number of independent random variables. *Acta Math. Acad. Sci. Hungar.*, *15*, 399–403.

Dubey, S. D. (1969). A new derivation of the logistic distribution. *Naval Res. Logist. Quart.*, *16*, 37–40.

Galambos, J. (1975). Characterizations of probability distributions by properties of order statistics. I, in *Statistical Distributions in Scientific Work*, Vol. 3, 71–88, Reidel, Dordrecht.

Galambos, J. (1976). A remark on the asymptotic theory of sums with random size. *Math. Proc. Cambridge Philos. Soc.*, *79*, 531–532.

Galambos, J. (1987). *The Asymptotic Theory of Extreme Order Statistics*, 2nd ed. Krieger, Melbourne, FL.

Galambos, J. (1988). *Advanced Probability Theory*. Marcel Dekker, New York.

Galambos, J. and Kotz, S. (1978). *Characterizations of Probability Distributions*, Lecture Notes in Math., Vol. 675, Springer-Verlag, Berlin.

George, E. O. and Mudholkar, G. S. (1981). Some relationships between the logistic and the exponential distributions, in *Statistical Distributions in Scientific Work*, Vol. 4, 401–409, Reidel, Dordrecht.

George, E. O. and Mudholkar, G. S. (1982). On the logistic and exponential laws. *Sankhya A*, *44*, 291–293.

Gnedenko, B. V. and Gnedenko, D. V. (1982). On Laplace and logistical distributions as limits in the theory of probability (in Russian). *Serdika Bolgarska Math.*, *8*, 229–234.

Govindarajulu, Z., Huang, J. S., and Saleh, A. K. M. E. (1975). Expected value of the spacings between order statistics, in *Statistical Distributions in Scientific Work*, Vol. 3, 143–147, Reidel, Dordrecht.

Hoeffding, W. (1953). On the distribution of the expected values of the order statistics. *Ann. Math. Statist.*, *24*, 93–100.

Huang, J. S. (1974). On a characterization of the exponential distribution by order statistics. *J. Appl. Prob.*, *11*, 605–608.

Huang, J. S. (1975). Characterizations of distributions by the expected values of the order statistics. *Ann. Inst. Statist. Math.*, *27*, 87–93.

Huang, J. S. (1989). Moment problem of order statistics: a review. *Intern. Statist. Rev.*, *57*, 59–66.

Huang, J. S. (1978). A note on Bernstein and Müntz-Szasz theorems with applications to the order statistics. *Ann. Inst. Statist. Math.*, *30*, 167–176.

Johnson, N. L. and Kotz, S. (1970). *Continuous Univariate Distributions*, Vol. 2. Wiley, New York.

Johnson, N. L. and Kotz, S. (1989). Characterization based on conditional distributions, Technical Report, Univ. North Carolina, Chapel Hill (3 pages).

Kadane, J. B. (1974). A characterization of triangular arrays which are expectations of order statistics. *J. Appl. Prob.*, *11*, 413–416.

Kakosyan, A. V. Klebanov, L. B., and Melamed, J. A. (1984). *Characterization of Distributions by the Method of Intensively Monotone Operators*, Lecture Notes in Math., Vol. 1088, Springer-Verlag, Berlin.

Lin, G. D. (1984). A note on equal distributions. *Ann. Inst. Statist. Math.*, *36*, 451–453.

Lin, G. D. (1988). Characterizations of distributions via relationships between two moments of order statistics. *J. Statist. Plan. Inf.*, *19*, 73–80.

Madreimov, I. and Petunin, Y. I. (1983). A characterization of the uniform distribution with the aid of order statistics. *Theor. Prob. Math. Statist.*, *27*, 105–110.

Mogyoródi, J. (1967). On the limit distribution of the largest term in the order statistics of a sample of random size (in Hungarian). *Magyar Tud. Akad. Mat. Fiz. Oszt. Kozl.*, *17*, 75–83.

Voorn, W. J. (1987). Characterization of the logistic and log-logistic distributions by extreme value related stability with random sample size. *J. Appl. Prob.*, *24*, 838–851.

8

Translated Families of Distributions

Norman L. Johnson
University of North Carolina at Chapel Hill, Chapel Hill, North Carolina

Pandu R. Tadikamalla
University of Pittsburgh, Pittsburgh, Pennsylvania

8.1 INTRODUCTION

Recall (from Chapter 2) that the standard logistic $L(0, \pi^2/3)$ variable Z has cdf

$$F_Z^*(z) = (1 + e^{-z})^{-1}, \qquad -\infty < z < \infty, \tag{8.1.1}$$

and the pdf is

$$f_Z^*(z) = e^{-z}(1 + e^{-z})^{-2}, \qquad -\infty < z < \infty. \tag{8.1.2}$$

The percentile function (inverse cdf) is

$$z = \ln\left\{\frac{F_Z^*(z)}{1 - F_Z^*(z)}\right\}. \tag{8.1.3}$$

If $Y = g(Z)$ is a monotonic increasing function of Z (and $Z = g^{-1}(Y)$ is also a monotonic increasing function of Y), the cdf of Y is

$$F_Y(y) = [1 + \exp\{-g^{-1}(y)\}]^{-1}. \tag{8.1.4}$$

So if it is required to form a random variable Y with any desired cdf, $H(y)$ say, this can be done by choosing $g(Z)$ such that

$$H(y) = [1 + \exp\{-g^{-1}(y)\}]^{-1}; \tag{8.1.5}$$

that is,

$$g^{-1}(y) = \ln\left\{\frac{H(y)}{1 - H(y)}\right\}. \tag{8.1.6}$$

Conversely, given a random variable Y with cdf $H(y)$, it is transformed to a standard logistic $L(0, \pi^2/3)$ variable by the transformation

$$Z = \ln\left\{\frac{H(Y)}{1 - H(Y)}\right\} \tag{8.1.7}$$

The Johnson systems of S-distributions (Johnson, 1949) were developed by ascribing the unit normal distribution to

$$Z = \gamma + \delta \ln Y \quad (0 < Y) \qquad \text{for } S_L \text{ (log normal) system;} \tag{8.1.8}$$

$$Z = \gamma + \delta \ln\{Y(1 - Y)\} \quad (0 < Y < 1) \qquad \text{for } S_B \text{ system;} \tag{8.1.9}$$

$$Z = \gamma + \delta \sinh^{-1} Y \quad (-\infty < Y < \infty) \qquad \text{for } S_U \text{ system.} \tag{8.1.10}$$

Analogous to these S-systems, the Tadikamalla and Johnson systems of L-distributions (Tadikamalla and Johnson, 1984) were developed by ascribing the standard logistic distribution to the Z's defined in (8.1.8)–(8.1.10). These new systems were denoted by L_L, L_B, and L_U, respectively.

The parameter δ is always positive; both γ and δ are shape parameters. Two other parameters, ξ (location) and λ (scale), are introduced by writing

$$Y = \frac{X - \xi}{\lambda}.$$

In view of the closeness in shapes of the logistic and normal distributions, it is to be expected that the new systems will exhibit some similarity in shape to S_L, S_B, and S_U. In fact, since (Johnson and Kotz, 1970, p. 6)

$$\left|\left\{1 + \exp\left(\frac{-\pi x}{\sqrt{3}}\right)\right\}^{-1} - \Phi\left(\frac{16x}{15}\right)\right| < 0.01 \qquad \text{for all } x, \tag{8.1.11}$$

where $\Phi(u) = (\sqrt{2\pi})^{-1} \int_{-\infty}^{u} e^{-t^2/2}\, dt$ the difference between the cdf's of $S_{L,B,U}$ with parameters γ, δ, ξ, λ and those of $L_{L,B,U}$ with parameters $\psi\gamma$, $\psi\delta$, ξ, λ, where $\psi = (\pi/\sqrt{3})(15/16) \doteq 1.7$, cannot exceed 0.01. Although this gives good agreement in the central parts of the distributions, there can be gross disparities in the tails. Sections 8.3–8.5 contain examples showing that when an S-curve and an L-curve are fitted to a given data set; we do not necessarily find that (i) the ratios L/S for γ and δ are each about 1.7 or (ii) the values of ξ and λ are about the same.

The very simple formula (8.1.3) for z in terms of $F_Z^*(z)$ is an obvious practical advantage of the L-system. Percentile points of fitted distributions can be obtained very simply.

8.2 L_L (LOG-LOGISTIC) DISTRIBUTIONS

There are *log-logistic* distributions, which were studied by Shah and Dave (1963). The pdf is

$$f_Y(y) = \delta e^{\gamma} y^{\delta-1}(1 + e^{\gamma} y^{\delta})^{-2} \qquad (y \geq 0; \delta > 0). \tag{8.2.1}$$

The distributions belong to Burr's (1942) Type XII family of distributions. Dubey (1966) called them *Weibull-exponential* distributions and fitted them to business failure data presented by Lomax (1954).

The pdf is unimodal. If $\delta \geq 1$, the mode is at $y = 0$ (giving a reversed J-shaped curve); if $\delta > 1$, the mode is at $y = e^{-\gamma}(\delta - 1)/(\delta + 1)$. The cdf is

$$F_Y(y) = (1 + e^{-\gamma} y^{-\delta})^{-1}. \tag{8.2.2}$$

This equation can be inverted, giving y as an explicit function of $F_Y(y)$:

$$y = e^{-\Omega}\left\{\frac{F_Y(y)}{1 - F_Y(y)}\right\}^{1/\delta}, \tag{8.2.3}$$

where $\Omega = \gamma/\delta$.

In particular, the median is at $y = e^{-\Omega}$, and the lower and upper quartiles are at $y = 3^{-1/\delta}e^{-\Omega}$ and $y = 3^{1/\delta}e^{-\Omega}$, respectively.

The rth moment of Y about zero is

$$\mu_r'(Y) = \int_{-\infty}^{\infty} e^{r(z-\gamma)/\delta} e^{z}(1 + e^{z})^{-2}\, dz.$$

Putting $u = e^{z}(1 + e^{z})^{-1}$ so that $du/dz = e^{z}(1 + e^{z})^{-2}$ and $e^{z} = u/(1 - u)$ gives

$$\begin{aligned} \mu_r'(Y) &= e^{-r\Omega} \int_0^1 u^{r/\delta}(1 - u)^{-r/\delta}\, du \\ &= e^{-r\Omega} B(1 + r\delta^{-1}, 1 - r\delta^{-1}) \\ &= e^{-r\Omega} r\theta \csc r\theta, \end{aligned} \tag{8.2.4}$$

with $\theta = \pi/\delta$, provided $r < \delta$. (If $r \geq \delta$, $\mu_r'(Y)$ is infinite.)

In particular,

$$E[Y] = e^{-\Omega}\theta \csc \theta \tag{8.2.5}$$

and

$$\text{var}(Y) = e^{-2\Omega}\left(2\int \csc 2\theta - \theta^2 \csc^2 \theta\right)$$
$$= e^{-2\Omega}\theta(\tan \theta - \theta) \csc^2\theta. \qquad (8.2.6)$$

The rth central moment $\mu_r(Y)$ is a multiple (depending on θ but not on Ω) of $\exp(-r\Omega)$, and so the moment ratios $\mu_r/\mu_2^{r/2}$, particularly $\sqrt{\beta_1} = \mu_3/\mu_2^{3/2}$ and $\beta_2 = \mu_4/\mu_2^2$, do not depend on Ω, but only on θ. The expression for $\sqrt{\beta_1}(Y)$ is

$$\sqrt{\beta_1}(Y) = \frac{3 \csc 3\theta - 6\theta \csc 3\theta \csc \theta + 2\theta^2 \csc^3 \theta}{\theta^{1/2}(\tan \theta - \theta)^{3/2} \csc^3 \theta} \qquad (8.2.7)$$

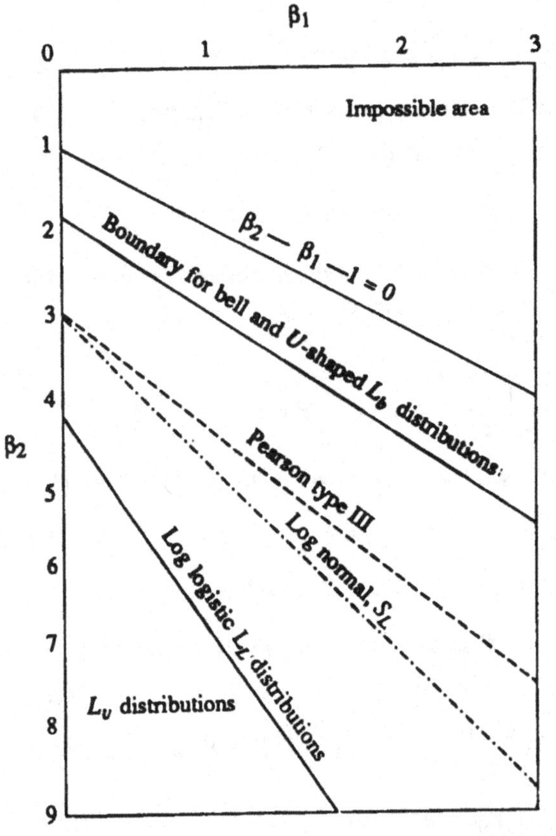

Fig. 8.2.1 β_1, β_2 region for L_U-, L_L-, L_B-distributions.

Table 8.2.1 The Log-Logistic Line

$\sqrt{\beta_1}$	β_2	δ	$\sqrt{\beta_1}$	β_2	δ	$\sqrt{\beta_1}$	β_2	δ
0.05	4.21	174.2	0.55	4.96	16.26	1.1	7.47	8.73
0.10	4.22	87.14	0.60	5.11	14.98	1.2	8.16	8.13
0.15	4.26	58.16	0.65	5.27	13.90	1.3	8.94	7.63
0.20	4.30	43.69	0.70	5.45	12.98	1.4	9.81	7.21
0.25	4.35	35.02	0.75	5.65	12.18	1.5	10.79	6.86
0.30	4.42	29.26	0.80	5.86	11.49	1.6	11.88	6.55
0.35	4.50	25.15	0.85	6.08	10.88	1.7	13.11	6.28
0.40	4.60	22.08	0.90	6.32	10.35	1.8	14.47	6.04
0.45	4.71	19.70	0.95	6.58	9.87	1.9	16.00	5.84
0.50	4.83	17.80	1.00	6.86	9.45	2.0	17.71	5.65

(For $\sqrt{\beta_1} = 0$, $\beta_2 = 4.2$ and $\delta = \infty$).

The (β_1, β_2) points $(\beta_1 = (\sqrt{\beta_1})^2)$ lie on a line through the "logistic point" (0, 4.2), which is approached as $\theta \to 0$ $(\delta \to \infty)$. Figure 8.2.1 shows the L_L(log-logistic) line and the S_L (log-normal) line.

Table 8.2.1 gives numerical values of β_2 for selected values of β_1 for points on the L_L line.

The (β_1, β_2) points for $L_B(L_U)$ distributions lie "above" ("below") the L_L line in Figure 8.2.1. This is analogous to the situation for S_B and S_U relative to the S_L line. However, the L_L line differs from the S_L line in an important respect. For the S_L line, $\beta_1 \to \infty$ as $\beta_2 \to \infty$, but this is not so for the L_L line, for which

$$\beta_1 \to \frac{4(12 - 6\pi + \pi^2)^2}{\pi(4 - \pi)^3} = 18.36 \qquad \text{as } \beta_2 \to \infty.$$

(This is the limiting value of β_1 as $\theta \to \pi/4$.)

8.3 THE L_B SYSTEM

The pdf of Y corresponding to transformation (8.1.10), with Z a standard logistic variable, is

$$f_Y(y) = e^{\gamma} y^{\delta-1}(1 - y)^{\delta-1}\{(1 - y)^{\delta} + e^{\gamma} y^{\delta}\}^{-2} \qquad (0 < y < 1) \quad (8.3.1)$$

If $\delta = 1$, this is a *power function* distribution, with pdf

$$f_Y(y) = e^{\gamma}(1 - y + e^{\gamma} y)^{-2} \qquad (0 < y < 1). \qquad (8.3.2)$$

If, in addition, $\gamma = 0$ we have a standard uniform distribution with

$$f_Y(y) = 1 \qquad (0 < y < 1). \tag{8.3.3}$$

In contrast to S_B-distributions, no L_B-curves are multimodal. There is a single mode (if $\delta > 1$) or antimode (if $\delta < 1$) at the unique value of y between 0 and 1 satisfying the equation

$$e^{\gamma} = \left(\frac{1-y}{y}\right)^{\delta} \frac{\delta - 1 + 2y}{\delta + 1 - 2y}. \tag{8.3.4}$$

The right-hand side of (8.3.4) is equal to 1 when $y = 1/2$, whatever the value of δ. The derivative of the logarithm of the right-hand side is

$$\delta(1 - \delta^2)y^{-1}(1 - y)^{-1}\{\delta^2 - (1 - 2y)^2\}. \tag{8.3.5}$$

If $\delta > 1$, this is negative, and so the mode is at $y \lesseqgtr 1/2$ according as $\gamma \lesseqgtr 0$.

For $\delta < 1$, the curve is U-shaped, and the antimode is between $(1 - \delta)/2$ and $(1 + \delta)/2$. Since (8.3.5) is positive for $(1 - \delta)/2 < y \times (1 + \delta)/2$, the antimode is between $(1 - \delta)/2$ and $1/2$ for $\gamma < 0$, and between $1/2$ and $(1 + \delta)/2$ for $\gamma > 0$.

The (β_1, β_2) line corresponding to $\delta = 1$ (power function distributions), which is the lower boundary of the region with U-shaped curves, is shown in Figure 8.2.1. Apart from the power function distributions, there are no J- or reverse J-shaped curves in the L_B-system.

The cdf corresponding to (8.3.1) is

$$F_Y(y) = \left\{1 + e^{-\gamma}\left(\frac{y}{1-y}\right)^{-\delta}\right\}^{-1} \qquad (0 < y < 1). \tag{8.3.6}$$

This equation can be inverted, giving y as an explicit function of $F_Y(y)$, leading to

$$y = \left[1 + e^{\Omega}\left\{\frac{F_Y(y)}{1 - F_Y(y)}\right\}^{-1/\delta}\right]^{-1}, \tag{8.3.7}$$

with $\Omega = \gamma/\delta$ as in Section 8.2.

In particular, the median is at $y = (1 + e^{\Omega})^{-1}$, and the lower and upper quartiles are $(1 + 3^{1/\delta}e^{\Omega})^{-1}$ and $(1 + 3^{-1/\delta}e^{\Omega})^{-1}$, respectively.

The rth moment about zero is

$$\mu_r'(Y) = \int_{-\infty}^{\infty} \{1 + e^{-(z-\gamma)/\delta}\}^{-r} e^{z}(1 + e^{z})^{-2}\, dz$$

$$= \int_0^1 \{1 + e^{\Omega}(1 - u)^{1/\delta}u^{-1/\delta}\}^{-r}\, du. \tag{8.3.8}$$

Generally, the integral in (8.3.8) must be evaluated by quadrature. (Of course, for some special values of δ, such as $\delta = 1$, explicit solutions can be obtained.) Note that expansion of the integrand as a power series in $(1 - u^{-1})^{1/\delta}$ cannot be valid over the whole range of integration. Dichotomy of the interval (0, 1) according as

$$(1 - u^{-1})^{1/\delta} \gtrless e^{\Omega} \qquad (\text{i.e., } u \gtrless (1 + e^{-\Omega})^{-1})$$

leads to valid expansions, but the resulting incomplete beta functions must themselves (in general) be evaluated by quadrature.

8.4 THE L_U-SYSTEM

The pdf of Y is

$$f_Y(y) = \frac{\delta e^{\gamma}}{(y^2 + 1)^{1/2}} \frac{\{y + \sqrt{y^2 + 1}\}^{\delta}}{[1 + e^{\gamma}\{y + \sqrt{y^2 + 1}\}^{\delta}]^2}. \qquad (8.4.1)$$

This curve is unimodal with mode at the unique value of y satisfying the equation

$$\delta[1 - e^{\gamma}\{y + \sqrt{y^2 + 1}\}] = y(y^2 + 1)^{-1/2}. \qquad (8.4.2)$$

The cdf is

$$F_Y(y) = [1 + \exp\{-\gamma - \delta \sinh^{-1} y\}]^{-1}. \qquad (8.4.3)$$

This equation can be inverted, giving y as an explicit function of $F_Y(y)$, namely

$$y = \frac{1}{2}\left[e^{-\Omega}\left\{\frac{F_Y(y)}{1 - F_Y(y)}\right\}^{1/\delta} - e^{\Omega}\left\{\frac{1 - F_Y(y)}{F_Y(y)}\right\}^{1/\delta}\right]. \qquad (8.4.4)$$

The median is at $y = -\sinh \Omega$, and the lower and upper quartiles are

$$\tfrac{1}{2}(3^{-1/\delta}e^{-\Omega} - 3^{1/\delta}e^{\Omega}) = \sinh(-\Omega - \delta^{-1} \log 3)$$

and

$$\sinh(-\Omega + \delta^{-1} \log 3),$$

respectively.

The rth moment of Y about zero is

$$\begin{aligned}
\mu_r' &= 2^{-r} \int_{-\infty}^{\infty} \{e^{(z-\gamma)/\delta} - e^{-(z-\gamma)/\delta}\}^r e^z(1 + e^z)^{-2}\, dz \\
&= 2^{-r} \sum_{j=0}^{r} (-1)^j \binom{r}{j} e^{-(r-2j)\Omega} \int_{-\infty}^{\infty} e^{(r-2j)z/\delta} e^z(1 + e^z)^{-2}\, dz \\
&= 2^{-r} \sum_{j=0}^{r} (-1)^j \binom{r}{j} e^{-(r-2j)\Omega}(r - 2j)\theta \csc(r - 2j)\theta, \qquad (8.4.5)
\end{aligned}$$

provided $r < \delta$. If $r \geq \delta$, $\mu'_r(Y)$ is infinite. When $r = 2j$, $\theta \csc \theta$ is interpreted as

$$\lim_{\theta \to 0} \theta \csc \theta = 1.$$

(As in Section 8.2, $\Omega = \gamma/\delta$ and $\theta = \pi/\delta$.)

Since $(-\alpha)\csc(-\alpha) = \alpha \csc \alpha$, (8.4.5) can be simplified.

For r even,

$$\mu'_r(Y) = 2^{-r}(-1)^{r/2}\binom{r}{r/2} + 2^{-(r-1)} \sum_{j=0}^{r/2-1} (-)^j \binom{r}{j} (r-2j)\theta \csc(r-2j)\theta \cosh(r-2j)\Omega. \quad (8.4.6)$$

For r odd,

$$\mu'_r(Y) = -2^{-(r-1)} \sum_{j=0}^{(r-1)/2} (-1)^j \binom{r}{j} \times (r-2j)\theta \csc(r-2j)\theta \sinh(r-2j)\Omega. \quad (8.4.7)$$

In particular,

$$\begin{aligned}
\mu'_1 &= -\theta \csc \theta \sinh \Omega, \\
\mu'_2 &= \theta \csc 2\theta \cosh 2\Omega - \tfrac{1}{2}, \\
\mu'_3 &= -\tfrac{3}{4}\theta(\csc 3\theta \sinh 3\Omega - \csc \theta \sinh \Omega), \\
\mu'_4 &= \tfrac{1}{2}\theta(\csc 4\theta \cosh 4\Omega - 2 \csc 2\theta \cosh 2\Omega) + \tfrac{3}{8},
\end{aligned} \quad (8.4.8)$$

so that

$$\begin{aligned}
E[Y] &= -\theta \csc \theta \sinh \Omega, \\
\operatorname{var}(Y) &= \tfrac{1}{2}\{(\theta \csc \theta)^2 - 1\} + \tfrac{1}{2}\theta(\tan \theta - \theta)\csc^2 \theta \cosh 2\Omega.
\end{aligned}$$

For β_2 to be finite we must have $\theta < \pi/4$; that is, $\delta > 4$. As $\theta \to \pi/4$,

$$\beta_1(Y) \to \frac{2\pi^2\{(\pi^2 - 6\pi + 12)\sinh 3\Omega + 3(\pi^2 - 2\pi - 4)\sinh \Omega\}^2}{\{(4\pi - \pi^2)\cosh 2\omega + \pi^2 - 8\}^3}.$$

This function increases with Ω, and, as $\Omega \to \infty$, β_1 tends to the limiting value $4(\pi^2 - 6\pi + 12)\pi^{-1}(4 - \pi)^{-3}$ of the L_L line. Thus, L_U distributions also cannot have β_1 in excess of this amount.

8.5 FITTING THE DISTRIBUTIONS

The methods of fitting described by Johnson (1949) for the S-systems are also applicable, with suitable modifications, for the L-systems, with the advantage that the cdf of the logistic is simpler than that of the normal distribution.

Introducing the location and scale parameters ξ and λ by the transformation $Y = (X - \xi)/\lambda$, we obtain a three-parameter family for L_L:

$$Z = \gamma + \delta \ln(X - \xi) \qquad (X > \xi), \tag{8.5.1}$$

and four-parameter systems:

$$Z = \gamma + \delta \ln\left(\frac{X - \xi}{\xi + \lambda - X}\right) \qquad (\xi < X < \xi + \lambda), \tag{8.5.2}$$

$$Z = \gamma + \delta \sinh^{-1}\left(\frac{X - \xi}{\lambda}\right) \qquad (-\infty < X < \infty), \tag{8.5.3}$$

for L_B, L_U, respectively.

We consider fitting by the methods of moments, percentile points, and maximum likelihood. Moments are usually more reliable for distributions with bounded support (such as L_B); order statistics (percentile points) are more reliable when support is unbounded (as for L_U). Unfortunately, moments are more difficult technically for L_B, but order statistics can be used, and they are easier for L_B than for L_U.

8.5.1 Moments

If four parameters are to be fitted, the first four moments of the distribution of X may be equated to those of the fitted curve. It is convenient to use them in the form of the mean (μ_1'), the variance (μ_2), and the moment ratios ($\sqrt{\beta_1}$ and β_2). Since $\sqrt{\beta_1}$ and β_2 are determined by γ, δ, and conversely (for L_B and L_U, as for S_B and S_U), the first step is to determine the values of γ and δ from the specified values of $\sqrt{\beta_1}$ and β_2. This can be done by using special tables (Table 8.2.1 for L_L; Tadikamalla and Johnson, 1982, for L_U) or, for L_U, the very accurate approximation formulas of Bowman and Shenton (1981).

Once γ and δ have been determined, values of ξ and λ are obtained from the equations

$$\begin{aligned} \text{mean } X &= \xi + \lambda E[Y|\gamma, \delta], \\ \text{var}(X) &= \lambda^2 \text{ var}(Y|\gamma, \sigma). \end{aligned} \tag{8.5.4}$$

8.5.2 Percentile Points

If four parameters are to be fitted, and if $f(\cdot)$ is monotonic increasing (and $\delta > 0$) and

$$Z = \gamma + \delta f\left(\frac{X - \xi}{\lambda}\right) \qquad (\lambda > 0),$$

then we use four equations of the form

$$Z_{P_i} = \gamma + \delta f\left(\frac{\hat{X}_{P_i} - \xi}{\lambda}\right) \qquad (i = 1, 2, 3, 4), \qquad (8.5.5)$$

where Z_{P_i} is the $100P_i\%$ point of the distribution of Z and $\hat{X}_{P_i}$ is the estimated $100P_i\%$ point of the distribution of X. When the distribution of Z is symmetric about zero (as for both L- and S-systems), solutions of especially simple form are obtained by taking symmetric percentiles (i.e., $P_3 = 1 - P_2$; $P_4 = 1 - P_1$; $P_1 < P_2 < P_3 < P_4$) so that $Z_{P_2} = -Z_{P_3}$ and $Z_{P_1} = -Z_{P_4}$. Then the solutions $\tilde{\xi}$, $\tilde{\lambda}$ of Eqs. (8.5.5) for ξ and λ must be linked by the same equation

$$f\left(\frac{\hat{X}_{P_1} - \tilde{\xi}}{\tilde{\lambda}}\right) + f\left(\frac{\hat{X}_{P_4} - \tilde{\xi}}{\tilde{\lambda}}\right) = f\left(\frac{\hat{X}_{P_2} - \tilde{\xi}}{\tilde{\lambda}}\right) + f\left(\frac{\hat{X}_{P_3} - \tilde{\xi}}{\tilde{\lambda}}\right), \qquad (8.5.6)$$

whatever the distribution of Z. The common value of each side of (8.5.6) is $-2\tilde{\Omega}$.

Another relation is

$$\frac{f\left(\frac{\hat{X}_{1-P_2} - \tilde{\xi}}{\tilde{\lambda}}\right) - f\left(\frac{\hat{X}_{P_2} - \tilde{\xi}}{\tilde{\lambda}}\right)}{f\left(\frac{\hat{X}_{1-P_1} - \tilde{\xi}}{\tilde{\lambda}}\right) - f\left(\frac{\hat{X}_{P_1} - \tilde{\xi}}{\tilde{\lambda}}\right)} = \frac{Z_{1-P_2}}{Z_{1-P_1}}. \qquad (8.5.7)$$

This relation does depend on the actual distribution of Z. Equations (8.5.6) and (8.5.7) suffice to determine $\tilde{\xi}$ and $\tilde{\lambda}$. As will be seen, for L_B-distributions it is possible to find explicit solutions for $\tilde{\xi}$ and $\tilde{\lambda}$.

If ξ and λ are known, two percentile points suffice to determine γ and δ. Knowledge of ξ and/or λ is more usual when a distribution with bounded support (e.g., L_B) is being fitted. Fitting of L_B in these cases is discussed in Section 8.7.

8.5.3 Maximum Likelihood

If ξ and λ are known, the transformed values of $(X - \xi)/\lambda$ can be used, and methods appropriate to fitting logistic distributions (e.g., Chapter 3 or Johnson and Kotz, 1970, Chapter 23) can be applied. If ξ and λ are not known, a series of pairs of values of these parameters can be tried, and the corresponding maximized likelihoods calculated, the maximum likelihood estimates being found by trial and error. Of course, a computer program which maximizes the likelihood function directly can be used. Except in special cases, maximum likelihood calculations are rather lengthy, but not impracticably so.

In the following sections, observed values of X are denoted by $X_1, X_2, \ldots, X_n$. Owen (1988) gives a useful general discussion of points arising in fitting transformed distributions.

8.6 FITTING THE L_L-SYSTEM

If ξ is known, the transformed values $Y_i = \log(X_i - \xi)$ can be used, and the parameters γ and δ fitted by methods appropriate to a logistic distribution.

When ξ is not known, there are three parameter values to be fitted (ξ, γ, and δ). With the method of *percentile points*, it is natural to use the sample median $\hat{X}_{0.5}$ and the lower and upper sample $100P\%$ points, $\hat{X}_P$ and $\hat{X}_{1-P}$, respectively ($P < 0.5$).

From (8.1.3) and (8.1.8),

$$-\ln\left(\frac{1-P}{P}\right) = \tilde{\gamma} + \tilde{\delta}\ln(\hat{X}_P - \tilde{\xi}),$$
$$0 = \tilde{\gamma} + \tilde{\delta}\ln(\hat{X}_{0.5} - \tilde{\xi}), \tag{8.6.1}$$
$$\ln\left(\frac{1-P}{P}\right) = \tilde{\gamma} + \tilde{\delta}\ln(\hat{X}_{1-P} - \tilde{\xi}),$$

which have the unique solutions

$$\tilde{\xi} = \frac{\hat{X}_{0.5}^2 - \hat{X}_{1-P}\hat{X}_P}{2\hat{X}_{0.5} - \hat{X}_{1-P}\hat{X}_P}, \qquad \tilde{\delta} = \frac{-2\ln\{P/(1-P)\}}{\ln\{(\hat{X}_{1-P} - \tilde{\xi})(\hat{X}_P - \tilde{\xi})\}},$$
$$\tilde{\gamma} = -\tilde{\delta}\ln(\hat{X}_{0.5} - \tilde{\xi}), \tag{8.6.2}$$

provided $\tilde{\xi} < \hat{X}_P$.

Of course, several values of P can be used. The resulting estimates $\tilde{\xi}$, $\tilde{\gamma}$, and $\tilde{\delta}$ should be reasonably consistent with each other if an L_L-distribution is suitable. If only one set of values is to be used, $P = 0.10$ (i.e., the sample median, and lower and upper deciles) would seem to be a good choice. Example 8.6.1 contains an illustration of this approach.

For the *method of moments*, the sample value of $\sqrt{\beta_1}$ is equated to Eq. (8.2.7), and the resulting equation solved for θ. Provided the sample value of $\sqrt{\beta_1}$ is positive and does not exceed the limiting vaue of 18.36 (see end of Section 8.2) there will be a unique solution $\theta = \bar{\theta}$ of the equation (otherwise an L_L-curve is not appropriate), where $\bar{\theta}$ is the moment estimator of θ, and $\bar{\delta} = \pi/\bar{\theta}$ is the moment estimator of δ. Table 8.2.1 gives values of δ corresponding to selected values of δ, as well as the corresponding values of β_2 on the L_L line.

Example 8.6.1 As can be seen from Figure 8.1.1, the L_L line is in the Pearson Type IV region. Suppose we want to use an L_L distribution to approximate a standardized Type IV with $\sqrt{\beta_1} = 0.4$, $\beta_2 = 4.6$. From Table 8.2.1 (with extra decimal places), $\delta = 22.080$ yields $\sqrt{\beta_1} = 0.4$, $\beta_2 = 4.5991$. In order to standardize the fitted L_L distribution we have to make the expected value zero and the variance 1. From (8.2.5) and (8.2.6) we obtain the equations

$$\bar{\xi} + e^{-\bar{\gamma}/\bar{\delta}}\bar{\theta} \csc \bar{\theta} = 0 \tag{8.6.3}$$

and

$$e^{-2\bar{\gamma}/\bar{\delta}}\bar{\theta}(\tan \bar{\theta} - \bar{\theta}) \csc^2 \bar{\theta} = 1. \tag{8.6.4}$$

Inserting $\bar{\delta} = 22.08$, $\bar{\theta} = \pi/\bar{\delta} = 0.14228$, we obtain from (8.6.4) that

$$\frac{\bar{\gamma}}{\bar{\delta}} = \frac{1}{2} \ln\{\bar{\theta}(\tan \bar{\theta} - \bar{\theta}) \csc^2 \bar{\theta}\} = -2.49075,$$

whence $\bar{\gamma} = -55.0$ and, from (8.6.3),

$$\bar{\xi} = -12.118.$$

Fitting by percentile points ($\hat{X}_{0.10} = -1.186$, $\hat{X}_{0.5} = -0.046$, $\hat{X}_{0.90} = 1.237$) leads to $\bar{\xi} = -10.245$, $\bar{\gamma} = -43.06$, $\bar{\delta} = 18.54$. Standardized percentiles of the fitted L_L-distributions are compared with the standardized percentiles of the Type IV distribution (obtained by interpolation in the tables of Bouver and Bargmann, 1974) in Table 8.6.1.

Shoukri, Mian, and Tracey (1988) considered probability-weighted moment (PWM) estimators and compared them with maximum likelihood estimators for the case when it is known that $\xi = 0$. The biases and variances

Table 8.6.1 Comparison of Standardized Type IV and L_L Percentiles

%	Type IV	L_L (a)	L_L (b)	%	Type IV	L_L (a)	L_L (b)
0.1	−3.315	−3.285	−3.218	99.9	4.346	4.395	4.557
0.5	−2.574	−2.615	−2.579	99.5	3.182	3.231	3.324
1	−2.262	−2.310	−2.285	99	2.717	2.753	2.882
2.5	−1.854	−1.887	−1.875	97.5	2.127	2.139	2.182
5	−1.532	−1.548	−1.544	95	1.686	1.682	1.709
10	−1.186	−1.185	−1.186	90	1.237	1.223	1.237
25	−0.641	−0.627	−0.633	75	0.587	0.575	0.576
50	−0.046	−0.041	−0.046				

($\sqrt{\beta_1} = 0.4$, $\beta_2 = 4.6$) fitted by (a) moments (b) percentiles $\hat{X}_{0.10}$, $\hat{X}_{0.5}$, $\hat{X}_{0.90}$

of the PWM estimators (of γ and δ) are smaller than those of ML estimators for small sample sizes.

Best linear unbiased estimation of parameters of log-logistic distributions is discussed by Balakrishnan, Malik, and Puthenpura (1987). Ragab and Green (1987) derived nearly best linear unbiased (NBLU) estimators for the logistic distribution based on doubly and singly censored (as well as complete) random samples, and applied them to fitting two-parameter log-logistic distributions.

Applications of order statistics from log-logistic and truncated log-logistic populations are described by Ali and Khan (1987) and Balakrishnan and Malik (1987).

8.7 FITTING THE L_B-SYSTEM

8.7.1 Moments

If all four parameters (ξ, λ, γ, δ) have to be estimated, it is necessary to have special tables (i) to determine γ, δ from $\sqrt{\beta_1}$ and β_2 and (ii) giving $E[Y|\gamma, \delta]$ and $\text{var}(Y|\gamma, \delta)$ (or $\text{S.D.}(Y|\gamma, \delta)$) for given values of γ and δ. Computer programs could be constructed for use instead of tables. (Such programs exist for fitting S_B-curves.)

8.7.2 Percentile Points

Introducing the notation

$$\tilde{\xi}_P = \hat{X}_P - \tilde{\xi},$$

Eqs. (8.5.6) and (8.5.7) become, for L_B-distributions with $P_3 = 1 - P_2$ and $P_4 = 1 - P_1$,

$$\frac{\tilde{\xi}_{P_1}\tilde{\xi}_{1-P_1}}{(\tilde{\lambda} - \tilde{\xi}_{P_1})(\tilde{\lambda} - \tilde{\xi}_{1-P_1})} = \frac{\tilde{\xi}_{P_2}\tilde{\xi}_{1-P_2}}{(\tilde{\lambda} - \tilde{\xi}_{P_2})(\tilde{\lambda} - \tilde{\xi}_{1-P_2})} \tag{8.7.1}$$

and

$$Z_{1-P_1} \ln\left\{\frac{\tilde{\xi}_{P_2}(\tilde{\lambda} - \tilde{\xi}_{1-P_2})}{\tilde{\xi}_{1-P_2}(\tilde{\lambda} - \tilde{\xi}_{P_2})}\right\} = Z_{1-P_2} \ln\left\{\frac{\tilde{\xi}_{P_1}(\tilde{\lambda} - \tilde{\xi}_{1-P_1})}{\tilde{\xi}_{1-P_1}(\hat{\lambda} - \tilde{\xi}_{P_1})}\right\}. \tag{8.7.2}$$

From (8.7.1),

$$\hat{\lambda} = \frac{\tilde{\xi}_{P_1}\tilde{\xi}_{1-P_1}(\tilde{\xi}_{P_2} + \tilde{\xi}_{1-P_2}) - \tilde{\xi}_{P_2}\tilde{\xi}_{1-P_2}(\tilde{\xi}_{P_1} + \tilde{\xi}_{1-P_1})}{\tilde{\xi}_{P_1}\tilde{\xi}_{1-P_1} - \tilde{\xi}_{P_2}\tilde{\xi}_{1-P_2}}. \tag{8.7.3}$$

Trial values of $\tilde{\xi}$ are used to calculate the $\tilde{\xi}_p$'s and, hence, $\hat{\lambda}$ from (8.7.3) and then the ratios

$$g(\tilde{\xi}) = \ln\left\{\frac{\tilde{\xi}_{P_1}(\tilde{\lambda} - \tilde{\xi}_{1-P_1})}{\tilde{\xi}_{1-P_1}(\hat{\lambda} - \tilde{\xi}_{P_1})}\right\} \Big/ \ln\left\{\frac{\tilde{\xi}_{P_2}(\tilde{\lambda} - \tilde{\xi}_{1-P_2})}{\tilde{\xi}_{1-P_2}(\hat{\lambda} - \tilde{\xi}_{P_2})}\right\}.$$

It is desired to find $\bar{\xi}$ so that $g(\bar{\xi}) = Z_{1-P_1}/Z_{1-P_2} = Z_{P_1}/Z_{P_2}$. Example 8.7.1 illustrates this method.

In connection with fitting S_B-curves, Bukač (1972), Mage (1980), and Slifker and Shapiro (1980) have obtained simpler equations by appropriate choices of values of the P's.

In particular, if Z_{1-P_1}/Z_{1-P_2} $(=Z_{P_1}/Z_{P_2}) = r$ is an integer, then (8.7.2) becomes

$$\frac{\bar{\xi}_{P_1}(\bar{\lambda} - \bar{\xi}_{1-P_1})}{\bar{\xi}_{1-P_1}(\bar{\lambda} - \bar{\xi}_{P_1})} = \left\{\frac{\bar{\xi}_{P_2}(\bar{\lambda} - \bar{\xi}_{1-P_2})}{\bar{\xi}_{1-P_2}(\bar{\lambda} - \bar{\xi}_{P_2})}\right\}^r. \tag{8.7.3}$$

This is a polynomial equation in $\bar{\lambda}$ of order $r + 1$, but since $\bar{\lambda} = 0$ is one solution, it is equivalent to a polynomial equation of order r.

If $Z_{1-P_1}/Z_{1-P_2} = 3$, then

$$Z_{P_2} - Z_{P_1} = Z_{1-P_2} - Z_{P_2} = Z_{1-P_1} - Z_{1-P_2}, \tag{8.7.4}$$

and explicit solutions for $\bar{\xi}$ and $\bar{\lambda}$ can be obtained, as shown by Mage (1980), who pointed out that this result applies to *any* set of values P_1, P_2, P_3, P_4 (not necessarily symmetrical) such that

$$Z_{P_2} - Z_{P_1} = Z_{P_3} - Z_{P_2} = Z_{P_4} - Z_{P_3}. \tag{8.7.5}$$

For the symmetrical case, since

$$1 + \exp(-Z_P) = P^{-1},$$

the condition Z_{P_1}/Z_{P_2} is equivalent to

$$P_1^{-1} = 1 + \exp(-3Z_{P_2}) = 1 + (P_2^{-1} - 1)^3;$$

that is,

$$P_1^{-1} - 1 = (P_2^{-1} - 1)^3. \tag{8.7.6}$$

From the background of order statistics in fitting logistic distributions (see David, 1981, p. 200) suitable choices might be $P_1 = 0.05$, $P_2 = 0.273$ (with $Z_{P_1} = -2.9444$; $Z_{P_2} = -0.9815$).

Mage points out that, in the general case with (8.7.5) satisfied, we have

$$\bar{\xi}_{P_{4-i}}\bar{\xi}_{P_{2-i}}(\bar{\lambda} - \bar{\xi}_{P_{3-i}})^2 = (\bar{\lambda} - \bar{\xi}_{P_{4-i}})(\bar{\lambda} - \bar{\xi}_{P_{2-i}})\bar{\xi}^2_{P_{3-i}} \qquad (i = 0, 1), \tag{8.7.6}$$

leading to the following explicit solutions for $\bar{\xi}$ and $\bar{\xi} + \bar{\lambda}$. They are roots of the equation (in α)

$$\alpha^2 + \phi\alpha + \theta = 0, \tag{8.7.7}$$

where

$$\phi = \frac{bf - ce}{bd - ae}; \qquad \theta = \frac{af - cd}{bd - ae},$$

with

$$a = \hat{X}_{P_3}^2 - \hat{X}_{P_2}\hat{X}_{P_4}, \quad b = \hat{X}_{P_2} + \hat{X}_{P_4} - 2\hat{X}_{P_3},$$
$$c = 2\hat{X}_{P_2}\hat{X}_{P_3}\hat{X}_{P_4} - (\hat{X}_{P_2} + \hat{X}_{P_4})\hat{X}_{P_3}^2, \quad d = \hat{X}_{P_2}^2 - \hat{X}_{P_1}\hat{X}_{P_3},$$
$$e = \hat{X}_{P_1} + \hat{X}_{P_3} - 2\hat{X}_{P_2}, \quad f = 2\hat{X}_{P_1}\hat{X}_{P_2}\hat{X}_{P_3} - (\hat{X}_{P_1} + \hat{X}_{P_3})\hat{X}_{P_2}^2.$$

Example 8.7.1 We use the data in Table 6-4 of Hahn and Shapiro (1967). For fitting four parameters we use, in addition to the values $\hat{X}_{0.09} = 0.84$ and $\hat{X}_{0.91} = 1.42$ employed by Hahn and Shapiro (for fitting S_B), the values $\hat{X}_{0.3162} = 0.97$ and $\hat{X}_{0.6838} = 1.18$. These values are chosen to make

$$Z_{0.91} = \ln \frac{0.91}{0.09} = 2.3136 = 3 \ln \frac{0.6836}{0.3162} = 3Z_{0.6838}.$$

The trial-and-error calculations are set out briefly below.

$\tilde{\xi}$	$\tilde{\lambda}(\tilde{\xi})$ (from (8.7.3))	$g(\tilde{\xi})$
0.5	6.32	2.59
0.75	0.95	3.10
0.72	1.12	2.97
0.73	1.06	3.01

Taking the values $\tilde{\xi} = 0.73$ and $\tilde{\lambda} = 1.06$, we obtain estimators $\tilde{\gamma} = 1.276$ and $\tilde{\delta} = 1.665$, for γ and δ, respectively, from (8.5.6) and (8.5.7). The fitted distribution is shown in Table 8.7.1.

Using Mage's approach, we obtain

$$a = (1.18)^2 - (0.97 \times 1.42) = 0.0150,$$
$$b = 0.97 + 1.42 - (2 \times 1.18) = 0.03,$$
$$c = (2 \times 0.97 \times 1.18 \times 1.42) - (0.97 + 1.42)(1 \cdot 18)^2 = -0.077172,$$
$$d = -0.0503,$$
$$e = 0.08,$$
$$f = 0.022362.$$

Equation (8.7.7) is now

$$\alpha^2 - 2.62661\alpha + 1.30904 = 0,$$

so

$$\tilde{\xi}, \tilde{\xi} + \tilde{\lambda} = \tfrac{1}{2}[2.62661 \pm \sqrt{6.38376 - 5.23616}]$$
$$= \tfrac{1}{2}[2.52661 \pm 1.07126] = [0.728, 1.799].$$

Table 8.7.1 Fitted L_B-Distributions.

Production time (min)	Observed %	Fitted L_B-distributions (a)	(b)	(c)
≤0.695	0.9	—	1.5	0.9
0.695–0.795	3.7	3.7	4.2	4.2
0.795–0.895	12.6	14.0	8.8	11.0
0.895–0.995	18.4	18.8	14.6	18.4
0.995–1.095	18.8	18.6	18.4	20.6
1.095–1.195	15.8	15.3	18.1	16.8
1.195–1.295	12.2	11.3	14.3	11.2
1.295–1.395	7.6	7.8	9.4	6.8
1.395–1.495	5.0	5.1	5.5	4.0
1.495–1.595	2.8	3.1	2.9	2.4
1.595–1.695	1.1	1.7	1.4	1.4
1.695–1.795	0.9	0.6	0.6	0.9
≥1.795	0.2	—	0.2	1.5
	ξ	0.73	0.5[a]	0.5[a]
	λ	1.06	1.5[a]	4.36
	γ	1.276	1.049	7.614
	δ	1.665	2.740	4.019

[a]Value assumed known. Other values estimated.
(a) $P_1 = 0.09$; $P_2 = 0.3162$; (b) $P = 0.09$; (c) $P = 0.09$.
Source: Hahn and Shapiro (1967).

Therefore,

$$\tilde{\xi} = 0.728, \qquad \tilde{\lambda} = 1.071.$$

The discrepancies with the earlier values (0.73, 1.06) can be attributed to round-off and to omission of further steps in the trial-and-error procedure.

If only one end point, say the lower (ξ), is known, then three sample percentiles are needed from which to estimate λ, γ, and δ. It is convenient to take the sample median $\hat{X}_{0.5}$ in addition to $\hat{X}_P$ and $\hat{X}_{1-P}$. We have

$$\left(\frac{\bar{\xi}_{0.5}}{\tilde{\lambda} - \bar{\xi}_{0.5}}\right)^2 = \frac{\bar{\xi}_P\bar{\xi}_{1-P}}{(\tilde{\lambda} - \bar{\xi}_P)(\tilde{\lambda} - \bar{\xi}_{1-P})},$$

whence

$$\tilde{\lambda} = \frac{\bar{\xi}_{0.5}\{(\bar{\xi}_P + \bar{\xi}_{1-P})\bar{\xi}_{0.5} - 2\bar{\xi}_P\bar{\xi}_{1-P}\}}{\bar{\xi}_{0.5}^2 - \bar{\xi}_P\bar{\xi}_{1-P}}. \tag{8.7.8}$$

Since ξ is known, this is an explicit formula for $\tilde{\lambda}$.

If both ξ and λ (and so both end points) are known, then $\tilde{\gamma}$ and $\tilde{\delta}$ can be determined from the equations

$$\tilde{\gamma} + \tilde{\delta} \log \frac{\hat{X}_P - \xi}{\xi + \lambda - \hat{X}_P} = Z_P = -\hat{\gamma} - \hat{\delta} \log \frac{\hat{X}_{1-P} - \xi}{\xi_{1-\lambda} - \hat{X}_{1-P}}. \quad (8.7.9)$$

Example 8.7.2 With the data used in Example 8.7.1 and $\hat{X}_{0.5} = 1.07$ with $P = 0.09$, if we take $\xi = 0.5$ then (8.7.8) gives $\tilde{\lambda} = 4.36$, $\tilde{\gamma} = 7.61$, and $\tilde{\delta} = 4.02$.

The value of $\tilde{\lambda}$ (4.36) is markedly different from the value (1.06 or 1.07) obtained in Example 8.7.1, but the fitted distributions (see Table 8.7.1) are not remarkably different.

If we take $\xi = 0.5$ and $\lambda = 1.5$, we find (from (8.7.9)), with $P = 0.09$,

$$\hat{\gamma} = 1.05, \qquad \tilde{\delta} = 2.74.$$

Again from Table 8.7.1 we see that the different sets of values of the parameters do not result in comparably great differences among fitted distributions.

8.8 FITTING L_U-DISTRIBUTIONS

Moments. Tadikamalla and Johnson (1982) give tables of δ and Ω $(=\gamma/\delta)$ corresponding to specified values of $\sqrt{\beta_1}$ and β_2, together with values of $E[Y]$ and S.D.(Y), for $\sqrt{\beta_1} = 0.00(0.05)1.00(0.1)2.0$ combined (for each $\sqrt{\beta_1}$) with 20 values of β_2 increasing by intervals of 0.2, starting from a value just "below" the L_L line.

Bowman and Shenton (1981) give formulas from which δ and γ can be calculated, given $\sqrt{\beta_1}$ and β_2. These formulas give accurate values over a much wider range than that covered by the tables.

The use of the tables is illustrated in the following example.

Example 8.8.1 We will compare a standardized Pearson Type IV curve having $\sqrt{\beta_1} = 1$, $\beta_2 = 8$ with an L_U-curve having the same first four moments.

From the table we find that the L_U has parameter values $\delta = 6.8469$, $\Omega = 0.7849$ $(\gamma = 5.3820)$ with $E[Y] = 0.8591$, S.D.$(Y) = 0.3731$.

The standardized percentage points are calculated from the formula

$$\frac{\sinh\left(\dfrac{Z_P - \gamma}{\delta}\right) - E[Y]}{\text{S.D.}(Y)} = \frac{\sinh(0.14584 Z_P + 0.7849) - 0.8991}{0.3731}$$

Table 8.8.1 Comparison of Standardized Type IV and L_U Distributions with $\sqrt{\beta_1} = 1, \beta_2 = 8$

% (100 P)	Type IV	L_U	% (100 P)	Type IV	L_U
0.1	−3.053	−3.011	99.9	5.343	5.411
0.5	−2.348	−2.340	99.5	3.637	3.665
1	−2.064	−2.102	99	3.009	3.019
2.5	−1.697	−1.730	97.5	2.252	2.244
5	−1.414	−1.436	95	1.720	1.706
10	−1.112	−1.120	90	1.209	1.194
25	−0.634	−0.624	75	0.522	0.518
50	−0.093	−0.048			

with

$$Z_P = \ln\left(\frac{P}{1 - P}\right).$$

The resulting values are shown in Table 8.8.1, compared with corresponding values for the Type IV curve, taken from Bouver and Bargmann (1974).

In fitting L_U-distributions it is less likely that the values of ξ and/or λ will be known than is the case for L_B or L_L. If they are both known, then, as noted before, the transformed variables $\sinh^{-1}((X_i - \xi)/\lambda)$ can be used as logistic variables, and maximum likelihood can be applied.

8.9 CHOICE AMONG L_L, L_B, AND L_U

Choice among L_L-, L_B-, and L_U-distributions can be based on values of $\sqrt{\beta_1}$ and β_2. Alternatively, order statistics can be used. The methods described by Slifker and Shapiro (1980) and by Bowman and Shenton (1989) for *S*-systems can be applied, *pari passu*, to *L*-systems.

With $Z_P = \ln\{P/(1 - P)\}$, if $P_1 < P_2 < 1 - P_2 < 1 - P_1$, and (as in (8.7.4)) $Z_{P_1} = 3Z_{P_2}$ (and so $Z_{1-P_1} = 3Z_{1-P_2}$), L_L-, L_B-, or L_U-curves are appropriate, according as

$$(X_{P_2} - X_{P_1})(X_{1-P_1} - X_{1-P_2}) =, <, \text{or} > (X_{1-P_2} - X_{P_2})^2$$

(Slifker and Shapiro, 1980). (See also Wheeler, 1980.)

Bowman and Shenton's (1989) extensions of Slifker and Shapiro's results to general values of the ratio Z_{P_1}/Z_{P_2} also apply (with $Z_P =$

$\ln\{P/(1 - P)\}$) to choice among L_L, L_B, and L_U. Of course, in applying these results, the X_P's have to be replaced by their sample estimates $\{\hat{X}_P\}$.

REFERENCES

Ali, M. M. and Khan, A. H. (1987). On order statistics from the log-logistic distribution, *J. Statist. Plann. Inf.*, *17*, 103–108.

Balakrishnan, N. and Malik, H. J. (1987). Moments of order statistics from truncated log-logistic distribution, *J. Statist. Plann. Inf.*, *17*, 251–267.

Balakrishnan, N., Malik, H. J., and Puthenpura, S. (1987). Best linear unbiased estimation of location and scale parameters of the log-logistic distribution, *Commun. Statist.—Theor. Meth.*, *16*, 3477–3495.

Bouver, H. and Bargmann, R. E. (1974). Tables of the standardized percentage points of the Pearson systems in terms of β_1 and β_2. *THEMIS Tech. Rep.*, *32*, Department of Statistics, University of Georgia, Athens, GA.

Bowman, K. O. and Shenton, L. R. (1981). Explicit accurate approximations for fitting the parameters of L_U, in *Statistical Distributions in Scientific Work*, *5*, 231–240, Reidel, Dordrecht, Netherlands.

Bowman, K. O. and Shenton, L. R. (1988). Solutions to Johnson's S_B and S_U, *Commun. Statist.—Simul. Comp.*, *17*, 343–348.

Bowman, K. O. and Shenton, L. R. (1989). S_B and S_U distributions fitted by percentiles. A general criterion, *Commun. Statist.—Simul. Comp.*, *18*, 1–13.

Bukač, J. (1972). Fitting S_B curves using symmetrical percentile points, *Biometrika*, *59*, 688–690.

Burr, I. W. (1942). Cumulative frequency functions. *Ann. Math. Statist.*, *13*, 215–232.

David, H. A. (1981). *Order Statistics*, 2nd ed., Wiley, New York.

Dubey, S. D. (1966). Transformations for estimation of parameters, *J. Indian Statist. Assoc.*, *4*, 109–124.

Hahn, G. J. and Shapiro, S. (1967). *Statistical Models in Engineering*, Wiley, New York.

Johnson, N. L. (1949). Systems of frequency curves generated by methods of translation, *Biometrika*, *36*, 149–176.

Johnson, N. L. (1954). Systems of frequency curves derived from the first law of Laplace, *Trab. Estad.*, *5*, 285–291.

Johnson, N. L. and Kotz, S. (1970). *Distributions in Statistics: Continuous Univariate Distributions—2*, Wiley, New York.

Lomax, K. S. (1954). Business failures: Another example of the analysis of failure data, *J. Amer. Statist. Assoc.*, *49*, 847–852.

Mage, D. T. (1980). An explicit solution for S_B parameters using four percentile points, *Technometrics*, *22*, 247–251.

Owen, D. B. (1988). The starship, *Commun. Statist.—Simul. Comp.*, *17*, 315–323.

Pearson, E. S., Johnson, N. L., and Burr, I. W. (1979). Comparisons of the percentage points of distributions with the same first four moments, chosen from eight different systems of frequency curves, *Commun. Statist.—Simul. Comp.*, *8*, 191–229.

Ragab, A. and Green, J. (1987). Estimation of the parameters of the log logistic distribution based on order statistics, *Amer. J. Math. Mgmt. Sci.*, *7*, 307–323.

Shah, B. K. and Dave, P. H. (1963). A note on log-logistic distribution, *J. M.S. Univ. Baroda* (*Sci. Number*), *12*, 21–22.

Shoukri, M. M., Mian, I. U. H., and Tracey, D. S. (1988). Sampling properties of estimates of the log-logistic distribution, with application to Canadian precipitation data, *Canad. J. Statist.*, *16*, 223–226.

Slifker, J. F. and Shapiro, S. S. (1980). The Johnson system: Selection and parameter estimation, *Technometrics*, *22*, 239–246.

Tadikamalla, P. R. and Johnson, N. L. (1982). Systems of frequency curves generated by transformation of logistic variables, *Biometrika*, *69*, 461–465.

Tadikamalla, P. R. and Johnson, N. L. (1982). Tables to facilitate fitting L_U distribution, *Commun. Statist.—Simul. Comp.*, *11*, 249–271.

Wheeler, R. E. (1980). Quantile estimators of Johnson curve parameters, *Biometrika*, *67*, 725–728.

9

Univariate Generalized Distributions

Daniel Zelterman
University of Minnesota, Minneapolis, Minnesota
N. Balakrishnan
McMaster University, Hamilton, Ontario, Canada

9.1 INTRODUCTION

Generalizations of the logistic distribution have appeared in the literature. There are several different distributions that are referred to as "*the* generalized distribution" by their authors. We will describe these and give some characterization of each. Most of these distributions are of theoretical interest, but Type I (given below) has received additional attention with regard to estimating its parameters for practical usage and the behavior of its order statistics. Some generalized logistic distributions were developed as alternatives to the usual logistic model for binomial regression data. Prentice (1976) proposed such a model, denoted Type IV, discussed below. Another generalized logistic regression model was recently proposed by Stukel (1988), who fitted a different parameter to each of the tails. Balakrishnan and Leung (1988a,b) defined three types of generalized logistic distributions. We will use their notation in this chapter in addition to the Type IV. Table 9.1.1 provides a brief summary of the four major univariate generalized logistic distributions discussed in this chapter.

The Type I distribution has cumulative distribution function (cdf)

$$G_\alpha^{\mathrm{I}}(y) = (1 + e^{-y})^{-\alpha}$$

for a parameter $\alpha > 0$. This cdf can be written as $[F^*(y)]^\alpha$, where $F^*(\cdot)$

Table 9.1.1 Four Major Types of Generalized Logistic Distributions

Type	Parameters	Density function $g(y)$	Moment generating function $m(t)$	Remarks
I	$\alpha > 0$	$\dfrac{\alpha e^{-y}}{(1 + e^{-y})^{\alpha+1}}$	$\dfrac{\Gamma(1 - t)\Gamma(\alpha + t)}{\Gamma(\alpha)}$	Logistic $L(0, \pi^2/3)$ when $\alpha = 1$.
II	$\alpha > 0$	$\dfrac{\alpha e^{-\alpha y}}{(1 + e^{-y})^{\alpha+1}}$	$\dfrac{\Gamma(1 + t)\Gamma(\alpha + t)}{\Gamma(\alpha)}$	Negative Type I $L(0, \pi^2/3)$ for $\alpha = 1$.
III	$\alpha > 0$	$\dfrac{\Gamma(2\alpha)}{[\Gamma(\alpha)]^2} e^{-\alpha y} \Big/ (1 + e^{-y})^{2\alpha}$	$\dfrac{\Gamma(\alpha - t)\Gamma(\alpha + t)}{[\Gamma(\alpha)]^2}$	Symmetric about zero. Logistic $L(0, \pi^2/3)$ when $\alpha = 1$.
IV	$p > 0$ $q > 0$	$\dfrac{\Gamma(p + q)}{\Gamma(p)\Gamma(q)} e^{-qy} \Big/ (1 + e^{-y})^{p+q}$	$\dfrac{\Gamma(p + t)\Gamma(q - t)}{\Gamma(p)\Gamma(q)}$	Type I when $q = 1$, $p = \alpha$. Type III for $p = q = \alpha$. $L(0, \pi^2/3)$ when $p = q = 1$.

is the $L(0, \pi^2/3)$ cdf given in (1.6). The Type I distribution coincides with the usual logistic distribution when $\alpha = 1$. This distribution has received considerable attention with respect to its characterizations, moments, order statistics, as well as parameter estimates in a (location, scale, shape) model $G_\alpha^{\text{I}}[(y - \mu)/\sigma]$. This literature is summarized in Sections 9.3 and 9.4.

The Type II generalized logistic distribution has cdf

$$G_\alpha^{\text{II}}(y) = 1 - \frac{e^{-\alpha y}}{(1 + e^{-y})^\alpha}$$

defined for $\alpha > 0$. If Y is a random variable distributed as Type I, then $-Y$ has a Type II distribution. Hence the Type II model has many properties of Type I, and will only be briefly mentioned here.

The Type III generalized logistic distribution has density function

$$g_\alpha^{\text{III}}(y) = \frac{\Gamma(2\alpha)}{[\Gamma(\alpha)]^2} e^{-\alpha y} \Big/ (1 + e^{-y})^{2\alpha}.$$

When $\alpha = 1$, this is the usual logistic density function. Of all the generalizations of the logistic distribution discussed in this chapter, only the Type III is symmetric about zero; i.e., $g_\alpha^{\text{III}}(y) = g_\alpha^{\text{III}}(-y)$ for all values of y. The Type III distribution has proved useful as an approximation to other symmetric distributions, most notably Student's t.

The generalized logistic model studied by Prentice (1976) and Kalbfleisch and Prentice (1980) will be called Type IV. The Type IV density with parameters $p > 0$, $q > 0$ can be written as

$$g_{pq}^{\mathrm{IV}}(y) = \Gamma(p + q)[\Gamma(p)\Gamma(q)]^{-1}[F^*(y)]^p[1 - F^*(y)]^q,$$

where $F^*(\cdot)$ is the logistic $L(0, \pi^2/3)$ cdf given in (1.6). This distribution is the most general one discussed in this chapter: it includes Types I and III as special cases. The Type I model is obtained by setting $p = \alpha$ and $q = 1$, and the usual logistic is obtained by setting $p = q = 1$ in the Type IV distribution. The Type III model can be obtained by setting $p = q = \alpha$ in the Type IV distribution.

The following are other generalizations of the logistic distribution but will only be mentioned. Perks (1932) proposed a very general distribution that includes the logistic and has density function

$$g^P(y) = \sum_k a_k e^{-k\theta y} \Big/ \sum_k b_k e^{-k\theta y},$$

where parameters a_k, b_k, θ are chosen so that this function is never negative and integrates to unity. See Johnson and Kotz (1970, pp. 14–15) for further discussion.

The Bayesian analysis of the Type I model's parameter α gives rise to a compound Type I generalized logistic distribution. Suppose Y given $\alpha > 0$ has the Type I distribution and suppose that the (conjugate) prior distribution of α is gamma with density function $\alpha^{\theta-1}e^{-\alpha}/\Gamma(\theta)$ with parameter $\theta > 0$. The marginal density function of Y is then

$$\theta e^{-y}(1 + e^{-y})^{-1}[1 + \log(1 + e^{-y})]^{-\theta-1}.$$

The posterior distribution of α is gamma and is discussed in Section 9.3.

The following section describes characterizations of the Type I–Type IV distributions. Sections 9.3 and 9.4 treat only Type I: methods of fitting it to data and its order statistics.

9.2 CHARACTERIZATIONS

One characterization of the Type I generalized logistic distribution was found by Dubey (1969). Let Y given η have the generalized extreme value distribution with density

$$L(y|\eta) = \eta e^{-y} \exp[-\eta e^{-y}].$$

If $\eta > 0$ has a gamma distribution with density $\eta^{\alpha-1}e^{-\eta}/\Gamma(\alpha)$ and parameter $\alpha > 0$, then the marginal distribution of Y is Type I generalized logistic with density function

$$g_\alpha^{\rm I}(y) = \frac{\alpha e^{-y}}{(1 + e^{-y})^{\alpha+1}}.$$

Dubey (1969) noted that in a sample of size α from the logistic $L(0, \pi^2/3)$, the largest order statistic has Type I distribution. Zelterman (1987b) described an estimator of the quantiles of $g_\alpha^{\rm I}(\cdot)$, using a multisample technique. Other methods of fitting the Type I distribution are given in the following section.

Balakrishnan and Leung (1988a), derived the moment generating function of the Type I distribution:

$$\mathcal{E}e^{tY} = m(t) = \frac{\Gamma(1 - t)\Gamma(\alpha + t)}{\Gamma(\alpha)}$$

and found the moments

$$\text{mean:} \quad \mathcal{E}(Y) = m'(0) = \psi(\alpha) - \psi(1), \tag{9.2.1}$$

$$\text{variance:} \quad V(Y) = \mathcal{E}(Y - \mathcal{E}Y)^2 = \psi'(\alpha) + \psi'(1), \tag{9.2.2}$$

$$\mu_3: \quad \mathcal{E}(Y - \mathcal{E}Y)^3 = \psi''(\alpha) - \psi''(1), \tag{9.2.3}$$

$$\mu_4: \quad \mathcal{E}(Y - \mathcal{E}Y)^4 = \psi'''(\alpha) + \psi'''(1) + 3[V(Y)]^2.$$

The functions $\psi(\cdot)$, $\psi'(\cdot)$, etc., are the successive derivatives of the function $\log \Gamma(\cdot)$. Most noticeably, the third moment (μ_3) is an increasing function of α. The Type I distribution is left (negatively)-skewed for $0 < \alpha < 1$ and right-skewed for $1 < \alpha < \infty$. Ahuja and Nash (1967) showed that if Y has the Type I distribution then $-\alpha Y$ behaves as a standard exponential random variable when α is near zero, and $Y - \log(\alpha)$ behaves as a Gumbel extreme value variate with cdf $\exp(-e^{-y})$ when α is large.

Other properties of the Type I distribution will be discussed in the following two sections. The Type II distribution is simply a negative Type I and can also be obtained via Dubey's gamma mixture of extreme value distributions and will not be discussed further.

The Type III distribution has been studied by Davidson (1980), who described the two characterizations given below. The moment generating function is $\Gamma(\alpha - t)\Gamma(\alpha + t)/[\Gamma(\alpha)]^2$, and the Type III distribution is symmetric about zero. The mean is zero (as are all moments of odd orders), the variance is $2\psi'(\alpha)$, and the fourth central moment is $2\psi'''(\alpha) + 12[\psi'(\alpha)]^2$, indicating longer tails than the normal distribution. If Y has the Type III distribution with parameter $\alpha > 0$, then $(2/\alpha)^{1/2}Y$ behaves approximately as standard normal when α is large.

Mudholkar and George (1978) noted the close approximation the logistic density exhibited to Student's t with 9 degrees of freedom. Continuing this idea, George and Ojo (1980) and George, El-Saidi, and Singh (1986) developed an approximation to Student's t with ν degrees of freedom, using the symmetric Type III distribution. By matching the kurtosis of these distributions, they recommend the use of $\alpha = (\nu - 3.25)/5.5$ as the appropriate Type III parameter when approximating Student's t with $\nu >$ 9 degrees of freedom.

Gumbel (1944) characterized the Type III as the limiting distribution of the αth midrange from certain symmetric distributions. If $X_{1:n} \leq \cdots \leq X_{n:n}$ are the order statistics from some population, then the αth midrange (for integer $\alpha \geq 1$) is $[X_{\alpha:n} + X_{n-\alpha+1:n}]/2$. Davidson (1980) showed that Type III can also be expressed as a difference of two independent and identically distributed (iid) random variables. In particular, he proved

Lemma

1. Let R_1R_2 be iid generalized extreme random variables with density function

$$f(r|\theta, \alpha) = \frac{\theta^\alpha}{\Gamma(\alpha)} e^{-\alpha r} \exp(-\theta e^{-r}),$$

$$-\infty < r < \infty, \alpha > 0, \theta > 0.$$

   The difference $R_1 - R_2$ behaves as a Type III generalized logistic with parameter $\alpha > 0$.
2. Let S_1S_2 be iid with the distribution of $-\log V$, where V given $T = t$ behaves as gamma $(\theta t, \alpha + \eta)$ and T behaves as a beta random variable with parameters (α, η). The difference $S_1 - S_2$ behaves as a Type III generalized logistic with parameter $\alpha > 0$.

Notice that the Type III distributions in this lemma are independent of the choices of $\theta > 0$ and $\eta > 0$.

The Type IV generalized logistic distribution can be obtained as the log-odds (logit) of a beta random variable. In particular, if T behaves as a beta random variable with parameters $p > 0$ and $q > 0$, then the random variable $S = \log\{T/(1 - T)\}$ behaves as a Type IV generalized logistic with parameters (p, q). This characterization should make the Type IV a good model for the comparison of log-odds of an event (such as cancer incidence) across populations. Similarly, Type IV is the same distribution as $-\log V$, where V has an F-distribution with $2p$ and $2q$ degrees of freedom. If S has a Type IV distribution with parameters (p, q), then $-S$ has a Type IV distribution with parameters (q, p). The Type IV cdf was proposed by Prentice (1976) as an alternative to modeling binary response data with the usual symmetric logistic cdf. Kalbfleisch and Prentice (1980) suggested using a scale and location model for log survival times where the error distribution has the Type IV generalized logistic distribution. In Sections 2.2.7 and 3.9 of their volume, they described the limiting forms of the Type IV distribution: lognormal if $p \to \infty$; Weibull if $p = 1$ and $q \to \infty$. The Type IV distribution was also used to model data in Farewell and Prentice (1977).

The moment generating function of the Type IV distribution is

$$m^{\text{IV}}(t) = \frac{\Gamma(p + t)\Gamma(q - t)}{\Gamma(p)\Gamma(q)}.$$

The first four central moments are

$$\begin{aligned} \mathcal{E}S &= \psi(p) - \psi(q), \\ V(S) &= \psi'(p) + \psi'(q), \\ \mu_3: \quad \mathcal{E}(S - \mathcal{E}S)^3 &= \psi''(p) - \psi''(q), \\ \mu_4: \quad \mathcal{E}(S - \mathcal{E}S)^4 &= \psi'''(p) + \psi'''(q) + 3[V(S)]^2. \end{aligned}$$

George and Ojo (1980) and George and Singh (1987) expanded the cumulant generating function ($\log m^{\text{IV}}(t)$) and expressed the first four Type IV cumulants in infinite series. This distribution is right-skewed if $p > q$, left-skewed when $p < q$, and symmetric when $p = q$, corresponding to the Type II model.

9.3 PARAMETER ESTIMATES FOR TYPE I

In order to fully utilize any generalized logistic model, we must be able to fit it to data. We begin with a discussion of estimating the single parameter $\alpha > 0$ in $g_\alpha^{\text{I}}(x)$ and then discuss the three-parameter (μ, σ, α) = (location, scale, shape) model $g_\alpha^{\text{I}}((x - \mu)/\sigma)$.

Given a sample $x_1 \cdots x_n$ from $g_\alpha^{\text{I}}(x)$ the maximum likelihood estimator $\hat{\alpha}$ of α is

$$\hat{\alpha} = n \Big/ \sum_{i=1}^{n} \log(1 + e^{-x_i}).$$

The denominator of $\hat{\alpha}$ behaves as the sum of n iid standard exponential random variables divided by α. That is, $\hat{\alpha}$ has the same distribution as $n\alpha/Z$ where Z is a gamma random variable with density function $z^{n-1}e^{-z}/\Gamma(n)$ for $z > 0$. For $n > 1$, $\mathcal{E}\hat{\alpha} = \alpha n/(n - 1)$ and $V(\hat{\alpha}) = \alpha^2(n - 2)^{-1}(n/(n - 1))^2$ for $n > 2$. For large samples, $\hat{\alpha}$ is approximately unbiased and attains the Cramér-Rao-Frêchet lower bound of α^2/n on the variance of an unbiased estimator of α. The approximate maximum likelihood estimator $\tilde{\alpha}$ defined by

$$\tilde{\alpha} = \frac{\hat{\alpha}(n-1)}{n} = (n-1)\Big/\sum_{i=1}^{n} \log(1 + e^{-x_i})$$

is unbiased for α when $n > 1$ and has variance $\alpha^2/(n-2)$ when $n > 2$, which is uniformly smaller than that of $\hat{\alpha}$ but is asymptotically equivalent. Estimators of α based on one and two order statistics are given in the following section.

Given a random sample $x_1 \cdots x_n$ from the three-parameter density $g_\alpha^1((x-\mu)/\sigma)$, we can obtain method of moments estimates by equating the sample moments with the theoretical distribution's moments. Define $\bar{x} = n^{-1}\Sigma x_i$. From (9.2.1) to (9.2.3), the expected values of the first three sample moments are

$$\mathcal{E}\bar{x} = \mu + \sigma[\psi(\alpha) + \gamma], \tag{9.3.1}$$

$$\mathcal{E}(n-1)^{-1}\Sigma(x_i - \bar{x})^2 = \sigma^2\left[\psi'(\alpha) + \frac{\pi^2}{6}\right], \tag{9.3.2}$$

and

$$\mathcal{E}(n-3)^{-1}\Sigma(x_i - \bar{x})^3 = \sigma^3[\psi''(\alpha) - \psi''(1)] + O(n^{-1}),$$

where $\gamma = 0.5772\ldots$. The sample skewness coefficient given by

$$b_1(x) = \frac{(n-3)^{-1}\Sigma(x_i - \bar{x})^3}{\{(n-1)^{-1}\Sigma(x_i - \bar{x})^2\}^{3/2}}$$

satisfies

$$b_1(x) = \frac{\psi''(\alpha) - \psi''(1)}{[\psi'(\alpha) + \pi^2/6]^{3/2}} \tag{9.3.3}$$

plus terms that are $o(1)$. The method of moments estimate α_m of α solves (9.3.3). The method of moments estimates for μ_m and σ_m are also found by equating sample moments (9.3.1) and (9.3.2), giving

$$\sigma_m^2 = \left[\psi'(\alpha_m) + \frac{\pi^2}{6}\right]^{-1}(n-1)^{-1}\Sigma(x_i - \bar{x})^2$$

and

$$\mu_m = \bar{x} - \sigma_m[\psi(\alpha_m) + \gamma].$$

Method of moments estimation was at one time a popular statistical technique, but it has fallen into disuse. Zelterman (1987a) asserted that Eq. (9.3.3) has a solution in α if and only if $-2 < b_1 < 1.1396$, so that the method of moments may not prove useful if the data are highly skewed. The method of moments estimates $(\mu_m, \sigma_m, \alpha_m)$ have not been studied further, but it is likely that standard δ-methods could be used to approximate their asymptotic mean-squared errors.

Maximum likelihood is a popular technique for many situations, but Zelterman (1987a) proved that maximum likelihood estimates do not exist. The problem is that the log likelihood $\Sigma \log g_\alpha^!((x_i - \mu)/\sigma)$ exhibits a "ridge" in its three parameters and becomes unbounded as $\alpha \to \infty$ and $\mu \to -\infty$. Zelterman (1987a) also derived the Fisher information for (μ, σ, α) and showed that it is nonsingular.

For a known value of the shape parameter α, Balakrishnan (1990) approximated the likelihood equations for a doubly censored sample and obtained explicit expressions for estimators of μ and σ. Let $Y_{1:n} \le \cdots \le Y_{n:n}$ denote the order statistics of a noncensored sample from $g_\alpha^!((y - \mu)/\sigma)$. Balakrishnan's approximate maximum likelihood estimates are

$$\bar{\mu} = B - \bar{\sigma}C, \tag{9.3.4}$$

$$\bar{\sigma} = \frac{-D + (D^2 + 4nE)^{1/2}}{2n}, \tag{9.3.5}$$

where

$$B = \Sigma\, b_i Y_{i:n} / \Sigma\, b_i,$$

$$C = \Sigma\, a_i / \Sigma\, b_i,$$

$$D = \Sigma\, a_i (Y_{i:n} - B),$$

$$E = \Sigma\, b_i (Y_{i:n} - B)^2,$$

$$b_i = (\alpha + 1)(1 - p_i^{1/\alpha}) p_i^{1/\alpha},$$

$$a_i = \alpha - (\alpha + 1) p_i^{1/\alpha} + b_i \log\left\{\frac{p_i^{1/\alpha}}{1 - p_i^{1/\alpha}}\right\},$$

$$p_i = \frac{i}{n + 1}.$$

Balakrishnan (1990) also found expressions for the approximate bias, variance, and covariance of these estimates.

Balakrishnan and Leung (1988b) found the best linear unbiased estimators (BLUEs) of μ and σ when the shape parameter α is known. Let $\tau = \{\tau_i\}$, where $\tau_i = \mathcal{E} Y_{i:n}$, and $\xi = \{\xi_{ij}\}$, where $\xi_{ij} = \text{cov}(Y_{i:n}, Y_{j:n})$. The

BLUEs of μ and σ are

$$\mu^* = \sum_i C_{1i} Y_{i:n} \tag{9.3.6}$$

and

$$\sigma^* = \sum_i C_{2i} Y_{i:n}, \tag{9.3.7}$$

where

$$\mathbf{C} = (\mathbf{G}^T \boldsymbol{\xi}^{-1} \mathbf{G})^{-1} \mathbf{G}^T \boldsymbol{\xi}^{-1}$$

is a $2 \times n$ matrix, $\mathbf{G} = (\mathbf{1}, \boldsymbol{\tau})$ is $n \times 2$, and $\mathbf{1}^T = (1, 1, \ldots, 1)$. Balakrishnan (1990) showed that his approximate likelihood estimates given in (9.3.4) and (9.3.5) are fully efficient relative to the BLUEs given in (9.3.6) and (9.3.7). Both of these sets of estimates are functions of the order statistics which are studied further in the following section.

This section ends with a description of an E-M–like algorithm derived by Zelterman (1987a) and studied in a simulation by Hwang and Zelterman (1986). To describe this algorithm, first notice that the gamma distribution is a conjugate Bayes prior distribution for the shape parameter $\alpha > 0$. Let $h(\alpha | r, \lambda) = \lambda^r \alpha^{r-1} e^{-\lambda\alpha} / \Gamma(r)$ denote the gamma prior density for $\alpha > 0$ with parameters $r > 0$ and $\lambda > 0$. The posterior distribution of α given observations $x_1 \cdots x_n$ is also gamma with parameters $r + n$ and

$$\lambda + \sum_i \log\left\{1 + \exp\left[-\frac{(x_i - \mu)}{\sigma}\right]\right\}.$$

The algorithm alternatively replaces the current estimate of α by its posterior mean and then maximizes the likelihood of (μ, σ) at this trial value of α. Specifically, starting with estimates $(\mu_0, \sigma_0, \alpha_0)$, such as the method of moments estimates, the E-M algorithm alternates between the two steps:

1. $\alpha_{j+1} \leftarrow \dfrac{r + n}{\lambda + \Sigma_i \log\{1 + \exp[-(x_i - \mu_j)/\sigma_j]\}}$.
2. Find values $(\mu_{j+1}, \sigma_{j+1})$ of (μ, σ) that maximize

$$-\log \sigma - \frac{\bar{x} - \mu}{\sigma} - (\alpha_{j+1} + 1) \sum_i \log\left\{1 + \exp\left[-\frac{x_i - \mu}{\sigma}\right]\right\}$$

until the values of $(\mu_j, \sigma_j, \alpha_j)$ converge or until a fixed number of iterations occur.

Step 1 replaces α by its posterior mean given (μ, σ), and step 2 maximizes the log likelihood at the current value of α. Hwang and Zelterman (1986) simulated this algorithm and found that, if the prior gamma variance r/λ^2 of α is small, then the algorithm is more likely to converge. The final estimates appeared to have a normal distribution.

9.4 ORDER STATISTICS OF THE TYPE I DISTRIBUTION

The Type I generalized logistic distribution possesses a functional form that facilitates the study of its order statistics. We described the estimation of the parameters of this distribution from order statistics in the previous section and will only briefly mention estimation here. Most of the results in this section were derived independently and at about the same time by Balakrishnan and Leung (1988a,b) and Zelterman (1989).

Let $G_\alpha(x) = (1 + e^{-x})^{-\alpha}$ denote the distribution function and $g_\alpha = \alpha e^{-x}(1 + e^{-x})^{-\alpha-1}$ the density of the Type I generalized logistic distribution. The density function of the rth order statistic $Y_{r:n}$ of sample of size n is denoted

$$g_{r:n,\alpha}(y) = \frac{n!}{(r-1)!(n-r)!} G_\alpha^{r-1}(y)[1 - G_\alpha(y)]^{n-r} g_\alpha(y)$$

and can be written as

$$g_{r:n,\alpha}(y) = \sum_{j=0}^{n-r} d_j g_{\alpha(r+j)}(y),$$

where

$$d_j = d_j(n, r) = \frac{(-1)^j n!}{j!(n-r-j)!(r-1)!(r+j)}. \tag{9.4.1}$$

That is, the density of $Y_{r:n}$ is a weighted average of other Type I generalized logistic density functions, but not a convex combination because the weights d_j are alternately negative and are usually greater than unity in absolute value. In particular, Dubey (1969) noted that the density function of the largest order statistic $Y_{n:n}$ is $g_{n\alpha}(y)$.

The mean of $Y_{r:n}$ is

$$\mathcal{E}Y_{r:n} = \gamma + \sum_{j=0}^{n-r} d_j \psi(\alpha(r+j)),$$

where $\gamma = 0.5772\ldots$, and the variance of $Y_{r:n}$ is

$$V(Y_{r:n}) = \frac{\pi^2}{6} + \sum_{j=0}^{n-r} d_j \psi'(\alpha(r+j)),$$

where d_j are given in (9.4.1). The mode of $Y_{r:n}$ is the root (in y) of the equation

$$[1 + e^{-y}]^{\alpha}[\alpha r e^{-y} - 1] - (\alpha n e^{-y} - 1) = 0.$$

Balakrishnan and Leung (1988a) derived the joint moment generating function for pairs of order statistics and found expressions for their joint product moments. Their (1988b) paper included extensive tables of means, variances, and covariances of the order statistics of the Type I generalized logistic distribution.

For large samples $Y_{n:n}$ and $-Y_{1:n}$ are in the domain of attraction of the Gumbel extreme value distribution with cdf $T(y) = \exp(-e^{-y})$. Specifically, for all real values of y,

$$\lim_{n\to\infty} P[Y_{n:n} - \log \alpha n \leq y] = T(y) \tag{9.4.2}$$

and

$$\lim_{n\to\infty} P[-\alpha Y_{1:n} - \log n \leq y] = T(y). \tag{9.4.3}$$

The Type I generalized logistic sample range $Z = Y_{n:n} - Y_{1:n}$ has a density function expressible as a convex combination of density functions of the largest order statistics from an exponential distribution (Zelterman, 1989). The expected range can be approximated for large n by using (9.4.2) and (9.4.3), giving

$$\mathcal{E}Z \approx (\alpha + 1)\alpha^{-1}[\gamma + \log n] + \log \alpha,$$

with $\gamma = 0.5772. \ldots$

We conclude with a discussion of the estimation of α from functions of single order statistics. Estimation of location and scale parameters from order statistics was discussed in the previous section. To estimate α from the order statistics $Y_{1:n} \leq \cdots \leq Y_{n:n}$ of $g_{\alpha}^{\mathrm{I}}(\cdot)$, define

$$\alpha_r = \frac{\log[(n + 1)/r]}{\log(1 + e^{-Y_{r:n}})}$$

for $r = 1, \ldots, n$ (Zelterman, 1989). Each α_r estimates α and has the same distribution as $\alpha \log[r/(n + 1)]/\log U_{r:n}$, where $U_{1:n} \leq \cdots \leq U_{n:n}$ are order statistics from a uniform distribution on (0, 1). When n is large and $r = np$ for $0 < p < 1$, we have

$$\mathcal{E}\alpha_r = \alpha, \qquad nV(\alpha_r) = \frac{\alpha^2(1 - p)}{p(\log p)^2}$$

plus terms that are $o(1)$. The asymptotic variance of α_r is minimized when $r = r^* = p^*n$ and p^* is the root of $\log p = 2(p - 1)$ near 0.2032, in which case $nV(\alpha_{r\cdot}) = 1.54\alpha^2$. The estimator $\alpha_{r\cdot}$ has approximately the same efficiency as the median relative to the mean in a normally distributed sample. The estimator based on two order statistics and defined as

$$\alpha_{rs} = \beta\alpha_r + (1 - \beta)\alpha_s,$$

with $\beta = 0.4675$, $r = 0.0735n$, and $s = 0.3614n$ has approximate variance $1.22\alpha^2/n$.

REFERENCES

Ahuja, J. C. and Nash, S. W. (1967). The generalized Gompertz-Verhulst family of distributions, *Sankhya A*, *29*, 141–156.

Balakrishnan, N. (1990). Approximate maximum likelihood estimation for a generalized logistic distribution, *J. Statist. Plan. and Inference*, *26*, 221–236.

Balakrishnan, N. and Leung, M. Y. (1988a). Order statistics from the Type I generalized logistic distribution, *Commun. Statist.—Simula.*, *17*, 25–50.

Balakrishnan, N. and Leung, M. Y. (1988b). Means, variances, and covariances of order statistics, BLUE's for the Type I generalized logistic distribution, and some applications, *Commun. Statist.—Simula.*, *17*, 51–84.

Davidson, R. R. (1980). Some properties of a family of generalized logistic distributions, *Statistical Climatology*, *Developments in Atmospheric Science 13*, S. Ikeda et al., (ed.), Elsevier, New York.

Dubey, S. D. (1969). A new derivation of the logistic distribution, *Naval Res. Logist. Quart.*, *16*, 37–40.

Farewell, V. T. and Prentice, R. L. (1977). A study of distributional shape in life testing, *Technometrics*, *19*, 69–76.

George, E. O. and Ojo, M. O. (1980). On a generalization of the logistic distribution, *Ann. Inst. Statist. Math. A*, *32*, 161–169.

George, E. O., El-Saidi, M., and Singh, K. (1986). A generalized logistic approximation of the Student *t* distribution, *Commun. Statist.—Simula.*, *15*, 1199–1208.

George, E. O. and Singh, K. (1987). An approximation of *F* distribution by binomial probabilities, *Statistics & Probability Letters*, *5*, 169–173.

Grumbel, E. J. (1944). Ranges and midranges, *Ann. Math. Statist.* *15*, 414–422.

Hwang, S. and Zelterman, D. (1986). On the distribution of EM estimates, *Proceedings of the Joint Statistical Meetings—Statistical Computing Section*, 144–146.

Johnson, N. L. and Kotz, S. (1970). *Distributions in Statistics: Continuous Univariate Distributions*, Vol. 2, Wiley, New York.

Kalbfleisch, J. D. and Prentice, R. L. (1980). *The Statistical Analysis of Failure Time Data*, Wiley, New York.

Mudholkar, G. S. and George, E. O. (1978). A remark on the shape of the logistic distribution, *Biometrika*, *65*, 667–668.

Perks, W. F. (1932). On some experiments in the graduation of mortality statistics, *Journal of the Institute of Actuaries*, *58*, 12–57.

Prentice, R. L. (1976). A generalization of the probit and logit methods for dose response curves, *Biometrics*, *32*, 761–768.

Stukel, T. (1988). Generalized logistic models, *J. Am. Statist. Assoc.*, *83*, 426–431.

Zelterman, D. (1987a). Parameter estimation in the generalized logistic distribution, *Computational Statistics and Data Analysis*, *5*, 177–184.

Zelterman, D. (1987b). Estimation of percentage points by simulation, *J. Statist. Comput. Simulation*, *27*, 107–125.

Zelterman, D. (1989). Order statistics of the generalized logistic distribution, *Computational Statistics and Data Analysis*, *7*, 69–77.

10
Some Related Distributions

E. Olusegun George and Meenakshi Devidas
Memphis State University, Memphis, Tennessee

10.1 INTRODUCTION

The logistic distribution, also known as the sech squared distribution, has a distribution function

$$F(x; \mu, \sigma) = \left[1 + \exp\left\{-\frac{(x-\mu)}{\sigma}\right\}\right]^{-1},$$
$$-\infty < x < \infty, \quad -\infty < \mu < \infty, \sigma > 0. \quad (10.1.1)$$

Its density function is simply related to its distribution function by

$$f(x; \mu, \sigma) = \frac{1}{\sigma} F(x; \mu, \sigma)[1 - F(x; \mu, \sigma)]. \quad (10.1.2)$$

For the most part, we shall present the results of this section in terms of the standard forms of the distribution and density functions

$$F^*(z) = [1 + e^{-z}]^{-1}, \quad -\infty < z < \infty, \quad (10.1.3)$$

and

$$f^*(z) = F^*(z)[1 - F^*(z)], \quad -\infty < z < \infty. \quad (10.1.4)$$

The relationship given by (10.1.4), or alternatively by (10.1.2), is the simplest characterization of the logistic distribution. It is equivalent to the logit transformation

$$\log\left\{\frac{F^*(z)}{1 - F^*(z)}\right\} = z. \tag{10.1.5}$$

Consequently, this simple characterization of the logistic forms the basis of the wide applications of the logistic function as a response function in various binary and quantal response problems, as well as in the modeling of hazard function in survival analysis. However, beyond this result, other results that meaningfully characterize the logistic have apparently remained elusive; also see Chapter 7.

In this chapter, we give a review of the characterizations and properties of the logistic distribution. These results invariably provide relationships between the logistic random variable, order statistics of the logistic, and functions of order statistics of the logistic, on the one hand, and the exponential, double exponential, and geometric distributions, on the other.

In Section 10.2, characterizations and other properties of the logistic in terms of the exponential, double exponential, and extreme value distribution are given, while in Section 10.3 joint characterizations with geometric distribution are described. In Section 10.4 we present a brief description of the folded form of the logistic distribution and its properties.

10.2 RELATIONSHIPS WITH THE EXPONENTIAL, DOUBLE EXPONENTIAL DISTRIBUTIONS, AND EXTREME VALUE DISTRIBUTION

Let $Z \sim L(0, \pi^2/3)$. The characteristic function, $\phi^*(t)$, of Z is

$$\phi^*(t) = \int_{-\infty}^{\infty} e^{itz} F^*(z)[1 - F^*(z)]\, dz.$$

Since $z = \text{logit}\,(F^*(z))$, we get

$$\phi^*(t) = \int_0^1 e^{it\,\text{logit}(u)}\, du.$$

Hence,

$$\phi^*(t) = \Gamma(1 - it)\Gamma(1 + it). \tag{10.2.1}$$

Alternatively, we may write

$$\phi^*(t) = \frac{\pi i t}{\sin \pi i t}. \tag{10.2.2}$$

From (10.2.1), we get the following relationships of the logistic to the double exponential (or Laplace) distribution.

Theorem 10.2.1 Let $\{W_j\}$ be a sequence of independent random variables with density

$$f_{W_j}(w) = \frac{j}{2} e^{-j|w|}, \qquad -\infty < w < \infty, \quad j = 1, 2, \ldots.$$

Then

$$Z \overset{\mathcal{D}}{=} \sum_{j=1}^{\infty} W_j. \tag{10.2.3}$$

Note that (10.2.3) obviously describes a relationship between the logistic and the exponential, namely

$$Z \overset{\mathcal{D}}{=} \sum_{j=1}^{\infty} (E_{ij} - E_{2j}), \tag{10.2.4}$$

where E_{ij}'s are independent exponential random variables with

$$f_{E_{ij}}(x) = je^{-jx}, \qquad x > 0, \tag{10.2.5}$$

$i = 1, 2$ and $j = 1, 2, \ldots$. Moreover, from the relationship (10.2.3) or (10.2.4), we may conclude immediately that the logistic distribution is infinitely divisible.

The characteristic function (10.2.1) or the relationship (10.2.4) may be used to show a relationship between the logistic and the extreme value maximum distribution.

Theorem 10.2.2 Let $Z \sim L(0, \pi^2/3)$ and let Y_1, Y_2 be independent random variables with density

$$h(x) = e^{-x} \exp(-e^{-x}), \qquad -\infty < x < \infty.$$

Then

$$Z \overset{\mathcal{D}}{=} Y_1 - Y_2. \tag{10.2.6}$$

Proof.

$$\phi_{Y_1}(t) = Ee^{itY_1} = \int_{-\infty}^{\infty} e^{-itx} e^{-x} \exp(-e^{-x})\, dx$$

$$= \int_0^{\infty} u^{it} e^{-u}\, du = \Gamma(1 + it).$$

Similarly, $\phi_{Y_2}(t) = \Gamma(1 - it)$. Hence, $Y_1 - Y_2 \overset{\mathcal{L}}{=} Z$.

Since the maximum order statistic of a large family of symmetric distributions converges to the extreme value random variable, Theorem 10.2.2 provides a basis for the fact that, under suitable conditions, the midrange

$$M_n = \frac{X_{1:n} + X_{n:n}}{2}$$

of a large family of distributions has asymptotic logistic distribution. Specifically, let $X_1, X_2, \ldots, X_n$ be iid with cdf F, and let

$$\Omega_F = \sup\{x : F(x) < 1\}.$$

The following is given by de Haan (1975).

Theorem 10.2.3 If there exists $x_0 < \Omega_F$ such that for all $x \in (x_0, \Omega_F)$, $f(x) > 0$; and if

$$\lim_{x \to \Omega_F} f(x) \int_x^{\Omega_F} [1 - F(t)]\, dt/[1 - F(x)]^2 \to 1, \tag{10.2.7}$$

then

$$\lim_{n \to \infty} P\{a_n[X_{n:n} - b_n] \le x\} = \exp(-e^{-x}), \tag{10.2.8}$$

where $a_n > 0$ and b_n are sequences of constants. Furthermore, if f is nonincreasing in a small neighborhood of Ω_F and

$$[F(a_n^{-1}x + b_n)]^n \to \exp(-e^{-x}),$$

as $n \to \infty$, then F satisfies (10.2.7).

Condition (10.2.7) is a stronger version of the von Mises (1936) condition

$$\lim_{x \uparrow \Omega_F} \frac{d}{dx}\left[\frac{1 - F(x)}{f(x)}\right] = 0, \tag{10.2.9}$$

which requires the existence of $f'(x)$ for $x \in (x_0, \Omega_F)$. However, it is clear that von Mises' condition is usually easier to verify.

It is easy to show that if (10.2.8) or (10.2.9) is satisfied and if F is symmetric, then for a suitable choice of sequence of constants c_n,

$$c_n M_n \xrightarrow{\mathcal{L}} L\left(0, \frac{\pi^2}{3}\right). \tag{10.2.10}$$

George and Rousseau (1990) extend Galambos' (1978) results in this respect. They also give detailed discussion for specific distributions such as the family of power distributions defined by densities

$$f_p(x) = \frac{1}{2\Gamma(1 + 1/p)} \exp(-|x|^p), \qquad -\infty < x < \infty, \qquad (10.2.11)$$

$p > 0$, and the double Weibull defined by Balakrishnan and Kocherlakota (1985) as

$$f_p(x) = \tfrac{1}{2}p|x|^{p-1} \exp(-|x|^p), \qquad -\infty < x < \infty, \quad p > 0. \qquad (10.2.12)$$

In view of the decomposition of the logistic into extreme value, double exponential, and exponential distributions, we may regard the exponential r.v.'s as constituting building blocks for the logistic. It would thus seem appropriate to look for characterizations of the logistic in terms of the exponential, double exponential, and, perhaps, extreme value distributions.

The earliest result (known to the authors) which provides a joint characterization of the logistic and the exponential distributions is given by Galambos and Kotz (1979).

Theorem 10.2.4 Let X be a continuous random variable with cdf F_X, which is symmetric about 0. Then

$$P\{-x < X | X < x\} = 1 - e^{-\lambda x}, \qquad x > 0, \qquad (10.2.13)$$

if and only if $F_X(x) = F^*(\lambda x)$.

George and Mudholkar (1981a,b) and (1982) provide joint characterizations of the logistic and the exponential and double exponential distributions. Their results use the observation that the characteristic functions of the order statistics of the logistic distribution can be written in the form $P_m(it)\phi^*(t)$, where $P_m(t)$ is a polynomial of degree m and ϕ^* is given by (10.2.1).

Specifically, let $Z_{1:n} \le Z_{2:n} \le \cdots \le Z_{n:n}$ be order statistics of a random sample from $L(0, \pi^2/3)$. Then it can be shown that the characteristic function of $Z_{r:n}$ is

$$\phi_{r:n}(t) = \prod_{j=1}^{r-1}\left(1 + \frac{it}{j}\right) \prod_{k=1}^{n-r}\left(1 - \frac{it}{k}\right) \phi^*(t). \qquad (10.2.14)$$

Consequently,

$$Z_{r:n} + \sum_{k=1}^{n-r} E_{1k} - \sum_{j=1}^{r-1} E_{2j} \stackrel{\mathcal{D}}{=} Z_1, \qquad (10.2.15)$$

where the E_{ij}'s are independent random variables with densities given by (10.2.5), $i = 1, 2$, $j = 1, \ldots, n - 1$.

The question posed by (10.2.14) is the following: Let $\phi_{r:n}$ denote the characteristic function of the rth order statistic and ϕ_Z the characteristic function of the population distribution. Under what condition does the relationship

$$\phi_{r:n}(t) = P_{n-1}(it)\phi_Z(t) \tag{10.2.16}$$

characterize the logistic distribution?

From the work of George and Mudholkar, it appears that the necessary conditions are that $t^k\phi_Z(t)$ is integrable for all k, P_m is a polynomial with zeros at $\pm ik$, and f has derivatives of order k, $k = 1, \ldots, m$.

To illustrate we consider the simplest of such results. Let X_1, X_2, X_3 be iid random variables with absolutely continuous cdf F pdf f, and characteristic function ϕ such that ϕ and $t\phi$ are integrable. Assume that $f'(x)$ and $f''(x)$ both exist for all x, and let $F(0) = 1/2$.

Theorem 10.2.5

$$\phi_{2:3}(t) = (1 + t^2)\phi(t), \qquad -\infty < t < \infty,$$

if and only if $F(x) = F^*(x)$.

Proof. It is easy to show that if $F = F^*$ then

$$\phi_{2:3}(t) = \Gamma(2 - it)\Gamma(2 + it) = (1 + t^2)\phi^*(t).$$

To prove the converse, we use the fact that $\phi(t)$ and $t\phi(t)$ are integrable. Hence,

$$\begin{aligned} 6F(x)[1 - F(x)]f(x) &= \frac{1}{2\pi}\int_{-\infty}^{\infty} (1 + t^2)e^{-itx}\phi(t)\,dt \\ &= f(x) - \frac{1}{2\pi}\int_{-\infty}^{\infty} \frac{\partial^2}{\partial x^2} e^{-itx}\phi(t)\,dt \\ &= f(x) - f''(x). \end{aligned}$$

Letting $y = F(x)$, we get the boundary value problem

$$6y(1 - y)\frac{dy}{dx} = \frac{dy}{dx} - \frac{d^2y}{dx^2}, \qquad y(0) = \frac{1}{2}. \tag{10.2.17}$$

Rewriting (10.2.17) leads to

$$\frac{d}{dx}(3y^2 - 2y^3) = \frac{d}{dx}\left(y - \frac{d^2y}{dx^2}\right).$$

By the Riemann-Lebesgue lemma

$$\lim_{y\to-\infty} y(x) = \lim_{|x|\to\infty} y'(x) = \lim_{|x|\to\infty} y''(x) = 0. \tag{10.2.18}$$

Hence,

$$\frac{d^2y}{dx^2} = y - 3y^2 + 2y^3. \tag{10.2.19}$$

Multiplying both sides of (10.2.19) by $2(dy/dx)$ and simplifying, we get

$$\frac{d}{dx}\left(\frac{dy}{dx}\right)^2 = \frac{d}{dx}(y^2 - 2y^3 + y^4).$$

This reduces to

$$\left(\frac{dy}{dx}\right)^2 = [y(1 - y)]^2. \tag{10.2.20}$$

Thus $y(x) = [1 + e^{-x}]^{-1}$.

From Theorem 10.2.5 we get the following:

Corollary 10.2.5 Let X_1, X_2, X_3 be iid random variables with cdf F and density f, where $F(0) = 1/2$. Let W be an independent double exponential random variable with density

$$g(w) = \tfrac{1}{2}e^{-|w|}, \qquad -\infty < w < \infty.$$

Then

$$X_{2:3} + W \overset{d}{=} X_1$$

if and only if $F(x) = F^*(x)$.

Extensions of these results to the case $n = 2p$, $p > 1$, are given by George and Mudholkar (1981).

Based on the order statistics $Z_{1:2}$ and $Z_{2:2}$, George and Mudholkar (1982) give the following joint characterization of the logistic and the exponential distributions:

Theorem 10.2.6 Let Z_1, Z_2, V_1, V_2 be independent random variables with $Z_1, Z_2 \sim F$, while $V_1 \sim G_1$ and $V_2 \sim G_2$, where $G_1(0) = G_2(0) = 0$ and at least one of G_1 and G_2 is non-lattice. If $Z_{1:2} + V_1 \overset{d}{=} Z_{2:2} - V_2 \overset{d}{=} Z_1$, then $F = F^*(\gamma + \sigma\cdot)$ for some real γ and $\sigma > 0$, and $G(x) = G_2(x) = 1 + e^{-\sigma x}$, and conversely.

Further interesting relationships between the logistic and the exponential distributions are found by studying the distribution of the logistic mid-range. We summarize some of the results obtained in this regard by George and Rousseau (1987).

Let $M_n = (Z_{1:n} + Z_{n:n})/2$. The characteristic function of M_n is given by

$$\varphi_n(t) = G_n(it) \times {}_3F_2\left[\begin{matrix} n + it, & it/2, & 1 + it/2 \\ & n + 1 + \frac{it}{2}, & n + \frac{it}{2}, \end{matrix} ; 1\right], \tag{10.2.21}$$

where

$$G_n(it) = n(n-1)B\left(n + it, 1 - \frac{it}{2}\right) B\left(1 + \frac{it}{2}, n - 1\right)$$

and $B(x, y)$ denotes the beta function.

By observing that the generalized hypergeometric function $3F_2$ in (10.2.21) is well poised, Dixon's theorem can be used to get

$$\varphi_n(t) = \begin{cases} \prod_{j=1}^{p-1}\left[1 + \frac{t^2}{4j^2}\right]\left(\phi^*\left(\frac{t}{2}\right)\right)^2, & \text{if } n = 2p, \\ \prod_{j=1}^{p}\left[1 + \frac{t^2}{(2j-1)^2}\right]\phi^*(t), & \text{if } n = 2p + 1. \end{cases} \tag{10.2.22}$$

Using (10.2.22), a number of relationships between the logistic midrange and the exponential and double exponential distributions can be obtained. We give a few of these results here.

As before, let W_j denote a double exponential random variable with density given by Theorem 10.2.1. From (10.2.22), we immediately get

$$M_{2p} + \sum_{j=1}^{p-1} W_{2j} \stackrel{r}{=} \frac{Z_1 + Z_2}{2}, \tag{10.2.23}$$

and

$$M_{2p+1} + \sum_{j=1}^{p} W_{2j-1} \stackrel{r}{=} Z. \tag{10.2.24}$$

To obtain the next relationship, first observe that the characteristic function of $Z_{p:2p-1}$ is

$$\phi_{p:2p-1}(t) = \prod_{j=1}^{p-1}\left(1 + \frac{t^2}{j^2}\right)\phi^*(t). \tag{10.2.25}$$

From (10.2.22),

$$\varphi_{2p}(2t) = \prod_{j=1}^{p-1}\left(1 + \frac{t^2}{j^2}\right)(\phi^*(t))^2. \tag{10.2.26}$$

Hence, from (10.2.25) and (10.2.26) we get

$$Z_{p:2p-1} + \sum_{j=1}^{p-1} W_j \stackrel{r}{=} Z_1 \tag{10.2.27}$$

and

$$2M_{2p} \stackrel{r}{=} Z_{p:2p-1} + Z', \tag{10.2.28}$$

where Z' is independent of $Z_1, \ldots, Z_{2p-1}$. When $n = 3$, (10.2.22) gives

$$\varphi_3(t) = (1 + t^2)\phi^*(t). \tag{10.2.29}$$

Consequently, from (10.2.25), the following result is obtained for the logistic:

$$\frac{Z_{1:3} + Z_{3:3}}{2} \stackrel{r}{=} Z_{2:3}. \tag{10.2.30}$$

10.3 RELATIONSHIPS WITH THE GEOMETRIC DISTRIBUTION

As with the exponential distribution, the joint characterization of the logistic and the geometric distributions involve the order statistics of the logistic. Let $Z_1, Z_2, \ldots$ be iid random variables with nondegenerate cdf F, and let N be a positive integer-valued random variable which is independent of the Z's. Let ρ denote the generating function of N. Baringhaus (1980) gives the following result:

Theorem 10.3.1 Let F be symmetric about 0, and let γ be a real-valued function of $\theta \in (0, 1)$. Then

$$\frac{\rho(\theta F(x))}{\rho(\theta)} = F(x + \gamma(\theta)), \qquad -\infty < x < \infty, \tag{10.3.1}$$

iff $F(x) = F^*(ax)$, for some $a > 0$, and ρ is the generating function of a geometric distribution.

Voorn (1987) generalizes Baringhaus result and gives an interpretation of this characterization in terms of what he calls "maximum stability" of the logistic distribution.

Definition 10.3.1 F is said to be maximum stable with respect to a random sample size N or with respect to the probability distribution of N if

$$\frac{Z_{N:N} - a}{b} \stackrel{r}{=} Z_1 \tag{10.3.2}$$

for some $-\infty < a < \infty$ and $b > 0$.

Equation (10.3.2) can be expressed by the equation

$$\sum_{j=1}^{\infty} p_j (F(z))^j = F\left[\frac{z - a}{b}\right], \qquad (10.3.3)$$

where $P(N = j) = p_j, j = 1, 2, \ldots$.

Definition 10.3.2 A distribution function F_1 is called extended logistic if

$$F_1(x) = \left\{F^*\left[\frac{x - a}{b}\right]\right\}^{-1/k},$$

where $-\infty < a < \infty$, $b > 0$, and k is a positive integer.

Note that Voorn's extended logistic is a special case of what is usually referred to as a generalized logistic; see Chapter 9 for details. Voorn gives the following theorems.

Theorem 10.3.2 Let $\{p_{n,m}\}$ be a sequence of nondegenerate distributions on the positive integers, such that

$$\lim_{n \to 0} p_{n,i} = 1$$

and

$$\lim_{n \to 0} \frac{p_{n,i}}{p_{n,k+1}} = 0, \qquad 1 < i \neq k + 1.$$

Then F is maximum stable with respect to $\{p_{n,m}\}$ iff the distributions $\{p_{n,m}\}$ are extended geometric and F is extended logistic or extended log-logistic or backward log-logistic with same parameter k.

Voorn gives definitions for extended geometric, extended log-logistic, and extended backward log-logistic. When $k = 1$, the result may be written as follows:

Theorem 10.3.3

$$Z_{N:N} + \log p \stackrel{r}{=} Z_1$$

iff N has a geometric distribution with parameter p and Z has logistic distribution function F^*.

10.4 RELATIONSHIPS WITH THE HALF-LOGISTIC DISTRIBUTION

Let $H = |Z|$, where Z is the standard logistic random variable. H is called the folded logistic or half-logistic random variable. Its distribution is given by the pdf

$$g(h) = 2f^*(h) = \frac{2e^{-h}}{(1 + e^{-h})^2}, \qquad h \geq 0, \tag{10.4.1}$$

and cdf

$$G(h) = 2F^*(h) - 1 = \frac{1 - e^{-h}}{1 + e^{-h}}, \qquad h \geq 0. \tag{10.4.2}$$

The density $g(h)$ is a monotonic decreasing function of h in $[0, \infty)$ and has an increasing hazard rate. Consequently, as illustrated by Balakrishnan and Cohen (1990), the half-logistic distribution is very useful for modeling lifetime distributions.

From (10.4.1) and (10.4.2) we observe that

$$g(h) = G(h)[1 - G(h)] + \tfrac{1}{2}[1 - G(h)]^2 \tag{10.4.3}$$

or, equivalently,

$$g(h) = [1 - G(h)] - \tfrac{1}{2}[1 - G(h)]^2. \tag{10.4.4}$$

Equation (10.4.3) (or (10.4.4)) characterizes the half-logistic distribution and is used by Balakrishnan (1985) to obtain several recurrence relationships between the moments of the order statistics of this distribution. Specifically, let $H_{i:n}(i = 1, \ldots, n)$ denote the order statistics of the half-logistic distribution. Balakrishnan (1985) gives the following results.

Theorem 10.4.1 For $k \geq 0$,

$$E(H_{1:n+1}^{k+1}) = 2\left[E(H_{1:n}^{k+1}) - \frac{k+1}{n} E(H_{1:n}^{k})\right], \qquad n \geq 1,$$

and

$$E(H_{i+1:n+1}^{k+1}) = \frac{1}{i}\left[\frac{(n+1)(k+1)}{n-i+1} E(H_{i:n}^{k}) + \frac{n+1}{2} E(H_{i:n+1}^{k+1})\right],$$

$$1 \leq i \leq n.$$

Theorem 10.4.2

$$E(H_{i:n+1}H_{i+1:n+1}) = E(H_{i:n+1}^2) + \frac{2(n+1)}{n-i+1}$$

$$\times \left[E(H_{i:n}H_{i+1:n}) - E(H_{i:n}^2) - \frac{1}{n-i} E(H_{i:n})\right], \qquad 1 \leq i \leq n-1,$$

$$E(H_{2:n+1}H_{3:n+1}) = E(H_{3:n+1}^2)$$

$$+ (n+1)\left[E(H_{2:n}) - \frac{n}{2} E(H_{1:n-1}^2)\right], \qquad n \geq 2,$$

and

$$E(H_{i+1:n+1}H_{i+2:n+1}) = E(H^2_{i+2:n+1}) + \frac{n+1}{i(i+1)}$$
$$\times \, [2E(H_{i+1:n}) + n[E(H_{i-1:n-1}H_{i:n-1}) - E(H^2_{i:n-1})]], \qquad 2 \le i \le n-1.$$

Further, since $H \stackrel{r}{=} |Z|$, we may invoke the results of Govindarajulu (1963) to obtain some relationships between the moments of order statistics from the logistic and those from the half-logistic distribution. These are given by the following theorem.

Theorem 10.4.3 For $k \ge 1$ and $1 \le i \le n$,

$$E(Z^k_{i:n}) = 2^{-n}\left[\sum_{r=0}^{i-1}\binom{n}{r} E(H^k_{i-r:n-r}) + (-1)^k \sum_{r=i}^{n}\binom{n}{r} E(H^k_{r-i+1:r})\right],$$

and for $1 \le i < j \le n$,

$$E(Z_{i:n}Z_{j:n}) = 2^{-n}\left[\sum_{r=0}^{i-1}\binom{n}{r} E(H_{i-r:n-r}H_{j-r:n-r})\right.$$
$$+ \sum_{r=j}^{n}\binom{n}{r} E(H_{r-j+1:r}H_{r-i+1:r})$$
$$\left. - \sum_{r=i}^{j-1}\binom{n}{r} E(H_{r-i+1:r}H_{j-r:n-r})\right].$$

Arguments similar to those used in the above results can be applied to give relationships in distribution between the order statistics of the logistic and the half-logistic.

REFERENCES

Balakrishnan, N. (1985). Order Statistic from the half logistic distribution. *J. Statist. Comput. Simul.*, *20*, 287–309.

Balakrishnan, N. and Kocherlakota, S. (1985). On the double Weibull distribution: Order statistics and estimation, Sankhya B, *47b*, 161–178.

Balakrishnan, N. and Cohen, A. C. (1991). *Order Statistics and Inference: Estimation Methods*, Academic Press, Boston.

Baringhaus, von, L. (1980). Eine simultane Charakterisierung der geometrischen Verteilung und der logistischen Verteilung, *Metrika*, *27*, 237–242.

de Haan, L. (1975). *On Regular Variation and Its Application to Weak Convergence of Sample Extremes*, 3rd. ed., Mathematical Center Tracts, *32*, Amsterdam.

Galambos, J. (1978). *The Asymptotic Theory of Extreme Order Statistics*, John Wiley, New York.

Galambos, J. and Kotz, S. (1979). *Characterizations of Probability Distributions: A Unified Approach with Emphasis on Exponential and Related Models*, Lecture Notes in Mathematics, *675*, Springer-Verlag, New York.

George, E. O. and Mudholkar, G. S. (1981a). A characterization of the logistic distribution by a sample median, *Ann. Inst. Stat. Math.*, 125–129.

George, E. O. and Mudholkar, G. S. (1981b), Some relationships between the logistic and the exponential distribution, in *Statistical Distributions in Scientific Work*, C. Tallie et al., (eds.) Vol. 4, 401–409. D. Reidel Publishing Co.

George, E. O. and Mudholkar, G. S. (1982). On the logistic and exponential laws, Sankhya A, *44*, 291–293.

George, E. O. and Rousseau, C. C. (1987). On the logistic midrange, *Ann. Inst. Stat. Math.*, *39*, 627–635.

George, E. O. and Rousseau, C. C. (1991). On the asymptotics of range and midrange, (Accepted for publication in Sankya).

Govindarajulu, Z. (1963). Relationships among moments of order statistics in samples from two related populations, *Technometrics*, *5*, 514–518.

von Mises, R. (1936). La distribution de la plus grande de *n* values, reprinted in Selected Papers II, *Amer. Math. Soc.* (1954), 271–294.

Voorn, W. J. (1987). Characterization of the logistic and loglogistic distributions by extreme value related stability with random sample size, *J. Appl. Prob.*, *24*, 838–851.

11

Multivariate Logistic Distributions

Barry C. Arnold
University of California, Riverside, California

11.1 INTRODUCTION

Although multivariate data sets with logistic-like marginals have always been around, it was not until 1961 that a bivariate logistic model was proposed by Gumbel. The natural multivariate extension of Gumbel's model had to wait 12 more years until it was provided by Malik and Abraham (1973). Gumbel actually provided three candidate bivariate logistic models. One of these (Eq. (11.7.1)) can be identified as an early example of the Farlie-Gumbel-Morgenstern paradigm for constructing bivariate distributions with specified marginals. The second (Eq. (11.6.8)) has been largely ignored, but it is a close relative of a recently introduced class of bivariate survival distributions involving frailty (Oakes, 1989, for example). The third bivariate logistic density introduced by Gumbel (the first one in his paper and the one analyzed in most detail) takes the simple form

$$F_{X,Y}(x, y) = [1 + e^{-x} + e^{-y}]^{-1}, \qquad x, y \in \mathbf{R}. \qquad (11.1.1)$$

Evidently this distribution does have logistic marginals (let $y \to \infty$ in (11.1.1) to get $F_X(x)$). Location and scale parameters may be introduced, but quite clearly the model is very specific with regard to the nature of the dependence between X and Y. In fact, by direct computation, Gumbel verified that the correlation coefficient is 1/2. The multivariate extension proposed by Malik and Abraham (1973) suffers from a similar restriction in that all its pairwise correlations are 1/2. A simple representation in terms of linear functions of independent extreme value random variables allows one to verify this correlation phenomenon without computation and suggests techniques for developing analogous models with a more flexible range of possible correlation structures. These are introduced in Section 11.4. The fact that geometrically distributed random minima and maxima of i.i.d. logistic random variables are again logistic can be exploited to generate broad classes of bivariate (and multivariate) logistic distributions incorporating many of the well-known models as special cases. These models are surveyed in Section 11.5. The problems of multivariate logistic simulation and estimation are then addressed. The chapter closes with brief comments on a few other multivariate logistic or generalized multivariate logistic models which have appeared in the literature. Included is some discussion of the class of distributions with logistic conditional distributions.

11.2 THE CLASSICAL MULTIVARIATE LOGISTIC DISTRIBUTION (Gumbel, 1961; Malik and Abraham, 1973)

A random vector $(Z_1, Z_2, \ldots, Z_k)$ has the classical k-variate logistic distribution if its joint distribution function is of the form

$$F_{Z_1,\ldots,Z_k}(z_1, \ldots, z_k) = \left(1 + \sum_{i=1}^{k} e^{-z_i}\right)^{-1}. \tag{11.2.1}$$

Real location parameters ($\boldsymbol{\mu}$) and positive scale parameters ($\boldsymbol{\sigma}$) may be introduced to obtain the full classical k-variate logistic family

$$F_{\mathbf{X}}(\mathbf{x}) = \left\{1 + \sum_{i=1}^{k} \exp\left[-\left(\frac{x_i - \mu_i}{\sigma_i}\right)\right]\right\}^{-1}, \tag{11.2.2}$$

If $\mathbf{X}$ has distribution (11.2.2) we write $\mathbf{X} \sim L^k(I)(\boldsymbol{\mu}, \boldsymbol{\sigma})$.

The marginal distributions in (11.2.1) are $L(0, \pi^2/3)$ variables in the nomenclature of Chapter 1. The marginals in (11.2.2) are $L(\mu_i, \sigma_i^2(\pi^2/3))$, $i = 1, 2, \ldots, k$. Gumbel provided an extensive discussion of distributional properties in the bivariate case. The bivariate density corresponding to

(11.2.2) is of the form

$$f_{X_1,X_2}(x_1, x_2) = 2\exp\left\{-\left[\frac{x_1-\mu_1}{\sigma_1}+\frac{x_2-\mu_2}{\sigma_2}\right]\right\}$$
$$\div\ \sigma_1\sigma_2\left(1+\exp\left\{-\left(\frac{x_1-\mu_1}{\sigma_1}\right)\right\}\right.$$
$$\left.+\exp\left\{-\left(\frac{x_2-\mu_2}{\sigma_2}\right)\right\}\right)^3 \qquad (11.2.3)$$

Contours of the density (11.2.3) for the case $\boldsymbol{\mu} = \mathbf{0}$, $\boldsymbol{\sigma} = \mathbf{1}$ are displayed in Fig. 11.2.1, reproduced from Gumbel (1961). Positive correlation is clearly indicated in the figure by the "neolithic arrowhead"-shaped contours.

Still focusing on the bivariate density with $\boldsymbol{\mu} = \mathbf{0}$ and $\boldsymbol{\sigma} = \mathbf{1}$, we may write the conditional density in the form

$$f_{X_1|X_2}(x_1|x_2) = \frac{2e^{-x_1}(1+e^{-x_2})^2}{(1+e^{-x_1}+e^{-x_2})^3}. \qquad (11.2.4)$$

The conditional moment generating function is

$$E(e^{t_1X_1}|X_2 = x_2) = (1+e^{-x_2})^{-t_1}\Gamma(2+t_1)\Gamma(1-t_1). \qquad (11.2.5)$$

From this we may compute conditional means and variances. Thus,

$$E(X_1|X_2 = x_2) = 1 - \ln(1+e^{-x_2}) \qquad (11.2.6)$$

and

$$\operatorname{var}(X_1|X_2 = x_2) = \frac{\pi^2}{3} - 1. \qquad (11.2.7)$$

Analogous expressions are appropriate for the conditional moments of X_2 given X_1. The resulting curvilinear regression curves are displayed in Fig. 11.2.2 (again reproduced from Gumbel's original paper). In the figure they are compared with the corresponding regression lines for a bivariate normal distribution with correlation matching that of the bivariate logistic.

The joint moment generating function of the k-variate distribution (11.2.1) is not difficult to compute. One finds, for $|t_i| < 1$, $i = 1, 2, \ldots, k$,

$$M_{\mathbf{X}}(\mathbf{t}) = \Gamma\left(1+\sum_{i=1}^{k} t_i\right)\prod_{i=1}^{k}\Gamma(1-t_i). \qquad (11.2.8)$$

Inspection of this expression suggests the following representation of the

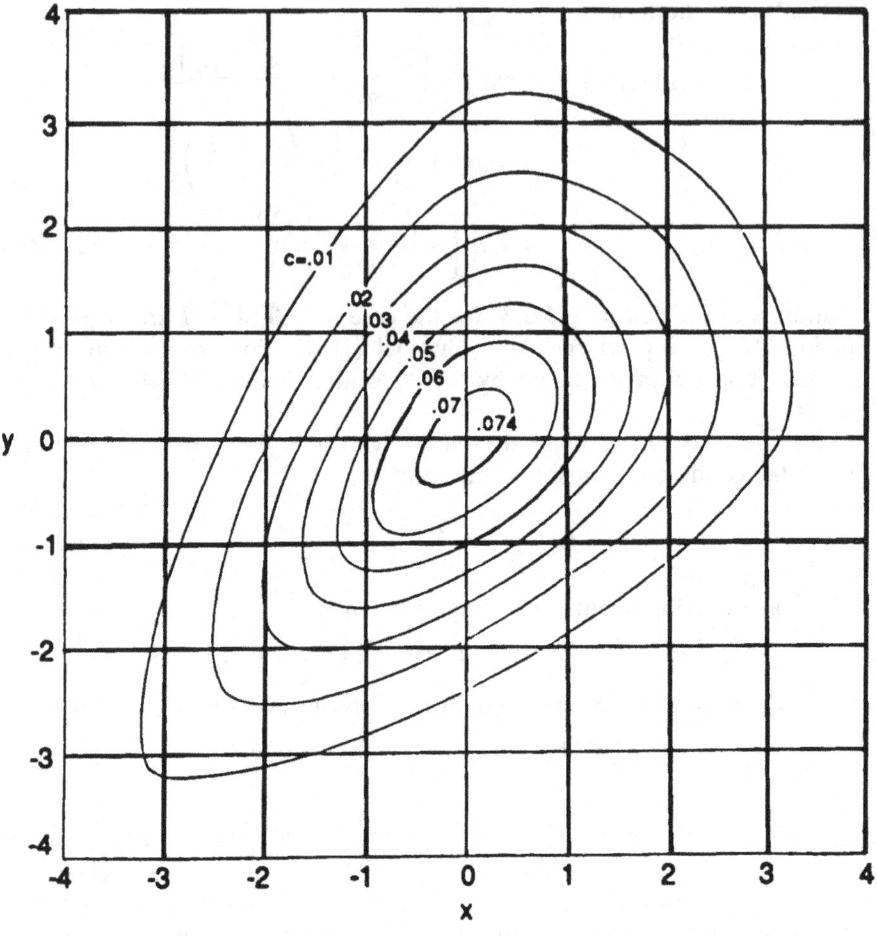

Fig. 11.2.1 Equi-Density Curves for Bivariate Logistic Distribution $c = f(x, y)$

k-variate logistic distribution. Let $Y_0, Y_1, \ldots, Y_k$ be i.i.d. random variables each having the extreme value distribution with density

$$f_Y(y) = e^{-y} \exp(e^{-y}), \qquad -\infty < y < \infty, \tag{11.2.9}$$

and moment generating function

$$M_Y(t) = \Gamma(1 - t),$$

where $|t| < 1$. Define $(X_1, \ldots, X_k)$ by

$$X_i = Y_i - Y_0, \qquad i = 1, 2, \ldots, k, \tag{11.2.10}$$

then $\mathbf{X}$ has the k-variate logistic distribution (11.2.1).

The representation (11.2.10) immediately permits determination of the correlations between the X_i's. Clearly

$$\operatorname{cov}(X_i, X_j) = \operatorname{var}(Y_0) = \frac{\pi^2}{6}$$

and, since $\operatorname{var}(X_i) = \pi^2/3$, the correlation is

$$\rho(X_i, X_j) = \tfrac{1}{2}. \tag{11.2.11}$$

Evidently a family of k-variate logistic distributions with a more flexible

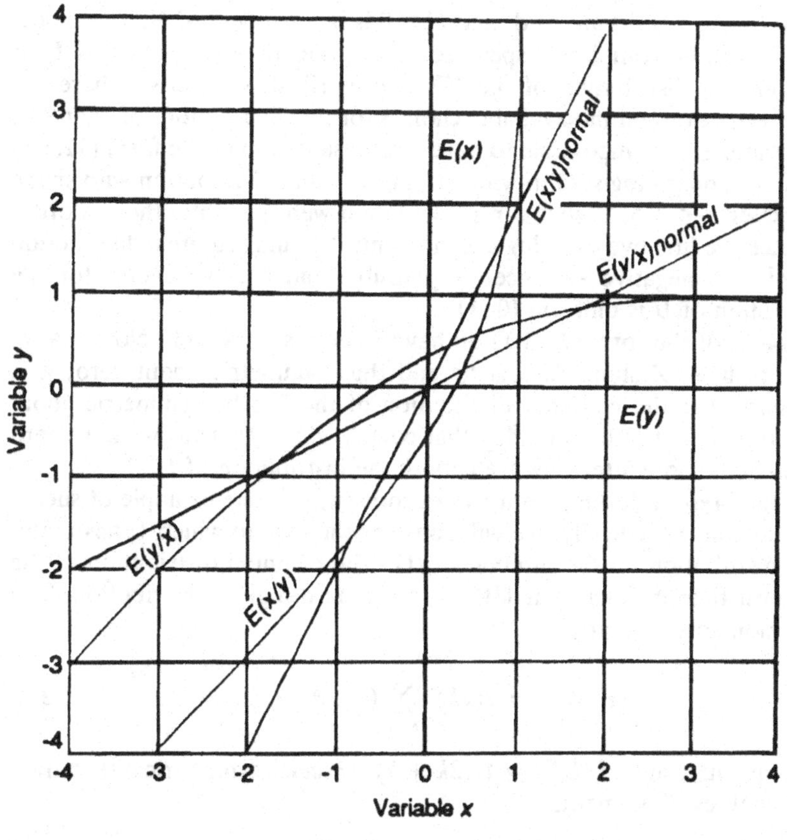

Fig. 11.2.2 Regression Curves for the Bivariate Normal and Logistic Distributions

correlation structure would be desirable. Such distributions will be encountered in later sections.

11.3 MIXTURE REPRESENTATIONS

A popular method for developing dependent k-dimensional distributions with prescribed marginals involves scale mixtures of suitably chosen distributions with independent marginals. Thus we wish to construct a model for our random vector $(Z_1, Z_2, \ldots, Z_k)$ such that each marginal distribution is standard logistic (i.e., $F_{Z_i}(z) = (1 + e^{-z})^{-1}$) and $Z_1, \ldots, Z_k$ is expressible in the form

$$Z_i = UV_i, \qquad i = 1, 2, \ldots, k, \tag{11.3.1}$$

where $(V_1, \ldots, V_k)$ are i.i.d. and $U \geq 0$ is independent of $\mathbf{V}$. The model (11.3.1) will be completely specified once either the distribution of U or the common distribution of the V_i's is prescribed. Of course, there are some restrictions on the available choices for the distributions of U and V. For example, it is impossible to have a representation of the form (11.3.1) (with Z_i standard logistic) in which U has a gamma distribution with shape parameter $\alpha < 1.5$. If such a representation were possible, then in order to match the first two even logistic moments it would require a distribution for V^2 with negative variance! A plausible and feasible choice for the distribution of U is uniform (0, 1).

Models of the form (11.3.1) do have a serious drawback. Note that in order to have Z standard logistic and thus symmetric about zero, it is necessary that the common distribution of the V_i's be symmetric about zero. But then it follows readily that $\operatorname{cov}(Z_i, Z_j) = 0$. Thus we have zero correlations, no matter how we choose the distribution of U.

In the present development we will present just one example of such a construction. Specifically, we will consider the case in which U has a uniform distribution on the interval (0, 1). The common distribution of the V_i's must then be such that UV has a standard logistic distribution. The even moments of Z are

$$E(Z^{2k}) = 2(2k)! \sum_{n=1}^{\infty} (-1)^{n-1} n^{-2k}. \tag{11.3.2}$$

Consequently, since $E(U^{2k}) = 1/(2k + 1)$ we seek a symmetric distribution for V with even moments

$$E(V^{2k}) = 2(2k + 1)! \sum_{n=1}^{\infty} (-1)^{n-1} n^{-2k}. \tag{11.3.3}$$

It is readily verified that a suitable density for V is of the form

$$f_V(v) = \frac{1}{2}\sum_{n=1}^{\infty} (-1)^{n-1}n^2|v|e^{-n|v|}, \qquad -\infty < v < \infty. \tag{11.3.4}$$

An alternative expression for the density is obtainable by summing the series

$$f_V(v) = \frac{1}{4}\left\{\frac{|v|}{2}\tanh\left(\frac{|v|}{2}\right)\operatorname{sech}^2\left(\frac{|v|}{2}\right)\right\}, \qquad -\infty < v < \infty. \tag{11.3.5}$$

The resulting k-dimensional joint density is then

$$f_{\mathbf{Z}}(\mathbf{z}) = \frac{1}{4^k}\int_0^1 \prod_{i=1}^{k}\left\{\frac{|v_i|}{2u}\tanh\left(\frac{|v_i|}{2u}\right)\operatorname{sech}^2\left(\frac{|v_i|}{2u}\right)\right\}\frac{du}{u^k}. \tag{11.3.6}$$

We are assured that (11.3.6) has standard logistic marginals, but a closed-form expression for the joint density appears to be elusive.

Alternative scale mixture constructions are possible. For example, we can generalize (11.3.1) to a form

$$Z_i = U_iV_i, \tag{11.3.7}$$

where $\mathbf{U} \geq 0$ and $\mathbf{V}$ are independent vectors but otherwise arbitrary, save for the marginal requirements that U_iV_i should be standard logistic for each i. A model in which $\mathbf{U}$ has some dependent multivariate uniform distribution might be worthy of study. Note that in such a case, if the V_i's were independent, we would still have zero correlations among the Z_i's. A special case of the model (11.3.7) will be encountered in the last section of this chapter. In that case, $U_1 = U_2 = \cdots = U_k$ while the V_i's are dependent. The model in question yields a spherically symmetric joint distribution for $\mathbf{Z}$.

Analogous additive models can be postulated; i.e., we might seek a representation of the form

$$Z_i = U_i + V_i, \tag{11.3.8}$$

where $\mathbf{Z}$ has standard logistic marginals and $\mathbf{U}$ and $\mathbf{V}$ are independent random vectors. Note that, since the standard logistic distribution is symmetric about zero, some representations of the form (11.3.8) are possible. Recall that a logistic random variable can be modeled as a difference between two i.i.d. extreme value random variables (see Eq. (11.2.9)), so $\mathbf{U}$ and $-\mathbf{V}$ could be chosen to be multivariate extreme value random variables in (11.3.8). Other models involving less intimate relationships between the distribution of U_i and V_i await investigation. The differences between independent extremes paradigm will be discussed in the following section.

11.4 DIFFERENCES OF EXTREMES

If U and V are independent extreme value random variables (density (11.2.9)), then $U - V$ has a standard logistic distribution. Immediately we have available a means of constructing multivariate logistic random vectors as differences of independent multivariate extreme random vectors. Thus if $\mathbf{U}$ and $\mathbf{V}$ are independent random vectors with extreme marginals, then $\mathbf{Z} = \mathbf{U} - \mathbf{V}$ is a multivariate logistic random vector. The classic example in this genre is the Gumbel-Malik-Abraham distribution (11.2.10), in which, in our present notation, the V_i's are identical ($V_1 = V_2 \cdots V_k$) extreme random variables while the U_i's are independent extremes. A reversal of roles, taking the V_i's to be independent and the U_i's identical, yields a variant of the Gumbel-Malik-Abraham distribution, actually a random vector whose negative has the Gumbel-Malik-Abraham distribution. In the bivariate case the density contours retain the neolithic arrowhead shape of Fig. 11.2.1 but now point up and to the right. This distribution is also encountered in Lindley and Singpurwalla (1986) in a reliability context.

More interesting variants are possible. Strauss (1979), in a random utility context, considers the one-parameter multivariate logistic distribution constructed by using the Gumbel (1958) multivariate extreme distribution, namely,

$$P(\mathbf{U} \leq \mathbf{u}) = \exp - \left\{ \sum_{i=1}^{k} e^{-\alpha u_i} \right\}^{1/\alpha}, \tag{11.4.1}$$

where $\alpha \geq 1$. If we then choose $\mathbf{V}$ to have identical extreme coordinates and be independent of $\mathbf{U}$, then the vector $\mathbf{Z} = \mathbf{U} - \mathbf{V}$ has a distribution given by

$$F_{\mathbf{Z}}(\mathbf{z}) = \left[1 + \left(\sum_{i=1}^{k} e^{-\alpha z_i} \right)^{1/\alpha} \right]^{-1}. \tag{11.4.2}$$

This of course includes the classic k-variate logistic, corresponding to the choice $\alpha = 1$. Since in Gumbel's distribution the correlations are $\rho(U_i, U_j) = 1 - \alpha^{-2}$ (Oliveira, 1961, who derives this result utilizing the fact that $U_i - U_j$ itself has an extreme distribution), an elementary computation yields the correlations for the multivariate logistic (11.4.2). Thus

$$\rho(Z_i, Z_j) = 1 - (2\alpha^2)^{-1}. \tag{11.4.3}$$

Strong correlation is always present (always at least 0.5).

It may be observed that if we let α in Gumbel's k-variate extreme distribution become arbitrarily large, the limiting distribution has identical extreme components. It follows that the distribution (11.4.2) can be viewed as a limiting case of a more flexible family of k-variate logistic distributions. For this assume that $\mathbf{U}$ has the k-variate Gumbel extreme distribution (11.4.1) with parameter α and that $\mathbf{V}$ independent of $\mathbf{U}$ has the same distribution but with parameter α'. If we define $\mathbf{Z} = \mathbf{U} - \mathbf{V}$, then we have a k-variate logistic distribution. For this distribution the correlation is more flexible. One finds (again using Oliveira's result on the correlation for k-variate Gumbel extremes) that

$$\rho(Z_i, Z_j) = \frac{1}{2}\left[\left(1 - \frac{1}{\alpha^2}\right) + \left(1 - \frac{1}{\alpha'^2}\right)\right]. \tag{11.4.4}$$

Thus any nonnegative correlation is possible in this model (model (11.4.2) is recoverable by letting $\alpha' \to \infty$). Unfortunately a closed-form expression for the distribution of this more flexible model is lacking.

If we are willing to forgo the existence of a closed form for the joint distribution, then quite arbitrary correlation structures can be developed. We will illustrate the construction in the bivariate case; the potential multivariate extension will be apparent.

We make use of the following representation of the standard logistic distribution in terms of independent exponential variates (see, e.g., George and Mudholkar, 1981):

$$Z \stackrel{d}{=} \sum_{j=1}^{\infty} \frac{E_j}{j} - \sum_{j=1}^{\infty} \frac{E_j'}{j}, \tag{11.4.5}$$

where the E_j's and E_j''s are independent standard exponential variates. To construct our bivariate logistic model, we consider an infinite sequence of independent four-dimensional exponential variates

$$\begin{pmatrix} E_1 \\ F_1 \\ G_1 \\ H_1 \end{pmatrix}, \begin{pmatrix} E_2 \\ F_2 \\ G_2 \\ H_2 \end{pmatrix}, \ldots . \tag{11.4.6}$$

Now define

$$I_1 = \{j : E_j, F_j, G_j, H_j \text{ are independent}\},$$
$$I_2 = \{j : E_j = G_j;\ E_j, F_j, H_j \text{ independent}\},$$
$$I_3 = \{j : E_j = H_j;\ E_j, F_j, G_j \text{ independent}\},$$
$$I_4 = \{j : F_j = G_j;\ E_j, F_j, H_j \text{ independent}\},$$
$$I_5 = \{j : F_j = H_j;\ E_j, F_j, G_j \text{ independent}\},$$
$$I_6 = \{j : E_j = G_j,\ F_j = H_j;\ E_j, F_j \text{ independent}\},$$
$$I_7 = \{j : E_j = H_j,\ F_j = G_j;\ E_j, F_j \text{ independent}\},$$

and define our bivariate logistic random variable (Z_1, Z_2) by

$$Z_1 = \sum_{j=1}^{\infty} \frac{E_j}{j} - \sum_{j=1}^{\infty} \frac{F_j}{j}$$

$$Z_2 = \sum_{j=1}^{\infty} \frac{G_j}{j} - \sum_{j=1}^{\infty} \frac{H_j}{j}. \tag{11.4.7}$$

Provided that every j is a member of one of the sets $I_1, \ldots, I_7$ (i.e., $\cup_{k=1}^{7} I_k = \mathbf{Z}^+$), then (Z_1, Z_2) will indeed have logistic marginals (using the representation (11.4.5)). However, the covariance is quite arbitrary. One finds

$$\text{cov}(Z_1, Z_2) = \sum_{j \in I_2 \cup I_6} \frac{1}{j^2} + \sum_{j \in I_3 \cup I_6} \frac{1}{j^2} - \sum_{j \in I_4 \cup I_7} \frac{1}{j^2} - \sum_{j \in I_5 \cup I_7} \frac{1}{j^2}. \tag{11.4.8}$$

Suitable choices for the I_i's will yield any desired covariance between $-\pi^2/3$ and $+\pi^2/3$, i.e., any correlation between -1 and 1. For example, to achieve perfect positive correlation put every j in I_6.

11.5 GEOMETRIC MINIMA AND MAXIMA

Suppose that $Z_1, Z_2, \ldots$ are i.i.d. standard logistic variables and that N, independent of the Z_i's, has a geometric distribution with parameter p (i.e., $P(N = n) = p(1 - p)^{n-1}, n = 1, 2, \ldots$). Define

$$Z^*(p) = \left(\min_{i \le N} Z_i\right) - \log p. \tag{11.5.1}$$

An elementary computation verifies that $Z^*(p)$ has again a standard logistic distribution. We may say that the logistic distribution is closed under the operation of geometric minimization. Analogous computations, or introspection on the consequences of the symmetry of the logistic distribution, indicate that the logistic is closed under geometric maximization also. Thus if we define

$$Z^{**}(p) = \left(\max_{i \le N} Z_i\right) + \log p, \tag{11.5.2}$$

then $Z^{**}(p)$ is again standard logistic. It is consequently clear that any multivariate geometric distribution can be used in conjunction with geometric minimization and/or maximization to generate a multivariate distribution with logistic marginals.

For example, consider a sequence of independent trials with $k + 1$ possible outcomes $0, 1, 2, \ldots, k$ and associated probabilities p_0,

$p_1, \ldots, p_k$ ($\Sigma_{i=0}^k p_i = 1$). Define a random vector $\mathbf{N} = (N_1, \ldots, N_k)$ such that N_i represents the number of outcomes of type i which precede the first outcome of type 0. The generating function of the random vector $\mathbf{N}$ is

$$P_{\mathbf{N}}(\mathbf{s}) = E\left(\prod_{j=1}^{k} s_j^{N_j}\right) = p_0\left(1 - \sum_{j=1}^{k} p_j s_j\right)^{-1}. \tag{11.5.3}$$

Note that, for each j, $N_j + 1$ has a geometric distribution with support 1, 2, ... Now consider k independent sequences of independent standard logistic variables: $Y_i^{(j)}$, $j = 1, 2, \ldots, k$, $i = 1, 2, \ldots$. Define a k-dimensional random vector $\mathbf{Z} = (Z_1, Z_2, \ldots, Z_k)$ by

$$Z_j = \min_{i \le N_j + 1} Y_i^{(j)}. \tag{11.5.4}$$

Evidently each Z_j will have a logistic distribution. The joint survival function of the Z_j's is readily derived. For $\mathbf{z} \in \mathbf{R}^k$,

$$\begin{aligned} P(\mathbf{Z} \ge \mathbf{z}) &= E(P(\mathbf{Z} \ge \mathbf{z}|\mathbf{N})) \\ &= \sum_{\mathbf{n}} \left\{\prod_{j=1}^{k} (1 + e^{z_j})^{-(n_j+1)}\right\} P(\mathbf{N} = \mathbf{n}) \\ &= \prod_{j=1}^{k} (1 + e^{z_j})^{-1} E\left(\prod_{j=1}^{k} (1 + e^{z_j})^{-N_j}\right) \\ &= p_0 \Big/ \left[\left(1 - \sum_{j=1}^{k} p_j(1 + e^{z_j})^{-1}\right) \prod_{j=1}^{k} (1 + e^{z_j})\right], \end{aligned} \tag{11.5.5}$$

where the last step follows from (11.5.3). Translation and reparameterization in (11.5.5) leads to the following k-variate survival function with standard logistic marginals

$$P(\mathbf{Z} \ge \mathbf{z}) = \left[1 + \sum_{j=1}^{k} e^{z_j} + \sum\sum_{j_1 \ne j_2} c_{j_1 j_2} e^{z_{j_1} + z_{j_2}} + \sum\sum\sum_{j_1 \ne j_2 \ne j_3} c_{j_1 j_2 j_3} e^{z_{j_1} + z_{j_2} + z_{j_3}} + \cdots + c_{1 \cdots k} e^{z_1 + z_2 + \cdots + z_k}\right]^{-1}. \tag{11.5.6}$$

In order for (11.5.6) to be a genuine survival function, certain restrictions must be placed on the c's (see Arnold, 1990 for more details).

If maximization rather than minimization were invoked in (11.5.5), an analogous development leads to a k-variate random vector with standard logistic marginals whose joint distribution function (instead of survival function) $P(\mathbf{Z} \le \mathbf{z})$ is of the form (11.5.6) with e^{z_i}'s replaced by e^{-z_i}'s throughout the right-hand side. Rather than rewrite (11.5.6) with such editorial amendment, we will switch for the remainder of this section to consideration of the bivariate case only. The reader will be able to envision straightforward, albeit notationally cumbersome, extensions to higher dimensions.

Thus in two dimensions, geometric minimization using (11.5.3) leads to the model

$$P(Z_1 \ge z_1, Z_2 \ge z_2) = [1 + e^{z_1} + e^{z_2} + \theta e^{z_1+z_2}]^{-1}, \qquad (11.5.7)$$

where $0 \le \theta \le 2$. Geometric maximization leads to the distinct model

$$P(Z_1 \le z_1, Z_2 \le z_2) = [1 + e^{-z_1} + e^{-z_2} + \theta e^{-z_1-z_2}]^{-1}, \qquad (11.5.8)$$

where again $0 \le \theta \le 2$. The classic Gumbel-Malik-Abraham model is recognized as a special case ($\theta = 0$) of the geometric maximization model (11.5.8). Ali, Mikhail, and Haq (1978) also discussed the bivariate model (11.5.8). They began by defining the function

$$H(z_1, z_2) = [P(Z_1 \le z_1, Z_2 \le z_2)]^{-1} - 1$$

and then solving the differential equation

$$\frac{\partial^2 H(z_1, z_2)}{\partial H(z_1, \infty)\partial H(\infty, z_2)} = \theta.$$

Of course, any bivariate geometric minimum (or maximum) of bivariate random vectors with any bivariate logistic distribution will again be bivariate logistic. The hierarchy of bivariate geometric distributions described in Arnold (1975) could be used to generate even more complicated models. The general construction is as follows. We envision a sequence of i.i.d. bivariate logistic random variables (Y_{1i}, Y_{2i}), $i = 1, 2, \ldots$, and an independent bivariate geometric random variable $\mathbf{N} = (N_1, N_2)$ (i.e., the marginals are of the form $P(N_j = i) = p_j(1 - p_j)^{i-1}$, $i = 1, 2, \ldots$). Then we define $\mathbf{Z}$ by

$$Z_j = \min_{i \le N_j} Y_{ji}. \qquad (11.5.9)$$

The resulting joint distribution for $\mathbf{Z}$ is then suitably translated to yield a bivariate distribution with standard logistic marginals. Sometimes the parameter space can subsequently be augmented, even though the resulting distribution no longer retains an interpretation as a bivariate geometric minimum of independent bivariate logistics. For example, (11.5.7), which is a result of such an operation, is a genuine survival function for any $\theta \in [0, 2]$, but it could have arisen by bivariate geometric minimization only if $\theta \in [0, 1]$. Such niceties are of some concern if we wish to simulate samples from these distributions.

Some other examples of the bivariate geometric minimization routine (using (11.5.9)) are listed below.

A. Assume $N_1 = N_2$ with $P(N_1 = n) = p(1 - p)^{n-1}$ and that, for each i, Y_{1i}, Y_{2i} are independent standard logistic variables. This leads to the model

$$P(Z_1 \geq z_1, Z_2 \geq z_2) = [1 + e^{z_1} + e^{z_2} + pe^{z_1+z_2}]^{-1}. \quad (11.5.10)$$

This reproduces the model (11.5.7), and, of course, parameter space augmentation is possible (p can be allowed to range over the interval $[0, 2]$). The k-dimensional extension of the paradigm A is predictably not as flexible as (11.5.6) since it would still involve only one parameter.

B. Assume $N_1 = N_2$ with $P(N_1 = n) = p(1 - p)^{n-1}$ and that for each i we have

$$P(Y_{1i} \geq y_1, Y_{2i} \geq y_2) = [1 + (e^{\alpha y_1} + e^{\alpha y_2})^{1/\alpha}]^{-1}. \quad (11.5.11)$$

The resulting translated distribution $\mathbf{Z}$ (defined by (11.5.9)) is again of the form (11.5.11). Thus the distribution (11.5.11) is closed under geometric minimization. Multiplying the coordinates of $\mathbf{Y}$ by -1 leads to the distribution (11.4.2) which is thus closed under geometric maximization.

C. Assume $Y_{1i} = Y_{2i}$ for each i and that (N_1, N_2) is bivariate geometric with survival function

$$\begin{aligned} P(N_1 \geq n_1, N_2 > n_2) &= p_{00}^{n_1}(p_{10} + p_{00})^{n_2-n_1}, \qquad n_1 \leq n_2 \\ &= p_{00}^{n_2}(p_{01} + p_{00})^{n_1-n_2}, \qquad n_2 < n_1 \quad (11.5.12) \end{aligned}$$

(see Block, 1975 for further details on this model). Application of (11.5.9) leads to the model.

$$\begin{aligned} P(Z_1 \geq z_1, Z_2 \geq z_2) &= [1 - p_{00} + \max\{(p_{10} + p_{11})e^{z_1}, (p_{01} + p_{11})e^{z_2}\}]^{-1} \\ &\quad \times \left\{\frac{p_{01}}{1 + e^{z_1}} + \frac{p_{10}}{1 + e^{z_2}} + p_{11}\right\}. \quad (11.5.13) \end{aligned}$$

A potentially unattractive feature of this distribution is the presence of a singular component.

D. Assume that for each i, Y_{1i} and Y_{2i} are independent standard logistic and the $\mathbf{N}$ is bivariate geometric with distribution (11.5.12). The resulting distribution is of the form

$$P(Z_1 \geq z_1, Z_2 \geq z_2) = \frac{p_{11} + \dfrac{p_{01}}{1 + e^{z_1}} + \dfrac{p_{01}}{1 + e^{z_2}}}{[1 + (p_{10} + p_{11})e^{z_1}][1 + (p_{01} + p_{11})e^{z_2}] - p_{00}}. \tag{11.5.14}$$

11.6 FRAILTY AND ARCHIMEDEAN MODELS

The concept of frailty is typically of interest in survival studies (see Oakes, 1989) for an interesting recent discussion of the topic). However, the model extends easily to random variables whose support is the entire real line, even though the motivation for such models may be lacking. Suppose Z is a standard logistic random variable. A frailty representation of Z corresponding to a given distribution F on $(0, \infty)$ is of the form

$$(1 + e^z)^{-1} = P(Z \geq z) = \int_0^\infty [P(U \geq z)]^\theta \, dF(\theta). \tag{11.6.1}$$

Since we may write $[P(U \geq z)]^\theta$ in the form $\exp[-\theta[-\log P(U \geq z)]]$, it is clear that once F is selected then the distribution U is determined by the uniqueness of Laplace transforms. In fact we must have

$$P(U \geq z) = \exp[-M_F^{-1}((1 + e^z)^{-1})], \tag{11.6.2}$$

where $M_F(\theta) = \int_0^\infty e^{-t\theta} \, dF(\theta)$ is the Laplace transform of the distribution F. However, there are no restrictions on the choice of the distribution F, except that its support be $(0, \infty)$. Now fix attention on a specific distribution F and consider a k-variate logistic distribution with survival function given by

$$P(\mathbf{Z} \geq \mathbf{z}) = \int_0^\infty \left[\prod_{i=1}^{k} P(U \geq z_i)\right]^\theta dF(\theta), \tag{11.6.3}$$

where the distribution of U is related to F by (11.6.2). The resulting distribution will have logistic marginals since (11.6.1) is true when U and F are related by (11.6.2). Note that (11.6.3) can be written completely in terms of M_F, the Laplace transform of F. Thus

$$\begin{aligned} P(\mathbf{Z} \geq \mathbf{z}) &= M_F\left(-\sum_{i=1}^{k} \log P(U \geq z_i)\right) \\ &= M_F\left(\sum_{i=1}^{k} M_F^{-1}((1 + e^{z_i})^{-1})\right). \end{aligned} \tag{11.6.4}$$

Actually the model could be extended. The function M_F in (11.6.4) does not have to be a Laplace transform. All that is needed is that it be non-negative and decreasing, have nonnegative second derivative, and satisfy $M_F(0) = 1$. In such generality (11.6.4) becomes what Genest and MacKay (1986) call an Archimedean distribution.

For example, if we choose F to be a gamma $(\alpha, 1)$ distribution, then its Laplace transform is $M_F(t) = (1 + t)^{-\alpha}$. The corresponding inverse function is $M_F^{-1}(u) = u^{-1/\alpha} - 1$, and the corresponding multivariate logistic distribution takes the form

$$P(\mathbf{Z} \geq \mathbf{z}) = \left[\sum_{i=1}^{k} (1 + e^{z_i})^{1/\alpha} - k + 1\right]^{-\alpha}. \qquad (11.6.5)$$

In the above development, survival functions could have been replaced by distribution functions, leading to related but distinct k-variate distributions. Alternatively, they can be viewed as being simply of the form $-\mathbf{Z}$, where $\mathbf{Z}$ has a k-variate logistic distribution of the type derived above. Thus we are led to k-variate logistic distributions whose joint distribution functions are of the form

$$P(\mathbf{Z} \leq \mathbf{z}) = M_F\left(\sum_{i=1}^{k} M_F^{-1}((1 + e^{-z_i})^{-1})\right), \qquad (11.6.6)$$

which includes as a special case

$$P(\mathbf{Z} \leq \mathbf{z}) = \left[\sum_{i=1}^{k} (1 + e^{-z_i})^{1/\alpha} - k + 1\right]^{-\alpha}. \qquad (11.6.7)$$

Note that the choice $\alpha = 1$ in (11.6.7) yields the Gumbel-Malik-Abraham distribution, so that we can view (11.6.7) as a one-parameter ($\alpha > 0$) extension of that classical distribution.

A second interesting special case of the distribution (11.6.6) is provided by the choice $M_F(t) = \exp\{-t^\alpha\}$, where $\alpha \leq 1$. The resulting joint distribution function is

$$P(\mathbf{Z} \leq \mathbf{z}) = \exp - \left\{\sum_{i=1}^{k} [\log(1 + e^{-z_i})]^{1/\alpha}\right\}^{\alpha}. \qquad (11.6.8)$$

This may be recognized as the second (and largely ignored) kind of multivariate logistic distribution introduced by Gumbel (1961).

11.7 FARLIE-GUMBEL-MORGENSTERN DISTRIBUTIONS

The one parameter family of k-variate logistic distributions of the Gumbel-Morgenstern type is

$$P(\mathbf{Z} \le \mathbf{z}) = \left[\prod_{i=1}^{k} F^*(z_i)\right]\left[1 + \alpha \prod_{i=1}^{k} (1 - F^*(z_i))\right], \quad (11.7.1)$$

where $F^*(z) = (1 + e^{-z})^{-1}$ and $-1 < \alpha < 1$. The Farlie generalization would allow the terms $1 - F^*(z_i)$ to be replaced by suitable functions of the form $g_i(z_i)$ (subject only to the constraint that the resulting expression be a genuine k-dimensional distribution function). The correlation in (11.7.1) is relatively easy to evaluate; one finds

$$\rho(Z_i, Z_j) = \frac{3\alpha}{\pi^2}, \quad (11.7.2)$$

which is always less than 0.304 in absolute value. This restricted range for the correlation (a characteristic feature of F-G-M distributions) has limited the use of the model (11.7.1). Perhaps a more serious drawback is the lack of a plausible stochastic model which might account for data with such a distribution.

Generalized F-G-M models can be developed. For example, let S_k denote the class of all possible vectors of 0's and 1's of length k with at least two coordinates equal to 1 (S_k thus contains $2^k - k - 1$ members, and we may denote a generic element of S_k by $\mathbf{s} = (s_1, \ldots, s_k)$). Define a k-dimensional distribution function by

$$P(\mathbf{Z} \le \mathbf{z}) = \left[\prod_{i=1}^{k} F^*(z_i)\right]\left[1 + \sum_{\mathbf{s} \in S_k} \alpha_{\mathbf{s}} \left\{\prod_{i=1}^{k} [1 - F^*(z_i)]^{s_i}\right\}\right]. \quad (11.7.3)$$

Clearly we have standard logistic marginals. Indeed the functions $F^*(z_i)$ in the second term on the right of (11.7.3) can be replaced by other distribution functions $G_i(z_i)$ to obtain an even more general class of k-variate logistic distributions. Details, including discussion of the relevant constraints on the $\alpha_{\mathbf{s}}$'s, are found in Johnson and Kotz (1975, 1977).

11.8 A FLEXIBLE MODEL

Several models which include the Gumbel-Malik-Abraham family have been encountered. Two of the most appealing in terms of mathematical simplicity were the difference of extremes model (Section 11.4) and the geometric maxima model (Section 11.5). Can they both be embedded in a relatively tractable "super" model. An affirmative answer is appropriate. For notational convenience we will focus on the bivariate case. Extensions to higher dimension are not intrinsically difficult, although there may be some difficulty in delineating the exact limits on the resulting parameter spaces.

In the two-dimensional case, the differences of extremes model is of the form

$$F_{Z_1,Z_2}(z_1, z_2) = [1 + (e^{-\alpha z_1} + e^{-\alpha z_2})^{1/\alpha}]^{-1}, \qquad (11.8.1)$$

and the geometric maxima model is

$$F_{Z_1,Z_2}(z_1, z_2) = [1 + e^{-z_1} + e^{-z_2} + \theta e^{-z_1-z_2}]^{-1}. \qquad (11.8.2)$$

A natural model which embraces the above two is provided by

$$F_{Z_1,Z_2}(z_1, z_2) = [1 + (e^{-\alpha z_1} + e^{-\alpha z_2} + \theta e^{-\alpha z_1-\alpha z_2})^{1/\alpha}]^{-1}, \qquad (11.8.3)$$

where α, θ are chosen so that (11.8.3) represents a genuine two-dimensional distribution function. In other words, α, θ must be chosen such that

$$\frac{\partial}{\partial z_1}\frac{\partial}{\partial z_2} F_{Z_1,Z_2}(z_1, z_2) \geq 0 \quad \forall z_1, z_2.$$

An equivalent question (which requires a little less algebraic bookkeeping) is, for which values of θ and α is it true that

$$\frac{\partial}{\partial x}\frac{\partial}{\partial y}[1 + (x + y + \theta xy)^{1/\alpha}]^{-1} \geq 0 \quad \forall x > 0, y > 0?$$

It is not difficult to verify that we must have

$$\alpha \geq 1, \qquad (11.8.4)$$

and for a given value of α we must have

$$0 \leq \theta \leq [(\alpha^2 - 1)^{1/\alpha} + (1 + \alpha)(\alpha^2 - 1)^{(1-\alpha)/\alpha}]^{\alpha}. \qquad (11.8.5)$$

Note that the right-hand side of (11.8.5) is always ≥ 2 (and, for large α, can be quite large).

11.9 EXTREMES

Gumbel, with his well-known profound interest in extreme values, predictably discussed the limiting distribution of coordinatewise minima and maxima of random samples from his bivariate logistic distribution. There is no difficulty in extending his result to the case in which the data are assumed to be a sample from the more general flexible bivariate distribution (11.8.3). To this end, we consider $(Z_{11}, Z_{21}), \ldots, (Z_{1n}, Z_{2n})$ i.i.d. with distribution (11.8.3) and define

$$Z_1^*(n) = \max_{j=1,2,\ldots,n} Z_{1j} - \log n,$$
$$Z_2^*(n) = \max_{j=1,2,\ldots,n} Z_{2j} - \log n. \qquad (11.9.1)$$

It is clear that

$$\begin{aligned}\lim_{n\to\infty} & P(Z_1^*(n) \le z_1, Z_2^*(n) \le z_2)\\ &= \lim_{n\to\infty}\left(1 + \left(\frac{e^{-\alpha z_1}}{n} + \frac{e^{-\alpha z_2}}{n} + \theta\,\frac{e^{-\alpha z_1 - \alpha z_2}}{n^2}\right)^{1/\alpha}\right)^{-n}\\ &= \exp[-(e^{-\alpha z_1} + e^{-\alpha z_2})^{1/\alpha}],\end{aligned} \qquad (11.9.2)$$

which can be recognized as a Gumbel bivariate extreme limiting distribution. Independence is encountered only in the case $\alpha = 1$.

If we consider minima rather than maxima, analogous computations are feasible. The limiting distribution turns out to depend on θ. We define

$$\tilde{Z}_1^*(n) = \min_{j=1,2,\ldots,n} Z_{1j} + \log n,$$
$$\tilde{Z}_2^*(n) = \min_{j=1,2,\ldots,n} Z_{2j} + \log n. \qquad (11.9.3)$$

We then find that, if $\theta = 0$,

$$\lim_{n\to\infty} P(\tilde{Z}_1^*(n) \ge z_1, \tilde{Z}_2^*(n) \ge z_1) = \exp[-e^{z_1} - e^{z_2} + (e^{-\alpha z_1} + e^{-\alpha z_2})^{-1/\alpha}] \qquad (11.9.4)$$

(as Gumbel observed, even in the case $\alpha = 1$, the asymptotic minima are not independent). If, however, $\theta \neq 0$, then asymptotic independence is encountered. One finds

$$\lim_{n\to\infty} P(\tilde{Z}_1^*(n) \ge z_1, \tilde{Z}_2^*(n) \ge z_2) = \exp[-e^{z_1} - e^{z_2}] \qquad (11.9.5)$$

(no matter what value α assumes).

If we consider extremes of samples from the bivariate "frailty" related distribution (11.6.7) (with $k = 2$) analogous results are obtainable. Retaining the same definitions for $(Z_1^*(n), Z_2^*(n))$ and $(\tilde{Z}_1^*(n), \tilde{Z}_2^*(n))$, we find, for maxima,

$$\lim_{n\to\infty} P(Z_1^*(n) \le z_1, Z_2^*(n) \le z_2) = \exp(-e^{-z_1} - e^{-z_2}) \qquad (11.9.6)$$

no matter what value α assumes in (11.6.7), while for minima,

$$\lim_{n\to\infty} P(\tilde{Z}_1^*(n) \geq z_1, \tilde{Z}_2^*(n) \geq z_2) = \exp[-e^{z_1} - e^{z_2} + (e^{-z_1/\alpha} + e^{-z_2/\alpha})^{-\alpha}]. \tag{11.9.7}$$

Note that if a suitable reparameterization were made in the frailty related distribution, the asymptotic distribution of the minima would be the same as that obtained in the case of the flexible distribution (11.8.3) with $\theta = 0$ (which is the difference of extremes model); i.e. (11.9.7) and (11.9.4) differ only by parameterization.

11.10 SIMULATION

In order to compare the quality and efficiency of various inferential techniques, it is desirable to be able to simulate samples from the spectrum of competing multivariate logistic models encountered in the present chapter. Since it is possible in a straightforward fashion to simulate univariate logistic variables, univariate extreme value variables, and k-dimensional geometric random variables, etc. (a good reference for such simulation is Devroye, 1986), there is no problem simulating data from the classic model (11.2.1), the difference of extremes model (11.4.2), geometric minima and maxima models (11.5.7), (11.5.8), (11.5.11), etc., and the frailty related models (11.6.5), (11.6.7), (11.6.8). The only families of distributions encountered in the chapter, which do not have a simple stochastic model associated with them are the Farlie-Gumbel-Morgenstern model (11.7.1) and the flexible bivariate model (11.8.3). Simulation is possible for the flexible model. By straightforward marginal transformations we will be able to simulate data from the model (11.8.3) if we can simulate bivariate data with joint survival function

$$P(X \geq x, Y \geq y) = [1 + (x + y + \theta xy)^{1/\alpha}]^{-1}. \tag{11.10.1}$$

The corresponding density on $(0, \infty) \times (0, \infty)$ may be denoted by $g(x, y; \theta, \alpha)$. One may verify that

$$g(x, y; \theta, \alpha) \leq \left(\frac{\theta}{\alpha - 1} + 5\right) g(x, y; 0, \alpha), \tag{11.10.2}$$

so that by using a rejection scheme it will be enough to simulate data in the case $\theta = 0$. However, substitution of $\theta = 0$ in (11.10.1) yields a distribution which may be marginally transformed to the model (11.8.1) which, as a difference of extremes, is easily simulated.

Finally, data from the Farlie-Gumbel-Morgenstern (FGM) distribution (11.7.1) can be simulated by marginally transforming FGM data with uniform marginals. A technique for generating such uniform marginal data is described (in the bivariate case) in Devroye (1986, pp. 580–581).

11.11 ESTIMATION

If location and scale parameters are introduced in the multivariate models, marginal sample means and sample variances provide convenient consistent estimates of these parameters. Typically the remaining structural parameters in the models can be consistently estimated by some correlation-like measure. In the difference of extremes model (11.4.2), the correlation took the form

$$\rho(Z_i, Z_j) = 1 - (2\alpha^2)^{-1}. \tag{11.11.1}$$

A suitable combination of all pairwise sample correlations will then provide a consistent estimate of $1 - (2\alpha^2)^{-1}$ from which a consistent estimate of α is readily obtained.

For the frailty related models which are Archimedean in the sense of Genest and Mackay (1986), the "right" measure of correlation is Kendall's coefficient of concordance, $\tau_{ij} = E(\text{sign}((Z_i - Z_i^*)(Z_j - Z_j^*)))$, where $\mathbf{Z}$, $\mathbf{Z}^*$ are i.i.d. vectors. For Archimedean distributions the τ_{ij}'s are always expressible in integral form and sometimes are easily calculated. For example, in the case of distribution (11.6.7) one finds

$$\tau_{ij} = (1 + 2\alpha)^{-1}. \tag{11.11.2}$$

For the distribution (11.6.8) an even simpler expression is encountered

$$\tau_{ij} = 1 - \alpha \tag{11.11.3}$$

(note that $0 \leq \alpha \leq 1$ in (11.6.8)). In both of these cases sample versions of the τ_{ij}'s can be sensibly combined to yield consistent estimates of the dependence parameter α.

The hybrid nature of the flexible model (11.8.3) can be expected to lead to more difficult estimation problems. The likelihood function associated with the model (11.8.3) is not simple. The moments are not available in closed form. However, consistent estimates of θ and α are possible by considering

$$W = \exp[-\max(Z_1, Z_2)]. \tag{11.11.4}$$

The corresponding density for W is

$$f_W(w) = \frac{2(2u^\alpha + \theta u^{2\alpha})^{1/\alpha - 1}[u^{\alpha-1} + \theta u^{2\alpha-1}]}{[1 + (2w^\alpha + \theta w^{2\alpha})^{1/\alpha}]^2}, \qquad 0 < w < \infty. \qquad (11.11.5)$$

Assuming we have n bivariate observations from (11.8.3), we can compute the corresponding W_i's using (11.11.4) and develop suitable likelihood equations (using (11.11.5)) which can be solved iteratively.

11.12 LOGISTIC CONDITIONALS

Under certain circumstances it may be easier to visualize and model conditional distributions rather than marginal distributions. In the present context, restricting attention to the bivariate case for simplicity, we may be interested in joint densities $f_{Z_1,Z_2}(z_1, z_2)$ such that

1. For every z_2, the conditional density of Z_1 given $Z_2 = z_2$ is logistic with parameters which may depend on z_2.
2. For every z_1, the conditional density of Z_2 given $Z_1 = z_1$ is logistic with parameters which may depend on z_1.

It is natural to admit the possibility that the conditional distributions of Z_1 given $Z_2 = z_2$ may have both location and scale parameters which depend on z_2. Analogously, the location and scale of $f_{Z_2|Z_1}(z_2|z_1)$ might both depend on z_1. In this generality, the determination of the class of all bivariate distributions with logistic conditionals is an open question. If we only allow the location parameters to depend on the value of the conditioning variable, the problem can be reduced to one already solved.

Thus we seek all bivariate densities $f_{Z_1,Z_2}(z_1, z_2)$ with the property that

1. For each z_2,

$$P(Z_1 \geq z_1|Z_2 = z_2) = [1 + e^{z_1 - a(z_2)}]^{-1} \qquad (11.12.1)$$

for some function $a(z_2)$.

2. For each z_1,

$$P(Z_2 \geq z_2|Z_1 = z_1) = [1 + e^{z_2 - b(z_1)}]^{-1} \qquad (11.12.2)$$

for some function $b(z_1)$.

If one makes the change of variable $U_i = e^{Z_i}$ in the above formulation, (11.12.1) and (11.12.2) indicate that U_1, U_2 should have Pareto conditionals in the sense discussed by Arnold (1987) (with $\alpha = 1$). It follows from that paper that if (11.12.1) and (11.12.2) are to hold then the joint density of

(Z_1, Z_2) must be of the form

$$f_{Z_1,Z_2}(z_1, z_2) = \left[\frac{1-\phi}{-\log\phi}\right] e^{(z_1-\mu_1)+(z_2-\mu_2)} \times [1 + e^{z_1-\mu_1} + e^{z_2-\mu_2} + \phi e^{z_1-\mu_1+z_2-\mu_2}]^{-2}, \quad (11.12.3)$$

where $\mu_1, \mu_2 \in \mathbf{R}$ and $\phi > 0$. When $\phi = 1$ (in which case $(1 - \phi)/[-\log \phi]$ is defined to be 1) we have independent logistic marginals. Scale parameters σ_1 and σ_2 can be introduced in (11.12.3) without upsetting the logistic conditionals property. Parameter estimation in the family (11.12.3) can be implemented numerically by maximum likelihood (see Arnold, 1987).

Note that (except in trivial cases where independent marginals occur) none of the bivariate logistic models discussed earlier have logistic conditional distributions.

11.13 NOTES

1. Satterthwaite and Hutchinson (1978) discuss a generalization of Gumbel's bivariate logistic distribution. Their model is

$$P(Z_1 \le z_1, Z_2 \le z_2) = [1 + e^{-z_1} + e^{-z_2}]^{-\gamma}, \quad (11.13.1)$$

where $\gamma > 0$. This distribution has more flexible correlation structure than Gumbel's distribution, but it does not have logistic marginals.

2. Cook and Johnson (1986) introduce the following family of bivariate distributions:

$$\begin{aligned} P(Z_1 \le z_1, Z_2 \le z_2) &= (1 + \beta)[1 + e^{-z_1} + e^{-z_2}]^{-\alpha} + \beta[1 + 2e^{-z_1} + 2e^{-z_2}]^{-\alpha} \\ &\quad - \beta[1 + 2e^{-z_1} + e^{-z_2}]^{-\alpha} - \beta[1 + e^{-z_1} + 2e^{-z_2}]^{-\alpha}, \quad (11.13.2) \end{aligned}$$

where $\alpha > 0$ and $-1 \le \beta \le 1$. Only when $\alpha = 1$ are logistic marginals encountered. The resulting one-parameter bivariate logistic family is flexible, but it is not clear why the particular linear combination of distribution functions exhibited in (11.14.2) should be preferred over some other linear combination.

Symanowski and Koehler (1989) introduce a variation on the Cook-Johnson model that always has standard logistic marginals. Their model takes the form

$$
\begin{aligned}
P(Z_1 \le z_1, Z_2 \le z_2) = {} & (1 + \beta)[(1 + e^{-z_1})^{\alpha^{-1}} + (1 + e^{-z_2})^{-\alpha^{-1}} - 1]^{-\alpha} \\
& + \beta[2(1 + e^{-z_1})^{\alpha^{-1}} + 2(1 + e^{-z_2})^{\alpha^{-1}} - 3]^{-\alpha} \\
& - \beta[2(1 + e^{-z_1})^{\alpha^{-1}} + 2(1 + e^{-z_2})^{\alpha^{-1}} - 2]^{-\alpha} \\
& - \beta[(1 + e^{-z_1})^{\alpha^{-1}} + 2(1 + e^{-z_2})^{\alpha^{-1}} - 2]^{-\alpha}.
\end{aligned}
$$

Further discussion of this model is found in Chapter 16.

3. Recent papers dealing with stationary logistic processes implicitly include descriptions of a variety of multivariate logistic distributions.

For example, from Arnold and Robertson (1989a) we might consider a stationary process $Z_1, Z_2, \ldots$ defined as follows. Let Z_1 be a standard logistic random variable, and let $\epsilon_1, \epsilon_2, \ldots$ be a sequence of i.i.d. standard logistic variables. Define, for $n = 2, 3, \ldots,$

$$
\begin{aligned}
Z_n &= Z_{n-1} - \log p \quad \text{w.p. } p \\
&= \min(Z_{n-1} - \log p, \epsilon_n) \quad \text{w.p. } 1 - p,
\end{aligned} \tag{11.13.3}
$$

where $p \in (0, 1)$. The resulting joint survival function for (Z_{n-1}, Z_n) (with logistic marginals) is then

$$
P(Z_{n-1} \ge z_{n-1}, Z_n \ge z_n) = \frac{1 + pe^{z_n}}{(1 + e^{z_n})(1 + \max(e^{z_{n-1}}, pe^{z_n}))}. \tag{11.13.4}
$$

This distribution has a singular component. A related absolutely continuous distribution is described in Arnold and Robertson (1989a, see Eqs. (4.4) and (4.5)).

In the context of a logistic process involving geometric minimization, Arnold (1989) discusses the following bivariate survival function with logistic marginals

$$
\begin{aligned}
P(Z_{n-1} \ge z_{n-1}, Z_n \ge z_n) = {} & (1 + pe^{z_{n-1}})^{-1} \\
& \times \left\{ \frac{p}{1 + e^{z_n}} + \frac{1 - p}{1 + \exp(\max(z_{n-1}, z_n))} \right\},
\end{aligned} \tag{11.13.5}
$$

where $p \in (0, 1)$.

4. Elliptically contoured k-variate logistic distributions exist. Such distributions correspond to random vectors of the form

$$
\mathbf{Z} = A\mathbf{Z}^* + \mathbf{b},
$$

where $\mathbf{Z}^*$ is a spherically symmetric random vector with standard logistic marginals. Such a random vector $\mathbf{Z}^*$ does exist (see Arnold and Robertson, 1989b for details). One may represent $\mathbf{Z}^*$ in the form

$$\mathbf{Z}^* = R\mathbf{U}^*,$$

where $\mathbf{U}^*$ is a random vector uniformly distributed over the unit k-sphere and R, independent of $\mathbf{U}^*$, has density

$$f_R(r) = \sum_{n=1}^{\infty} (-1)^{n-1} \frac{r^{k/2} n^{k/2+1} 2^{2-k/2}}{\Gamma(k/2)} K_{k/2-1}(nr), \qquad r > 0, \quad (11.13.6)$$

where K_ν is a modified Bessel function of the third kind. Curiously (11.13.6) simplifies considerably in the case where $k = 3$. In that case, the resulting spherically contoured three-dimensional distribution with logistic marginals has the following simple expression for its density:

$$f_{Z_1,Z_2,Z_3}(z_1, z_2, z_3) = \frac{\tanh\left(\dfrac{\sqrt{z_1^2 + z_2^2 + z_3^2}}{2}\right) \operatorname{sech}^2\left(\dfrac{\sqrt{z_1^2 + z_2^2 + z_3^2}}{2}\right)}{8\pi\sqrt{z_1^2 + z_2^2 + z_3^2}}. \quad (11.13.7)$$

In dimensions other than three no simple expression seems to be available.

REFERENCES

Ali, M. M., Mikhail, N. N., and Haq, M. S. (1978). A class of bivariate distributions including the bivariate logistic, *J. Multivariate Anal.*, *8*, 405–412.

Arnold, B. C. (1975). Multivariate exponential distributions based on hierarchical successive damage, *J. Appl. Probab.*, *12*, 142–147.

Arnold, B. C. (1987). Bivariate distributions with Pareto conditionals, *Statist. Probab. Lett.*, *5*, 263–266.

Arnold, B. C. (1989). A logistic process constructed using geometric minimization, *Statist. Probab. Lett.*, *7*, 253–257.

Arnold, B. C. (1990). A flexible family of multivariate Pareto distributions, *J. Statist. Planning Inference*, *24*, 249–258.

Arnold, B. C. and Robertson, C. A. (1989a). Autoregressive logistic processes, *J. Appl. Probab.*, *26*, 524–531.

Arnold, B. C. and Robertson, C. A. (1989b). Elliptically contoured distributions with logistic marginals, *Technical Report #180*, Department of Statistics, University of California, Riverside, CA.

Block, H. W. (1975). Physical models leading to multivariate exponential and negative binomial distributions, *Modeling and Simulation*, *6*, 445–450.

Cook, R. D. and Johnson, M. E. (1986). Generalized Burr-Pareto-logistic distributions with applications to a uranium exploration data set, *Technometrics*, *28*, 123–131.

Devroye, L. (1986). *Non-uniform Random Variate Generation*, Springer-Verlag, New York.

Genest, C. and MacKay, J. (1986). The joy of copulas: Bivariate distributions with uniform marginals, *The American Statistician*, *40*, 280–283.

George, E. O. and Mudholkar, G. S. (1981). Some relationships between the logistic and the exponential distributions, in *Statistical Distributions in Scientific Work*, C. Taillie, G. P. Patil, and B. Baldessari, (eds.), *4*, 401–409, Reidel, Dordrecht-Holland.

Gumbel, E. J. (1958). *Statistics of Extremes*, Columbia University Press, New York.

Gumbel, E. J. (1961). Bivariate logistic distributions, *J. Am. Statist. Assoc.*, *56*, 335–349.

Johnson, N. L. and Kotz, S. (1975). On some generalized Farlie-Gumbel-Morgenstern distributions, *Commun. Statist.*, *4*, 415–427.

Johnson, N. L. and Kotz, S. (1977). On some generalized Farlie-Gumbel-Morgenstern distributions. II: Regression, correlation and further generalizations, *Commun. Statist.*, *6*, 485–496.

Lindley, D. V. and Singpurwalla, N. D. (1986). Multivariate distributions for the life lengths of components of a system sharing a common environment, *J. Appl. Probab.*, *23*, 418–431.

Malik, H. J. and Abraham, B. (1973). Multivariate logistic distributions, *Ann. Statist.*, *1*, 588–590.

Oakes, D. (1989). Bivariate survival models induced by frailties, *J. Am. Statist. Assoc.*, *84*, 487–493.

de Oliveira, J. T. (1961). La representation des distributions extremales bivariees, *Bull. Int. Statist. Inst.*, *33*, 477–480.

Satterthwaite, S. P. and Hutchinson, T. P. (1978). A generalization of Gumbel's bivariate logistic distribution, *Metrika*, *25*, 163–170.

Strauss, D. J. (1979). Some results on random utility, *J. Math. Psychology*, *20*, 35–52.

Symanowski, J. T. and Koehler, K. J. (1989). A bivariate logistic distribution with applications to categorical responses, *Technical Report #89-29*, Department of Statistics, Iowa State University, Ames, Iowa.

[illegible]

12
Outlier and Robustness of Estimators

N. Balakrishnan
McMaster University, Hamilton, Ontario, Canada

12.1 INTRODUCTION

In Chapter 3 we discussed the maximum likelihood and the approximate maximum likelihood estimation methods for the location and scale parameters of a logistic population. In Chapter 4 we presented a detailed account of various linear estimation methods for the location and scale parameters based on complete as well as Type II censored samples. Further, there are some omnibus robust estimators like the trimmed, Winsorized estimators, modified maximum likelihood estimators, the median, the linearly weighted mean, Gastwirth mean, etc., available in the literature for the estimation of the location and scale parameters. In this chapter, we study the performance of these estimators by considering a single outlier logistic model. Similar work has been carried out by Shu (1978), David and Shu (1978), and Arnold and Balakrishnan (1989) for the normal distribution, Kale and Sinha (1971) and Joshi (1972) for the exponential distribution, and Balakrishnan and Ambagaspitiya (1988) for the double exponential or Laplace distribution.

Density functions and joint density functions of order statistics arising from a sample containing a single outlier have been given by David and Shu (1978) and Arnold and Balakrishnan (1989), and have been made use of by David et al. (1977) and Balakrishnan et al. (1991) in tabulating the means, variances, and covariances of order statistics from a single outlier normal and logistic models, respectively. One may also refer to Vaughan and Venables (1972) for more general expressions of distributions of order statistics when the sample, in fact, includes k outliers. The importance of a systematic study of order statistics from an outlier model and the usefulness of the tables of means, variances, and covariances of these order statistics in the context of robustness have been well demonstrated by several authors, including Andrews et al. (1972), David (1981), and Tiku et al. (1986).

We first present the density function and the joint density function of order statistics in Section 12.2. In Section 12.3 we describe the single (location as well as scale) outlier logistic model considered in this study and explain the evaluation of the means, variances, and covariances of order statistics under these models as carried out by Balakrishnan et al. (1991). The functional behavior of order statistics from a single location— and scale—outlier models is discussed in detail in Section 12.4. Finally, in Section 12.5 we make use of these results to study the robustness properties (through bias and mean square error) of various estimators of the location and scale parameters of a logistic population.

12.2 DISTRIBUTIONS OF ORDER STATISTICS

We present the distributions of order statistics obtained from a sample of size n when an unidentified outlier is present in the sample. For the sake of convenience, let us represent the sample by n independent absolutely continuous random variables Y_i ($i = 1, 2, \ldots, n - 1$) and X, such that Y_i has pdf $f(x)$ and cdf $F(x)$, and X has pdf $g(x)$ and cdf $G(x)$. Further, let

$$O_{1:n} \le O_{2:n} \le \cdots \le O_{n:n} \tag{12.2.1}$$

be the order statistics obtained by arranging the n independent observations in increasing order of magnitude.

The cumulative distribution function of $O_{n:n}$, denoted by $H_{n:n}(x)$, is obtained as

$$\begin{aligned} H_{n:n}(x) &= \Pr\{\text{all of } Y_1, Y_2, \ldots, Y_{n-1}, X \le x\} \\ &= \{F(x)\}^{n-1}G(x), \qquad -\infty < x < \infty. \end{aligned} \tag{12.2.2}$$

Similarly, the cumulative distribution function of $O_{i:n}$ ($1 \le i \le n - 1$), denoted by $H_{i:n}(x)$, is

$$\begin{aligned} H_{i:n}(x) &= \Pr\{\text{at least } i \text{ of } Y_1, Y_2, \ldots, Y_{n-1}, X \le x\} \\ &= \Pr\{\text{exactly } i - 1 \text{ of } Y_1, Y_2, \ldots, Y_{n-1} \le x \text{ and } X \le x\} \end{aligned}$$

$$+ \Pr\{\text{at least } i \text{ of } Y_1, Y_2, \ldots, Y_{n-1} \le x\}$$

$$= \binom{n-1}{i-1} \{F(x)\}^{i-1}\{1 - F(x)\}^{n-i}G(x) + F_{i:n-1}(x), \quad (12.2.3)$$

where $F_{i:n-1}(x)$ is the distribution function of the ith order statistic in a sample of size $n - 1$ drawn from a population with pdf $f(x)$ and cdf $F(x)$. Differentiating the expressions in (12.2.2) and (12.2.3), we derive the density function of $O_{i:n}$ $(1 \le i \le n)$ as

$$\begin{aligned} h_{i:n}(x) &= \frac{(n-1)!}{(i-2)!(n-i)!} \{F(x)\}^{i-2}\{1 - F(x)\}^{n-i}G(x)f(x) \\ &+ \frac{(n-1)!}{(i-1)!(n-i)!} \{F(x)\}^{i-1}\{1 - F(x)\}^{n-i}g(x) \\ &+ \frac{(n-1)!}{(i-1)!(n-i-1)!} \{F(x)\}^{i-1} \\ &\times \{1 - F(x)\}^{n-i-1}\{1 - G(x)\}f(x), \qquad -\infty < x < \infty, \end{aligned} \quad (12.2.4)$$

where the first term drops out when $i = 1$, and the last when $i = n$.

The above density function may also be derived in an alternative way as given below, which lends itself to further extensions. The event $x < O_{i:n} \le x + \delta x$ may be seen as follows:

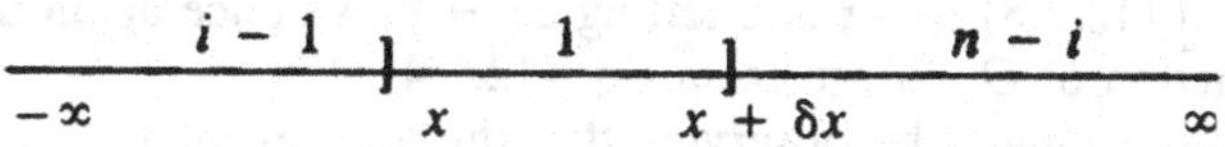

$O_r \le x$ for $i - 1$ of the O_r, $x < O_r \le x + \delta x$ for exactly one O_r, and $O_r > x + \delta x$ for the remaining $n - i$ of the O_r. Realizing now that the outlying variable X could belong to any one of three intervals, we have the following three possibilities:

1. $Y_r \le x$ for $i - 2$ of the Y_r and $X \le x$, $x < Y_r \le x + \delta x$ for exactly one Y_r, and $Y_r > x + \delta x$ for the remaining $n - i$ of the Y_r, with a probability

$$\frac{(n-1)!}{(i-2)!(n-i)!} \{F(x)\}^{i-2}G(x)\{F(x + \delta x) - F(x)\} \times \{1 - F(x + \delta x)\}^{n-i}, \qquad -\infty < x < \infty. \quad (12.2.5)$$

2. $Y_r \le x$ for $i - 1$ of the Y_r, $x < X \le x + \delta x$, and $Y_r > x + \delta x$ for the remaining $n - i$ of the Y_r, with a probability

$$\frac{(n-1)!}{(i-1)!(n-i)!} \{F(x)\}^{i-1}\{G(x + \delta x) - G(x)\} \times \{1 - F(x + \delta x)\}^{n-i}, \qquad -\infty < x < \infty. \quad (12.2.6)$$

3. $Y_r \le x$ for $i - 1$ of the Y_r, $x < Y \le x + \delta x$ for exactly one Y_r, and $Y_r > x + \delta x$ for the remaining $n - i - 1$ of the Y_r and $X > x + \delta x$, with a probability

$$\frac{(n-1)!}{(i-1)!(n-i-1)!}\{F(x)\}^{i-1}\{F(x+\delta x) - F(x)\}\{1 - F(x+\delta x)\}^{n-i-1} \times \{1 - G(x+\delta x)\}, \qquad -\infty < x < \infty. \quad (12.2.7)$$

By taking δx small, we can write

$$\begin{aligned}\Pr(x < O_{i:n} \le x + \delta x) &= \frac{(n-1)!}{(i-2)!(n-i)!}\{F(x)\}^{i-2}G(x)\{1 - F(x+\delta x)\}^{n-i}f(x)\,\delta x \\ &+ \frac{(n-1)!}{(i-1)!(n-i)!}\{F(x)\}^{i-1}\{1 - F(x+\delta x)\}^{n-i}g(x)\,\delta x \\ &+ \frac{(n-1)!}{(i-1)!(n-i-1)!}\{F(x)\}^{i-1}\{1 - F(x+\delta x)\}^{n-i-1} \\ &\times \{1 - G(x+\delta x)\}f(x)\,\delta x + O((\delta x)^2), \end{aligned} \quad (12.2.8)$$

where $O((\delta x)^2)$ denotes the probability of more than one O_r falling in the interval $(x, x + \delta x)$ and, therefore, is a term of order $(\delta x)^2$. By dividing both sides of (12.2.8) by δx and letting $\delta x \to 0$, we once again derive the density function of $O_{i:n}$ as given in Eq. (12.2.4).

Proceeding similarly by observing that the joint event $x < O_{i:n} \le x + \delta x$, $y < O_{j:n} \le y + \delta y$ is obtained by the configuration (by neglecting terms of lower order of probability)

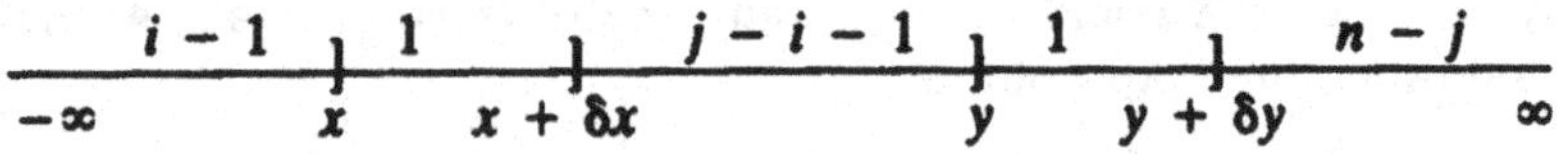

we derive the joint density function of $O_{i:n}$ and $O_{j:n}$ $(1 \le i < j \le n)$ as

$$\begin{aligned}h_{i,j:n}(x, y) &= \frac{(n-1)!}{(i-2)!(j-i-1)!(n-j)!}\{F(x)\}^{i-2} \\ &\times \{F(y) - F(x)\}^{j-i-1}\{1 - F(y)\}^{n-j}G(x)f(x)f(y) \\ &+ \frac{(n-1)!}{(i-1)!(j-i-1)!(n-j)!} \\ &\times \{F(x)\}^{i-1}\{F(y) - F(x)\}^{j-i-1}\{1 - F(y)\}^{n-j}g(x)f(y) \\ &+ \frac{(n-1)!}{(i-1)!(j-i-2)!(n-j)!}\{F(x)\}^{i-1}\end{aligned}$$

$$\times \{F(y) - F(x)\}^{j-i-2}\{1 - F(y)\}^{n-j}\{G(y) - G(x)\}f(x)f(y)$$
$$+ \frac{(n-1)!}{(i-1)!(j-i-1)!(n-j)!}\{F(x)\}^{i-1}$$
$$\times \{F(y) - F(x)\}^{j-i-1}\{1 - F(y)\}^{n-j}f(x)g(y)$$
$$+ \frac{(n-1)!}{(i-1)!(j-i-1)!(n-j-1)!}\{F(x)\}^{i-1}$$
$$\times \{F(y) - F(x)\}^{j-i-1}\{1 - F(y)\}^{n-j-1}\{1 - G(y)\}f(x)f(y),$$
$$-\infty < x < y < \infty, \quad (12.2.9)$$

where the first term drops out if $i = 1$, the middle term if $j = i + 1$, and the last term if $j = n$. The density functions in (12.2.4) and (12.2.9) are special cases of the general results of Vaughan and Venables (1972), who have expressed densities of order statistics from nonidentically distributed variables in terms of permanents. By using the density functions in (12.2.4) and (12.2.9), Balakrishnan (1988a,b) has established several recurrence relations and identities satisfied by the single and the product moments of order statistics from a general single-outlier model. These results have also been subsequently generalized by Balakrishnan (1988c, 1989a,b) for moments of order statistics arising from nonidentically distributed random variables. Many of these results are presented in Chapter 5 of Arnold and Balakrishnan (1989).

12.3 SINGLE-OUTLIER LOGISTIC MODELS AND MOMENTS OF ORDER STATISTICS

With the density function of $O_{i:n}$ and the joint density function of $O_{i:n}$ and $O_{j:n}$ as given in Eqs. (12.2.4) and (12.2.9), respectively, let us denote the single moments by

$$\nu_{i:n}^{(k)} = E(O_{i:n}^k) = \int_{-\infty}^{\infty} x^k h_{i:n}(x)\,dx, \qquad 1 \le i \le n,\ k \ge 1, \quad (12.3.1)$$

and the product moments by

$$\nu_{i,j:n} = E(O_{i:n}O_{j:n})$$
$$= \iint\limits_{-\infty<x<y<\infty} xyh_{i,j:n}(x, y)\,dy\,dx, \qquad 1 \le i < j \le n. \quad (12.3.2)$$

Let us also denote the covariance between $O_{i:n}$ and $O_{j:n}$ by

$$\eta_{i,j:n} = \text{Cov}(O_{i:n}, O_{j:n}) = \nu_{i,j:n} - \nu_{i:n}\nu_{j:n}, \qquad 1 \le i \le j \le n. \quad (12.3.3)$$

Further, let us use $f_{i:n}(x)$ and $f_{i,j:n}(x, y)$ to denote the density of the ith order statistic and the joint density of the ith and jth order statistics, respectively, in a sample of size n drawn from a population with pdf $f(x)$ and cdf $F(x)$ (without an outlier) and denote the single moments, product moments, and covariances of order statistics in this case by

$$\alpha_{i:n}^{(k)} = \int_{-\infty}^{\infty} x^k f_{i:n}(x)\, dx, \qquad 1 \le i \le n,\ k \ge 1, \tag{12.3.4}$$

$$\alpha_{i,j:n} = \iint\limits_{-\infty<x<y<\infty} xy f_{i,j:n}(x, y)\, dy\, dx, \qquad 1 \le i < j \le n, \tag{12.3.5}$$

and

$$\beta_{i,j:n} = \alpha_{i,j:n} - \alpha_{i:n}\alpha_{j:n}, \qquad 1 \le i \le j \le n, \tag{12.3.6}$$

respectively.

12.3.1 Location-Outlier Logistic Model

In the location-outlier logistic model, we assume that the single outlier in the sample has a location shift from the remaining variables. That is, we assume that $Y_1, Y_2, \ldots, Y_{n-1}$ are independent logistic $L(\mu, \sigma^2)$ variables with pdf $f(x; \mu, \sigma)$ and cdf $F(x; \mu, \sigma)$ as given in Eqs. (1.1) and (1.2), respectively, and the single outlier X is an independent logistic $L(\mu + \lambda\sigma, \sigma^2)$ variable. By considering the standardized form of this outlier model (with $\mu = 0$ and $\sigma = 1$), Balakrishnan et al. (1991) have tabulated the values of $\nu_{i:n}^{(1)}$ $(1 \le i \le n)$ and $\eta_{i,j:n}$ $(1 \le i \le j \le n)$ for sample size n up to 20 and for $\lambda = 0.5(0.5)3, 4$. Of course, for the general location-outlier case, the means and covariances of order statistics may be obtained as $\mu + \sigma\nu_{i:n}^{(1)}$ $(1 \le i \le n)$ and $\sigma^2\eta_{i,j:n}$ $(1 \le i \le j \le n)$ by using these tabulated values.

12.3.2 Scale-Outlier Logistic Model

Under the scale-outlier logistic model, we assume that the single outlier in the sample has a scale shift from the remaining variables. In other words, we assume that $Y_1, Y_2, \ldots, Y_{n-1}$ are independent logistic $L(\mu, \sigma^2)$ variables with pdf $f(x; \mu, \sigma)$ and cdf $F(x; \mu, \sigma)$ as given in Eqs. (1.1) and (1.2), respectively, and the single outlier X is an independent logistic $L(\mu, \sigma^2\tau^2)$ variable. Balakrishnan et al. (1991) have tabulated the values of $\nu_{i:n}^{(1)}$ $(1 \le i \le n)$ and $\eta_{i,j:n}$ $(1 \le i \le j \le n)$ for sample size n up to 20 and for $\tau = 0.5(0.5)2(1)4$ by considering the standardized form of this outlier model (with $\mu = 0$ and $\sigma = 1$). Once again, the means and covariances of order statistics for the general scale-outlier case may be obtained from these tables as $\mu + \sigma\nu_{i:n}^{(1)}$ $(1 \le i \le n)$ and $\sigma^2\eta_{i,j:n}$ $(1 \le i \le j \le n)$.

Expected values, variances, and covariances of order statistics in the nonoutlier case (corresponding to either $\lambda = 0$ in the location-outlier model or $\tau = 1$ in the scale-outlier model), as mentioned already in Chapter 2, have been tabulated rather extensively by Gupta and Shah (1965), Shah (1966), Gupta et al. (1967), and Balakrishnan and Malik (1991). The values of means and covariances tabulated by Balakrishnan et al. (1991) in the above described outlier cases may also be supplemented by results for $\lambda = \infty$ in the location-outlier case and for $\tau = \infty$ in the scale-outlier case which are obtainable from the nonoutlier case for sample size $n - 1$ as shall be explained in the following section. Similar tables of means, variances, and covariances of order statistics from a single-outlier normal model have been prepared by David et al. (1977), which have been subsequently utilized by David and Shu (1978) in robustness studies.

12.4 FUNCTIONAL BEHAVIOR OF ORDER STATISTICS

Let us first consider the single location-outlier model, that is, $G(x) = F(x - \lambda)$ for all x. We may then write the outlying variable X as $X = Y_n + \lambda$, where Y_n is a random variable with pdf $f(x)$ and cdf $F(x)$ and independent of the remaining $n - 1$ variables $Y_1, Y_2, \ldots, Y_{n-1}$. For the sake of clarity in the forthcoming discussion, let us denote the ith order statistic in this case by $O_{i:n}(\lambda)$, and its pdf and cdf by $h_{i:n}(x; \lambda)$ and $H_{i:n}(x; \lambda)$, respectively. We then observe directly from Eqs. (12.2.2) and (12.2.3) that $H_{i:n}(x; \lambda)$, for $1 \le i \le n$, is a decreasing function of λ. The behaviour of $O_{i:n}(\lambda)$ as a function of λ has been studied by David and Shu (1978); interested readers may also refer to Hampel (1974). Let us now denote the realizations of the random variables Y, X, and O by y, x, and o. Then, by inserting the outlier $x = y_n + \lambda$ into the ordered sample of size $n - 1$ denoted by $y_{1:n-1} \le y_{2:n-1} \le \cdots \le y_{n-1:n-1}$, we realize for fixed values of $y_1, y_2, \ldots, y_n$ that

$$
\begin{aligned}
o_{1:n}(\lambda) &= y_n + \lambda && \text{if } y_n + \lambda \le y_{1:n-1}, \\
&= y_{1:n-1} && \text{if } y_n + \lambda > y_{1:n-1},
\end{aligned}
\tag{12.4.1}
$$

$$
\begin{aligned}
o_{i:n}(\lambda) &= y_{i-1:n-1} && \text{if } y_n + \lambda \le y_{i-1:n-1}, \\
&= y_n + \lambda && \text{if } y_{i-1:n-1} < y_n + \lambda \le y_{i:n-1}, \\
&= y_{i:n-1} && \text{if } y_n + \lambda > y_{i:n-1},
\end{aligned}
\tag{12.4.2}
$$

for $i = 2, 3, \ldots, n-1$, and

$$o_{n:n}(\lambda) = y_{n-1:n-1} \qquad \text{if } y_n + \lambda \le y_{n-1:n-1},$$
$$= y_n + \lambda \qquad \text{if } y_n + \lambda > y_{n-1:n-1}. \tag{12.4.3}$$

From Eqs. (12.4.1) to (12.4.3), we see that $o_{i:n}(\lambda)$ is a nondecreasing function of λ. Further, we also observe that

$$\lim_{\lambda \to -\infty} o_{1:n}(\lambda) = o_{1:n}(-\infty) = -\infty, \tag{12.4.4}$$

$$\lim_{\lambda \to -\infty} o_{i:n}(\lambda) = o_{i:n}(-\infty) = y_{i-1:n-1}, \qquad 2 \le i \le n, \tag{12.4.5}$$

$$\lim_{\lambda \to \infty} o_{i:n}(\lambda) = o_{i:n}(\infty) = y_{i:n-1}, \qquad 1 \le i \le n-1, \tag{12.4.6}$$

and

$$\lim_{\lambda \to \infty} o_{n:n}(\lambda) = o_{n:n}(\infty) = \infty. \tag{12.4.7}$$

By using the monotone convergence theorem (Loeve, 1977), we also have for $2 \le i \le n$,

$$\lim_{\lambda \to -\infty} E(O_{i:n}(\lambda)) = E\left(\lim_{\lambda \to -\infty} O_{i:n}(\lambda)\right)$$

implying that

$$\nu_{i:n}(-\infty) = E(Y_{i-1:n-1}) = \alpha_{i-1:n-1}, \tag{12.4.8}$$

and similarly for $1 \le i \le n-1$,

$$\lim_{\lambda \to \infty} E(O_{i:n}(\lambda)) = E\left(\lim_{\lambda \to \infty} O_{i:n}(\lambda)\right),$$

implying that

$$\nu_{i:n}(\infty) = E(Y_{i:n-1}) = \alpha_{i:n-1}, \tag{12.4.9}$$

and also that

$$\lim_{\lambda \to -\infty} E(O_{1:n}(\lambda)) = \nu_{1:n}(-\infty) = -\infty \tag{12.4.10}$$

and

$$\lim_{\lambda \to \infty} E(O_{n:n}(\lambda)) = \nu_{n:n}(\infty) = \infty. \tag{12.4.11}$$

Furthermore, for fixed values of x and y, upon noting that

$$f(x-\lambda) \to 0, \quad F(x-\lambda) \to 1, \quad F(y-\lambda) - F(x-\lambda) \to 0 \tag{12.4.12}$$

as $\lambda \to -\infty$, and that

$$f(x - \lambda) \to 0, \quad F(x - \lambda) \to 0, \quad F(y - \lambda) - F(x - \lambda) \to 0 \tag{12.4.13}$$

as $\lambda \to \infty$, we observe from Eqs. (12.2.4) and (12.2.9) that

$$\lim_{\lambda \to -\infty} h_{i:n}(x; \lambda) = h_{i:n}(x; -\infty) = f_{i-1:n-1}(x), \qquad 2 \le i \le n, \tag{12.4.14}$$

$$\lim_{\lambda \to -\infty} h_{i,j:n}(x, y; \lambda) = h_{i,j:n}(x, y; -\infty)$$
$$= f_{i-1,j-1:n-1}(x, y), \qquad 2 \le i < j \le n, \tag{12.4.15}$$

$$\lim_{\lambda \to \infty} h_{i:n}(x; \lambda) = h_{i:n}(x; \infty) = f_{i:n-1}(x), \qquad 1 \le i \le n - 1, \tag{12.4.16}$$

and

$$\lim_{\lambda \to \infty} h_{i,j:n}(x, y; \lambda) = h_{i,j:n}(x, y; \infty) = f_{i,j:n-1}(x, y), \qquad 1 \le i < j \le n - 1. \tag{12.4.17}$$

Now upon using the Lebesgue dominated convergence theorem (Loeve, 1977), we obtain

$$\lim_{\lambda \to -\infty} \eta_{i,j:n}(\lambda) = \eta_{i,j:n}(-\infty) = \mathrm{Cov}(Y_{i-1:n-1}, Y_{j-1:n-1})$$
$$= \beta_{i-1,j-1:n-1}, \qquad 2 \le i < j \le n, \tag{12.4.18}$$

and

$$\lim_{\lambda \to \infty} \eta_{i,j:n}(\lambda) = \eta_{i,j:n}(\infty) = \mathrm{Cov}(Y_{i:n-1}, Y_{j:n-1})$$
$$= \beta_{i,j:n-1}, \qquad 1 \le i < j \le n - 1. \tag{12.4.19}$$

Next, let us consider the single scale-outlier model, that is, $G(x) = F(x/\tau)$ for all x. We may then write the outlying variable X as $X = \tau Y_n$, where Y_n is a random variable with pdf $f(x)$ and cdf $F(x)$ and independent of the remaining $n - 1$ variables $Y_1, Y_2, \ldots, Y_{n-1}$. In this case, let us denote the ith order statistic by $O^*_{i:n}(\tau)$, and its pdf and cdf by $h^*_{i:n}(x; \tau)$ and $H^*_{i:n}(x; \tau)$, respectively. From Eqs. (12.2.2) and (12.2.3), we then observe that $H^*_{i:n}(x; \tau)$, for $1 \le i \le n$, is a decreasing function of τ for fixed positive values of x and an increasing function of τ for fixed negative values of x. The behavior of $O^*_{i:n}(\tau)$ as a function of τ has been examined by David and Shu (1978). As before, let us denote the realizations of the variables Y, X, and O^* by y, x, and o^*, and insert an outlier $x = \tau y_n$ into the ordered sample of size $n - 1$ given by $y_{1:n-1} \le y_{2:n-1} \le \cdots \le y_{n-1:n-1}$. We then observe for fixed values of $y_1, y_2, \ldots, y_n$ that

$$\begin{aligned} o^*_{1:n}(\tau) &= \tau y_n && \text{if } \tau y_n \le y_{1:n-1}, \\ &= y_{1:n-1} && \text{if } \tau y_n > y_{1:n-1}, \end{aligned} \tag{12.4.20}$$

$$\begin{aligned} o^*_{i:n}(\tau) &= y_{i-1:n-1} && \text{if } \tau y_n \le y_{i-1:n-1}, \\ &= \tau y_n && \text{if } y_{i-1:n-1} < \tau y_n \le y_{i:n-1}, \\ &= y_{i:n-1} && \text{if } \tau y_n > y_{i:n-1}, \end{aligned} \tag{12.4.21}$$

for $i = 2, 3, \ldots, n-1$, and

$$\begin{aligned} o^*_{n:n}(\tau) &= y_{n-1:n-1} && \text{if } \tau y_n \le y_{n-1:n-1}, \\ &= \tau y_n && \text{if } \tau y_n > y_{n-1:n-1}. \end{aligned} \tag{12.4.22}$$

From Eqs. (12.4.20) to (12.4.22), we see that $o^*_{i:n}(\tau)$ is a nondecreasing function of τ if $y_n > 0$ and a nonincreasing function of τ if $y_n < 0$. Further, for $2 \le i \le n-1$ we observe that

$$\begin{aligned} \lim_{\tau\to\infty} o^*_{i:n}(\tau) = o^*_{i:n}(\infty) &= y_{i-1:n-1} && \text{if } y_n < 0, \\ &= y_{i:n-1} && \text{if } y_n > 0. \end{aligned} \tag{12.4.23}$$

Since the logistic $L(0, 1)$ distribution is symmetric about zero, we obtain from Eq. (12.4.23) that for $2 \le i \le n-1$,

$$\begin{aligned} \lim_{\tau\to\infty} E(O^*_{i:n}(\tau)) &= E\left(\lim_{\tau\to\infty} O^*_{i:n}(\tau)\right) \\ &= E(O^*_{i:n}(\infty)) \\ &= \Pr(Y_n < 0)E(Y_{i-1:n-1}) + \Pr(Y_n > 0)E(Y_{i:n-1}), \end{aligned}$$

which implies that

$$\nu^*_{i:n}(\infty) = \tfrac{1}{2}(\alpha_{i-1:n-1} + \alpha_{i:n-1}), \qquad 2 \le i \le n-1. \tag{12.4.24}$$

We also observe that

$$\lim_{\tau\to\infty} E(O^*_{1:n}(\tau)) = \nu^*_{1:n}(\infty) = -\infty \tag{12.4.25}$$

and

$$\lim_{\tau\to\infty} E(O^*_{n:n}(\tau)) = \nu^*_{n:n}(\infty) = \infty. \tag{12.4.26}$$

All these results will be used in the next section to determine the limiting bias and mean square error of various estimators of the location and scale parameters of the logistic distribution when an unidentified single outlier is present in a sample of size n.

12.5 ROBUSTNESS OF ESTIMATORS OF μ AND σ

The robust estimation of the location and scale parameters of symmetric distributions, in particular, the normal distribution, has received much attention in the recent years; for example, see Crow and Siddiqui (1967), Gastwirth and Cohen (1970), and David and Shu (1978). Andrews et al. (1972) have used Monte Carlo simulations to study a rather large class of robust estimators of the location parameter. Interested readers may also refer to Tiku et al. (1986) for a detailed comparison of various robust estimators of the location and scale parameters.

In this section, we will consider various estimators of μ and σ of the logistic $L(0, 1)$ distribution and study their bias and mean square error when an unidentified single outlier (location or scale) is present in the sample. When the estimators are linear functions of order statistics, the bias and mean square error of these estimators were determined exactly by using the tables of means, variances and covariances prepared by Balakrishnan et al. (1991) and when the estimators are nonlinear functions of order statistics we resorted to Monte Carlo simulations to determine their bias and mean square error.

12.5.1 Estimators of the Mean μ

Here, we restrict our attention to the following estimators of μ that are based on a sample of size n.

1. Sample mean

$$\overline{O}_n = \frac{1}{n}\sum_{i=1}^{n} O_{i:n}. \tag{12.5.1}$$

2. Sample median

$$\begin{aligned}\mathrm{Med}_n &= O_{(n+1)/2:n} && \text{for odd } n,\\ &= \tfrac{1}{2}(O_{n/2:n} + O_{n/2+1:n}) && \text{for even } n.\end{aligned} \tag{12.5.2}$$

3. Best linear unbiased estimator

$$\mu_r^* = \sum_{i=r+1}^{n-r} a_i O_{i:n} \tag{12.5.3}$$

is the BLUE of μ based on a Type II censored sample where the smallest r and the largest r observations in the sample are censored. Formulas for the coefficients a_i are presented in Section 4.2, and tables have been prepared by Gupta et al. (1967) and Balakrishnan (1991);

4. Raghunandanan and Srinivasan's (1970) estimator of μ based on the sample quasi-midranges has been discussed in Section 4.6.

5. Approximate maximum likelihood estimator of $\mu(\hat{\mu}_r)$, based on a symmetrically Type II censored sample with r smallest and largest observations in the sample censored, has been discussed in detail in Section 3.3.

In addition to the above estimators, we also included the following omnibus robust estimators of μ in this study.

6. Trimmed means

$$T_n(r) = \frac{1}{n - 2r} \sum_{i=r+1}^{n-r} O_{i:n} \tag{12.5.4}$$

is the trimmed estimator of μ with r smallest and largest observations in the sample trimmed as proposed by Dixon (1960), Dixon and Tukey (1968), and Tukey and McLaughlin (1963).

7. Winsorized means

$$W_n(r) = \frac{1}{n}\left\{\sum_{i=r+1}^{n-r} O_{i:n} + r(O_{r+1:n} + O_{n-r:n})\right\} \tag{12.5.5}$$

is the Winsorized estimator of μ with r smallest and largest observations in the sample winsorized as proposed by Dixon and Tukey (1968) and Tukey and McLaughlin (1963).

8. Modified maximum likelihood estimators

$$M_n(r) = \frac{1}{m}\left\{\sum_{i=r+1}^{n-r} O_{i:n} + r\beta(O_{r+1:n} + O_{n-r:n})\right\}, \tag{12.5.6}$$

where $m = n - 2r + 2r\beta$ is the modified maximum likelihood estimator of μ based on a symmetrically Type II censored sample with r smallest and largest observations in the sample censored as proposed by Tiku (1967, 1980). The expression for β has been given by Tiku (1967, 1980), while Tiku et al. (1986) have tabulated the values of β for various choices of n and r.

9. Gastwirth mean

$$G_n = 0.3(O_{[n/3]+1:n} + O_{n-[n/3]:n}) + 0.2(O_{n/2:n} + O_{n/2+1:n}) \tag{12.5.7}$$

for even values of n, where $[n/3]$ denotes the integral part of $n/3$.

10. Linearly weighted means

$$L_n(r) = \frac{1}{2(n/2 - r)^2} \sum_{i=1}^{n/2-r} (2i - 1)(O_{r+i:n} + O_{n-r-i+1:n})$$
(12.5.8)

is the linearly weighted mean for even values of n, as proposed by Gastwirth and Cohen (1970), based on a censored sample where the r smallest and largest observations in the sample have been censored.

For the single-location-outlier logistic model, we may write all the estimators considered above as

$$M(\lambda) = \sum_{i=1}^{n} c_i O_{i:n}(\lambda). \tag{12.5.9}$$

From Eqs. (12.4.1) to (12.4.3), we see that $\Sigma_{i=1}^{n} c_i O_{i:n}(\lambda)$ is a nondecreasing function of λ; in addition, $E(M(\lambda)) = \Sigma_{i=1}^{n} c_i \nu_{i:n}(\lambda)$ is an increasing function of λ with $E(M(-\infty)) = -\infty$ except when $c_1 = 0$ and $E(M(\infty)) = \infty$ except when $c_n = 0$. From Eqs. (12.4.8) and (12.5.9), we further get when $c_1 = 0$ that

$$E(M(-\infty)) = \sum_{i=2}^{n} c_i \alpha_{i-1:n-1}, \tag{12.5.10}$$

and similarly when $c_n = 0$, we obtain from Eq. (12.4.9) that

$$E(M(\infty)) = \sum_{i=1}^{n-1} c_i \alpha_{i:n-1}. \tag{12.5.11}$$

By making use of the expected values of order statistics from a single-location-outlier logistic model given by Balakrishnan et al. (1991), we computed the bias of all the estimators of μ of a logistic $L(0, 1)$ population when a single-location-outlier from a logistic $L(\lambda, 1)$ population is present in the sample. These values of bias are presented in Tables 12.5.1 and 12.5.4 for sample sizes 10 and 20, respectively, and for $\lambda = 0.5(0.5)3.0, 4.0, \infty$. From Table 12.5.1 we observe the orderings

$$\overline{O}_{10} < \text{BLUE } 0\% < \text{AMLE } 10\% < \text{BLUE } 10\% < \text{AMLE } 20\% < \text{BLUE } 20\% < \text{RSE} < \text{AMLE } 30\% < \text{Med}_{10}$$

and

$$W_{10}(1) < M_{10}(1) < T_{10}(1) < W_{10}(2) < M_{10}(2) < L_{10}(1) < T_{10}(2) < L_{10}(2) < G_{10},$$

Table 12.5.1 Bias of Various Estimators of μ for $n = 10$ When a Single Outlier Is from $L(\lambda, 1)$ and the Others from $L(0, 1)$

	λ							
Estimator	0.5	1.0	1.5	2.0	2.5	3.0	4.0	∞
Sample mean	0.0500	0.1000	0.1500	0.2000	0.2500	0.3000	0.4000	∞
Sample median	0.0476	0.0836	0.1051	0.1158	0.1206	0.1226	0.1238	0.1240
BLUE 0%	0.0490	0.0927	0.1283	0.1560	0.1777	0.1954	0.2251	∞
BLUE 10%	0.0489	0.0915	0.1242	0.1466	0.1604	0.1683	0.1746	0.1766
BLUE 20%	0.0483	0.0886	0.1167	0.1334	0.1421	0.1462	0.1488	0.1498
R & S estimator	0.0482	0.0869	0.1123	0.1261	0.1328	0.1357	0.1375	0.1378
Appr. MLE 10%	0.0490	0.0926	0.1270	0.1512	0.1667	0.1756	0.1830	0.1854
Appr. MLE 20%	0.0486	0.0893	0.1180	0.1352	0.1442	0.1485	0.1512	0.1518
Appr. MLE 30%	0.0480	0.0859	0.1101	0.1230	0.1291	0.1318	0.1334	0.1337
Trimmed mean 10%	0.0491	0.0929	0.1278	0.1525	0.1682	0.1774	0.1849	0.1873
Trimmed mean 20%	0.0484	0.0883	0.1156	0.1315	0.1397	0.1435	0.1459	0.1464
Winsorized mean 10%	0.0495	0.0957	0.1351	0.1651	0.1854	0.1977	0.2084	0.2119
Winsorized mean 20%	0.0488	0.0906	0.1211	0.1400	0.1501	0.1549	0.1581	0.1587
MML estimator 10%	0.0494	0.0954	0.1343	0.1636	0.1834	0.1954	0.2057	0.2091
MML estimator 20%	0.0487	0.0903	0.1204	0.1388	0.1487	0.1534	0.1564	0.1570
Gastwirth mean	0.0480	0.0856	0.1094	0.1220	0.1279	0.1305	0.1320	0.1323
Linearly weighted mean 10%	0.0483	0.0881	0.1156	0.1322	0.1412	0.1458	0.1491	0.1500
Linearly weighted mean 20%	0.0480	0.0859	0.1102	0.1233	0.1297	0.1325	0.1342	0.1346

Table 12.5.2 Mean Square Error of Various Estimators of μ for $n = 10$ When a Single Outlier Is from $L(\lambda, 1)$ and the Others from $L(0, 1)$

	λ							
Estimator	0.5	1.0	1.5	2.0	2.5	3.0	4.0	∞
Sample mean	0.1025	0.1100	0.1225	0.1400	0.1625	0.1900	0.2600	∞
Sample median	0.1158	0.1241	0.1316	0.1363	0.1386	0.1396	0.1402	0.1404
BLUE 0%	0.0962	0.1035	0.1128	0.1222	0.1306	0.1381	0.1515	∞
BLUE 10%	0.0967	0.1041	0.1133	0.1219	0.1285	0.1329	0.1371	0.1387
BLUE 20%	0.0982	0.1057	0.1142	0.1209	0.1252	0.1275	0.1291	0.1304
R & S estimator	0.1069	0.1149	0.1233	0.1294	0.1328	0.1344	0.1354	0.1356
Appr. MLE 10%	0.0969	0.1043	0.1141	0.1235	0.1311	0.1364	0.1416	0.1437
Appr. MLE 20%	0.0990	0.1068	0.1153	0.1224	0.1269	0.1293	0.1310	0.1314
Appr. MLE 30%	0.1044	0.1122	0.1199	0.1253	0.1282	0.1296	0.1305	0.1306
Trimmed mean 10%	0.0971	0.1045	0.1144	0.1241	0.1320	0.1374	0.1428	0.1450
Trimmed mean 20%	0.0992	0.1067	0.1150	0.1214	0.1253	0.1274	0.1288	0.1291
Winsorized mean 10%	0.0998	0.1076	0.1190	0.1316	0.1428	0.1512	0.1601	0.1641
Winsorized mean 20%	0.1002	0.1080	0.1175	0.1256	0.1309	0.1339	0.1360	0.1365
MML estimator 10%	0.0993	0.1071	0.1183	0.1305	0.1413	0.1493	0.1578	0.1615
MML estimator 20%	0.0998	0.1075	0.1168	0.1246	0.1298	0.1326	0.1346	0.1350
Gastwirth mean	0.1046	0.1123	0.1199	0.1251	0.1279	0.1292	0.1300	0.1302
Linearly weighted mean 10%	0.0992	0.1066	0.1146	0.1209	0.1250	0.1273	0.1291	0.1297
Linearly weighted mean 20%	0.1034	0.1110	0.1185	0.1238	0.1267	0.1281	0.1290	0.1292

Table 12.5.3 Variances of Various Estimators of μ for $n = 10$ When a Single Outlier is from $L(0, \tau^2)$ and the Others from $L(0, 1)$

	τ						
Estimator	0.5	1.0	1.5	2.0	3.0	4.0	∞
Sample mean	0.0925	0.1000	0.1125	0.1300	0.1800	0.2500	∞
Sample median	0.0954	0.1120	0.1202	0.1249	0.1298	0.1324	0.1404
BLUE 0%	0.0837	0.0935	0.1021	0.1096	0.1227	0.1346	∞
BLUE 10%	0.0839	0.0939	0.1022	0.1083	0.1164	0.1213	0.1387
BLUE 20%	0.0842	0.0952	0.1031	0.1083	0.1146	0.1180	0.1304
R & S estimator	0.0907	0.1035	0.1117	0.1168	0.1226	0.1257	0.1356
Appr. MLE 10%	0.0846	0.0941	0.1027	0.1093	0.1182	0.1237	0.1437
Appr. MLE 20%	0.0852	0.0960	0.1040	0.1094	0.1159	0.1195	0.1314
Appr. MLE 30%	0.0879	0.1011	0.1089	0.1137	0.1189	0.1218	0.1306
Trimmed mean 10%	0.0850	0.0943	0.1029	0.1097	0.1188	0.1245	0.1450
Trimmed mean 20%	0.0847	0.0962	0.1040	0.1091	0.1151	0.1184	0.1291
Winsorized mean 10%	0.0889	0.0971	0.1067	0.1150	0.1270	0.1348	0.1641
Winsorized mean 20%	0.0872	0.0973	0.1056	0.1115	0.1188	0.1229	0.1365
MML estimator 10%	0.0883	0.0966	0.1061	0.1142	0.1259	0.1333	0.1615
MML estimator 20%	0.0866	0.0968	0.1051	0.1109	0.1179	0.1219	0.1350
Gastwirth mean	0.0907	0.1035	0.1117	0.1168	0.1226	0.1257	0.1302
Linearly weighted mean 10%	0.0842	0.0961	0.1038	0.1089	0.1150	0.1184	0.1297
Linearly weighted mean 20%	0.0866	0.1001	0.1077	0.1124	0.1176	0.1204	0.1292

Table 12.5.4 Bias of Various Estimators of μ for $n = 20$ When a Single Outlier Is from $L(\lambda, 1)$ and the Others from $L(0, 1)$

	λ							
Estimator	0.5	1.0	1.5	2.0	2.5	3.0	4.0	∞
Sample mean	0.0250	0.0500	0.0750	0.1000	0.1250	0.1500	0.2000	∞
Sample median	0.0236	0.0407	0.0503	0.0549	0.0569	0.0576	0.0581	0.0582
BLUE 0%	0.0244	0.0457	0.0623	0.0743	0.0827	0.0889	0.0978	∞
BLUE 10%	0.0244	0.0453	0.0606	0.0704	0.0759	0.0785	0.0803	0.0807
BLUE 20%	0.0242	0.0440	0.0572	0.0645	0.0681	0.0697	0.0706	0.0708
R & S estimator	0.0241	0.0433	0.0555	0.0620	0.0650	0.0663	0.0670	0.0671
Appr. MLE 10%	0.0244	0.0456	0.0616	0.0719	0.0778	0.0807	0.0827	0.0831
Appr. MLE 20%	0.0242	0.0441	0.0575	0.0650	0.0687	0.0703	0.0713	0.0714
Appr. MLE 30%	0.0239	0.0424	0.0537	0.0594	0.0621	0.0631	0.0638	0.0639
Trimmed mean 10%	0.0245	0.0459	0.0622	0.0728	0.0787	0.0817	0.0836	0.0841
Trimmed mean 20%	0.0241	0.0434	0.0559	0.0626	0.0658	0.0672	0.0681	0.0682
Winsorized mean 10%	0.0248	0.0479	0.0673	0.0812	0.0897	0.0942	0.0974	0.0981
Winsorized mean 20%	0.0244	0.0452	0.0598	0.0683	0.0726	0.0745	0.0756	0.0759
MML estimator 10%	0.0247	0.0477	0.0666	0.0801	0.0883	0.0926	0.0956	0.0962
MML estimator 20%	0.0243	0.0449	0.0592	0.0675	0.0716	0.0735	0.0746	0.0748
Gastwirth mean	0.0239	0.0423	0.0535	0.0591	0.0617	0.0628	0.0634	0.0636
Linearly weighted mean 10%	0.0240	0.0432	0.0556	0.0624	0.0658	0.0673	0.0682	0.0684
Linearly weighted mean 20%	0.0238	0.0420	0.0529	0.0585	0.0610	0.0620	0.0626	0.0628

where < denotes "inferior to." Similarly, from Table 12.5.4 we observe the orderings

$$\overline{O}_{20} < \text{BLUE } 0\% < \text{AMLE } 10\% < \text{BLUE } 10\% < \text{AMLE } 20\% \\ < \text{BLUE } 20\% < \text{RSE} < \text{AMLE } 30\% < \text{Med}_{20}$$

and

$$W_{20}(2) < M_{20}(2) < T_{20}(2) < W_{20}(4) < M_{20}(4) \\ < L_{20}(2) < T_{20}(4) < G_{20} < L_{20}(4).$$

We note that the trimmed means have a smaller bias than the corresponding modified maximum likelihood estimators which, in turn, have a smaller bias than the corresponding Winsorized estimators. We also note that the estimators based on 20% censoring is less subject to bias than those based on 10% censoring, and this is to be expected since an estimator based on 20% censoring excludes the single outlier present in the sample with a larger probability than an estimator based on only 10% censoring. Also, as rightly mentioned by David and Shu (1978), the sample median is observed to be more biased than what we may have naively thought it to be. We also see that the Gastwirth mean has the smallest bias among all omnibus robust estimators for a sample of size 10 and has only a slightly larger bias than the linearly weighted mean based on 20% censoring for a sample of size 20.

By considering the expression of $E(M^2(\lambda))$ from (12.5.9) and integrating by parts, we get

$$E(M^2(\lambda)) = 2\int_0^\infty x \Pr\{|M(\lambda)| > x\}\, dx. \qquad (12.5.12)$$

Since the logistic $L(0, 1)$ distribution is symmetric about 0 and taking accordingly $c_i = c_{n-i+1}$ for $i = 1, 2, \ldots, [(n + 1)/2]$, we have $M(0)$ to be symmetrically distributed about 0 and also $M(-\lambda)$ to be distributed as $-M(\lambda)$. Then we may expect

$$\Pr\{|M(\lambda)| > x\} = \Pr\{M(\lambda) > x\} + \Pr\{M(-\lambda) > x\}$$

to be an increasing function of $|\lambda|$ and, therefore, $E(M^2(\lambda))$ in (12.5.12) may be expected to be an increasing function of $|\lambda|$. If, further, $c_1 = c_n = 0$, we have

$$\begin{aligned} E(M^2(\pm\infty)) &= \lim_{\lambda\to\infty} E(M^2(\lambda)) \\ &= E\left\{\lim_{\lambda\to\infty} \sum_{i=2}^{n-1} c_i O_{i:n}(\lambda)\right\}^2 \end{aligned}$$

$$= E\left\{\sum_{i=2}^{n-1} c_i \lim_{\lambda\to\infty} O_{i:n}(\lambda)\right\}^2$$

$$= E\left\{\sum_{i=2}^{n-1} c_i Y_{i:n-1}\right\}^2$$

$$= (E(M(\infty)))^2 + \sum_{i=2}^{n-1}\sum_{j=2}^{n-1} c_i c_j \beta_{i,j:n-1}. \qquad (12.5.13)$$

By making use of the variances and covariances of order statistics from a single-location-outlier logistic model given by Balakrishnan et al. (1991), we computed the mean square error of all the estimators of μ of a logistic $L(0, 1)$ population when a single location-outlier from a logistic $L(\lambda, 1)$ population is present in the sample. These values of mean square error are presented in Tables 12.5.2 and 12.5.5 for sample sizes 10 and 20, respectively, and for $\lambda = 0.5(0.5)3.0, 4.0, \infty$. From Tables 12.5.2 and 12.5.5 we see that the ordering of the estimators based on the mean square error is almost the reverse of their ordering based on the bias. We note that the sample median has the largest mean square error values among all the estimators (excluding sample mean and the BLUE based on complete sample) for a sample of size 20 and nearly the largest mean square error value for a sample of size 10.

Table 12.5.5 Mean Square Error of Various Estimators of μ for $n = 20$ When a Single Outlier Is from $L(\lambda, 1)$ and the Others from $L(0, 1)$

	λ							
Estimator	0.5	1.0	1.5	2.0	2.5	3.0	4.0	∞
Sample mean	0.0506	0.0525	0.0556	0.0600	0.0661	0.0725	0.0900	∞
Sample median	0.0591	0.0612	0.0629	0.0639	0.0643	0.0644	0.0646	0.0646
BLUE 0%	0.0469	0.0487	0.0509	0.0529	0.0547	0.0559	0.0578	∞
BLUE 10%	0.0470	0.0489	0.0510	0.0529	0.0542	0.0549	0.0554	0.0556
BLUE 20%	0.0479	0.0497	0.0517	0.0532	0.0540	0.0544	0.0546	0.0547
R & S estimator	0.0524	0.0544	0.0564	0.0578	0.0586	0.0588	0.0591	0.0591
Appr. MLE 10%	0.0471	0.0489	0.0512	0.0532	0.0546	0.0554	0.0560	0.0561
Appr. MLE 20%	0.0479	0.0498	0.0518	0.0533	0.0541	0.0545	0.0548	0.0548
Appr. MLE 30%	0.0502	0.0521	0.0538	0.0549	0.0555	0.0557	0.0559	0.0559
Trimmed mean 10%	0.0472	0.0491	0.0514	0.0535	0.0549	0.0557	0.0564	0.0565
Trimmed mean 20%	0.0482	0.0500	0.0519	0.0532	0.0539	0.0542	0.0544	0.0545
Winsorized mean 10%	0.0489	0.0509	0.0538	0.0568	0.0592	0.0607	0.0619	0.0622
Winsorized mean 20%	0.0488	0.0507	0.0530	0.0548	0.0559	0.0564	0.0567	0.0568
MML estimator 10%	0.0486	0.0505	0.0533	0.0562	0.0584	0.0598	0.0609	0.0612
MML estimator 20%	0.0484	0.0504	0.0526	0.0543	0.0553	0.0558	0.0561	0.0562
Gastwirth mean	0.0506	0.0525	0.0542	0.0553	0.0559	0.0561	0.0563	0.0563
Linearly weighted mean 10%	0.0483	0.0502	0.0520	0.0532	0.0540	0.0543	0.0545	0.0546
Linearly weighted mean 20%	0.0504	0.0523	0.0540	0.0550	0.0555	0.0558	0.0559	0.0559

For the single-scale-outlier logistic model, we may similarly write all the estimators of μ as

$$M^*(\tau) = \sum_{i=1}^{n} c_i O^*_{i:n}(\tau). \tag{12.5.14}$$

When $c_i = c_{n-i+1}$ for $i = 1, 2, \ldots, [(n+1)/2]$, it is clear that $M^*(\tau)$ in (12.5.14) is symmetrically distributed about 0 for all values of τ. Hence, we have

$$E(M^*(\tau)) = 0 \tag{12.5.15}$$

for all values of τ; therefore, under the single-scale-outlier model all the estimators of μ discussed earlier are unbiased for all values of τ. Furthermore, from the discussions presented in Section 12.4 we see that the limiting behavior of $M^*(\tau)$ as $\tau \to \infty$ corresponds exactly to that of $M(\lambda)$ as $\lambda \to -\infty$ or ∞ accordingly as $Y_n < 0$ or $Y_n > 0$. So, we have

$$\begin{aligned}\lim_{\tau\to\infty} E(M^{*2}(\tau)) &= E(M^{*2}(\infty))\\ &= \Pr(Y_n < 0)E(M^2(-\infty)) + \Pr(Y_n > 0)E(M^2(\infty))\\ &= \tfrac{1}{2}\{E(M^2(-\infty)) + E(M^2(\infty))\}\\ &= E(M^2(\infty)).\end{aligned} \tag{12.5.16}$$

By making use of the variances and covariances of order statistics from a single-scale-outlier logistic model given by Balakrishnan et al. (1991), we computed the variance of all the estimators of μ of a logistic $L(0, 1)$ population when a single scale-outlier from a logistic $L(0, \tau^2)$ population is present in the sample. The values of variance are presented in Tables 12.5.3 and 12.5.6 for sample sizes 10 and 20, respectively, and for $\tau = 0.5(0.5)2.0(1.0)4.0, \infty$. From Tables 12.5.3 and 12.5.6 we see that the variance of the sample mean and that of the BLUE based on the complete sample both tend to ∞ as $\tau \to \infty$. We also note that the sample median has the largest variance for all values of τ among all other estimators. We once again observe that the trimmed means have smaller variance than the modified maximum likelihood estimators, which, in turn, have smaller variance than the corresponding Winsorized means. We also observe that the R & S estimator has its variance to be larger than those of the approximate MLE based on 10%, 20%, and 30% censored samples for all values of τ. Furthermore, we observe that both the Gastwirth mean and the linearly weighted mean based on 20% censoring have their variances to be larger than those of the trimmed mean, Winsorized mean, and the modified maximum likelihood estimator based on 20% censoring.

Table 12.5.6 Variance of Various Estimators of μ for $n = 20$ When a Single Outlier Is from $L(0, \tau^2)$ and the Others from $L(0, 1)$

	τ						
Estimator	0.5	1.0	1.5	2.0	3.0	4.0	∞
Sample mean	0.0481	0.0500	0.0531	0.0575	0.0700	0.0875	∞
Sample median	0.0532	0.0581	0.0602	0.0612	0.0623	0.0629	0.0646
BLUE 0%	0.0436	0.0462	0.0482	0.0498	0.0522	0.0540	∞
BLUE 10%	0.0438	0.0463	0.0483	0.0497	0.0514	0.0523	0.0556
BLUE 20%	0.0443	0.0471	0.0490	0.0502	0.0516	0.0523	0.0547
R & S Estimator	0.0485	0.0516	0.0536	0.0548	0.0562	0.0569	0.0591
Appr. MLE 10%	0.0439	0.0464	0.0484	0.0498	0.0516	0.0526	0.0561
Appr. MLE 20%	0.0444	0.0472	0.0491	0.0503	0.0517	0.0524	0.0548
Appr. MLE 30%	0.0460	0.0494	0.0512	0.0523	0.0534	0.0540	0.0559
Trimmed mean 10%	0.0441	0.0465	0.0485	0.0500	0.0519	0.0529	0.0565
Trimmed mean 20%	0.0444	0.0474	0.0492	0.0504	0.0516	0.0523	0.0545
Winsorized mean 10%	0.0463	0.0483	0.0506	0.0525	0.0551	0.0567	0.0622
Winsorized mean 20%	0.0456	0.0480	0.0501	0.0514	0.0530	0.0539	0.0568
MML estimator 10%	0.0458	0.0479	0.0502	0.0520	0.0545	0.0560	0.0612
MML estimator 20%	0.0452	0.0477	0.0497	0.0510	0.0526	0.0534	0.0562
Gastwirth mean	0.0463	0.0498	0.0516	0.0527	0.0538	0.0544	0.0563
Linearly weighted mean 10%	0.0444	0.0476	0.0494	0.0505	0.0518	0.0524	0.0546
Linearly weighted mean 20%	0.0459	0.0496	0.0514	0.0525	0.0536	0.0541	0.0559

In conclusion, we would like to highlight the following points observed from Tables 12.5.1 to 12.5.6. The sample mean and the BLUE based on the complete sample are both quite nonrobust, since their bias as well as mean square error tend to ∞ as $\lambda \to \infty$ under a single-location-outlier model and also that their variance tend to ∞ as $\tau \to \infty$ under a single-scale-outlier model. Even though the sample median has the smallest bias among all estimators under a single-location-outlier model, it has the largest mean square error under a single-location-outlier model and also the largest variance under a single-scale-outlier model. The approximate maximum likelihood estimator based on 10% and 20% censoring both are seen to perform very well overall under both location-outlier and scale-outlier models. The approximate maximum likelihood estimator, unlike Raghunandanan and Srinivasan's estimator, is an explicit estimator which does not require the construction of any special tables. Furthermore, another appealing aspect of the approximate maximum likelihood estimator of μ is that it has a natural matching estimator of σ which is also explicit in nature and is efficient and robust under both location- and scale-outlier models, as will be shown in the discussion to follow. Finally, among all the omnibus robust estimators the 20% trimmed mean seems to have overall superiority based on the mean square error and the variance under the location- and scale-outlier models, respectively.

12.5.2 Estimators of the Standard Deviation σ

In this study, we consider the following estimators of σ that are based on a sample of size n.

1. Sample standard deviation

$$S_n = \left\{\frac{1}{n-1}\sum_{i=1}^{n}(O_{i:n} - \overline{O}_n)^2\right\}^{1/2}. \tag{12.5.17}$$

2. Best linear unbiased estimator

$$\sigma_r^* = \sum_{i=r+1}^{n-r} b_i O_{i:n} \tag{12.5.18}$$

is the BLUE of σ based on a Type II censored sample where the smallest r and the largest r observations in the sample are censored. Formulas for the coefficients b_i are presented in Section 4.2, and tables have been prepared by Gupta et al. (1967) and Balakrishnan (1991).
3. Raghunandanan and Srinivasan's (1970) estimator of σ based on the sample quasi-ranges, with r smallest and largest observations in the sample being censored, has been discussed in Section 4.6.
4. Approximate maximum likelihood estimator of σ ($\tilde{\sigma}_r$), based on a symmetrically Type II censored sample with r smallest and largest observations in the sample censored, has been discussed in detail in Section 3.3.

In addition to the above estimators, we also considered in this study the following omnibus robust estimators of σ:

5. Trimmed estimators
The trimmed mean $T_n(r)$ is defined in Eq. (12.5.4); the matching trimmed estimator of σ is given by

$$\hat{\sigma}_T(r) = \left[\frac{1}{n-2r-1}\left\{\sum_{i=r+1}^{n-r}(O_{i:n} - T_n(r))^2 + r(O_{r+1:n} - T_n(r))^2 + r(O_{n-r:n} - T_n(r))^2\right\}\right]^{1/2}, \tag{12.5.19}$$

where, as before, the r smallest and largest observations in the sample are trimmed for the estimator $\hat{\sigma}_T(r)$ in (12.5.19); see Huber (1970).
6. Modified maximum likelihood estimators.
The modified maximum likelihood estimator $M_n(r)$ of μ is defined in Eq. (12.5.6); the associated modified maximum likelihood esti-

mator of σ is given by (Tiku et al., 1986)

$$\hat{\sigma}_M(r) = \frac{B + (B^2 + 4AC)^{1/2}}{2\{A(A - 1)\}^{1/2}}, \tag{12.5.20}$$

where

$$A = n - 2r,$$
$$m = n - 2r + 2r\beta,$$
$$B = r\alpha(O_{n-r:n} - O_{r+1:n}),$$
$$C = \sum_{i=r+1}^{n-r} (O_{i:n} - M_n(r))^2 + r\beta(O_{r+1:n} - M_n(r))^2 + r\beta(O_{n-r:n} - M_n(r))^2.$$

Expressions for the constants α and β have been given by Tiku (1967, 1980) while Tiku et al. (1986) have tabulated the values of α and β for various choices of n and r.

By making use of the means, variances, and covariances of order statistics from a single-location-outlier logistic model given by Balakrishnan et al. (1991), we computed the bias and the mean square error of the BLUE of σ (based on 0%, 10%, and 20% symmetric censoring) and also Raghunandanan and Srinivasan's estimator of σ (with $r = 0$, 1, and 2) of a logistic $L(0, 1)$ population when a single location-outlier from a logistic $L(\lambda, 1)$ population is present in the sample. The values of bias are presented in Tables 12.5.7 and 12.5.11 for $n = 10$ and 20 and $\lambda = 0.5(0.5)3.0, 4.0$, and the values of mean square error are presented in Tables 12.5.8 and 12.5.12 for $n = 10$ and 20 and $\lambda = 0.5(0.5)3.0, 4.0$. Also given in these tables are the values of bias and mean square error of all other estimators (which are nonlinear functions of order statistics) determined by Monte Carlo simulations (based on 10,000 Monte Carlo runs). From these tables, we first of all observe that the sample standard deviation, the BLUE based on complete sample and Raghunandanan and Srinivasan's estimator with $r = 0$ all have their bias as well as mean square error tending to ∞ as $\lambda \to \infty$. We further observe that the approximate maximum likelihood estimators of σ with 10% and 20% censoring both seem to be overall highly efficient and robust under the single-location-outlier model.

Similarly, by making use of the means, variances, and covariances of order statistics from a single-scale-outlier logistic model given by Balakrishnan et al. (1991), we computed the bias and the mean square error of the BLUE of σ (based on 0%, 10%, and 20% symmetric censoring) and also Raghunandanan and Srinivasan's estimator of σ (with $r = 0$, 1,

Table 12.5.7 Bias of Various Estimators of σ for $n = 10$ When a Single Outlier Is from $L(\lambda, 1)$ and the Others from $L(0, 1)$

Estimator	λ 0.5	1.0	1.5	2.0	2.5	3.0	4.0
Sample std. dev.	−0.0282	0.0077	0.0656	0.1425	0.2348	0.3396	0.5766
BLUE 0%	0.0138	0.0523	0.1090	0.1767	0.2502	0.3264	0.4820
BLUE 10%	0.0151	0.0545	0.1034	0.1479	0.1809	0.2022	0.2211
BLUE 20%	0.0163	0.0548	0.0958	0.1261	0.1440	0.1530	0.1589
R & S estimator (r=0)	0.0137	0.0523	0.1098	0.1789	0.2542	0.3325	0.4920
R & S estimator (r=1)	0.0155	0.0549	0.1030	0.1459	0.1770	0.1966	0.2137
R & S estimator (r=2)	0.0162	0.0549	0.0966	0.1278	0.1464	0.1559	0.1621
Appr. MLE 10%	−0.0520	−0.0167	0.0303	0.0738	0.1058	0.1267	0.1456
Appr. MLE 20%	−0.1060	−0.0734	−0.0371	−0.0090	0.0079	0.0163	0.0207
Appr. MLE 30%	−0.2045	−0.1741	−0.1476	−0.1285	−0.1197	−0.1145	−0.1111
Trimmed estimator 10%	−0.1534	−0.1218	−0.0790	−0.0390	−0.0092	0.0105	0.0286
Trimmed estimator 20%	−0.2700	−0.2435	−0.2138	−0.1906	−0.1767	−0.1697	−0.1660
Trimmed estimator 30%	−0.4132	−0.3909	−0.3713	−0.3572	−0.3507	−0.3468	−0.3443
MML estimator 10%	−0.0861	−0.0520	−0.0058	0.0375	0.0696	0.0908	0.1102
MML estimator 20%	−0.1309	−0.0993	−0.0639	−0.0363	−0.0197	−0.0114	−0.0071
MML estimator 30%	−0.1899	−0.1589	−0.1319	−0.1125	−0.1035	−0.0982	−0.0948

and 2) of a logistic $L(0, 1)$ population when a single scale-outlier from a logistic $L(0, \tau^2)$ population is present in the sample. The values of bias are presented in Tables 12.5.9 and 12.5.13 for $n = 10$ and 20 and $\tau = 0.5(0.5)2.0(1.0)4.0$, and the values of mean square error are presented in Tables 12.5.10 and 12.5.14 for $n = 10$ and 20 and $\tau = 0.5(0.5)2.0(1.0)4.0$.

Table 12.5.8 Mean Square Error of Various Estimators of σ for $n = 10$ When a Single Outlier Is from $L(\lambda, 1)$ and the Others from $L(0, 1)$

Estimator	λ 0.5	1.0	1.5	2.0	2.5	3.0	4.0
Sample std. dev.	0.0762	0.0779	0.0855	0.1054	0.1442	0.2081	0.4311
BLUE 0%	0.0781	0.0835	0.0958	0.1177	0.1508	0.1957	0.3222
BLUE 10%	0.1002	0.1082	0.1232	0.1430	0.1631	0.1798	0.1994
BLUE 20%	[illegible]	[illegible]	[illegible]	[illegible]	[illegible]	[illegible]	[illegible]

MML estimator 20%	0.1242	0.1245	0.1263	0.1314	
MML estimator 30%	0.2036	0.2046	0.2091	0.2123	

.1404 0.1436
.2158 0.2170

R & S estimator (r=2)	476	0.1597	0.1773	0.1949	0.2081	0.2160	0.2221
Appr. MLE 10%	902	0.0928	0.1004	0.1134	0.1281	0.1413	0.1578
Appr. MLE 20%	244	0.1266	0.1306	0.1373	0.1436	0.1481	0.1515
Appr. MLE 30%	034	0.2033	0.2067	0.2090	0.2110	0.2119	0.2129
Trimmed estimator 10%	938	0.0891	0.0864	0.0889	0.0954	0.1029	0.1140
Trimmed estimator 20%	485	1.1402	0.1320	0.1281	0.1274	0.1279	0.1290
Trimmed estimator 30%	587	0.2468	0.2383	0.2323	0.2300	0.2285	0.2277
MML estimator 10%	893	0.0893	0.0934	0.1033	0.1159	0.1280	0.1437
MML estimator 20%	242	0.1245	0.1263	0.1314	0.1366	0.1404	0.1436
MML estimator 30%	036	0.2046	0.2091	0.2123	0.2147	0.2158	0.2170

Table 12.5.9 Bias of Various Estimators of σ for $n = 10$ When a Single Outlier Is from $L(0, \tau^2)$ and the Others from $L(0, 1)$

	τ					
Estimator	0.5	1.0	1.5	2.0	3.0	4.0
Sample std. dev.	−0.0787	−0.0399	0.0163	0.0848	0.2466	0.4299
BLUE 0%	−0.0449	0.0000	0.0526	0.1084	0.2237	0.3412
BLUE 10%	−0.0534	0.0000	0.0436	0.0758	0.1172	0.1417
BLUE 20%	−0.0624	0.0000	0.0398	0.0649	0.0939	0.1096
R & S estimator (r=0)	−0.0439	0.0000	0.0531	0.1099	0.278	0.3481
R & S estimator (r=1)	−0.0539	0.0000	0.0434	0.0748	0.1147	0.1381
R & S estimator (r=2)	−0.0614	0.0000	0.0401	0.0657	0.0953	0.1114
Appr. MLE 10%	−0.1133	−0.0647	−0.0243	0.0064	0.0444	0.0677
Appr. MLE 20%	−0.1714	−0.1161	−0.0814	−0.0600	−0.0353	−0.0215
Appr. MLE 30%	−0.2670	−0.2136	−0.1860	−0.1692	−0.1500	−0.1401
Trimmed estimator 10%	−0.2073	−0.1648	−0.1284	−0.1003	−0.0653	−0.0437
Trimmed estimator 20%	−0.3234	−0.2784	−0.2500	−0.2324	−0.2121	−0.2008
Trimmed estimator 30%	−0.4594	−0.4199	−0.3996	−0.3872	−0.3730	−0.3657
MML estimator 10%	−0.1443	−0.0984	−0.0591	−0.0288	0.0089	0.0322
MML estimator 20%	−0.1943	−0.1407	−0.1069	−0.0860	−0.0619	−0.0485
MML estimator 30%	−0.2535	−0.1991	−0.1711	−0.1539	−0.1343	−0.1243

Also given in these tables are the values of bias and mean square error of all other estimators (which are nonlinear functions of order statistics) determined by Monte Carlo simulations (based on 10,000 Monte Carlo runs). From these tables, we observe once again that the sample standard deviation, the BLUE based on a complete sample, and Raghunandanan and

Table 12.5.10 Mean Square Error of Various Estimators of σ for $n = 10$ When a Single Outlier Is from $L(0, \tau^2)$ and the Others from $L(0, 1)$

	τ					
Estimator	0.5	1.0	1.5	2.0	3.0	4.0
Sample std. dev.	0.0778	0.0761	0.0901	0.1298	0.3111	0.6517
BLUE 0%	0.0743	0.0768	0.0903	0.1158	0.1987	0.3082
BLUE 10%	0.0933	0.0979	0.1097	0.1219	0.1416	0.1550
BLUE 20%	0.1310	0.1419	0.1554	0.1662	0.1802	0.1885
R & S estimator (r=0)	0.0752	0.0774	0.0913	0.1178	0.2047	0.3202
R & S estimator (r=1)	0.0952	0.1002	0.1118	0.1235	0.1418	0.1541
R & S estimator (r=2)	0.1333	0.1437	0.1577	0.1690	0.1840	0.1929
Appr. MLE 10%	0.0928	0.0910	0.0951	0.1029	0.1155	0.1252
Appr. MLE 20%	0.1270	0.1224	0.1261	0.1300	0.1349	0.1380
Appr. MLE 30%	0.2057	0.2011	0.2018	0.2025	0.2043	0.2054
Trimmed estimator 10%	0.1078	0.0963	0.0926	0.0935	0.0971	0.1011
Trimmed estimator 20%	0.1699	0.1502	0.1423	0.1386	0.1345	0.1325
Trimmed estimator 30%	0.2842	0.2609	0.2507	0.2445	0.2381	0.2349
MML estimator 10%	0.0964	0.0903	0.0923	0.0980	0.1081	0.1164
MML estimator 20%	0.1303	0.1228	0.1245	0.1272	0.1306	0.1329
MML estimator 30%	0.2037	0.2009	0.2027	0.2040	0.2066	0.2082

Table 12.5.11 Bias of Various Estimators of σ for $n = 20$ When a Single Outlier Is from $L(\lambda, 1)$ and the Others from $L(0, 1)$

	λ						
Estimator	0.5	1.0	1.5	2.0	2.5	3.0	4.0
Sample std. dev.	−0.0140	0.0044	0.0344	0.0752	0.1256	0.1844	0.3228
BLUE 0%	0.0069	0.0263	0.0543	0.0872	0.1225	0.1586	0.2315
BLUE 10%	0.0079	0.0277	0.0513	0.0713	0.0844	0.0917	0.0968
BLUE 20%	0.0082	0.0273	0.0464	0.0594	0.0663	0.0695	0.0714
R & S estimator (r=0)	0.0070	0.0263	0.0542	0.0870	0.1222	0.1585	0.2322
R & S estimator (r=1)	0.0078	0.0276	0.0508	0.0698	0.0821	0.0887	0.0933
R & S estimator (r=2)	0.0082	0.0275	0.0472	0.0609	0.0683	0.0717	0.0738
Appr. MLE 10%	−0.0247	−0.0064	0.0157	0.0349	0.0485	0.0564	0.0622
Appr. MLE 20%	−0.0500	−0.0331	−0.0164	−0.0045	0.0023	0.0055	0.0071
Appr. MLE 30%	−0.1024	−0.0876	−0.0735	−0.0651	−0.0601	−0.0581	−0.0574
Trimmed estimator 10%	−0.1463	−0.1302	−0.1102	−0.0925	−0.0797	−0.0723	−0.0666
Trimmed estimator 20%	−0.2567	−0.2434	−0.2303	−0.2209	−0.2156	−0.2130	−0.2117
Trimmed estimator 30%	−0.3891	−0.3789	−0.3694	−0.3636	−0.3602	−0.3588	−0.3584
MML estimator 10%	−0.0764	−0.0589	−0.0372	−0.0180	−0.0042	0.0039	0.0100
MML estimator 20%	−0.1099	−0.0940	−0.0783	−0.0670	−0.0606	−0.0576	−0.0560
MML estimator 30%	−0.1493	−0.1352	−0.1218	−0.1138	−0.1090	−0.1071	−0.1065

Srinivasan's estimator with $r = 0$ all have their bias as well as mean square error tending to ∞ as $\tau \rightarrow \infty$. We also observe in this case that both the trimmed estimator and the MML estimator develop a rather serious bias and also possess larger mean square error than the corresponding approximate maximum likelihood estimator. Finally, we note that the approximate

Table 12.5.12 Mean Square Error of Various Estimators of σ for $n = 20$ When a Single Outlier Is from $L(\lambda, 1)$ and the Others from $L(0, 1)$

	λ						
Estimator	0.5	1.0	1.5	2.0	2.5	3.0	4.0
Sample std. dev.	0.0376	0.0380	0.0401	0.0458	0.0572	0.0767	0.1494
BLUE 0%	0.0369	0.0382	0.0412	0.0465	0.0542	0.0646	0.0933
BLUE 10%	0.0462	0.0482	0.0518	0.0562	0.0601	0.0627	0.0649
BLUE 20%	0.0651	0.0678	0.0715	0.0748	0.0768	0.0779	0.0785
R & S estimator (r=0)	0.0370	0.0383	0.0413	0.0464	0.0542	0.0646	0.0935
R & S estimator (r=1)	0.0472	0.0492	0.0528	0.0570	0.0605	0.0628	0.0647
R & S estimator (r=2)	0.0667	0.0696	0.0737	0.0773	0.0797	0.0809	0.0817
Appr. MLE 10%	0.0429	0.0440	0.0459	0.0488	0.0518	0.0540	0.0563
Appr. MLE 20%	0.0601	0.0605	0.0620	0.0633	0.0643	0.0650	0.0652
Appr. MLE 30%	0.0964	0.0969	0.0975	0.0978	0.0984	0.0987	0.0987
Trimmed estimator 10%	0.0541	0.0510	0.0476	0.0457	0.0452	0.0454	0.0462
Trimmed estimator 20%	0.1012	0.0957	0.0910	0.0877	0.0860	0.0853	0.0849
Trimmed estimator 30%	0.1912	0.1850	0.1792	0.1757	0.1737	0.1730	0.1727
MML estimator 10%	0.0441	0.0433	0.0429	0.0348	0.0455	0.0471	0.0490
MML estimator 20%	0.0628	0.0611	0.0605	0.0602	0.0604	0.0606	0.0606
MML estimator 30%	0.0994	0.0984	0.0976	0.0970	0.0970	0.0971	0.0970

Table 12.5.13 Bias of Various Estimators of σ for $n = 20$ When a Single Outlier Is from $L(0, \tau^2)$ and the Others from $L(0, 1)$

	τ					
Estimator	0.5	1.0	1.5	2.0	3.0	4.0
Sample std. dev.	−0.0399	−0.0200	0.0103	0.0490	0.1452	0.2591
BLUE 0%	−0.0228	0.0000	0.0260	0.0530	0.1080	0.1634
BLUE 10%	−0.0271	0.0000	0.0214	0.0362	0.0541	0.0643
BLUE 20%	−0.0327	0.0000	0.0192	0.0308	0.0436	0.0504
R & S estimator (r=0)	−0.0229	0.0000	0.0260	0.0530	0.1083	0.1643
R & S estimator (r=1)	−0.0276	0.0000	0.0210	0.0354	0.0527	0.0624
R & S estimator (r=2)	−0.0315	0.0000	0.0195	0.0314	0.0448	0.0519
Appr. MLE 10%	−0.0586	−0.0318	−0.0102	0.0043	0.0211	0.0305
Appr. MLE 20%	−0.0892	−0.0579	−0.0398	−0.0292	−0.0180	−0.0117
Appr. MLE 30%	−0.1427	−0.1088	−0.0934	−0.0856	−0.0764	−0.0714
Trimmed estimator 10%	−0.1753	−0.1526	−0.1332	−0.1199	−0.1045	−0.0959
Trimmed estimator 20%	−0.2873	−0.2630	−0.2487	−0.2403	−0.2315	−0.2265
Trimmed estimator 30%	−0.4163	−0.3933	−0.3828	−0.3775	−0.3713	−0.3679
MML estimator 10%	−0.1077	−0.0831	−0.0621	−0.0478	−0.0312	−0.0218
MML estimator 20%	−0.1465	−0.1174	−0.1003	−0.0902	−0.0797	−0.0738
MML estimator 30%	−0.1874	−0.1553	−0.1407	−0.1333	−0.1246	−0.1199

maximum likelihood estimators of σ with 10% and 20% censoring are overall highly efficient and robust under the single-scale-outlier model. The points made above, together with the facts that the approximate maximum likelihood estimators of μ and σ are a pair of naturally matching estimators and that they are explicit estimators which do not need the construction

Table 12.5.14 Mean Square Error of Various Estimators of σ for $n = 20$ When a Single Outlier Is from $L(0, \tau^2)$ and the Others from $L(0, 1)$

	τ					
Estimator	0.5	1.0	1.5	2.0	3.0	4.0
Sample std. dev.	0.0381	0.0375	0.419	0.0553	0.1224	0.2576
BLUE 0%	0.0361	0.0366	0.0398	0.0456	0.0629	0.0793
BLUE 10%	0.0447	0.0457	0.0483	0.0509	0.0546	0.0570
BLUE 20%	0.0619	0.0642	0.0671	0.0694	0.0721	0.0736
R & S estimator (r=0)	0.0361	0.0367	0.0398	0.0457	0.0632	0.0800
R & S estimator (r=1)	0.0457	0.0467	0.0493	0.0517	0.0552	0.0574
R & S estimator (r=2)	0.0638	0.0658	0.0690	0.0714	0.0744	0.0761
Appr. MLE 10%	0.0438	0.0429	0.0444	0.0461	0.0487	0.0503
Appr. MLE 20%	0.0616	0.0600	0.0607	0.0615	0.0624	0.0630
Appr. MLE 30%	0.0985	0.0970	0.0971	0.0974	0.0975	0.0973
Trimmed estimator 10%	0.0621	0.0557	0.0522	0.0504	0.0488	0.0481
Trimmed estimator 20%	0.1155	0.1039	0.0982	0.0950	0.0917	0.0900
Trimmed estimator 30%	0.2096	0.1942	0.1876	0.1843	0.1803	0.1781
MML estimator 10%	0.0484	0.0449	0.0442	0.0445	0.0454	0.0460
MML estimator 20%	0.0688	0.0637	0.0622	0.0616	0.0610	0.0609
MML estimator 30%	0.1053	0.1006	0.0992	0.0987	0.0978	0.0972

of any special tables, unlike the best linear unbiased estimators or Raghunandanan and Srinivasan's (1970) estimators, lead us to recommend the usage of the approximate maximum likelihood estimators of μ and σ based on either 10% or 20% symmetric censoring whenever an outlier is suspected to be present in the data.

REFERENCES

Andrews, D. F., Bickel, P. J., Hampel, F. R., Huber, P. J., Rogers, W. H., and Tukey, J. W. (1972). *Robust Estimators of Location*, Princeton University Press, Princeton, New Jersey.

Arnold, B. C. and Balakrishnan, N. (1989). *Relations, Bounds and Approximations for Order Statistics*, Lecture Notes in Statistics No. 53, Springer-Verlag, New York.

Balakrishnan, N. (1988a). Relations and identities for the moments of order statistics from a sample containing a single outlier, *Commun. Statist.—Theor. Meth.*, *17*(7), 2173–2190.

Balakrishnan, N. (1988b). Recurrence relations among moments of order statistics from two related outlier models, *Biomet. J.*, *30*, 741–746.

Balakrishnan, N. (1988c). Recurrence relations for order statistics from n independent and non-identically distributed random variables, *Ann. Inst. Statist. Math.*, *40*, 273–277.

Balakrishnan, N. (1989a). Recurrence relations among moments of order statistics from two related sets of independent and non-identically distributed random variables, *Ann. Inst. Statist. Math.*, *41*, 323–329.

Balakrishnan, N. (1989b). A relation for the covariances of order statistics from n independent and non-identically distributed random variables, *Statist. Hefte*, *30*, 141–146.

Balakrishnan, N. (1991). Best linear unbiased estimates of the location and scale parameters of logistic distribution for complete and censored samples of sizes 2(1)25(5)40. Submitted for publication.

Balakrishnan, N. and Ambagaspitiya, R. S. (1988). Relationships among moments of order statistics in samples from two related outlier models and some applications, *Commun. Statist.—Theor. Meth.*, *17*(7), 2327–2341.

Balakrishnan, N., Chan, P. S., Ho, K. L., and Lo, K. K. (1991). Means, variances, and covariances of logistic order statistics in the presence of an outlier, to appear in *Selected Tables in Mathematical Statistics*.

Balakrishnan, N. and Malik, H. J. (1991). Means, variances, and covariances of logistic order statistics for sample sizes up to fifty, to appear in *Selected Tables in Mathematical Statistics*.

Crow, E. L. and Siddiqui, M. M. (1967). Robust estimation of location, *J. Amer. Statist. Assoc.*, *62*, 353–389.

David, H. A. (1981). *Order Statistics*, 2nd ed., Wiley, New York.

David, H. A., Kennedy, W. J., and Knight, R. D. (1977). Means, variances, and covariances of normal order statistics in the presence of an outlier, *Selected Tables in Mathematical Statistics*, *5*, 75–204.

David, H. A. and Shu, V. S. (1978). Robustness of location estimators in the presence of an outlier, in *Contributions to Survey Sampling and Applied Statistics: Papers in Honor of H. O. Hartley*, H. A. David, (ed.), 235–250, Academic Press, Boston.

Dixon, W. J. (1960). Simplified estimation from censored normal samples, *Ann. Math. Statist.*, *31*, 385–391.

Dixon, W. J. and Tukey, J. W. (1968). Approximate behavior of the distribution of Winsorized *t* (trimming/Winsorization 2), *Technometrics*, *10*, 83–98.

Gastwirth, J. L. and Cohen, M. L. (1970). Small sample behavior of robust linear estimators of location, *J. Amer. Statist. Assoc.*, *65*, 946–973.

Gupta, S. S., Qureishi, A. S., and Shah, B. K. (1967). Best linear unbiased estimators of the parameters of the logistic distribution using order statistics, *Technometrics*, *9*, 43–56.

Gupta, S. S. and Shah, B. K. (1965). Exact moments and percentage points of the order statistics and the distribution of the range from the logistic distribution, *Ann. Math. Statist.*, *36*, 907–920.

Hampel, F. R. (1974). The influence curve and its role in robust estimation, *J. Amer. Statist. Assoc.*, *69*, 383–393.

Huber, P. J. (1970). Studentizing robust estimates, in *Nonparametric Techniques in Statistical Inference*, M. L. Puri, (ed.), Cambridge University Press, Cambridge, England.

Joshi, P. C. (1972). Efficient estimation of the mean of an exponential distribution when an outlier is present, *Technometrics*, *14*, 137–144.

Kale, B. K. and Sinha, S. K. (1971). Estimation of expected life in the presence of an outlier observation, *Technometrics*, *13*, 755–759.

Loeve, M. (1977). *Probability Theory I*, 4th ed., Springer-Verlag, New York.

Raghunandanan, K. and Srinivasan, R. (1970). Simplified estimation of parameters in a logistic distribution, *Biometrika*, *57*, 677–678.

Shah, B. K. (1966). On the bivariate moments of order statistics from a logistic distribution, *Ann. Math. Statist.*, *37*, 1002–1010.

Shu, V. S. (1978). Robust estimation of a location parameter in the presence of outliers, Ph.D. Thesis, Iowa State University, Ames, Iowa.

Tiku, M. L. (1967). Estimating the mean and standard deviation from a censored normal sample, *Biometrika*, *54*, 155–165.

Tiku, M. L. (1980). Robustness of MML estimators based on censored samples and robust test statistics, *J. Statist. Plann. Inf.*, *4*, 123–143.

Tiku, M. L., Tan, W. Y., and Balakrishnan, N. (1986). *Robust Inference*, Marcel Dekker, New York.

Tukey, J. W. and McLaughlin, D. H. (1963). Less vulnerable confidence and significance procedures for location based on a single sample: trimming/Winsorization 1, *Sankhyā* A, *25*, 331–352.

Vaughan, R. J. and Venables, W. N. (1972). Permanent expressions for order statistics densities, *J. Roy. Statist. Soc., Ser. B*, *34*, 308–310.

13

Goodness-of-Fit Tests

Ralph B. D'Agostino and Joseph M. Massaro
Boston University, Boston, Massachusetts

13.1 INTRODUCTION

In this chapter we present goodness-of-fit techniques for the logistic distribution with cumulative distribution function (cdf)

$$F(x; \mu, \sigma) = \frac{1}{1 + e^{-\pi(x-\mu)/\sigma\sqrt{3}}}. \qquad (13.1.1)$$

Here $-\infty < x < \infty$ and μ and σ are, respectively, the mean and standard deviation. By goodness-of-fit techniques, we mean procedures for evaluating if sample data were obtained from a random variable X with cdf of (13.1.1). We distinguish two types of goodness-of-fit techniques: (1) informal techniques involving plots or graphs of the data; (2) formal procedures consisting of hypotheses tests with null hypothesis that the logistic distribution of (13.1.1) is the underlying distribution of the data versus the alternative that it is not.

The graphical techniques give us a feel for the data and indicate graphically the "closeness" of the data to the logistic distribution. They also supply information on how the data deviate from the logistic distribution. The formal testing techniques quantify with the appropriate level of significance the goodness or deviation from the logistic. Used jointly they can supply a substantial evaluation of the data.

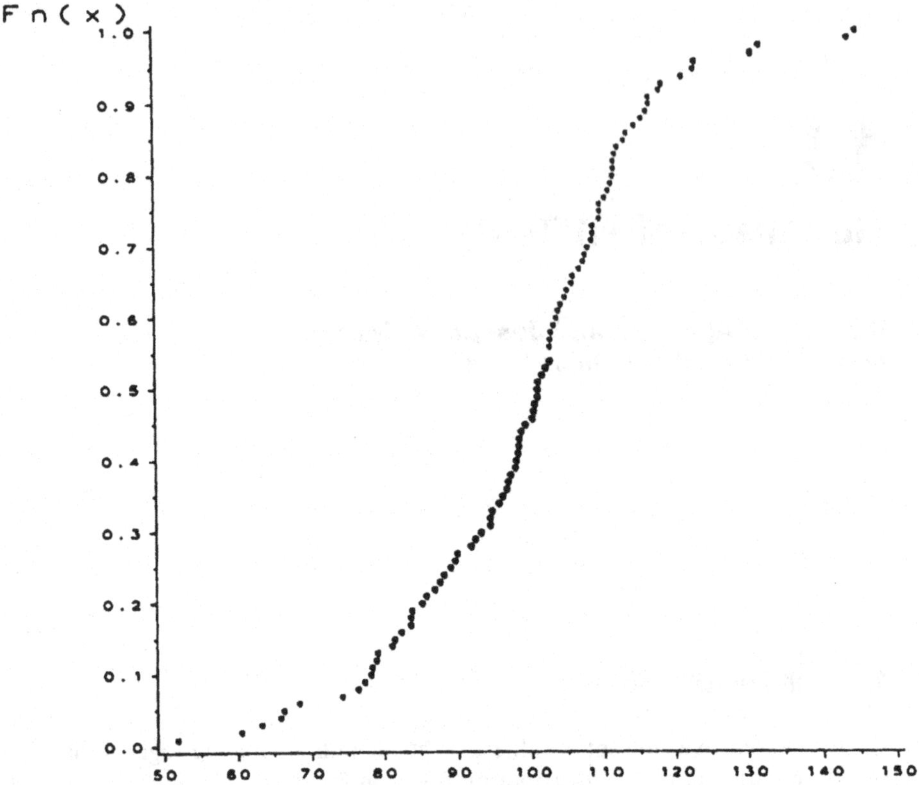

Fig. 13.2.1 (a) Plot of edf for LOG data set.

In Section 13.2 we begin the discussion of the graphical, informal analysis techniques by defining the empirical distribution function (edf) of a random sample and discussing the uses of its graph for the goodness-of-fit evaluation. In Section 13.3 we develop the logistic probability plot. This is the preferred device for graphical analysis. Sections 13.4 to 13.7 present examples and various uses of it. In particular Section 13.6 contains its use for censored data.

Section 13.8 begins the discussion of the formal tests. We include chi-squared tests (Section 13.9), empirical distribution function procedures (Section 13.10), normalized spacing procedures (Section 13.11), and tests based on regression and correlation statistics (Section 13.12). Some of these procedures are not well developed in the literature. This chapter develops them more completely than elsewhere.

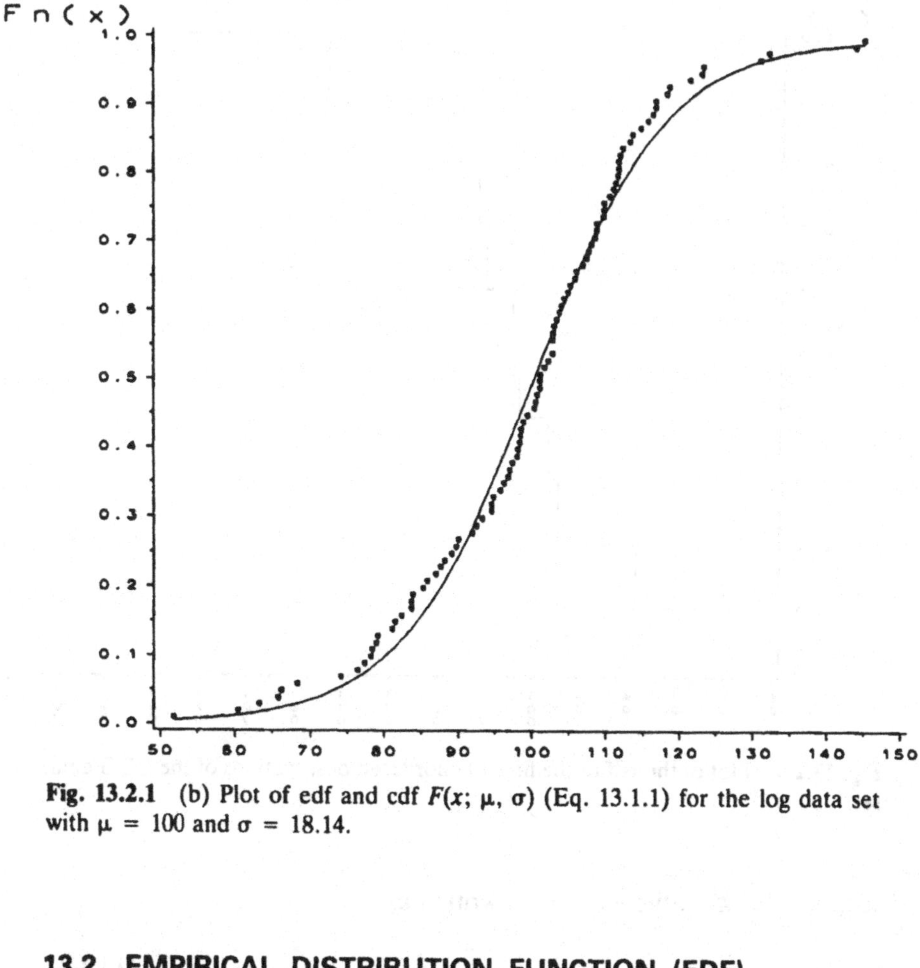

Fig. 13.2.1 (b) Plot of edf and cdf $F(x; \mu, \sigma)$ (Eq. 13.1.1) for the log data set with $\mu = 100$ and $\sigma = 18.14$.

13.2 EMPIRICAL DISTRIBUTION FUNCTION (EDF)

13.2.1 Definition

Let $X_1, X_2, X_3, \ldots, X_n$ represent a random sample of size n from the logistic distribution with cumulative distribution function (cdf) given by Eq. (13.1.1). The empirical cumulative function or empirical distribution function (edf) is defined as

$$F_n(x) = \frac{\#(X_j \leq x)}{n}, \tag{13.2.1}$$

where $j = 1, 2, \ldots, n$. In terms of the ordered observations $X_{1:n} \leq$

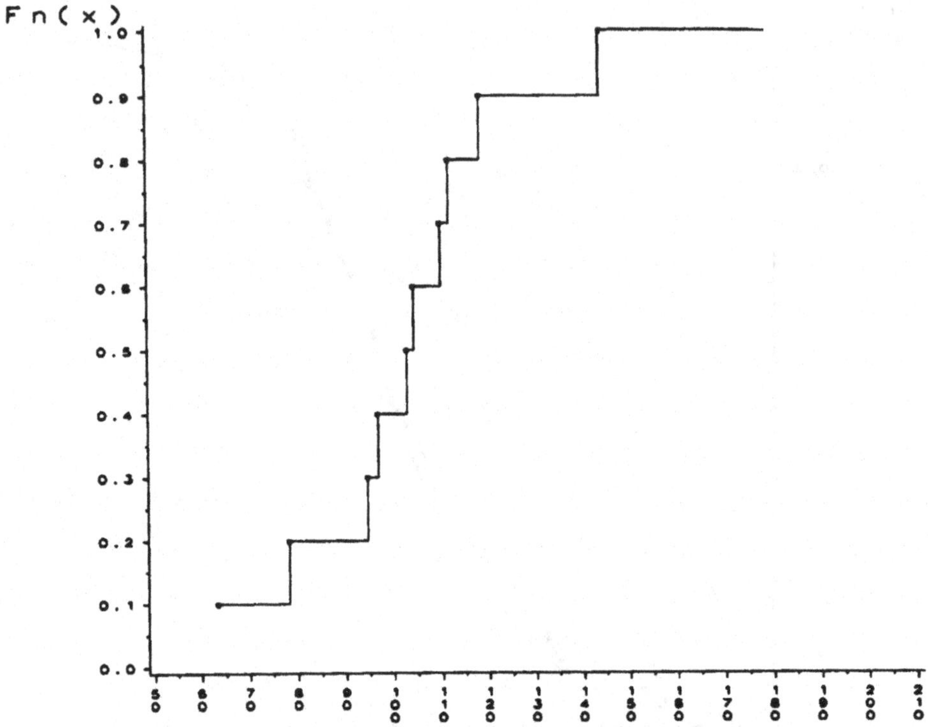

Fig. 13.2.2 Plot of the edf of the first 10 unordered observations of the LOG data set.

$X_{2:n} \leq \cdots \leq X_{n:n}$, the edf can be written as

$$F_n(x) = \frac{i}{n}, \tag{13.2.2}$$

where $i = \#(X_{j:n} \leq x)$. So the value of $F_n(x)$ for $X_{i:n}$ is i/n, where i is the rank of observation $X_{i:n}$ for $i = 1, 2, \ldots, n$. A plot of the edf can be displayed on arithmetic graph paper by plotting i/n versus $X_{i:n}$ (i/n is the ordinate, $X_{i:n}$ is the abscissa). For all x, $F_n(x)$ converges for large n to (13.1.1). This convergence is strong convergence uniformly for all x (Rényi, 1970). Thus, the edf plot does supply a "good picture" of the population or underlying cdf as n increases. Figure 13.2.1(a) displays the edf for the LOG data set, which is a random sample of size 100 from a logistic distribution with mean $\mu = 100$ and standard deviation $\sigma = 18.14$. (A copy of the data set is given in the appendix to this chapter.) Figure 13.2.1(b)

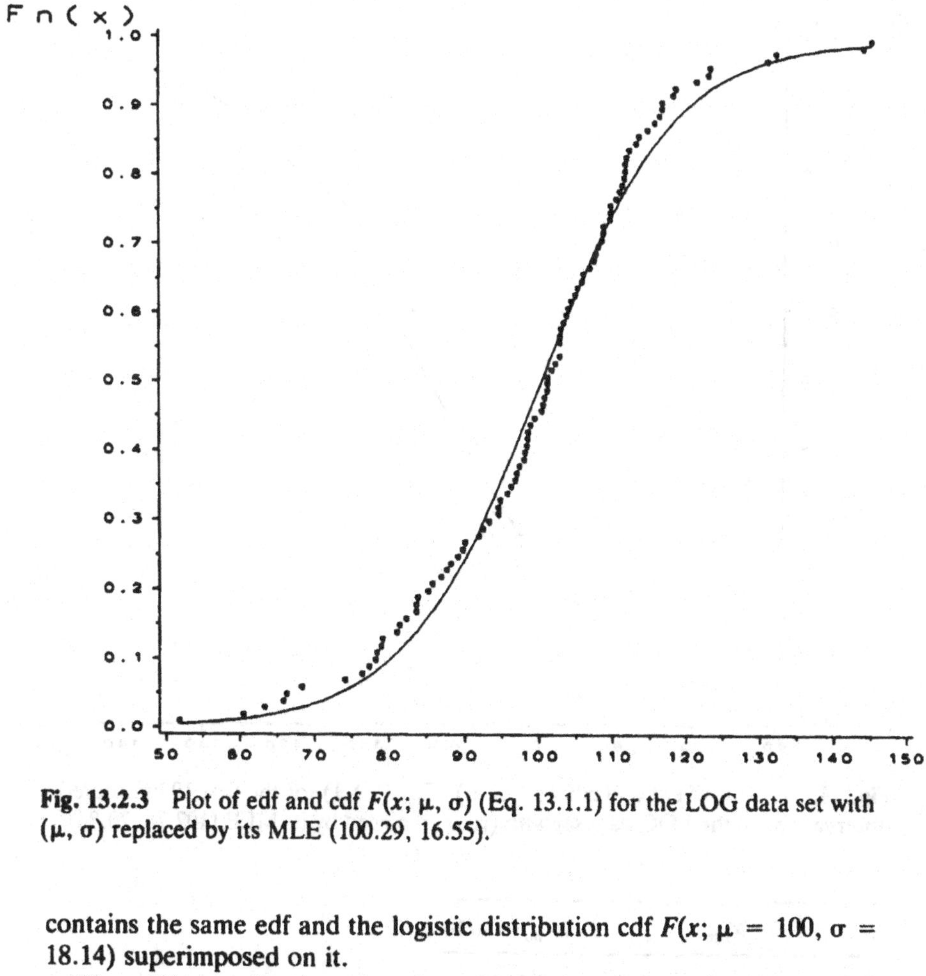

Fig. 13.2.3 Plot of edf and cdf $F(x; \mu, \sigma)$ (Eq. 13.1.1) for the LOG data set with (μ, σ) replaced by its MLE (100.29, 16.55).

contains the same edf and the logistic distribution cdf $F(x; \mu = 100, \sigma = 18.14)$ superimposed on it.

The edf is actually a step function with steps at the values of the random variable occurring in the data. It is not often displayed as such especially when the sample size is large, such as in Fig. 13.2.1. Figure 13.2.2 displays the edf as a step function using only the first 10 unordered observations of the LOG data set as the sample. The ordered values of these 10 observations and their corresponding edf values are as follows:

i	$X_{i:10}$	$F_n(X_{i:10}) = i/10$
1	63.35	0.1
2	78.32	0.2

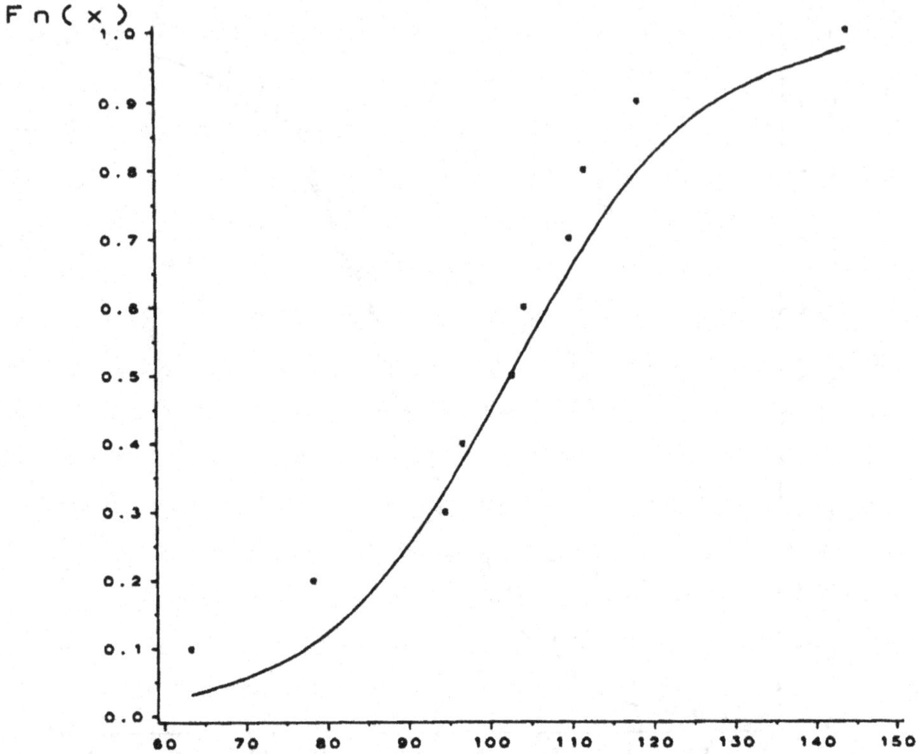

Fig. 13.2.4 Plot of edf and cdf $F(x; \mu, \sigma)$ (Eq. 13.1.1) for the first 10 unordered observations of the LOG data set with (μ, σ) replaced by its MLE (102.74, 20.84).

i	$X_{i:10}$	$F_n(X_{i:10}) = i/10$
3	94.63	0.3
4	96.91	0.4
5	102.97	0.5
6	104.47	0.6
7	109.99	0.7
8	111.81	0.8
9	118.54	0.9
10	144.28	1.0

Note, if there were ties, they would receive the highest of the corresponding ranks.

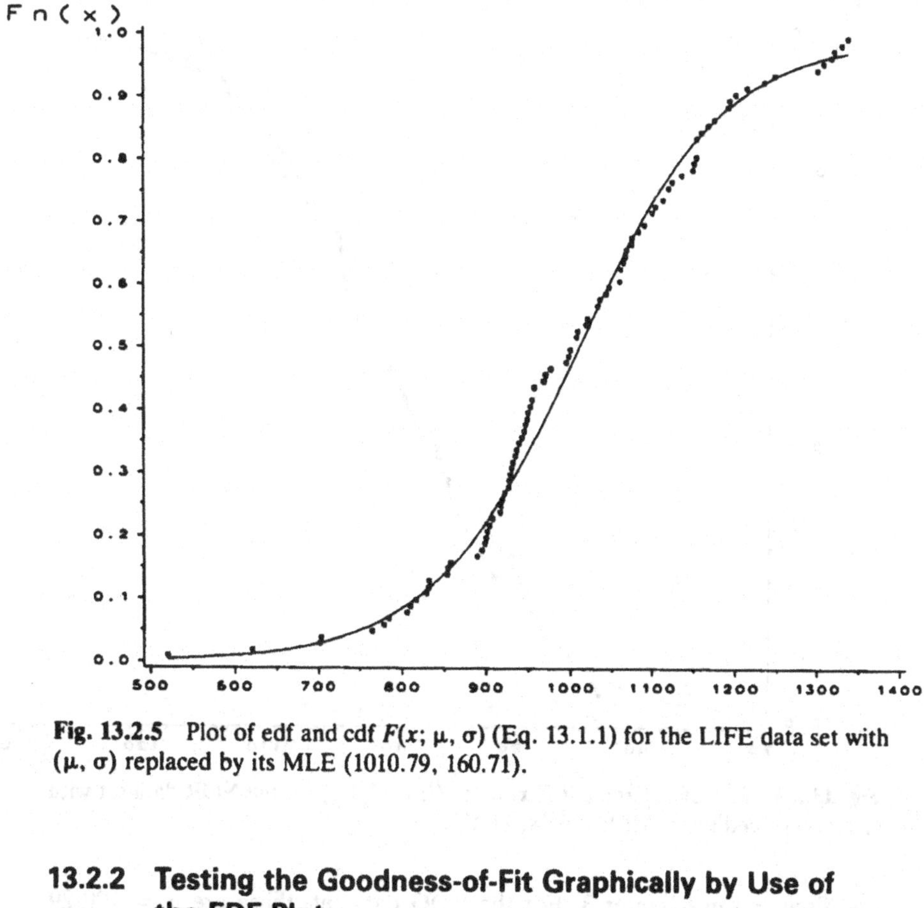

Fig. 13.2.5 Plot of edf and cdf $F(x; \mu, \sigma)$ (Eq. 13.1.1) for the LIFE data set with (μ, σ) replaced by its MLE (1010.79, 160.71).

13.2.2 Testing the Goodness-of-Fit Graphically by Use of the EDF Plot

The edf plot can be used to assess how well the data fits the logistic distribution by imposing on the edf plot the graph of the theoretical logistic cdf and informally judging the closeness of the edf plot to the cdf graph. When the values of μ and σ are specified, this is achieved simply by using these values in the formula for the cdf and producing the resulting graph. Figure 13.2.1(b) is an example of such a plot. Except for some deviation in the upper tail the agreement of the edf plot and the cdf curve appears to be very good.

When, as is the usual case, μ and σ are unknown, estimates of them must be produced and the cdf plot is attained by the substitution of these into the formula for the logistic cdf. One procedure is to use the maximum likelihood estimates (MLE) of μ and σ. Procedures for obtaining the MLE

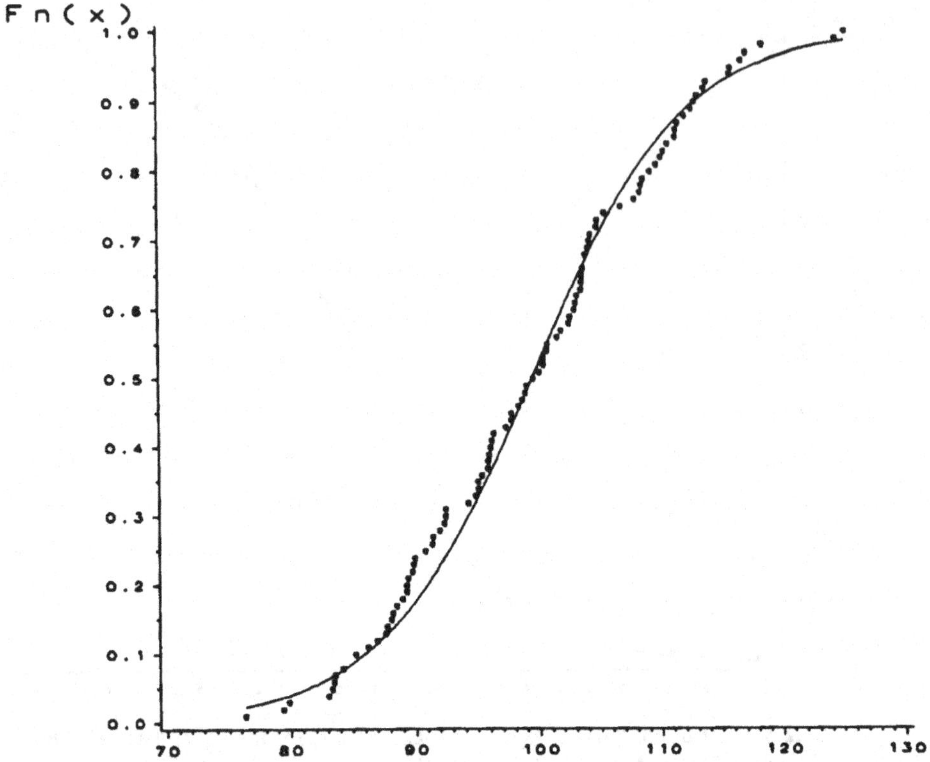

Fig. 13.2.6 Plot of edf and cdf $F(x; \mu, \sigma)$ (Eq. 13.1.1) for the NOR data set with (μ, σ) replaced by its MLE (99.48, 11.10).

are discussed in Chapter 3. For the LOG data set, these are $\hat{\mu} = 100.29$ and $\hat{\sigma} = 16.55$. Figure 13.2.3 contains the edf plot for the full LOG data set and a graph of the logistic cdf $F(x; \hat{\mu} = 100.29, \hat{\sigma} = 16.55)$. There is little difference between this graph and Fig. 13.2.1(b), where the true μ and σ were used. Often the plot with the estimated values of the parameters will give better fits (agreements of edf and cdf) than the plot with the prespecified parameters.

Figures 13.2.4 and 13.2.5 supply two further illustrations of the edf and cdf comparisons. Figure 13.2.4 contains the first 10 unordered observations from the LOG data set. The MLE of (μ, σ) is (102.74, 20.84), and these are used for the cdf graph. Figure 13.2.5 contains the edf from the first 100 of the 417 lifetimes of incandescent light bulbs given by Davis (1952) and the logistic cdf graph in which the MLE of $(\mu, \sigma) = (1010.79, 160.71)$ were used. (We refer to this data set as the LIFE data set, and it is given in the appendix to this chapter.) For both these examples the logistic distribution provides, at least by an informal examination of the figures, good fits to the data. In Section 13.10 we discuss the edf goodness-of-fit tests which test formally the closeness of the edf and cdf. These tests summarize in a single statistic the agreement of the two.

Figure 13.2.6 supplies an illustration of the edf of the NOR data set, a random sample of size 100 from a normal distribution with mean 100 and standard deviation 10, with the cdf (13.1.1), with (μ, σ) replaced by its MLE based on the logistic distribution, (99.48, 11.10). (The NOR data set is also given in the appendix to this chapter.) Unfortunately, this nonlogistic distribution seems to approximate very well the logistic distribution. A better assessment of the deviation of these data from the logistic distribution is supplied by the logistic probability plot, which we now will discuss.

13.3 LOGISTIC PROBABILITY PLOTTING

13.3.1 Definition and Plot

The difficulty with the use of the edf to judge informally its agreement or closeness to the cdf is that one has to judge deviations from nonlinear curves. For most it is easier to judge deviations from a straight line. A probability plot is a useful way of judging informally the fit of data to a distribution where the judgment is based on deviations from a straight line. The key element in the probability plot for the logistic distribution is to transform the edf in such a way that it will be, within random fluctuations, a straight line if the underlying distribution is logistic. We now develop this probability plot and show its many uses.

The cdf of the logistic distribution is given as $F(x; \mu, \sigma)$ of (13.1.1), where μ and σ are the mean and standard deviation. If we standardize x with the transformation

$$Y = \frac{X - \mu}{\sigma} \tag{13.3.1}$$

and use $G(y)$ as the cdf of the standardized distribution, we have $F(x; \mu, \sigma) = G(y)$, which, upon taking the inverse of $G(\cdot)$, produces

$$G^{-1}(F(x; \mu, \sigma)) = y = \frac{x - \mu}{\sigma}.$$

From this we have

$$x = \mu + \sigma y = \mu + \sigma G^{-1}(F(x; \mu, \sigma)).$$

Since y is linearly related to x, the plot of x on $y = G^{-1}(F(x; \mu, \sigma))$ is a straight line. Specifically, for the logistic distribution we have

$$y = G^{-1}(F(x; \mu, \sigma)) = \frac{\sqrt{3}}{\pi} \log\left(\frac{F(x; \mu, \sigma)}{1 - F(x; \mu, \sigma)}\right) \tag{13.3.2}$$

where log represents the natural logarithm. For a random sample we replace the cdf in (13.3.2) with the edf $F_n(x)$, where in order to avoid taking logs of infinity, $F_n(x)$ is defined as

$$F_n(x) = \frac{i - 0.5}{n}, \tag{13.3.3}$$

where $i = \#(X_j \leq x)$. From this we obtain a plot of

$$x \text{ vs. } y = \frac{\sqrt{3}}{\pi} \log\left(\frac{F_n(x)}{1 - F_n(x)}\right). \tag{13.3.4}$$

This plot is the *logistic probability plot*. If the underlying distribution is logistic, this plot will be, within random fluctuations, a straight line. D'Agostino in Chapter 2 of D'Agostino and Stephens (1986) demonstrated that (13.3.4) can be used for all sample sizes $n \geq 10$.

In the above we redefined $F_n(x)$ from i/n of (13.2.2) to $(i - 0.5)/n$ in (13.3.3). Barnett (1975) discussed the general question of modifying the definition of the edf. He investigated the general form $F_n(x) = (i - c)/(n - 2c + 1)$ for $0 \leq c \leq 1$. In (13.3.3) we use $c = 0.5$. In our experience it is a useful selection. We suggest the reader review Barnett (1975) for further discussion of this.

Figure 13.3.1 is the logistic probability plot for the LOG data set.

Presence of Ties

Due to rounding, even data from continuous distributions such as logistic will contain tied observations. For example, for the LOG data set there are two values of 102.90, and these occupy the 55th and 56th positions in the order sample. The corresponding y values for these ranks are 0.100 and 0.122. In the probability plot the mean of these, namely 0.111, should be plotted as the y corresponding to $x = 102.90$. In general, for tied observations the mean y of the tied observations is plotted.

13.3.2 Informal Goodness-of-Fit and Estimation of Parameters

Informal examination of the plot in Fig. 13.3.1 clearly shows that a straight line fits the data well. This indicates that the logistic distribution is a reasonable and acceptable distribution for the underlying distribution of the LOG data set.

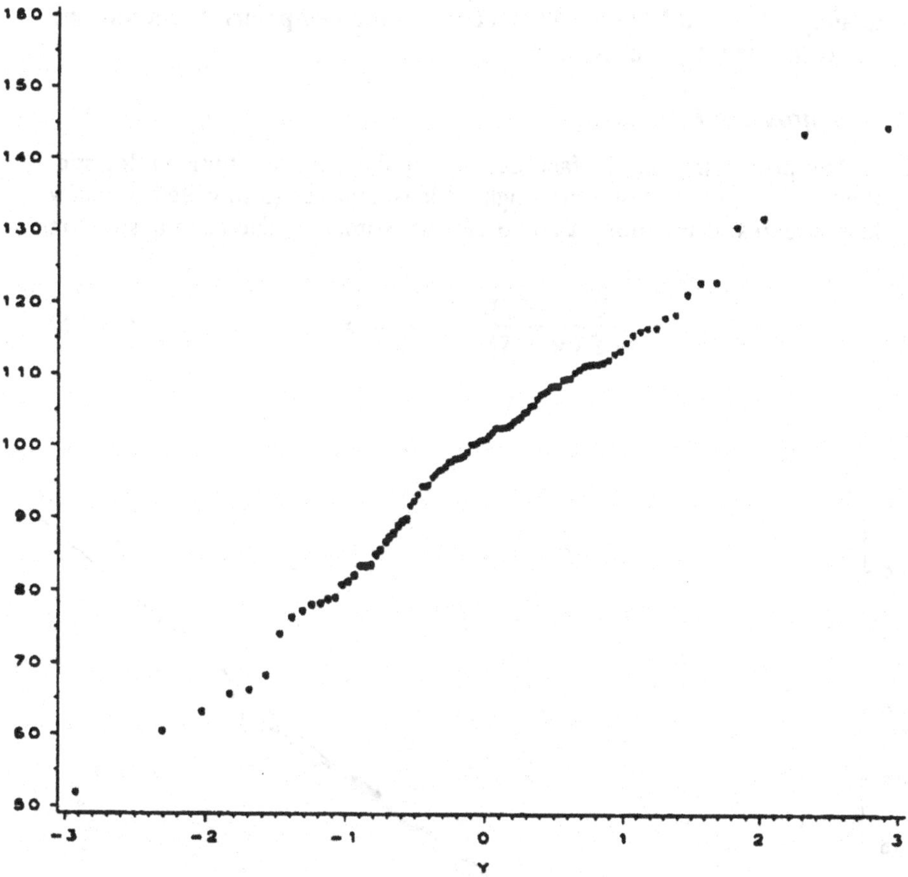

Fig. 13.3.1 Logistic probability plot of the LOG data set.

Given a logistic probability plot of an arbitrary data set, there are several ways of obtaining a straight line to "fit" to the data. Informal examination of the deviation of the probability plot from the straight line can be used to assess if the fit is good. Four techniques for obtaining the straight line will be discussed below. Formal goodness-of-fit techniques also exist (see regression techniques, Section 13.12).

Drawing a Line by Eye

One way of fitting informally a straight line is to do so "by eye." We have found it useful to locate points on the plot corresponding to approximately $F_n(x) = 0.10$ and 0.90 and connect the two points. From this one can determine if the fit is good.

Regression Estimates

This procedure uses the fact that if the underlying distribution is logistic, then $x = \mu + \sigma y$. Using unweighted least-squares (also called ordinary least squares) estimation, μ and σ can be estimated. The general solution is

$$\tilde{\sigma} = \frac{\Sigma\,(y - \bar{y})x}{\Sigma\,(y - \bar{y})^2}, \qquad \tilde{\mu} = \bar{x} - \tilde{\sigma}\bar{y}, \tag{13.3.5}$$

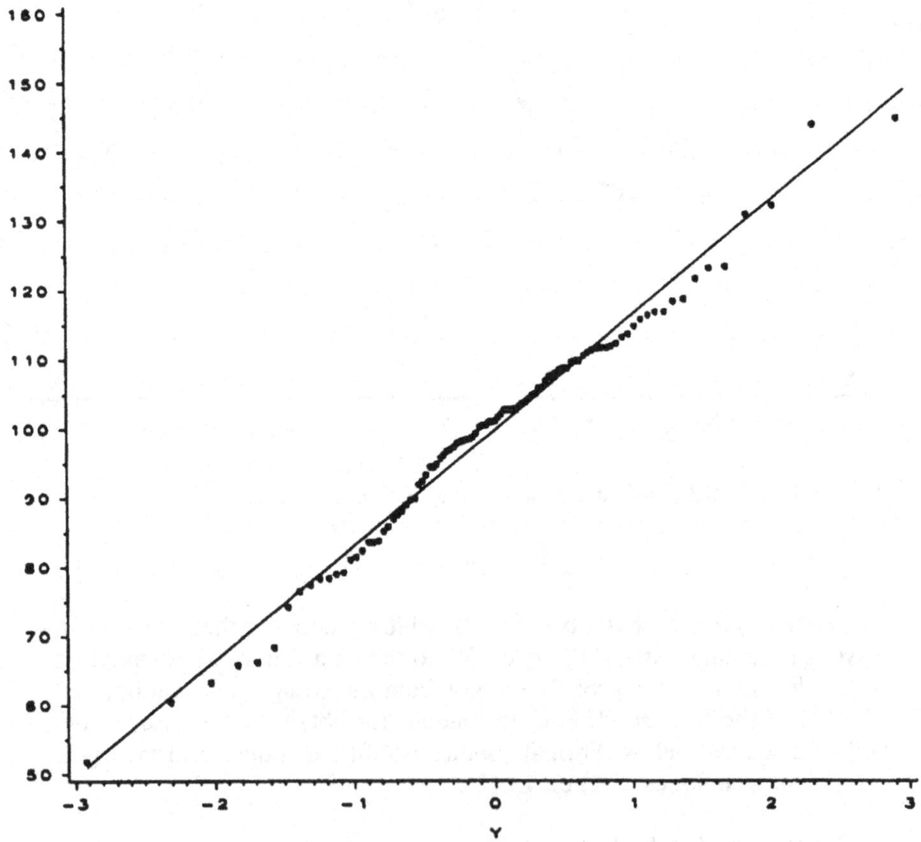

Fig. 13.3.2 Logistic probability plot of the LOG data set with straight-line approximation obtained from ordinary least squares ($\tilde{\mu}$ = 99.68, $\tilde{\sigma}$ = 16.69).

where the summations run from 1 to n. Note since $\bar{y} = 0$, $\tilde{\mu} = \bar{x}$. For the LOG data set, $\tilde{\mu} = 99.68$ and $\tilde{\sigma} = 16.69$. Figure 13.3.2 shows the line $x = 99.68 + 16.69y$ imposed on the probability plot of the LOG data set. Two points are important to make here. First, because the ordered observations are dependent the nonrandom pattern of points about the straight line as shown in Fig. 13.3.2 is to be expected. Second, it is the vertical distance from the points to the line that should be used in judging deviations, since x is now plotted on the vertical axis. Figure 13.3.2 indicates a very good fit of the logistic distribution to the LOG data set.

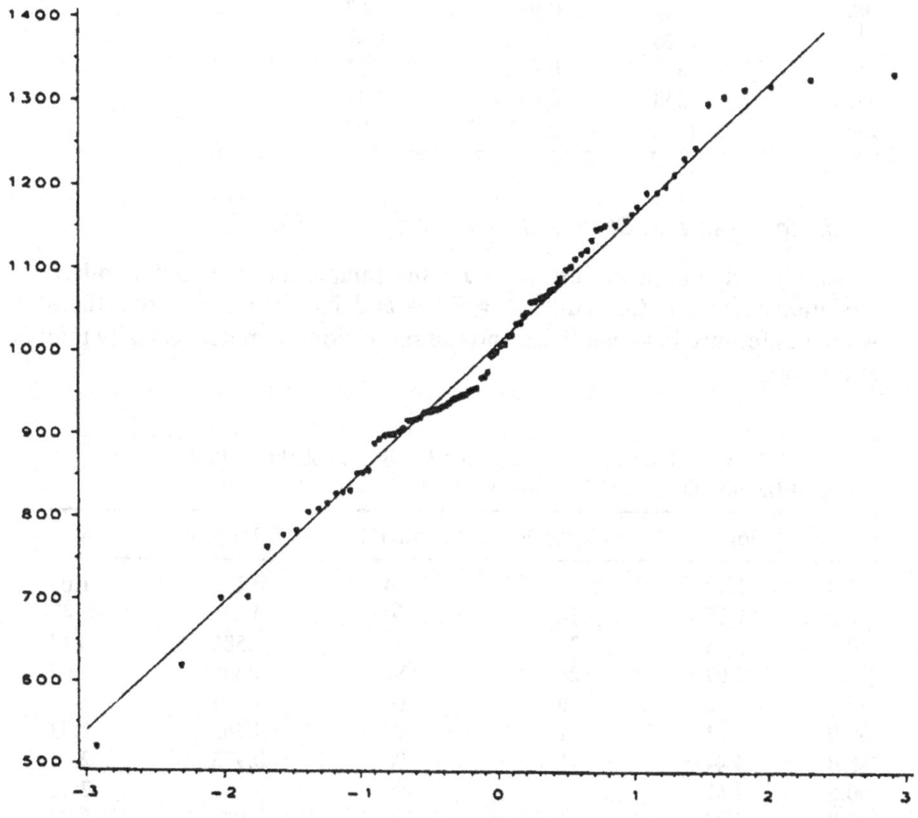

Fig. 13.4.1 Logistic probability plot of the LIFE data set with straight-line approximation obtained from ordinary least squares ($\tilde{\mu} = 1011.05$, $\tilde{\sigma} = 157.33$).

Table 13.4.1 Calculation of y Used in the Logistic Probability Plot of the LIFE Data Set

Rank (I)	x	$F_n(x)$	y
1	521	0.005	−2.92
2	621	0.015	−2.31
3	702	0.025	−2.02
4	704	0.035	−1.83
5	765	0.045	−1.68
6	780	0.055	−1.57
7	785	0.065	−1.47
94	1250	0.935	1.47
95	1303	0.945	1.57
96	1311	0.955	1.68
97	1320	0.965	1.83
98	1324	0.975	2.02
99	1333	0.985	2.31
100	1340	0.995	2.92

Using Mean and Standard Deviation of the Sample

Another simple procedure is to use the sample mean $\overline{X}$ and standard deviation S. For the LOG data $\overline{X} = 99.68$ and $S = 16.71$. Note that there is little difference between these and the regression estimates. Usually that is the case.

Table 13.4.2 Calculation of y Used in the Logistic Probability Plot for the Grouped Income Data for Fisk (1961)

x	log(x)	Frequency	Rank(i)	$F_n(\log(x))$	y
3.0	1.10	14	14	0.135	−1.03
7.0	1.95	16	30	0.295	−0.48
10.5	2.35	29	59	0.585	0.19
21.5	3.07	28	87	0.865	1.03
35.5	3.56	9	96	0.955	1.69
50.0	3.91	1	97	0.965	1.83
85.0	4.44	1	98	0.975	2.02
100.5	4.61	1	99	0.985	2.31
120.0	4.79	1	100	0.995	2.92

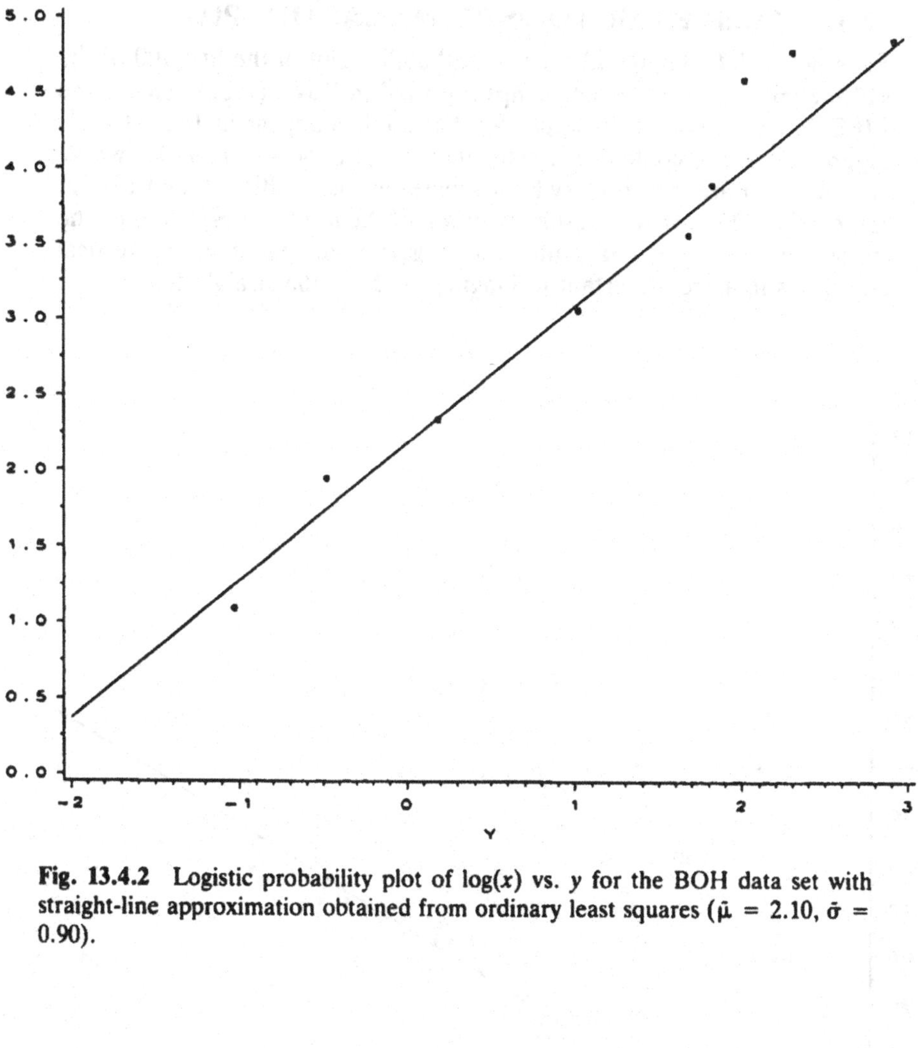

Fig. 13.4.2 Logistic probability plot of log(x) vs. y for the BOH data set with straight-line approximation obtained from ordinary least squares ($\tilde{\mu}$ = 2.10, $\tilde{\sigma}$ = 0.90).

Maximum Likelihood Estimates

A fourth technique is to use the maximum likelihood estimates. For the LOG data $\hat{\mu}$ = 100.29 and $\hat{\sigma}$ = 16.55. A plot using these would look similar to Fig. 13.3.2.

13.4 EXAMPLES OF LOGISTIC PROBABILITY PLOTS

Example 13.4.1 Figure 13.4.1 is a probability plot of the first 100 of the 417 lifetimes of incandescent lamps reported in Davis (1952). This is the LIFE data set given in the appendix. Partial data appear in Table 13.4.1. Figure 13.4.1 also contains the estimated straight line $x = \tilde{\mu} + \tilde{\sigma}y$, where $\tilde{\mu} = \overline{X}$ and $\tilde{\sigma}$ are the ordinary least-squares estimates 1011.05 and 157.33, respectively. Note that with the exception of the last few observations, the straight line fits the data quite well. Again note that it is the vertical deviations that are important in judging the fit to the straight line.

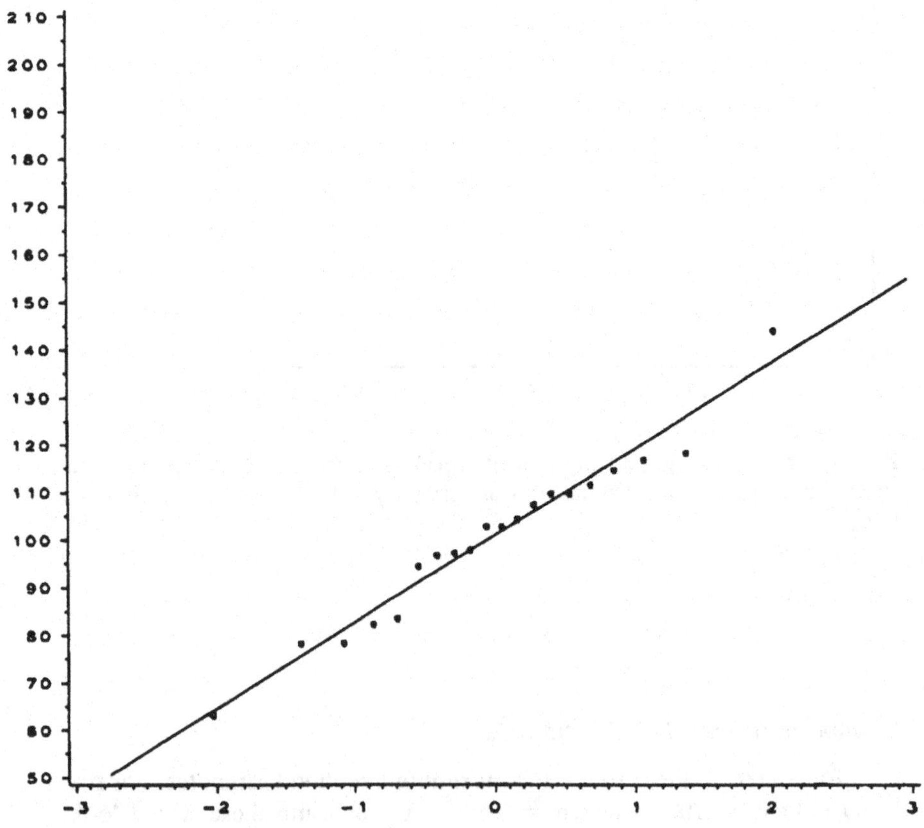

Fig. 13.5.1 (a) Logistic probability plot for detecting outliers: first 20 unordered observations of LOG data set.

Example 13.4.2 Grouped Data In Section 13.3.1 we recommended the use of mean rankings or y values in the case of ties in the data. For grouped data, instead of using average ranking of the observation for each group in the calculation of $F_n(x)$, the highest ranking is used.

The data in Table 13.4.2 is based on data from Fisk (1961). It is a random sample of size 100 from his data of the distribution of incomes in Bohemia in 1933 in units of 1000 kroner. (We refer to this data set as the BOH data set, and it is also given in the appendix to this chapter.) Fisk claims that the natural log of the data has the logistic distribution as its underlying distribution. Note that the data are grouped by their midpoints; i.e., 3 represents $0 \le x \le 6$, 7 represents $6 < x \le 8$, 10.5 represents $8 < x \le 13$, etc. Also, $\log(x)$ represents the natural logarithm of x.

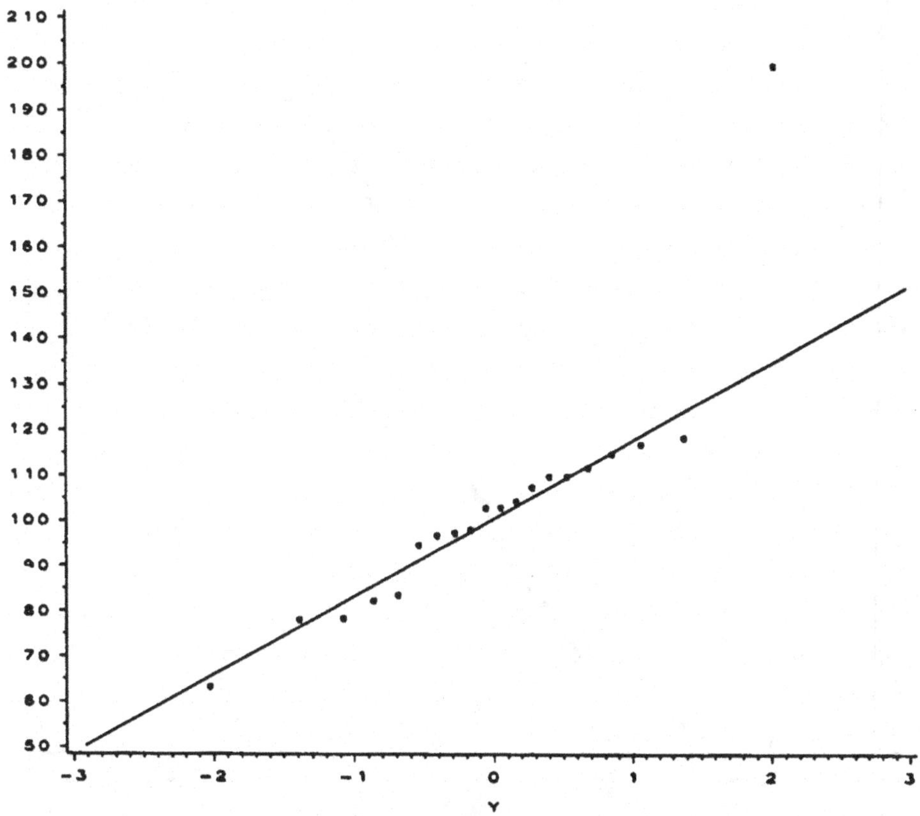

Fig. 13.5.1 (b) Logistic probability plot for detecting outliers: first 20 unordered observations of LOG data set with largest value replaced with an outlier.

Figure 13.4.2 shows the probability plot on which the theoretical line of log(x) versus y is imposed. Again the estimates of (μ, σ) used to draw the line were the least-squares estimates, (2.18, 0.90). Note that the overall fit of log(x) to the logistic distribution is good.

13.5 JUDGING DEVIATIONS FROM LINEARITY IN PROBABILITY PLOTS

13.5.1 Detection of Outliers

The logistic probability plot can be used to detect outliers. Figure 13.5.1 illustrates its use. Figure 13.5.1(a) shows the logistic probability plot of the

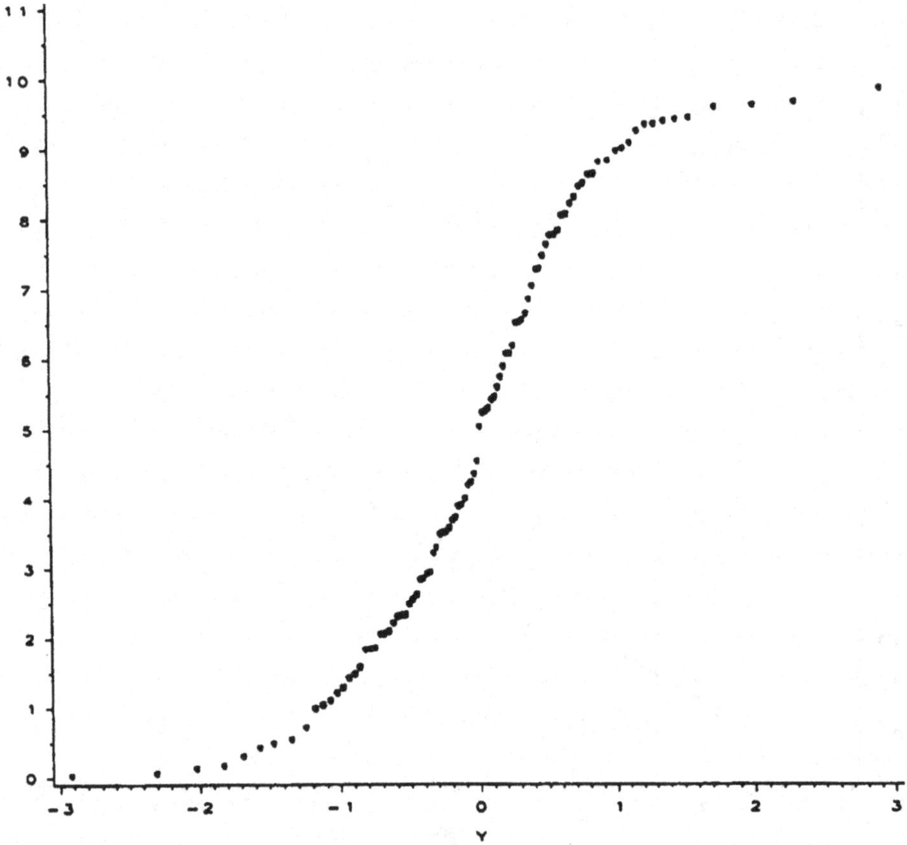

Fig. 13.5.2 (a) Logistic probability plot of a data set of size 100 simulated from the uniform (0, 10) distribution (β_2 = 1.8).

first 20 unordered observations of the LOG data set. Figure 13.5.1(b) gives the probability plot of the first 20 unordered observations of the LOG data set with the largest of these, 144.28, replaced with an outlier value of 200.28. We deliberately selected the large outlier so the reader could see the impact. Outliers appear as large vertical deviations.

13.5.2 Unimodal Distributions

D'Agostino in Chapter 2 of D'Agostino and Stephens (1986) reports that one way to distinguish unimodal nonnormal distributions from the normal is in terms of the skewness (measuring symmetry) and kurtosis (measuring

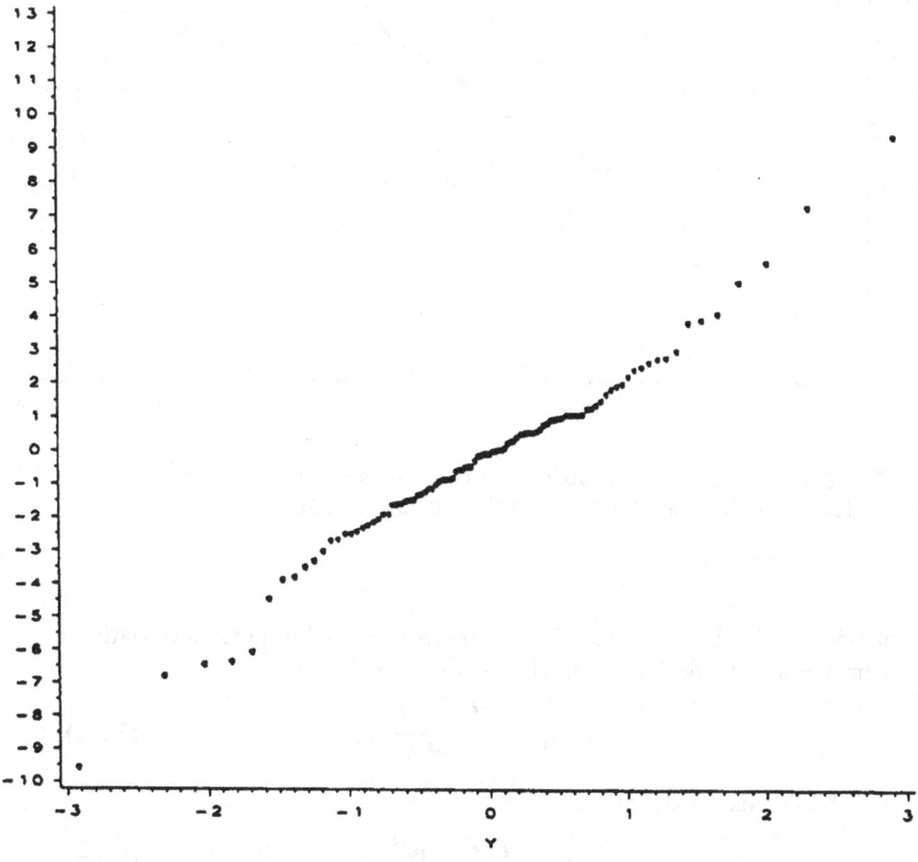

Fig. 13.5.2 (b) Logistic probability plot of a data set of size 100 simulated from the Laplace (0, 2) distribution ($\beta_2 = 6.0$).

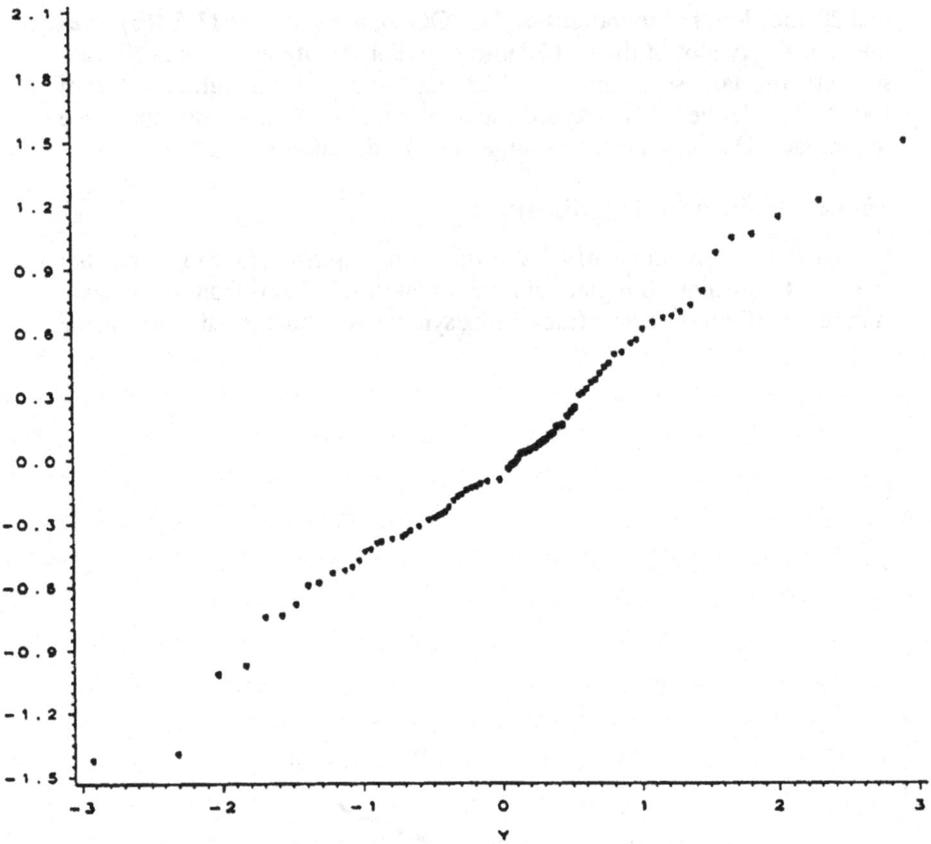

Fig. 13.5.2 (c) Logistic probability plot of a data set of size 100 simulated from the Johnson unbounded $SU(0, 2)$ distribution ($\beta_2 = 4.51$).

heaviness of tails) measures. These can also be used in judging deviations from the logistic distribution. The skewness is defined as

$$\sqrt{\beta_1} = \frac{E(x - \mu)^3}{\sigma^3}, \tag{13.5.1}$$

and the kurtosis is defined as

$$\beta_2 = \frac{E(x - \mu)^4}{\sigma^4}, \tag{13.5.2}$$

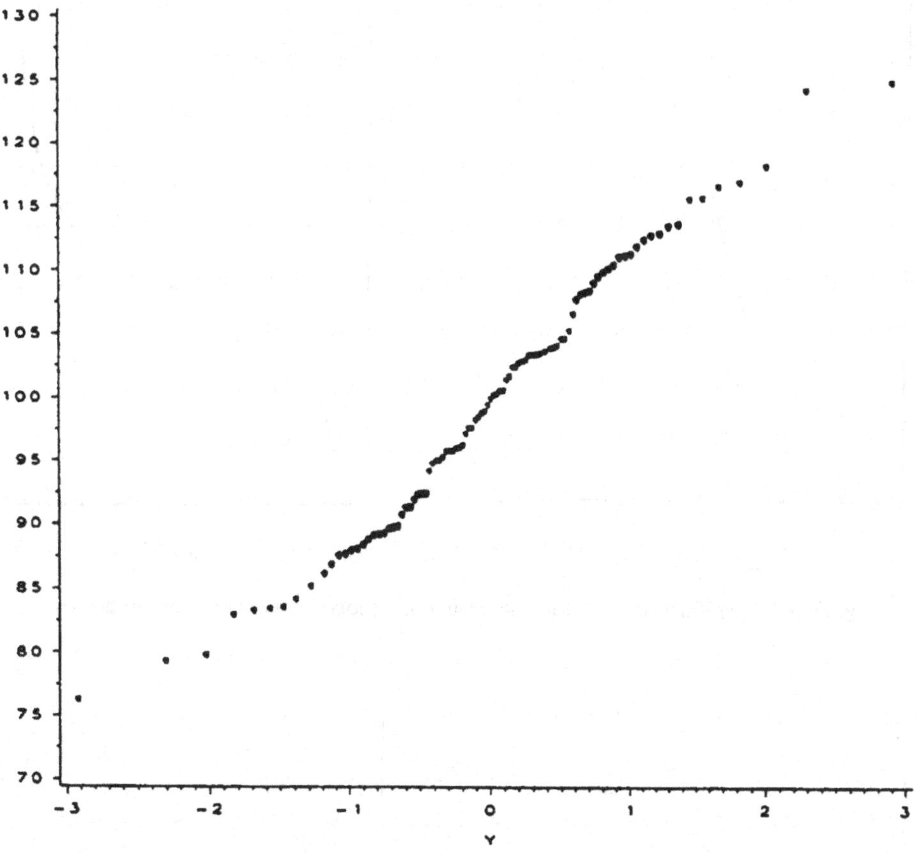

Fig. 13.5.2 (d) Logistic probability plot of a data set of size 100 simulated from the normal (100, 100) distribution ($\beta_2 = 3.0$).

where E represents the expected value operator. For the logistic distribution $\sqrt{\beta_1} = 0$ (thus implying symmetry) and $\beta_2 = 4.2$. Figures 13.5.2 (a) through (d) show the logistic probability plot for simulated sets of data from four *symmetric* distributions with different values of kurtosis. The UNI data set was generated from a uniform (0, 10) distribution. The SU(0, 2) data set was generated from a Johnson unbounded SU(0, 2) distribution. The LAP data set was generated from the Laplace distribution with location parameter 0 and scale parameter 2 and NOR from a normal (100, 100) distribution.

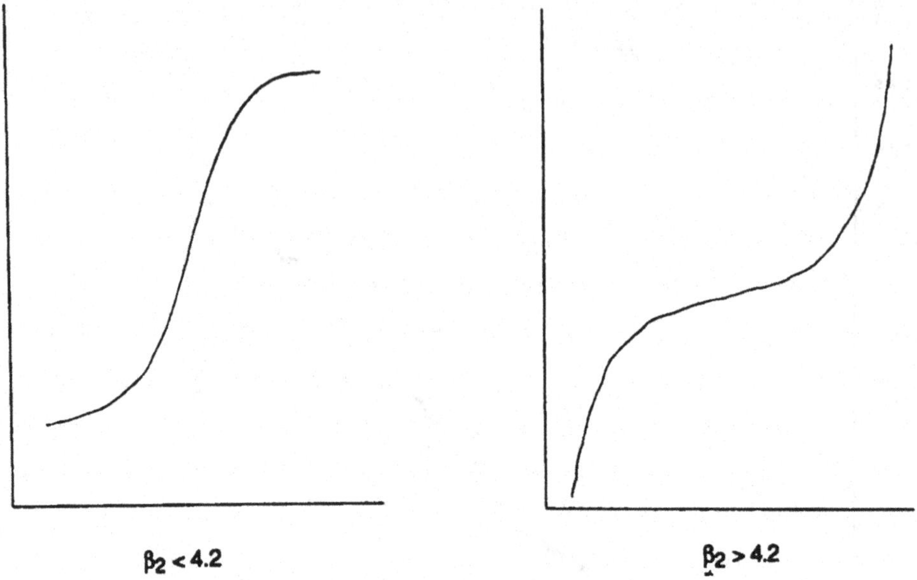

Fig. 13.5.3 (a) Illustration of logistic probability plots for symmetric distributions.

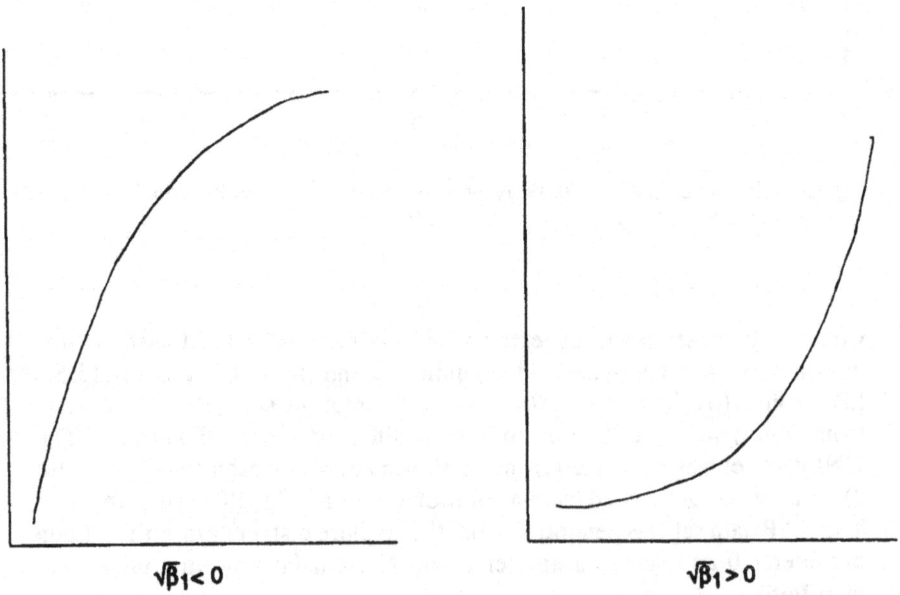

Fig. 13.5.3 (b) Illustration of logistic probability plots for skewed distributions.

Figure 13.5.3(a) illustrates the general appearance of a logistic probability plot for symmetric data sets. The reader should note that the simulated data sets of Figure 13.5.2 conform to these general forms. The imposition of straight lines on these plots by the methods of Section 13.3.2 would illustrate this even more clearly. Figure 13.5.4 contains probability plots for simulated data sets from skewed distributions: the negative exponential distribution (EXP) with mean equal to 5 and the Johnson unbounded with parameters 1 and 2 (SU(1, 2)). Figure 13.5.3(b) illustrates the general appearance of the logistic probability plots for skewed distributions. The data sets used for Figs. 13.5.2 and 13.5.4 are also given in the appendix to this chapter.

13.6 PROBABILITY PLOTTING FOR CENSORED SAMPLES

13.6.1 Singly Censored Sample

Often life-test data is censored; that is not all observations are available. Censoring can occur, for example, when measurements are taken until a certain prespecified magnitude has been reached. For example, in testing the life of light bulbs, only the lifetimes of all light bulbs failing before a three-month time limit might be recorded. This is called singly Type I censoring (or time censoring). For the above example, the resulting data would be said to be censored *on the right*. The number of censored observations, r, is a random variable. The second type of singly censoring, Type II censoring (or failure censoring) occurs when the number of ordered observations to be measured is prespecified. For example, interest may lie in only the 15 hottest days of the year. With respect to temperature, this example is of data censored *on the left*. The number of censored observations, s, is *not* a random variable.

The probability plotting of a singly censored sample is exactly the same as plotting a full sample, except, of course, that the censored observations are not plotted. Also, be sure that in the denominator of $F_n(x)$ of (13.3.3), that n, the number of total observation, is still used, and not $n - r$ or $n - s$, the number of uncensored observations.

Example 13.6.1 Suppose only the 40 lowest observations of the LOG data set (with total $n = 100$) are available. The data set is then singly right censored with $r = 60$. The observations are then ordered with i being the ordered observation number, $F_n(x)$ of (13.3.3) is calculated with $n = 100$, and y is calculated as usual using this $F_n(x)$. Then the plot of x versus y is plotted. Figure 13.6.1 contains the probability plot along with the theoretical line of x versus y for this data, with μ and σ being estimated by the

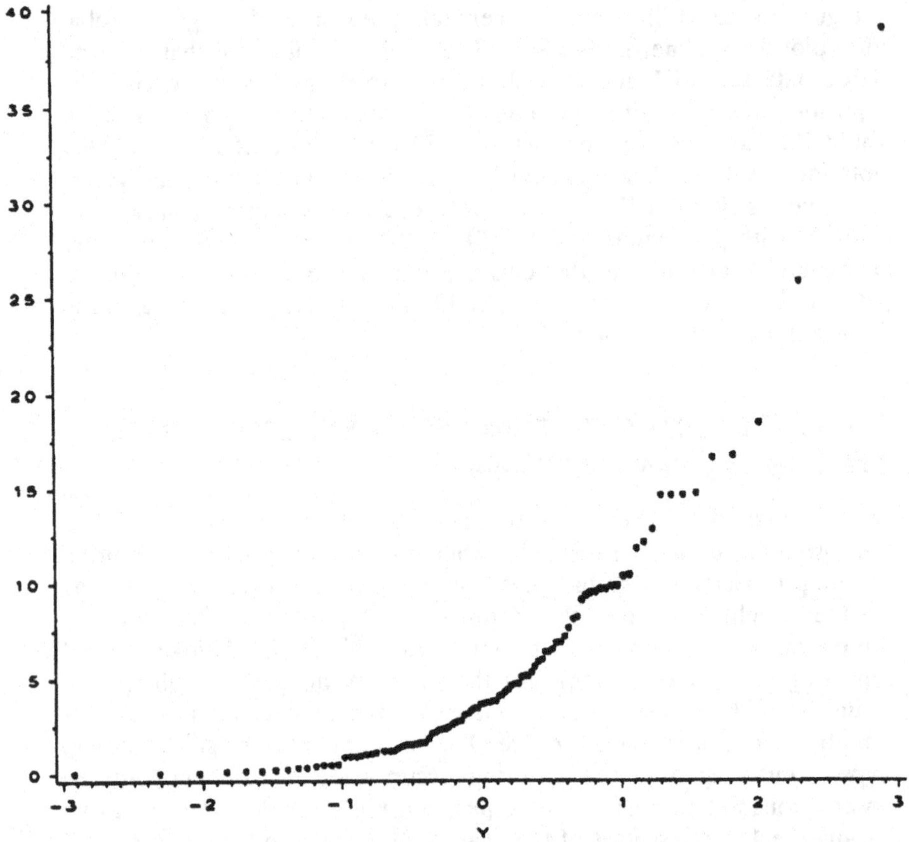

Fig. 13.5.4 (b) Logistic probability plot of a data set of size 100 simulated from the Johnson unbounded $SU(1, 2)$ distribution ($\sqrt{\beta_1} = -0.87$, $\beta_2 = 5.59$).

ordinary least-squares regression estimates 101.32 and 10.32, respectively. The procedure for plotting singly left-censored data is the same. Note, for probability plotting it does not matter whether the censoring is Type I or Type II.

13.6.2 Multiply Censored Sample

Censoring can be extended to both tails simultaneously and also to censoring of observations not in the tails. Censoring beyond just a single tail is called multiple censoring. Probability plotting can be extended to this. *Random time censoring* occurs when each subject's censoring time is an

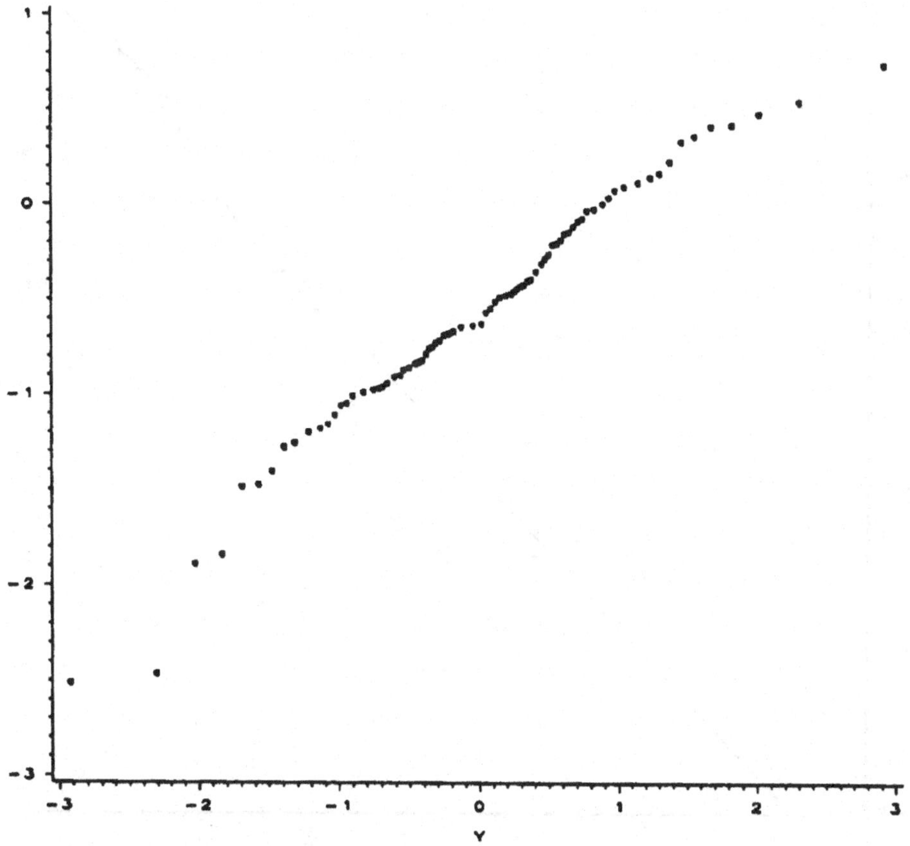

Fig. 13.5.4 (b) Logistic probability plot of a data set of size 100 simulated from the Johnson unbounded $SU(1, 2)$ distribution ($\sqrt{\beta_1} = -0.87$, $\beta_2 = 5.59$).

independent variable and is independent of the measurement taken. For example, if the survival times of individuals who have received a certain drug is the variable of interest, some subjects may leave the study before the event occurs or the study may end before the event occurs. In the latter case the larger values of interest may not be observed, which is why this censoring is also referred to as *arbitrary right censoring*.

If the different censoring times of each subject are known beforehand, the data are *progressively censored* (Type I). This occurs, for example, when subjects are placed into a life test at different starting times, but each one has the same fixed termination point (Michael and Schucany in Chapter 11 of D'Agostino and Stephens, 1986).

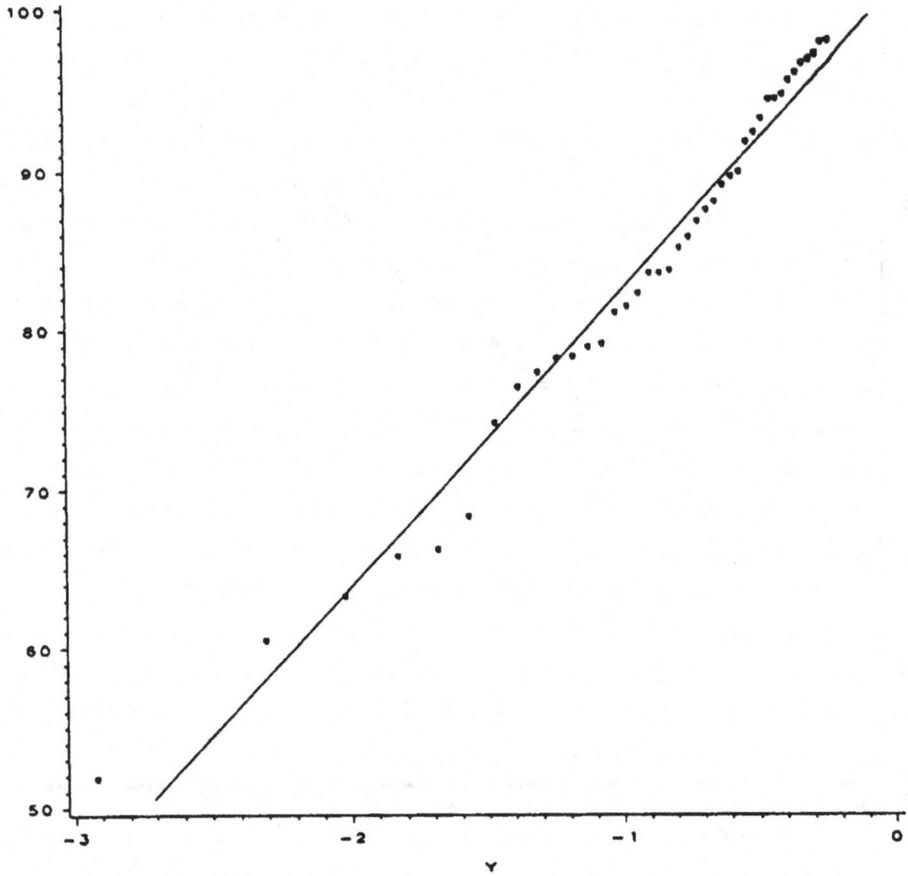

Fig. 13.6.1 Logistic probability plot of the LOG data set with 60 right-censored observations.

To graph a logistic probability plot of multiply right-censored data, a new definition of the empirical distribution function $F_n(x)$ is needed. There are a number of alternatives. Say that there are n subjects whose lifetimes are to be observed. Let $X_{1:n}, X_{2:n}, \ldots, X_{n:n}$ be the n lifetimes, where some are failure times and some are censoring times. Let

$$\delta_i = \begin{cases} 1 & \text{if } X_{i:n} \text{ is a failure time,} \\ 0 & \text{if } X_{i:n} \text{ is a censoring time.} \end{cases}$$

The Kaplan-Meier (1958) edf is defined as follows:

$$_{\text{K-M}}F_n(x) = 1 - \prod_{\substack{j \\ X_j \le x}} \left(\frac{n - j}{n - j + 1}\right)^{\delta_j}. \tag{13.6.1}$$

Note that this is undefined for $X > X_{n:n}$ if $X_{n:n}$ is a censoring time. Kaplan-Meier showed this is the maximum likelihood nonparametric estimator of the cdf $F(x; \mu, \sigma)$ for a multiply right-censored (Type I) sample. Equation (13.6.1) can be redefined at each $X_{i:n}$ as

$$_{\text{K-M}}F_n(x) = {}_{\text{K-M}}p_i = 1 - \prod_{\substack{j \\ j \le i}} \left(\frac{n - j}{n - j + 1}\right)^{\delta_j} \tag{13.6.2}$$

for all i such that $\delta_i = 1$. Note we use p_i here to reflect that the Kaplan-Meier edf depends on the index i for the ordered observation. Herd (1960) and Johnson (1964) adjust the Kaplan-Meier estimate to

$$_{\text{H-J}}p_i = 1 - \prod_{\substack{j \\ j \le i}} \left(\frac{n - j + 1}{n - j + 2}\right)^{\delta_j} \tag{13.6.3}$$

for all i such that $\delta_i = 1$. Nelson (1972) defines

$$_{\text{N}}p_i = 1 - \prod_{\substack{j \\ j \le i}} e^{-1/(n-j+1)} \tag{13.6.4}$$

for all i such that $\delta_i = 1$. Nelson's method is called the cumulative hazard property. It can be shown that $_{\text{K-M}}p_i \ge {}_{\text{N}}p_i \ge {}_{\text{H-J}}p_i$. Now note that for complete samples, Eq. (13.6.2) reduces to i/n, Eq. (13.6.3) reduces to $i/(n + 1)$, and Eq. (13.6.4) reduces to $1 - e^{-s_i}$, where

$$s_i = \sum_{j=1}^{i} (n - j + 1)^{-1}.$$

It is desired to find a p_i that reduces to $(i - c)/(n - 2c + 1)$ for complete samples. Michael and Schucany in D'Agostino and Stephens (1986) define

$$_{c}p_i = 1 - \frac{n - c + 1}{n - 2c + 1} \prod_{\substack{j \\ j \le i}} \left(\frac{n - j - c + 1}{n - j - c + 2}\right)^{\delta_j}. \tag{13.6.5}$$

In the following we will use (13.6.5) for plotting with $c = 0.5$.

Example 13.6.2 LOG Data Set The LOG data set will be censored as follows, following an algorithm used by Michael and Schucany in

Table 13.6.1 EDFs for the LOG Data Set with Multiply Censored Observations

j	x	Delta	K-Mp_i	Np_i	H-Jp_i	$_cp_i$, where c=0.5
1	51.90	1	0.010	0.010	0.010	0.005
2	60.57	1	0.020	0.020	0.020	0.015
3	63.35	1	0.030	0.030	0.030	0.025
4	65.87	1	0.040	0.040	0.040	0.035
5	66.35	1	0.050	0.050	0.050	0.045
6	68.44	1	0.060	0.060	0.059	0.055
7	74.29	1	0.070	0.070	0.069	0.065
8	76.52	1	0.080	0.080	0.079	0.075
9	77.45	1	0.090	0.090	0.089	0.085
10	78.32	1	0.100	0.100	0.099	0.095
11	78.48	1	0.110	0.109	0.109	0.105
12	79.07	1	0.120	0.119	0.119	0.115
13	79.23	1	0.130	0.129	0.129	0.125
14	81.17	1	0.140	0.139	0.139	0.135
15	81.61	1	0.150	0.149	0.149	0.145
16	82.45	1	0.160	0.159	0.158	0.155
17	83.73	1	0.170	0.169	0.168	0.165
18	83.74	1	0.180	0.179	0.178	0.175
19	83.89	1	0.190	0.189	0.188	0.185
20	85.30	1	0.200	0.199	0.198	0.195
21	85.94	1	0.210	0.209	0.208	0.205
22	86.98	1	0.220	0.219	0.218	0.215
23	87.71	1	0.230	0.229	0.228	0.225
24	88.22	1	0.240	0.239	0.238	0.235
25	89.22	1	0.250	0.249	0.248	0.245
26	89.81	1	0.260	0.259	0.257	0.255
27	90.00	0	0.260	0.259	0.257	0.255
28	90.00	0	0.260	0.259	0.257	0.255
29	90.00	0	0.260	0.259	0.257	0.255
30	90.00	0	0.260	0.259	0.257	0.255
31	90.00	0	0.260	0.259	0.257	0.255
32	90.00	0	0.260	0.259	0.257	0.255
33	90.00	0	0.260	0.259	0.257	0.255
34	90.00	0	0.260	0.259	0.257	0.255
35	90.00	0	0.260	0.259	0.257	0.255
36	90.00	0	0.260	0.259	0.257	0.255
37	90.00	0	0.260	0.259	0.257	0.255
38	90.00	0	0.260	0.259	0.257	0.255
39	90.00	0	0.260	0.259	0.257	0.255
40	90.00	0	0.260	0.259	0.257	0.255
41	90.00	0	0.260	0.259	0.257	0.255
42	90.00	0	0.260	0.259	0.257	0.255
43	90.00	0	0.260	0.259	0.257	0.255
44	90.00	0	0.260	0.259	0.257	0.255
45	90.00	0	0.260	0.259	0.257	0.255
46	90.09	1	0.273	0.272	0.271	0.268
47	91.97	1	0.287	0.285	0.284	0.282
48	92.55	1	0.300	0.299	0.297	0.295
49	93.44	1	0.314	0.312	0.310	0.309
50	94.66	1	0.327	0.325	0.324	0.322

Table 13.6.1 (*continued*)

j	x	Delta	$_{K\text{-}M}p_i$	$_Np_i$	$_{H\text{-}J}p_i$	$_cp_i$, where c=0.5
51	94.95	1	0.341	0.339	0.337	0.336
52	95.00	0	0.341	0.339	0.337	0.336
53	95.00	0	0.341	0.339	0.337	0.336
54	95.00	0	0.341	0.339	0.337	0.336
55	95.00	0	0.341	0.339	0.337	0.336
56	95.00	0	0.341	0.339	0.337	0.336
57	95.00	0	0.341	0.339	0.337	0.336
58	95.00	0	0.341	0.339	0.337	0.336
59	95.00	0	0.341	0.339	0.337	0.336
60	95.00	0	0.341	0.339	0.337	0.336
61	95.00	0	0.341	0.339	0.337	0.336
62	95.00	0	0.341	0.339	0.337	0.336
63	95.00	0	0.341	0.339	0.337	0.336
64	95.00	0	0.341	0.339	0.337	0.336
65	95.00	0	0.341	0.339	0.337	0.336
66	95.00	0	0.341	0.339	0.337	0.336
67	95.00	0	0.341	0.339	0.337	0.336
68	95.00	0	0.341	0.339	0.337	0.336
69	96.30	1	0.361	0.359	0.357	0.356
70	97.12	1	0.382	0.380	0.377	0.376
71	98.21	1	0.403	0.400	0.397	0.397
72	98.51	1	0.423	0.420	0.417	0.417
73	98.71	1	0.444	0.441	0.437	0.438
74	98.95	1	0.464	0.461	0.458	0.458
75	100.00	0	0.464	0.461	0.458	0.458
76	100.00	0	0.464	0.461	0.458	0.458
77	100.00	0	0.464	0.461	0.458	0.458
78	100.00	0	0.464	0.461	0.458	0.458
79	100.00	0	0.464	0.461	0.458	0.458
80	100.00	0	0.464	0.461	0.458	0.458
81	100.00	0	0.464	0.461	0.458	0.458
82	100.00	0	0.464	0.461	0.458	0.458
83	100.00	0	0.464	0.461	0.458	0.458
84	100.00	0	0.464	0.461	0.458	0.458
85	100.00	0	0.464	0.461	0.458	0.458
86	100.00	0	0.464	0.461	0.458	0.458
87	100.00	0	0.464	0.461	0.458	0.458
88	100.00	0	0.464	0.461	0.458	0.458
89	101.32	1	0.509	0.504	0.499	0.502
90	102.86	1	0.554	0.547	0.541	0.545
91	105.00	0	0.554	0.547	0.541	0.545
92	105.00	0	0.554	0.547	0.541	0.545
94	105.00	0	0.554	0.547	0.541	0.545
95	105.00	0	0.554	0.547	0.541	0.545
96	105.00	0	0.554	0.547	0.541	0.545
97	105.00	0	0.554	0.547	0.541	0.545
98	105.00	0	0.554	0.547	0.541	0.545
99	105.00	0	0.554	0.547	0.541	0.545
100	105.00	0	0.554	0.547	0.541	0.545

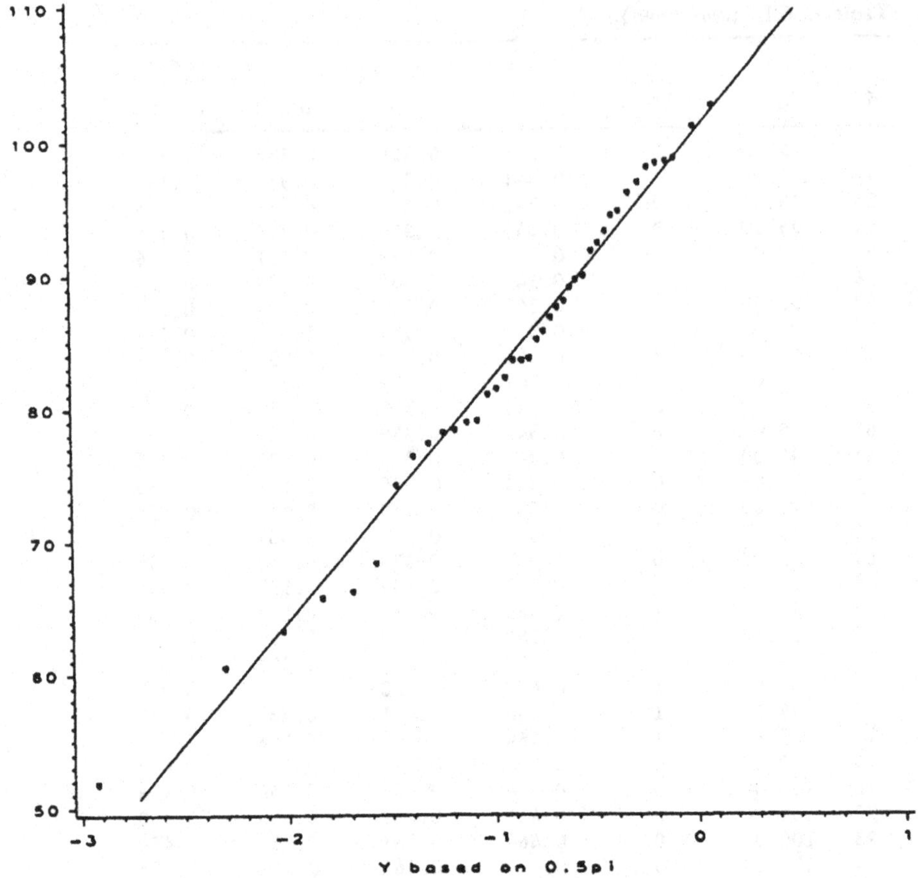

Fig. 13.6.2 Logistic probability plot of the LOG data set with random censoring.

D'Agostino and Stephens (1986). Among the first 25 unordered observations, all those greater than 90 will be set equal to 90 (so unordered observations among the first 25 that are greater than 90 will be censored at 90). Observations among the second, third, and fourth sets of 25 unordered observations that are greater than 95, 100, and 105, respectively, will be set equal to 95, 100, and 105, respectively. Values are then ranked from smallest to largest, with the observations appearing in Table 13.6.1. Note that the DELTA column of Table 13.6.1 is the indication of a failure time (DELTA = 1) or a censoring time (DELTA = 0). Figure 13.6.2 contains the probability plot of x versus y (where ${}_cp_i$ with $c = 0.05$ in the last column of Table 13.6.1 is used in the calculation of y), along with the theoretical line of x versus y using 100.67 and 18.33 (the least-squares estimates based solely on the uncensored observations) to estimate μ and σ.

Example 13.6.3 UNI Data Set Here, the UNI data will be multiply censored as follows. Observations among the first, second, third, and fourth sets of 25 unordered observations that are greater than 3, 4, 5, and 6, respectively will be set equal to 3, 4, 5, and 6, respectively. Table 13.6.2 contains this data ranked from smallest to largest. Figure 13.6.3 contains the probability plot of x versus y (where again ${}_cp_i$ with $c = 0.05$ is used in the calculation of y). Figure 13.6.3 also shows the theoretical line of x

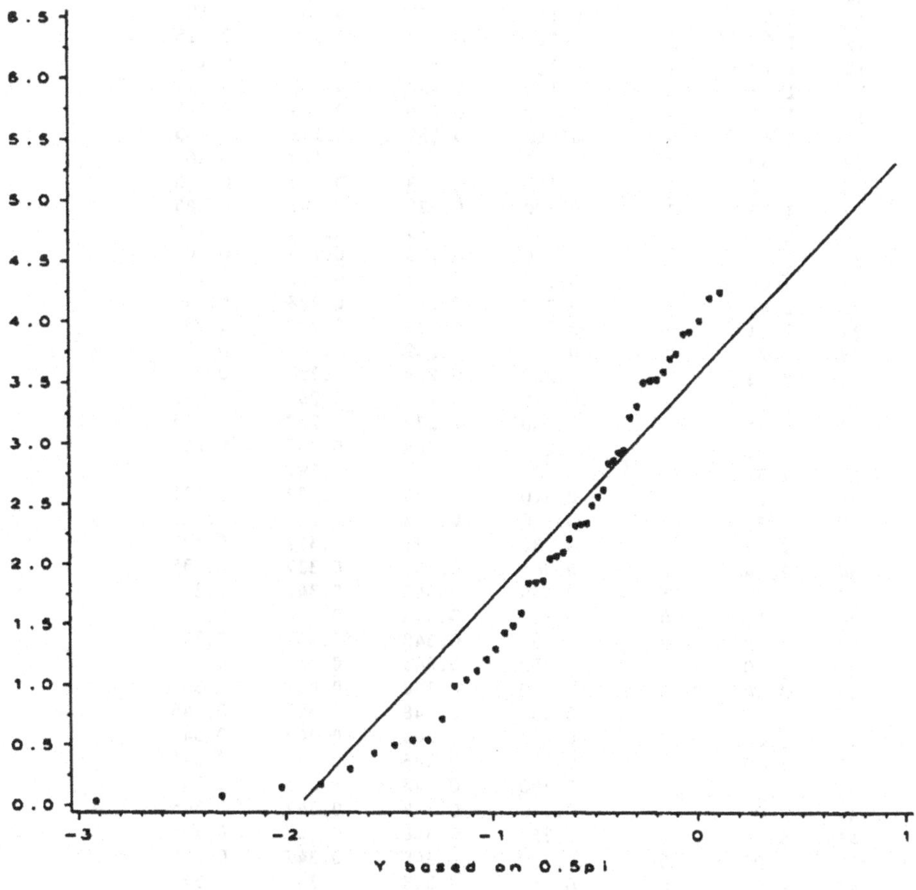

Fig. 13.6.3 Logistic probability plot of the UNI data set with random censoring.

Table 13.6.2 EDFs for the UNI Data Set with Multiply Censored Observations

j	x	Delta	K-MP_i	NP_i	H-JP_i	$_cP_i$, where c=0.5
1	0.04	1	0.010	0.010	0.010	0.005
2	0.08	1	0.020	0.020	0.020	0.015
3	0.15	1	0.030	0.030	0.030	0.025
4	0.18	1	0.040	0.040	0.040	0.035
5	0.31	1	0.050	0.050	0.050	0.045
6	0.44	1	0.060	0.060	0.059	0.055
7	0.50	1	0.070	0.070	0.069	0.065
8	0.55	1	0.080	0.080	0.079	0.075
9	0.55	1	0.090	0.090	0.089	0.085
10	0.72	1	0.100	0.100	0.099	0.095
11	1.00	1	0.110	0.109	0.109	0.105
12	1.05	1	0.120	0.119	0.119	0.115
13	1.12	1	0.130	0.129	0.129	0.125
14	1.22	1	0.140	0.139	0.139	0.135
15	1.30	1	0.150	0.149	0.149	0.145
16	1.44	1	0.160	0.159	0.158	0.155
17	1.50	1	0.170	0.169	0.168	0.165
18	1.60	1	0.180	0.179	0.178	0.175
19	1.85	1	0.190	0.189	0.188	0.185
20	1.86	1	0.200	0.199	0.198	0.195
21	1.87	1	0.210	0.209	0.208	0.205
22	2.06	1	0.220	0.219	0.218	0.215
23	2.08	1	0.230	0.229	0.228	0.225
24	2.11	1	0.240	0.239	0.238	0.235
25	2.22	1	0.250	0.249	0.248	0.245
26	2.33	1	0.260	0.259	0.257	0.255
27	2.34	1	0.270	0.269	0.267	0.265
28	2.35	1	0.280	0.279	0.277	0.275
29	2.50	1	0.290	0.289	0.287	0.285
30	2.57	1	0.300	0.299	0.297	0.295
31	2.63	1	0.310	0.308	0.307	0.305
32	2.85	1	0.320	0.318	0.317	0.315
33	2.87	1	0.330	0.328	0.327	0.325
34	2.94	1	0.340	0.338	0.337	0.335
35	2.96	1	0.350	0.348	0.347	0.345
36	3.00	0	0.350	0.348	0.347	0.345
37	3.00	0	0.350	0.348	0.347	0.345
38	3.00	0	0.350	0.348	0.347	0.345
39	3.00	0	0.350	0.348	0.347	0.345
40	3.00	0	0.350	0.348	0.347	0.345
41	3.00	0	0.350	0.348	0.347	0.345
42	3.00	0	0.350	0.348	0.347	0.345
43	3.00	0	0.350	0.348	0.347	0.345
44	3.00	0	0.350	0.348	0.347	0.345
45	3.00	0	0.350	0.348	0.347	0.345
46	3.00	0	0.350	0.348	0.347	0.345
47	3.00	0	0.350	0.348	0.347	0.345
48	3.00	0	0.350	0.348	0.347	0.345
49	3.00	0	0.350	0.348	0.347	0.345
50	3.00	0	0.350	0.348	0.347	0.345

Table 13.6.2 (*continued*)

j	x	Delta	$_{K-M}P_i$	$_{N}P_i$	$_{H-J}P_i$	$_{c}P_i$, where c=0.5
51	3.00	0	0.350	0.348	0.347	0.345
52	3.00	0	0.350	0.348	0.347	0.345
53	3.00	0	0.350	0.348	0.347	0.345
54	3.00	0	0.350	0.348	0.347	0.345
55	3.23	1	0.364	0.362	0.360	0.359
56	3.32	1	0.378	0.376	0.374	0.373
57	3.51	1	0.392	0.390	0.388	0.387
58	3.53	1	0.407	0.404	0.402	0.401
59	3.54	1	0.421	0.418	0.416	0.415
60	3.60	1	0.435	0.432	0.430	0.430
61	3.71	1	0.449	0.446	0.444	0.444
62	3.75	1	0.463	0.460	0.458	0.458
63	3.91	1	0.477	0.474	0.472	0.472
64	3.93	1	0.491	0.488	0.486	0.486
65	4.00	0	0.491	0.488	0.486	0.486
66	4.00	0	0.491	0.488	0.486	0.486
67	4.00	0	0.491	0.488	0.486	0.486
68	4.00	0	0.491	0.488	0.486	0.486
69	4.00	0	0.491	0.488	0.486	0.486
70	4.00	0	0.491	0.488	0.486	0.486
71	4.00	0	0.491	0.488	0.486	0.486
72	4.00	0	0.491	0.488	0.486	0.486
73	4.00	0	0.491	0.488	0.486	0.486
74	4.00	0	0.491	0.488	0.486	0.486
75	4.00	0	0.491	0.488	0.486	0.486
76	4.00	0	0.491	0.488	0.486	0.486
77	4.00	0	0.491	0.488	0.486	0.486
78	4.00	0	0.491	0.488	0.486	0.486
79	4.02	1	0.514	0.511	0.508	0.509
80	4.21	1	0.538	0.534	0.530	0.532
81	4.26	1	0.561	0.557	0.553	0.554
82	5.00	0	0.561	0.557	0.553	0.554
83	5.00	0	0.561	0.557	0.553	0.554
84	5.00	0	0.561	0.557	0.553	0.554
85	5.00	0	0.561	0.557	0.553	0.554
86	5.00	0	0.561	0.557	0.553	0.554
87	5.00	0	0.561	0.557	0.553	0.554
88	5.00	0	0.561	0.557	0.553	0.554
89	6.00	0	0.561	0.557	0.553	0.554
90	6.00	0	0.561	0.557	0.553	0.554
91	6.00	0	0.561	0.557	0.553	0.554
92	6.00	0	0.561	0.557	0.553	0.554
93	6.00	0	0.561	0.557	0.553	0.554
94	6.00	0	0.561	0.557	0.553	0.554
95	6.00	0	0.561	0.557	0.553	0.554
96	6.00	0	0.561	0.557	0.553	0.554
97	6.00	0	0.561	0.557	0.553	0.554
98	6.00	0	0.561	0.557	0.553	0.554
99	6.00	0	0.561	0.557	0.553	0.554
100	6.00	0	0.561	0.557	0.553	0.554

versus y based on the logistic distribution using 3.56 and 1.84 (the least-squares estimates based solely on the uncensored observations) as estimates of μ and σ in (13.3.2). Notice the deviation of the probability plot from the straight line.

If the data is multiply left-censored, simply multiply the observations by minus 1 and treat it as multiply right-censored.

13.7 SUMMARY PROBABILITY PLOTTING

Logistic probability plotting can be a tremendous informal aid to determine the fit of a given sample to the logistic distribution. These plots can be drawn by hand on arithmetic graph paper, semilog graph paper, or even logistic paper which is available commercially. They can also be done very easily on computer using software packages such as SAS and SASGRAPH. They not only can aid in determining the appropriateness of the logistic distribution, they can also be helpful to decide upon what type of deviation from the logistic distribution is present (Section 13.5). They do have limitations. A serious one is that they can overemphasize chance fluctuations. This is especially true for small n. In order to avoid this, formal testing procedures should be used in conjunction with them. We now turn to these formal tests.

13.8 FORMAL TESTS OF GOODNESS-OF-FIT

We now begin our discussion of formal goodness-of-fit tests for the logistic distribution. In all cases we will be testing a null hypothesis that the underlying distribution is the logistic distribution with cdf given by (13.1.1). The alternative hypothesis will be that the underlying distribution is not logistic.

We present the case where the null hypothesis is simple (complete specification of μ and σ) and the more realistic cases where the null hypothesis is composite and the parameters are unspecified. In all cases the alternative hypothesis will be composite and will state the true underlying distribution is not logistic. We will not consider the situations where the alternative hypothesis is directional (e.g., skewed to the right). Tests for full samples and censored samples will be presented.

While there are many varieties of goodness-of-fit tests, in fact only limited attention has been paid to the logistic distribution. The most complete presentation to date is given in D'Agostino and Stephens (1986). This chapter reviews that material and extends much of it. Four classes of tests are reviewed, the chi-squared test, the empirical distribution function tests (edf tests), tests based on normalized spacings, and tests based on regression and correlation.

13.9 CHI-SQUARED TESTS

13.9.1 Simple Hypothesis

Suppose we have a random sample $X_1, X_2, \ldots, X_n$ and we desire to test the simple null hypothesis

$$H_0\text{: } X_1, X_2, \ldots, X_n \text{ is from the logistic distribution with cdf } F(\cdot) \text{ of (13.1.1) where } \mu, \sigma \text{ are completely specified.} \tag{13.9.1}$$

This is the situation where the chi-squared test applies immediately. First partition the data into M cells, $C_1, C_2, \ldots, C_M$. Let $n_1, n_2, \ldots, n_M$ be the *observed* number of X_j's in the ith cell, for $i = 1, 2, \ldots, M$. Then, under H_0, n_i is distributed as a binomial random variable with parameters n and p_i. Here

$$p_i = P(X_j \text{ falls in } C_i) = \int_{C_i} dF(x; \mu, \sigma)$$

$$= \int_{C_i} \frac{\pi}{\sqrt{3}\sigma} \frac{e^{-\pi(x-\mu)/\sqrt{3}\sigma}}{(1 + e^{-\pi(x-\mu)/\sqrt{3}\sigma})^2}\, dx.$$

Let the bounds of C_i be (a_{i-1}, a_i), where $a_0 = -\infty$, $a_M = \infty$. Now, under H_0, $E(n_i) = np_i$, $V(n_i) = np_i(1 - p_i)$, where E is the expected value operator and V is the variance operator. Therefore by the central limit theorem,

$$\frac{n_i - np_i}{\sqrt{np_i(1 - p_i)}}$$

is asymptotically normally distributed with mean zero and standard deviation one. The joint distribution of $n_i, n_2, \ldots, n_M$ is asymptotically jointly normal, and by simple normal distribution theory

$$X^2 = \sum \frac{(n_i - np_i)^2}{np_i} \tag{13.9.2}$$

is distributed asymptotically as a chi-squared variable with $M - 1$ degrees of freedom; i.e., $X^2 \sim \chi^2(M - 1)$. To test the simple null hypothesis

(13.9.1), the statistic (13.9.2) can be employed. Rejection of the null hypothesis results when X^2 of (13.9.2) exceeds the upper $100(1 - \alpha)$ percentile (for example, the 95th percentile with $\alpha = .05$) of the $\chi^2(M - 1)$ distribution. X^2 is called the Pearson chi-squared statistic.

13.9.2 Composite Null Hypothesis

Here, we have

H_0: $X_1, X_2, \ldots, X_n$ is from the logistic distribution with cdf $F(\cdot)$ of (13.1.1) where μ, σ are not specified. (13.9.3)

Let $\dot{\theta} = (\dot{\mu}, \dot{\sigma})$ be an estimator of (μ, σ). Then we will test the goodness-of-fit of the data to $F(x; \dot{\mu}, \dot{\sigma})$, that is, to the logistic distribution with mean $\mu = \dot{\mu}$ and standard deviation $\sigma = \dot{\sigma}$. The partitioning of (13.9.1) is repeated here with

$$p_i(\dot{\mu}, \dot{\sigma}) = \int_{C_i} dF(x; \dot{\mu}, \dot{\sigma}),$$

$$X^2(\dot{\mu}, \dot{\sigma}) = \sum \frac{(n_i - np_i(\dot{\mu}, \dot{\sigma}))^2}{np_i(\dot{\mu}, \dot{\sigma})}. \qquad (13.9.4)$$

This is the chi-squared statistic. The problem is to determine the appropriate degrees of freedom. Fisher (1924) showed that as $M \to \infty$, $X^2(\dot{\mu}, \dot{\sigma})$ does not necessarily converge to the chi-squared distribution with $M - 1$ degrees of freedom even if $(\dot{\mu}, \dot{\sigma})$ is consistent for (μ, σ). The distribution of $X^2(\dot{\mu}, \dot{\sigma})$ depends on the method of estimation. Fisher suggests the grouped data maximum likelihood estimate (MLE) of (μ, σ) as an appropriate estimator. The estimators of μ and σ are the solutions of the equations

$$\sum_{i=1}^{M} \frac{n_i}{p_i(\mu, \sigma)} \frac{\delta p_i(\mu, \sigma)}{\delta \mu} = 0,$$

$$\sum_{i=1}^{M} \frac{n_i}{p_i(\mu, \sigma)} \frac{\delta p_i(\mu, \sigma)}{\delta \sigma} = 0.$$

If the grouped data MLE of (μ, σ) is employed, then the appropriate degrees of freedom are $M - 3$. Fisher also mentioned several other estimators (see Chapter 3 by David S. Moore in D'Agostino and Stephens, 1986). Due to the complexity of these and the complexity of the logistic cdf, these are of little practical use.

13.9.3 Estimation of (μ, σ) by the Ungrouped MLE $(\hat{\mu}, \hat{\sigma})$

If $\dot{\theta} = \hat{\theta} = (\hat{\mu}, \hat{\sigma})$, the ungrouped MLE of (μ, σ), then under H_0,

$$X^2(\hat{\mu}, \hat{\sigma}) \xrightarrow{d} \chi^2(M - 2 - 1) + \sum_{k=1}^{2} \lambda_k(\theta)\chi_k^2(1)$$

(Chernoff and Lehmann, 1954), where $0 \leq \lambda_k(\theta) \leq 1$, for $k = 1, 2$, and $\chi^2(M - 2 - 1)$, $\chi_1^2(1)$ and $\chi_2^2(1)$ are all independent random variables. The critical value of $X^2(\hat{\mu}, \hat{\sigma})$ for testing the composite null hypothesis H_0 of (13.9.3) falls between $\chi^2(M - 3)$ and $\chi^2(M - 1)$.

13.9.4 Choosing the Number of Cells and Cell Sizes

The best choice for M, the number of cells, has been researched in great detail. The following recommendation has been shown to be optimal for the simple hypothesis, but David Moore in D'Agostino and Stephens (1986) indicates that it appears useful also for composite hypotheses. The recommendation is that the number of cells to use should be

$$M = 2n^{2/5} \tag{13.9.5}$$

for $\alpha \geq 0.05$. More generally, Mann and Wald (1942) recommended

$$M = 4\left\{\frac{2n^2}{C(\alpha)^2}\right\}^{1/5},$$

where $C(\alpha)$ is the upper α percentage point of the $N(0, 1)$ distribution. Schorr (1974), however, demonstrated that the optimal M is actually smaller.

For cell sizes (i.e., probability in each cell), Mann and Wald (1942) recommended that cells equiprobable under H_0 should be used. The advantages of this are summarized by Moore in D'Agostino and Stephens (1986) and include (1) X^2 is unbiased (Mann and Wald, Cohen and Sackrowitz, 1975); and (2) the chi-squared distribution is a more accurate approximation to the exact null distribution of X^2 (Larntz, 1978; Read, 1984).

13.9.5 Estimation of (μ, σ) by the Sample Mean and Standard Deviation

Instead of using the MLE $(\hat{\mu}, \hat{\sigma})$ as estimates of (μ, σ), it would be computationally easier if $(\overline{X}, S)$ could be used in the calculation of X^2, where

$$\overline{X} = \frac{\sum_{j=1}^{n} X_i}{n}, \qquad S^2 = \frac{\sum_{j=1}^{n} (X_i - \overline{X})^2}{n}.$$

Dahiya and Gurland (1972) showed that $X^2(\overline{X}, S)$ is invariant with respect to (μ, σ) as long as the cell boundaries are of the form $\overline{X} + c_i S$, $i = 0, 1, 2, \ldots, M$, and that its distribution is of the form

$$\sum_{i=1}^{M} \lambda_i \chi_i^2(1).$$

It is not necessary that the cells be equiprobable, but here we will only consider that case. Massaro and D'Agostino (1991) have shown for M up to 16 that $\lambda_1, \lambda_2 > 1$; $\lambda_3, \lambda_4 < 1$; $\lambda_i = 1$, for $i = 5, 6, \ldots, M - 1$; and $\lambda_M = 0$. Therefore,

$$X^2(\overline{X}, S) \sim \chi^2(M - 5) + \sum_{i=1}^{4} \lambda_i \chi_i^2(1). \qquad (13.9.6)$$

The values for λ_i, $i = 1, 2, 3, 4$, along with the critical points of (13.9.6) at $\alpha = 0.10$, 0.05, and 0.01 have been calculated by Massaro and D'Agostino (1991) and are given in Table 13.9.1.

13.9.6 Examples

Simple Null Hypothesis for LOG Data

Here, the LOG data set will be taken (with $\mu = 100$, $\sigma = 18.14$) and the following *simple* hypothesis will be tested at the $\alpha = 0.05$ level:

$$H_0: X_1, X_2, \ldots, X_n \text{ have cdf(13.1.1) } F(x; 100, 18.14).$$

Table 13.9.1 Critical Points for $X^2(\overline{X}, S)$ of (13.9.6) along with λ_i, $i = 1, 2, 3, 4$

M	0.10	0.05	0.10	λ_1	λ_2	λ_3	λ_4
3	2.81	3.82	6.32	—	—	0.901	0.277
4	4.42	5.70	8.68	—	1.009	0.812	0.201
5	5.93	7.40	10.74	1.018	1.003	0.738	0.157
6	7.38	9.00	12.62	1.026	1.006	0.676	0.128
7	8.78	10.53	14.41	1.033	1.010	0.624	0.108
8	10.15	12.03	16.11	1.039	1.015	0.581	0.094
9	11.49	13.47	17.77	1.043	1.019	0.544	0.082
10	12.81	14.90	19.38	1.047	1.022	0.512	0.073
11	14.11	16.29	20.94	1.051	1.026	0.484	0.066
12	15.40	17.67	22.47	1.054	1.029	0.460	0.060
13	16.68	19.03	23.99	1.057	1.032	0.438	0.055
14	17.94	20.37	25.47	1.059	1.035	0.418	0.051
15	19.19	21.70	26.94	1.061	1.038	0.401	0.047
16	20.44	23.01	28.39	1.063	1.041	0.385	0.044

The number of cells used by (13.9.5) is $M = 2(100^{2/5}) \approx 13$. For convenience, however, will use $M = 16$. Letting $p_i = 1/16$, $i = 1, 2, \ldots, 16$, guarantees equiprobable cells. Then

$$X^2 = \sum_{i=1}^{16} \frac{(n_i - 100(1/16))^2}{100(1/16)} = 0.16 \sum_{i=1}^{16} (n_i - 6.25)^2. \quad (13.9.7)$$

To find n_i the boundary points of each cell are needed. Let Y be from the standardized logistic distribution $L(0, 1)$. Then to obtain the boundary points of Y, let

$$P(Y < y_i) = 1 - d_i = \frac{1}{1 + e^{-\pi y_i/\sqrt{3}}}.$$

Here, $d_i = (16 - i)/16$; i.e., $d_1 = 15/16 = 0.9375$, $d_2 = 14/16 = 0.8750, \ldots, d_{15} = 1/16 = 0.0625$, $d_{16} = 0$. Solving for y_i, we have

$$y_i = \frac{\sqrt{3}}{\pi} \ln\left(\frac{1 - d_i}{d_i}\right).$$

The cell boundary points for x based on $F(x; 100, 18.14)$ are then $100 + 18.14y_i$. The computation for the cells C_i and the determination of the n_i, the observed value for cell C_i, now follow immediately and are given in Table 13.9.2.

Table 13.9.2 Calculation of the Cells C_i for Testing Fit of the LOG Data Set to the Logistic $L(100, 18.14)$ Distribution

i	d_i	y_i	C_i	n_i
1	0.9375	−1.493	(−∞, 72.917]	6
2	0.8750	−1.073	(72.917, 80.536]	7
3	0.8125	−0.808	(80.536, 85,343]	7
4	0.7500	−0.606	(85.343, 89.007]	4
5	0.6875	−0.435	(89.007, 92.109]	4
6	0.6250	−0.282	(92.109, 94.884]	4
7	0.5625	−0.139	(94.884, 97.479]	6
8	0.5000	0.000	(97.479, 100.000]	7
9	0.4375	0.139	(100.000, 102.521]	8
10	0.3750	0.282	(102.521, 105.115]	10
11	0.3125	0.435	(105.115, 107.891]	5
12	0.2500	0.606	(107.891, 110.993]	9
13	0.1875	0.808	(110.993, 114.657]	9
14	0.1250	1.073	(114.657, 119.464]	7
15	0.0625	1.493	(119.464, 127.083]	3
16	0.0000	∞	(127.083, ∞]	4

Intermediate computation for X^2 of (13.9.2) and (13.9.7) is given in Table 13.9.3. Summing up the last column of this table and multiplying the sum by 0.16 as in (13.9.7) produces $X^2 = 10.72$. For $M - 1 = 16 - 1 = 15$ degrees of freedom, the 0.05 critical value is 25.00. Since $10.72 < 25.00$, we do not reject the null hypothesis that the underlying distribution is a logistic distribution with $\mu = 100$ and $\sigma = 18.14$.

Composite Null Hypothesis for LOG Data Using $(\hat{\mu}, \hat{\sigma})$ as Parameter Estimates

Again, the LOG data set will be used and the composite null hypothesis (13.9.3) will be tested. Again we use $M = 16$. Here, the ungrouped MLE $(\hat{\mu}, \hat{\sigma})$ of (μ, σ) will be used. Recall that $(\hat{\mu}, \hat{\sigma}) = (100.29, 16.55)$ for the LOG data set based on the logistic distribution. With $p_i(\hat{\mu}, \hat{\sigma}) \equiv 1/16$, $i = 1, 2, \ldots, 16$, the test statistic will be

$$X^2(\hat{\mu}, \hat{\sigma}) = 0.16 \sum_{i=1}^{16} (n_i - 6.25)^2. \tag{13.9.8}$$

The upper boundaries for cell C_i here are $100.29 + 16.55y_i$ (y_i defined as before). Note that the cells are now data-dependent. Table 13.9.4 gives

Table 13.9.3 Calculation of X^2 for Testing Fit of the LOG Data Set to the Logistic $L(100, 18.14)$ Distribution

i	n_i	$n_i - 6.25$	$(n_i - 6.25)^2$
1	6	−0.25	0.0625
2	7	0.75	0.5625
3	7	0.75	0.5625
4	4	−2.25	5.0625
5	4	−2.25	5.0625
6	4	−2.25	5.0625
7	6	−0.25	0.0625
8	7	0.75	0.5625
9	8	1.75	3.0625
10	10	3.75	14.0625
11	5	−1.25	1.5625
12	9	2.75	7.5625
13	9	2.75	7.5625
14	7	0.75	0.5625
15	3	−3.25	10.5625
16	4	−2.25	5.0625

Table 13.9.4 Calculation of Cells C_i and $X^2(\hat{\mu}, \hat{\sigma})$ for Testing Fit of the LOG Data Set to the Logistic Distribution with (μ, σ) Unspecified

i	C_i	n_i	$n_i - 6.25$	$(n_i - 6.25)^2$
1	$(-\infty, 75.581]$	7	0.75	0.5625
2	(75.581, 82.532]	9	2.75	7.5625
3	(82.532, 86.918]	5	−1.25	1.5625
4	(86.918, 90.261]	6	−0.25	0.0625
5	(90.261, 93.091]	2	−4.25	18.0625
6	(93.091, 96.623]	4	−2.25	5.0625
7	(95.623, 97.990]	5	−1.25	1.5625
8	(97.990, 100.290]	7	0.75	0.5625
9	(100.290, 102.590]	8	1.75	3.0625
10	(102.590, 104.957]	9	2.75	7.5625
11	(104.957, 107.489]	5	−1.25	1.5625
12	(107.489, 110.319]	9	2.75	7.5625
13	(110.319, 113.662]	9	2.75	7.5625
14	(113.662, 118.048]	6	−0.25	0.0625
15	(118.048, 125.000]	5	−1.25	1.5625
16	$(125.000, \infty]$	4	−2.25	5.0625

the steps for the computation of $X^2(\hat{\mu}, \hat{\sigma})$ and the determination of the cells. From these data we obtain $X^2(\hat{\mu}, \hat{\sigma}) = 11.04$. The 0.05 critical values for $\chi^2(M - 1) = \chi^2(15)$ and $\chi^2(M - 3) = \chi^2(13)$ are, respectively, 25.00 and 22.36. Since $X^2(\hat{\mu}, \hat{\sigma}) < 22.36$, we do not reject the composite null hypothesis (13.9.3).

Composite Null Hypothesis for NOR Data Using $(\hat{\mu}, \hat{\sigma})$ as Parameter Estimates

The procedure of using the ungrouped MLEs was also applied to the NOR data set. Based on the logistic distribution, $(\hat{\mu}, \hat{\sigma})$ for the NOR data set is the vector (99.48, 11.10). Computing $X^2(\hat{\mu}, \hat{\sigma})$ of (13.9.4) we obtain $X^2(\hat{\mu}, \hat{\sigma}) = 31.20$. For these data the null hypothesis is rejected since $31.20 > 25.00$. The reader should recall the edf for the NOR data (Fig. 13.2.6) did not indicate a deviation from the logistic distribution. The logistic probability plot (Fig. 13.5.2(d)) did indicate a deviation. Now formal tests support this.

Composite Null Hypothesis for LOG Data Using $(\overline{X}, S)$ as Parameter Estimates

Here, the above composite null hypothesis will be tested for the LOG data set using $(\overline{X}, S)$ as estimates of (μ, σ). Again let $M = 16$ and let

Table 13.9.5 Calculation of Cells C_i and $X^2(\overline{X}, S)$ for Testing Fit of the LOG Data Set to the Logistic Distribution with (μ, σ) Unspecified

i	C_i	n_i	$n_i - 6.25$	$(n_i - 6.25)^2$
1	$(-\infty, 74.941]$	7	0.75	0.5625
2	(74.941, 81.900]	8	1.75	3.0625
3	(81.900, 86.291]	6	−0.25	0.0625
4	(86.291, 89.639]	4	−2.25	5.0625
5	(89.639, 92.472]	3	−3.25	10.5625
6	(92.472, 95.007]	5	−1.25	1.5625
7	(95.007, 97.377]	4	−2.25	5.0625
8	(97.377, 99.680]	8	1.75	3.0625
9	(99.680, 101.983]	7	0.75	0.5625
10	(101.983, 104.353]	9	2.75	7.5625
11	(104.353, 106.888]	5	−1.25	1.5625
12	(106.888, 109.721]	7	0.75	0.5625
13	(109.721, 113.069]	11	4.75	22.5625
14	(113.069, 117.460]	7	0.75	0.5625
15	(117.460, 124.419]	5	−1.25	1.5625
16	$(124.419, \infty]$	4	−2.25	5.0625

$p_i(\overline{X}, S) \equiv 1/16$, guaranteeing equiprobable cells. For the LOG data set, $\overline{X} = 99.68$, $S = 16.57$. The form of the chi-squared test statistic is the same as in (13.9.7). The cell boundaries for cell C_i are now of the form $99.68 + 16.57y_i$. See Table 13.9.5 for the computation of $X^2(\overline{X}, S)$ in the usual way based on these cell boundaries. The results show by adding up the last column of Table 13.9.5 and multiplying the sum by 0.16 that $X^2(\overline{X}, S) = 11.04$. Comparing this value to the row for $M = 16$ in Table 13.9.1 shows that the null hypothesis can be accepted at all three levels of significance given since $X^2(\overline{X}, S)$ is less than all three respective values.

Bohemian Income Data Set (Fisk, 1961)

The Bohemian data set is grouped data, and so it is necessary to analyze the data in that form; that is, choose cells based on the groupings. Table 13.9.6 shows the computation of $X^2(\hat{\mu}, \hat{\sigma})$ where (2.51, 0.86), the ***grouped*** maximum likelihood estimate of (μ, σ) are employed. The lower cell boundary for cell C_i is the midpoint between $\log(x_i)$ and $\log(x_{i-1})$, $i = 1, 2, \ldots, 9$, where the log taken is the natural log and where $\log(x_0)$ is set to 0. Similarly, the upper cell boundary is the midpoint between $\log(x_i)$

Table 13.9.6 Computation of the Cells C_i and $X^2(\hat{\mu}, \hat{\sigma})$ for Testing Fit of the Bohemian Income Data Set to Logistic Distribution with (μ, σ) Unspecified

i	log(x)	C_i	n_i	p_i	$(n_i - np_i)^2/np_i$
1	1.10	(0.55, 1.53]	14	0.0955	2.074
2	1.95	(1.53, 2.15]	16	0.2075	1.087
3	2.35	(2.15, 2.71]	29	0.2851	0.008
4	3.07	(2.71, 3.32]	28	0.2413	0.620
5	3.56	(3.32, 3.74]	9	0.0846	0.035
6	3.91	(3.74, 4.18]	1	0.0412	2.364
7	4.44	(4.18, 4.53]	1	0.0149	0.162
8	4.61	(4.53, 4.70]	1	0.0043	0.756
9	4.79	(4.70, 4.90]	1	0.0033	1.383

and $\log(x_{i+1})$, $1 = 1, 2, \ldots, 9$, where $\log(x_{10})$ is set to 5. The observed values of the test statistic is 8.49 (sum up the last column of the table to obtain this). With 9 cells the degrees of freedom are $M - 3 = 9 - 3 = 6$. The 0.05 critical value is 12.59. Because $8.49 < 12.59$, we do not reject the null hypothesis that the underlying distribution is the logistic distribution.

Table 13.9.7 shows the computation of X^2 using the mean and standard deviation of log(x) (2.51 and 0.83, respectively) as estimates of μ and σ. The sum of the last column is 9.42, which is also less than 12.59, so the composite null hypothesis of (13.9.3) is not rejected again.

Table 13.9.7 Computation of the Cells C_i and X^2 Using the Mean and Standard Deviation of Log(x) for Testing Fit of the Bohemian Income Data Set to the Logistic Distribution with (μ, σ) Unspecified

i	log(x)	C_i	n_i	p_i	$(n_i - np_i)^2/np_i$
1	1.10	(0.55, 1.53]	14	0.0905	2.709
2	1.95	(1.53, 2.15]	16	0.2087	1.140
3	2.35	(2.15, 2.71]	29	0.2947	0.007
4	3.07	(2.71, 3.32]	28	0.2455	0.483
5	3.56	(3.32, 3.74]	9	0.0826	0.067
6	3.91	(3.74, 4.18]	1	0.0387	2.131
7	4.44	(4.18, 4.53]	1	0.0135	0.092
8	4.61	(4.53, 4.70]	1	0.0038	1.005
9	4.79	(4.70, 4.90]	1	0.0029	1.786

13.10 EDF STATISTICS

13.10.1 Introduction

Recall that the edf is given by

$$F_n(x) = \frac{\#(X_j \leq x)}{n}$$

for a random sample of observations $X_1, X_2, \ldots, X_n$. Any statistic measuring the differences between $F_n(x)$ and (13.1.1) is called an edf statistic and falls into one of two classes, *supremum* or *quadratic* statistics. We first give a brief description of these followed by their calculation and use.

13.10.2 Supremum EDF Statistics

The first two supremum statistics are defined as follows:

$$D^+ = \sup_x \{F_n(x) - F(x; \mu, \sigma)\}, \tag{13.10.1}$$

$$D^- = \sup_x \{F(x; \mu, \sigma) - F_n(x)\}, \tag{13.10.2}$$

where $F(x; \mu, \sigma)$ is the logistic cdf (13.1.1). Note that D^+ is the largest vertical difference for all points x such that $F_n(x) > F(x; \mu, \sigma)$. Similarly, D^- is the largest vertical distance for all points x such that $F_n(x) < F(x; \mu, \sigma)$. See Fig. 13.10.1.

From D^+ and D^-, the Kolmogorov (1933) statistic is defined as

$$D = \sup_x |F_n(x) - F(x; \mu, \sigma)| = \max(D^+, D^-). \tag{13.10.3}$$

Another supremum statistic is $V = D^+ + D^-$ given by Kuiper (1960). V is used for observations on a circle but is also used to test the goodness of fit of a sample to the logistic distribution with cdf (13.1.1).

13.10.3 Quadratic EDF Statistics

Quadratic statistics arise from the Cramér–von Mises family

$$Q = n \int_{-\infty}^{\infty} \{F_n(x) - F(x; \mu, \sigma)\}^2 \psi(x)\, dF(x; \mu, \sigma), \tag{13.10.4}$$

where $\psi(x)$ is a weight function. If $\psi(x) \equiv 1$ in (13.10.4) then W^2, the Cramér–von Mises statistic is obtained. If

$$\psi(x) = \frac{1}{F(x; \mu, \sigma)\{1 - F(x; \mu, \sigma)\}},$$

the resulting statistic is the Anderson-Darling (1954) statistic, denoted A^2.

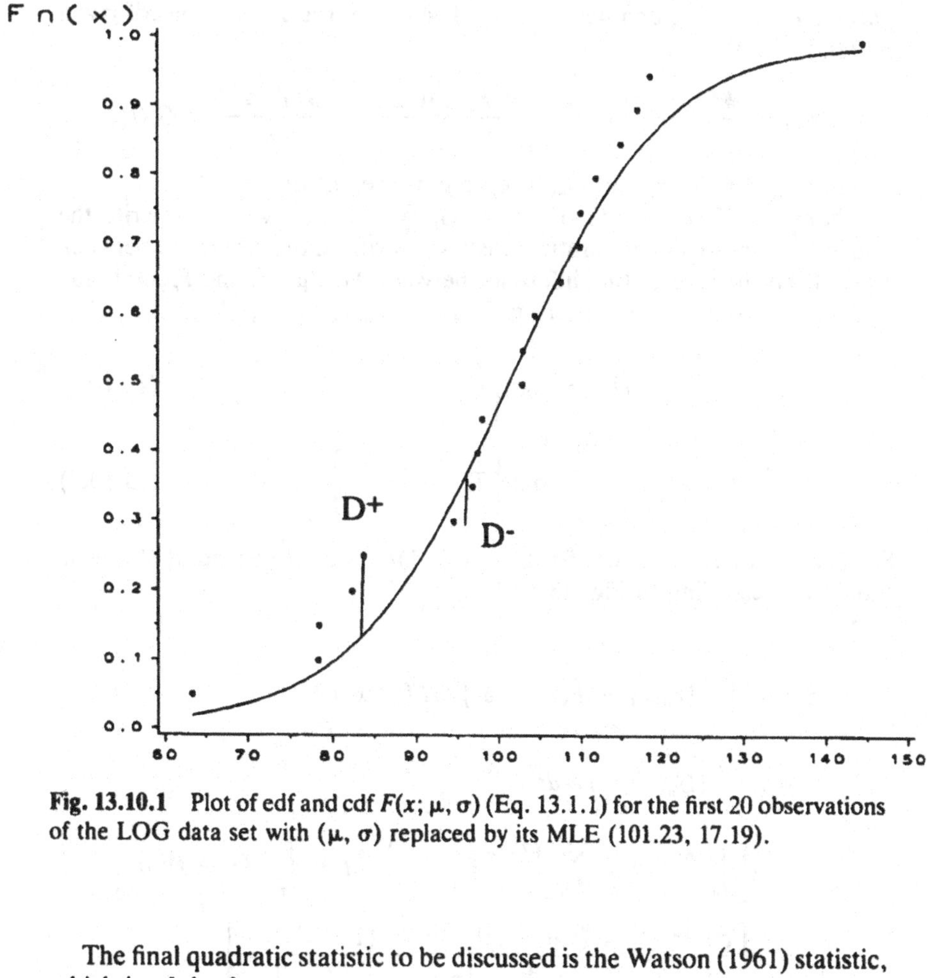

Fig. 13.10.1 Plot of edf and cdf $F(x; \mu, \sigma)$ (Eq. 13.1.1) for the first 20 observations of the LOG data set with (μ, σ) replaced by its MLE (101.23, 17.19).

The final quadratic statistic to be discussed is the Watson (1961) statistic, which is of the form

$$U^2 = n \int_{-\infty}^{\infty} [F_n(x) - F(x; \mu, \sigma) - \int_{-\infty}^{\infty} \{F_n(x) - F(x; \mu, \sigma)\} \, dF(x; \mu, \sigma)]^2 \, dF(x; \mu, \sigma). \quad (13.10.5)$$

13.10.4 Computation of the Statistics

Let the sample be $X_1, X_2, \ldots, X_n$. Let $T_i = F(X_i; \mu, \sigma)$ where $F(x; \mu, \sigma)$ is the cdf (13.1.1) of the logistic(μ, σ) distribution. If the logistic distribution is indeed the underlying distribution of the given sample, then $T_1, T_2, \ldots, T_n$ is a random sample from the uniform distribution with parameters 0 and 1, denoted $U(0, 1)$. Therefore, the T_i have the cdf given by $G(t) = t, 0 \le t \le 1$. So

$$F_n(x) = \frac{\#(X_j \le x)}{n} = \frac{\#\{T_j \le F(x; \mu, \sigma)\}}{n} = \frac{\#(T_j \le t)}{n} \equiv G_n(t),$$

where $t = F(x; \mu, \sigma)$ and $G_n(t)$ represents the edf of t.

Therefore, $F_n(x) - F(x; \mu, \sigma) = G_n(t) - t$; i.e., we can rewrite the above supremum and quadratic statistics in terms of the difference between t and its edf instead of the difference between $F(x; \mu, \sigma)$ and $F_n(x)$. It can easily be seen that, after ordering $T_1, T_2, \ldots, T_n$ as $T_{1:n}, T_{2:n}, \ldots, T_{n:n}$,

$$D^+ = \max_i \left\{\frac{i}{n} - T_{i:n}\right\}, \tag{13.10.6}$$

$$D^- = \max_i \left\{T_{i:n} - \frac{i-1}{n}\right\}. \tag{13.10.7}$$

See Fig. 13.10.2. Note that for D^-, $(i - 1)/n$ is used instead of i/n, as it should be according to Fig. 13.10.1.

Now,

$$\begin{aligned}
W^2 &= n \int_{-\infty}^{\infty} \{F_n(x) - F(x; \mu, \sigma)\}^2 \, dF(x; \mu, \sigma) \\
&= n \int_0^1 (G_n(t) - t)^2 \, dt \\
&= n\left[\int_0^{T_{1:n}} t^2 \, dt + \sum_{i=1}^{n-1} \int_{T_{i:n}}^{T_{i+1:n}} \left(\frac{i}{n} - t\right)^2 dt + \int_{T_{n:n}}^1 (1 - t)^2 \, dt\right] \\
&= n\left[\frac{t^3}{3}\bigg|_0^{T_{1:n}} - \sum_{i=1}^{n-1} \frac{(i/n - t)^3}{3}\bigg|_{T_{i:n}}^{T_{i+1:n}} - \frac{(1-t)^3}{3}\bigg|_{T_{n:n}}^1\right] \\
&= \sum_{i=1}^{n} \left(T_{i:n} - \frac{2i - 1}{2n}\right)^2 + \frac{1}{12n}.
\end{aligned} \tag{13.10.8}$$

Similarly, it can be shown that

$$U^2 = W^2 - n(0.5 - \overline{T})^2, \tag{13.10.9}$$

where $\overline{T}$ is the mean of the n T's. Also,

$$A^2 = n \int_0^1 \frac{(G_n(t) - t)^2}{t(1 - t)} \, dt$$

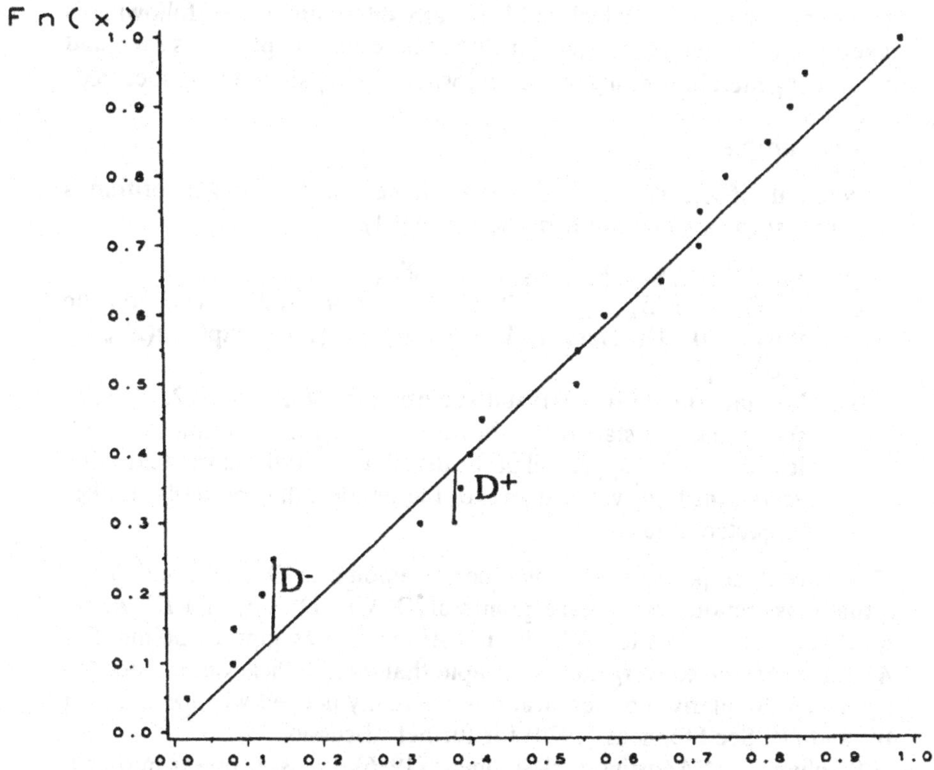

Fig. 13.10.2 Plot of edf and cdf for the uniform transformations of the first 20 observations of the LOG data set.

$$= -n - \frac{1}{n}\sum_{i=1}^{n}(2i-1)\{\log(T_{i:n}) + \log(1 - T_{n+1-i:n})\}$$

$$= -n - \frac{1}{n}\sum_{i=1}^{n}[(2i-1)\log(T_{i:n})$$

$$+ (2n+1-2i)\log(1 - T_{i:n})], \qquad (13.10.10)$$

where $\log(T_{i:n})$ is the natural log of $T_{i:n}$.

13.10.5 Simple Hypothesis

We will now use the edf statistics to test the simple null hypothesis (13.9.1). Stephens in Chapter 4 of D'Agostino and Stephens (1986) gives a complete discussion for this case which holds for any distribution. The following is taken from that reference and is included here for completeness sake and because it generalizes easily to the case where (μ, σ) are not fully specified.

Full Sample

When all of $X_1, X_2, \ldots, X_n$ are available, the following algorithm is used to test the simple null hypothesis (13.9.1):

1. Order the X_i's as $X_{1:n}, X_{2:n}, \ldots, X_{n:n}$.
2. Let $T_{i:n} = F(X_{i:n}; \mu, \sigma)$. Then $T_{i:n}$ are ordered uniforms on the interval (0, 1). Here $F(X_{i:n}; \mu, \sigma) = [1 + \exp(-\pi(X_{i:n} - \mu)/\sigma\sqrt{3})]^{-1}$.
3. Calculate the desired edf statistic using the $T_{i:n}$, $i = 1, 2, \ldots, n$.
4. Modify the test statistic as shown in Table 13.10.1 using the modifications for the upper tail of the distribution. If the calculated value exceeds the table value at level α, the simple null hypothesis (13.9.1) is rejected at level α.

The percentage points in the table corresponding to D^+, D^-, and D are actually asymptotic percentage points of $D^+\sqrt{n}$, $D^-\sqrt{n}$, and $D\sqrt{n}$, respectively. The points for W^2, U^2, and A^2 are also asymptotic points. For A^2, however, the convergence is so rapid that no modifications are needed for $n \geq 5$. Similarly, no modifications are really needed when $n \geq 20$ for W^2 and U^2. See Stephens (1970) for further discussion.

Stephens in D'Agostino and Stephens (1986) discusses power in detail, but generally the edf statistics are more powerful than the Pearson chi-squared statistics discussed earlier since no grouping of the data is involved; i.e., no information is lost. A^2 and W^2 behave similarly, but A^2 is more

Table 13.10.1 Modifications and Percentage Points for EDF Statistics for Testing a Completely Specified Distribution for a Full Sample

		Significance level α				
Statistic	Modification	0.15	0.10	0.05	0.025	0.01
D^+ (or D^-)	$D^+(\sqrt{n} + 0.12 + 0.11/\sqrt{n})$	0.973	1.073	1.224	1.358	1.518
D	$D(\sqrt{n} + 0.12 + 0.11/\sqrt{n})$	1.138	1.224	1.358	1.480	1.628
V	$V(\sqrt{n} + 0.155 + 0.24/\sqrt{n})$	1.537	1.620	1.747	1.862	2.001
W^2	$(W^2 - 0.4/n + 0.6/n^2)(1 + 1/n)$	0.284	0.347	0.461	0.581	0.743
U^2	$(U^2 - 0.1/n + 0.1/n^2)(1 + 0.8/n)$	0.131	0.152	0.187	0.222	0.268
A^2	None needed for $n \geq 5$	1.610	1.933	2.492	3.070	3.857

Source: Stephens and D'Agostino (1986).

powerful when (13.1.1) departs from the actual underlying distribution in the tails.

Censored Sample

Edf statistics exist when μ and σ are known (case 0 in D'Agostino and Stephens 1986 vocabulary) for all types of censoring (Type I, Type II, singly right, singly left, double, and random). Here, Type I and Type II single censoring will be discussed. For doubly censored data, refer to Pettitt and Stephens (1976). For random censoring, refer to Kaplan and Meier (1958), Koziol and Green (1976), Koziol (1980), and Csörgõ and Horvath (1981).

To start, the $T_{i:n}$ are calculated as above for all available X_i. Suppose the sample is Type I right-censored with r censored observations; in other words, $X_{1:n}, X_{2:n}, \ldots, X_{n-r:n}$ are available, where $X_{n-r:n}$ is less than some fixed value X^*. Then $T_{n-r:n} < t^*$, where $t^* = F(X^*; \mu, \sigma)$. The Kolmogorov statistic (13.10.3) is calculated as

$$
\begin{aligned}
{}_1D_{t,n} &= \sup_{0 \le t \le t^*} |G_n(t) - t| \\
&= \max_{1 \le i \le n-r} \left\{ \frac{i}{n} - T_{i:n}, T_{i:n} - \frac{i-1}{n}, t^* - \frac{n-r}{n} \right\}. \qquad (13.10.11)
\end{aligned}
$$

If the sample is Type II right-censored with r censored observations, then $T_{1:n}, T_{2:n}, \ldots, T_{n-r:n}$ are available, where r is fixed. The Kolmogorov statistic is then

$$
\begin{aligned}
{}_2D_{r,n} &= \sup_{0 \le t \le T_{n-r,n}} |G_n(t) - t| \\
&= \max_{1 \le i \le n-r} \left\{ \frac{i}{n} - T_{i:n}, T_{i:n} - \frac{i-1}{n} \right\}. \qquad (13.10.12)
\end{aligned}
$$

${}_1D_{t,n}$ and ${}_2D_{r,n}$ converge to one asymptotic distribution, given by Koziol and Byar (1975). Asymptotic percentage points for this distribution are given in Table 13.10.2. Dufour and Maag (1978) give modifications of ${}_1D_{t,n}$ and ${}_2D_{r,n}$ so that this asymptotic distribution could be used with finite samples. Their algorithm to test the simple null hypothesis of (13.9.1) is as follows:

1. Calculate $T_{i:n}$ as above.
2. If censoring is Type I, then
   a. Calculate ${}_1D_{t,n}$ as in (13.10.11).
   b. Modify ${}_1D_{t:n}$ to D_t^*, where $D_t^* = \sqrt{n}\,{}_1D_{t,n} + 0.19/\sqrt{n}$ for $n \ge 25$ and $t^* \ge 0.25$.

Table 13.10.2 Upper Tail Asymptotic Percentage Points for $\sqrt{n}D$, W^2, and A^2 for Testing a Completely Specified Distribution for Type I or Type II Singly Censored Data

	Significance Level α				
p	0.15	0.10	0.05	0.025	0.01
$\sqrt{n}D$					
.2	0.7443	0.8155	0.9268	1.0282	1.1505
.3	0.8784	0.9597	1.0868	1.2024	1.3419
.4	0.9746	1.0616	1.1975	1.3209	1.4696
.5	1.0438	1.1334	1.2731	1.3997	1.5520
.6	1.0914	1.1813	1.3211	1.4476	1.5996
.7	1.1208	1.2094	1.3471	1.4717	1.6214
.8	1.1348	1.2216	1.3568	1.4794	1.6272
.9	1.1379	1.2238	1.3581	1.4802	1.6276
1.0	1.1379	1.2238	1.3581	1.4802	1.6276
W^2					
.2	0.033	0.041	0.057	0.074	0.094
.3	0.066	0.083	0.115	0.147	0.194
.4	0.105	0.136	0.184	0.231	0.295
.5	0.153	0.186	0.258	0.330	0.427
.6	0.192	0.241	0.327	0.417	0.543
.7	0.231	0.286	0.386	0.491	0.633
.8	0.259	0.321	0.430	0.544	0.696
.9	0.278	0.341	0.455	0.573	0.735
1.0	0.284	0.347	0.461	0.581	0.743
A^2					
.2	0.333	0.436	0.588	0.747	0.962
.3	0.528	0.649	0.872	1.106	1.425
.4	0.700	0.857	1.150	1.455	1.872
.5	0.875	1.062	1.419	1.792	2.301
.6	1.028	1.260	1.676	2.112	2.707
.7	1.184	1.451	1.920	2.421	3.083
.8	1.322	1.623	2.146	2.684	3.419
.9	1.467	1.798	2.344	2.915	3.698
1.0	1.610	1.933	2.492	3.070	3.857

Source: Stephens and D'Agostino (1986).

   c. Let $p = t^*$ and enter Table 13.10.2 at that row and at the desired significance level α. If D_r^* exceeds the tabulated percentage point, reject the simple null hypothesis of (13.9.1).
3. If censoring is Type II, then
   a. Calculate ${}_2D_{r,n}$ as in (13.10.12).
   b. Modify ${}_2D_{r,n}$ to D_r^*, where $D_r^* = \sqrt{n}\,{}_2D_{r,n} + 0.24/\sqrt{n}$ for $n \geq 25$ and $(n - r)/n \geq 0.4$.
   c. Refer to Table 13.10.2. Let $p = (n - r)/n$ and enter the table at that row at the desired significance level α. If D_r^* exceeds the tabulated percentage point, reject the simple null hypothesis of (13.9.1).

For $n \leq 25$ and for higher censoring ratios, consult Barr and Davidson (1973) or Dufour and Maag (1978).

For the quadratic statistics, versions of W^2, A^2, and U^2 for singly right-censored data were found by Pettitt and Stephens (1976). Given $T_{1:n}$, $T_{2:n}, \ldots, T_{n-r:n}$, the Type II censoring versions are

$$
\begin{aligned}
{}_2W^2_{n-r:n} &= \sum_{i=1}^{n-r}\left(T_{i:n} - \frac{2i-1}{2n}\right)^2 + \frac{n-r}{12n^2} \\
&\quad + \frac{n}{3}\left(T_{n-r:n} - \frac{n-r}{n}\right)^3, \qquad (13.10.13) \\
{}_2U^2_{n-r:n} &= {}_2W^2_{n-r:n} - nT_{n-r:n} \\
&\quad \times \left[\frac{n-r}{n} - \frac{T_{n-r:n}}{2} - \frac{(n-r)\overline{T}}{nT_{n-r:n}}\right]^2, \qquad (13.10.14)
\end{aligned}
$$

where $\overline{T}$ is defined as above.

$$
\begin{aligned}
{}_2A^2_{n-r:n} &= -\frac{1}{n}\sum_{i=1}^{n-r}(2i-1)[\log(T_{i:n}) - \log(1 - T_{i:n})] \\
&\quad -2\sum_{i=1}^{n-r}\log(1 - T_{i:n}) - \frac{1}{n}[r^2\log(1 - T_{n-r:n}) \\
&\quad - (n-r)^2\log(T_{n-r:n}) + n^2T_{n-r:n}]. \qquad (13.10.15)
\end{aligned}
$$

To test the simple null hypothesis of (13.9.1), calculate ${}_2W^2_{n-r:n}$ or ${}_2A^2_{n-r:n}$ as above and refer to Table 13.10.3 entering at the level of n and $p = (n - r)/n$. Reject the simple null hypothesis of (13.9.1) at level α if ${}_2W^2_{n-r:n}$ or ${}_2A^2_{n-r:n}$ exceeds its respective tabulated percentage point at the desired significance level α.

For Type I right-censored data, let t^* be the fixed censoring value of T. Add this point to the available sample, thereby giving $n - r + 1$ available

Table 13.10.3 Upper Tail Percentage Points for ${}_2W^2_{r,n}$ and ${}_2A^2_{r,n}$ for Testing a Completely Specified Distribution for Type II Right-Censored Data

p	n	Significance Level α 0.15	0.10	0.05	0.025	0.01
${}_2W^2_{r,n}$						
	20	0.038	0.058	0.099	0.152	0.243
	40	0.032	0.046	0.084	0.128	0.198
	60	0.031	0.044	0.074	0.107	0.154
.2	80	0.031	0.043	0.069	0.097	0.136
	100	0.031	0.043	0.066	0.092	0.127
	∞	0.031	0.041	0.057	0.074	0.094
	10	0.101	0.144	0.229	0.313	0.458
	20	0.095	0.132	0.209	0.297	0.419
	40	0.100	0.128	0.191	0.267	0.381
.4	60	0.102	0.130	0.189	0.256	0.354
	80	0.103	0.132	0.187	0.251	0.342
	100	0.103	0.132	0.187	0.248	0.335
	∞	0.105	0.135	0.184	0.236	0.307
	10	0.159	0.205	0.297	0.408	0.547
	20	0.172	0.216	0.302	0.408	0.538
	40	0.180	0.226	0.306	0.398	0.522
.6	60	0.184	0.231	0.313	0.404	0.528
	80	0.186	0.233	0.316	0.407	0.531
	100	0.187	0.235	0.318	0.409	0.532
	∞	0.192	0.241	0.327	0.417	0.539
	10	0.217	0.266	0.354	0.453	0.593
	20	0.235	0.289	0.389	0.489	0.623
	40	0.247	0.303	0.401	0.508	0.651
.8	60	0.251	0.308	0.410	0.520	0.667
	80	0.253	0.311	0.415	0.526	0.675
	100	0.254	0.313	0.418	0.529	0.680
	∞	0.259	0.320	0.430	0.544	0.700
	10	0.246	0.301	0.410	0.502	0.645
	20	0.263	0.322	0.431	0.536	0.675
	40	0.271	0.330	0.437	0.546	0.701
.9	60	0.273	0.333	0.442	0.553	0.713
	80	0.274	0.335	0.445	0.558	0.718
	100	0.275	0.336	0.447	0.561	0.722
	∞	0.278	0.341	0.455	0.573	0.735

Table 13.10.3 *(continued)*

p	n	Significance Level α 0.15	0.10	0.05	0.025	0.01
$_2W^2_{r,n}$						
.95	10	0.266	0.324	0.430	0.534	0.676
	20	0.275	0.322	0.444	0.551	0.692
	40	0.280	0.329	0.448	0.557	0.715
	60	0.280	0.338	0.451	0.562	0.724
	80	0.281	0.340	0.453	0.566	0.729
	100	0.282	0.341	0.454	0.569	0.735
	∞	0.283	0.346	0.460	0.579	0.742
1.0	10	0.288	0.349	0.456	0.564	0.709
	20	0.288	0.350	0.459	0.572	0.724
	40	0.288	0.350	0.461	0.576	0.731
	100	0.288	0.351	0.462	0.578	0.736
	∞	0.284	0.347	0.461	0.581	0.743
$_2A^2_{r,n}$						
.2	20	0.337	0.435	0.626	0.887	1.278
	40	0.337	0.430	0.607	0.804	1.111
	60	0.341	0.432	0.601	0.785	1.059
	80	0.344	0.433	0.598	0.775	1.034
	100	0.345	0.434	0.596	0.769	1.019
	∞	0.351	0.436	0.588	0.747	0.962
.4	10	0.627	0.803	1.127	1.483	2.080
	20	0.653	0.824	1.133	1.513	2.011
	40	0.681	0.843	1.138	1.460	1.903
	60	0.686	0.848	1.142	1.458	1.892
	80	0.688	0.850	1.144	1.457	1.887
	100	0.689	0.851	1.145	1.457	1.884
	∞	0.695	0.857	1.150	1.455	1.872
.6	10	0.944	1.174	1.577	2.055	2.774
	20	0.984	1.207	1.650	2.098	2.688
	40	1.001	1.229	1.635	2.071	2.671
	60	1.011	1.239	1.649	2.084	2.683
	80	1.017	1.244	1.655	2.091	2.689
	100	1.020	1.248	1.659	2.095	2.693
	∞	1.033	1.260	1.676	2.112	2.707

(continued)

Table 13.10.3 (*continued*)

p	n	Significance Level α 0.15	0.10	0.05	0.025	0.01
${}_2A^2_{r,n}$						
	10	1.237	1.498	2.021	2.587	3.254
	20	1.280	1.558	2.068	2.570	3.420
	40	1.321	1.583	2.088	2.574	3.270
.8	60	1.330	1.596	2.107	2.610	3.319
	80	1.335	1.603	2.117	2.629	3.344
	100	1.338	1.607	2.123	2.640	3.359
	∞	1.350	1.623	2.146	2.684	3.419
	10	1.435	1.721	2.281	2.867	3.614
	20	1.457	1.765	2.295	2.858	3.650
	40	1.478	1.778	2.315	2.860	3.628
.9	60	1.482	1.784	2.325	2.878	3.648
	80	1.485	1.788	2.330	2.888	3.661
	100	1.486	1.790	2.332	2.893	3.668
	∞	1.492	1.798	2.344	2.915	3.698
	10	1.525	1.842	2.390	2.961	3.745
	20	1.533	1.853	2.406	2.965	3.750
	40	1.543	1.860	2.416	2.968	3.743
.95	60	1.545	1.863	2.421	2.977	3.753
	80	1.546	1.865	2.423	2.982	3.760
	100	1.547	1.866	2.424	2.984	3.763
	∞	1.550	1.870	2.430	2.995	3.778
1.0	All n	1.610	1.933	2.492	3.070	3.857

observations in this sample. Let $T_{n-r+1:n} = t^*$. Recalculate the above statistics replacing $n - r$ with $n - r + 1$ and denote the statistics as ${}_1W^2_{r^*,n}$, ${}_1U^2_{r^*,n}$, and ${}_1A^2_{r^*,n}$. When n is finite, percentage points have been generated for various levels of t^*. However, these converge so rapidly to the asymptotic percentage points (which are the same as those for their Type II counterparts) given in Table 13.10.2 that a table for finite n is not needed. Reject the simple null hypothesis of (13.9.1) at the significance level α if ${}_1A^2_{r^*,n}$ or ${}_1W^2_{r^*,n}$ exceeds the tabulated percentage point at level α for a given level of $p = t^*$.

If the data is left censored with s censored observations; i.e., if the ordered uniform transformations of the available observations are $T_{s+1:n}$, $T_{s+2:n}, \ldots, T_{n:n}$, then transform it into a right-censored sample by letting $T^*_{i:n} = 1 - T_{n+1-i:n}$, $i = 1, 2, \ldots, n - s$. In addition, if the censoring is Type I, let $t^{**} = 1 - t^*$. The above statistics are then calculated, depending on the type of censoring, using $T^*_{i:n}$, $i = 1, 2, \ldots, n - s$ and with s replacing r.

13.10.6 Composite Hypothesis

Stephens in D'Agostino and Stephens (1986) gives the test for full samples of the usual H_0 where μ, σ, or both are unknown (cases 1, 2, and 3, respectively). The algorithm is straightforward:

1. Find the MLEs of any unknown parameters.
2. Let $T_{i:n} = F(X_{i:n}; \hat{\mu}, \hat{\sigma})$, where $F(x; \hat{\mu}, \hat{\sigma})$ is (13.1.1) with the parameters replaced by their corresponding MLEs.
3. Calculate the edf statistics using $T_{i:n}$ from Step 2.
4. For W^2, U^2, and A^2, modify the statistic as in Table 13.10.4. If the modified statistic exceeds the percentage point given for the chosen significance level α, reject the composite null hypothesis of (13.9.3).
5. For D^+, D^-, D, and V, multiply the statistic by $\sqrt{n}$ and refer to Table 13.10.5. If the modified statistic exceeds the percentage point at the given n and the chosen α, then reject the composite null hypothesis of (13.9.3).

Table 13.10.4 actually gives the asymptotic distribution for quadratic sta-

Table 13.10.4 Modifications and Upper Tail Percentage Points for W^2, A^2, and U^2 in Tests for the Logistic Distribution with One or Two Unknown Parameters

			Significance level α			
Statistic	Case	Modification	0.10	0.05	0.025	0.01
W^2	1	$(1.9nW^2 - 0.15)/(1.9n - 1)$	0.119	0.148	0.177	0.218
	2	$(0.95n^2 - 0.45)/(0.95n - 1)$	0.323	0.438	0.558	0.721
	3	$(nW^2 - 0.08)/(n - 1)$	0.081	0.098	0.114	0.136
U^2	2	$(U^2 - 0.16)/(1.6n - 1.0)$	0.116	0.145	0.174	0.214
A^2	1	$A^2 + 0.15/n$	0.857	1.046	1.241	1.505
	2	$(0.6nA^2 - 1.8)/(0.6n - 1)$	1.725	2.290	2.880	3.685
	3	$A^2(1 + 0.25/n)$	0.563	0.660	0.769	0.906

For U^2, cases 1 and 3, use modifications and percentage points for W^2, cases 1 and 3, respectively.
Source: D'Agostino and Stephens (1986).

Table 13.10.5 Upper Tail Percentage Points for the Supremum Statistics Multiplied by $\sqrt{n}$ in Tests for the Logistic Distribution with One or Two Unknown Parameters

Statistic and Case	n	Significance Level α 0.10	0.05	0.025	0.01
$D^+\sqrt{n}$					
1	5	0.702	0.758	0.805	0.854
	10	0.730	0.792	0.846	0.913
	20	0.744	0.809	0.867	0.944
	50	0.752	0.819	0.880	0.962
	∞	0.757	0.826	0.888	0.974
2	5	0.971	1.120	1.239	1.380
	10	0.990	1.143	1.268	1.423
	20	0.999	1.150	1.282	1.444
	50	1.005	1.161	1.290	1.456
	∞	1.009	1.166	1.297	1.464
3	5	0.603	0.650	0.690	0.735
	10	0.636	0.687	0.736	0.789
	20	0.653	0.705	0.758	0.816
	50	0.663	0.716	0.773	0.832
	∞	0.669	0.723	0.781	0.842
$D\sqrt{n}$					
1	5	0.736	0.791	0.845	0.883
	10	0.777	0.837	0.895	0.653
	20	0.800	0.865	0.926	0.997
	50	0.808	0.874	0.937	1.011
	∞	0.816	0.883	0.947	1.025
2	5	1.108	1.236	1.349	1.474
	10	1.148	1.274	1.388	1.521
	20	1.167	1.294	1.406	1.545
	50	1.179	1.305	1.419	1.559
	∞	1.187	1.313	1.427	1.568
3	5	0.643	0.679	0.723	0.751
	10	0.679	0.730	0.774	0.823
	20	0.968	0.755	0.800	0.854
	50	0.708	0.770	0.817	0.873
	∞	0.715	0.780	0.827	0.886

Table 13.10.5 *(continued)*

Statistic and Case	n	Significance Level α 0.10	0.05	0.025	0.01
$V\sqrt{n}$					
	5	1.369	1.471	1.580	1.658
	10	1.410	1.520	1.630	1.741
	20	1.433	1.550	1.659	1.790
	50	1.447	1.564	1.675	1.815
	∞	1.454	1.574	1.685	1.832
	5	1.314	1.432	1.547	1.674
	10	1.372	1.483	1.587	1.711
	20	1.400	1.510	1.607	1.730
	50	1.417	1.525	1.619	1.741
	∞	1.429	1.535	1.627	1.748
	5	1.170	1.246	1.299	1.373
	10	1.230	1.311	1.381	1.466
	20	1.260	1.344	1.422	1.514
	50	1.277	1.364	1.448	1.542
	∞	1.289	1.376	1.463	1.560

Source: Stephens and D'Agostino (1986).

tistics. See Stephens (1979). Modifications for the quadratic statistics along with percentage points for $D^-\sqrt{n}$, $D\sqrt{n}$, and $V\sqrt{n}$ were derived from Monte Carlo studies. The percentage points for $D^+\sqrt{n}$ and $D^-\sqrt{n}$ are identical.

13.10.7 Examples of Testing the Composite Hypothesis—LOG Data Set

Calculations of the edf statistics using the first 20 (unordered) observations of the LOG data set as our full sample is shown in Table 13.10.6. Here we assume both μ and σ are unknown (case 3). The MLE of (μ, σ) in this case is (101.23, 17.19). For the supremum edf statistics:

$$D^+ = \max(i/n - T_{i:n}) = 0.114,$$

$$D^- = \max(T_{i:n} - (i - 1)/n) = 0.096,$$

$$D = \max(D^+, D^-) = 0.114,$$

$$V = D^+ + D^- = 0.210.$$

Table 13.10.6 Calculation of the EDF Statistics Using the First 20 Observations of the LOG data set

i	$X_{i:n}$	i/n	$(i-1)/n$	$T_{i:n}$	$(i/n) - T_{i:n}$	$T_{i:n} - (i-1)/n$	$(T_{i:n} - (2i-1)/2n)^2$	$\log(T_{i:n})$	$\log(1 - T_{i:n})$	$(2i-1)\log(T_{i:n}) + (2n+1-2i)\log(1 - T_{i:n})$
1	63.35	0.05	0.00	0.018	0.032	0.018	0.0000484	−4.015	−0.018	−4.73
2	78.32	0.10	0.05	0.082	0.018	0.032	0.0000471	−2.503	−0.085	−10.67
3	78.48	0.15	0.10	0.083	0.067	−0.017	0.0017525	−2.487	−0.087	−15.47
4	82.45	0.20	0.15	0.121	0.079	−0.029	0.0028996	−2.111	−0.129	−19.04
5	83.73	0.25	0.20	0.136	0.114	−0.064	0.0078707	−1.993	−0.147	−22.48
6	94.63	0.30	0.25	0.333	−0.033	0.083	0.0033191	−1.101	−0.404	−23.84
7	96.91	0.35	0.30	0.388	−0.038	0.088	0.0039661	−0.947	−0.491	−25.57
8	97.45	0.40	0.35	0.402	−0.002	0.052	0.0007070	−0.912	−0.514	−26.52
9	98.07	0.45	0.40	0.417	0.033	0.017	0.0000577	−0.874	−0.540	−27.28
10	102.97	0.50	0.45	0.546	−0.046	0.096	0.0050085	−0.606	−0.789	−28.08
11	103.08	0.55	0.50	0.549	0.001	0.049	0.0005591	−0.600	−0.796	−27.72
12	104.47	0.60	0.55	0.585	0.015	0.035	0.0000930	−0.537	−0.879	−27.28
13	107.64	0.65	0.60	0.663	−0.013	0.063	0.0014382	−0.411	−1.087	−26.59
14	109.91	0.70	0.65	0.714	−0.014	0.064	0.0015366	−0.337	−1.253	−25.37
15	109.99	0.75	0.70	0.716	0.034	0.016	0.0000825	−0.334	−1.259	−23.53
16	111.81	0.80	0.75	0.753	0.047	0.003	0.0004704	−0.283	−1.400	−21.38
17	114.97	0.85	0.80	0.810	0.040	0.010	0.0002261	−0.211	−1.661	−18.58
18	116.99	0.90	0.85	0.841	0.059	−0.009	0.0011814	−0.174	−1.837	−15.26
19	118.54	0.95	0.90	0.861	0.089	−0.039	0.0040526	−0.149	−1.976	−11.45
20	144.28	1.00	0.95	0.989	0.011	0.039	0.0002092	−0.011	−4.553	−4.97
							0.036			−405.79

$D^+ = 0.114$
$D = 0.096$
$D = \max(D^+, D^-) = 0.114$
$V = D^+ + D^- = 0.210$
$W^2 = 0.036 + 1/240 = 0.040$
$A^2 = -20 - (1/20)(-405.79) = 0.290$

To use these to test the composite null hypothesis (13.9.3) by using Table 13.10.5, they must first be multiplied by the square root of 20: $D^+\sqrt{20} = 0.114\sqrt{20} = 0.510$. The tabulated value in Table 13.10.5 corresponding to case 3 (both parameters estimated) with $n = 20$ and $\alpha = 0.05$ is 0.705; since $0.510 < 0.705$, we indeed do not reject the composite null hypothesis of (13.9.3). Similarly $D\sqrt{20} = 0.510$. This does not exceed 0.755, the tabulated percentage point for $D\sqrt{20}$ at $\alpha = 0.05$ and $n = 20$ given in Table 13.10.5, so again, do not reject the composite null hypothesis (13.9.3) at level α. Finally, $V\sqrt{20} = 0.939$. The tabulated percentage point is 1.344, thereby leading us to not reject the composite null hypothesis (13.9.3) once again.

For the quadratic statistics, from Table 13.10.6 we have

$$W^2 = \sum_{i=1}^{n}\left(T_{i:n} - \frac{2i-1}{2n}\right)^2 + \frac{1}{12n} = 0.036 + \frac{1}{240} = 0.040,$$

$$A^2 = -n - \frac{1}{n}\sum_{i=1}^{n}[(2i-1)\log(T_{i:n}) + (2n+1-2i)\log(1-T_{i:n})]$$

$$= -20 - \frac{1}{20}(-405.79) = 0.290.$$

To use W^2 to test the composite null hypothesis (13.9.3), look at the row in Table 13.10.4 corresponding to case 3. Modify W^2 as shown; i.e., in this example calculate $(20W^2 - 0.08)/(20 - 1) = (20(0.04) - 0.08)/19 = 0.038$. Since this does not exceed the tabulated point of 0.098 corresponding to $\alpha = 0.05$, we do not reject (13.9.3) at a significance level of 0.05.

Similarly, to use A^2 to test the composite null hypothesis (13.9.3), modify it as shown in Table 13.10.4 for case 3; i.e., calculate $A^2(1 + 0.25/20)$, which for this example equals $0.290(1 + 0.25/20) = 0.294$. Since this does not exceed 0.660, the tabulated percentage point at $\alpha = 0.05$, we do not have significant evidence to reject (13.9.3) at a significance level of 0.05.

13.11 TESTS BASED ON NORMALIZED SPACINGS

13.11.1 Normalized Spacing Statistics

Suppose the following observations are available after Type II censoring: $X_{s+1:n}, X_{s+2:n}, \ldots, X_{n-r:n}$, where again s is the number of observations censored on the left and r is the number of observations censored on the right. We wish to test the composite null hypothesis (13.9.3).

Let $E_i = X_{i+1:n} - X_{i:n}$, $i = s+1, s+2, \ldots, n-r-1$; then E_i is the spacing between successive observations. Suppose that $X_{s+1:n}$,

$X_{s+2:n}, \ldots, X_{n-r:n}$ are order statistics from the logistic $L(\mu, \sigma)$ distribution. Then $Y_{s+1:n}, Y_{s+2:n}, \ldots, Y_{n-r:n}$ can be considered order statistics from the $L(0, 1)$ distribution where $X_{i:n} = \mu + \sigma Y_{i:n}$. Here μ and σ are the underlying location (mean) and scale (standard deviation) parameters, respectively, of the sample. Furthermore, let $m_i = E(Y_{i:n})$ and define G_i as

$$G_i = \frac{E_i}{m_{i+1} - m_i}. \tag{13.11.1}$$

These are *normalized* spacings for $i = s + 1, s + 2, \ldots, n - r - 1$. Pyke (1965) has shown that the G_i are asymptotically distributed as exponential(0, σ) random variables with cdf $H(x; 0, \sigma) = -e^{(-x/\sigma)}$.

For the logistic distribution, $m_{i+1} - m_i = n/i(n - i)$ for $i = 1, 2, \ldots, n - 1$. Therefore

$$G_i = \frac{i(n - i)E_i}{n} \tag{13.11.2}$$

for $i = s + 1, s + 2, \ldots, n - r - 1$. To use G_i in a goodness-of-fit test, first let

$$T_{i:n} = \sum_{j=s+1}^{i} G_j \Big/ \sum_{j=s+1}^{n-r-1} G_j \tag{13.11.3}$$

for $i = s + 1, s + 2, \ldots, n - r - 2$. These are then used in the following version of the Anderson-Darling statistic:

$$\begin{aligned} A_s^2 = &-(n - r - s - 2) \\ &- \frac{1}{n - r - s - 2} \\ &\times \sum_{i=s+1}^{n-r-2} \{[2(i - s) - 1][\log(T_{i:n}) + \log(1 - T_{n-r+s-1-i:n})]\}. \end{aligned}$$

Lockhart, O'Reilly, and Stephens (1986) discuss the asymptotic distribution of A_s^2 for tests of the composite null hypothesis of (13.9.3). Asymptotic percentage points for various levels of the censoring ratio and for various levels of significance are given in Table 13.11.1. For an approximate test of the composite null hypothesis of (13.9.3) when $n \geq 40$, calculate A_s^2 as described above, enter Table 13.11.1 at the appropriate α-level and censoring ratio level (where $q = s/n$ and $p = (n - r)/n$). If A_s^2 is greater than the tabulated point, reject the composite null hypothesis of (13.9.3) at that significance level. Note the table can be used for complete samples (i.e., $q = 0, p = 1$).

Table 13.11.1 Asymptotic Percentage Points for A_s^2 for Samples from Logistic Populations

Left censoring point $q = s/n$	Right censoring point $p = (n - r)/n$	Significance level α 0.15	0.10	0.05	0.025	0.01
0	1	1.448	1.720	2.206	2.716	3.413
0	0.75	1.468	1.741	2.230	2.741	3.441
0	0.50	1.521	1.806	2.318	2.852	3.584
0	0.25	1.574	1.873	2.409	2.969	3.736
0.25	0.75	1.517	1.801	2.308	2.838	3.561
0.25	0.50	1.584	1.885	2.424	2.989	3.761

Source: Lockhart, O'Reilly, and Stephens (1986).

Another test statistic using G_i that is given by Tiku (1980) is

$$Z^* = \frac{2 \sum_{i=s+1}^{n-r-2} (n - r - 1 - i)G_i}{(n - s - r - 2)(X_{n-r} - X_{s+1})}. \tag{13.11.4}$$

Tiku (1980) has shown that when the censored sample $X_{s+1:n}, X_{s+2:n}, \ldots, X_{n-r:n}$ comes from the logistic distribution, Z^* is approximately normally distributed with mean 1 and variance $(n - r - s)/3(n - r - s - 1)(n - r - s - 2)$.

Percentage points calculated by Tiku for various levels of α, n, and censoring ratio based on this normal approximation and on Monte Carlo studies are reproduced in Table 13.11.2. To test (13.9.3), calculate Z^* from the available observations. If Z^* exceeds the tabulated percentage point, reject (13.9.3) at the corresponding significance level.

13.11.2 Example

Here we take the LOG data set with $n = 100$. Suppose the last 50 and the first 25 ordered observations are Type II censored (i.e., $r = 50$, $s = 25$). We wish to calculate A_s^2 to test the composite null hypothesis of (13.9.3). Tables 13.11.3, 13.11.4, and 13.11.5 show the calculation of G_i, $T_{i:n}$, and A_s^2, respectively, for this example. Note that the sum of the last column of Table 13.11.5 is -567.32, so $A_s^2 = -(n - r - s - 2) - (1/(n - r - s - 2))(-567.32) = 1.667$. This value does not exceed 2.424, the tabulated percentage point in Table 13.11.1 with $q = 0.25$ and $p = 0.50$ at a

Table 13.11.2 Simulated and Approximate Probabilities $P(Z^* \leq C)$ for Logistic Populations

n	s	r	C	Simulated Probability	Approximate Probability
10	0	0	0.65	0.025	0.031
			0.71	0.046	0.058
			1.29	0.946	0.942
			1.35	0.977	0.969
20	0	0	0.77	0.024	0.026
			0.81	0.047	0.051
			1.19	0.956	0.949
			1.23	0.979	0.974
20	0	4	0.74	0.026	0.028
			0.78	0.048	0.051
			1.22	0.953	0.949
			1.26	0.974	0.972
20	4	4	0.66	0.024	0.028
			0.72	0.052	0.058
			1.28	0.947	0.942
			1.34	0.975	0.972
50	0	0	0.86	0.026	0.026
			0.88	0.047	0.046
			1.12	0.951	0.954
			1.14	0.972	0.974
50	0	10	0.84	0.025	0.026
			0.87	0.052	0.054
			1.13	0.947	0.946
			1.16	0.975	0.974
50	10	10	0.80	0.026	0.027
			0.83	0.048	0.052
			1.17	0.948	0.948
			1.20	0.973	0.973

Source: Tiku (1980).

Table 13.11.3 Calculation of G_i

i	$X_{i:n}$	$X_{i+1:n}$	E_i	$m_{i+1}-m_i$	G_i
26	89.81	90.09	0.28	0.0519751	5.387
27	90.09	91.97	1.88	0.0507357	37.055
28	91.97	92.55	0.58	0.0496032	11.693
29	92.55	93.44	0.89	0.0485673	18.325
30	93.44	94.63	1.19	0.0476190	24.990
31	94.63	94.66	0.03	0.0467508	0.642
32	94.66	94.95	0.29	0.0459559	6.310
33	94.95	95.86	0.91	0.0452284	20.120
34	95.86	96.30	0.44	0.0445633	9.874
35	96.30	96.91	0.61	0.0439560	13.877
36	96.91	97.12	0.21	0.0434028	4.838
37	97.12	97.45	0.33	0.0429000	7.692
38	97.45	98.07	0.62	0.0424448	14.607
39	98.07	98.21	0.14	0.0420345	3.331
40	98.21	98.51	0.30	0.0416667	7.200
41	98.51	98.58	0.07	0.0413394	1.693
42	98.58	98.71	0.13	0.0410509	3.167
43	98.71	98.95	0.24	0.0407997	5.882
44	98.95	99.52	0.57	0.0405844	14.045
45	99.52	100.46	0.94	0.0404040	23.265
46	100.46	100.62	0.16	0.0402576	3.974
47	100.62	100.81	0.19	0.0401445	4.733
48	100.81	101.18	0.37	0.0400641	9.235
49	101.18	101.19	0.01	0.0400160	0.250
50	101.19	.	.	0.0400000	.

Note: $G_i = \dfrac{E_i}{m_{i+1} - m_i}$.

significance level of 0.05, so we do not reject the composite null hypothesis of (13.9.3).

13.12 TESTS BASED ON REGRESSION AND CORRELATION

Recall the probability plotting section 13.3. Take a sample of observations $X_1, X_2, \ldots, X_n$, for which it is desired to test the composite null hypothesis of (13.9.3). A plot of

$$X_{i:n} \text{ versus } Y_i = \frac{\sqrt{3}}{\pi} \ln\left(\frac{F_n(X_{i:n})}{1 - F_n(X_{i:n})}\right), \tag{13.12.1}$$

Table 13.11.4 Calculation of $T_{i:n}$

i	G_i	Numerator of $T_{i:n}$	Denominator of $T_{i:n}$ (Sum of all G_i)	$T_{i:n}$
26	5.3872	5.387	252.186	0.021362
27	37.0548	42.442	252.186	0.168296
28	11.6928	54.135	252.186	0.214662
29	18.3251	72.460	252.186	0.287327
30	24.9900	97.450	252.186	0.386420
31	0.6417	98.092	252.186	0.388965
32	6.3104	104.402	252.186	0.413987
33	20.1201	124.522	252.186	0.493770
34	9.8736	134.396	252.186	0.532922
35	13.8775	148.273	252.186	0.587951
36	4.8384	153.112	252.186	0.607137
37	7.6923	160.804	252.186	0.637639
38	14.6072	175.411	252.186	0.695561
39	3.3306	178.742	252.186	0.708768
40	7.2000	185.942	252.186	0.737319
41	1.6933	187.635	252.186	0.744033
42	3.1668	190.802	252.186	0.756590
43	5.8824	196.684	252.186	0.779916
44	14.0448	210.729	252.186	0.835608
45	23.2650	233.994	252.186	0.927861
46	3.9744	237.968	252.186	0.943621
47	4.7329	242.701	252.186	0.962389
48	9.2352	251.936	252.186	0.999009
49	0.2499	.	.	.
50	.	.	.	.
.				

Note: $T_{i:n} = \sum_{j=s+1}^{i} G_j \Big/ \sum_{j=s+1}^{n-r-1} G_j$

where $F_n(X_{i:n}) = (i - 0.5)/n$, is the probability plot. If it approximates a straight line, then as an informal test, the underlying distribution of the sample appears to be logistic.

The informal "by eye" test can be formalized by use of regression techniques where the regression model is

$$X_{i:n} = \alpha + \beta Y_i + \epsilon_i. \tag{13.12.2}$$

Here the expected value of ϵ_i, denoted $E(\epsilon_i)$, is 0. Much of the work for

Table 13.11.5 Calculation of A_s^2

i	$T_{i:n}$	$\log(T_{i:n})$	$n-r+s-1-i$	$T_{n-r+s-1-i:n}$	$\log(1-T_{n-r+s-1-i:n})$	$(2(i-s)-1)(\log(T_{i:n}) + \log(1-T_{n-r+s-1-i:n}))$
26	0.021362	−3.8461	48	0.999009	−6.9169	−10.76
27	0.168296	−1.7820	47	0.962389	−3.2804	−15.19
28	0.214662	−1.5387	46	0.943621	−2.8757	−22.07
29	0.287327	−1.2471	45	0.927861	−2.6292	−27.13
30	0.386420	−0.9508	44	0.835608	−1.8055	−24.81
31	0.388965	−0.9443	43	0.779916	−1.5137	−27.04
32	0.413987	−0.8819	42	0.756590	−1.4130	−29.83
33	0.493770	−0.7057	41	0.744033	−1.3627	−31.03
34	0.532922	−0.6294	40	0.737319	−1.3368	−33.43
35	0.587951	−0.5311	39	0.708768	−1.2336	−33.53
36	0.607137	−0.4990	38	0.695561	−1.1893	−35.45
37	0.637639	−0.4500	37	0.637639	−1.0151	−33.70
38	0.695561	−0.3630	36	0.607137	−0.9343	−32.43
39	0.708768	−0.3442	35	0.587951	−0.8866	−33.23
40	0.737319	−0.3047	34	0.532922	−0.7613	−30.91
41	0.744033	−0.2957	33	0.493770	−0.6808	−30.27
42	0.756590	−0.2789	32	0.413987	−0.5344	−26.84
43	0.779916	−0.2486	31	0.388965	−0.4926	−25.94
44	0.835608	−0.1796	30	0.386420	−0.4884	−24.72
45	0.927861	−0.0749	29	0.287327	−0.3387	−16.13
46	0.943621	−0.0580	28	0.214662	−0.2416	−12.29
47	0.962389	−0.0383	27	0.168296	−0.1843	−9.57
48	0.999009	−0.0010	26	0.021362	−0.0216	−1.02
						−567.32

Note: $A_s^2 = -(n-r-s-2) - (1/(n-r-s-2))(-567.32) = 1.667$.

this model still needs to be developed with respect to the logistic distribution; however, Stephens in Chapter 5 of D'Agostino and Stephens (1986) provides a procedure and statistic based on this approach that is used to test the composite null hypothesis of (13.9.3).

Suppose we have a full sample of observations available, namely X_1, $X_2, \ldots, X_n$. Slightly change the definition of $F_n(X_{i:n})$ to

$$F_n(X_{i:n}) = \frac{i}{n+1} \tag{13.12.3}$$

in the definition of Y_i and perform the regression of $X_{i:n}$ on Y_i using ordinary least squares to estimate α and β in (13.12.2). Then calculate

a. SSR = regression sum of squares = $\sum_{i=1}^{n} (\hat{X}_{i:n} - \overline{X})^2$,

b. SSE = error sum of squares = $\sum_{i=1}^{n} (\hat{X}_{i:n} - X_{i:n})^2$,

c. SST = total sum of squares = $\sum_{i=1}^{n} (X_{i:n} - \overline{X})^2$.

Here $\hat{X}_{i:n}$ is the predicted value of $X_{i:n}$ using the regression estimates of α and β in (13.12.2); $\overline{X}$ is the mean of all $X_{i:n}$. Let $R^2(X, Y) = SSR/SST$ (Note, $R^2(X, Y) = r^2(X, Y)$, the Pearson correlation coefficient between $X_{i:n}$ and Y_i.) Let $Z(X, Y) = n[1 - R^2(X, Y)]$. If the fit of the plot to a straight line is good, then $R^2(X, Y)$ is large and close to 1. (If the fit is exact, $R^2(X, Y)$ is 1.) Therefore, if $Z(X, Y)$ is "large," then $R^2(X, Y)$ is "small," and (13.9.3) should be rejected.

If the sample is Type II right-censored, where the available ordered observations are $X_{1:n}, X_{2:n}, \ldots, X_{n-r:n}$, then $F_n(X_{i:n})$ and Y_i are still calculated as above. The regression between available $X_{i:n}$ and their corresponding Y_i is run, and the above sums of squares are calculated summing from 1 to $n - r$. $R^2(X, Y)$ and $Z(X, Y)$ are defined as before, and (13.9.3) is rejected if $Z(X, Y)$ is "large."

Table 13.12.1 gives the upper tail percentage points of $Z(X, Y)$ for a test of (13.9.3) if the sample is complete or Type II right-censored with a censoring ratio of $p = (n - r)/n$. Once $Z(X, Y)$ is calculated from the available data, the table is entered at the desired significance level α, the censoring ratio p, and the sample size n. If the calculated $Z(X, Y)$ exceeds the tabulated percentage point, the composite null hypothesis (13.9.3) is rejected.

Table 13.12.1 Upper Tail Percentage Points for $Z(X, Y)$ for a Test for the Logistic Distribution, Parameters Unknown, for Complete or Type II Censored Data

		Significance Level α				
p	n	0.15	0.10	0.05	0.025	0.01
	20	3.57	4.03	4.78	6.09	7.35
	40	5.70	6.65	8.29	9.73	11.13
0.2	60	6.92	8.21	10.37	14.48	15.32
	80	8.08	9.71	12.67	15.40	19.35
	100	9.03	10.95	14.70	17.96	22.78
	10	1.78	2.00	2.37	3.01	3.76
	20	2.78	3.26	4.10	4.90	5.71
	40	4.01	4.79	6.25	7.93	9.87
0.4	60	4.75	5.73	7.66	9.68	12.99
	80	5.43	6.66	9.01	11.70	15.61
	100	5.97	7.43	10.12	13.48	17.68
	10	1.57	1.84	2.26	2.68	3.16
	20	2.24	2.63	3.41	4.19	5.02
	40	3.07	3.72	5.07	6.37	8.38
0.6	60	3.56	4.40	5.99	7.72	10.43
	80	4.00	4.95	6.86	9.07	12.14
	100	4.35	5.38	7.57	10.20	13.48
	10	1.36	1.58	1.99	2.34	2.81
	20	1.85	2.20	2.86	3.54	4.43
	40	2.42	2.94	3.99	5.15	6.99
0.8	60	2.76	3.39	4.62	6.01	8.15
	80	3.05	3.76	5.21	6.90	9.27
	100	3.27	4.05	5.68	7.64	10.18
	10	1.33	1.54	1.96	2.32	2.82
	20	1.71	2.05	2.63	3.29	4.20
	40	2.18	2.64	3.50	4.59	6.14
0.9	60	2.48	3.02	4.02	5.31	7.17
	80	2.70	3.30	4.47	5.97	8.07
	100	2.86	3.51	4.82	6.50	8.79
	10	1.31	1.52	1.94	2.31	2.82
	20	1.71	2.03	2.57	3.17	4.00
	40	2.12	2.51	3.30	4.27	5.71
0.95	60	2.38	2.84	3.76	4.89	6.49
	80	2.56	3.10	4.12	5.42	7.28
	100	2.89	3.28	4.39	5.83	7.92

Table 13.12.1 (*continued*)

p	n	Significance Level α 0.15	0.10	0.05	0.025	0.01
	10	1.29	1.51	1.93	2.31	2.84
	20	1.84	2.19	2.78	3.42	4.20
	40	2.46	2.94	3.76	4.64	5.94
1.0	60	2.88	3.40	4.38	5.37	6.99
	80	3.15	3.74	4.83	5.93	7.73
	100	3.35	3.99	5.16	6.34	8.26

Source: D'Agostino and Stephens (1986).

Example LOG Data Set

We will take all 100 observations of the LOG data set and test the composite null hypothesis (13.9.3) for this sample using $Z(X, Y)$. Table 13.12.2 shows for several observations the calculation of Y_i and $F_n(X_{i:n})$ using (13.12.3). The results of the regression of $X_{i:n}$ vs. Y_i are

$$SSR = 27{,}027.28,$$

$$SST = 27{,}458.83,$$

$$R^2 = \frac{27{,}027.28}{27{,}458.83} = 0.9843,$$

$$Z(X, Y) = 100(1 - 0.9843) = 1.57.$$

Since 1.57 is less than 5.16, the percentage point obtained from Table 13.12.1 with $n = 100$, $p = 1.0$, and $\alpha = 0.05$, we do not reject the composite null hypothesis of (13.9.3).

Table 13.12.2 Calculation of Y_i in (13.12.1)

i	$X_{i:n}$	$F_n(X_{i:n}) = i/(n+1)$	$Y_{i:n}$
1	51.90	0.0099	−2.54
2	60.57	0.0198	−2.15
3	63.35	0.0297	−1.92
4	65.87	0.0396	−1.76
97	131.24	0.9604	1.76
98	132.40	0.9703	1.92
99	144.28	0.9802	2.15
100	145.33	0.9901	2.54

Example UNI Data Set

Now we will take all 100 observations of the UNI data set and test the composite null hypothesis of (13.9.3) using $Z(X, Y)$. The regression of $X_{i:n}$ versus Y_i gives

$$SSR = 875.66,$$
$$SST = 955.44,$$
$$R^2 = \frac{875.66}{955.44} = 0.9165,$$
$$Z(X, Y) = 100(1 - 0.9165) = 8.35.$$

Since 8.35 exceeds the tabulated percentage point, 5.16, the composite null hypothesis of (13.9.3) is rejected at a significance level of 0.05.

13.13 RECOMMENDATIONS

Much needs to be done before firm recommendations can be made regarding which of the formal numerical tests should be employed in practice. As of yet no substantial, systematic power studies for the logistic distribution have been undertaken. Given what is available (see D'Agostino and Stephens, 1986), it appears that the edf tests followed by the chi-squared tests are the tests of choice. This should be considered only as tentative.

REFERENCES

Anderson, T. W. and Darling, D. A. (1954). A test of goodness-of-fit, *J. Amer. Stat. Assoc.*, *49*, 765–769.

Barnett, V. (1975). Probability plotting methods and order statistics, *J. R. Stat. Soc. C*, *24*, 95–108.

Barr, D. R. and Davidson, T. (1973). A Kolmogorov-Smirnov test for censored samples, *Technometrics*, *15*, 739–757.

Chernoff, H. and Lehmann, E. L. (1954). The use of maximum-likelihood estimates of χ^2 test for goodness of fit, *Ann. Math. Stat.*, *25*, 579–586.

Cohen, A. and Sackrowitz, H. B. (1975). Unbiasedness of the chi-square, likelihood ratio, and other goodness of fit tests for the equal cell case, *Ann. Stat.*, *4*, 959–964.

Csorgo, S. and Horvath, L. (1981). On the Koziol-Green model for random censorship, *Biometrika*, *68*, 391–401.

D'Agostino and Stephens (1986). *Goodness-of-Fit Techniques*, Marcel Dekker, New York.

Dahiya, R. C. and Gurland, J. (1972). Pearson chi-square test of fit with random intervals, *Biometrika*, *59*, 147–153.

Davis, D. J. (1952). An analysis of some failure data, *J. Amer. Stat. Assoc.*, *47*, 113–150.

Dufour, R. and Maag, U. R. (1978). Distribution results for modified Kolmogorov-Smirnov statistics for truncated or censored data, *Technometrics*, *20*, 29–32.

Fisher, R. A. (1924). The conditions under which χ^2 measures the discrepancy between observation and hypothesis, *J. R. Stat. Soc.*, *87*, 442–450.

Fisk, P. R. (1961). The graduation of income distributions, *Econometrica*, *29*, 171–185.

Herd, G. R. (1960). Estimation of reliability from incomplete data, in *Proceedings of the Sixth National Symposium on Reliability and Quality Control*, 202–217.

Kaplan, E. L. and Meier, P. (1958). Non-parametric estimation from incomplete observations, *J. Amer. Stat. Assoc.*, *53*, 457–481.

Kolmogorov, A. N. (1933). Sulla determinazione empirica di una legge di distibuziane, *Giorna. Ist. Attuari.*, *4*, 83–91.

Koziol, J. A. (1980). Goodness-of-fit tests for randomly censored data. *Biometrika*, *67*, 693–696.

Koziol, J. A. and Byar, D. P. (1975). Percentage points of the asymptotic distributions of one and two sample *k-s* statistics for truncated or censored data, *Technometrics*, *17*, 507–510.

Koziol, J. A. and Green, S. B. (1976). A Cramer-von Mises statistic for randomly censored data, *Biometrika*, *63*, 465–474.

Kuiper, N. H. (1960). Tests concerning random points on a circle, *Proc. Koninkl. Neder. Akad. van. Wetenschappen. A*, *63*, 38–47.

Larntz, K. (1978). Small-sample comparisons of exact levels for chi-squared goodness-of-fit statistics, *J. Amer. Stat. Assoc.*, *73*, 253–263.

Lockhart, R. A., O'Reilly, F. J., and Stephens, M. A. (1986). Tests of fit based on normalized spacings, *J. R. Stat. Soc. B*, *48*, 344–352.

Mann, H. B. and Wald, A. (1942). On the choice of the number of class intervals in the application of the chi-square test, *Ann. Math. Stat.*, *13*, 306–317.

Massaro, J. M. and D'Agostino, R. B. (1992). To appear.

Nelson, W. (1972). Theory and applications of hazard plotting for censored failure data, *Technometrics*, *14*, 945–966.

Pettitt, A. N. and Stephens, M. A. (1976). Modified Cramer-von Mises statistics for censored data, *Biometrika*, *63*, 291–298.

Pettitt, A. N. and Stephens, M. A. (1977). The Kolmogorov-Smirnov goodness-of-fit statistic with discrete and grouped data, *Technometrics*, *19*, 205–210.

Pyke, R. (1965). Spacings, *J. R. Stat. Soc. B*, *27*, 395–449.

Read, T. R. C. (1984). Small sample comparisons for the power divergence goodness-of-fit statistics, *J. Amer. Stat. Assoc.*, *79*, 929–935.

Renyi, A. (1970). *Probability Theory*, North-Holland, Amsterdam.

Schorr, B. (1974). On the choice of the class intervals in the application of the chi-square test, *Math. Operations Forsch. Stat.*, *5*, 357–377.

Stephens, M. A. (1970). Use of the Kolmogorov-Smirnov, Cramer-von Mises and related statistics without extensive tables, *J. R. Stat. Soc. B*, *32*, 115–122.

Stephens, M. A. (1979). Tests of fit for the logistic distribution based on the empirical distribution function, *Biometrika*, *66*, 591–595.

Tiku, M. L. (1980). Goodness of fit statistics based on the spacings of complete or censored samples, *Austral. J. Stat.*, *22*, 260–275.

Watson, G. S. (1961). Goodness-of-fit tests on a circle, *Biometrika*, *48*, 109–114.

APPENDIX I.
LOG Data Set

No.	Observation	No.	Observation	No.	Observation	No.	Observation
1	96.91	26	86.98	51	112.50	76	98.51
2	109.99	27	79.23	52	109.82	77	107.91
3	102.97	28	110.70	53	94.66	78	132.40
4	118.54	29	98.58	54	107.08	79	101.32
5	63.35	30	76.52	55	108.22	80	116.01
6	94.63	31	93.44	56	81.61	81	111.18
7	144.28	32	89.81	57	102.90	82	65.87
8	104.47	33	100.62	58	85.94	83	96.30
9	111.81	34	108.75	59	66.35	84	83.74
10	78.32	35	101.91	60	97.12	85	91.97
11	109.91	36	87.71	61	90.09	86	94.95
12	98.07	37	145.33	62	111.92	87	98.95
13	82.45	38	121.83	63	83.89	88	98.21
14	114.97	39	99.52	64	77.45	89	98.71
15	103.08	40	116.58	65	74.29	90	108.88
16	78.48	41	106.05	66	102.90	91	68.44
17	97.45	42	92.55	67	113.41	92	118.92
18	107.64	43	79.07	68	102.37	93	117.01
19	83.73	44	111.59	69	100.46	94	89.22
20	116.99	45	101.18	70	104.14	95	123.39
21	103.82	46	105.03	71	51.90	96	85.30
22	131.24	47	101.19	72	105.34	97	123.58
23	95.86	48	100.81	73	108.94	98	113.79
24	111.90	49	106.17	74	103.43	99	102.86
25	60.57	50	112.12	75	81.17	100	88.22

APPENDIX II. LAP Data Set

No.	Observation	No.	Observation	No.	Observation	No.	Observation
1	1.11	26	1.41	51	7.37	76	−2.41
2	0.59	27	−0.05	52	−0.81	77	1.03
3	1.54	28	−2.22	53	−6.42	78	5.13
4	−0.99	29	−3.86	54	0.07	79	−2.02
5	0.96	30	1.98	55	−0.43	80	0.33
6	0.64	31	0.84	56	0.02	81	1.11
7	3.87	32	1.01	57	−3.78	82	−1.89
8	−1.47	33	−1.26	58	−0.10	83	−0.06
9	−2.48	34	−2.67	59	2.54	84	−0.53
10	−1.11	35	−0.27	60	−3.27	85	−0.13
11	1.72	36	−3.48	61	−0.43	86	3.03
12	−4.44	37	−1.58	62	−1.19	87	−2.48
13	−0.57	38	2.68	63	1.11	88	0.05
14	1.87	39	2.80	64	1.12	89	4.15
15	0.35	40	−0.89	65	0.55	90	−6.05
16	−6.79	41	−0.05	66	−6.34	91	1.33
17	1.14	42	2.05	67	2.24	92	−0.79
18	−0.46	43	3.95	68	1.31	93	−1.89
19	−9.56	44	0.87	69	−0.84	94	−2.12
20	0.56	45	2.48	70	1.00	95	2.83
21	−1.43	46	−1.59	71	0.58	96	−0.84
22	0.04	47	0.15	72	−1.44	97	0.59
23	−2.64	48	0.44	73	9.52	98	−2.29
24	0.28	49	−1.29	74	5.71	99	−1.10
25	−3.01	50	0.72	75	−0.52	100	−1.54

APPENDIX III.
LIFE Data Set

No.	Observation	No.	Observation	No.	Observation	No.	Observation
1	1067	26	938	51	830	76	1250
2	919	27	970	52	1063	77	1203
3	1196	28	1237	53	930	78	1078
4	785	29	956	54	807	79	890
5	1126	30	1102	55	954	80	1303
6	936	31	1022	56	1063	81	1011
7	918	32	978	57	1002	82	1102
8	1156	33	832	58	909	83	996
9	920	34	1009	59	1077	84	780
10	948	35	1157	60	1021	85	900
11	855	36	1151	61	1062	86	1106
12	1092	37	1009	62	1157	87	704
13	1162	38	765	63	999	88	621
14	1170	39	958	64	932	89	854
15	929	40	1311	65	1035	90	1178
16	950	41	1037	66	944	91	1138
17	905	42	702	67	1049	92	951
18	972	43	521	68	940	93	902
19	1035	44	933	69	1122	94	923
20	1045	45	928	70	1115	95	1333
21	1157	46	1153	71	833	96	811
22	1195	47	946	72	1320	97	1217
23	1195	48	858	73	901	98	1085
24	1340	49	1071	74	1324	99	896
25	1122	50	1069	75	818	100	958

APPENDIX IV.
BOH Data Set

Observation	Frequency
1.10	14
1.95	16
2.35	29
3.07	28
3.56	9
3.91	1
4.44	1
4.61	1
4.79	1

APPENDIX V.
SU12 Data Set

No.	Observation	No.	Observation	No.	Observation	No.	Observation
1	−0.41	26	−0.10	51	−1.00	76	−0.47
2	−0.90	27	0.11	52	−0.89	77	−0.82
3	−0.63	28	−0.93	53	−0.34	78	−1.88
4	−1.25	29	−0.47	54	−0.78	79	−0.64
5	0.50	30	0.18	55	−0.83	80	−1.15
6	−0.34	31	−0.30	56	0.05	81	−0.95
7	−2.46	32	−0.19	57	−0.63	82	0.44
8	−0.68	33	−0.54	58	−0.07	83	−0.39
9	−0.97	34	−0.85	59	0.43	84	−0.01
10	0.13	35	−0.66	60	−0.42	85	−0.25
11	−0.90	36	−0.13	61	−0.20	86	0.35
12	−0.45	37	−2.51	62	−0.98	87	−0.48
13	0.02	38	−1.40	63	−0.02	88	−0.46
14	−1.10	39	−0.50	64	0.16	89	−0.48
15	−0.63	40	−1.17	65	0.24	90	−0.85
16	0.13	41	−0.74	66	−0.63	91	0.38
17	−0.43	42	−0.27	67	−1.04	92	−1.27
18	−0.81	43	0.11	68	−0.68	93	−1.19
19	−0.01	44	−0.96	69	−0.54	94	−0.17
20	−1.19	45	−0.64	70	−0.67	95	−1.47
21	−0.66	46	−0.71	71	0.76	96	−0.06
22	−1.83	47	−0.56	72	−0.72	97	−1.48
23	−0.38	48	−0.62	73	−0.86	98	−1.05
24	−0.98	49	−0.75	74	−0.64	99	−0.62
25	0.56	50	−0.98	75	0.09	100	−0.14

APPENDIX VI.
NOR Data Set

No.	Observation	No.	Observation	No.	Observation	No.	Observation
1	92.55	26	102.56	51	111.38	76	88.13
2	96.20	27	79.43	52	103.22	77	102.98
3	84.27	28	105.48	53	113.17	78	103.71
4	90.87	29	85.29	54	108.39	79	95.14
5	101.58	30	83.53	55	103.60	80	87.71
6	106.82	31	104.21	56	103.90	81	103.56
7	98.70	32	100.75	57	89.35	82	89.44
8	113.75	33	92.02	58	124.60	83	88.26
9	98.98	34	100.10	59	104.34	84	97.80
10	100.42	35	87.83	60	85.29	85	97.33
11	118.52	36	89.00	61	97.78	86	103.90
12	89.90	37	108.67	62	109.76	87	96.38
13	92.45	38	103.09	63	94.92	88	94.33
14	115.92	39	99.12	64	95.12	89	99.62
15	103.61	40	91.46	65	88.56	90	95.95
16	96.13	41	125.28	66	115.96	91	104.89
17	95.45	42	91.45	67	100.79	92	83.34
18	108.52	43	92.56	68	104.87	93	87.04
19	112.69	44	102.66	69	95.89	94	89.80
20	90.03	45	101.91	70	110.72	95	83.07
21	111.56	46	76.35	71	86.28	96	112.14
22	109.26	47	111.30	72	107.97	97	113.90
23	83.67	48	89.33	73	117.23	98	100.46
24	112.97	49	79.89	74	104.12	99	110.39
25	116.87	50	110.17	75	95.97	100	98.43

APPENDIX VII.
EXP Data Set

No.	Observation	No.	Observation	No.	Observation	No.	Observation
1	8.15	26	11.89	51	1.27	76	5.19
2	4.69	27	7.26	52	1.56	77	0.26
3	2.17	28	14.71	53	16.81	78	9.46
4	0.37	29	0.23	54	6.07	79	0.95
5	16.69	30	1.21	55	3.89	80	0.51
6	0.06	31	0.18	56	9.60	81	1.39
7	6.48	32	1.24	57	3.12	82	3.74
8	2.63	33	12.94	58	4.16	83	4.37
9	0.44	34	4.78	59	0.07	84	3.87
10	0.89	35	18.53	60	1.67	85	5.40
11	6.96	36	9.20	61	3.80	86	2.41
12	5.15	37	1.65	62	1.52	87	5.93
13	9.78	38	2.20	63	2.79	88	39.12
14	6.47	39	1.13	64	0.36	89	1.05
15	0.99	40	5.20	65	4.49	90	0.47
16	7.70	41	14.74	66	9.76	91	9.57
17	1.61	42	2.86	67	2.37	92	8.29
18	1.68	43	0.19	68	9.91	93	3.79
19	0.92	44	0.08	69	6.60	94	2.35
20	1.87	45	3.22	70	0.17	95	1.09
21	14.80	46	1.21	71	14.68	96	4.19
22	9.96	47	3.51	72	3.72	97	12.21
23	25.92	48	5.67	73	6.92	98	1.57
24	3.37	49	10.50	74	2.53	99	3.52
25	2.76	50	10.45	75	4.77	100	0.48

APPENDIX VIII.
UNI Data Set

No.	Observation	No.	Observation	No.	Observation	No.	Observation
1	8.10	26	6.54	51	3.93	76	4.26
2	2.06	27	8.24	52	0.08	77	3.32
3	1.60	28	9.12	53	3.51	78	9.29
4	8.87	29	0.31	54	0.44	79	2.57
5	9.90	30	2.63	55	1.22	80	0.55
6	6.58	31	6.20	56	1.12	81	6.53
7	8.68	32	5.47	57	2.34	82	2.33
8	7.31	33	7.80	58	1.86	83	9.01
9	2.85	34	1.30	59	8.35	84	7.86
10	6.09	35	9.39	60	3.53	85	7.06
11	6.10	36	8.67	61	5.05	86	8.54
12	2.94	37	1.87	62	5.28	87	9.71
13	1.85	38	6.67	63	6.87	88	8.49
14	9.04	39	5.90	64	2.96	89	2.08
15	9.38	40	0.15	65	2.35	90	0.50
16	7.30	41	3.91	66	4.02	91	3.54
17	2.11	42	8.87	67	1.44	92	3.75
18	4.55	43	2.50	68	9.63	93	9.46
19	7.66	44	7.49	69	9.44	94	0.04
20	9.63	45	0.55	70	5.44	95	7.79
21	9.48	46	5.25	71	3.71	96	8.08
22	5.31	47	5.61	72	4.21	97	3.60
23	5.76	48	1.00	73	2.22	98	8.85
24	9.66	49	3.23	74	2.87	99	1.50
25	4.37	50	1.05	75	0.72	100	0.18

APPENDIX IX.
SU02 Data Set

No.	Observation	No.	Observation	No.	Observation	No.	Observation
1	0.10	26	0.41	51	-0.39	76	0.05
2	-0.31	27	0.65	52	-0.31	77	-0.25
3	-0.09	28	-0.33	53	0.17	78	-1.01
4	-0.58	29	0.04	54	-0.22	79	-0.10
5	1.15	30	0.73	55	-0.26	80	-0.50
6	0.17	31	0.21	56	0.57	81	-0.35
7	-1.39	32	0.32	57	-0.09	82	1.07
8	-0.14	33	-0.02	58	0.44	83	0.12
9	-0.37	34	-0.27	59	1.05	84	0.51
10	0.68	35	-0.12	60	0.09	85	0.25
11	-0.31	36	0.38	61	0.31	86	0.16
12	0.06	37	-1.42	62	-0.37	87	0.03
13	0.55	38	-0.68	63	0.50	88	0.06
14	-0.47	39	0.01	64	0.70	89	0.04
15	-0.10	40	-0.52	65	0.80	90	-0.28
16	0.67	41	-0.19	66	-0.09	91	0.98
17	0.08	42	0.23	67	-0.42	92	-0.59
18	-0.24	43	0.65	68	-0.14	93	-0.53
19	0.51	44	-0.36	69	-0.01	94	0.34
20	-0.53	45	-0.10	70	-0.13	95	-0.73
21	-0.12	46	-0.16	71	1.51	96	0.46
22	-0.97	47	-0.04	72	-0.17	97	-0.74
23	0.13	48	-0.09	73	-0.28	98	-0.43
24	-0.37	49	-0.19	74	-0.11	99	-0.09
25	1.23	50	-0.38	75	0.62	100	0.37

APPENDIX X.
SU12 Data Set

No.	Observation	No.	Observation	No.	Observation	No.	Observation
1	-0.41	26	-0.10	51	-1.00	76	-0.47
2	-0.90	27	0.11	52	-0.89	77	-0.82
3	-0.63	28	-0.93	53	-0.34	78	-1.88
4	-1.25	29	-0.47	54	-0.78	79	-0.64
5	0.50	30	0.18	55	-0.83	80	-1.15
6	-0.34	31	-0.30	56	0.05	81	-0.95
7	-2.46	32	-0.19	57	-0.63	82	0.44
8	-0.68	33	-0.54	58	-0.07	83	-0.39
9	-0.97	34	-0.85	59	0.43	84	-0.01
10	0.13	35	-0.66	60	-0.42	85	-0.25
11	-0.90	36	-0.13	61	-0.20	86	0.35
12	-0.45	37	-2.51	62	-0.98	87	-0.48
13	0.02	38	-1.40	63	-0.02	88	-0.46
14	-1.10	39	-0.50	64	0.16	89	-0.48
15	-0.63	40	-1.17	65	0.24	90	-0.85
16	0.13	41	-0.74	66	-0.63	91	0.38
17	-0.43	42	-0.27	67	-1.04	92	-1.27
18	-0.81	43	0.11	68	-0.68	93	-1.19
19	-0.01	44	-0.96	69	-0.54	94	-0.17
20	-1.19	45	-0.64	70	-0.67	95	-1.47
21	-0.66	46	-0.71	71	0.76	96	-0.06
22	-1.83	47	-0.56	72	-0.72	97	-1.48
23	-0.38	48	-0.62	73	-0.86	98	-1.05
24	-0.98	49	-0.75	74	-0.64	99	-0.62
25	0.56	50	-0.98	75	0.09	100	-0.14

APPENDIX XI.
LAP Data Set

No.	Observation	No.	Observation	No.	Observation	No.	Observation
1	1.11	26	1.41	51	7.37	76	-2.41
2	0.59	27	-0.05	52	-0.81	77	1.03
3	1.54	28	-2.22	53	-6.42	78	5.13
4	-0.99	29	-3.86	54	0.07	79	-2.02
5	0.96	30	1.98	55	-0.43	80	0.33
6	0.64	31	0.84	56	0.02	81	1.11
7	3.87	32	1.01	57	-3.78	82	-1.89
8	-1.47	33	-1.26	58	-0.10	83	-0.06
9	-2.48	34	-2.67	59	2.54	84	-0.53
10	-1.11	35	-0.27	60	-3.27	85	-0.13
11	1.72	36	-3.48	61	-0.43	86	3.03
12	-4.44	37	-1.58	62	-1.19	87	-2.48
13	-0.57	38	2.68	63	1.11	88	0.05
14	1.87	39	2.80	64	1.12	89	4.15
15	0.35	40	-0.89	65	0.55	90	-6.05
16	-6.79	41	-0.05	66	-6.34	91	1.33
17	1.14	42	2.05	67	2.24	92	-0.79
18	-0.46	43	3.95	68	1.31	93	-1.89
19	-9.56	44	0.87	69	-0.84	94	-2.12
20	0.56	45	2.48	70	1.00	95	2.83
21	-1.43	46	-1.59	71	0.58	96	-0.84
22	0.04	47	0.15	72	-1.44	97	0.59
23	-2.64	48	0.44	73	9.52	98	-2.29
24	0.28	49	-1.29	74	5.71	99	-1.10
25	-3.01	50	0.72	75	-0.52	100	-1.54

14

Tolerance Limits and Sampling Plans Based on Censored Samples

N. Balakrishnan
McMaster University, Hamilton, Ontario, Canada

Karen Y. Fung
University of Windsor, Windsor, Ontario, Canada

14.1 INTRODUCTION

The best linear unbiased estimation of the location and scale parameters (μ and σ) based on complete and Type II censored samples has been discussed in Chapter 4. Hall (1975) has used these BLUEs to determine one-sided tolerance limits based on right-censored samples for some selected sample sizes up to 25. Similar work has been carried out by Mann and Fertig (1973) and Hall and Sampson (1974) for the extreme value and normal distributions, respectively. In this chapter we first describe the work of Hall (1975) in Section 14.2 and extend his table of one-sided tolerance limits for sample sizes up to 40. Next, we present in Section 14.3 the large-sample normal approximation for the one-sided tolerance limits based on right-censored samples; by comparing these approximate values with the simulated values for sample sizes 30, 35, and 40, we show that this normal approximation provides a very good approximation to the tolerance limits for large sample sizes. In Section 14.4 we describe the two-sided tolerance limits and present tables based on doubly censored samples for sample sizes 5, 7, 10 and 15. We show in Section 14.5 that acceptance sampling plans may be set up by using the appropriate tolerance limits; one may refer to Owen (1967) for similar comments while discussing the tolerance limits and acceptance sampling plans for the normal distribution. Finally, we present two examples in Section 14.6 and illustrate the method of determining the tolerance limits developed in the earlier sections.

14.2 ONE-SIDED TOLERANCE LIMITS

Let $X_{1:n} \leq X_{2:n} \leq \cdots \leq X_{n-s:n}$ be a Type II right-censored sample available from the logistic $L(\mu, \sigma^2)$ population with pdf $f(x; \mu, \sigma)$ and cdf $F(x; \mu, \sigma)$ as given in Eqs. (1.1) and (1.2), respectively. Let μ^* and σ^* be the best linear unbiased estimators of μ and σ based on the above right-censored sample. These BLUEs can be computed by using the tables set up by Gupta, Qureishi, and Shah (1967) for sample sizes up to 25 and the tables recently set up by Balakrishnan (1991) for sample sizes up to 40. Both these tables, as mentioned in Chapter 4, give the coefficients required for the computation of the best linear unbiased estimates and also the variances and covariance of these estimates for general Type II doubly censored samples.

Then, $L(\underset{\sim}{X})$ is said to be a lower γ tolerance limit for proportion β if

$$\Pr\{1 - F(L(\underset{\sim}{X}); \mu, \sigma) \geq \beta\} = \gamma. \tag{14.2.1}$$

Since the pth percentile of the distribution $F(\mu, \sigma)$ is $\mu - r\sigma$, where

$$r = \frac{\sqrt{3}}{\pi} \ln\left(\frac{1-p}{p}\right),$$

a logical estimate of the pth percentile, as pointed out by Hall (1975), will be $\mu^* - r\sigma^*$. This estimator is easily seen to be unbiased for the pth percentile of the population distribution; for more details, refer to Balakrishnan and Cohen (1990). Hence, a reasonable lower $100\gamma\%$ confidence limit for the pth percentile which is a (γ, β) lower tolerance limit for the distribution $F(\mu, \sigma)$ would seem to be

$$L(\underset{\sim}{X}) = \mu^* - r_1\sigma^*, \tag{14.2.2}$$

where r_1 is chosen such that (14.2.1) is satisfied.

Let us now write

$$\mu^* = \sum_{i=1}^{n-s} a_i X_{i:n} \quad \text{and} \quad \sigma^* = \sum_{i=1}^{n-s} b_i X_{i:n}. \tag{14.2.3}$$

Using (14.2.2) in (14.2.1), we need to find r_1 such that

$$\gamma = \Pr\left[\Pr\left\{X \geq \sum_{i=1}^{n-s} (a_i - r_1 b_i) X_{i:n}\right\} \geq \beta\right]$$

$$= \Pr\left[\Pr\left\{Y \geq \sum_{i=1}^{n-s} (a_i - r_1 b_i) Y_{i:n}\right\} \geq \beta\right]$$

$$= \Pr\left[\sum_{i=1}^{n-s} (a_i - r_1 b_i) Y_{i:n} \leq F^{-1}(1 - \beta) = -\frac{\sqrt{3}}{\pi} \ln\left(\frac{\beta}{1 - \beta}\right)\right]$$

$$= \frac{n!}{s!} \int \int_{A_1} \cdots \int \{1 - F(y_{n-s:n})\}^s$$

$$\times f(y_{1:n}) \cdots f(y_{n-s:n})\, dy_{1:n} \cdots dy_{n-s:n}, \tag{14.2.4}$$

where $Y = (X - \mu)/\sigma$, $Y_{i:n} = (X_{i:n} - \mu)/\sigma$, $f(y)$ and $F(y)$ are as defined in Eqs. (1.3) and (1.4), respectively, and A_1 is the region in an $(n - s)$-dimensional space defined by

$$A_1 = \left\{(y_1, \ldots, y_{n-s}) : y_1 < y_2 < \cdots < y_{n-s} \quad \text{and} \quad \sum_{i=1}^{n-s} (a_i - r_1 b_i) y_i \leq -\frac{\sqrt{3}}{\pi} \ln\left(\frac{\beta}{1 - \beta}\right)\right\}. \tag{14.2.5}$$

Hall (1975) used numerical procedures to determine the exact values of r_1 which satisfy Eq. (14.2.4) for large values of s (or small values of $n - s$). Thus, for n = 5, 7, 10, 12, 15, 20, and 25 and $s = n - 2$, $n - 3$, and $n - 4$, Hall (1975) numerically computed the values of r_1 when $\gamma = 0.90$, 0.95, and $\beta = 0.90$, 0.95. These values are presented in Table 14.2.1. For other choices of n and s in the table, the values of r_1 were determined by Monte Carlo simulations. For this purpose, Eq. (14.2.4) was first rewritten as

$$\gamma = \Pr\left\{\frac{\sum_{i=1}^{n-s} a_i Y_{i:n} + \frac{\sqrt{3}}{\pi} \ln\left(\frac{\beta}{1 - \beta}\right)}{\sum_{i=1}^{n-s} b_i Y_{i:n}} \leq r_1\right\}. \tag{14.2.6}$$

Then, after simulating the necessary censored sample from the logistic $L(0, 1)$ population by using the method of Lurie and Hartley (1972), we computed the value of

$$T_1 = \frac{\left\{\sum_{i=1}^{n-s} a_i Y_{i:n} + \frac{\sqrt{3}}{\pi} \ln\left(\frac{\beta}{1 - \beta}\right)\right\}}{\sum_{i=1}^{n-s} b_i Y_{i:n}}. \tag{14.2.7}$$

Table 14.2.1 Values of r_1 (r_2) for Lower (Upper) Tolerance Limits Based on BLUEs

(γ, β)		(0.90, 0.90)		(0.95, 0.90)		(0.90, 0.95)		(0.95, 0.95)	
n	s	r_1	r_2	r_1	r_2	r_1	r_2	r_1	r_2
5	0	2.60	2.60	3.23	3.23	3.35	3.35	4.17	4.17
5	1	2.84	3.07	3.69	4.09	3.73	3.97	4.85	5.25
5	2	3.41	4.57	4.86	6.85	4.64	5.84	6.64	8.72
5	3	5.81	13.87	11.24	28.95	8.93	17.20	16.00	35.72
7	0	2.24	2.24	2.66	2.66	2.90	2.90	3.42	3.42
7	1	2.34	2.45	2.80	2.96	3.05	3.10	3.67	3.84
7	2	2.47	2.69	3.04	3.47	3.26	3.52	4.05	4.47
7	3	2.67	3.40	3.44	4.63	3.59	4.36	4.65	5.87
7	4	3.10	5.22	4.35	7.97	4.36	6.56	6.19	9.93
7	5	4.74	16.03	8.90	33.64	7.84	19.51	15.21	40.72
10	0	1.98	1.98	2.28	2.28	2.59	2.59	2.95	2.95
10	1	2.01	2.06	2.35	2.38	2.65	2.68	3.03	3.10
10	2	2.07	2.16	2.43	2.54	2.72	2.81	3.15	3.29
10	3	2.13	2.30	2.52	2.80	2.81	3.00	3.30	3.59
10	4	2.20	2.55	2.64	3.20	2.93	3.28	3.51	4.04
10	5	2.30	2.95	2.81	3.87	3.10	3.76	3.84	4.88
10	6	2.45	3.82	3.12	5.27	3.40	4.81	4.35	6.59
10	7	2.73	5.93	3.74	9.16	4.01	7.32	5.60	11.21
10	8	3.63	18.19	6.47	38.33	6.56	21.78	12.43	45.64
12	0	1.89	1.89	2.13	2.13	2.45	2.45	2.75	2.75
12	1	1.91	1.96	2.15	2.22	2.49	2.52	2.77	2.85
12	2	1.94	2.00	2.21	2.33	2.54	2.60	2.88	2.98
12	3	1.97	2.09	2.25	2.46	2.59	2.71	2.97	3.14
12	4	2.01	2.20	2.32	2.64	2.65	2.86	3.07	3.40
12	5	2.06	2.36	2.39	2.90	2.73	3.05	3.23	3.76
12	6	2.13	2.48	2.51	3.35	2.85	3.36	3.41	4.24
12	7	2.22	3.21	2.70	4.10	3.05	3.92	3.70	5.14
12	8	2.33	4.04	2.93	5.61	3.28	5.05	4.17	6.95
12	9	2.54	6.29	3.42	9.76	3.80	7.70	5.27	11.85
12	10	3.14	19.35	5.38	40.82	5.92	22.99	11.04	48.27
15	0	1.79	1.79	2.00	2.00	2.34	2.34	2.60	2.60
15	1	1.80	1.82	2.03	2.05	2.36	2.37	2.63	2.65
15	2	1.82	1.86	2.05	2.10	2.38	2.42	2.67	2.71
15	3	1.85	1.90	2.08	2.17	2.42	2.47	2.71	2.80
15	4	1.86	1.95	2.11	2.28	2.45	2.55	2.76	2.90
15	5	1.88	2.03	2.15	2.39	2.49	2.64	2.82	3.04
15	6	1.92	2.14	2.20	2.53	2.54	2.76	2.89	3.29
15	7	1.95	2.28	2.25	2.72	2.60	2.94	2.98	3.56
15	8	1.99	2.48	2.33	3.04	2.68	3.21	3.12	3.94

(continued)

Table 14.2.1 (*continued*)

(γ, β)		(0.90, 0.90)		(0.95, 0.90)		(0.90, 0.95)		(0.95, 0.95)	
n	s	r_1	r_2	r_1	r_2	r_1	r_2	r_1	r_2
15	9	2.05	2.75	2.42	3.52	2.76	3.56	3.28	4.47
15	10	2.11	3.27	2.54	4.37	2.91	4.14	3.52	5.44
15	11	2.18	4.31	2.79	6.02	3.12	5.34	3.93	7.40
15	12	2.31	6.72	3.03	10.48	3.55	8.16	4.84	12.62
15	13	2.62	20.66	4.18	43.67	5.13	24.35	9.31	51.22
20	0	1.69	1.69	1.85	1.85	2.21	2.21	2.42	2.42
20	1	1.70	1.70	1.87	1.88	2.22	2.22	2.44	2.44
20	2	1.71	1.72	1.88	1.90	2.24	2.24	2.46	2.47
20	3	1.72	1.74	1.89	1.94	2.25	2.27	2.48	2.51
20	4	1.73	1.77	1.91	1.98	2.27	2.30	2.50	2.56
20	5	1.74	1.80	1.93	2.02	2.29	2.35	2.53	2.62
20	6	1.76	1.84	1.95	2.07	2.30	2.40	2.56	2.70
20	7	1.77	1.90	1.97	2.14	2.32	2.46	2.59	2.78
20	8	1.79	1.95	1.99	2.23	2.34	2.54	2.63	2.90
20	9	1.80	2.04	2.01	2.37	2.37	2.64	2.66	3.04
20	10	1.82	2.13	2.05	2.54	2.41	2.76	2.71	3.27
20	11	1.84	2.28	2.09	2.70	2.46	2.91	2.76	3.50
20	12	1.87	2.45	2.13	2.97	2.52	3.12	2.84	3.81
20	13	1.89	2.68	2.17	3.34	2.58	3.42	2.96	4.24
20	14	1.91	2.96	2.24	3.83	2.66	3.80	3.07	4.84
20	15	1.95	3.52	2.31	4.70	2.75	4.39	3.24	5.79
20	16	1.99	4.65	2.43	6.55	2.92	5.74	3.62	7.94
20	17	2.04	7.26	2.57	11.37	3.21	8.73	4.28	13.55
20	18	2.12	22.30	3.00	47.23	4.18	26.05	7.22	54.90
25	0	1.65	1.65	1.79	1.79	2.15	2.15	2.32	2.32
25	1	1.65	1.66	1.79	1.80	2.15	2.16	2.33	2.33
25	2	1.66	1.67	1.80	1.81	2.16	2.17	2.35	2.35
25	3	1.66	1.68	1.81	1.82	2.16	2.18	2.36	2.37
25	4	1.67	1.69	1.82	1.84	2.17	2.20	2.37	2.40
25	5	1.67	1.70	1.83	1.86	2.18	2.23	2.38	2.42
25	6	1.68	1.73	1.84	1.90	2.19	2.25	2.40	2.47
25	7	1.69	1.76	1.85	1.94	2.20	2.28	2.41	2.52
25	8	1.70	1.79	1.86	1.99	2.21	2.32	2.43	2.58
25	9	1.70	1.82	1.87	2.04	2.22	2.36	2.45	2.65
25	10	1.71	1.86	1.88	2.11	2.24	2.42	2.48	2.73
25	11	1.72	1.91	1.90	2.18	2.26	2.48	2.50	2.81
25	12	1.73	1.98	1.91	2.26	2.28	2.56	2.53	2.90
25	13	1.74	2.06	1.92	2.37	2.30	2.66	2.56	3.05

Table 14.2.1 (*continued*)

(γ, β)		(0.90, 0.90)		(0.95, 0.90)		(0.90, 0.95)		(0.95, 0.95)	
n	s	r_1	r_2	r_1	r_2	r_1	r_2	r_1	r_2
25	14	1.75	2.14	1.94	2.48	2.33	2.77	2.60	3.18
25	15	1.76	2.23	1.96	2.62	2.35	2.88	2.64	3.36
25	16	1.77	2.39	1.98	2.84	2.40	3.10	2.70	3.66
25	17	1.79	2.62	2.01	3.14	2.44	3.34	2.77	4.00
25	18	1.81	2.87	2.05	3.55	2.51	3.81	2.84	4.43
25	19	1.82	3.26	2.08	4.10	2.56	4.05	2.93	5.00
25	20	1.84	3.76	2.12	5.00	2.63	4.63	3.07	6.11
25	21	1.85	4.90	2.19	6.95	2.76	5.97	3.37	8.37
25	22	1.87	7.67	2.27	12.07	2.95	9.16	3.85	14.27
25	23	1.88	23.54	2.39	49.92	3.54	27.32	5.78	57.66
30	0	1.59	1.59	1.72	1.72	2.08	2.08	2.24	2.24
30	1	1.60	1.60	1.73	1.72	2.09	2.09	2.26	2.24
30	2	1.60	1.61	1.74	1.73	2.10	2.10	2.27	2.26
30	3	1.61	1.61	1.75	1.74	2.11	2.11	2.28	2.28
30	4	1.61	1.61	1.75	1.75	2.12	2.12	2.28	2.30
30	5	1.62	1.63	1.76	1.77	2.12	2.14	2.30	2.32
30	6	1.62	1.64	1.77	1.80	2.13	2.15	2.32	2.34
30	7	1.63	1.66	1.77	1.82	2.14	2.17	2.33	2.36
30	8	1.63	1.67	1.77	1.87	2.15	2.20	2.34	2.43
30	9	1.64	1.70	1.79	1.88	2.17	2.23	2.36	2.45
30	10	1.65	1.73	1.80	1.90	2.17	2.26	2.37	2.47
30	11	1.66	1.74	1.80	1.93	2.19	2.27	2.37	2.51
30	12	1.66	1.78	1.81	1.97	2.19	2.32	2.40	2.56
30	13	1.66	1.82	1.82	2.03	2.21	2.36	2.42	2.63
30	14	1.67	1.86	1.83	2.10	2.22	2.42	2.44	2.72
30	15	1.68	1.90	1.85	2.16	2.24	2.47	2.46	2.79
30	16	1.68	1.94	1.86	2.23	2.25	2.52	2.49	2.87
30	17	1.70	2.00	1.87	2.33	2.27	2.59	2.52	3.00
30	18	1.70	2.09	1.88	2.43	2.29	2.69	2.55	3.09
30	19	1.71	2.17	1.91	2.56	2.31	2.80	2.58	3.27
30	20	1.73	2.31	1.93	2.75	2.35	2.95	2.63	3.48
30	21	1.74	2.48	1.94	2.98	2.37	3.15	2.66	3.77
30	22	1.74	2.64	1.96	3.35	2.40	3.35	2.71	4.05
30	23	1.76	2.91	1.98	3.73	2.45	3.67	2.79	4.64
30	24	1.76	3.39	2.00	4.38	2.48	4.21	2.86	5.35
30	25	1.77	4.02	2.03	5.35	2.54	4.93	3.02	6.47
30	26	1.78	5.22	2.07	7.40	2.65	6.30	3.25	8.86
30	27	1.78	8.09	2.10	13.36	2.80	9.57	3.56	15.79
30	28	1.78	25.55	2.10	55.55	3.23	29.62	4.88	63.51

(*continued*)

Table 14.2.1 (*continued*)

(γ, β)		(0.90, 0.90)		(0.95, 0.90)		(0.90, 0.95)		(0.95, 0.95)	
n	*s*	r_1	r_2	r_1	r_2	r_1	r_2	r_1	r_2
35	0	1.56	1.56	1.67	1.67	2.04	2.04	2.19	2.19
35	1	1.56	1.56	1.68	1.67	2.05	2.04	2.20	2.19
35	2	1.56	1.56	1.68	1.68	2.06	2.05	2.20	2.19
35	3	1.57	1.56	1.69	1.68	2.06	2.05	2.20	2.20
35	4	1.57	1.57	1.70	1.69	2.07	2.06	2.21	2.21
35	5	1.58	1.58	1.70	1.71	2.07	2.07	2.22	2.23
35	6	1.58	1.59	1.70	1.72	2.08	2.08	2.23	2.25
35	7	1.58	1.60	1.71	1.74	2.08	2.10	2.24	2.27
35	8	1.59	1.62	1.72	1.77	2.09	2.12	2.25	2.30
35	9	1.60	1.63	1.73	1.78	2.10	2.13	2.27	2.33
35	10	1.60	1.64	1.73	1.81	2.11	2.16	2.27	2.36
35	11	1.61	1.66	1.74	1.83	2.12	2.18	2.29	2.38
35	12	1.61	1.68	1.75	1.85	2.13	2.20	2.30	2.41
35	13	1.61	1.69	1.75	1.88	2.13	2.22	2.31	2.45
35	14	1.62	1.71	1.75	1.90	2.14	2.24	2.32	2.47
35	15	1.63	1.74	1.77	1.93	2.15	2.27	2.33	2.51
35	16	1.63	1.77	1.78	1.97	2.16	2.30	2.35	2.56
35	17	1.64	1.80	1.78	2.03	2.18	2.35	2.37	2.62
35	18	1.64	1.83	1.79	2.07	2.18	2.38	2.39	2.68
35	19	1.65	1.87	1.80	2.12	2.20	2.43	2.41	2.74
35	20	1.65	1.92	1.81	2.20	2.21	2.49	2.43	2.83
35	21	1.66	2.00	1.82	2.31	2.24	2.59	2.45	2.95
35	22	1.66	2.08	1.83	2.39	2.24	2.69	2.47	3.06
35	23	1.66	2.17	1.84	2.51	2.25	2.79	2.50	3.21
35	24	1.67	2.25	1.85	2.66	2.27	2.88	2.50	3.37
35	25	1.68	2.40	1.86	2.86	2.29	3.04	2.54	3.59
35	26	1.69	2.55	1.87	3.04	2.32	3.24	2.58	3.81
35	27	1.69	2.73	1.89	3.39	2.35	3.44	2.64	4.20
35	28	1.70	3.04	1.89	3.85	2.36	3.79	2.69	4.74
35	29	1.70	3.49	1.92	4.48	2.41	4.31	2.78	5.49
35	30	1.70	4.09	1.93	5.53	2.46	5.00	2.85	6.73
35	31	1.70	5.24	1.93	7.72	2.51	6.30	2.98	9.21
35	32	1.70	8.27	1.93	13.34	2.61	9.79	3.32	15.61
35	33	1.75	25.33	1.93	56.24	2.80	29.14	4.12	64.22
40	0	1.53	1.53	1.63	1.63	2.01	2.01	2.13	2.13
40	1	1.53	1.53	1.64	1.63	2.01	2.01	2.14	2.14
40	2	1.54	1.53	1.64	1.64	2.02	2.02	2.15	2.15

Table 14.2.1 (*continued*)

(γ, β)		(0.90, 0.90)		(0.95, 0.90)		(0.90, 0.95)		(0.95, 0.95)	
n	s	r_1	r_2	r_1	r_2	r_1	r_2	r_1	r_2
40	3	1.54	1.53	1.65	1.64	2.02	2.02	2.16	2.15
40	4	1.54	1.54	1.65	1.65	2.03	2.02	2.16	2.16
40	5	1.55	1.54	1.66	1.66	2.03	2.03	2.17	2.17
40	6	1.55	1.55	1.66	1.66	2.03	2.03	2.17	2.18
40	7	1.55	1.56	1.66	1.68	2.04	2.04	2.18	2.20
40	8	1.56	1.57	1.67	1.69	2.05	2.06	2.18	2.21
40	9	1.56	1.57	1.68	1.70	2.05	2.07	2.19	2.22
40	10	1.57	1.59	1.68	1.72	2.06	2.08	2.21	2.25
40	11	1.57	1.60	1.69	1.73	2.06	2.10	2.22	2.27
40	12	1.57	1.61	1.69	1.75	2.06	2.11	2.23	2.29
40	13	1.57	1.63	1.70	1.78	2.08	2.13	2.24	2.32
40	14	1.58	1.64	1.71	1.79	2.09	2.15	2.25	2.33
40	15	1.59	1.65	1.71	1.81	2.10	2.17	2.25	2.37
40	16	1.59	1.67	1.71	1.84	2.10	2.18	2.25	2.40
40	17	1.59	1.69	1.71	1.85	2.10	2.21	2.26	2.42
40	18	1.60	1.71	1.72	1.88	2.11	2.24	2.27	2.45
40	19	1.60	1.73	1.74	1.93	2.12	2.26	2.30	2.51
40	20	1.60	1.76	1.74	1.97	2.12	2.30	2.30	2.56
40	21	1.60	1.79	1.75	2.02	2.13	2.33	2.32	2.60
40	22	1.60	1.83	1.75	2.05	2.14	2.40	2.33	2.64
40	23	1.61	1.86	1.76	2.10	2.15	2.41	2.35	2.70
40	24	1.61	1.91	1.77	2.18	2.17	2.48	2.36	2.80
40	25	1.62	1.97	1.77	2.26	2.17	2.54	2.37	2.89
40	26	1.62	2.03	1.78	2.32	2.18	2.61	2.40	2.97
40	27	1.62	2.09	1.78	2.43	2.19	2.69	2.41	3.10
40	28	1.62	2.18	1.79	2.57	2.21	2.78	2.44	3.26
40	29	1.63	2.29	1.79	2.71	2.22	2.91	2.44	3.42
40	30	1.64	2.41	1.80	2.89	2.24	3.06	2.47	3.64
40	31	1.64	2.60	1.81	3.17	2.27	3.28	2.53	3.96
40	32	1.64	2.79	1.82	3.47	2.28	3.49	2.54	4.30
40	33	1.64	3.04	1.83	3.89	2.30	3.78	2.59	4.79
40	34	1.65	3.52	1.83	4.61	2.33	4.33	2.65	5.60
40	35	1.65	4.23	1.84	5.64	2.39	5.14	2.75	6.81
40	36	1.65	5.56	1.85	7.92	2.42	6.68	2.88	9.38
40	37	1.66	8.73	1.85	13.58	2.47	10.27	3.02	15.86
40	38	1.74	26.74	1.87	57.67	2.55	30.58	3.48	64.98

This Monte Carlo process was repeated 8001 times and the values of T_1 were then ordered. The 8001γth value of T_1 was taken as an estimate of r_1 that satisfies Eq. (14.2.4) for specified values of γ and β. These values are also reported in Table 14.2.1. Thus, the values of r_1 required for computing the (γ, β) lower tolerance limit in (14.2.2) for $n = 5, 7, 10, 12, 15(5)40$, $s = 0(1)n - 2$, $\gamma = 0.90, 0.95$, and $\beta = 0.90, 0.95$ are presented in Table 14.2.1.

Now, $U(\underset{\sim}{X})$ is said to be an upper γ tolerance limit for proportion β if

$$\Pr\{F(U(\underset{\sim}{X}); \mu, \sigma) \geq \beta\} = \gamma. \tag{14.2.8}$$

We may set

$$U(\underset{\sim}{X}) = \mu^* + r_2\sigma^* \tag{14.2.9}$$

and determine r_2 such that (14.2.8) is satisfied. In other words, we need to find r_2 such that

$$\begin{aligned}
\gamma &= \Pr\left[\Pr\left\{X \leq \sum_{i=1}^{n-s} (a_i + r_2 b_i) X_{i:n}\right\} \geq \beta\right] \\
&= \Pr\left[\Pr\left\{Y \leq \sum_{i=1}^{n-s} (a_i + r_2 b_i) Y_{i:n}\right\} \geq \beta\right\} \\
&= \Pr\left[\sum_{i=1}^{n-s} (a_i + r_2 b_i) Y_{i:n} \geq F^{-1}(\beta) = \frac{\sqrt{3}}{\pi} \ln\left(\frac{\beta}{1-\beta}\right)\right] \\
&= \frac{n!}{s!} \int\int_{A_2} \cdots \int \{1 - F(y_{n-s:n})\}^s \\
&\quad \times f(y_{1:n}) \cdots f(y_{n-s:n})\, dy_{1:n} \cdots dy_{n-s:n},
\end{aligned} \tag{14.2.10}$$

where A_2 is the region in an $(n - s)$-dimensional space defined by

$$A_2 = \left\{(y_1, \ldots, y_{n-s}) : y_1 < y_2 < \cdots y_{n-s} \quad \text{and} \quad \sum_{i=1}^{n-s} (a_i + r_2 b_i) y_i \geq \frac{\sqrt{3}}{\pi} \ln\left(\frac{\beta}{1-\beta}\right)\right\}. \tag{14.2.11}$$

Hall (1975) employed some numerical procedures to determine the exact values of r_2 which satisfy Eq. (14.2.10) for large values of s (or small values of $n - s$). For $n = 5, 7, 10, 12, 15, 20$, and 25, $s = n - 2, n - 3$, and $n - 4$, $\gamma = 0.90, 0.95$, and $\beta = 0.90, 0.95$, Hall (1975) numerically computed the values of r_2 and these are presented in Table 14.2.1. For other choices of n and s, the values of r_2 were determined once again by

Monte Carlo simulations. For this purpose, we first rewrote Eq. (14.2.10) as

$$\gamma = \Pr\left\{ \frac{-\sum_{i=1}^{n-s} a_i Y_{i:n} + \frac{\sqrt{3}}{\pi} \ln\left(\frac{\beta}{1-\beta}\right)}{\sum_{i=1}^{n-s} b_i Y_{i:n}} \le r_2 \right\}. \tag{14.2.12}$$

The value of

$$T_2 = \frac{\left\{-\sum_{i=1}^{n-s} a_i Y_{i:n} + \frac{\sqrt{3}}{\pi} \ln\left(\frac{\beta}{1-\beta}\right)\right\}}{\sum_{i=1}^{n-s} b_i Y_{i:n}} \tag{14.2.13}$$

was computed for each of 8001 simulated Type II censored samples from the logistic $L(0, 1)$ population. The values of T_2 were then ordered and the 8001γth value of T_2 was taken as an estimate of r_2 that satisfies Eq. (14.2.10) for specified values of γ and β. These values are also given in Table 14.2.1. Thus, the values of r_2 required for computing the (γ, β) upper tolerance limit in (14.2.9) for $n = 5, 7, 10, 12, 15(5)40$, $s = 0(1)n - 2$, $\gamma = 0.90, 0.95$, and $\beta = 0.90, 0.95$ are presented in Table 14.2.1.

14.3 LARGE-SAMPLE NORMAL APPROXIMATION

For approximating the factors r_1 and r_2 in case of large sample sizes, we may use the result that the BLUEs μ^* and σ^* in (14.2.3) have asymptotically a bivariate normal distribution

$$\mathrm{BVN}\left(\begin{pmatrix}\mu \\ \sigma\end{pmatrix}, \sigma^2\begin{pmatrix}\sigma_{11} & \sigma_{12} \\ & \sigma_{22}\end{pmatrix}\right), \tag{14.3.1}$$

so that

$$\sum_{i=1}^{n-s} (a_i - r_1 b_i) Y_{i:n} \xrightarrow{\text{asy}} N(-r_1, \sigma_{11} + r_1^2\sigma_{22} - 2r_1\sigma_{12}). \tag{14.3.2}$$

Then, from Eq. (14.2.4) we have

$$\gamma = \Pr\left\{ Z \le \frac{F^{-1}(1-\beta) + r_1}{(\sigma_{11} + r_1^2\sigma_{22} - 2r_1\sigma_{12})^{1/2}} \right\}, \tag{14.3.3}$$

where Z is a standard normal variable. Now, by denoting the upper α

percentage point of the standard normal distribution by $z_{(\alpha)}$, we obtain from (14.3.3) that

$$\frac{F^{-1}(1-\beta)+r_1}{(\sigma_{11}+r_1^2\sigma_{22}-2r_1\sigma_{12})^{1/2}} = z_{(1-\gamma)}. \tag{14.3.4}$$

Equation (14.3.4) may be rewritten by the quadratic equation

$$r_1^2(1-z_{(1-\gamma)}^2\sigma_{22}) + r_1(2F^{-1}(1-\beta)+2z_{(1-\gamma)}^2\sigma_{12}) + \{F^{-1}(1-\beta)\}^2 - z_{(1-\gamma)}^2\sigma_{11} = 0. \tag{14.3.5}$$

From (14.3.5), we get the large-sample approximation for r_1 to be

$$r_1 = \frac{t_1 + (t_1^2 - uv)^{1/2}}{v}, \tag{14.3.6}$$

where

$$v = 1 - z_{(1-\gamma)}^2\sigma_{22},$$

$$u = \{F^{-1}(1-\beta)\}^2 - z_{(1-\gamma)}^2\sigma_{11} = \frac{3}{\pi^2}\left\{\ln\left(\frac{\beta}{1-\beta}\right)\right\}^2 - z_{(1-\gamma)}^2\sigma_{11},$$

and

$$t_1 = -F^{-1}(1-\beta) - z_{(1-\gamma)}^2\sigma_{12} = \frac{\sqrt{3}}{\pi}\ln\left(\frac{\beta}{1-\beta}\right) - z_{(1-\gamma)}^2\sigma_{12}.$$

By using the formula in (14.3.6), the large-sample approximate values of r_1 were computed for $n = 30, 35$ and 40, $s = 0(1)n-6$, $\gamma = 0.90, 0.95$, and $\beta = 0.90, 0.95$. These values are presented in Table 14.3.1. The values of σ_{11}, σ_{12}, and σ_{22} required in the above computation were taken from the tables of Balakrishnan (1991). We observe that the formula in (14.3.6) yields very close approximation to r_1 whenever s is small or moderately large; but, if s is large the normal approximation in (14.3.6) does not give satisfactory results. However, in this case one may use the integral form in Eq. (14.2.4) directly to solve for r_1.

In order to derive a similar large-sample approximation for r_2, we first note that

$$\sum_{i=1}^{n-s}(a_i + r_2 b_i)Y_{i:n} \xrightarrow{\text{asy}} N(r_2, \sigma_{11} + r_2^2\sigma_{22} + 2r_2\sigma_{12}). \tag{14.3.7}$$

Table 14.3.1 Large-Sample Approximate Values of r_1 (r_2) for Lower (Upper) Tolerance Limits Based on BLUEs

(γ, β)		(0.90, 0.90)		(0.95, 0.90)		(0.90, 0.95)		(0.95, 0.95)	
n	s	r_1	r_2	r_1	r_2	r_1	r_2	r_1	r_2
30	0	1.60	1.60	1.74	1.74	2.10	2.10	2.27	2.27
30	1	1.60	1.61	1.75	1.75	2.10	2.10	2.28	2.28
30	2	1.61	1.61	1.75	1.76	2.11	2.11	2.29	2.29
30	3	1.61	1.62	1.76	1.77	2.12	2.12	2.30	2.31
30	4	1.62	1.63	1.77	1.78	2.13	2.14	2.31	2.33
30	5	1.63	1.64	1.78	1.80	2.13	2.15	2.33	2.35
30	6	1.63	1.65	1.79	1.82	2.14	2.17	2.34	2.38
30	7	1.64	1.67	1.80	1.84	2.15	2.19	2.36	2.41
30	8	1.64	1.69	1.81	1.87	2.17	2.21	2.38	2.44
30	9	1.65	1.71	1.82	1.90	2.18	2.24	2.40	2.48
30	10	1.66	1.73	1.83	1.94	2.19	2.27	2.42	2.53
30	11	1.67	1.76	1.85	1.98	2.20	2.30	2.44	2.58
30	12	1.67	1.79	1.86	2.03	2.22	2.34	2.46	2.65
30	13	1.68	1.83	1.88	2.09	2.23	2.39	2.49	2.72
30	14	1.69	1.87	1.89	2.16	2.25	2.44	2.52	2.80
30	15	1.70	1.93	1.91	2.25	2.27	2.51	2.55	2.90
30	16	1.71	1.99	1.93	2.35	2.29	2.58	2.59	3.03
30	17	1.72	2.06	1.95	2.47	2.31	2.67	2.63	3.17
30	18	1.73	2.15	1.98	2.62	2.34	2.78	2.69	3.36
30	19	1.74	2.26	2.00	2.82	2.36	2.91	2.74	3.59
30	20	1.76	2.40	2.04	3.07	2.40	3.07	2.82	3.89
30	21	1.77	2.59	2.08	3.42	2.44	3.29	2.91	4.30
30	22	1.79	2.84	2.12	3.92	2.49	3.59	3.03	4.90
30	23	1.80	3.20	2.18	4.72	2.55	4.01	3.21	5.83
30	24	1.82	3.77	2.27	6.15	2.63	4.67	3.51	7.51
35	0	1.56	1.56	1.69	1.69	2.05	2.05	2.21	2.21
35	1	1.57	1.57	1.70	1.70	2.06	2.06	2.21	2.22
35	2	1.57	1.57	1.70	1.70	2.06	2.07	2.22	2.22
35	3	1.58	1.58	1.71	1.71	2.07	2.07	2.23	2.24
35	4	1.58	1.59	1.71	1.72	2.08	2.08	2.24	2.25
35	5	1.58	1.59	1.72	1.73	2.08	2.09	2.25	2.26
35	6	1.59	1.60	1.73	1.75	2.09	2.10	2.26	2.28
35	7	1.59	1.61	1.73	1.76	2.10	2.12	2.27	2.30
35	8	1.60	1.62	1.74	1.78	2.10	2.13	2.28	2.32
35	9	1.60	1.64	1.75	1.80	2.11	2.15	2.30	2.35
35	10	1.61	1.65	1.76	1.82	2.12	2.17	2.31	2.38
35	11	1.61	1.67	1.76	1.84	2.13	2.19	2.32	2.41
35	12	1.62	1.69	1.77	1.87	2.14	2.21	2.34	2.44
35	13	1.62	1.71	1.78	1.91	2.15	2.24	2.35	2.48

Table 14.3.1 (*continued*)

(γ, β)		(0.90, 0.90)		(0.95, 0.90)		(0.90, 0.95)		(0.95, 0.95)	
n	s	r_1	r_2	r_1	r_2	r_1	r_2	r_1	r_2
35	14	1.63	1.73	1.79	1.94	2.16	2.27	2.37	2.53
35	15	1.64	1.76	1.80	1.99	2.17	2.30	2.39	2.58
35	16	1.64	1.79	1.81	2.03	2.18	2.34	2.41	2.64
35	17	1.65	1.83	1.83	2.09	2.19	2.39	2.43	2.71
35	18	1.65	1.87	1.84	2.16	2.21	2.44	2.46	2.79
35	19	1.66	1.92	1.85	2.23	2.22	2.49	2.48	2.88
35	20	1.67	2.00	1.87	2.32	2.24	2.56	2.51	2.99
35	21	1.68	2.04	1.88	2.43	2.26	2.64	2.55	3.12
35	22	1.68	2.12	1.90	2.56	2.28	2.73	2.59	3.28
35	23	1.69	2.22	1.92	2.73	2.30	2.85	2.63	3.47
35	24	1.70	2.34	1.94	2.94	2.32	2.99	2.68	3.72
35	25	1.71	2.48	1.96	3.21	2.35	3.16	2.75	4.04
35	26	1.72	2.68	1.99	3.58	2.38	3.39	2.83	4.47
35	27	1.73	2.94	2.03	4.11	2.43	3.70	2.94	5.10
35	28	1.74	3.32	2.07	4.95	2.48	4.14	3.09	6.09
35	29	1.75	3.92	2.12	6.48	2.55	4.82	3.35	7.87
40	0	1.54	1.54	1.65	1.65	2.02	2.02	2.16	2.16
40	1	1.54	1.54	1.66	1.66	2.02	2.02	2.17	2.17
40	2	1.54	1.55	1.66	1.66	2.03	2.03	2.17	2.17
40	3	1.55	1.55	1.66	1.67	2.03	2.03	2.18	2.18
40	4	1.55	1.55	1.67	1.67	2.04	2.04	2.18	2.19
40	5	1.55	1.56	1.67	1.68	2.04	2.05	2.19	2.20
40	6	1.56	1.57	1.68	1.69	2.05	2.06	2.20	2.21
40	7	1.56	1.57	1.68	1.70	2.05	2.07	2.21	2.23
40	8	1.56	1.58	1.69	1.71	2.06	2.08	2.22	2.24
40	9	1.57	1.59	1.69	1.73	2.06	2.09	2.22	2.26
40	10	1.57	1.60	1.70	1.74	2.07	2.10	2.23	2.28
40	11	1.57	1.61	1.71	1.76	2.08	2.12	2.24	2.30
40	12	1.58	1.62	1.71	1.78	2.08	2.13	2.25	2.32
40	13	1.58	1.64	1.72	1.80	2.09	2.15	2.27	2.35
40	14	1.59	1.65	1.73	1.82	2.10	2.17	2.28	2.38
40	15	1.59	1.67	1.73	1.85	2.10	2.19	2.29	2.41
40	16	1.59	1.69	1.74	1.87	2.11	2.21	2.30	2.44
40	17	1.60	1.71	1.75	1.91	2.12	2.24	2.32	2.48
40	18	1.60	1.73	1.76	1.94	2.13	2.27	2.33	2.53
40	19	1.61	1.76	1.76	1.98	2.14	2.30	2.35	2.58
40	20	1.61	1.79	1.77	2.03	2.15	2.34	2.37	2.64
40	21	1.62	1.83	1.78	2.08	2.16	2.38	2.38	2.70

Table 14.3.1 *(continued)*

(γ, β)		(0.90, 0.90)		(0.95, 0.90)		(0.90, 0.95)		(0.95, 0.95)	
n	s	r_1	r_2	r_1	r_2	r_1	r_2	r_1	r_2
40	22	1.62	1.86	1.79	2.14	2.17	2.43	2.40	2.77
40	23	1.63	1.91	1.80	2.22	2.18	2.48	2.43	2.86
40	24	1.63	1.96	1.81	2.30	2.20	2.54	2.45	2.95
40	25	1.64	2.02	1.83	2.39	2.21	2.61	2.48	3.07
40	26	1.65	2.09	1.84	2.51	2.23	2.69	2.51	3.21
40	27	1.65	2.17	1.85	2.65	2.24	2.79	2.54	3.37
40	28	1.66	2.27	1.87	2.82	2.26	2.91	2.58	3.57
40	29	1.66	2.40	1.88	3.04	2.28	3.05	2.63	3.83
40	30	1.67	2.55	1.90	3.32	2.31	3.23	2.68	4.16
40	31	1.68	2.76	1.92	3.71	2.34	3.47	2.75	4.62
40	32	1.68	3.03	1.94	4.28	2.37	3.79	2.85	5.27
40	33	1.69	3.43	1.97	5.16	2.41	4.24	2.98	6.30
40	34	1.69	4.04	2.00	6.77	2.47	4.96	3.20	8.17

Then, from Eq. (14.2.10) we have

$$\gamma = \Pr\left\{Z \geq \frac{F^{-1}(\beta) - r_2}{(\sigma_{11} + r_2^2\sigma_{22} + 2r_2\sigma_{12})^{1/2}}\right\}. \tag{14.3.8}$$

From (14.3.8) we obtain

$$\frac{F^{-1}(\beta) - r_2}{(\sigma_{11} + r_2^2\sigma_{22} + 2r_2\sigma_{12})^{1/2}} = z_{(\gamma)}, \tag{14.3.9}$$

where $z_{(\gamma)}$ is the upper γ percentage point of the standard normal distribution. Equation (14.3.9) may be rewritten by the quadratic equation

$$r_2^2(1 - z_{(\gamma)}^2\sigma_{22}) - r_2(2F^{-1}(\beta) + 2z_{(\gamma)}^2\sigma_{12}) + \{F^{-1}(\beta)\}^2 - z_{(\gamma)}^2\sigma_{11} = 0. \tag{14.3.10}$$

From (14.3.10), we get the large sample approximation for r_2 to be

$$r_2 = \frac{t_2 + (t_2^2 - uv)^{1/2}}{v}, \tag{14.3.11}$$

where

$$t_2 = F^{-1}(\beta) + z_{(\gamma)}^2\sigma_{12} = \frac{\sqrt{3}}{\pi}\ln\left(\frac{\beta}{1 - \beta}\right) + z_{(\gamma)}^2\sigma_{12}.$$

We should mention here that the formula for r_2 in (14.3.11) has been given by Hall (1975) wrongly with a positive sign inside the square root. By using the formula in (14.3.11), the large-sample approximate values of r_2 were computed for $n = 30, 35$, and 40, $s = 0(1)n - 6$, $\gamma = 0.90, 0.95$, and $\beta = 0.90, 0.95$. These values are presented in Table 14.3.1. As mentioned earlier, the values of σ_{11}, σ_{12}, and σ_{22} required in the above computation were taken from the tables of Balakrishnan (1991). We observe that the formula in (14.3.11) yields very close approximation to r_2 whenever s is small or moderately large; but if s is large, then the normal approximation in (14.3.11) does not give satisfactory results. In this case, however, one may use the integral form in Eq. (14.2.10) directly to solve for r_2.

It should be mentioned here that Hall (1975), when presenting the approximate values of r_1 and r_2 (in Table 3), has wrongly switched γ and δ; (γ, δ) in Table 3 should be (δ, γ). In addition, the simulated values of r_1 and r_2 for the case $n = 7$, $s = 3$, $\delta = 0.90$, and $\gamma = 0.95$ have been wrongly reported in Table 3 as 3.04 and 3.47, respectively. They should actually be 3.44 and 4.63, respectively, as may be verified from Table 2 of Hall (1975). Finally, we would also like to point out an error in one of the entries in Table 2. The value of r_2 for the case $n = 15$, $s = 4$, $\delta = 0.90$, and $\gamma = 0.90$ has been wrongly reported as 2.95 when it actually is 1.95.

14.4 TWO-SIDED TOLERANCE LIMITS

Let $X_{s_1+1:n} \le X_{s_1+2:n} \le \cdots \le X_{n-s_2:n}$ be a Type II doubly censored sample available from the logistic $L(\mu, \sigma^2)$ population with pdf $f(x; \mu, \sigma)$ and cdf $F(x; \mu, \sigma)$ as given in Eqs. (1.1) and (1.2), respectively. Let μ^* and σ^* be the best linear unbiased estimators of μ and σ based on the above doubly censored sample which may be computed from the tables of Gupta, Qureishi, and Shah (1967) and Balakrishnan (1991).

Then $(L(\underset{\sim}{X}), U(\underset{\sim}{X}))$ are said to be two-sided γ tolerance limits for proportion β if

$$\Pr\{1 - F(L(\underset{\sim}{X}); \mu, \sigma) \ge \beta_1 \text{ and } F(U(\underset{\sim}{X}); \mu, \sigma) \ge \beta_2\} = \gamma, \qquad (14.4.1)$$

where $\beta = \beta_1 + \beta_2 - 1$ with β_1 and β_2 being the proportions controlled by the lower and upper tolerance limits $L(\underset{\sim}{X})$ and $U(\underset{\sim}{X})$, respectively. As before, we shall take $L(\underset{\sim}{X}) = \mu^* - r_1\sigma^*$ and $U(\underset{\sim}{X}) = \mu^* + r_2\sigma^*$. If we now choose the proportions β_1 and β_2 to be equal, i.e., $\beta_1 = \beta_2 = (1 + \beta)/2$, then we may take $r_1 = r_2 = r$ due to the symmetry of the logistic distribution in which case we get symmetric tolerance limits. Equation

(14.4.1) can be rewritten in this case as

$$\gamma = \Pr\left[\Pr\left\{X \geq \sum_{i=s_1+1}^{n-s_2} (a_i - rb_i)X_{i:n}\right\} \geq \frac{1+\beta}{2} \quad \text{and} \right.$$

$$\left. \Pr\left\{X \leq \sum_{i=s_1+1}^{n-s_2} (a_i + rb_i)X_{i:n}\right\} \geq \frac{1+\beta}{2}\right]$$

$$= \Pr\left[\Pr\left\{Y \geq \sum_{i=s_1+1}^{n-s_2} (a_i - rb_i)Y_{i:n}\right\} \geq \frac{1+\beta}{2} \quad \text{and} \right.$$

$$\left. \Pr\left\{Y \leq \sum_{i=s_1+1}^{n-s_2} (a_i + rb_i)Y_{i:n}\right\} \geq \frac{1+\beta}{2}\right]$$

$$= \Pr\left[\sum_{i=s_1+1}^{n-s_2} (a_i - rb_i)Y_{i:n} \leq F^{-1}\left(\frac{1-\beta}{2}\right) \quad \text{and} \right.$$

$$\left. \sum_{i=s_1+1}^{n-s_2} (a_i + rb_i)Y_{i:n} \geq F^{-1}\left(\frac{1+\beta}{2}\right)\right]$$

$$= \Pr\left[\sum_{i=s_1+1}^{n-s_2} a_i Y_{i:n} - r\sum_{i=s_1+1}^{n-s_2} b_i Y_{i:n} \leq F^{-1}\left(\frac{1-\beta}{2}\right) \quad \text{and} \right.$$

$$\left. -\sum_{i=s_1+1}^{n-s_2} a_i Y_{i:n} - r\sum_{i=s_1+1}^{n-s_2} b_i Y_{i:n} \leq F^{-1}\left(\frac{1-\beta}{2}\right)\right]$$

$$= \frac{n!}{s_1!\, s_2!} \int\int_{\cdots A \cdots}\int \{1 - F(y_{n-s_2:n})\}^{s_2}\{F(y_{s_1+1:n})\}^{s_1}$$

$$\times f(y_{s_1+1:n}) \cdots f(y_{n-s_2:n})\, dy_{s_1+1:n} \cdots dy_{n-s_2:n}, \tag{14.4.2}$$

where $Y = (X - \mu)/\sigma$, $Y_{i:n} = (X_{i:n} - \mu)/\sigma$, $f(y)$ and $F(y)$ are as defined in Eqs. (1.3) and (1.4), respectively, and A is the region in an $(n - s_1 - s_2)$-dimensional space defined by

$$A = \left\{(y_{s_1+1}, \ldots, y_{n-s_2}) : y_{s_1+1} < y_{s_1+2} < \cdots < y_{n-s_2} \quad \text{and} \right.$$

$$\sum_{i=s_1+1}^{n-s_2} (a_i - rb_i)y_i \leq \frac{\sqrt{3}}{\pi} \ln\left(\frac{1-\beta}{1+\beta}\right) \quad \text{and}$$

$$\left. -\sum_{i=s_1+1}^{n-s_2} (a_i + rb_i)y_i \leq \frac{\sqrt{3}}{\pi} \ln\left(\frac{1-\beta}{1+\beta}\right)\right\}. \tag{14.4.3}$$

One may employ numerical procedures to determine the exact values of r which satisfy Eq. (14.4.3) for large values of $s_1 + s_2$ (or small values of

$n - s_1 - s_2$). The values of r may also be determined by Monte Carlo simulations by rewriting Eq. (14.4.2) as

$$\gamma = \Pr\left[\frac{\sum_{i=s_1+1}^{n-s_2} a_i Y_{i:n} - F^{-1}\left(\frac{1-\beta}{2}\right)}{\sum_{i=s_1+1}^{n-s_2} b_i Y_{i:n}} \le r \quad \text{and} \right.$$

$$\left. \frac{-\sum_{i=s_1+1}^{n-s_2} a_i Y_{i:n} - F^{-1}\left(\frac{1-\beta}{2}\right)}{\sum_{i=s_1+1}^{n-s_2} b_i Y_{i:n}} \le r \right]. \tag{14.4.4}$$

The values of r were determined from Eq. (14.4.4) by Monte Carlo process (based on 10,000 runs) for $n = 5, 7, 10$, and 15 over various choices of s_1 and s_2, $\gamma = 0.90, 0.95$, and $\beta = 0.90, 0.95$. These values are presented in Table 14.4.1.

Table 14.4.1 Values of r for Two-Sided Tolerance Limits Based on BLUEs

			(γ, β)			
n	s_1	s_2	(0.90, 0.90)	(0.95, 0.90)	(0.90, 0.95)	(0.95, 0.95)
5	0	0	3.80	4.60	4.70	5.70
5	0	1	4.53	5.82	5.45	7.02
5	0	2	6.47	9.86	7.72	11.65
5	0	3	18.40	36.00	21.70	42.50
5	1	1	6.15	8.99	7.40	10.80
5	1	2	16.00	34.00	19.20	40.00
7	0	0	3.24	3.78	3.92	4.56
7	0	1	3.47	4.18	4.18	5.03
7	0	2	3.86	4.80	4.65	5.78
7	0	3	4.68	6.18	5.61	7.47
7	0	4	6.86	10.35	8.17	12.30
7	0	5	20.50	40.05	23.68	47.80
7	1	1	3.82	4.82	4.64	5.80
7	1	2	4.58	6.00	5.52	7.20
7	1	3	6.44	9.50	7.75	11.40
7	1	4	17.50	36.80	20.90	43.90
7	2	2	6.30	9.30	7.63	11.20
7	2	3	16.00	34.30	19.50	41.40

Table 14.4.1 (*continued*)

n	s_1	s_2	(γ, β) (0.90, 0.90)	(0.95, 0.90)	(0.90, 0.95)	(0.95, 0.95)
10	0	0	2.81	3.20	3.40	3.88
10	0	1	2.89	3.34	3.49	4.02
10	0	2	3.01	3.50	3.65	4.24
10	0	3	3.22	3.78	3.89	4.59
10	0	4	3.49	4.32	4.22	5.20
10	0	5	4.08	5.20	4.90	6.25
10	0	6	5.01	6.80	6.01	8.07
10	0	7	7.40	11.32	8.74	13.30
10	0	8	22.00	46.90	25.50	53.52
10	1	1	3.02	3.50	3.66	4.23
10	1	2	3.17	3.71	3.82	4.48
10	1	3	3.42	4.12	4.14	5.00
10	1	4	3.86	4.95	4.67	5.98
10	1	5	4.74	6.36	5.72	7.64
10	1	6	6.95	10.60	8.28	12.55
10	1	7	19.30	39.10	22.50	45.00
10	2	2	3.38	4.10	4.08	4.96
10	2	3	3.78	4.75	4.58	5.75
10	2	4	4.66	6.37	5.65	7.71
10	2	5	6.68	9.77	8.03	11.73
10	2	6	19.00	39.50	22.70	47.00
10	3	3	4.54	5.95	5.50	7.20
10	3	4	6.43	9.60	7.75	11.60
10	3	5	16.40	33.50	19.80	40.40
10	4	4	16.40	34.40	19.65	40.50
15	0	0	2.50	2.75	3.25	3.52
15	0	1	2.55	2.80	3.30	3.55
15	0	2	2.60	2.90	3.34	3.68
15	0	3	2.65	2.94	3.39	3.78
15	0	4	2.71	3.05	3.46	3.88
15	0	5	2.79	3.20	3.60	4.10
15	0	6	2.98	3.37	3.78	4.30
15	0	7	3.08	3.53	3.95	4.55
15	0	8	3.28	4.00	4.20	5.00
15	0	9	3.65	4.69	4.62	5.75
15	0	10	4.30	5.52	5.48	6.90
15	0	11	5.35	7.52	6.65	9.30
15	0	12	8.30	13.00	10.00	15.00
15	0	13	24.30	49.00	29.00	59.00
15	1	1	2.59	2.87	3.32	3.68
15	1	2	2.65	2.95	3.38	3.80

(*continued*)

Table 14.4.1 (*continued*)

n	s_1	s_2	(0.90, 0.90)	(0.95, 0.90)	(0.90, 0.95)	(0.95, 0.95)
15	1	3	2.71	3.03	3.47	3.88
15	1	4	2.78	3.14	3.56	4.02
15	1	5	2.89	3.30	3.70	4.26
15	1	6	3.03	3.55	3.90	4.52
15	1	7	3.25	3.85	4.15	4.95
15	1	8	3.60	4.40	4.60	5.63
15	1	9	4.13	5.40	5.25	6.80
15	1	10	5.23	7.00	6.50	8.50
15	1	11	7.50	11.50	9.70	14.80
15	1	12	21.50	43.00	27.00	52.00
15	2	2	2.72	3.07	3.47	3.92
15	2	3	2.80	3.17	3.58	4.05
15	2	4	2.89	3.30	3.72	4.22
15	2	5	2.96	3.45	3.82	4.48
15	2	6	3.15	3.70	4.05	4.75
15	2	7	3.50	4.35	4.50	5.45
15	2	8	4.00	5.00	5.10	6.45
15	2	9	5.10	6.80	6.50	8.50
15	2	10	7.40	10.80	9.40	14.20
15	2	11	21.00	45.00	25.00	55.00
15	3	3	2.90	3.30	3.70	4.25
15	3	4	3.00	3.50	3.90	4.50
15	3	5	3.20	3.75	4.15	4.90
15	3	6	3.43	4.15	4.40	5.32
15	3	7	3.85	4.90	4.95	6.40
15	3	8	4.70	6.00	6.00	8.10
15	3	9	7.20	10.50	9.50	13.50
15	3	10	20.00	36.00	25.00	47.00
15	4	4	3.20	3.80	4.15	4.90
15	4	5	3.48	4.10	4.50	5.50
15	4	6	3.85	4.70	4.95	6.20
15	4	7	4.60	6.20	5.90	8.00
15	4	8	6.40	9.30	8.20	12.50
15	4	9	18.00	37.00	23.00	44.00
15	5	5	3.90	4.90	5.00	6.30
15	5	6	4.70	6.10	6.00	7.80
15	5	7	6.50	9.20	8.00	13.00
15	5	8	16.00	33.00	20.00	40.00
15	6	6	6.30	9.50	8.00	12.00
15	6	7	16.50	35.00	20.00	42.00

(γ, β) applies to the four value columns.

14.5 ACCEPTANCE SAMPLING PLANS

In this section, we consider the acceptance sampling plans based on Type II censored samples from the logistic $L(\mu, \sigma^2)$ distribution, and show that the factors involved in the acceptance sampling plans are exactly the same as the factors involved in the tolerance limits.

To see this, let us first start with the acceptance sampling plan with a lower specification limit L which would control the proportion below L so that it is no more than a specified proportion $1 - \beta$. The lot is accepted if $\mu^* - r_1^* \sigma^* \geq L$, where the factor r_1^* is chosen in such a way that the consumer's risk, for example, is at a specified level. So, we require the constant r_1^* such that

$$\Pr\{\mu^* - r_1^* \sigma^* \geq L\} = 1 - \gamma, \tag{14.5.1}$$

where $1 - \gamma$ is usually taken to be 0.05 or 0.10 for sampling plans as mentioned by Owen (1967). That is,

$$\Pr\left\{\sum_{i=1}^{n-s} a_i Y_{i:n} - r_1^* \sum_{i=1}^{n-s} b_i Y_{i:n} \geq F^{-1}(1 - \beta)\right\} = 1 - \gamma,$$

which may be rewritten as

$$\Pr\left\{\frac{\sum_{i=1}^{n-s} a_i Y_{i:n} - F^{-1}(1 - \beta)}{\sum_{i=1}^{n-s} b_i Y_{i:n}} \leq r_1^*\right\} = \gamma. \tag{14.5.2}$$

We note here that this is precisely the same as Eq. (14.2.6), the equation for the factor r_1 of the lower γ tolerance limit for proportion β. Hence we simply have $r_1^* = r_1$ and, therefore, the factor r_1^* in acceptance sampling plan with a lower specification limit may be taken from Table 14.2.1.

Next, let us consider an acceptance sampling plan with an upper specification limit U which would control the proportion above U so that it is no more than a specified proportion $1 - \beta$. The lot is accepted if $\mu^* + r_2^* \sigma^* \leq U$, where the factor r_2^* is chosen in such a way that the consumer's risk is at a specified level. So, we need to find the constant r_2^* such that

$$\Pr\{\mu^* + r_2^* \sigma^* \leq U\} = 1 - \gamma. \tag{14.5.3}$$

That is,

$$\Pr\left\{\sum_{i=1}^{n-s} a_i Y_{i:n} + r_2^* \sum_{i=1}^{n-s} b_i Y_{i:n} \leq F^{-1}(\beta)\right\} = 1 - \gamma,$$

which may be rewritten as

$$\Pr\left\{\frac{-\sum_{i=1}^{n-s} a_i Y_{i:n} + F^{-1}(\beta)}{\sum_{i=1}^{n-s} b_i Y_{i:n}} \le r_2^*\right\} = \gamma. \tag{14.5.4}$$

We note that this is precisely the same as Eq. (14.2.12), the equation for the factor r_2 of the upper γ tolerance limit for proportion β. Hence, we have $r_2^* = r_2$, and, therefore, the factor r_2^* in acceptance sampling plan with an upper specification limit may be taken from Table 14.2.1.

Finally, let us consider a two-sided acceptance sampling plan with a lower specification limit L and an upper specification limit U which would control the proportions below L and above U so that they are no more than some specified proportions $1 - \beta_1$ and $1 - \beta_2$, respectively. The lot is accepted if $\mu^* - r_1^*\sigma^* \ge L$ and $\mu^* + r_2^*\sigma^* \le U$, where the factors r_1^* and r_2^* are chosen in such a way that the consumer's risk is at a specified level. So, we need to find the constants r_1^* and r_2^* such that

$$\Pr\{\mu^* - r_1^*\sigma^* \ge L \quad \text{and} \quad \mu^* + r_2^*\sigma^* \le U\} = 1 - \gamma. \tag{14.5.5}$$

As Owen (1967) pointed out, the situation which calls for the largest constants r_1^* and r_2^* is the one where one tail has no defectives, i.e., $1 - \beta_1 = 0$, and the other tail has its full defectiveness. This corresponds to a one-sided sampling plan. Hence, to control $1 - \beta_1$ in the lower tail, look up r_1^* from the one-sided sampling plan table (i.e., factor r_1 of the lower tolerance limit from Table 14.2.1) and simultaneously, to control $1 - \beta_2$ in the upper tail, look up for r_2^* (i.e. factor r_2 of the upper tolerance limit from Table 14.2.1) Accept the lot if $\mu^* - r_1^*\sigma^* \ge L$ and $\mu^* + r_2^*\sigma^* \le U$, and reject the lot otherwise. This will guarantee the consumer with probability at least γ that no more than the proportion defictiveness $1 - \beta_1$ will be in the lower tail and no more than the proportion defectiveness $1 - \beta_2$ will be in the upper tail.

Therefore, if the consumer wants to accept lots with $100(1 - \beta_1)\%$ or more defectiveness in the lower tail and $100(1 - \beta_2)\%$ or more defectiveness in the upper tail $100(1 - \gamma)\%$ of the time or less, the consumer should choose r_1^* from a one-sided plan which controls $100(1 - \beta_1)\%$ defectiveness in the lower tail and should choose r_2^* from a one-sided plan which controls $100(1 - \beta_2)\%$ defectiveness in the upper tail.

14.6 ILLUSTRATIVE EXAMPLES

In this section, we shall consider two examples and illustrate the determination of tolerance limits that have been developed in this chapter.

Example 14.6.1 Bliss (1967) has given the average number of hours of sleep of 10 patients suffering from insomnia. The average number of hours of sleep for the patients after taking 0.6 mg of L-hyoscine were recorded. For the purpose of illustration, let us consider the following Type II right-censored sample where the largest observation is censored:

2.5, 3.8, 4.4, 5.6, 5.7, 5.8, 6.1, 6.3, 7.6, —

Then, by using the tables of Gupta, Qureishi, and Shah (1967), we obtain the best linear unbiased estimates of μ and σ to be

$$\begin{aligned}\mu^* &= 0.0230(2.5) + 0.0713(3.8) + 0.1096(4.4) + 0.1361(5.6)\\ &\quad + 0.1501(5.7) + 0.1512(5.8) + 0.1393(6.1)\\ &\quad + 0.1145(6.3) + 0.1048(7.6)\\ &= 5.67\end{aligned}$$

and

$$\begin{aligned}\sigma^* &= -0.1755(2.5) - 0.1743(3.8) - 0.1404(4.4) - 0.0893(5.6)\\ &\quad - 0.0292(5.7) + 0.0334(5.8) + 0.0928(6.1)\\ &\quad + 0.1425(6.3) + 0.3400(7.6)\\ &= 1.85\end{aligned}$$

Suppose we are interested in the 95% lower tolerance limit with proportion $\beta = 0.90$. Then, from Table 14.2.1 we determine the desired tolerance limit to be

$$L = \mu^* - r_1\sigma^* = 5.67 - 2.35(1.85) = 1.32 \text{ hours.}$$

Similarly, from Table 14.2.1 we determine the 95% upper tolerance limit with proportion 0.90 to be

$$U = \mu^* + r_2\sigma^* = 5.67 + 2.38(1.85) = 10.07 \text{ hours.}$$

Instead, suppose we are interested in 90% two-sided tolerance limits with the proportions controlled by the lower and upper tolerance limits being 0.95 each. In this case, we have $\gamma = 0.90$, $\beta_1 = \beta_2 = 0.95$, and $\beta = \beta_1 + \beta_2 - 1 = 0.90$; from Table 14.4.1, we then get the simultaneous lower and upper tolerance limits to be

$$L = \mu^* - r\sigma^* = 5.67 - 2.89(1.85) = 0.32 \text{ hours}$$

and

$$U = \mu^* + r\sigma^* = 5.67 + 2.89(1.85) = 11.02 \text{ hours,}$$

respectively.

Example 14.6.2 Davis (1952) has presented lifetimes in hours of 417 forty-watt incandescent lamps taken from 42 weekly forced life-test samples. Suppose 50% censoring had occurred on the right in the first 20 observations and the following Type II right-censored sample is available:

785, 855, 905, 918, 919, 920, 929, 936, 948, 950.

This example was considered earlier in Chapter 5.

Based on the above censored sample, we compute the best linear unbiased estimates of μ and σ by using the tables of Gupta, Qureishi, and Shah (1967) and Balakrishnan (1991) to be

$$\begin{aligned}\mu^* &= -0.04336(785) - 0.02968(855) - 0.01300(905) + 0.00454(918) \\ &\quad + 0.02200(919) + 0.03882(920) + 0.05460(929) + 0.06901(936) \\ &\quad + 0.08176(948) + 0.81530(950) \\ &= 956.28\end{aligned}$$

and

$$\begin{aligned}\sigma^* &= -0.16913(785) - 0.17466(855) - 0.16574(905) - 0.14860(918) \\ &\quad -0.12598(919) - 0.09961(920) - 0.07070(929) - 0.04027(936) \\ &\quad -0.00916(948) + 1.00385(950) \\ &= 65.67\end{aligned}$$

Suppose we are interested in the 95% lower tolerance limit with the proportion controlled by it being 0.90. Then, from Table 14.2.1 we determine the desired tolerance limit to be

$$L = \mu^* - r_1\sigma^* = 956.28 - 2.05(65.67) = 821.66 \text{ hours.}$$

This may be compared to the value of 814.66 calculated by Bain et al. (1991) in Chapter 5 for the 95% lower tolerance limit with proportion 0.90 based on the maximum likelihood estimates.

Similarly, from Table 14.2.1 we compute the 95% upper tolerance limit with the proportion controlled by it being 0.90 to be

$$U = \mu^* + r_2\sigma^* = 956.28 + 2.54(65.67) = 1123.08 \text{ hours.}$$

REFERENCES

Bain, L. J., Balakrishnan, N., Eastman, J., Engelhardt, M., and Antle, C. (1991). Reliability estimation based on MLE's for complete and censored samples, *Chapter 5*.

Balakrishnan, N. (1991). Best linear unbiased estimates of the location and scale parameters of logistic distribution for complete and censored samples of sizes 2(1) 25(5) 40, Submitted for publication.

Balakrishnan, N. and Cohen, A. C. (1990). *Order Statistics and Inference: Estimation Methods*, Academic Press, Boston.

Bliss, C. I. (1967). *Statistics in Biology*, Vol. 1, McGraw-Hill, New York.

Davis, D. J. (1952). An analysis of some failure data, J. Amer. Statist. Assoc., *47*, 113–150.

Gupta, S. S., Qureishi, A. S., and Shah, B. K. (1967). Best linear unbiased estimators of the parameters of the logistic distribution using order statistics, *Technometrics*, *9*, 43–56.

Hall, I. J. (1975). One-sided tolerance limits for a logistic distribution based on censored samples, *Biometrics*, *31*, 873–879.

Hall, I. J. and Sampson, C. B. (1974). One-sided tolerance limits for a normal population based on censored samples, *J. Statist. Comput. Simul.*, *2*, 317–324.

Lurie, D. and Hartley, H. O. (1972). Machine-generation of order statistics for Monte Carlo computations, *Amer. Statist.*, *26*, 26–27.

Mann, N. R. and Fertig K. W. (1973). Tables for obtaining Weibull confidence bounds and tolerance bounds based on best linear invariant estimates of parameters of the extreme-value distribution, *Technometrics*, *15*, 87–102.

Owen, D. B. (1967). Variables sampling plans based on the normal distribution, *Technometrics*, *9*, 417–423.

15

Logistic Stochastic Growth Models and Applications

Wai-Yuan Tan
Memphis State University, Memphis, Tennessee

15.1 INTRODUCTION

In the studies of growth of biological populations such as bacteria or cancer cell populations, usually exponential growth has been used. This growth law specifies that the changing rate of the population size N at any fixed time is proportional to the population size N at that time. When N is small, such a mechanism appears to fit the observed size quite well in most of the cases. However, when N is large, the above law appeared to be inadequate to describe the actual growth of the population due to limitations of food and space (Pielou, 1977; Eisen, 1979). To take into account the limitations of food and space, Verhulst (1838) introduced the logistic growth law, which is specified by the following ordinary differential equation:

$$\frac{dN(t)}{dt} = a(t)N(t) - b(t)N^2(t), \qquad t \geq 0, \tag{15.1.1}$$

where $N(t)$ satisfies $0 \leq N(t) \leq M$, with M being a positive integer representing the maximum population size.

In Eq. (15.1.1), M is usually very large and $b(t)$ is proportional to M^{-1}. It follows that if $N(t)$ is small, Eq. (15.1.1) is well approximated by $(d/dt)N(t) \cong a(t)N(t)$, which is the exponential growth law. To interpret Eq. (15.1.1) intuitively, let $\lambda(N, t)$ and $\mu(N, t)$ be the birthrate and death rate of any individual at time t with population size $N(t) = N$ at time t. If $N(t) = N$ is small, and if one assumes $\lambda(N, t) \cong \lambda(t)$ and $\mu(N, t) \cong \mu(t)$, then one has exponential growth law for the population growth. If $N(t) = N$ is large, then competitions for food and space occur so that $\lambda(N, t)$ decreases with $N(t)$ as $N(t)$ increases. Suppose that the effects of competition are proportional to the population size $N(t)$. Since $\lambda(M, t) = 0$, where M is the maximum population size, one may then assume that $\lambda(N, t) \cong \lambda(t)\{1 - N(t)/M\}$. If one assumes $\mu(N, t) = \mu(t)\{1 + \beta(t)N(t)/M\}$, then one has a logistic growth law with $a(t) = \lambda(t) - \mu(t)$ and $b(t) = [\lambda(t) + \beta(t)\mu(t)]/M$. Note that in Eq. (15.1.1), $N(t)$, $a(t)$, and $b(t)$ are deterministic differentiable functions of t. Hence, Eq. (15.1.1) is referred to as a deterministic logistic growth law for the population size.

Let t_0 be the starting time and suppose $N(t_0) = N_0$. Then Eq. (15.1.1) can easily be solved to give

$$N(t) = N_0 \exp\left[\int_{t_0}^{t} a(x)\, dx\right]\left\{1 + N_0 \int_{t_0}^{t} b(x) \exp\left[\int_{t_0}^{x} a(y)\, dy\right] dx\right\}^{-1}, \tag{15.1.2}$$

If $a(t) = a$ and $b(t) = b$, then Eq. (15.1.2) reduces to

$$N(t) = N_0 \exp[a(t - t_0)]\left\{1 - \frac{N_0 b}{a} + \frac{N_0 b}{a} \exp[a(t - t_0)]\right\}^{-1} \tag{15.1.3}$$

Equations (15.1.2) and (15.1.3) are the well-known logistic growth equations. These equations have been used widely by biologists and mathematicians to describe the growth of many biological populations, including humans (Pearl and Reed, 1920), fish (Jensen, 1975), animals (Miller and Botkin, 1974), bacteria and cells (Tan, 1983), and tumor cells (Eisen, 1979), including breast tumors (Moolgavkar, 1986).

Note that if one assumes $\lambda(N, t) = \lambda(t)\{1 - N(t)/M\}$ and $\mu(N, t) = \mu(t)\{1 - N(t)/M\}$, then $a(t) = \lambda(t) - \mu(t) = \epsilon(t)$ and $b(t) = \epsilon(t)/M$. In this case, Eq. (15.1.2) reduces to

$$N(t) = N_0 \exp\left[\int_{t_0}^{t} \epsilon(x)\, dx\right]\left\{1 - \frac{N_0}{M} + \frac{N_0}{M} \exp\left[\int_{t_0}^{t} \epsilon(x)\, dx\right]\right\}^{-1} \tag{15.1.4}$$

If $a(t) = \epsilon$ and $b(t) = \epsilon/M$, then Eq. (15.1.1) is a homogeneous two-parameter logistic growth equation (see Jensen, 1975). In these cases, $N(t) = M$ is a stable point at which the population neither grows nor dies. This is the situation for women's breast tumors.

15.2 SOME STOCHASTIC LOGISTIC GROWTH MODELS

While the above deterministic logistic growth laws have been widely used by biologists to describe population growth of many biological populations, there are many problems which the above models appear to be inadequate to solve (see Norden, 1982, 1984). Specifically, the above deterministic models do not allow for the extinction of the population; nor do the models provide any avenue to assess effects of random changes of many factors on the population growth. To take into account the random nature of the biological populations and to provide an avenue to accommodate for the possibility of extinction of the population, this chapter develops a stochastic model for the logistic growth. Such an extension is further justified by the following comments.

1. In the real world, environments are changing with time and many factors are subjected to random disturbances. Hence, the $N(t)$'s are, in essence, stochastic processes rather than deterministic functions.
2. One may obtain more information by working with stochastic models than with deterministic models. For example, by working with stochastic models, one may derive formulas for the variances of $N(t)$ and for the probability of extinction of the population; these are certainly not possible by the deterministic approach.
3. Under some conditions, deterministic models are equivalent to working with the expected values of the random processes. In this sense, then, the deterministic models are special cases of the stochastic models.

It is obvious that the stochastic models are more complicated than deterministic models. As a consequence, the mathematics for stochastic models are also more difficult.

To define a stochastic logistic law, let $X(t)$, $t \geq t_0$, be a Markov stochastic process with state space $S = \{0, 1, \ldots, M\}$, where $M > 0$ is a large positive integer representing the maximum population size. The Markov condition implies that for any $m > 0$, for any $t_0 < t_1 < \cdots < t_m < t$ and for any $0 \leq i_0, i_1, \ldots, i_m, i \leq M$,

$$\Pr\{X(t) = i \mid X(t_0) = i_0, X(t_i) = i_1, \ldots, X(t_m) = i_m\}$$
$$= \Pr\{X(t) = i \mid X(t_m) = i_m\}, \qquad (15.2.1)$$

Definition 15.2.1 Let $\lambda_i(t) = i\lambda(t)[1 - i/M]$ and $\mu_i(t) = i\mu(t)[1 + \beta(t)i/M]$, where $\lambda(t) > 0$, $\mu(t) \geq 0$, and $\beta(t)$ are differentiable functions of t.

Then the Markov process $X(t)$ given above is defined as a logistic birth-death process with birthrate $\lambda_i(t)$ and death rate $\mu_i(t)$ if and only if the following conditions are satisfied:

a. For $0 \leq j < j + 1 \leq M$,

$$\Pr\{X(t + \Delta t) = j + 1 | X(t) = j\} = \lambda_j(t)\Delta t + o(\Delta t),$$

$$\text{where } \lim_{\Delta t \to 0} \frac{o(\Delta t)}{\Delta t} = 0. \quad (15.2.2a)$$

b. For $0 \leq j - 1 < j \leq M$,

$$\Pr\{X(t + \Delta t) = j - 1 | X(t) = j\} = \mu_j(t)\Delta t + o(\Delta t). \quad (15.2.2b)$$

c. For $0 \leq k, j \leq M$ and $|k - j| \geq 2$,

$$\Pr\{X(t + \Delta t) = k | X(t) = j\} = o(\Delta t). \quad (15.2.2c)$$

The above definition for stochastic logistic growth has been used by Pielou (1977), Norden (1982, 1984), and Tan and Piantadosi (1988). In Tan and Piantadosi (1988), it is assumed that $\beta(t) = -1$. This assumption is supported by the fact that in many tumor cell populations and in women's breasts, the population is quite stable when it reaches the maximum size. For the homogeneous model, Jensen (1975) has also advocated the use of this assumption in many cases. For simplicity, in what follows I thus assume $\beta(t) = -1$ so that the above process may be applied to the growth of tumor cells in cancer research and women's breasts. In the homogeneous case, this is then a stochastic analogue of the deterministic two-parameter logistic model considered by Jensen (1975).

Given $\beta(t) = -1$, $\lambda(t) > 0$, and $\mu(t) > 0$, then the states 0 and M are absorbing states while the states j for $j = 1, \ldots, M - 1$ are transient states. Thus, starting at t_0 with any state j for $1 \leq j \leq M - 1$, with probability 1 the process will eventually reach 0 or M. The state 0 corresponds to the extinction of the population, while the state M corresponds to the maximum size of the population. In cancer research, the state 0 corresponds to the state of regression of cancer tumors.

15.2.1 Transition Probabilities and Absolute Probabilities

For $0 \leq i_0, i, j \leq M$ and for $t \geq s \geq t_0$, put

$$P_{ij}(s, t) = \Pr\{X(t) = j | X(s) = i\},$$

$$P_j(t) = \Pr\{X(t) = j | X(t_0) = i_0\},$$

$$P(s, t) = [P_{ij}(s, t)], \text{ and } \mathbf{P}(t) = [P_0(t), P_1(t), \ldots, P_M(t)]^T,$$

(The superscript T denotes transposition of the vector or matrix.)

In the above, $P(s, t)$ is an $(M + 1) \times (M + 1)$ matrix whose $(i + 1, j + 1)$th element is $P_{ij}(s, t)$, $i, j = 0, 1, \ldots, M$, and $\mathbf{P}(t)$ is $(1 + M) \times 1$ column whose $(i + 1)$st element is $P_i(t)$, $i = 0, 1, \ldots, M$; $P_{ij}(s, t)$ is the probability of transition from state i at time s to the state j at time t, and $P_i(t)$ is the probability that $X(t) = i$ given $X(t_0) = i_0$. If $\lambda(t) = \lambda$ and $\mu(t) = \mu$ are independent of t (i.e., time homogeneous), then $P_{ij}(s, t) = P_{ij}(t - s)$ depends on s and t through $t - s$ so that $P(s, t) = P(t - s)$.

To obtain $P(s, t)$ and $\mathbf{P}(t)$, observe that from the Chapman-Kolmogorov equation,

$$P(s, t) = P(s, r)P(r, t) \tag{15.2.3}$$

for any $s \le r \le t$.

It follows that for any $s \le r_1 < \cdots < r_m \le t$, and for any $m \ge 1$,

$$P(s, t) = P(s, r_1)\left\{\prod_{j=1}^{m-1} P(r_j, r_{j+1})\right\} P(r_m, t).$$

If $\lambda(t) = \lambda$ and $\mu(t) = \mu$ and if $r_1 - s = t - r_m = t_{j+1} - t_j = (t - s)/(m + 1) = \Delta$, $j = 1, \ldots, m - 1$, then

$$P(s, t) = P(t - s) = \{P(\Delta)\}^{m+1}.$$

For obtaining $\mathbf{P}(t)$, denoted by

$$\exp(-tC) = \sum_{j=0}^{\infty} \frac{1}{j!}(-t)^j C^j$$

for any positive regular matrix C; further, let $\mathbf{e}_0 = (e_0, \mathbf{e}_1^{\mathrm{T}}, e_M)^{\mathrm{T}}$, where $\mathbf{e}_1^{\mathrm{T}} = (e_1, \ldots, e_{M-1})$, with $e_j = 1$ if $j = i_0$ and $e_j = 0$ if $j \ne i_0$, $0 \le j, i_0 \le M$.

Theorem 15.2.1 If $\lambda(t) = \lambda$ and $\mu(t) = \mu$, then

$$\mathbf{P}(t) = \exp[-(t - t_0)A_1^{\mathrm{T}}]\mathbf{e}_0, \tag{15.2.4}$$

where A_1 is a $(1 + M) \times (1 + M)$ matrix given by

$$A_1 = \begin{bmatrix} 0 & \mathbf{0}^{\mathrm{T}} & 0 \\ -\mathbf{h}_1 & A & -\mathbf{h}_2 \\ 0 & \mathbf{0}^{\mathrm{T}} & 0 \end{bmatrix},$$

and $\mathbf{h}_1 = [\mu(1 - 1/M), 0, \ldots, 0, 0]^{\mathrm{T}}$, $\mathbf{h}_2 = [0, 0, \ldots, 0, \lambda(1 - 1/M)]^{\mathrm{T}}$, $\mathbf{0}^{\mathrm{T}}$ is a $1 \times (M - 1)$ row of 0's and $A = (a_{ij})$, $i, j = 1, \ldots, M - 1$,

with

$$a_{ij} = -\mu i\left[1 - \frac{i}{M}\right] \qquad \text{if } j = i - 1$$
$$= (\lambda + \mu)i\left[1 - \frac{i}{M}\right] \qquad \text{if } j = i$$
$$= -\lambda i\left[1 - \frac{i}{M}\right] \qquad \text{if } j = i + 1$$
$$= 0 \qquad \text{if } j \neq i + 1, j \neq i - 1, \text{ and } j \neq i.$$

Proof. Since $1 \leq j \leq M - 1$ are transient states, $A^n \to 0$ as $n \to \infty$ and A (and hence A_1) is positive regular; in fact, if $\lambda = \mu$, the eigenvalues of A are $\mu_n = (\lambda/M)n(n + 1)$, $n = 1, 2, \ldots, M - 1$ $(0 < \mu_n < 1)$. To prove Eq. (15.2.4), partition the time interval $[t_0, t]$ by $[t_0, t] = U_{j=0}^{n-1}L_j$, where $L_j = [t_j, t_j + \Delta t)$, $j = 0, 1, \ldots, n - 2$, and $L_{n-1} = [t_{n-1}, t_n]$ with $t_j = t_0 + j\,\Delta t$, $j = 1, \ldots, n$ and $t = t_n$. Then, $P(t_j, t_{j+1}) = [I_{M+1} - A_1\,\Delta t]$, so that $P(t_0, t) = [I_{M+1} - A_1\,\Delta t]^n$.

Noting $t - t_0 = n\,\Delta t$ and letting $\Delta t \to 0$ (or $n \to \infty$), we have

$$\mathbf{P}^{\mathrm{T}}(t) = \mathbf{e}_0^{\mathrm{T}}\left\{\lim_{\Delta t \to 0} P(t_0, t)\right\} = \mathbf{e}_0^{\mathrm{T}}\left\{\lim_{n \to \infty}\left[I_{M+1} - \frac{1}{n}(t - t_0)A_1\right]^n\right\}$$
$$= \mathbf{e}_0^{\mathrm{T}} \exp\{-(t - t_0)A_1\}$$

or

$$\mathbf{P}(t) = \exp[-(t - t_0)A_1^{\mathrm{T}}]\mathbf{e}_0. \qquad \blacksquare$$

To simplify Eq. (15.2.4), note that for $N \geq 1$,

$$A_1^n = \begin{bmatrix} 0 & \mathbf{0}^{\mathrm{T}} & 0 \\ -A^{n-1}\mathbf{h}_1 & A^n & -A^{n-1}\mathbf{h}_2 \\ 0 & \mathbf{0}^{\mathrm{T}} & 0 \end{bmatrix};$$

$$\sum_{j=0}^{\infty} \frac{1}{j!}(-A)^j(t - t_0)^j = \exp[-(t - t_0)A],$$

and

$$\boldsymbol{\omega}_i(t) = -\sum_{j=1}^{\infty} \frac{1}{j!}(-1)^j(t - t_0)^j A^{j-1}\mathbf{h}_i$$
$$= -A^{-1}\left\{\sum_{j=0}^{\infty} \frac{1}{j!}(-1)^j(t - t_0)^j A^j - I_{M-1}\right\}\mathbf{h}_i$$
$$= A^{-1}\{I_{M-1} - \exp[-(t - t_0)A]\}\mathbf{h}_i$$
$$= \{I_{M-1} - \exp[-(t - t_0)A]\}A^{-1}\mathbf{h}_i, \qquad i = 1, 2.$$

Hence,

$$\exp[-(t - t_0)A_1] = I_{m+1} + \sum_{j=1}^{\infty} \frac{1}{j!}(-A_1)^j(t - t_0)^j$$

$$= \begin{bmatrix} 1 & \mathbf{0}^T & 0 \\ \boldsymbol{\omega}_1(t) & \exp[-(t - t_0)A] & \boldsymbol{\omega}_2(t) \\ 0 & \mathbf{0}^T & 1 \end{bmatrix};$$

or

$$\mathbf{P}(t) = \begin{bmatrix} 1 & \boldsymbol{\omega}_1^T & 0 \\ \mathbf{0} & \exp[-(t - t_0)A^T] & \mathbf{0} \\ 0 & \boldsymbol{\omega}_2^T(t) & 1 \end{bmatrix} \mathbf{e}_0.$$

Thus, if $\lambda(t) = \lambda$ and $\mu(t) = \mu$,

$$P_0(t) = e_0 + \mathbf{e}_1^T\boldsymbol{\omega}_1(t), \tag{15.2.5}$$

$$P_M(t) = e_M + \mathbf{e}_1^T\boldsymbol{\omega}_2(t), \tag{15.2.6}$$

and

$$\mathbf{P}_1(t) = [P_1(t), \ldots, P_{M-1}(t)]^T = \exp[-(t - t_0)A^T]\mathbf{e}_1, \tag{15.2.7}$$

In the above, $P_0(t)$ is the probability of absorption into the state 0 at time t given that at t_0 there are i_0 individuals, $P_M(t)$ is the probability of absorption into the state M at time t given that at t_0 there are i_0 individuals, and $\mathbf{P}_1(t)$ is the $(M - 1) \times 1$ vector of probabilities of staying in transient states at time t given that at t_0 there are i_0 individuals. Further, $\boldsymbol{\omega}_1(t)$ is the vector of absorption probabilities into the state 0 at or before time t given the transient states at t_0 and $\boldsymbol{\omega}_2$ the vector of absorption probabilities into the state M at or before time t given the transient states at t_0. (That is, the jth element of $\boldsymbol{\omega}_i(t)$ is the probability of absorption into the state 0 ($i = 1$) or M ($i = 2$) at or before time t given j individuals at t_0, $1 \leq j \leq M - 1$).

Since A is positive regular, $\exp[-(t - t_0)A^T] \to 0$ as $t \to \infty$, so that $\mathbf{P}_1(t) \to \mathbf{0}$ as $t \to \infty$ and $\boldsymbol{\omega}_i(t) \to \boldsymbol{\omega}_i = A^{-1}\mathbf{h}_i$, $i = 1, 2$, as $t \to \infty$, where $\mathbf{1}_{M-1}$ is an $(M - 1) \times 1$ column of 1's. Also, since $-(\mathbf{h}_1 + \mathbf{h}_2) + A\mathbf{1}_{M-1} = \mathbf{0}$, $\boldsymbol{\omega}_1 + \boldsymbol{\omega}_2 = A^{-1}(\mathbf{h}_1 + \mathbf{h}_2) = \mathbf{1}_{M-1}$. Note that the conditions $\mathbf{P}_1(t) \to \mathbf{0}$ as $t \to \infty$ and $\boldsymbol{\omega}_1 + \boldsymbol{\omega}_2 = \mathbf{1}_{M-1}$ are equivalent to stating that with probability 1 the transient states $(1 \leq j \leq N - 1)$ will eventually be absorbed into the state 0 or M.

If A is diagonalizable with real distinct eigenvalues π_j, $j = 1, \ldots, k$, then

$$A = \sum_{j=1}^{k} \pi_j E_j,$$

where

$$E_j = \prod_{i \neq j} (\lambda_j - \lambda_i)^{-1}(A - \lambda_i I_{M-1})$$

(Tan, 1977; Rao, 1973).

Since the E_j's satisfy the conditions $E_j^2 = E_j$, $E_iE_j = 0$ for $i \neq j$, and $\Sigma_{j=1}^{k} E_j = I_{M-1}$ (Tan, 1977), one has

$$A^n = \sum_{j=1}^{k} \pi_j^n E_j, \qquad A^{-1} = \sum_{j=1}^{k} \pi_j^{-1} E_j,$$

and

$$\exp[-(t - t_0)A] = \sum_{j=1}^{k} \exp[-(t - t_0)\pi_j]E_j.$$

It follows that

$$\boldsymbol{\omega}_i(t) = \sum_{j=1}^{k} \pi_j^{-1}\{1 - \exp[-(t - t_0)\pi_j]\}E_j\mathbf{h}_i, \qquad i = 1, 2,$$

$$\mathbf{P}_1(t) = \exp[-(t - t_0)A^{\mathrm{T}}]\mathbf{e}_1 = \sum_{j=1}^{k} \exp[-(t - t_0)\pi_j]E_j^{\mathrm{T}}\mathbf{e}_1.$$

15.2.2 Probability Distributions of First Absorption Times

Since the states 0 and M are the absorbing states while $1 \leq j \leq M - 1$ are the transient states, with probability 1 the transient states will eventually be absorbed into 0 or M. The vector of ultimate absorption probabilities into 0 is $\boldsymbol{\omega}_1 = A^{-1}\mathbf{h}_1$, while the vector of ultimate absorption probabilities into M is $\boldsymbol{\omega}_2 = A^{-1}\mathbf{h}_2$.

To obtain the probability distribution of first absorption into the absorbing states 0 and M when $\lambda(t) = \lambda$ and $\mu(t) = \mu$, let $\mathbf{f}(t)$ be the vector of probability density functions of first absorbing times into 0 or M given at t_0 the transient states $1 \leq j \leq M - 1$. (That is, the jth element $f_j(t)$ of $\mathbf{f}(t)$ is the probability density function of first absorption time into 0 or M given at t_0 there are j individuals.) For obtaining $\mathbf{f}(t)$ and moments for first absorption times, in what follows we shall also use the notation

$$d\mathbf{f}(t) = [df_1(t), df_2(t), \ldots, df_n(t)]^{\mathrm{T}},$$

$$\frac{d}{dt}\mathbf{f}(t) = \left[\frac{d}{dt} f_1(t), \frac{d}{dt} f_2(t), \ldots, \frac{d}{dt} f_n(t)\right]^{\mathrm{T}},$$

$$\int_0^\infty \mathbf{f}(t)\,dt = \left[\int_0^\infty f_1(t)\,dt, \int_0^\infty f_2(t)\,dt, \ldots, \int_0^\infty f_n(t)\,dt\right]^T,$$

$$\int_0^\infty g(t)\,d\mathbf{f}(x) = \left[\int_0^\infty g(t)\,df_1(t), \int_0^\infty g(t)\,df_2(t), \ldots, \int_0^\infty g(t)\,df_n(t)\right]^T.$$

Then, since $\boldsymbol{\omega}(t) = \boldsymbol{\omega}_1(t) + \boldsymbol{\omega}_2(t)$ is the vector of absorption probabilities into the state 0 or M at or before time t given the transient states $1 \le j \le M - 1$ at t_0, we have

$$\begin{aligned}\mathbf{f}(t) &= \frac{d}{dt}\boldsymbol{\omega}(t) = \frac{d}{dt}\{[I_{M-1} - \exp(-(t - t_0)A)]A^{-1}(\mathbf{h}_1 + \mathbf{h}_2)\}\\ &= \frac{d}{dt}\{I_{M-1} - \exp[-(t - t_0)A]\}\mathbf{1}_{M-1} = A\exp[-(t - t_0)A]\mathbf{1}_{M-1}\\ &= \exp[-(t - t_0)A]A\mathbf{1}_{M-1}, \qquad (15.2.8)\end{aligned}$$

In Eq. (15.2.8) $\mathbf{f}(t)$ is the matrix analog of the exponential distribution (Tan, 1976, 1984). If A is diagonable with distinct real eigenvalues π_j, $j = 1, \ldots, k$, $\mathbf{f}(t)$ reduces to

$$\mathbf{f}(t) = \sum_{j=1}^{k} \pi_j \exp[-(t - t_0)\pi_j]E_j^T\mathbf{1}_{M-1}, \qquad (15.2.9)$$

Note that $\mathbf{f}(t)$ in Eq. (5.2.9) is a vector of weighted exponential distributions with weights $E_j^T\mathbf{1}_{M-1}$. (Note $\Sigma_{j=1}^{k} E_j^T\mathbf{1}_{M-1} = \mathbf{1}_{M-1}$.)

Using $\mathbf{f}(t)$, one may readily obtain moments of first absorption times into the state 0 or M. In particular, one may obtain the vector $\mathbf{U}$ of mean absorption times and $\mathbf{V}$ the vector of variances of first absorption times for the homogeneous cases $\lambda(t) = \lambda$ and $\mu(t) = \mu$ (Tan, 1976). We have

$$\begin{aligned}\mathbf{U} &= \int_{t_0}^\infty (t - t_0)\exp[-(t - t_0)A]A\mathbf{1}_{M-1}\,dt\\ &= -\int_0^\infty t\,d[\exp(-tA)\mathbf{1}_{M-1}] = \int_0^\infty \exp(-tA)\mathbf{1}_{M-1}\,dt\\ &= -A^{-1}\int_0^\infty d[\exp(-tA)\mathbf{1}_{M-1}] = A^{-1}\mathbf{1}_{M-1}, \qquad (5.2.10)\end{aligned}$$

and

$$\mathbf{V} = \int_{t_0}^{\infty} (t - t_0)^2 \exp[-(t - t_0)A] A \mathbf{1}_{M-1}\, dt - \mathbf{U}_{sq}$$

$$= 2 \int_0^{\infty} t \exp(-tA) \mathbf{1}_{M-1}\, dt - \mathbf{U}_{sq}$$

$$= 2A^{-1}\mathbf{U} - \mathbf{U}_{sq} = 2A^{-2}\mathbf{1}_{M-1} - \mathbf{U}_{sq}, \qquad (5.2.11)$$

where $\mathbf{U}_{sq} = (u_1^2, u_2^2, \ldots, u_{M-1}^2)^{\mathrm{T}}$ and $\mathbf{U} = (u_1, u_2, \ldots, u_{M-1})^{\mathrm{T}}$.

If A is diagonalizable with real distinct eigenvalues π_j, $j = 1, \ldots, r$, then

$$A = \sum_{j=1}^{r} \pi_j E_j, \qquad \text{where } E_j = \prod_{i \neq j} (\pi_j - \pi_i)^{-1}(A - \pi_i I_{M-1}),$$

and, hence, $\mathbf{U}$ and $\mathbf{V}$ reduce to

$$\mathbf{U} = \sum_{j=1}^{r} \pi_j^{-1} E_j \mathbf{1}_{M-1}, \qquad (15.2.12)$$

and

$$\mathbf{V} = 2 \sum_{j=1}^{r} \pi_j^{-2} E_j \mathbf{1}_{M-1} - \mathbf{U}_{sq}; \qquad (15.2.13)$$

From above, to compute ω_i, $i = 1, 2$, $\mathbf{U}$ and $\mathbf{V}$, one would either need A^{-1} or the eigenvalues of A. When $\lambda = \mu$, the eigenvalues are real and available; but if $\lambda \neq \mu$, the eigenvalues are difficult to obtain. Fortunately, $A^{-1} = (c_{ij})$ is available from Glaz (1979).

In fact, one has

$$c_{ij} = \left(\sum_{l=1}^{\min(i,j)} \mu^{l-1} \lambda^{j-l} \right) \left(\sum_{l=1}^{M-j-\max(i-j,0)} \mu^{M-l-j} \lambda^{l-1} \right) \times \left[j\left(1 - \frac{j}{M}\right) \right]^{-1} \left[\sum_{l=1}^{M} \mu^{M-l} \lambda^{l-1} \right]^{-1},$$

$$1 \leq i, j \leq M - 1. \qquad (15.2.14)$$

Using Eq. (15.2.14), we obtain, for $\lambda(t) = \lambda$ and $\mu(t) = \mu$,

$$\omega_1^{\mathrm{T}} = [\omega_{11}, \omega_{12}, \ldots, \omega_{1,M-1}] \qquad (15.2.15)$$

$$\omega_2^{\mathrm{T}} = [\omega_{21}, \omega_{22}, \ldots, \omega_{2,M-1}], \qquad (15.2.16)$$

where

$$\omega_{1i} = \sum_{l=1}^{M-i} \mu^{M-l}\lambda^{l-1} \Big/ \sum_{l=1}^{M} \mu^{M-l}\lambda^{l-1},$$

$$\omega_{2i} = \sum_{l=1}^{i} \mu^{l-1}\lambda^{M-l} \Big/ \sum_{l=1}^{M} \mu^{M-l}\lambda^{l-1},$$

$i = 1, \ldots, M - 1.$

Similarly, we have for $\lambda(t) = \lambda$ and $\mu(t) = \mu$:

$$\mathbf{U}^{\mathrm{T}} = (u_1, u_2, \ldots, u_{M-1}) \tag{15.2.17}$$

and

$$\mathbf{V}^{\mathrm{T}} = 2(V_{10}, V_{20}, \ldots, V_{M-1,0}) - \mathbf{U}^{\mathrm{T}}_{\mathrm{sq}}, \tag{15.2.18}$$

where

$$u_i = \left[\sum_{l=1}^{M} \mu^{M-l}\lambda^{l-1}\right]^{-1} \left\{\sum_{j=1}^{i} \left[\sum_{l=1}^{j} \mu^{l-1}\lambda^{j-l}\right]\left[\sum_{l=1}^{M-i} \mu^{M-l-j}\lambda^{l-1}\right] + \sum_{j=i+1}^{M-1} \left[\sum_{l=1}^{i} \mu^{l-1}\lambda^{j-l}\right]\left[\sum_{l=1}^{M-j} \mu^{M-l-j}\lambda^{l-1}\right]\right\}\left[i\left(1 - \frac{i}{M}\right)\right]^{-1},$$

$$\mathbf{U}^{\mathrm{T}}_{\mathrm{sq}} = (u_1^2, u_2^2, \ldots, u_{M-1}^2),$$

$$V_{i0} = \left[\sum_{l=1}^{M} \mu^{M-l}\lambda^{l-1}\right]^{-1} \left\{\sum_{j=1}^{i} \left[\sum_{l=1}^{j} \mu^{l-1}\lambda^{j-l}\right]\left[\sum_{l=1}^{M-i} \mu^{M-l-j}\lambda^{l-1}\right] + \sum_{j=i+1}^{M-1} \left[\sum_{l=1}^{i} \mu^{l-1}\lambda^{j-l}\right]\left[\sum_{l=1}^{M-j} \mu^{M-l-j}\lambda^{l-1}\right]\right\} u_j\left[i\left(1 - \frac{i}{M}\right)\right]^{-1},$$

$$i = 1, \ldots, M - 1.$$

15.3 THE MOMENTS OF $X(t)$, $t \geq t_0$

Given $X(t_0) = i_0$, the rth moment around 0 of $X(t)$ is

$$\mu_r'(t) = E\{[X(t)]^r | X(t_0) = i_0\} = \sum_{j=0}^{M} j^r P_j(t), \tag{15.3.1}$$

While one may use Eqs. (15.2.7) to obtain $\mu_r'(t)$ theoretically, in this section we shall derive $\mu_r'(t)$ alternatively through the Kolmogorov forward equation. The Kolmogorov forward equation for $P_j(t)$ is

$$\begin{aligned}\frac{d}{dt}P_j(t) &= \lambda(t)(j-1)\left[1-\frac{j-1}{M}\right]P_{j-1}(t)\\ &\quad+\mu(t)(j+1)\left[1-\frac{j+1}{M}\right]P_{j+1}(t)\\ &\quad-[\lambda(t)+\mu(t)]j\left[1-\frac{j}{M}\right]P_j(t),\end{aligned} \tag{15.3.2}$$

for $j = 0, 1, \ldots, M-1$, and

$$\frac{d}{dt}P_M(t) = \lambda(t)\left[1-\frac{1}{M}\right]P_{M-1}(t),$$

where $P_j(t_0) = \delta_{i_0 j}$, the Kronecker's δ (i.e., $\delta_{ij} = 1$ if $i = j$ and $\delta_{ij} = 0$ if $i \neq j$).

Let $Q(z, t) = Q$ be the probability generating function (PFG) of $[P_i(t), i = 0, 1, \ldots, M]$ given $X(t_0) = i_0$. That is,

$$Q = \sum_{j=0}^{M} z^j P_j(t), \qquad \text{with } Q(z; t_0) = z^{i_0}.$$

Then, one has $(j!)P_j(t) = \{d^jQ(z, t)/dz^j\}_{(z=0)}$. Further, if one denotes by $k_j(t)$ the jth cumulant of $X(t)$ given $X(t_0) = i_0$, then $k_j(t)$ is obtained from

$$(j!)k_j(t) = \left\{\frac{d^j}{dz^j}\log Q[\exp(z), t]\right\}_{(z=0)}, \tag{15.3.3}$$

where $C(z, t) = \log Q[\exp(z), t]$ is the cumulant generating function of $X(t), t \geq t_0$.

To obtain $Q = Q(z, t)$, one multiplies Eq. (15.3.2) on both sides by z^j and sums over z from $j = 0$ to $j = M$. Since $P_{-1}(t) = 0$ and since

$$\begin{aligned}&\sum_{j=0}^{M} jz^j\left[1-\frac{j}{M}\right]P_j(t)\\ &= z\left\{\left[1-\frac{1}{M}\right]\sum_{j=0}^{M} jz^{j-1}P_j(t) - \frac{z}{M}\sum_{j=0}^{M} j(j-1)z^{j-2}P_j(t)\right\}\\ &= z\left\{\left[1-\frac{1}{M}\right]\frac{\partial}{\partial z}Q(z, t) - \frac{z}{M}\frac{\partial^2}{\partial z^2}Q(z, t)\right\},\end{aligned}$$

one obtains

$$\frac{\partial}{\partial t} Q(z, t) = (z - 1)[z\lambda(t) - \mu(t)]$$
$$\times \left\{ \left[1 - \frac{1}{M} \right] \frac{\partial}{\partial z} - \frac{z}{M} \frac{\partial^2}{\partial z^2} \right\} Q(z, t), \quad (15.3.4)$$

with $Q(z, t_0) = z^{i_0}$.

Equation (15.3.4) in general is very difficult to solve. However, one may always use Eq. (15.3.4) to derive equations for the cumulants (hence moments) of $X(t)$. We have in fact the following differential equations for the first four cumulants of $X(t)$:

$$\frac{d}{dt} \kappa_1(t) = \epsilon(t) \left\{ \kappa_1(t) - \frac{1}{M} [(\kappa_1(t))^2 + \kappa_2(t)] \right\}, \quad (15.3.5)$$

where $\kappa_1(t_0) = i_0$, and $\epsilon(t) = \lambda(t) - \mu(t)$;

$$\frac{d}{dt} \kappa_2(t) = \kappa_2(t) \left\{ 2\epsilon(t) \left[1 - \frac{2\kappa_1(t)}{M} \right] - \frac{\omega(t)}{M} \right\}$$
$$+ \kappa_1(t)\omega(t) \left\{ 1 - \frac{\kappa_1(t)}{M} \right\} - 2\epsilon(t) \frac{\kappa_3(t)}{M}, \quad (15.3.6)$$

where $\kappa_2(t_0) = 0$ and $\omega(t) = \lambda(t) + \mu(t)$;

$$\frac{d}{dt} \kappa_3(t) = 3\kappa_3(t)h_1(t) + \epsilon(t)h_2(t) + 3\omega(t)\kappa_2(t)$$
$$\times \left\{ 1 - \frac{2\kappa_1(t)}{M} \right\} - \frac{3\epsilon(t)\kappa_4(t)}{M}, \quad (15.3.7)$$

where $\kappa_3(t_0) = 0$, $h_1(t) = \epsilon(t)\{1 - 2\kappa_1(t)/M\} - \omega(t)/M$, and $h_2(t) = \kappa_1(t) - [\kappa_2(t) + \kappa_1^2 + 6\kappa_2^2(t)](1/M)$; and

$$\frac{d}{dt} \kappa_4(t) = \kappa_4(t) \left\{ 4h_1(t) - \frac{2\omega(t)}{M} \right\} + h_3(t) - \frac{4\epsilon(t)\kappa_5(t)}{M}, \quad (15.3.8)$$

where $\kappa_4(t_0) = 0$, and

$$h_3(t) = \omega(t) \left\{ h_2(t) + 6\kappa_3(t) - 6[\kappa_2^2(t) + 2\kappa_1(t)\kappa_3(t)] \frac{1}{M} \right\}$$
$$- 4\epsilon(t) \left\{ \kappa_2(t) - [6\kappa_3(t)\kappa_2(t) + \kappa_3(t) + 2\kappa_2(t)\kappa_1(t)] \frac{1}{M} \right\}.$$

If $\kappa_i(t) = O(M)$ for $i = 1, 2$, then

$$\frac{d}{dt}\kappa_1(t) = \epsilon(t)\left\{\kappa_1(t) - \frac{1}{M}[\kappa_1^2(t) + \kappa_2(t)]\right\}$$
$$\cong \epsilon(t)\kappa_1(t)\left[1 - \frac{1}{M}\kappa_1(t)\right], \qquad \kappa_1(t_0) = i_0.$$

It follows that if $\kappa_i(t) = O(M)$ for $i = 1, 2$, a close approximation to $\kappa_1(t)$ is given by

$$\kappa_1(t) = i_0 \exp\left\{\int_{t_0}^{t} \epsilon(x)\,dx\right\}\left\{1 - \frac{1}{M} + \frac{1}{M}\exp\left[\int_{t_0}^{t}\epsilon(x)\,dx\right]\right\}^{-1}, \tag{15.3.9}$$

Equation (15.3.9) is the nonhomogeneous logistic growth function. It is shown in Tan and Piantadosi (1988) that when $\lambda(t) = \lambda$ and $\mu(t) = \mu$, Eq. (15.3.9) provides a very close approximation for most of the situations which correspond to doubling time of bacteria and cell populations. Similarly, if $\kappa_i(t) = O(M)$, $i = 1, 2$, and $\epsilon(t)\kappa_3(t) = O(M)$,

$$\frac{d}{dt}\kappa_2(t) \cong \kappa_2(t)\left\{2h_1(t) + \frac{\omega(t)}{M}\right\} + \kappa_1(t)\omega(t)\left\{1 - \frac{\kappa_1(t)}{M}\right\}$$
$$= \kappa_2(t)f_1(t) + f_2(t), \tag{15.3.10}$$

where $\kappa_2(t_0) = 0$, $f_1(t) = 2h_1(t) + \omega(t)/M$, and $f_2(t) = \kappa_1(t)\omega(t)\{1 - \kappa_1(t)/M\}$. From (3.10),

$$\kappa_2(t) \cong \exp\left\{\int_{t_0}^{t} f_1(x)\,dx\right\}\int_{t_0}^{t} f_2(x)\exp\left[-\int_{t_0}^{x} f_1(s)\,ds\right]dx. \tag{15.3.11}$$

It is shown in Tan and Piantadosi (1988) that when $\lambda(t) = \lambda$ and $\mu(t) = \mu$, Eq. (15.3.11) provides a very close approximation for many plausible values of the doubling time of bacteria and cell populations.

15.4 DIFFUSION APPROXIMATION

In Section 15.2, we have demonstrated how to derive probabilities of first absorption times and the means and the variances of first absorption times. In this section we give an alternative approach to these problems by using diffusion approximation. These results are very useful since M is usually very large in most of the cases.

To illustrate, let $Y(t) = X(t)/M$. When M is very large, one may approximate $Y(t)$ by a continuous process with state space [0, 1] and parameter space $t \geq t_0$, the states 0 and 1 being absorbing states. In fact it will be shown that up to order $O(M^{-2})$, one may approximate $Y(t)$ be a diffusion process with state space [0, 1] and diffusion coefficients $m(y, t) = \epsilon(t)y(1 - y)$ and $V(y, t) = V(t)y(1 - y)$, where $\epsilon(t) = \lambda(t) - \mu(t)$ and $V(t) = \lambda(t) + \mu(t)$. The states 0 and 1 are absorbing states.

Let $x = u/M$, $y = j/M$, and $dt = M^{-1}$. For large M, write $P_{uj}(s, t)$ as $P_{uj}(s, t) \cong f(x, y; s, t)\, dt$, where $f(x, y; s, t)$ is a continuous function of x and y. Then, we have

Theorem 15.4.1 Up to order $O(M^{-2})$, $f(x, y; s, t)$ satisfies the partial differential equation

$$\frac{\partial}{\partial t} f(x, y; s, t) = -\frac{\partial}{\partial y}\{f(x, y; s, t)m(y, t)\} + (2M)^{-1}\frac{\partial^2}{\partial y^2}\{f(x, y; s, t)V(y, t)\}, \quad (15.4.1)$$

where

$$f(x, y; s, s) = \delta(y - x)$$

is the Dirac δ function defined by

$$\int_{-\infty}^{\infty} g(x)\delta(x)\, dx = g(0).$$

Proof. Let $\phi(z, t)$ be the characteristic function (c.f.) of $Y(t) = X(t)/M$ given $Y(t_0) = X(t_0)/M = u/M = x$. Then, with $i = \sqrt{-1}$

$$\phi(z, t) = \sum_{j=0}^{M} \exp\left(\frac{izj}{M}\right) P_{uj}(s, t).$$

Using the forward equation (15.3.2) with $t_0 = s$ and $i_0 = u$, we have, noting $P_{u,-1}(s, t) = P_{u,M+1}(s, t) = 0$,

$$\frac{\partial}{\partial t}\phi(z, t) = \sum_{j=0}^{M} \exp\left(\frac{izj}{M}\right)\frac{d}{dt}P_{uj}(s, t)$$

$$= \left\{\sum_{j=0}^{M} \exp\left[\frac{iz(j-1)}{M}\right]\lambda(t)(j-1)\left[1 - \frac{j-1}{M}\right]P_{u,j-1}(s, t)\right\}$$

$$\times \exp\left(\frac{iz}{M}\right) + \left\{\sum_{j=0}^{M} \exp\left[\frac{iz(j+1)}{M}\right]\mu(t)(j+1)\right.$$

$$\times \left[1 - \frac{j+1}{M}\right] P_{u,j+1}(s, t)\Bigg\} \exp\left(-\frac{iz}{M}\right)$$

$$- \sum_{j=0}^{M} \exp\left(\frac{izj}{M}\right) [\lambda(t) + \mu(t)]j\left[1 - \frac{j}{M}\right] P_{uj}(s, t)$$

$$= \sum_{j=0}^{M} \exp\left(\frac{izj}{M}\right) j\left[1 - \frac{j}{M}\right] P_{uj}(s, t)$$

$$\times \left\{\lambda(t)\left[\exp\left(\frac{iz}{M}\right) - 1\right] + \mu(t)\left[\exp\left(-\frac{iz}{M}\right) - 1\right]\right\}$$

$$= \sum_{j=0}^{M} \exp\left(\frac{izj}{M}\right)\left(\frac{j}{M}\right)\left[1 - \frac{j}{M}\right] P_{uj}(s, t)$$

$$\times \{[\lambda(t) - \mu(t)](iz) + (2M)^{-1}[\lambda(t) + \mu(t)](iz)^2 + O(M^{-2})\}$$

$$= (iz)\epsilon(t)\left\{\frac{\partial}{\partial(iz)} - \frac{\partial^2}{\partial(iz)^2}\right\} Q(z, t) + (2M)^{-1}(iz)^2 V(t)$$

$$\times \left\{\frac{\partial}{\partial(iz)} - \frac{\partial^2}{\partial(iz)^2}\right\} Q(z, t) + O(M^{-2}), \tag{15.4.2}$$

By the inversion formula

$$f(x, y; s, t) = (2\pi)^{-1} \int_{-\infty}^{\infty} \exp(-iyz)\phi(z, t)\, dz,$$

and by the Lemma given in the Appendix we have

$$(2\pi)^{-1} \int_{-\infty}^{\infty} (iz) \exp(-iyz) \left[\frac{\partial}{\partial(iz)} - \frac{\partial^2}{\partial(iz)^2}\right] Q(z, t)\, dz$$

$$= \left(-\frac{\partial}{\partial y}\right) (2\pi)^{-1} \int_{-\infty}^{\infty} \exp(-iyz) \left[\frac{\partial}{\partial(iz)} - \frac{\partial^2}{\partial(iz)^2}\right] Q(z, t)\, dz$$

$$= -\frac{\partial}{\partial y} \{y(1 - y)f(x, y; s, t)\};$$

$$(2\pi)^{-1} \int_{-\infty}^{\infty} (iz)^2 \exp(-iyz) \left[\frac{\partial}{\partial(iz)} - \frac{\partial^2}{\partial(iz)^2}\right] Q(z, t)\, dz$$

$$= \left(\frac{\partial^2}{\partial y^2}\right) (2\pi)^{-1} \int_{-\infty}^{\infty} \exp(-iyz) \left[\frac{\partial}{\partial(iz)} - \frac{\partial^2}{\partial(iz)^2}\right] Q(z, t)\, dz$$

$$= \frac{\partial^2}{\partial y^2} \{y(1 - y)f(x, y; s, t)\},$$

Multiplying Eq. (15.4.2) on both sides by $(2\pi)^{-1}\exp(-iyz)$ and integrating from $-\infty$ to ∞, we obtain

$$\frac{\partial}{\partial t} f(x, y; s, t) = -\frac{\partial}{\partial y}\{m(y, t)f(x, y; s, t)\} + (2M)^{-1}\frac{\partial^2}{\partial y^2}\{V(y, t)f(x, y; s, t)\},$$

$$f(x, y; s, s) = \delta(y - x). \quad ■$$

Note that $m(0, t) = m(1, t) = V(0, t) = V(1, t) = 0$ for all $t \geq t_0$, and Eq. (15.4.1) is the Kolmogorov forward equation for a diffusion process with state space [0, 1], diffusion coefficients $m(y, t)$ and $V(y, t)$, 0 and 1 being absorbing states.

Theorem 15.4.2 Up to order $O(M^{-2})$, $f(x, y; s, t)$ also satisfies the partial differential equation

$$-\frac{\partial}{\partial s} f(x, y; s, t) = m(x, s)\frac{\partial}{\partial x} f(x, y; s, t) + (2M)^{-1}V(x, s)\frac{\partial^2}{\partial x^2} f(x, y; s, t), \quad (15.4.3)$$

where $f(x, y; s, s) = \delta(y - x)$ is the Dirac δ-function.

Proof. By the Kolmogorov backward equation, one has

$$-\frac{\partial}{\partial s} P_{uj}(s, t) = u\left[1 - \frac{u}{M}\right] \times \{\lambda(s)P_{u+1,j}(s, t) + \mu(s)P_{u-1,j}(s, t) - [\lambda(s) + \mu(s)]P_{uj}(s, t)\},$$
$$u, j = 0, 1, \ldots, M. \quad (15.4.4)$$

Let $dx = M^{-1}$ so that $(u - 1)/M = x - dx$ and $(u + 1)/M = x + dx$, and write $\phi(z, t)$ as

$$\phi(z, t) = \phi(x; s, t) = \sum_{j=0}^{M} \exp\left(\frac{izj}{M}\right) P_{uj}(s, t).$$

Then, multiplying both sides of Eq. (15.4.4) by $\exp(izj/M)$ with $i =$

$\sqrt{-1}$ and summing over j from 0 to M, we obtain

$$-\frac{\partial}{\partial s}\phi(x; s, t) = u\left[1 - \frac{u}{M}\right]\left\{-[\lambda(s) + \mu(s)]\sum_{j=0}^{M}\exp\left(\frac{izj}{M}\right)P_{uj}(s, t)\right.$$
$$+ \lambda(s)\sum_{j=0}^{M}\exp\left(\frac{izj}{M}\right)P_{u+1,j}(s, t)$$
$$\left.+ \mu(s)\sum_{j=0}^{M}\exp\left(\frac{izj}{M}\right)P_{u-1,j}(s, t)\right\}$$
$$= u\left[1 - \frac{u}{M}\right]\{\lambda(s)[\phi(x + dx; s, t) - \phi(x; s, t)]$$
$$+ \mu(s)[\phi(x - dx; s, t) - \phi(x; s, t)]\}. \qquad (15.4.5)$$

Expanding $\phi(x + dx; s, t)$ and $\phi(x - dx; s, t)$ in Taylor series around s with $dx = M^{-1}$,

$$\phi(x + dx; s, t) = \phi(x; s, t) + M^{-1}\frac{\partial}{\partial x}\phi(x; s, t)$$
$$+ M^{-1}(2M)^{-1}\frac{\partial^2}{\partial x^2}\phi(x; s, t) + O(M^{-3});$$
$$\phi(x - dx; s, t) = \phi(x; s, t) - M^{-1}\frac{\partial}{\partial x}\phi(x; s, t)$$
$$+ M^{-1}(2M)^{-1}\frac{\partial^2}{\partial x^2}\phi(x; s, t) + O(M^{-3}).$$

On substituting these results into Eq. (15.4.5), we obtain

$$-\frac{\partial}{\partial s}\phi(x; s, t) = x(1 - x)$$
$$\times\left\{\epsilon(s)\frac{\partial}{\partial x}\phi(x; s, t) + (2M)^{-1}V(s)\frac{\partial^2}{\partial x^2}\phi(x; s, t)\right\}$$
$$+ O(M^{-2}), \qquad (15.4.6)$$

Multiplying both sides of Eq. (15.4.6) by $(2\pi)^{-1}\exp(-izy)$ and integrating over z from $-\infty$ to ∞ gives

$$-\frac{\partial}{\partial s}f(x, y; s, t) = m(x, s)\frac{\partial}{\partial x}f(x, y; s, t)$$
$$+ (2M)^{-1}V(x, s)\frac{\partial^2}{\partial x^2}f(x, y; s, t),$$
$$f(x, y; s, s) = \delta(y - x). \qquad \blacksquare$$

Equation (15.4.3) is the Kolmogorov backward equation for a diffusion process with state space [0, 1], diffusion coefficients $m(y, t)$ and $V(y, t)$, with 0 and 1 being absorbing states. Hence, Theorems 15.4.1 and 15.4.2 imply that up to order $O(M^{-2})$, $Y(t) = X(t)/M$ is approximated by a diffusion process with state space [0, 1], diffusion coefficients $m(y, t)$ and $V(y, t)$, the states 0 and 1 being the absorbing states.

If $\lambda(t) = \lambda$ and $\mu(t) = \mu$ are independent of t, then $f(x, y; s, t) = f(x, y; t - s)$ depends on the time s and t through $t - s$. In these cases, Eqs. (15.4.1) and (15.4.3) reduce respectively, to

$$\frac{\partial}{\partial t} f(x, y; t) = -\epsilon \frac{\partial}{\partial y}\{y(1 - y)f(x, y; t)\} + V(2M)^{-1}\frac{\partial^2}{\partial y^2}\{y(1 - y)f(x, y; t)\}, \quad (15.4.7)$$

$$\frac{\partial}{\partial t} f(x, y; t) = \epsilon x(1 - x)\frac{\partial}{\partial x} f(x, y; t) + V(2M)^{-1}x(1 - x)\frac{\partial^2}{\partial x^2} f(x, y; t), \quad (15.4.8)$$

where $\epsilon = \lambda - \mu$, $V = \lambda + \mu$, and $f(x, y; 0) = \delta(y - x)$, the Dirac δ-function.

15.4.1 The Stationary Distribution

Let $F(x, y; s, t) = \int_0^y f(x, z; s, t)\, dz$, $0 < y \le 1$. Integrating Eq. (15.4.1), we obtain

$$\frac{\partial}{\partial t} F(x, y; s, t) = -m(y, t)f(x, y; s, t) + (2M)^{-1}\frac{\partial}{\partial y}[V(y, t)f(x, y; s, t)], \quad (15.4.9)$$

Theorem 15.4.3 Put $P(x, y; s, t) = -(\partial/\partial t)F(x, y; s, t)$. Then

$$P(x, y; s, t) = m(y, t)f(x, y; s, t) - (2M)^{-1}\frac{\partial}{\partial y}[V(y, t)f(x, y; s, t)]$$

is the total probability mass that $Y(t)$ passes y at time t given $Y(s) = x$.

Proof. Given $Y(t) = z$, let $g(z, u; \delta t)$ be the conditional probability density of a change u during $[t, t + \delta t]$. Then

$$m(z, t)\,\delta t = \int_{-\infty}^{\infty} u g(z, u; \delta t)\, du + o(\delta t),$$

$$M^{-1}V(z, t)\,\delta t = \int_{-\infty}^{\infty} u^2 g(z, u; \delta t)\, du + o(\delta t),$$

$$\int_{-\infty}^{\infty} u^r g(z, u; \delta t)\, du = o(\delta t) \qquad \text{if } r \geq 3.$$

Further, the probability that the process $Y(t)$ flows over the point y from the left during $[t, t + \delta t]$ is

$$\int_{u>y-z>0} g(z, u; \delta t)\, du.$$

Thus, the total probability mass flowing across y from the left during $[t, t + \delta t]$ is

$$\begin{aligned} P_+(x, y; s, t)\,\delta t &= \int_{y>z} f(x, z; s, t)\left\{\int_{u>y-z} g(z, u; \delta t)\, du\right\} dz \\ &= \int_{u>0}\left\{\int_{y-u}^{y} f(x, z; s, t) g(z, u; \delta t)\, dz\right\} du. \end{aligned}$$

Similarly, the total probability mass flowing across y from the right during $[t, t + \delta t]$ is

$$P_-(x, y; s, t)\,\delta t = \int_{u<0}\left\{\int_{y}^{y-u} f(x, z; s, t) g(z; u; \delta t)\, dz\right\} du.$$

Hence the total probability mass flowing across y during $[t, t + \delta t]$ is

$$\begin{aligned} &P_+(x, y; s, t)\,\delta t - P_-(x, y; s, t)\,\delta t = P(x, y; s, t)\,\delta t \\ &= \int_{-\infty}^{\infty}\left\{\int_{y-u}^{y} f(x, z; s, t) g(z; u; \delta t)\, dz\right\} du. \qquad (15.4.10) \end{aligned}$$

Expanding $f(x, z; s, t)g(z; u; \delta t)$ in a Taylor series with respect to z around y and noting

$$\frac{1}{j!}\int_{y-u}^{y} (y - z)^j\, dz = \frac{1}{j!}\int_{0}^{u} z^j\, dz = \frac{1}{(j + 1)!} u^{j+1},$$

Eq. (15.4.10) simplifies to

$$P(x, y; s, t)\delta t = f(x, y; s, t) \int_{-\infty}^{\infty} ug(y, u; \delta t)\, du$$

$$- \frac{1}{2}\frac{\partial}{\partial y}\left\{f(x, y; s, t) \int_{-\infty}^{\infty} u^2 g(y, u; \delta t)\, du\right\} + o(\delta t).$$

It follows that

$$P(x, y; s, t) = m(y, t)f(x, y; s, t) - (2M)^{-1}\frac{\partial}{\partial y}\{f(x, y; s, t)V(y, t)\}. \qquad \blacksquare$$

The above result is due to Kimura (1964).

If $P(x, y; s, t) = 0$, then

$$0 = m(y, t)f(x, y; s, t) - (2M)^{-1}\frac{\partial}{\partial y}\{V(y, t)f(x, y; s, t)\}. \qquad (15.4.11)$$

Since $P(x, y; s, t) = 0$ implies that the net flow of probability mass crossing y at time t is 0, the solution of the above equation provides the conditional stationary distribution of $Y(t)$ given no absorption.

To solve Eq. (15.4.11), let $h = h(x, y; s, t) = V(y, t)f(x, y; s, t)$. Then Eq. (15.4.11) becomes

$$2M\theta(t) = h^{-1}\frac{\partial}{\partial y}h = \frac{\partial}{\partial y}\log h(x, y; s, t),$$

where $\theta(t) = \epsilon(t)/V(t)$. Hence,

$$h(x, y; s, t) = h(y, t) = C_0 \exp[2M\theta(t)y].$$

Or the solution of Eq. (15.4.11) is

$$f_0(y, t) = [V(t)y(1 - y)]^{-1}C_0 \exp[2M\theta(t)y], \qquad 0 < y < 1,$$

where

$$C_0^{-1} = [V(t)]^{-1}\int_0^1 [y(1 - y)]^{-1} \exp[2M\theta(t)y]\, dy.$$

If $\lambda(t) = \lambda$ and $\mu(t) = \mu$, then $f_0(y, t) = f_0(y) = C_1[y(1 - y)]^{-1} \exp[2M(\epsilon/V)y]$, $0 < y < 1$, where

$$C_1^{-1} = \int_0^1 [y(1 - y)]^{-1} \exp\left[2M\left(\frac{\epsilon}{V}\right)y\right] dy.$$

15.4.2 Absorption Probability Distribution and Moments of First Absorption Times

Since the states j $(1 \le j \le M - 1)$ are transient states while 0 and M are absorbing states for $X(t)$, so for the diffusion process $Y(t) = X(t)/M$, the states $0 < x < 1$ are transient states while the two boundaries 0 and 1 are absorbing states. Thus, starting with $0 < x < 1$, with probability 1 the diffusion process $Y(t)$ will eventually be absorbed into the boundaries 0 or 1.

To obtain the absorption probability distributions and the absorption probabilities for the diffusion process $Y(t)$, put $x = u/M$, $y = j/M$ and for large M, write $P_{u0}(s, t) \cong U_0(x; s, t)$ and $P_{uM}(s, t) \cong U_1(x; s, t)$. Then for large M, $\xi(x; s\, t) = U_0(x; s, t) + U_1(x; s, t)$ is the absorption probability into 0 or 1 at or before time t. Let $\zeta(x; s, t)$ be the probability density function of first absorption time of $Y(t)$ given $Y(s) = x$. Then

$$\xi(x; s, t) = \int_s^t \zeta(x; s, v)\, dv.$$

If $\lambda(t) = \lambda$ and $\mu(t) = \mu$ are independent of t, then

$$U_i(x; s, t) = U_i(x; t - s), \qquad \zeta(x; s, t) = \zeta(x; t - s)$$

and (15.4.12)

$$\xi(x; s, t) = \xi(x; t - s) = \int_0^{t-s} \zeta(x; v)\, dv.$$

Theorem 15.4.4 $U_0(x; s, t)$, $U_1(x; s, t)$, and $\zeta(x; s, t)$ satisfy, respectively, the following partial differential equations:

$$-\frac{\partial}{\partial s} U_0(x; s, t) = m(x, s) \frac{\partial}{\partial x} U_0(x; s, t) + (2M)^{-1} V(x, s) \frac{\partial^2}{\partial x^2} U_0(x; s, t), \tag{15.4.13}$$

$U_0(x; s, s) = 0$ if $0 < x \le 1$ and $U_0(x; s, s) = 1$ if $x = 0$;

$$-\frac{\partial}{\partial s} U_1(x; s, t) = m(x, s) \frac{\partial}{\partial x} U_1(x; s, t) + (2M)^{-1} V(x, s) \frac{\partial^2}{\partial x^2} U_1(x; s, t), \tag{15.4.14}$$

$U_1(x; s, s) = 0$ if $x \neq 1$ and $U_1(x; s, s) = 1$ if $x = 1$;

$$-\frac{\partial}{\partial s}\zeta(x; s, t) = m(x, s)\frac{\partial}{\partial x}\zeta(x; s, t) + (2M)^{-1}V(x, s)\frac{\partial^2}{\partial x^2}\zeta(x; s, t). \quad (15.4.15)$$

Proof. To prove Eq. (15.4.13), observe first that for $r = 1, 2, \ldots,$

$$\sum_{j=0}^{M}\left[\frac{j-u}{M}\right]^r P_{uj}(s - ds, s) = \sum_{j=0}^{M}(y - x)^r P_{uj}(s - ds, s)$$
$$\cong \int_{-\infty}^{\infty} z^r g(x, z; ds)\, dz$$
$$= \begin{cases} m(x, s)\, ds + o(ds) & \text{if } r = 1, \\ M^{-1}V(x, s)\, ds + o(ds) & \text{if } r = 2, \\ o(ds) & \text{if } r \geq 3, \end{cases}$$

where $g(x, u; ds)$ is the probability density function of a change u during a time interval of length ds given $Y(s) = x$. Now, by the Chapman-Kolmogorov equation,

$$P_{u0}(s - ds, t) = \sum_{j=0}^{M} P_{uj}(s - ds, s)P_{j0}(s, t)$$
$$\cong \sum_{j=0}^{M} P_{uj}(s - ds, s)f(y, 0; s, t), \qquad y = \frac{j}{M}.$$

Expanding $f(y, 0; s, t)$ in Taylor series around $x = u/M$, we have

$$f(y, 0; s, t) = f(x, 0; s\ t) + (y - x)\frac{\partial}{\partial x}f(x, 0; s, t) + \frac{1}{2}(y - x)^2\frac{\partial^2}{\partial x^2}f(x, 0; s, t) + O((y - x)^3).$$

Hence,

$$P_{u0}(s - ds, t) \cong f(x, 0; s - ds, t)$$
$$\cong f(x, 0; s, t) + m(x, s)\, ds\frac{\partial}{\partial x}f(x, 0; s, t) + (2M)^{-1}V(x, s)\, ds\frac{\partial^2}{\partial x^2}f(x, 0; s, t) + o(ds).$$

Subtracting $f(x, 0; s, t)$ and then dividing by ds on both sides and letting $ds \to 0$, we obtain

$$-\frac{\partial}{\partial s} f(x, 0; s, t) = m(x, s) \frac{\partial}{\partial x} f(x, 0; s, t) + (2M)^{-1}V(x, s) \frac{\partial^2}{\partial x^2} f(x, 0; s, t).$$

Obviously the initial conditions are satisfied. This proves Eq. (15.4.13). Similarly, one proves Eq. (15.4.14).

Summing Eq. (15.4.13) and Eq. (15.4.14), one has

$$-\frac{\partial}{\partial s} \int_s^t \zeta(x; u, t)\, du = \zeta(x; s, t)$$

$$= m(x, s) \frac{\partial}{\partial x} \int_s^t \zeta(x, 0; u, t)\, du + (2M)^{-1}V(x, s) \frac{\partial^2}{\partial x^2} \int_s^t \zeta(x, 0; u, t)\, du.$$

Interchanging $\partial/\partial x$ (or $\partial^2/\partial x^2$) with $\int_s^t$ and taking the derivative with respect to s, we obtain

$$-\frac{\partial}{\partial s} \zeta(x; s, t) = m(x, s) \frac{\partial}{\partial x} \zeta(x; s, t) + (2M)^{-1}V(x, s) \frac{\partial^2}{\partial x^2} \zeta(x; s, t). \quad \blacksquare$$

If $\lambda(t) = \lambda$ and $\mu(t) = \mu$ so that $U_i(x; s, t) = U_i(x; t - s)$, $i = 0, 1$, and $\zeta(x; s, t) = \zeta(x; t - s)$, then Eqs. (15.4.13), (15.4.14), and (15.4.15) reduce respectively to

$$\frac{\partial}{\partial t} U_0(x; t) = \epsilon x(1 - x) \frac{\partial}{\partial x} U_0(x; t) + (2M)^{-1}Vx(1 - x) \frac{\partial^2}{\partial x^2} U_0(x; t), \quad (15.4.16)$$

$$U_0(x; 0) = 0 \quad \text{if } 0 < x \le 1 \quad \text{and} \quad U_0(x; 0) = 1 \quad \text{if } x = 0;$$

$$\frac{\partial}{\partial t} U_1(x; t) = \epsilon x(1 - x) \frac{\partial}{\partial x} U_1(x; t) + (2M)^{-1}Vx(1 - x) \frac{\partial^2}{\partial x^2} U_1(x; t), \quad (15.4.17)$$

$$U_1(x; 0) = 0 \quad \text{if } 0 \le x < 1 \quad \text{and} \quad U_1(x; 0) = 1 \quad \text{if } x = 1;$$

$$\frac{\partial}{\partial t} \zeta(x; t) = \epsilon x(1 - x) \frac{\partial}{\partial x} \zeta(x; t) + (2M)^{-1}Vx(1 - x) \frac{\partial^2}{\partial x^2} \zeta(x; t). \quad (15.4.18)$$

Let $U_i(x) = \lim_{t\to\infty} U_i(x; t), i = 0, 1$. Then $U_i(x)$ is the ultimate absorption probability into the state i ($i = 0, 1$) given $Y(0) = x$. From Eqs. (15.4.16) and (15.4.17), we have

$$\epsilon x(1 - x)\frac{d}{dx} U_i(x) + (2M)^{-1}Vx(1 - x)\frac{d^2}{dx^2} U_i(x) = 0, \quad (15.4.19)$$

with $U_1(1) = 1$ and $U_1(x) = 0$ for $0 \le x < 1$, and $U_0(0) = 1$ and $U_0(x) = 0$ for $0 < x \le 1$. Solving Eqs. (15.4.19), we obtain

$$\begin{aligned} U_0(x) &= \left\{\exp\left[-2M\left(\frac{\epsilon}{V}\right)x\right] - \exp\left[-2M\left(\frac{\epsilon}{V}\right)\right]\right\} \\ &\quad \times \left\{1 - \exp\left[-2M\left(\frac{\epsilon}{V}\right)\right]\right\}^{-1}, \\ U_1(x) &= \left\{1 - \exp\left[-2M\left(\frac{\epsilon}{V}\right)x\right]\right\}\left\{1 - \exp\left[-2M\left(\frac{\epsilon}{V}\right)\right]\right\}^{-1}. \end{aligned}$$

Note that $U_0(x) + U_1(x) = 1$. That is, the ultimate absorption probabilities are 1. This is equivalent to stating that starting with any state x ($0 < x < 1$), with probability 1 the process $Y(t)$ will eventually be absorbed into the absorbing boundaries 0 or 1.

To obtain the probability distribution of first absorption time, one would need to solve Eq. (15.4.15) or (15.4.18). The solution of Eq. (15.4.15) or (15.4.18) is in general very difficult; however, in the homogeneous cases $\lambda(t) = \lambda$ and $\mu(t) = \mu$, one may use Eq. (15.4.18) to obtain moments of first absorption times. In particular, one may obtain the mean absorption time $T(x, s)$ of $Y(t)$ given $Y(s) = x$ ($0 < x < 1$) and the variance $\sigma^2(x, s) = W(x, s) - [T(x, s)]^2$ of first absorption time of $Y(t)$ given $Y(s) = x$ ($0 < x < 1$). Note that if $\lambda(t) = \lambda$ and $\mu(t) = \mu$, then

$$T(x, s) = \int_s^\infty (t - s)\zeta(x; s, t)\, dt = \int_0^\infty t\zeta(x; t)\, dt = T(x),$$

$$W(x, s) = \int_s^\infty (t - s)^2\zeta(x; s, t)\, dt = \int_0^\infty t^2\zeta(x; t)\, dt = W(x).$$

Thus, if $\lambda(t) = \lambda$ and $\mu(t) = \mu$, both $T(x, s) = T(x)$ and $\sigma^2(x, s) = \sigma^2(x) = W(x) - [T(x)]^2$ are independent of s.

Theorem 15.4.5 If $\lambda(t) = \lambda$ and $\mu(t) = \mu$ and if $\lim_{t\to\infty} \zeta(x, t) = 0$ for all $0 < x < 1$, then $T(x)$ and $W(x)$ satisfy, respectively, the following differential equations:

$$\epsilon x(1-x)\frac{d}{dx}T(x) + (2M)^{-1}Vx(1-x)\frac{d^2}{dx^2}T(x) = -1, \quad (15.4.19)$$

with $T(0) = T(1) = 0$ and

$$\epsilon x(1-x)\frac{d}{dx}W(x) + (2M)^{-1}Vx(1-x)\frac{d^2}{dx^2}W(x) = -2T(x), \quad (15.4.20)$$

with $W(0) = W(1) = 0$.

Proof. Obviously the initial conditions are satisfied. To prove Eqs. (15.4.19) and (15.4.20), observe that

$$T(x) = \int_0^\infty t\zeta(x, t)\, dt$$

and

$$W(x) = \int_0^\infty t^2\zeta(x, t)\, dt.$$

Hence, using Eq. (15.4.18), we have

$$\begin{aligned}\epsilon x(1-x)&\frac{d}{dx}T(x) + (2M)^{-1}Vx(1-x)\frac{d^2}{dx^2}T(x)\\ &= \int_0^\infty t\left\{\epsilon x(1-x)\frac{\partial}{\partial x}\zeta(x,t) + (2M)^{-1}Vx(1-x)\frac{\partial^2}{\partial x^2}\zeta(x,t)\right\} dt\\ &= \int_0^\infty t\frac{\partial}{\partial t}\zeta(x,t)\, dt = \{t\zeta(x,t)\}_0^\infty - \int_0^\infty \zeta(x,t)\, dt = -1.\end{aligned}$$

Similarly, using Eq. (15.4.18), we get

$$\begin{aligned}\epsilon x(1-x)&\frac{d}{dx}W(x) + (2M)^{-1}Vx(1-x)\frac{d^2}{dx^2}W(x)\\ &= \int_0^\infty t^2\left\{\epsilon x(1-x)\frac{\partial}{\partial x}\zeta(x,t) + (2M)^{-1}Vx(1-x)\frac{\partial^2}{\partial x^2}\zeta(x,t)\right\} dt\\ &= \int_0^\infty t^2\frac{\partial}{\partial t}\zeta(x,t)\, dt = -2\int_0^\infty t\zeta(x,t)\, dt = -2T(x)\end{aligned}$$

by integration by parts. ■

By using the method of variation of parameters, one can solve Eqs.

(15.4.19) and (15.4.20) to obtain

$$T(x) = \epsilon^{-1}U_0(x) \int_0^x [y(1 - y)\psi(y)]^{-1}[1 - \psi(y)]\, dy$$
$$+ \epsilon^{-1}U_1(x) \int_x^1 [y(1 - y)\psi(y)]^{-1}[\psi(y) - \psi(1)]\, dy, \quad (15.4.21)$$

$$W(x) = \frac{2}{\epsilon} U_0(x) \int_0^x T(y)[y(1 - y)\psi(y)]^{-1}[1 - \psi(y)]\, dy$$
$$+ \frac{2}{\epsilon} U_1(x) \int_x^1 T(y)[y(1 - y)\psi(y)]^{-1} \times [\psi(y) - \psi(1)]\, dy, \quad (15.4.22)$$

where

$$\psi(y) = \exp\left[-2M\left(\frac{\epsilon}{V}\right) y\right], \qquad U_0(y) = \frac{\psi(y) - \psi(1)}{1 - \psi(1)},$$

and

$$U_1(y) = \frac{1 - \psi(y)}{1 - \psi(1)}.$$

15.5 CONCLUSIONS

In this chapter we have developed a general theory for the stochastic logistic growth process. In addition, approximations to the first two moments of the stochastic logistic process have been derived. Also, we have developed approximations for the absorption probabilities and the moments of the first absorption times using a diffusion approximation. The theory outlined in the chapter should be useful for building stochastic models in many areas of biomedical research.

REFERENCES

Eisen, M. (1979). *Mathematical Models in Cell Biology and Cancer Chemotherapy*, Lecture Notes in Biomathematics, *30*, Springer-Verlag, Berlin and New York.

Glaz, J. (1979). Probabilities and moments for absorption in finite homogeneous birth-death processes, *Biometrics*, *35*, 813–816.

Jensen, A. L. (1975). Comparison of logistic equations for population growth, *Biometrics*, *31*, 853–862.

Kimura, M. (1964). Diffusion models in population genetics, *J. Appl. Prob.*, *1*, 177–232.

Miller, R. S. and Botkin, D. B. (1974). Endangered species: models and predictions, *American Scientist*, *62*, 172–181.

Moolgavkar, S. H. (1986). Carcinogenesis modeling: from molecular biology to epidemiology, *Ann. Rev. Public Health*, *7*, 151–169.

Norden, R. H. (1982). On the distribution of the time to extinction in the stochastic logistic population model, *Adv. Appl. Prob.*, *14*, 687–708.

Norden, R. H. (1984). On the numerical evaluation of the moments of the distribution of states at time *t* in the stochastic logistic process, *J. Statist. Comput. Simul.*, *20*, 1–20.

Pearl, R. and Reed, L. J. (1920). On the rate of growth of the population of the United States since 1790 and its mathematical representation, *Proceedings of the National Academy of Sciences*, *6*, 275–288.

Pielou, E. C. (1977). *Mathematical Ecology*, Wiley, New York.

Rao, C. R. (1973). *Linear Statistical Inferences and Its Applications*, Wiley, New York.

Tan, W. Y. (1975). On the absorption probability and absorption time to finite homogeneous birth-death processes by diffusion approximation, *Metron*, *33*, 389–401.

Tan, W. Y. (1976). On the absorption probability and absorption time of finite homogeneous birth-death processes, *Biometrics*, *32*, 745–752.

Tan, W. Y. (1977). On the distribution of second degree polynomial in normal random variables, *Canadian J. Statistics*, *5*, 241–250.

Tan, W. Y. (1983). On the distribution of number of mutants at the hypoxanthine-quanine phosphoribosal transferase locus in Chinese hamster ovary cells, *Math. Biosciences*, *67*, 175–192.

Tan, W. Y. (1984). On the first passage probability distributions in continuous time Markov processes, *Utilitas Mathematica*, *26*, 89–102.

Tan, W. Y. and Piantadosi, S. (1988). On stochastic growth process with application to stochastic logistic growth, *Technical Report 88-4*, Department of Mathematical Sciences, Memphis State University, Memphis, TN. To appear in Statistica Sinica, 1990.

Verhulst, P. F. (1838). Notice sur la loi que la population suti dans son accroissement, *Correspondence mathematique et physique*, *10*, 113–121.

APPENDIX

Lemma If $f(x, y; s, t) = (2\pi)^{-1} \int_{-\infty}^{\infty} \exp(-izy)\phi(z, t)\, dz$, $i = \sqrt{-1}$ and if $\lim_{z\to\pm\infty}\{\partial^r\phi(z, t)/\partial z^r\} = 0$ for $r = 0, 1, \ldots$, then for any polynomial $P(z, t)$ in z, we have

$$(2\pi)^{-1} \int_{-\infty}^{\infty} \exp(-izy)P\left[\left(\frac{\partial}{\partial(iz)}\right), t\right] \phi(z, t)\, dz = P(y, t)f(x, y; s, t).$$

Proof. See Tan (1975).

16

Logistic Growth Models and Related Problems

Dieter Rasch
Institute of Biometry, Rostock, Germany

16.1 INTRODUCTION

The logistic function is one of the oldest growth functions. It is used to describe both population and organismic growth. First proposed as a growth function by Verhulst (1838) in the form

$$f(x, \theta) = \frac{\alpha}{1 + \beta e^{\gamma x}}, \tag{16.1.1}$$

the parameters α, β, and γ are elements of the parameter vector $\theta^T = (\alpha, \beta, \gamma)$ and can be interpreted as follows: α is the limiting (for $\gamma < 0$ the maximal) value of $f(x, \theta)$ if x tends to infinity. Since this chapter discusses the use of the function (16.1.1) to describe growth processes, the assumptions $\gamma < 0$ and $\beta > 0$ will be made throughout. The inflection point of f is $x_I = (1/\gamma)\ln(1/\beta)$, where the value of f is $f(x_I, \theta) = \alpha/2$. The parameter γ characterizes the curvature of the function f, which increases as the module of γ (i.e., under the above assumption $\gamma < 0$ the curvature is increasing with decreasing γ). For modeling population growth the logistic model was used and popularized by Pearl (1925) and Pearl and Reed (1920), who modeled the population of the United States. Yule (1925) used the logistic model to forecast human populations, and Leach (1981) reevaluated the use of the logistic function for modeling the growth of human populations (see also Lotka, 1931; Feller, 1940).

The fixed value $\alpha/2$ of the function at the point of inflection seems to be a drawback for the general use of the logistic function as a growth function, and several generalizations leading to a more flexible value of the function at the inflection point were proposed. Since these will not be discussed, we give only a few examples. Count (1943) proposed the use of

$$f_c(x) = \left(\frac{\alpha}{1 + \beta e^{\gamma x}}\right)^{-1/\alpha},$$

and Shohoji and Sasaki (1985) added linear and logarithmic terms to f in (16.1.1) to obtain a seven-parametric function. For forecasting purposes, Meade (1988) introduced a so-called local logistic model.

If the link to growth problems and the assumption $\gamma < 0$ are ignored, the function f is called a logistic regression function in connection with model (16.2.1). Logistic regression nowadays also has a slightly different meaning, being regarded as a special case of a generalized linear regression model with a logistic link function f usually with two parameters (i.e., with $\beta = 1$), as described by McCullagh and Nelder (1989). This kind of logistic regression for dichotomic random variables (odds ratio model) is mainly used in medical applications.

The logistic function can also be transformed into the three-parametric tanh function, which is nothing more than a reparametrization of f. This has no effect on most properties of the function such as its graph or its corresponding D-optimal experimental design. For this reparametrization we write f in (16.1.1) as

$$\begin{aligned} f(x, \theta) &= \frac{\alpha}{2}\left(1 + \frac{1/\beta e^{-\gamma x} - 1}{1/\beta e^{-\gamma x} + 1}\right) \\ &= \alpha_T\left(1 + \frac{e^{-2\beta_T\gamma_T + 2\gamma_T x} - 1}{e^{-2\beta_T\gamma_T + 2\gamma_T x} + 1}\right) \\ &= \alpha_T\{1 + \tanh[\gamma_T(x - \beta_T)]\} \\ &= f_T(x, \theta_T) \qquad \text{with } \theta_T^T = (\alpha_T, \beta_T, \gamma_T), \\ \alpha_T &= \alpha/2, \beta_T = -1/\gamma \ln \beta, \gamma_T = -\gamma/2. \end{aligned}$$

The vector $\nabla_\theta f$ of the derivatives of f in (16.1.1) with respect to the components of θ is, with

$$z = z(x, \theta) = (1 + \beta e^{\gamma x})^{-1}, \tag{16.1.2}$$

given by

$$\nabla_\theta f^T(x, \theta) = \{f_1(x, \theta), f_2(x, \theta), f_3(x, \theta)\}$$
$$= \{z, -\alpha e^{\gamma x} z^2, -\alpha\beta e^{\gamma x} x z^2\}. \tag{16.1.3}$$

If

$$C(\theta) = \begin{pmatrix} 1 & 0 & 0 \\ 0 & -\alpha & 0 \\ 0 & 0 & -\alpha \end{pmatrix},$$

we can write

$$\nabla_\theta f(x, \theta) = C(\theta) \begin{pmatrix} z \\ e^{\gamma x} z^2 \\ \beta e^{\gamma x} z^2 \end{pmatrix},$$

which means (Rasch, 1991) that f is partially nonlinear with the linearity parameter α and nonlinearity parameters β and γ.

16.2 PARAMETER ESTIMATION

Let us now consider a character y which depends on the time x in average by f in (16.1.1) with $\gamma < 0$. The growth model for $n > 3$ measurements y_i at the points x_i, respectively, is then given by

$$y_i = \frac{\alpha}{1 + \beta e^{\gamma x_i}} + e_i; \quad i = 1, \ldots, n > 3; \text{ at least three } x_i \text{ different},$$
$$\theta^T = (\alpha, \beta, \gamma) \in \Omega \tag{16.2.1}$$

The e_i are assumed to be i.i.d. random variables with expectation zero and variance $\sigma^2 = 1$. The set $X = \{x_1, \ldots, x_n\}$ is called the experimental design, and $\mathfrak{y}^T = (y_1, \ldots, y_n)$ is the observation vector. We further assume for Section 16.3.3 on robustness the existence of the fourth moments of the distribution of the e_i. Using z from (16.1.2), we write

$$z_i = z(x_i, \theta) \tag{16.2.2}$$

and introduce the following abbreviations (all sums from 1 to n):

$$A = \Sigma z_i^2, \quad B = \Sigma z_i^3 e^{\gamma x_i}, \quad C = \Sigma z_i^3 x_i e^{\gamma x_i},$$
$$D = \Sigma z_i^4 e^{2\gamma x_i}, \quad E = \Sigma z_i^4 x_i e^{2\gamma x_i}, \quad G = \Sigma z_i^4 x_i^2 e^{2\gamma x_i}. \tag{16.2.3}$$

The least squares estimator $\hat{\theta}^T = (\hat{\alpha}, \hat{\beta}, \hat{\gamma}) = (a, b, c)$ of θ^T, which for normally distributed e_i is identical to the maximum likelihood estimator, is defined by

$$\hat{\theta} = \arg \inf_{\Omega} \Sigma \{y_i - f(x_i, \theta)\}^2$$
$$\stackrel{\text{def}}{=} \arg \inf_{\Omega} R(\mathfrak{y}, X, \theta). \tag{16.2.4}$$

The three simultaneous equations (16.2.4) can only be solved numerically by iteration starting with some initial values a_0, b_0, and c_0. Good initial or starting values are important for successful iteration. The various iteration procedures such as gradient methods, the Gauss-Newton method, or the Marquardt algorithm have different degrees of sensitivity to starting values that differ greatly from the solution of (16.2.4). Experience has shown that good starting values can be derived from the geometrical behavior of the n observations y_i (the empirical growth curve) by roughly estimating the asymptotes and inflection point and using a modification of the Marquardt algorithm for iteration.

We now give a short geometrical interpretation of (16.2.4). The vector $\mathfrak{y}^T = (y_1, \ldots, y_n)$ of the n observations is an element of the euklidean space R^n. For a given θ the set of all $\mathfrak{y}^* \in R^n$ fulfilling the condition

$$\text{SL} = \{\mathfrak{y}^* | \exists \theta: \mathfrak{y}^* = f(x, \theta)\} \tag{16.2.5}$$

is called the solution locus of $f(x, \theta)$. Usually (with probability zero) $\mathfrak{y}$ is not an element of SL. The solution of (16.2.4) is the value $\theta = \hat{\theta}$ in (16.2.5) for that value $\mathfrak{y}^*$ which is closest to the observed $\mathfrak{y}$, and the infimum of (16.2.4) is exactly the square of that distance.

Example 16.2.1 In (16.1.1) let $\alpha = 10$, $\beta = 2$, and let only γ be unknown. Then

$$f(x, \gamma) = \frac{10}{1 + 2e^{\gamma x}}. \tag{16.2.6}$$

Because only one component of θ is unknown, γ can be estimated even if n, despite the side condition in (16.2.1), is equal to 2. A graphic demonstration is simple for $n = 2$. We therefore assume $n = 2$, $x_1 = 1$, $x_2 = 2$.

Table 16.2.1 presents a few values of the corresponding solution locus, which is the line in the (y_1, y_2)-plane shown in Fig. 16.2.1. In this figure four observation vectors $\mathfrak{y}_1$, $\mathfrak{y}_2$, $\mathfrak{y}_3$, $\mathfrak{y}_4$ are also shown together with their corresponding least distances from the solution locus. The scale on the line of the solution locus in Fig. 16.2.1 bears the values of γ corresponding to the $\mathfrak{y}$-values of SL. The observation vectors are $\mathfrak{y}_1^T = (4, 8)$, $\mathfrak{y}_2^T = (7, 3)$, $\mathfrak{y}_3^T = (1.25, 2.5)$, and $\mathfrak{y}_4^T = (1.25, 1.5)$. The least-squares estimators for

Table 16.2.1 Some Coordinates of the Solution Locus of Example 16.2.1

θ	$\frac{10}{1 + 2e^{\theta}}$	$\frac{10}{1 + 2e^{2\theta}}$
−1.9	7.70	9.57
−1.7	7.32	9.37
−1.5	6.91	9.09
−1.3	6.47	8.71
−1.1	6.00	8.18
−0.9	5.52	7.52
−0.82	5.32	7.20
−0.7	5.02	6.70
−0.5	4.52	5.76
−0.3	4.03	4.77
−0.24	3.89	4.47
−0.1	3.56	3.79
0	3.33	3.33
0.2	2.90	2.51
0.4	2.51	1.83
0.44	2.44	1.72
0.6	2.15	1.31
0.78	1.86	0.95
1.0	1.55	0.63
2.0	0.63	0.09

these four vectors are $\hat{\gamma}_1 = -0.82$, $\hat{\gamma}_2 = -0.24$, $\hat{\gamma}_3 = 0.44$, and $\hat{\gamma}_4 = 0.78$, respectively. For $\mathfrak{y}_4$, for instance, we have (calculated with the more exact value of $\hat{\gamma}_4 = 0.4358$)

$$\sum_{i=1}^{2} \left\{ y_i - \frac{10}{1 + 2e^{0.4958x_i}} \right\}^2 = 2.018.$$

The square root, 1.421, of this value is the distance of $\mathfrak{y}_4$ from the solution locus.

Example 16.2.2 This example demonstrates the computation of the estimation $\hat{\theta}$ of θ from (16.2.4) using real-world data taken from Baráth and Rasch (1989). Table 16.2.2 presents 14 measurements of hemp growth (plant heights). The x_i are the ages of the hemp plants in weeks, and the y_i are the plant heights in centimeter's.

Using the program LOWA (logistic growth curve analysis) of the module complex "Analysis of Growth Curves" in the expert system CADEMO

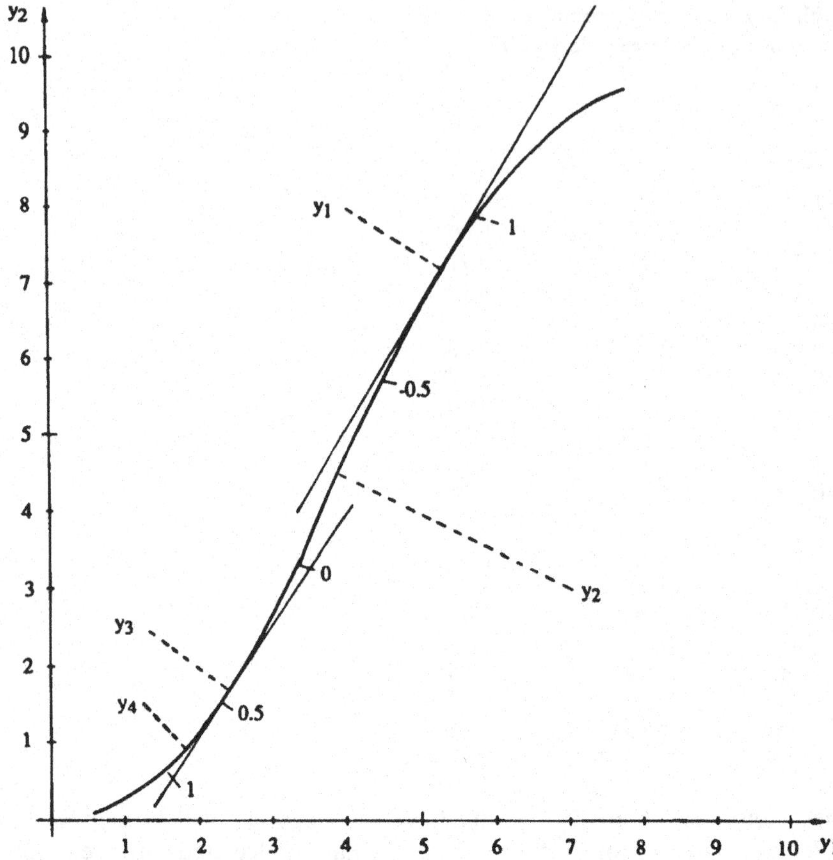

Fig. 16.2.1 Solution locus and four observation vectors of Example 16.2.1.

(for further information see Rasch et al., 1987, 1988), we obtained the starting values a_0, b_0, and c_0 and the values of the Marquardt iteration presented in Table 16.2.3.

Figure 16.2.2 shows the points of the 14 pairs (x_i, y_i) in Table 16.2.2 and the fitted curve defined by

$$f(x, a, b, c) = \frac{126.19}{1 + 12.449e^{-0.4607x}}.$$

To indicate the dependence of the estimator $\hat{\theta}$ in (16.2.4) on the size n of the vector $\mathfrak{y}_n = \mathfrak{y}$ of observations, we can also write $\hat{\theta}_n = \hat{\theta}$. Jennrich (1969) showed that for $N\{f(x_i, \theta), \sigma^2\}$ independently distributed compo-

Table 16.2.2 Results of Hemp Growth Measurements

Age x_i (weeks minus 1)	Plant height, y_i (cm)
0	8.3
1	15.2
2	24.7
3	32.0
4	39.3
5	55.4
6	69.0
7	84.4
8	98.1
9	107.7
10	112.0
11	116.9
12	119.9
13	121.1

Source: Barath and Rasch (1989).

nents of $\mathfrak{y}_n$,

$$\sqrt{n}(\hat{\theta}_n - \theta) \tag{16.2.7}$$

is asymptotically $N\{0_3, \sigma^2 I^{-1}(X, \theta)\}$ distributed $\{0_3^T = (0, 0, 0)\}$ with

$$I(X, \theta) = \lim_{n \to \infty} \frac{1}{n} \sum_{i=1}^{n} F_i^T(X, \theta) F_i(X, \theta)$$

$$= \lim_{n \to \infty} \frac{1}{n} F^T(X, \theta) F(X, \theta), \tag{16.2.8}$$

$$F_i(X, \theta) = \nabla_\theta f(x_i, \theta) \qquad \text{from (16.1.3)},$$

Table 16.2.3 Starting Values ($j = 0$) and Values after Five Iterations ($j = 5$) of the Numerical Solution of (16.2.4) Using the Data in Table 16.2.2 and the Corresponding Sums of Squared Deviations $R(\mathfrak{y}, X, \hat{\theta}_j) = R_j$ and $s_j^2 = R_j/11$

j	a_j	b_j	c_j	R_j	s_j^2
0	125.21	12.358	−0.4657	43.1607	3.9237
5	126.19	12.449	−0.4607	40.7473	3.7043

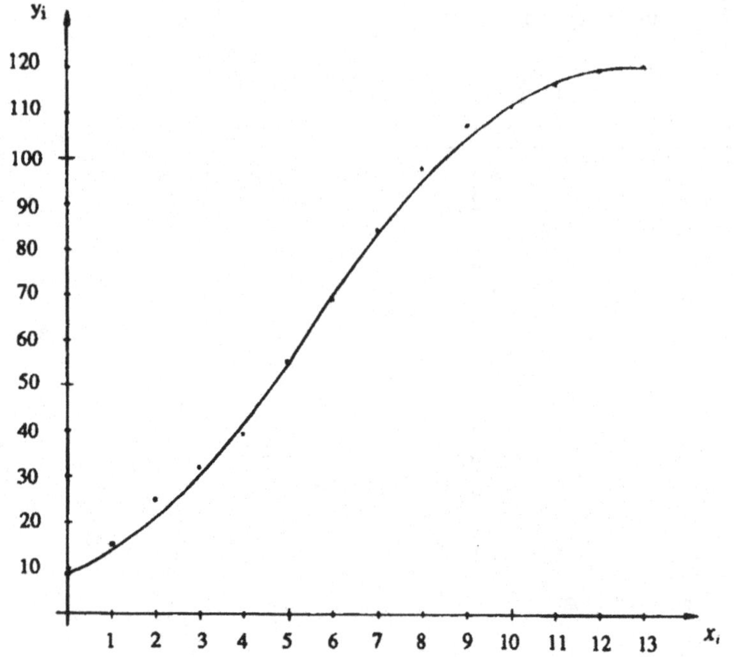

Fig. 16.2.2 Data of Example 16.2.2 and the graph of the fitted logistic function.

and

$$F^{T}(X, \theta) = \{F_1(X, \theta), \ldots, F_n(X, \theta)\}. \qquad (16.2.9)$$

The matrix $M(X, \theta) = 1/nF^{T}(X, \theta)F(X, \theta)$ in (16.2.8) is called the information matrix and

$$V_A(X, \theta) = \sigma^2\{F^{T}(X, \theta)F(X, \theta)\}^{-1} \qquad (16.2.10)$$

is the asymptotic covariance matrix of $\hat{\theta}_n$. Using the abbreviations in (16.2.3), we obtain

$$F^{T}(X, \theta)F(X, \theta) = \begin{pmatrix} A & -\alpha B & -\alpha\beta C \\ -\alpha B & \alpha^2 D & \alpha^2\beta E \\ \alpha\beta C & \alpha^2\beta E & \alpha^2\beta^2 G \end{pmatrix} \qquad (16.2.11)$$

and

$$|F^{T}(X, \theta)F(X, \theta)| = \alpha^4\beta^2\Delta,$$

with

$$\Delta = ADG + 2BCE - C^2D - AE^2 - B^2G. \qquad (16.2.12)$$

Hence, the asymptotic covariance matrix is

$$V_A(X, \theta) = \frac{\sigma^2}{\Delta}$$

$$\times \begin{pmatrix} DG - E^2 & -1/\alpha(EC - BG) & -1/(\alpha\beta)(BE - CD) \\ -1/\alpha(EC - BG) & 1/\alpha^2(AG - C^2) & 1/(\alpha^2\beta)(BC - AE) \\ 1/(\alpha\beta)(BE - CD) & 1/(\alpha^2\beta)(BC - AE) & 1/(\alpha^2\beta^2)(AD - B^2) \end{pmatrix}. \tag{16.2.13}$$

16.3 HYPOTHESIS TESTING AND INTERVAL ESTIMATION

16.3.1 Test Procedures Proposed

No exact method is known for testing any of the hypotheses

$$H_{0\alpha}: \quad \alpha = \alpha_0,$$
$$H_{0\beta}: \quad \beta = \beta_0,$$
$$H_{0\gamma}: \quad \gamma = \gamma_0. \tag{16.3.1}$$

In the case of exponential regression Rasch and Schimke (1985) proposed the use of a t-test based on the corresponding asymptotic covariance matrix to test hypotheses analogous to (16.3.1). They showed by an extensive simulation experiment that for risks of the first kind $\alpha_N = 0.05$ and $\alpha_N = 0.1$, respectively such tests are approximately α_N-tests with at most 20% deviation between the nominal and actual risks of the first kind over a wide range of the parameter space for samples of only $n = 4$. We now propose an analogous procedure for logistic regression, and report the results of a corresponding simulation experiment in Section 16.3.3.

It can be seen from (16.2.13) that the asymptotic variances of the parameter estimates a, b, and c of α, β, and γ, respectively, are

$$V(a) = \frac{\sigma^2}{\Delta}(DG - E^2)$$

$$V(b) = \frac{\sigma^2}{\Delta\alpha^2}(AG - C^2)$$

$$V(c) = \frac{\sigma^2}{\Delta\alpha^2\beta^2}(AD - B^2) \tag{16.3.2}$$

respectively. In analogy to the results for the exponential case we propose further that σ^2 be estimated by

$$s^2 = \frac{1}{n - 3} R(\mathfrak{y}, X, \hat{\theta}), \tag{16.3.3}$$

with R defined by (16.2.4). If we replace α, β, and γ in (16.3.2) by a, b, and c respectively, and write $\hat{A}$, $\hat{B}$, $\hat{C}$, $\hat{D}$, $\hat{E}$, $\hat{F}$, $\hat{G}$, and $\hat{\Delta}$ instead of A, B, C, D, E, F, G, and Δ, respectively, we obtain estimates (with unknown properties) of the asymptotic variances in (16.3.2) in the form

$$\hat{V}(a) = \frac{s^2}{\hat{\Delta}(\hat{D}\hat{G} - \hat{E}^2)} = s_a^2 = \frac{s^2}{\hat{\Delta}}(\hat{D}\hat{G} - \hat{E}^2) \tag{16.3.4}$$

$$\hat{V}(b) = s^2/\hat{\Delta}(\hat{A}\hat{G} - \hat{C}^2)/a^2 = s_b^2, \tag{16.3.5}$$

$$\hat{V}(c) = s^2/\hat{\Delta}(\hat{A}\hat{D} - \hat{B}^2)/a^2b^2 = s_c^2. \tag{16.3.6}$$

The following test statistics can now be defined:

For $H_{0\alpha}$: $\alpha = \alpha_0$ use

$$t_\alpha = \frac{a - \alpha_0}{s_a}. \tag{16.3.7}$$

For $H_{0\beta}$: $\beta = \beta_0$ use

$$t_\beta = \frac{b - \beta_0}{s_b}. \tag{16.3.8}$$

For $H_{0\gamma}$: $\gamma = \gamma_0$ use

$$t_\gamma = \frac{c - \gamma_0}{s_c}. \tag{16.3.9}$$

For $\kappa = \alpha$, β, or γ the proposed tests are defined by

$$k_\kappa = \begin{cases} 1, & \text{if } |t_\kappa| > t(n-3|1-\alpha_N/2) \\ 0, & \text{otherwise} \end{cases} \quad (\kappa = \alpha, \beta, \gamma). \tag{16.3.10}$$

Nothing is known theoretically about the tests defined by (16.3.10). Because $t(n - 3|1 - \alpha_N/2)$ is the $(1 - \alpha_N/2)$-quantile of a central t-distribution with $n - 3$ d.f., it is conjectured that even for small n the asymptotic results of Jennrich are useful for normally distributed error terms. We performed a simulation experiment to investigate the behavior of the tests in (16.3.10) (see section 16.3.3).

16.3.2 Confidence Estimation

Confidence regions for α, β, and γ, respectively, may be identified from the three statistical tests defined by (16.3.10) by defining the confidence region for κ ($\kappa = \alpha, \beta, \gamma$) as the set of κ_0-values ($\kappa_0 = \alpha_0, \beta_0, \gamma_0$) for which, for given sample estimators a, b, and c, $H_{0\kappa}$ is accepted. In this way the following confidence intervals can be constructed:

$$\text{For } \alpha\text{:}\quad \{a - t(n-3|1-\alpha_N/2)s_a,\ a + t(n-3|1-\alpha_N/2)s_a\},\tag{16.3.11}$$

$$\text{For } \beta\text{:}\quad \{b - t(n-3|1-\alpha_N/2)s_b,\ b + t(n-3|1-\alpha_N/2)s_b\},\tag{16.3.12}$$

$$\text{For } \gamma\text{:}\quad \{c - t(n-3|1-\alpha_N/2)s_c,\ c + t(n-3|1-\alpha_N/2)s_c\}.\tag{16.3.13}$$

16.3.3 Results of a Simulation Experiment

A series of simulation experiments were planned and performed with specific growth functions, parameter values, and experimental designs to investigate the difference between the nominal first-kind risk α_N and the estimated actual first-kind risk α_R of those tests or the corresponding difference between the actual confidence coefficient $1 - \alpha_R$ and the nominal confidence coefficient $1 - \alpha_N$. The sample size of the experiments (i.e., the number of runs) was chosen as 10,000. The estimated actual first kind risk α_R is defined as

$$\alpha_R = \frac{\text{number of false rejections}}{\text{number of runs}} = \frac{\text{NR}}{10{,}000}.\tag{16.3.14}$$

In each run, $n \geq 4$ pseudo-observations

$$y_i = f(x_i, \theta) + e_i \qquad (i = 1, \ldots, n;\ x_i \in [x_l, x_u])$$

were generated for different simulated error distributions with preassigned skewness γ_1 and kurtosis γ_2. Each pseudosample $\mathfrak{y}^T = (y_1, \ldots, y_n)$ led to a t_κ-value for each κ. We used $\alpha_N = 0.05$ and $\alpha_N = 0.1$, and for (γ_1, γ_2) the pairs

γ_1	0	0	0	1.5	0	2
γ_2	0	1.5	3.75	3.75	7	7

In other words, besides the normal distribution for $\gamma_1 = \gamma_2 = 0$, we included some nonnormal error distributions to investigate the robustness of the procedure. We planned to start with sample size $n = 4$ and to increase n until

$$|\alpha_N - \alpha_R| < 0.2\alpha_N.\tag{16.3.15}$$

Normal Error Distribution

Table 16.3.1 shows the values of $1 - \alpha_R$ (obtained by Lippert, 1986) for some parameter configurations, a normal error distribution ($\gamma_1 = \gamma_2 = 0$), and $\alpha_N = 0.05$ and $\alpha_N = 0.1$. Because the inequalities $|0.05 - \alpha_R| <$

Table 16.3.1 Results of Simulation Experiments for Sample Size $n = 4$ and Two Values of α, α_l (α_r) is the Percentage of Left (Right)-Sided Rejections, and $1 - \alpha_R$ is the Actual Confidence Coefficient

Number of experiment	Parameter value	$\alpha_N = 0.05$ α_l	α_r	$1 - \alpha_R$	$\alpha_N = 0.1$ α_l	α_r	$1 - \alpha_R$
	$\alpha_0 = 50$	2.38	2.62	95.00	4.95	5.12	89.93
1	$\beta_0 = 3.32$	2.82	2.51	94.67	5.79	4.45	89.76
	$\gamma_0 = -0.12$	2.33	2.41	95.26	4.62	5.06	90.32
	$\alpha_0 = 50$	2.52	2.18	95.30	5.32	4.48	90.20
2	$\beta_0 = 1.64$	1.96	2.73	95.31	4.30	5.58	90.12
	$\gamma_0 = -0.05$	2.20	2.52	95.28	4.93	4.97	90.10
	$\alpha_0 = 50$	2.59	2.34	95.07	5.24	4.73	90.03
3	$\beta_0 = 14.88$	2.78	2.07	95.15	5.67	3.99	90.34
	$\gamma_0 = -0.09$	2.32	2.51	95.17	4.63	5.16	90.21
	$\alpha_0 = 50$	2.64	2.83	94.53	5.02	5.42	89.56
4	$\beta_0 = 2.71$	2.68	2.53	94.79	5.35	4.77	89.88
	$\gamma_0 = -0.10$	2.52	2.56	94.92	5.15	5.02	89.83
	$\alpha_0 = 50$	2.63	2.49	94.88	5.02	5.24	89.74
5	$\beta_0 = 4.48$	2.91	2.21	94.88	5.74	4.79	89.47
	$\gamma_0 = -0.15$	2.47	2.58	94.95	5.00	4.93	90.07
	$\alpha_0 = 100$	2.81	1.84	95.35	5.45	3.74	90.81
6	$\beta_0 = 1.35$	2.53	1.95	95.52	4.95	3.98	91.07
	$\gamma_0 = -0.03$	2.31	2.47	95.22	4.67	4.85	90.48
	$\alpha_0 = 100$	2.81	2.33	94.86	5.81	4.43	89.76
7	$\beta_0 = 90.02$	3.61	1.86	94.53	6.73	3.58	89.69
	$\gamma_0 = -0.09$	2.27	3.03	94.70	4.34	5.59	90.07
	$\alpha_0 = 100$	2.37	2.46	95.17	4.89	4.83	90.28
8	$\beta_0 = 2.46$	2.91	2.34	94.75	5.34	4.75	89.91
	$\gamma_0 = -0.09$	2.46	2.61	94.93	4.90	5.14	89.96
	$\alpha_0 = 100$	2.43	2.79	94.78	4.68	5.31	90.01
9	$\beta_0 = 90.02$	3.05	2.00	94.95	5.91	4.16	89.93
	$\gamma_0 = -0.15$	2.17	2.89	94.94	4.41	5.67	89.92
	$\alpha_0 = 100$	2.43	2.62	94.95	4.92	5.28	89.90
10	$\beta_0 = 14.88$	2.72	2.33	94.95	5.50	4.56	89.94
	$\gamma_0 = -0.09$	2.35	2.65	95.00	4.66	5.36	89.98

Table 16.3.1 *(continued)*

Number of experiment	Parameter value	$\alpha_N = 0.05$			$\alpha_N = 0.1$		
		α_l	α_r	$1 - \alpha_R$	α_l	α_r	$1 - \alpha_R$
	$\alpha_0 = 200$	3.16	2.20	94.64	5.80	4.37	89.83
11	$\beta_0 = 90.02$	2.84	2.48	94.68	5.76	4.73	89.51
	$\gamma_0 = -0.09$	2.79	2.52	94.69	5.19	5.18	89.63
	$\alpha_0 = 200$	2.45	2.53	95.02	5.09	5.13	89.78
12	$\beta_0 = 90.02$	2.64	2.57	94.79	5.22	4.82	89.96
	$\gamma_0 = -0.15$	2.73	2.58	94.69	5.04	5.05	89.91
	$\alpha_0 = 200$	2.57	2.28	95.15	4.98	4.88	90.14
13	$\beta_0 = 14.88$	2.40	2.32	95.28	4.92	4.76	90.32
	$\gamma_0 = -0.09$	2.50	2.41	95.09	4.88	4.98	90.14
	$\alpha_0 = 200$	2.91	1.01	95.18	6.09	3.96	89.95
14	$\beta_0 = 4.48$	2.82	1.92	95.26	5.98	4.08	89.94
	$\gamma_0 = -0.03$	2.44	2.26	95.30	4.89	4.92	90.19
	$\alpha_0 = 200$	2.69	2.27	95.04	5.45	4.61	89.94
15	$\beta_0 = 1.35$	2.47	2.37	95.16	5.16	4.86	89.98
	$\gamma_0 = -0.03$	2.45	2.53	95.02	5.09	5.20	89.71

Table 16.3.2 Minimal Values n_0 for Different (γ_1, γ_2)-Pairs for Which (16.3.15) was Satisfied for $\alpha_N = 0.05$ and all Three Hypotheses (number of parameter combinations as defined in Table 16.3.2)

Number of parameter combinations	$\gamma_1 = 0$, $\gamma_2 = 1.5$	$\gamma_1 = 0$, $\gamma_2 = 3.75$	$\gamma_1 = 1.5$, $\gamma_2 = 3.75$	$\gamma_1 = 0$, $\gamma_2 = 7$	$\gamma_1 = 2$, $\gamma_2 = 7$
1	4	4	10	4	20
2	4	4	4	4	4
3	4	4	4	4	4
4	4	4	4	4	19
5	4	7	4	4	17
6	4	4	4	4	4
7	4	4	10	8	18
8	4	4	4	4	4
9	4	4	4	5	4
10	4	4	4	4	4
11	4	4	4	4	4
12	4	5	4	6	4
13	4	4	4	4	4
14	4	4	4	4	5
15	4	4	4	4	4

$0.2\alpha_N = 0.01$ and $|0.1 - \alpha_R| < 0.2\alpha_N = 0.02$ already for $n = 4$ are fulfilled for $\alpha_N = 0.05$ and $\alpha_N = 0.1$, respectively, the tests proposed in Section 16.3.1 and the corresponding confidence intervals proposed in Section 16.3.2 can be used even if n is as small as 4.

Nonnormal Error Distributions—Robustness

To investigate the robustness of the proposed tests against violation of the normality assumption, we simulated five nonnormal error distributions with the skewness-kurtosis pairs noted above. These distributions are elements of the Fleishman system of probability distributions (Fleishman, 1978). The corresponding random variables e_i were generated by the method described by Nürnberg (1986). Again 10,000 runs were made, starting with $n = 4$ and increasing n until for $\alpha_N = 0.05$ the inequality (16.3.15) was satisfied. Table 16.3.2 presents the smallest value n_0 of n for which (16.3.15) was satisfied.

16.4 EXPERIMENTAL DESIGN PROBLEMS

16.4.1 Results for General Nonlinear Regression Functions

Let $f(x, \theta)$ be any nonlinear regression function with parameter vector $\theta \in \Omega \subset R^p$ with p elements. The experimental design

$$X^{\mathrm{T}} = \{x_1, \ldots, x_n\}$$

may contain exactly m different values $(p \le m \le n)$ $s_1, \ldots, s_m$ which w.l.o.g. may be ordered $(s_1 < s_2, \ldots, < s_m)$. Let n_i be the frequency of s_i; i.e., $n_i = |\{j: x_j \in X, x_j = s_i\}|$.

Definition 16.4.1 A mapping D assigning a positive integer n_i to each s_i $(i = 1, \ldots, m)$ in its domain $\mathrm{Dom}(D) \in [x_l, x_u]$ is called an exact (experimental) design, $D_{n,m}$; $\mathrm{Dom}(D)$ is called the support, and $n = \Sigma n_i$ the size of the exact experimental design. The cardinality, $|\mathrm{Dom}(D)| = m$ of $\mathrm{Dom}(D)$, defines the design as an m-point design. Let $D(n, m)$ denote the set of all exact m-point designs of size n, and let $D(n) = D(n, p) \cup D(n, p + 1) \cup \cdots \cup D(n, n)$ be the set of all exact designs of a given size n. Each element $D_{n,m} \in D(n)$ can be written uniquely as

$$D_{n,m} = \begin{pmatrix} s_1, \ldots, s_m \\ n_1, \ldots, n_m \end{pmatrix}. \qquad (16.4.1)$$

The interval $B = [x_l, x_u]$ is called the experimental region.

If we use the representation (16.4.1) of an experimental design X to denote the dependence on $\theta_0 \in \Omega$, we write the matrix $V_A(X, \theta)$ in (16.2.10) as $V(D_{n,m}, \theta_0)$. Let $\mathfrak{M}_n = \{V(D_{n,m}, \theta_0): D_{n,m} \in D(n)\}$ be the set of these matrices for all $D_{n,m} \in D(n)$ for any fixed n. The optimality criteria for exact designs considered here are special cases of

Definition 16.4.2 Let $Z: \mathfrak{M}_n \to R^+$ be a real functional. Then

$$D_n^* = \arg\inf_{D_{n,m} \in D(n)} \{Z[V(D_{n,m}, \theta_0)],$$

$$\theta_0 \in \Omega \subset R^p, m = p, p + 1, \ldots, n\} \quad (16.4.2)$$

is called an exact locally Z-optimum (experimental) design of size n at $\theta = \theta_0$, and

$$D_{n,m}^* = \arg\inf_{D_{n,m} \in D(n,m)} \{Z[V(D_{n,m}, \theta_0)], \theta_0 \in \Omega \subset R^p\}$$

is called an exact locally Z-optimum m-point design of size n at $\theta = \theta_0$. If

$$Z[V(D_{n,m}, \theta_0)] = |V(D_{n,m}, \theta_0)|, \quad (16.4.3)$$

D_n^* ($D_{n,m}^*$) is called an exact locally D-optimum (m-point) design of size n at $\theta = \theta_0$ or, briefly, an EDD(n, θ_0) (EDD(n, m, θ_0)). If

$$Z[V(D_{n,m}, \theta_0)] = E_i^T V(D_{n,m}, \theta_0) E_i \quad (16.4.4)$$

with the ith unity vector E_i in R^p (i.e., E_i is a vector of length p with unity as its ith element and zeros otherwise), D_n^* ($D_{n,m}^*$) is called an exact locally $C_{\theta i}$-optimum (m-point) design of size n at $\theta = \theta_0$ or, briefly, an $EC_{\theta i}D(n, \theta_0)$ ($EC_{\theta i}D(n, m, \theta_0)$).

For an exact locally D-optimum m-point design $D_{n,m}$ the matrix $M(X, \theta)$ in (16.2.8) can be written

$$M(\theta, D_{n,m}) = \sum_{i=1}^{m} n_i/nk(s_i, \theta)k^T(s_i, \theta), \quad (16.4.5)$$

with $k^T(x, \theta) = \nabla_\theta f(x, \theta)$ as defined in (16.1.3).

Most papers on experimental design in nonlinear regression deal with D-optimality, for which also most of the analytical results were obtained. Rasch et al. (1985) present the historical development of experimental design in nonlinear regression. Box and Lucas (1959) obtained the first results for exact D-optimum designs in the special case $n = m = p$.

If $m = p$, the matrix $nM(\theta, D_{n,p})$ can be written

$$nM(\theta, D_{n,p}) = G^T(\theta, D_{n,p})\mathfrak{N}G(\theta, D_{n,p}), \quad (16.4.6)$$

with

$$G(\theta, D_{n,p}) = (f_j(s_i, \theta))_{i,j=1,\dots,p} \tag{16.4.7}$$

and

$$\mathfrak{N} = \text{diag}(n_1, \dots, n_p). \tag{16.4.8}$$

(If, in addition, $n = p$, we have $\mathfrak{N} = I_p$; i.e., $\mathfrak{N}$ is the identity matrix.) Minimizing $|V(D_{n,p}, \theta_0)|$ means maximizing $|M(D_{n,p}, \theta_0)|$ and, for $m = p$, $|G(\theta, D_{n,p})|$. This leads to

Theorem 16.4.1 The support of an EDD(n, p, θ_0) is independent of n.

Proof. For $m = p$ it follows from (16.4.6) that

$$D^*_{n,p} = \underset{D_{n,p}\in D(n,p)}{\arg\sup} |G^T(\theta, D_{n,p})\mathfrak{N}G(\theta, D_{n,p})|,$$

and, since $|ABC| = |A|\,|B|\,|C|$ for any three square matrices A, B, and C of the same order, we have

$$D^*_{n,p} = \underset{D_{n,p}\in D(n,p)}{\arg\sup} |G(\theta, D_{n,p}|^2|\mathfrak{N}|. \tag{16.4.9}$$

This completes the proof.

From Theorem 16.4.1 it follows that the support and the n_i of an EDD(n, p, θ_0) can be determined independently.

Corollary The EDD(n, p, θ_0) $D^*_{n,p}$ in (16.4.9) for a fixed n is given by a support which maximizes $|G(\theta, D_{n,p})|$ for $D_{n,p} \in D(n, p)$ and by n_i-values which are as equal as possible.

This statement follows from (16.4.9) because under $\Sigma\, n_i = n$ the determinant $|\mathfrak{N}| = \Pi_{i=1}^{p} n_i$ is maximized if the n_i are as equal as possible; i.e., if v is the greatest positive number for which $vp \le n$, we have either $n_i = v$ $(i = 1, \dots, p)$ if $vp = n$ or in the case $n - vp = t > 0$ the optimum set of the n_i is any permutation of $(v, \dots, v, v + 1, \dots, v + 1)$ with $p - t$ elements equal to v (and therefore t elements equal to $v + 1$).

Definition 16.4.3 Let η: $\Omega \to H$ be a parameter transformation with $\dim(\Omega) = \dim(H) = p$ and existing derivatives

$$\frac{\partial\eta(\theta)}{\partial\theta_j} \quad (j = 1, \dots, p).$$

Then

$$w(x, \eta) = w\{x, \eta(\theta)\} = f(x, \theta) \tag{16.4.10}$$

is called a reparametrization of $f(x, \theta)$.

Theorem 16.4.2 The support of an EDD(n, θ_0) for estimating θ in $f(x, \theta)$ is identical to the support of an EDD(n, η_0) at $\eta_0 = \eta(\theta_0)$ for estimating $\eta(\theta)$ in the reparametrization $w(x, \eta)$ of $f(x, \theta)$.

Proof. We write $\eta^T = (\eta_1, \ldots, \eta_p)$ and put

$$t_{i,j} = \frac{\partial \eta_i(\theta)}{\partial \theta_j} \quad (i, j = 1, \ldots, p), \quad T = (t_{i,j}),$$

$$\left(\frac{\partial w(x, \eta)}{\partial \eta}\right)^T = \left(\frac{\partial w(x, \eta)}{\partial \eta_1}, \ldots, \frac{\partial w(x, \eta)}{\partial \eta_p}\right),$$

$$\left(\frac{\partial \eta}{\partial \theta_j}\right) = \left(\frac{\partial \eta_1}{\partial \theta_j}, \ldots, \frac{\partial \eta_p}{\partial \theta_j}\right),$$

and

$$u_i(x_j, \theta) = \left(\frac{\partial w(x_i, \eta)}{\partial \eta}\right)^T \frac{\partial \eta}{\partial \theta_j} = \left(\frac{\partial \eta}{\partial \theta_j}\right)^T \left(\frac{\partial w(x_i, \eta)}{\partial \eta}\right).$$

It follows that

$$F^T(X, \theta)F(X, \theta) = T^T W^T(\eta, X) W(\eta, X) T$$

with

$$W(\eta, X) = \left(\frac{\partial w(x_i, \eta)}{\partial \eta_j}\right)$$

and, further, that

$$|F^T(X, \theta)F(X, \theta)| = |T^T||W^T(\eta, X)W(\eta, X)||T|.$$

The fact that T is independent of X completes the proof. Without proof (see Rasch, 1990) we state

Theorem 16.4.3 The support of an EDD(n, m, θ_0) is independent of the linearity parameters of f in (16.1.1).

16.4.2 D-Optimum Experimental Designs for the Logistic Function

From Theorem 16.4.2 it follows that the support of the EDD(n, θ_0) is the same for both the function (16.1.1) and its reparametrization as a tanh function. More specifically, this means that the EDD(n, 3, θ_0) of both representations are identical. It follows from Theorem 16.4.3 and from the form of the asymptotic covariance matrix of the parameter estimators of the logistic functions that the support (not the value of the determinant) of and EDD(n, m, θ_0) is independent of α_0. It further follows from Theorem 16.4.1 that only the support of a locally *D*-optimum 3-point design has to be determined because the *n* observations have to be distributed over the three support points as equally as possible. If it were certain that a *D*-optimum design is always a 3-point design, we could restrict ourselves to the case $m = 3$. Though all searches performed so far by our expert system CADEMO have led to *D*-optimum designs that are 3-point designs, no proof that the support size of a *D*-optimum design is equal to 3 for all possible parameter configurations has been found.

Example 16.4.1 For the parameters found for the hemp data in Example 16.2.2, namely $a = 126.19$, $b = 12.449$, and $c = -0.4607$, we will use the search procedure in the module "Logistic Growth" of the module complex "Growth Curve Analysis" of the expert system CADEMO (Rasch et al. 1987, 1988) to find the locally *D*-optimum design for $n = 14$ measurements at

$$\hat{\theta}_0^T = (126.19,\ 12.449,\ -0.4607).$$

This search procedure yields the design

$$\begin{pmatrix} 2.89453 & 7.24902 & 13 \\ 5 & 5 & 4 \end{pmatrix}$$

There are two other solutions with $n_1 = n_3 = 5$, $n_2 = 4$ and with $n_1 = 4$, $n_2 = n_3 = 5$, respectively, but the support is the same.

To illustrate the influence of the parameter vector θ_0 on the support of

Table 16.4.1 Values of x_1 and x_2 ($x_3 = 13$ remained constant) of an EDD($14, 3, \theta_0$) for $\alpha_0 = 126.191$ and Different Values of β_0 and γ_0

β_0	γ_0	x_1	x_2
12.449	−0.46074	2.89453	7.24902
12.449	−0.3	2.95801	8.89941
12.449	−0.4	3.07227	7.93457
12.449	−0.5	2.76758	6.83008
12.449	−0.6	2.39941	5.85254
5	−0.46074	1.07910	5.52246
10	−0.46074	2.47559	6.85547
12	−0.46074	2.83105	7.18555
13	−0.46074	2.98340	7.32520
15	−0.46074	3.26270	7.59180
20	−0.46074	3.80859	8.08691

the design, we changed the values of β_0 and γ_0. The results are given in Table 16.4.1.

In this example it therefore seems reasonable, even if the parameter values are totally unknown, to make the first five measurements between the second and the fourth week, five further measurements between the sixth and the ninth week, and four measurements after 13 weeks. Finally, we compare the value of the *D*-criterion of the optimal design with the value of the *D*-criterion of the design used in Example 16.2.2 with a support of 14 equidistant points between 0 and 13 for the parameter values estimated in Example 16.2.2, respectively. For the optimal design we obtained the value $0.34995 \times 10^{-6} \sigma^6$, and for the equidistant design the corresponding value (calculated by help of CADEMO) is $0.65371 \times 10^{-6} \sigma^6$; i.e., the latter design is much worse than the optimum one.

REFERENCES

Baráth, C. and Rasch, D. (1989). A CADEMO "szakértöi rendszer" alkalmazása növényi növekedésvizsgálatok modellezésére, *Növénytermelés*, *38*, 281–288.

Box, G. E. P. and Lucas, H. L. (1959). Design of experiments in nonlinear situations, *Biometrika*, *46*, 77–90.

Cook, R. D. and Nachtsheim, C. J. (1980). A comparison of algorithms for constructing exact *D*-optimal designs, *Technometrics*, *22*, 315–323.

Count, E. W. (1943). Growth pattern of human physique: an approach to kinetic anthropometry, *Human Biol.*, *15*, 1–32.

Duchrau, P. and Frischmuth, K. (1990). A numerical search algorithm for experimental design, *Rostock Math. Kolloq.*, 40, 83–94.

Duchrau, P. and Frischmuth, K. (1990). A numerical search algorithm for experimental design, *Rostock Math. Kolloqu.*, to appear.

Feller, W. (1940). On the logistic law of growth and its empirical verification in biology, *Acta Biomathematica*, *5*, 51–66.

Fleishman, A. I. (1978). A method for simulating non-normal distributions, *Psychometrika*, *43*, 521–532.

Jennrich, R. J. (1969). Asymptotic properties of nonlinear least-squares estimators, *Ann. Math. Statist.*, *40*, 633–643.

Leach, D. (1981). Re-evaluation of the logistic curve for human populations, *J. R. Statist. Soc. A*, *144*, 94–103.

Lippert, S. (1986). Untersuchungen des t-Tests bei Nichtlinearität und Nichtnormalität am Beispiel der logistischen Funktion, Master Thesis, University Rostock, Department of Mathematics.

Lotka, A. J. (1931). The structure of a growing population, *Hum. Biol.*, *3*, 459–493.

McCullagh, P. and Nelder, J. A. (1989). *Generalized Linear Models*, 2nd ed., London, New York.

Meade, N. (1988). A modified logistic model applied to human populations, *J. R. Statist. Soc. A*, *151*, 491–498.

Nürnberg, G. (1985, 1986). Robustness of two-sample tests for variances, in Rasch and Tiku (1985, 1986), 75–82.

Pearl, R. (1925) *The Biology of Population Growth*, New York, Knopf.

Pearl, R. and Reed, L. J. (1920). On the rate of growth of the population of the United States since 1790 and its mathematical representation, *Proc. Nat. Acad. Sci.*, *6*, 275–287.

Pohl, S. (1988). Lokal-optimale Versuchspläne für spezielle nichtlineare Regressionsfunktionen Diplomarbeit, Sekt. Mathematik, WPU Rostock.

Rasch, D. (1984). Robust confidence estimation and tests for parameters of growth functions, in Szamitasttechnikai es kibernetikai modszerek alkalmazasa az orvostodomanyban es a biologiaban Szeged, I. Gyoeri (ed.), 306–313.

Rasch, D. (1985). Finite sample behaviour of an asymptotic *t*-test in exponential regression, *Statistics*, *16*, 121–124.

Rasch, D. (1986). Optimum experimental design for fitting growth curves in cattle, *Arch. Tierzucht*, *29*, 85–91.

Rasch, D. (1988). Recent results in growth analysis, *Statistics*, *19*, 585–604.

Rasch, D. (1990). *Optimum Experimental Design in Nonlinear Regression, Commun. Statist.—Theor. Meth.*, *19*(12), 4789–4806.

Rasch, D. Nürnberg, G., and Busch, K. (1988). CADEMO—Ein Expertensystem zur Versuchsplanung, in *Fortschritte der Statistik-Software* 1, F. Faulbaum, und H. M. Ühlinger, (eds.), 193–201, Fischer Verlag Stuttgart, New York.

Rasch, D., Guiard, V., Nürnberg, G., Rudolph, P. E., and Teuscher, F. (1987). The expert system CADEMO. Computer aided design of experiments and modelling, *Statistical Software Newsletter*, *13*, 107–114.

Rasch, D., Rudolph, E., and Schimke, E. (1985). Optimale Versuchspläne in der nichtlinearen Regression, *Probleme der angewandten Statistik*, *13*, 93–121.

Rasch, D. and Schimke, E. (1985, 1986). A test for exponential regression, in D. Rasch and M. L. Tiku, (1985, 1986), 104–112.

Rasch, D. and Tiku, M. L. (eds.), (1985, 1986). *Robustness of Statistical Methods and Nonparametric Statistics*, VEB Deutscher Verlag der Wissenschaften (1985); Reidel, Dordrecht, (1986).

Shohoji, T. and Sasaki, H. (1985). An aspect of growth analysis of weight in savannah baboon, *Growth*, *49*, 300–309.

Verhulst, P. J. (1838). Notice sur la lois que la population suit dans sons accroissement, *Corr. Math. Phys.*, *10*, 113–121.

Yule, G. U. (1925). The growth of population and the factor which controls it, *J. R. Statist. Soc. A*, *88*, 1–58.

17

Applications in Health and Social Sciences

Chris P. Tsokos and Peter S. DiCroce
University of South Florida, Tampa, Florida

17.1 INTRODUCTION

The logistic growth function was first derived as a tool for use in demographic studies by Verhulst (1845), and was given its present name by Reed and Berkson (1929), two of the function's early enthusiasts. Verhulst assumed that the increase of the natural logarithm of a population size for a given geographic area as a function of the time is a constant minus a function which increases with the population. One of the solutions of his system of equations was

$$P_t = \frac{P_s}{1 + [P_s/P_0 - 1]\exp\{-at\}}, \tag{17.1.1}$$

where P_t is the population after time t, P_0 is the initial population, P_s is the asymptotic maximum or saturation level, and $a > 0$ is the growth rate coefficient. One of the more popular parameterizations of (17.1.1) is obtained by letting $b = P_s/P_0 - 1$, which yields the form

$$P_t = \frac{P_s}{1 + b\exp\{-at\}}. \tag{17.1.2}$$

In either form it is important to note that P_s is assumed to be independent of dynamic environmental conditions. In the case of human populations, this means technical and social changes. It could mean constant temperature, no further introduction of catalyst in a cell growth experiment, or no use of mass-media techniques in the measure of the spread of an innovation or rumor.

An illustrative example of the sensitivity of the logistic growth function to this assumption is the fitting to U.S. census population results, excluding Alaska and Hawaii, for the 170-year period from 1790 to 1960. Table 17.1.1 reports an extension of the results derived by Pearl and Reed (1940), i.e., actual U.S. census reports of population, the logistic estimates, and the error of the estimates.

The logistic estimates begin to diverge dramatically from the actual U.S. population beginning in 1950. The function could not take into account the post–World War II baby boom and the vastly improved health care services available.

Table 17.1.1 The Fit of the Logistic Curve to the U.S. Census Populations (excluding Alaska and Hawaii) for the 170-Year Period From 1790 to 1960

	Population (millions)		
Year	U.S. census	Logistic estimate	Error of the estimate
1790	3.9	3.7	0.2
1880	5.3	5.1	0.2
1810	7.2	7.0	0.2
1820	9.6	9.5	0.1
1830	12.9	12.8	0.1
1840	17.1	17.3	−0.2
1850	23.2	23.0	0.2
1860	31.4	30.3	1.1
1870	38.6	39.3	−0.7
1880	50.2	50.2	0.0
1890	62.9	62.8	0.1
1900	76.0	76.7	0.7
1910	92.0	91.4	0.6
1920	105.7	106.1	−0.4
1930	122.8	120.1	2.7
1940	131.4	132.8	−1.4
1950	150.7	143.8	6.9
1960	178.5	153.0	25.5

Another form is the general logistic distribution with cdf

$$F(x) = \frac{1}{1 + \exp\{-a(x - \mu)/\sigma\}}, \qquad -\infty < x < \infty, \qquad (17.1.3)$$

where $a = \pi/\sqrt{3}$, $-\infty < \mu < \infty$, is the mean, $\sigma > 0$ is the standard deviation, and μ and σ are parameters of location and scale, respectively. The general logistic distribution is described by the density function

$$f(x) = \frac{a\sigma^{-1}\exp\{-a(x - \mu)/\sigma\}}{[1 + \exp\{-a(x - \mu)/\sigma\}]^2}, \qquad -\infty < x < \infty, \qquad (17.1.4)$$

where a, μ and σ are as defined in (17.1.3).

Logistic regression analysis has emerged not only as the most used but as the most important application of the logistic distribution. This is especially true of multiple logistic regression. Initially, dose-response (bioassay) analysis was the primary use for the logistic regression model, and much of the early terminology remains today.

Let Y be a random variable that can take on two values, say 0 and 1. These two possibilities could represent, for example, life and death. Suppose also that $\Pr\{Y = 1\}$ depends on an independent (sometimes called explanatory) variable x, called dose (or stimulus). The logistic dose-response curve specifies that

$$\begin{aligned} P_i = \Pr\{Y = 1\} &= \frac{1}{1 + \exp\{-(\alpha + \beta x)\}} \\ &= \frac{\exp\{\alpha + \beta x\}}{1 + \exp\{\alpha + \beta x\}}, \end{aligned} \qquad (17.1.5)$$

and

$$Q = 1 - P_i = \Pr\{Y = 0\} = \frac{1}{1 + \exp\{\alpha + \beta x\}}, \qquad (17.1.6)$$

where $0 < \alpha = a/\sigma$ and $-\infty < \beta = a\mu/\sigma < \infty$ are parameters. Now, the linear transformation of P_i/Q_i, which is equivalent to the linear transformation of the density function of the general logistic distribution (17.1.4), is given by

$$\text{logit } P_i = \ln\left(\frac{P_i}{Q_i}\right) = \alpha + \beta x_i, \qquad (17.1.7)$$

where α and β are as defined in (17.1.5) and (17.1.6).

There are a multitude of extensions and generalizations of the logit (17.1.7), but the one we consider here is the logit generalized for k in-

Table 17.1.2 Summary of Different Logistic Model Forms

Name	Form	Conditions on parameters	Remarks
Growth function	$P_t = \dfrac{P_s}{1 + [P_s/P_0 - 1]\exp\{-at\}}$ $t \geq 0,\ P_s > 0$	$a > 0,\ b > 0$	
Logit	$\ln(P_i/Q_i) = \alpha + \beta x_i$ $\alpha = a/\sigma,\ \beta = a/\sigma,\ Q_i = 1 - P_i$	$\alpha > 0$ $-\infty < \beta < \infty$	Linear transformation of general logistic distribution
Generalized logit	$\ln(P_{ij}/Q_{ij}) = \alpha_j + \beta_j x_{ij}$ $j = 1, \ldots, k \quad Q_{ij} = 1 - P_{ij}$	$\alpha_j > 0$ $-\infty < \beta < \infty$	k sample form of logit
Dose-response	$P(x) = \dfrac{\exp\{\alpha + \beta\}}{1 + \exp\{\alpha + \beta x\}}$ $-\infty < x < \infty$	$0 < \alpha = a/\sigma$ $-\infty < \beta = a\mu/\sigma < \infty$	S-shaped curve
General logistic distribution	$f(x) = \dfrac{a\sigma^{-1}\exp\{-a(x - \mu)/\sigma\}}{[1 + \exp\{-a(x - \mu)/\sigma\}]^2}$ $-\infty < x < \infty,\ a = \pi/\sqrt{3}$	$-\infty < \mu < \infty$ $\sigma > 0$	Symmetrical

dependent samples. Suppose that $Y_{1j}, Y_{2j}, \ldots, Y_{nj}$ are independent binary random variables from the jth sample and x_{ij} is the covariate value associated with Y_{ij}. Then (17.1.7) becomes for the jth sample

$$\text{logit } P_{ij} = \ln \frac{P_{ij}}{1 - P_{ij}} = \alpha_j + \beta_j x_{ij}, \tag{17.1.8}$$

where α_j and β_j are parameters and

$$P_{ij} = \Pr\{Y_{ij} = 1 | x_{ij}\}. \tag{17.1.9}$$

Obviously, when $k = 1$, Eq. (17.1.8) reduces to (17.1.7).

A summary of some of the different forms the logistic function can assume are presented in Table 17.1.2.

17.2 ESTIMATION OF THE PARAMETERS OF THE LOGISTIC FUNCTION

We shall begin with the logistic function given by

$$P_i = 1 - Q_i = [1 + \exp\{-(\alpha + \beta x)\}]^{-1}, \tag{17.2.1}$$

where (response) P_i is the probability of an event at (dose/stimulus) X_i; α and β are its parameters. This function is also referred to as the logistic dose-response curve. The relative frequency, $P_i = r_i/n_i$, where r_i is the number of responses in n_i trials, is assumed to be a binomial variable. When the response, P_i, is measured in terms of the observed proportion,

p, affected out of n exposed, then the response is said to be binary (or quantal).

In applications, the reduced variate form of (17.2.1) is often used. This is defined as the linear transformation of the ratio of P_i and Q_i and is given by

$$\text{logit } P_i = \ln\left(\frac{P_i}{Q_i}\right) = \alpha + \beta x, \tag{17.2.2}$$

which is called the logit.

Berkson (1944, 1953) suggested that estimates of α and β may be obtained using the minimum logit chi-square method. This involves choosing $\bar{\alpha}$, $\bar{\beta}$, so as to minimize

$$X^2 = \Sigma\, W_i(l_i - \bar{\alpha} - \bar{\beta}x_i)^2, \tag{17.2.3}$$

where $l_i = \ln[r_i/(n_i - r_i)]$ and $W_i = (r_i/n_i)(n_i - r_i)$. The estimates are given by

$$\bar{\alpha} = \frac{\Sigma\, Wx^2\, \Sigma\, Wl - \Sigma\, Wx\, \Sigma\, Wlx}{\Sigma\, W\, \Sigma\, Wx^2 - (\Sigma\, Wx)^2}, \tag{17.2.4}$$

and

$$\bar{\beta} = \frac{\Sigma\, W\, \Sigma\, Wlx - \Sigma\, Wx\, \Sigma\, Wl}{\Sigma\, W\, \Sigma\, Wx^2 - (\Sigma\, Wx)^2}, \tag{17.2.5}$$

respectively.

Suppose that k different dosages are used and that a response is observed with relative frequency $P_i = r_i/n_i$ subjected to dose x_i, the method of estimation that is most useful and commonly used in this setting is the maximum likelihood estimation (MLE) procedure. The MLE of α and β, denoted by $\hat{\alpha}$ and $\hat{\beta}$ respectively, are obtained by solving the following system of equations:

$$\sum_{i=1}^{k} n_i p_i = \sum_{i=1}^{k} n_i \hat{p}_i, \tag{17.2.6}$$

$$\sum_{i=1}^{k} n_i p_i x_i = \sum_{i=1}^{k} n_i \hat{p}_i x_i, \tag{17.2.7}$$

where $\hat{p}_i$ is the estimate of P_i given by

$$\hat{p}_i = [1 + \exp\{-(\hat{\alpha} + \hat{\beta}x)\}]^{-1}. \tag{17.2.8}$$

Generally, the solutions of (17.2.6) and (17.2.7) cannot be obtained explicitly, and an iterative solution must be employed (Berkson, 1957). Thus, there is no closed-form solution for the MLE of α and β. The iterations would be continued until a desired tolerance is obtained.

The logistic function has a minimal set of sufficient statistics. When a minimal sufficient set of statistics exists, it is known that the MLE is a function of these sufficient statistics (Fisher, 1925; Neyman, 1935). It can be seen that the MLE, $\hat{\alpha}$ and $\hat{\beta}$, are functions of $\Sigma\, n_i p_i = \Sigma\, r_i$ and $\Sigma\, n_i p_i x_i = \Sigma\, r_i x_i$. We have $\Sigma\, n_i p_i$ minimally sufficient for α, with β fixed, and $\Sigma\, n_i p_i x_i$ is minimally sufficient for β when α is fixed (Berkson, 1955). Berkson (1957) developed tables for the MLE of the logistic function based on different configurations of the inherent parameter structure.

Another early method developed for the binomial logistic model was the transfer method, which simplified the work of fitting a logistic response curve to binary data by maximum likelihood (Hodges, 1958). The transfer method is based on two observations in addition to the above discussion of the existence of minimal sufficient statistics. That is, the following two conditions must hold:

1. The MLE are functions of $\Sigma\, r_i$ and $\Sigma\, r_i x_i$ and therefore have the same value for the observed r_i and for any other set r_i' satisfying

$$\sum_{i=1}^{k} r_i = \sum_{i=1}^{k} r_i', \sum_{i=1}^{k} r_i x_i = \sum_{i=1}^{k} r_i' x_i. \qquad (17.2.9)$$

   When this condition is satisfied it is said that r_i and r_i' are equivalent.
2. If we choose an equivalent set of r_i such that for $l_i' = \ln[r_i'/(n_i - r_i)]$, the points (x_i, l_i') are collinear, then the desired MLE may be obtained from the line $l_i' = \alpha + \beta x_i$ drawn through the points (x_i, l_i). The transfer method was an important and useful method of its time. However, it has apparently lost stature with the advent of faster, more accurate computers.

Worcester and Wilson (1943) noted in the case of three equally spaced dosages with equal n, that the MLE can be presented in table form related to two quantities which are in a one-to-one relation with $\Sigma\, p_i$ and $\Sigma\, p_i x_i$. These tables could be constructed for any specific dosage configuration (Cornfield and Mantel, 1950), but Berkson (1953) suggested nomographic presentation.

The nomograms presented by Berkson (1960) are based on the following facts. For any value of $\gamma = -\alpha/\beta$ with varying β, the corresponding values of $\Sigma\, p$, $\Sigma\, xp$ are the same values which, if found from a sample yield as the MLE the γ in question. Similarly, we can obtain the values of $\Sigma\, p$ and $\Sigma\, xp$ corresponding to a fixed β with varying α.

For large sample sizes Berkson (1960) gave expressions for the calculation of the variance of the estimates of β and γ. Namely,

$$s^2(\beta) = [n \, \Sigma \, \hat{p}_i \hat{q}_i (x_i - \bar{x})^2]^{-1} \tag{17.2.10}$$

and

$$s^2(\gamma) = \frac{1}{n\beta^2}\left[\frac{1}{\Sigma \, \hat{p}_i \hat{q}_i} + \frac{(\gamma - \bar{x})^2}{\Sigma \, \hat{p}_i \hat{q}_i (x_i - x)^2}\right], \tag{17.2.11}$$

where

$$\bar{x} = \frac{\Sigma \, n_i p_i q_i x_i}{\Sigma \, n_i \hat{p}_i \hat{q}_i}. \tag{17.2.12}$$

Finally, Berkson presents separate nomograms for both γ and β for doses 3, 4, 5, and 6. He also presents tables for the standard error of each estimate. It should be noted here that $l_i = \ln[r_i(n_i - r)]$ used by Berkson (1953, 1960) is a biased estimator of $\ln(P_i/Q_i)$, but Anscombe (1956) has shown that the bias can be practically eliminated by taking

$$l = \ln \frac{r + 0.5}{n - r + 0.5}, \tag{17.2.13}$$

provided n is large (>20) and neither P nor Q is close to zero.

Although the individual transformed values in (17.2.13) have expectations very close to their corresponding $\ln(P_i/Q_i)$, the forms (17.2.4) and (17.2.5) given by Berkson (1953) have greater bias in general for $k > 2$ when equation (17.2.13) is used than when $l = \ln[r_i/(n_i - r_i)]$ is used (Hitchcock, 1962). Hitchcock, in her calculations, assumed that all $n_i = n$ and the x_i were equally spaced. However, in general,

$$l_i = \ln \frac{r_i + \lambda_i}{n - r_i + \delta_i}, \tag{17.2.14}$$

though usually $\lambda_i = \delta_i = \epsilon$. Where its expected value is given by

$$E(l_i) = \alpha + \beta x_i + \frac{1}{n}\left[\frac{\lambda_i}{P_i} - \frac{\delta_i}{Q_i} - \frac{Q_i}{2P_i} + \frac{P_i}{2Q_i}\right] + o(n^{-2}). \tag{17.2.15}$$

Note that when $k = 2$, the expression (17.2.4) and (17.2.5) reduces to

$$\hat{\alpha} = \frac{l_1 x_2 - l_2 x_1}{x_2 - x_1}, \tag{17.2.16}$$

and

$$\hat{\beta} = \frac{l_2 - l_1}{x_2 - x_1}. \tag{17.2.17}$$

In this case, the estimates of α and β do not take the weights into account, and $E(\hat{\alpha})$ and $E(\hat{\beta})$ are simply linear functions of $E(l_i)$. Therefore, the best form would be the one suggested by Anscombe (1956), that is, $\epsilon = 0.5$, which would give zero terms of order n^{-1} in $E(\alpha)$ and $E(\beta)$. We note that, when $k = 3$, the forms of $E(\hat{\alpha})$, $E(\hat{\beta})$ are much more complicated and will not be presented here.

However, Hitchcock developed a table of the bias of order n^{-1} in $E(\hat{\alpha})$ when $k = 3$ for various values of P_i with $\epsilon = 0, 0.5$, as well as the expected value of $\hat{\alpha}$. Similarly, tables have been constructed for $\hat{\beta}$ when $k = 3, 5$ and $\epsilon = 0, 0.25, 0.5$. It was concluded that $E(\hat{\alpha})$ is much closer to α when $\epsilon = 0$ than when $\epsilon = 0.5$ over the whole range of the P_i considered. Therefore, for the estimation of α we usually take $\epsilon = 0$, or

$$l = \ln \frac{r}{n - r}. \tag{17.2.18}$$

For the estimation of β the bias is much larger and opposite in sign when $\epsilon = 0$ as opposed to when $\epsilon = 0.5$. Thus, Hitchcock proposed taking $\epsilon = 0.25$ when $k = 3$. When $k = 5$ she proposed taking $\epsilon = 0$ for both α and β. It was noted that as k gets large, the bias gets smaller when $\epsilon = 0$, as opposed to $\epsilon = 0.5$. Therefore for $k > 3$ it suffices to take l equal to (17.2.18).

Hamilton (1977) considered four estimation methods. Three of the methods require grouped data, and the fourth is the MLE computed from the original, ungrouped data. The four methods

1. MLE
2. MLE from grouped data (GMLE)
3. The two-cycle minimum logit chi-square estimates (TLCE)
4. The minimum logit chi-square estimates (LCE)

were compared through their comparative mean square error (MSE) from Monte Carlo simulations.

Specifically, the MLE method used to solve Eqs. (17.2.6) and (17.2.7) was the iterative method of Silverstone (1957); the initial approximation to the MLE suggested by Cox (1966) was used to start the iterative process.

The MLE from grouped data (GMLE) of α and β, denoted $\hat{\alpha}_G$ and $\hat{\beta}_G$ are (Hamilton, 1977) the solutions of the equations

$$\sum_1^k r_k - \sum_1^k n_i[1 + \exp\{-\hat{\alpha}_G - \hat{\beta}_G\}]^{-1} = 0 \tag{17.2.19}$$

and

$$\sum_1^k \bar{x}_k r_k - \sum_1^k \bar{x}_k n_k[1 + \exp\{-\hat{\alpha}_G - \hat{\beta}_G \bar{x}_k\}]^{-1} = 0, \tag{17.2.20}$$

where $k < n$ is the number of cells, and r_k is the number of observed successes out of n_k associated with the kth cell; also, the mean of the kth cell is given by

$$\bar{x}_k = \sum_{i=1}^{n_k} \frac{x_{k_i}}{n_k}. \tag{17.2.21}$$

The GMLE are of interest primarily in the case where the dose scale is coarse, i.e., when a coarse measuring device is used to determine dose. The LCE of α and β, denoted α_L and β_L are the intercept and slope, respectively, of a weighted least-squares regression line fit through the points (logit(p_k), $\bar{x}_k$), $k = 1, 2, \ldots, k$ developed by Berkson (1956). The TLCE of α and β, α_T and β_T, are the intercept and slope of a weighted least-squares regression line fit through the (logit(p_k), $\bar{x}_k$) points (Hamilton, 1977). The weights depend on α_L and β_L, hence the name "two-cycle." The individual dose levels are used in the calculations and TLCE cannot be used in situations of implicit grouping, i.e., grouping schemes implied by the collection and recording procedures of data, since exact individual doses would have never been recorded. The equations for the estimates are

$$\beta_T = \sum_{k=1}^{k} W_k L_k(\bar{x}_k - \bar{x}_W)\left[\sum_{k=1}^{k} W_k(\bar{x}_k - \bar{x}_W)^2\right]^{-1} \tag{17.2.22}$$

and

$$\alpha_T = \bar{L}_W - \beta_T \bar{x}_W, \tag{17.2.23}$$

where W_k is proportional to the reciprocal of an estimate of the variance of L_k,

$$\bar{L}_W = \sum_{1}^{k} W_k L_k, \quad \bar{x}_W = \sum_{1}^{k} W_k x_k. \tag{17.2.24}$$

Hamilton compared the MSE of the four estimators described above using Monte Carlo methods of 30 to 60 responses. The doses were grouped into equal width cells, each cell containing about 10 observations. There were two main results. First, the estimators based on grouped data usually yield smaller MSE's than the MLE based on original, ungrouped data. Second, the MLE and GMLE yield smaller MSE than the LCE when estimating to 50% or 90% effective dose. Finally, the estimators of the variance of the MLE, GMLE are more accurate than the LCE or TLCE. Hamilton (1977) summarized the above conclusions in a table by different types of grouping. Hitchcock (1962) apparently was the first to suggest that the value of δ should vary with the number k of design points and with the

Table 17.2.1 Some Optimal Values of δ When Estimating β for Various Numbers of Design Points

k	Optimal δ	k	Optimal δ
2	0.5000	7	−0.2500
3	0.2500	.	.
4	0.0000	.	.
5	−0.0625	.	.
6	−0.1875	∞	−0.5000

intercept α and the slope β. Following, Gart et al. (1985), on the basis of exact bias calculations, suggested some "optimal" values for δ when estimating β when the number of design points $k = 2, 3, 4, 5, 6, 7, \ldots, \infty$. These values are presented in Table 17.2.1.

Davis (1985) extended the work of Hitchcock to $k > 5$. Davis assumed that the x_i were equally spaced and all $n_i = n$; then she studied cases for $k = 3(1)20$. The method used by Davis to reduce the bias of Eq. (17.2.14) was to calculate $\delta_0 = (\delta_1, \delta_2)$, where δ_i ($i = 1, 2$) is the value of δ for which the n^{-1} order bias of the estimator of β_i is zero. For each case considered, δ_i is a strictly decreasing function of k, tending towards -0.5 as k increases. Since δ_0 is dependent on β, Davis averaged over all cases to give a single best δ, namely, $\delta^* = (\delta_1, \delta_2)$, for each k. She presented the values of the means for δ^* in tabular form by values of $k = 3, 4, \ldots, 20$.

The estimation of the parameters of the logistic distribution is discussed in Chapters 3, 4, 5, and 14.

17.3 ESTIMATION OF THE PARAMETERS OF THE LOGISTIC GROWTH FUNCTION

Pearl and Reed (1924) proposed a method to estimate the parameters of the logistic growth function that relies only on an initial estimate of a, say a_0. Estimate values $\hat{P}_s$, $\hat{b}$ and a correction to $\hat{a}$, $\Delta\hat{a}$ are obtained by minimizing the sum of squares

$$S = \Sigma\left[P_t \exp\{-\hat{a}_0 t\}(1 - t\,\Delta\hat{a}) - \frac{y - \hat{P}_s}{\hat{b}}\right]^2 \qquad (17.3.1)$$

with respect to $\hat{P}_s/\hat{b}$, $1/\hat{b}$, and $\Delta\hat{a}$. This method has the disadvantage that it relies solely on the initial value of a, for which the calculation is neither trivial nor unambiguous.

Schultz (1930) pointed out that the Pearl-Reed method is not the true least-squares procedure, and proposed the following iterative method. He chose initial values of P_s, b, and a, denoted $\hat{P}_{s_0}$, $\hat{b}_0$, and $\hat{a}_0$. Then, by minimizing

$$S = \Sigma\, P_t - f_0 - \left[\Delta\hat{a}\frac{\partial f}{\partial a_0} - \Delta\hat{b}\frac{\partial f}{\partial b_0} - \Delta\hat{P}_s\frac{\partial f}{\partial P_s}\right]^2, \qquad (17.3.2)$$

where f_0, $\partial f/\partial a_0$, $\partial f/\partial b_0$, and $\partial f/\partial P_s$ are, respectively, $f(P_s)$, $\partial f/\partial a$, $\partial f/\partial b$, and $\partial f/\partial P_s$ with $\hat{P}_{s_0}$, $\hat{b}_0$, and $\hat{a}_0$ substituted for the unknown parameters, the corrected values $\hat{P}_{s_0} + \Delta\hat{P}_s$, $\hat{b}_0 + \Delta\hat{b}$, and $\hat{a}_0 + \Delta\hat{a}$ are derived. This process may be repeated until sufficient accuracy is obtained. Also, it has the ordinary least-squares properties including maximum likelihood, the only drawback being the complexity of the calculations.

Davis (1941) developed the following estimation procedure, which is an adaptation of an earlier method by Hotelling (1927). Differentiating the logistic growth function logarithmically, we have

$$\frac{1}{P_t}\frac{dP_t}{dt} = a - \frac{a}{P_s}P_t. \qquad (17.3.3)$$

Approximating dP_t/dt by $\Delta P_t/\Delta t$ gives

$$\frac{1}{P_t}\frac{\Delta P_t}{\Delta t} = a - (a/P_s)P_t, \qquad (17.3.4)$$

which is fitted by linear least squares to the values of $(\Delta P_t/\Delta t,\ P_t)$. The parameters a and P_s follow directly. Solving (17.1.2) for b we have

$$b = P_t^{-1}(P_s - P_t)\exp\{at\}, \qquad (17.3.5)$$

from which we obtain each value of b. The estimate of b is the arithmetic mean of these values.

Tintner (1952) developed two methods for estimating b after obtaining estimates of a and P_s by fitting (17.3.4) by least squares. The geometric mean of the values of b can be obtained from (17.3.5). Unfortunately, the method does not work if at least one of the observations is above the estimated asymptote. In his second method, b is chosen so that the function passes through the bivariate mean $(\bar{t}, \bar{P}_t)$.

Oliver (1964) took six practical examples to compare the methods of Tintner (1952), Davis (1941), and Schultz (1930) versus true least-squares for the estimation of the parameters of the logistic growth function (17.1.2). His conclusion was that there is no substitute for full least squares in estimating the parameters of the logistic growth function. Presently, solving a system of three simultaneous nonlinear equations is a relatively simple task, and true least squares offers the following advantages:

1. It is often the ML estimator.
2. It does not depend on an arbitrary selection of points to estimates.
3. It gives least-squares estimates of functions of parameters usually needed in practice.

Suppose that a stochastic term ϵ_t is added to (17.1.2), so that

$$P_t = \frac{P_s}{1 + b\exp\{-at\}} + \epsilon_t. \tag{17.3.6}$$

Assume that ϵ_t is distributed normally with mean zero and variance σ^2, each successive value of ϵ_t is independent, and the values of t are predetermined. The MLE of the parameters are determined by the usual iterative solution. The estimates of these parameters are, for sufficiently large numbers of observations, distributed approximately normal, with mean (σ^2, P_s, b, a) and variance-covariance matrix $\mathbf{V}$ whose inverse, denoted $\sigma\mathbf{V}^{-1}$, is given by

$$\begin{bmatrix} \dfrac{n}{2\sigma^2} & 0 & 0 & 0 \\ 0 & \Sigma P^2 & -P_s \Sigma e^{-at}P^3 & P_s b \Sigma te^{-at}P^3 \\ 0 & -P_s \Sigma e^{-at}P^3 & P_s^2 \Sigma e^{-2at}P^4 & -P_s^2 b \Sigma te^{-2at}P^4 \\ 0 & P_s b \Sigma te^{-at}P^3 & -P_s^2 b \Sigma te^{-2at}P^4 & P_s^2 b^2 \Sigma t^2 e^{-2at}P^4 \end{bmatrix}, \tag{17.3.7}$$

where e is the base of the natural logarithm and $P = 1 + be^{-at}$. Although $\mathbf{V}^{-1}$ cannot be inverted directly, Oliver (1966) gives a numerical method to approximate $\mathbf{V}$.

In addition, Oliver did a Monte Carlo simulation of nine small sample cases and showed that the estimates are subject to some bias, decreasing with the number of observations.

17.4 ESTIMATION OF THE PARAMETERS AND DISCRIMINATION PROCEDURES OF THE LOGISTIC REGRESSION MODEL

The problem of estimating the parameters in a logistic regression model is related to the configuration of the data points. In this section we state the development of the types of data configurations, as well as some theoretical results on existence, uniqueness, and location of maximum likelihood estimators (MLEs).

There is a practical problem that can occur when attempting to estimate the parameters of the logistic regression model. Namely, there is sometimes a nonunique maximum or the boundary of the parameter space at infinity. Nonunique, infinite parameter estimates are not usually acceptable. Albert and Anderson (1984) show that estimates of this type are indicative of the pattern of data points. They define three types of data configuration:

1. Completely separated
2. Quasicompletely separated
3. Overlapped

If the data set X is not completely separable, then the vector $\alpha \in R$ is said to give quasicomplete separation of the sample points if for all $i \in E_j$ and for $j, t = 1, \ldots, k$ $(j \neq t)$ we have

$$(\alpha_j - \alpha_t)^{\mathrm{T}} x_i \geq 0, \tag{17.4.1}$$

with equality for at least one (i, j, t) triplet.

If neither complete nor quasicomplete separation of the data points exist, then they must overlap in the sense that for any vector $\alpha \in R$, there exists a triplet (i, j, t) where $j, t \in \{1, \ldots, k\}$, $j \neq t$, $i \in E_j$, and

$$(\alpha_j - \alpha_t)^{\mathrm{T}} x_i < 0. \tag{17.4.2}$$

Day and Kerridge (1967) considered the case of completely separated data points in the logistic regression. Anderson (1972) also made some preliminary observations into the case of completely separated data points based on posterior probabilities. Wedderburn (1976) proved some very general theorems regarding sufficient (but not necessary) conditions for existence, uniqueness, and location of the parameters of the generalized linear model defined by Nelder and Wedderburn (1972).

Logistic discrimination for binary data was considered by Cox (1966) using the likelihood ratio of Welch (1939). That is, if the vector $\mathbf{x} = (x_1, \ldots, x_n)$ has known density functions say $f_1(x)$ and $f_0(x)$, and if the prior probabilities are $\Pr\{I_1|x\}$, $\Pr\{I_0|x\}$, where I_i $(i = 0, 1)$ is an indicator variable, then the ratio of the posterior probabilities is

$$\ln\left(\frac{p_i}{q_i}\right) = \ln\frac{\Pr\{I_1|x\}}{\Pr\{I_0|x\}} + \ln\frac{f_1(x)}{f_0(x)}. \tag{17.4.3}$$

The general case of quasicomplete separation was introduced by Albert and Anderson (1984).

The function to be maximized is the log likelihood

$$\ln L(X, \alpha) = \sum_{j=1}^{k} \sum_{i \in E_j} \ln \sum_{t=1}^{k} \exp\{(\alpha_t - \alpha_j)^T x_i\} \qquad (17.4.4)$$

where $\mathbf{X}$ is the $n \times (p + 1)$ matrix with x_i^T as rows, E_j is the set of row identities for $\mathbf{X}$ from observations from I_i.

When the data is of type 1, 2, and 3, respectively, Albert and Anderson (1984) proved the following theorems:

Theorem 17.4.1 The MLE $\hat{\alpha}$ does not exist, and maximum of $L(\mathbf{X}, \alpha) = 1$ for all α in the observation space.

Theorem 17.4.2 The MLE $\hat{\alpha}$ does not exist, and maximum of $L(\mathbf{X}, \alpha) < 1$ for all α in the observation space.

Theorem 17.4.3 The MLE $\hat{\alpha}$ exists and is unique.

Albert and Anderson (1984) proved existence theorems that went beyond all previous work through classification of the different types of data sets. They also suggested criteria for recognizing the different data configurations.

Santner and Duffy (1986) gave some modifications in the statement and proof of Albert and Anderson's (1984) Theorem 17.4.2 and Theorem 17.4.3. They noted that $\alpha = 0$ must be excluded from Eq. (17.4.2) otherwise it is trivially false. Furthermore, Santner and Duffy developed a linear program which determined whether the data is completely separated, quasicompletely separated, or overlapping.

Suppose that we have k independent samples, where the independent, binary random variables from the jth sample are given by $X_{1j}, X_{2j}, \ldots, X_{nj}$ and x_{ij} is the covariate associated with y_{ij}. Then the linear logistic model for the jth sample is represented by Eq. (17.1.6). For sample j, the likelihood is given by

$$L_j = \Pr\{Y_j = y_j | x_{ij}\ (i = 1, \ldots, n_j)\}$$
$$= \frac{\exp\{S_j\alpha_j + \beta t_j\}}{\prod_{i=1}^{n_j} [1 + \exp\{\alpha_j + \beta x_{ij}\}]}, \qquad (17.4.5)$$

where y_j is the response vector for sample j, and $S_j = \Sigma\, y_{ij}$, $t_j = \Sigma\, y_{ij}x_{ij}$. Sufficient statistics are $\mathbf{s} = (s_1, s_2, \ldots, s_k)^T$ and $t = \Sigma\, t_i$, and the overall likelihood is the product of the L_j; i.e.,

$$L = \prod_{j=1}^{k} L_j = \frac{\exp \sum_{j=1}^{k} S_j\alpha_j + \beta \sum_{j=1}^{k} t_j}{\prod_{j=1}^{k} \prod_{i=1}^{n_j} [1 + \exp\{\alpha_j + \beta x_{ij}\}]}. \tag{17.4.6}$$

The conditional mass function is given by

$$\Pr\{T = t|S = s\} = \frac{C(s, t)\exp\{\beta t\}}{\sum_r C(s, r)\exp\{\beta r\}}. \tag{17.4.7}$$

Note that when we have only one random sample, that is, when $k = 1$ Eq. (17.1.6) reduces to Eq. (17.1.5), Eq. (17.4.5) reduces to

$$\Pr\{Y_1 = y_1, \ldots, Y_n = y_n|x_1, \ldots, x_n\} = \frac{\exp\{S\alpha + \beta t\}}{\prod_{i=1}^{n} [1 + \exp\{\alpha + \beta x_i\}]}, \tag{17.4.8}$$

where $s = \Sigma\, y_i$ and $t = \Sigma\, y_i x_i$.

Cox (1970) based his conditional inference on the sufficient statistics

$$S = \sum_{i=1}^{n} Y_i \quad \text{and} \quad T = \sum_{i=1}^{n} Y_i x_i, \tag{17.4.9}$$

whose joint distribution is obtained by summing over all binary sequences generating each (s, t). Thus,

$$\Pr\{S = s \text{ and } T = t\} = \frac{C(s, t)\exp\{S\alpha + \beta t\}}{\prod_{i=1}^{n} [1 + \exp\{\alpha + \beta x_i\}]}, \tag{17.4.10}$$

where $C(s, t)$ is the number of distinct binary sequences yielding the values s and t for S and T, respectively.

Exact inferences concerning β can be based (Cox, 1970) on the conditional distribution $\Pr\{T = t|S = s\}$. Furthermore, from the properties of the exponential family of distributions, the critical region defined by the upper tail values of t provide a uniformly most powerful unbiased (UMPU) test of

$$H_0: \beta = \beta_0 \quad \text{versus} \quad H_a: \beta > \beta_0 \tag{17.4.11}$$

(Lehmann, 1959). From Eq. (17.4.10) the conditional distribution used to test (17.4.11) is

$$\Pr\{T = t|S = s\} = \frac{C(s, t)\exp\{\beta t\}}{\sum_r C(s, r)\exp\{\beta r\}}, \tag{17.4.12}$$

where r is an index ranging over all values of T. Thus, for a one sided test of hypotheses, one needs $C(s, r)$ for fixed s, since α is not related to any covariate effect.

Tritchler (1984) developed an algorithm for the exact logistic analysis of a single parameter. The algorithm calculates estimates of the coefficient $C(s, \cdot)$, where $\cdot$ indicates that the corresponding agreement varies over all its values. The coefficient $C(s, t)$ is the number of distinct binary sequences of s 1's and $n - s$ 0's that yield the value t defined in (17.4.8). The algorithm estimates the coefficients by first calculating their Fourier transform and then inverting it.

Define

$$f(l, k, p) \equiv \sum_{\mathbf{l} \in G(k,l)} \exp\{ip\mathbf{x}^k\mathbf{l}\} \qquad (l, k < n), \tag{17.4.13}$$

where $G(k, l) = \{\mathbf{l}:\mathbf{l}$ is a column vector of r 1's and $k - r$ 0's$\}$, and $\mathbf{x}^k = (x_1, \ldots, x_k)$. Then the Fourier transform of $C(s, \cdot)$ is

$$\begin{aligned}\Psi(p) &= \sum_t C(s, t)\exp\{ipt\} = \sum_{\mathbf{l} \in G(n,s)} \exp\{ip\mathbf{x}\mathbf{l}\} \\ &= f(s, n, \theta).\end{aligned} \tag{17.4.14}$$

The $\Psi(p)$ may be obtained by the following

$$f(l, k, p) = \exp\{ip\mathbf{x}^k\}f(l - 1, k - 1, p) + f(l, k - 1, p). \tag{17.4.15}$$

The proof is given in Pagano and Tritchler (1983), Inverting the Fourier transform of Ψ requires computing the quantities $\Psi(p)$ for all frequencies

$$P = \frac{2\pi m}{M} \qquad (m = 0, 1, \ldots, M - 1), \tag{17.4.16}$$

where $M \geq \max T$.

We need to assume that **x** takes on only positive integer values to ensure that the discrete Fourier transform will yield the coefficient $C(s, \cdot)$ (Singleton, 1989) and these coefficients from the inversion process will be in a one-dimensional array, indexed by the value of the sufficient statistic T.

To calculate $C(\mathbf{s}, t)$, let $C_j(s_j, T_j)$ be the number of sequences of $\mathbf{Y}_j$ yielding s_j and T_j. Then

$$C(\mathbf{s}, T) = \sum_T \prod_{j=1}^{k} C_j(s_j, T_j). \tag{17.4.17}$$

The Fourier transform of $C(s, \cdot)$ can be obtained as the product of the transforms of $C_j(s_j, \cdot)$ with $j = 1, 2, \ldots, k$. Then using the fast Fourier transform, this can be inverted to give $f(s_j, n_j, p)$, which can be evaluated using equation (17.4.15).

For $k = 1$, we have two possible test of hypothesis; i.e., H_0: $\beta = \beta_0$ versus

$$H_a: \beta > \beta_0 \tag{17.4.18}$$

or

$$H_a: \beta < \beta_0. \tag{17.4.19}$$

The p-values for the testing of H_0 versus the alternative hypotheses given by (17.4.18) and (17.4.19), respectively, are (Tritchler, 1984)

$$p_+(t, \beta_0) = \sum_{v \geq t} \frac{C(\mathbf{s}, v) \exp\{\beta_0 v\}}{\sum_r C(\mathbf{s}, r) \exp\{\beta_0 r\}}, \tag{17.4.20}$$

and

$$p_-(t, \beta_0) = \sum_{v \leq t} \frac{C(\mathbf{s}, v) \exp\{\beta_0 v\}}{\sum_r C(\mathbf{s}, r) \exp\{\beta_0 r\}}, \tag{17.4.21}$$

where r is an index ranging over all values of T. Note that if $k = 1$, then **s** becomes s as defined in (17.4.8). Tritchler also developed a method for the calculation of a $(1 - \alpha)100\%$ confidence interval (CI) for β corresponding to the observed value t (see also Cox, 1970). The CI can be written

$$\Pr\{\beta_L < \beta < \beta_U\} = 1 - \alpha, \tag{17.4.22}$$

where β_L and β_U can be found from the solutions to

$$p_+(t, \beta_L) = \alpha \tag{17.4.23}$$

and

$$p_-(t, \beta_U) = \alpha, \tag{17.4.24}$$

respectively. The solutions to Eqs. (17.4.23) and (17.4.24) are solved numerically; Tritchler suggests the algorithm by Forsythe, Malcolm, and Moler (1977) ZEROIN.

Finally, a reasonable point estimate (Tritchler, 1984) can be obtained as the β that most nearly satisfies

$$p_-(t, \hat{\beta}) = p_+(t, \hat{\beta}) = 0.5. \tag{17.4.25}$$

17.5 GOODNESS-OF-FIT TECHNIQUES FOR THE LOGISTIC FUNCTION

17.5.1 An Overview of Techniques

This section is primarily concerned with the problem of goodness-of-fit techniques for use in logistic regression. Much of the work done in recent years has focused on adapting techniques from linear regression. Tremendous amounts of work have been done in linear regression diagnostic techniques providing useful guides for the advancement of similar techniques for logistic regression. That is, the detection of outliers and other diagnostics based on residuals have gained widespread use in logistic regression as a result of their success in linear regression. Some specific recent developments of goodness-of-fit techniques are outlined for the logit in Section 17.2.

A secondary effort has been made in tests of fit for the logistic distribution based on the empirical distribution function that will follow.

The remainder of this section provides an overview of techniques and motivation for the development of certain trends in the field. Some of the problems associated with the definition of residuals is discussed, as well as the necessity of including practical techniques for the detection of outliers in any exposition of assessment of fit procedures.

Several authors have contributed work in this vein, including Tsiatis (1980), Pregibon (1979, 1981), Landwehr, Pregibon, and Shoemaker (1984), and Cook and Wiseberg (1982). Jennings (1986) warned that direct application of linear regression techniques to logistic regression is not necessarily legitimate, and that each diagnostic tool must be considered individually.

For analysis based on the linear logistic transform given by Eq. (17.1.6), methods of examining goodness-of-fit used in Gaussian theory problems are applicable with some minor modifications. To this end, Cox (1970) outlined the following possibilities.

1. For grouped data, where it is assumed that (approximately) the probability of success is constant within groups, the residual sum of squares (RSS) can be scaled to correspond to theoretical variance of unity if the model is correct. Thus, the RSS will be distributed approximately chi-squared with $k - m$ degrees of freedom, where k is the number of groups and m is the number of parameters fitted.
2. The residuals can be defined to be the difference between the observed and fitted values. An analysis of the distribution of the residuals can indicate the existence of outliers. Cox and Snell (1968) have given a more general definition of residuals for any general linear model, that will give a more nearly normal distribution of the residuals, and hence a more nearly linear plot on probability paper.
3. A more complicated nonlinear model can be fitted and tested against the linear model. This is relatively easy if the fitting is done with MLE.

Methods using residuals are of particular importance in analyzing data where many types of departure from the initial model are of possible importance (Cox, 1970).

Pregibon (1981) developed model diagnostics for the detection of outliers and extreme points for the logistic regression model, as well as schemes for quantifying their effects. The results are designed for use with the standard output from a statistical computer package when attempting to fit the model using MLE techniques.

The standard output from a well-designed computer package to fit the logit model (17.1.7) with the intercept α considered unimportant, is typically a subset of the (Pregibon, 1981)

1. MLE of β, $\hat{\beta}$
2. Standard error of the individual $\hat{\beta}_i$
3. Estimated covariance matrix of $\hat{\beta}$:

$$\mathrm{Var}(\hat{\beta}) = (\mathbf{X}^{\mathrm{T}}\mathbf{V}\mathbf{X})^{-1}; \qquad (17.5.1)$$

4. Chi-squared goodness-of-fit statistic χ^2
5. Individual components of χ^2:

$$\chi_i = \frac{y_i - n_i\hat{p}_i}{\sqrt{np_iq_i}} \qquad (17.5.2)$$

6. Deviance:

$$D = \Sigma\, d_i^2. \tag{17.5.3}$$

The identification of outlying and extreme points is based on a residual vector and a projection matrix. We shall define residuals in the following two ways found most useful by Pregibon (1981). One is the individual components of the chi-squared goodness-of-fit statistic, given by (17.5.2), and the other is the components of deviance given by

$$d_i = \sqrt{2[\ln L(p_i, \hat{y}_i) - \ln L(x_i\beta, \hat{y}_i)]} \tag{17.5.4}$$

where + is used if $\hat{p}_i > x_i\hat{\beta}$, and − if $\hat{p}_i < x_i\hat{\beta}$. Both X^2 and D are measures of goodness-of-fit of the model. The latter (17.5.4) measures the (lack of) agreement between the maxima of the observed and fitted log-likelihood functions. When the jth component of χ_i^2 or d_i is large, the jth observation is not well accounted for by the proposed model.

The projection matrix for the logistic model, denoted by $\mathbf{M} = [m_{ii}]$, given in general form is

$$\mathbf{M} = \mathbf{I} - \mathbf{V}^{1/2}\mathbf{X}(\mathbf{X}^T\mathbf{V}\mathbf{X})^{-1}\mathbf{V}^T\mathbf{V}^{1/2} = \mathbf{I} - \mathbf{H}. \tag{17.5.5}$$

Pregibon showed that $\mathbf{M}$ is symmetric, idempotent, and spans the χ^2 residual space. Thus, small m_{ii} should be useful in detecting extreme points. This suggestion is supported by the hypothetical data set created by Pregibon (1981) and listed in Table 17.5.1. The 17th observation has the

Table 17.5.1 Pregibon's Hypothetical Data Set and the Basic Building Blocks Associated with the Maximum Likelihood Fit of a Logistic Regression Model

Data		Building blocks		
x_i	logit(y_i/n_i)	χ_i^2	d_i	m_{ii}
1	0.5	−0.506	−0.496	0.7192
2	0.5	−0.615	−0.600	0.7885
3	0.8	−0.247	−0.244	0.8409
4	1.0	−0.055	−0.054	0.8773
5	1.3	0.248	0.252	0.8989
6	1.3	0.157	0.159	0.9070
7	1.6	0.417	0.431	0.9030
8	1.8	0.535	0.561	0.8882
9	2.1	0.715	0.768	0.8640
10	2.1	0.642	0.686	0.8316
17	1.0	−1.383	−1.227	0.4813

smallest value of m_{ii}, and this will always be the case when a recorded observation is far removed from the others.

In most cases, the examination of χ_i^2, d_i, and m_{ii} will indicate potential outliers and influential points. Plots of χ_i^2, d_i, and m_{ii} against i are strongly suggested, especially when the order of the observations are important.

Consider also the matrix

$$\mathbf{H}^* = [h_{ii}^*], \tag{17.5.6}$$

where

$$h_{ii}^* = h_{ii} + \frac{\chi_i^2}{\chi^2} \tag{17.5.7}$$

determines the diagonal elements with $0 \le h_{ii}^* \le 1$, and h_{ii}, χ_i^2, and χ^2 are defined in Eqs. (17.5.5), (17.5.2) and statement (4), respectively. Values of h_{ii}^* near 1 correspond to one of three possible problems with the data. First, a large relative χ_i^2 in (17.5.7) indicates observations poorly fit by the proposed model. A second possibility is observations considered to be outlying which is indicated by large h_{ii} in (17.5.7). Finally, observations with both h_{ii} and χ_i^2 relatively large in (17.5.7) indicates observations both poorly fit and extreme in the design space. A scatter plot of χ_i^2/χ^2 versus h_{ii} will display any extreme values of h_{ii} or χ_i^2.

Finally, Chapter 13 offers a more detailed description of the various treatments of goodness-of-fit for the logistic distribution.

17.6 GOODNESS-OF-FIT TECHNIQUES FOR LOGISTIC REGRESSION

This section contains an outline of some of the more recent advances in goodness-of-fit procedures for the logistic regression model given by

$$\ln(P_i/q_i) = \alpha + \beta x_i. \tag{17.6.1}$$

It appears that assessing the fit of the model to the data in the design space has lagged far behind not only the theoretical development of the logit, but the applications as well. The 1980s saw the start of a great deal of literature on the subject, specifically the development of some specific and useful procedures. For a discussion of earlier, more general works refer to Section 17.5.

Tsiatis (1980) proposed a goodness-of-fit test for the logistic regression model given by (17.6.1) which is asymptotically chi-squared distributed. The space of covariates $x_1, \ldots, x_k$ can be partitioned into m distinct regions in k-dimensional space denoted by $R_1, \ldots, R_k$; and the indicator functions $I^{(j)}$ $(j = 1, \ldots, m)$ are defined by

$$I^{(j)} = \begin{cases} 1 & \text{if } (x_1, \ldots, x_k) \in R_j, \\ 0 & \text{otherwise.} \end{cases} \tag{17.6.2}$$

Consider the model

$$\ln\left(\frac{p_i}{q_i}\right) = \beta' X + \alpha' I', \tag{17.6.3}$$

where $\beta' = (\beta_0, \ldots, \beta_k)$, $X = (x_0, x_1, \ldots, x_k)$, $\alpha' = (\alpha_1, \ldots, \alpha_m)$, $I' = (I^{(1)}, \ldots, I^{(m)})$.

The goodness-of-fit test consists of testing the hypotheses

$$H_0\colon \alpha_1 = \cdots = \alpha_m = 0.$$

$$H_a\colon \text{at least one } \alpha_i \text{ is different from zero.} \tag{17.6.4}$$

The test proposed by Tsiatis (1980) is based on the efficient scores test statistic

$$\mathbf{T} = \mathbf{Z}^{\mathrm{T}}\mathbf{V}^{-1}\mathbf{Z}, \tag{17.6.5}$$

where the jth element of the m-dimensional column vector $\mathbf{Z}$ is given by

$$\sum_{i=1}^{n} X_i I_i^{(j)} - \sum_{i=1}^{n} \frac{I_i^{(j)} \exp(\hat{\beta} X_i)}{1 + \exp(\hat{\beta} X_i)}$$
$$= O_j - E_j, \tag{17.6.6}$$

and the $m \times m$ matrix $\mathbf{V}^{-1}$ is equal to the inverse of

$$\mathbf{V} = \mathbf{A} - \mathbf{B}\mathbf{C}^{-1}\mathbf{B}^{\mathrm{T}}, \tag{17.6.7}$$

with

$$A_{ij} = \begin{cases} \sum_{i'} \hat{p}_{i'} \hat{q}_{i'} & (i = j), \\ 0 & i \neq j;\ i, j = 1, \ldots, m, \end{cases} \tag{17.6.8}$$

$$B_{ij} = \sum_{i'} X_{i'j} \hat{p}_{i'} \hat{q}_{i'}, \qquad i = 1, \ldots, m;\ j = 1, \ldots, k, \tag{17.6.9}$$

$$C_{ij} = \sum_{i'=1}^{n} X_{i'i} X_{i'j} \hat{p}_{i'} \hat{q}_{i'} \qquad i, j = 0, 1, \ldots, k, \tag{17.6.10}$$

where i' denotes the set of indices such that

$$(X_{i1}, \ldots, X_{ik}) \in R_j \tag{17.6.11}$$

and

$$\hat{p}_i = \frac{\exp(\hat{\beta}X_i)}{1 + \exp(\hat{\beta}X_i)}, \qquad \hat{q}_i = 1 - \hat{p}_i. \tag{17.6.12}$$

Tsiatis chose the model (17.6.3) so that through the use of standard likelihood theory, the derived test statistic (17.6.6) would be a quadratic form of $O_i - E_i$ which is asymptotically distributed as a chi-square. This method appears better than the Pearson chi-squared test which has been suggested as a goodness-of-fit test for the logistic regression model. The Pearson test does not in general have a chi-squared distribution, hence making the calculation of significance levels difficult.

Cook and Weisberg (1982) considered a logistic outlier model of the form

$$y = X\beta + e_i\delta + \epsilon, \tag{17.6.13}$$

where $e' = \{0, \ldots, 0, 1, 0, \ldots, 0\}$ with the 1 (indicating outlier) in the ith position. The test of $\delta = 0$ yields a residual-like quantity given by

$$t_i = \frac{y_i - x_i'\beta}{s_i\sqrt{m_i}}, \tag{17.6.14}$$

where $m_i = 1 - x_i'(\mathbf{X}^T\mathbf{X})^{-1}x_i$ and s_i^2 is the estimated variance without the ith case. This is a measure of how the model explains a given observation by testing the need for an additional variable specifically to model that case.

Landweher, Pregibon and Shoemaker (1984) structured three graphical methods to analyze logistic regression models, specifically

1. Local mean deviance plots
2. Empirical probability plots
3. Partial residual plots

The local mean deviance plot assesses the overall adequacy of the fit of the model. The deviance measures the global disagreement between the observed data and the fitted values. This deviance is given by

$$\begin{aligned} D(\hat{p}; y) &= \sum_{i=1}^{n} d(\hat{p}_i; y_i) \\ &= -2\sum_{i=1}^{n} [y_i \ln \hat{p}_i + (1 - y_i)\ln(1 - \hat{p}_i)]. \end{aligned} \tag{17.6.15}$$

The procedure is based on a partition of $D(\hat{p}; y)$ into a pure-error component and a lack-of-fit component. If the model accounts for the variation in the data, then the lack-of-fit component will be small. The pure-error component is obtained by a method based on near neighbors and an approximate decomposition of the deviance. Namely, comparing running estimates of approximate pure-error paired with their degrees of freedom against the mean of the overall deviance.

The empirical probability plot is useful in detecting outliers. The plot has standardized residuals on the vertical axis, and quantiles from a reference distribution on the horizontal axis. The residuals used are the d_i from the first line of (17.6.15) standardized by its approximate standard error. Landweher et al. (1984) propose a simulation procedure to obtain the reference distribution. This procedure is based on p and estimates the distribution that the residuals would have if the fitted model were correct. As in the case of normal probability plots for a linear regression, this plot approximates a straight line when the model correctly fits the data. Any points straying too far from the line are possible outliers.

Jennings (1986) presented the case that outliers are necessary to logistic regression. Censoring outliers is equivalent to censoring in only one tail, since we do not observe $Y_t = i$ if $\Pr\{Y_t = i|x_t\} < \epsilon$. This changes the systematic part of the model. Fitting the true model to the censored data will produce estimates of β that are biased. Any a posteriori approach to censoring must consider the binary nature of the data. A better approach than censoring individual points might be to find groups of data that are not well fit by the model.

Jennings also points out that the residuals from the fitted models do not have the same properties as the residuals from the true model. Furthermore, the useful residuals-like quantities have distributions which depend on p. Specifically, $E(d_i)$ is not 0 and $E(d_i^2)$ is not constant; that is, they both rely on p_i.

Jennings (1986) proposed a standardization of d^2 so that $E(d^2) = 1$ for all p, namely,

$$t^2 = \frac{d^2}{E(d^2)} = \frac{y \ln p + (1 - y)\ln(1 - p)}{p \ln p + (1 - p)\ln(1 - p)}. \tag{17.6.16}$$

This proposal is based on the success of a similar technique used by Williams (1976) to provide a better approximation to a chi-squared distribution for the likelihood ratio statistic in log-linear models. The behavior of (17.6.16) is unknown when p is replaced by $\hat{p}$.

17.7 GOODNESS-OF FIT TESTS FOR THE LOGISTIC DISTRIBUTION

Goodness-of-fit tests for the logistic distribution given by

$$F(x) = \left[1 + \exp\frac{-a(x - \mu)}{\sigma}\right]^{-1}, \qquad (17.7.1)$$

are derived in the following section. Very little work has been done in this area; the major effort has been in assessing the adequacy of fit of the logit. This is due primarily to the popularity of the logistic regression model with the epidemiologists and biostatisticians.

Stephens (1979) derived three test statistics based on measurement of the discrepancy between the theoretical distribution function (17.7.1), with estimates inserted for any unknown parameters, and the empirical distribution function of the sample. The null hypothesis H_0 is that the sample of x-values is a random sample from the logistic distribution (17.7.1). The aim of the test is to indicate the goodness-of-fit of the observed values to Eq. (17.7.1).

Stephens (1979) investigated four cases of testing H_0 with respect to knowledge of the parameters. These cases are given as follows:

1. Both μ and σ are known so that $F(x)$ is completely specified.
2. σ is known; μ must be estimated.
3. μ is known; σ must be estimated.
4. Both μ and σ are unknown and must be estimated.

Once the parameters have been estimated, the steps in testing the null hypothesis are as follows.

First, calculate $z_i = F(x_i)$, where $F(x)$ is given in (17.7.1), with the appropriate estimates substituted for unknown parameters as per cases (i), (ii), or (iii) above.

Then, compute the desired test statistic:

$$W^2 = \sum_i \left[z_i - \frac{2i - 1}{2n}\right]^2 + (12n)^{-1}, \qquad (17.7.2)$$

$$U^2 = W^2 - n(\bar{z} - 1/2)^2, \qquad (17.7.3)$$

or

$$A^2 = -n^{-1}\left\{\sum_i (2i - 1)[\ln z_i + \ln(1 - z_{n+1-i})]\right\}, \qquad (17.7.4)$$

where z is the mean of the z_i.

Table 17.7.1 Modifications for Stephens' Goodness-of-Fit Test Statistics W^2, U^2, and A^2 for Different Cases

Test statistic	Case	Modification
W^2	1	$(W^2 - 0.4/n + 0.6/n^2)(1.0 + 1.0/n)$
	2	$(1.9nW^2 - 0.15)/(1.9n - 1.0)$
	3	$(0.95nW^2 - 0.45)(0.95n - 1.0)$
	4	$(nW^2 - 0.08)/(n - 1.0)$
U^2	1	$(U^2 - 0.1/n + 0.1/n^2)(1.0 + 0.8/n)$
	2	None
	3	$(1.6nU^2 - 0.16)/(1.6n - 1.0)$
	4	None
A^2	1	None
	2	$A^2 + 0.15/n$
	3	$(0.6nA^2 - 1.8)/(0.6n - 1.0)$
	4	$A^2(1.0 + 0.25/n)$

Finally, calculate the modified test statistic using Table 17.7.1; then refer to Table 17.7.2 comparing the modified result with the upper tail points given for the appropriate case. The modifications make only slight changes to the given value of the test statistic, but they make it possible to dispense with tables of points for each sample size n; that is, they can be used with small samples. The percentage points given in Table 17.7.2 are those of

Table 17.7.2 Percentage Points for Stephens' Modified Goodness-of-Fit Test Statistics W^2, U^2, and A^2

Modified statistic	Case	Upper tail percentage points, α				
		0.900	0.950	0.975	0.990	0.995
W^2	1	0.347	0.461	0.581	0.743	0.869
	2	0.119	0.148	0.177	0.218	0.249
	3	0.323	0.438	0.558	0.721	0.847
	4	0.081	0.098	0.114	0.136	0.152
U^2	1	0.152	0.187	0.221	0.267	0.304
	3	0.116	0.145	0.174	0.214	0.246
A^2	1	1.933	2.492	3.070	3.857	4.500
	2	0.857	1.046	1.241	1.505	1.710
	3	1.725	2.290	2.880	3.685	4.308
	4	0.563	0.660	0.769	0.906	1.010

the asymptotic distribution of the test statistics. They were determined from the theory of the empirical process discussed by Durbin (1973).

Stephens (1979) derived the asymptotic theory for the statistics W^2, U^2, and A^2, and supports the results by Monte Carlo runs. Also, some Monte Carlo results were given for the Kolmogorov-Smirnov D statistics and the Kuiper V statistic. These statistics are defined as follows:

$$D = \max\{D^+, D^-\}, \tag{17.7.5}$$

and

$$V = D^+ + D^-, \tag{17.7.6}$$

where

$$D^+ = \max_i\left\{\frac{i}{n - z_i}\right\}, \tag{17.7.7}$$

and

$$D^- = \max_i\left\{z_i - \frac{i-1}{n}\right\}. \tag{17.7.8}$$

The smoothed percentage points of $D^+\sqrt{n}$, $D\sqrt{n}$, and $V\sqrt{n}$ (points for $D^-\sqrt{n}$ can be obtained from $D^+\sqrt{n}$), were derived by Stephens (1979) for n = 5, 10, 20, 50, and ∞. The asymptotic values were found by extrapolation, and their accuracy is difficult to determine (Stephens, 1979).

17.8 APPLICATIONS OF THE LOGISTIC FUNCTION

The logistic function has been found useful in a variety of applications. Initially discovered and used as a model for the growth of human population by Verhulst (1845), it has since been used in this context by various authors. The logistic function has also been used in studies of physiochemical phenomenon by Pearl and Reed (1929), geological studies by Aitchison and Shen (1980), and psychological studies by Birnbaum and Dudman (1963), Lord (1965), Sanathanan (1974), and Formann (1982).

In this study we will pay particular attention to four areas of application of the logistic function:

1. Population growth
2. Bioassay
3. Medical diagnosis
4. Public health

Within the broad category of public health we shall concentrate in the areas of epidemiology, categorical data analysis, retrospective studies, and survival analysis. For each of the four categories above, we shall cite as completely as possible the relevant work given in the literature, as well as give closer examination to some selected applications.

17.8.1 Applications of the Logistic Function to Population Growth

The first recorded use of the logistic curve was for estimating the growth of human population by Verhulst (1845). The logistic function was also used in this context by Pearl and Reed (1920) and Pearl et al. (1940) to project U.S. population growth. Schultz (1930) applied Pearl and Reed's (1920) population data to the logistic function. More recently, Oliver (1982) applied the logistic curve to human population growth.

The logistic growth function has been applied to other areas of growth as well, most notedly in biology by Pearl and Reed (1924); Schultz (1930) applied it to yeast cell growth. Oliver (1964) used the yeast cell data of Schultz and compared several methods of estimation. That is, he compared the methods of Schultz (1930), Davis (1941), and Tintner (1952). Oliver (1969) successfully used the logistic growth function to model the spread of an innovation, where it fit substantially better than did the three-parameter logistic given by (17.8.1).

Viera and Hoffman (1977) applied the logistic growth functin to weight-gain data of Holstein cows. They employed a heteroscedastic model a priori since their data was only over the first 56 months of life and smaller variances seemed reasonable. Also, Glasbey (1979) applied the generalized logistic curve to the weight-gain analysis of Ayrshire steer calves, which were recorded weekly from birth to slaughter at 880 pounds. He presented a method of analysis that assumed the residuals to be the realization of a first-order autoregressive process.

It should also be noted that both Schultz (1930) and Oliver (1964) also applied the logistic growth function to the same agricultural production data.

Oliver (1982) extended the work of Leach (1981) in which he applied the logistic curve given by (17.1.1) as a model for the growth of human population. Oliver (1982) considered three particular aspects of the logistic function.

He derived and evaluated the variance-covariance matrix for the purpose of obtaining the standard errors of the parameters of the logistic function given by

$$P_t = \frac{P_s}{1 + (P_s/P_0)\exp\{-at\}}, \qquad (17.8.1)$$

where P_t is the population t years after the initial population P_0, P_s is the asymptotic maximum or saturation level, and a is the growth rate coefficient. The inverse of the variance-covariance matrix is given by

$$\mathbf{V}^{-1} = \sigma^{-2}\begin{bmatrix} \frac{n}{2\sigma^2} & 0 & 0 & 0 \\ 0 & \Sigma F^2 & \Sigma FG & \Sigma FH \\ 0 & \Sigma FG & \Sigma G^2 & \Sigma GH \\ 0 & \Sigma FH & \Sigma GH & \Sigma H^2 \end{bmatrix}, \tag{17.8.2}$$

where

$$F = \frac{P_s t(P_s/P_0 - 1)\exp\{-at\}}{[1 + (P_s/P_0 - 1)\exp\{-at\}]^2}, \tag{17.8.3}$$

$$G = \frac{(P_s/P_0)^2 \exp\{-at\}}{[1 + (P_s/P_0 - 1)\exp\{-at\}]^2} \tag{17.8.4}$$

and

$$H = (1 - \exp\{-at\})\left[1 + \left(\frac{P_s}{P_0} - 1\right)\exp\{-at\}\right]^2. \tag{17.8.5}$$

For $n = 18$ decennial estimates of the population of Great Britain from 1801 to 1971, as defined by Leach (1981), the least-squares estimates and their standard errors (SE) as defined from (17.8.2) are displayed in Table 17.8.1. Oliver (1982) compared the SE of the least-square estimates of (17.1.1) with those from a generalized logistic function originally put forth by Pearl (1924).

This generalization is by means of a vertical displacement and is given by

$$P_t = d + \frac{P_s}{1 + (P_s/P_0 - 1)\exp\{-at\}}. \tag{17.8.6}$$

Table 17.8.1 Estimates with Associated Standard Error of the Parameters of the Logistic Growth Function Using Leach's 1801–1971 Human Population Data as Derived by Oliver

Parameter	Estimate	Standard error
a	0.019625	0. 000689
P_0	10.003828	0.302281
P_s	63.646872	1.593014

Table 17.8.2 Estimates with Their Associated Standard Errors of the Parameters of the Generalized Logistic Growth Function Using Leach's 1801–1971 Human Population Data as Derived by Oliver

Parameter	Estimate	Standad error
a	0.023489	0.001990
d	4.211378	1.715073
P_0	6.453336	1.376688
P_s	63.646872	1.593014

This function increases from a lower asymptote of d to an upper asymptote or saturation level of $d + P_s$; and obviously, $P_{t=0} = d + P_0$.

Using the same notation as above, Oliver (1982) derived the inverse of the asymptotic variance-covariance matrix for the generalization of the logistic growth function (17.8.6), and is given by

$$\mathbf{V}^{-1} = \sigma^{-2}\begin{bmatrix} \frac{n}{2\sigma^2} & 0 & 0 & 0 & 0 \\ 0 & \Sigma F^2 & \Sigma F & \Sigma FG & \Sigma FH \\ 0 & \Sigma F & n & \Sigma G & \Sigma H \\ 0 & \Sigma FG & \Sigma G & \Sigma G^2 & \Sigma GH \\ 0 & \Sigma FH & \Sigma H & \Sigma GH & \Sigma H^2 \end{bmatrix}. \tag{17.8.7}$$

Oliver (1982) showed that $\mathbf{V}^{-1}$, and hence $\mathbf{V}$, does not depend on d. He estimated the parameters of (17.8.6) by direct least squares for the same data of the population of Great Britain, and used the results in (17.8.7) to obtain their SE. These results are presented in Table 17.8.2.

The logistic growth function (17.8.1) and the generalization given by (17.8.6) produced similar values in the middle decades, but they diverged increasingly as they were extrapolated. The results are presented in Table 17.8.3 along with the official projection of the Office of Population Censuses and Surveys (OPCS). It can be seen that results of the generalized function follow more closely the OPCS projections, which were prepared by the component method (OPCS, 1980).

Finally, Oliver (1982) outlined an alternative parameterization for the logistic growth function which involves the ratio of the saturation level to the initial population. The alternative function is given by

$$P_t = \frac{P_s}{1 + b\exp\{-at\}}, \tag{17.8.8}$$

Table 17.8.3 Comparison of Extrapolated Results From the Logistic Growth Function, Oliver's Generalized Growth Function, and the OPCS Official Projection of Population Growth in Great Britain

Year	Logistic	Generalized logistic	OPCS official projection
1981	55,022,000	54,340,000	54,387,000
1991	56,384,000	55,413,000	55,421,000
2001	57,554,000	56,295,000	56,428,000
2011	58,553,000	57,014,000	56,800,000
2021	59,400,000	57,597,000	NA
2031	60,115,000	58,067,000	NA
∞	63,647,000	59,918,000	NA

where $b = P_s/P_0 - 1$ is the location parameter. Applied to the same data, (17.8.8) gave the same estimates as (17.8.1) for P_s and a; and b was estimated as 5.362252 with a SE of 0.149802.

The generalized form of (17.8.8) gave the same estimates with the same SE as (17.8.6) for a, d, and P_s, and b was estimated at 7.632290 with SE 1.349320.

The use of P_0 rather than b does not affect the estimates of the other parameters of their standard errors, but the estimate of b has a smaller SE than does the estimate of P_0, and it is less strongly correlated with the other parameters. Thus, neither parameterization seems to have any particular advantage over the other, and the choice of function should be made based on preference for direct calculation of P_0 or b.

17.8.2 Applications of the Logistic Model to Bioassay

Some of the early proponents of the logistic function applied to bioassay were, among others, Emmens (1941) on endocrinology, Wilson and Worcester (1943) on bacteriology and Berkson (1944, 1951, 1953) on numerous bioassay applications. Specifically, Berkson (1944) applied the logit to dose mortality data. He compared logistic results to the normal curve in terms of the LD50, that is, the dose which is lethal to just 50% of the population exposed. Berkson (1951) again compared the logistic (logit) to the normal probit), i.e., the linear transformations of the reduced variate forms. He illustrated examples including autocatalysis, an oxidation-reduction reaction, a biomolecular reaction, hydrolysis, chemical concentration and hemolysis. Berkson (1953) used the minimum logit X^2 method to approximate the parameters of the logistic model applied to pharmacological data, comparative assays, and the relative toxicity of an unknown drug to a standard drug.

Gupta, Qureishi, and Shah (1967) used order statistics to estimate the mean and standard deviation of the logistic distribution. They applied their method to chemical concentration data.

Berkson (1968) considered five examples of the "no interaction" problem taken from Grizzle (1961), Woolf (1955), Norton (1945), Bartlett (1935), and Kastenbaum and Lamphiear (1959) in relation to estimation by minimum logit X^2, comparing these with estimation by MLE.

Ku and Kullback (1968) considered the same five data sets as Berkson (1968) above using the method of restricted minimum discrimination information (RMDI) estimation. They reached the same estimates as previously found by MLE in all cases.

Grizzle (1971) reanalyzed Ashford and Sowden's (1970) data using a logistic approach. He treated the data as two sets of binomial data, one for each of two reported respiratory symptoms in British coal miners. The data was fitted by a minimum chi-square approach, and since the data were not fit in its totality, Grizzle noted that he could not make a goodness-of-fit test for the adequacy of the model.

Berkson (1972) reconsidered the same five data sets as he did in 1968. He applied the RMDI estimate, which was obtained by adjustment of the marginal totals, and the unrestricted minimum discrimination information (UMDI) estimate.

Mantel and Brown (1973) used a model based on Mantel's (1966) generalization of the logistic curve to a polychotomous dose-response curve to reanalyze Ashford and Sowden's (1970) data on respiratory symptoms in coal miners. Various models with different assumptions were fit to the data by MLE and were compared with each other by chi-squared statistics.

Montfort and Otten (1976) extended the usual logistic to a more general type of symmetric distribution applied to quantal assay data. Their form contains a parameter related to kurtosis. They noted a loss of precision in estimating the parameters due to the addition of the kurtosis parameter when, in fact, the logistic distribution applies.

Morgan (1985) proposed and applied the cubic logistic model to quantal assay data. The usefulness of such a model would be to determine if the fit of a simpler model, such as the logit, could be improved. The problem lies in the difficulty of application to the data. Also, some alternative, nonlogistic models were suggested and compared to the logistic showing them to be at least as adequate.

Tsutakawa (1980) extended his previous work (Tsutakawa, 1972) by using the logistic quantal response curve to estimate the extreme percentage point (EDγ) when the location and scale parameters are unknown. The method was based on a prior distribution of the parameters and a predicted value of the posterior variance. The case when the scale parameter is known was presented by Tsutakawa (1972), where he presented approximate Bayesian solutions for estimating ED50 of the logistic curve.

17.8.3 Applications to Medical Diagnosis

The problem of medical diagnosis by the logistic discriminant function introduced by Cox (1966) and Day and Kerridge (1967) was extended by J. A. Anderson (1972 1973, 1974). Anderson and Richards (1979) made adjustments to small sample bias and also applied their results to medical diagnosis.

Wijesinha et al. (1983) applied the polychotomous logistic regression model to a large data set of patients where there were many distinct diagnostic categories. They encountered a variety of practical problems in their analysis. Consequently, Begg and Gray (1984) developed a simplified analytic method where each diagnostic category was individually compared with a normal baseline category using simple logistic models.

Begg and Gray (1984) also explored the possibility of using the method of individualized logistic regression to calculate the parameters of a polychotomous logistic regression model. That is, a series of separate simple logistic regression analyses are performed in place of a polychotomous logistic regression. They derived the asymptotic distribution of the estimates given by the individualized method; then they evaluated the asymptotic efficiency of the individualized method relative to a full polychotomous analysis.

However, Begg and Gray (1984) encountered some practical, unexpected problems in their medical diagnosis application. The first was a lack of computer storage space as a result of having a large number of parameters. The second was that only the SAS software package had the polychotomous logistic model. The third problem was that the SAS package would not accept sparse data. Finally, the logistic model does not accommodate symptoms with zero incidence in any disease category. Therefore, any software package that does not permit selective inclusion of a variable in some, but not necessarily all, of the regressions will require exclusion of that particular variable.

In order to compare, Begg and Gray (1984) examined the asymptotic relative efficiencies of three quantities:

1. Estimates of individual parameters
2. Estimates of predicted probabilities
3. Joint tests of the effect of a factor on several of the regressions

The factors of a particular application that might affect the outcome of the individualized logistic method are the

1. Relative prevalence of the chosen baseline category
2. Distance between the diagnostic categories with respect to the covariates
3. Marginal distribution of the covariates
4. Number of diagnostic categories
5. Number of covariates

To evaluate the performance of the individualized model under the influence of the five listed factors, Begg and Gray (1984) constructed situations with 4, 7, and 11 diagnostic categories, and 2 and 4 independent binary covariates. Table 17.8.4 illustrated the effect on efficiency by these factors. In general, the median asymptotic relative efficiencies (MARE) are high throughout, although there exists some inefficiencies among some individual parameters, predicted probabilities and tests. This is evidenced by the low mean of the minimum asymptotic relative efficiencies or average minimum efficiencies (AME). There appears to be a trend for the efficiencies to decrease as the number of parameters and diagnostic categories increase. Although, through the comparison of situations where the baseline probabilities are kept constant, it appears that the chief influence on efficiency is the size of the baseline category.

Begg and Gray (1984) stopped short of suggesting guidelines for use of the individualized method, but they feel that it is applicable to many situations.

Table 17.8.4 The Effect on Efficiency on the Outcome of the Individualized Logistic Model by 5 Different Factors as Shown by Begg and Gray (1984)

Parameter configurations			Asymptotic relative efficiencies					
			Parameters		Probabilities		Tests	
Number of diagnostic categories	Number of explanatory variables	Approximate size of baseline category	MARE	AME	MARE	AME	MARE	AME
4	2	1/4	99	96	99	96	98	95
	4	1/4	97	87	96	76	91	80
7	2	1/7	99	90	99	88	94	83
	4	1/7	95	83	93	70	88	77
11	2	1/11	99	88	99	81	95	77
	4	1/11	94	80	92	61	87	63
7	2	1/4	99	93	99	97	98	89
	4	1/4	97	85	96	79	93	82
11	2	1/4	100	91	99	90	97	86
	4	1/4	97	85	97	78	94	81

17.8.4 Applications to Public Health

Under the extremely broad heading of public health, we informally delineate five categories:

1. Retrospective/prospective studies
2. Survival analysis
3. Response-time data analysis
4. Epidemiology
5. Categorical data analysis

What follows is not intended to be a comprehensive outline of the applications of the logistic model in the public health sciences. Rather, the sample of articles presented here are hopefully representative of the major trends in this arena.

For example, Breslow and Powers (1978) compared two approaches to the logistic regression analysis of retrospective study data. Specifically, they compared the prospective model to the retrospective model. The prospective model has the dependent variable being a binary indicator of case-control status, while the regression variables represent both risk and confounding factors. Conversely, the retrospective logistic model, proposed by Prentice (1976), has as dependent variable a binary indicator of exposure to one particular risk factor. Here the regression variables represent disease (case-control) status, other risk factors, and confounding factors.

The goal in the analysis of retrospective data is to estimate the ratio of disease incidence among those exposed versus those not exposed to the risk factor(s) of interest. This is known as the relative risk (RR).

The two logistic models were compared by Breslow and Powers (1978) by relating both to a log-odds ratio regression model for a series of 2×2 tables. The two models yielded similar results for the estimate of the RR. The models were illustrated, and the results compared using data from the Oxford Childhood Cancer Survey reported by Kneale (1971).

Plackett (1959) was the first to use the logistic function in the analysis of survival data. His applications include survival after operations on cancer patients, labor turnover, and business failures. All applications to the data were based on order statistics. Bain, Eastman, and Engelhardt (1973) as well as Engelhardt (1975) used the logistic distribution as a life-testing model; refer also to Chapter 5. A log-logistic regression model for survival data was used by Bennett (1983) applied to lung cancer survival data; while a separate model was suggested by Paiva-Franco (1984). Furthermore, logistic regression was applied to estimate the survival time of diagnosed leukemia patients by Johnson (1985).

Meyers, Hankey, and Mantel (1973) and then later Mantel and Hankey

(1978) used a logistic-exponential regression model applied to a response-time analysis. The same breast cancer data was used in both experiments. In addition, Mantel and Hankey (1978) also applied the logistic-exponential regression model to the survival of patients diagnosed with malignant melanomas.

Dyke and Patterson (1952) used the logit form of the logistic to model proportions pertaining to good and bad knowledge of cancer facts relative to a linear combination of categorical predictors. Grizzle (1961) used the logistic model to compare the mothers of school children referred by their teachers as having behavioral problems to mothers of a comparable group of control children. The comparison was done relative to whether or not the mother had suffered any previous infant losses prior to the child in the study. McCullagh (1977) extended some simple odds ratio statistics of Clayton (1974) for comparing two independent samples of ordered categorical data to a related model for paired data. He gives an application concerning the degrees of pneumoconiosis in coal miners as measured radiologically.

Lachenbruch (1980) developed a simple method for combining risks based on repeated observations on an individual. This was applied to the estimation of the risk of a disease based on a given set of risk factors. Specifically, Lachenbruch (1980) illustrated the technique in a study in the detection of hemophilia carriers. The formulation of the problem was developed using a logistic discriminant function approach, and the combination of risks was shown to be a simple computation.

A major two-part work on categorical data analysis was done by Imrey, Koch, and Stokes (1981, 1982). They surveyed the literature, including both methodology and applications for the log-linear model as well as the logistic regression model relative to categorical data analysis.

Greenland (1985) illustrated some extensions of logistic models to the modeling of probabilities of ordinal responses. These are responses which involve only ranking of the response values without any specific distance between the values. Methods of estimation and assessment of fit for the ordinal models were extended from the binary logistic model. An application ensues with an analysis of the dependence of chronic obstructive respiratory disease prevalence on smoking and age.

Scott and Wild (1985) considered categorical response variables with independent samples drawn for each of k categories. Specifically, they looked at the problem of fitting logistic regression models for binary data from an unstratified case control study.

Scott and Wild (1986) continued their work on the special case of fitting logistic regression models under case control or choice-based sampling, where the population is stratified by values of the categorical response variable.

The regressive logistic model introduced by Bonney (1986) was designed to merge the goals and methodologies of both the epidemiologist and geneticist in the study of familial disease and other binary traits. The aim of the epidemiologist is to assess the significance of familiality as a risk factor for disease; while the aim of the geneticist is to interpret the underlying physiology as it relates to disease risk. Several relevant models were developed to fulfill the projected goals. That is, they could be used to summarize family data without reference to underlying biology, and they could be extended to include parameters relating to genetic transmission.

Kay and Little (1986) applied a method proposed by Copas (1983) to outcome of children diagnosed with hemolytic uremic syndrome. The primary aims of the study were to investigate the dependence of the probability of favorable outcome on specific covariates measured at diagnosis. Also of prime importance was the provision of a scoring scheme to aid discrimination between the outcome categories and the identification of high-risk children.

Their method requires the need for the computation of a smoothed value for the response variable for every subject which is a weighted average of the values of the outcome variable over all subject. For covariate x for the ith subject they compute

$$\hat{p}(x_i) = \sum_{j=1}^{n} w(x_j)y_j\left[\sum_{j=1}^{n} w(x_j)\right]^{-1}, \tag{17.8.9}$$

where

$$w(x_j) = \exp\left\{-\frac{(x_j - x_i)^2}{h}\right\} \tag{17.8.10}$$

and $h > 0$ in the smoothing parameter taken so that the function tails off at the desired rate; $p(x_i)$ is the observed proportion of good outcomes smoothed over neighboring values of x_i.

The covariates of interest are (x_i, $i = 1, \ldots, 4$): years of age, sex, existence of diarrhea, and season of admission (summer/winter), respectively. The ages of the children ranged between 0.1 and 13.5 years.

Different models based on the method described above were tried based on the distribution of the covariates. For example, the distribution of x_1 (years) was assumed to be Gamma. This implies that both x_1 and $\ln(x_1)$ should be included in the model. A series of tests for independence about

Table 17.8.5 Assessing the Fit of a Logistic Regression Model Applied to Children with the Haemolytic Uraemic Symdrome

	Model MLE of coefficient (standard error)			
Covariate	1	2	3	4
Constant	1.962 (0.477)	1.700 (0.548)	1.895	2.223 (0.508)
x_1(age)	−0.140 (0.077)	−0.164 (0.085)	−0.164 (0.085)	
x_2(sex)		1.173 (0.773)	1.73 (0.773)	
x_3(diarrhea)	−1.159 (0.603)	0.791 (1.055)	0.791 (1.055)	−1.155 (0.643)
x_4(season)	−1.383 (0.645)	−1.601 (0.689)	−1.601 (0.689)	−1.203 (0.677)
x_2x_3		−3.454 (1.405)	−3.453 (1.405)	
x_3x_4			−2.56	
$\ln(x_1)$				
$(\ln x_1)^2$				−0.498 (0.190)
$2 \ln(L)$	80.22	73.26	70.18	74.82
d.f.	71	69	68	70

the covariates for inclusion of certain interaction terms and cross-product terms were conducted. Kay and Little (1986) found no dependence between x_1 and x_2, x_3, x_4; also, they found no dependence between x_2 (sex) and x_4 (season). Dependence between any two covariates would indicate the need for a cross-product term in the model. Table 17.8.5 gives details of the fit of the various models.

All of the models did well in terms of predicting a good outcome patient, but not as well in predicting a poor outcome. To this end, Kay and Little (1986) used a variety of ways of assessing the reliability of the fitted models. For example, they used the average probability of correct allocation defined by Hilden, Habbema, and Bjerregaard (1978) given by

$$T = \frac{1}{n} \sum_{i=1}^{n} [y_i p_i - (1 - y_i)(1 - p_i)]. \qquad (17.8.11)$$

Furthermore, the techniques of Pregibon (1981) for assessing the fit of a logistic regression model were applied by Kay and Little (1986) to evaluate the fit of the various models.

The application put forth by Kay and Little (1986) is complete in the sense that one is taken through all of the calculations utilizing many important previous theoretical assertions. Their use of the logistic regression model to analyze the hemolytic uremic syndrome data is a relevant application in this field. This case study could be used as a model or guide for the evaluation of similar data.

REFERENCES

Aitchison, J. and Shen, S. M. (1980). Logistic-normal distributions: Some properties and uses, *Biometrika*, *67*, 261–272.

Albert, A. and Anderson, J. A. (1984). On the existence of maximum likelihood estimates in logistic regression models, *Biometrika*, *71*, 1–10.

Anderson, J. A. (1972). Separate sample logistic discrimination, *Biometrika*, *59*, 19–35.

Anderson, J. A. (1973). Logistic discrimination with medical applications, in *Discriminant Analysis and Applications*, T. Cacoullos (ed.), 1–15, Academic Press, New York.

Anderson, J. A. (1974). Diagnosis by logistic discriminant function: further practical problems and results, *Applied Statistics*, *23*, 397–404.

Anderson, J. A. and Richardson, S. C. (1979). Logistic discrimination and bias correction in maximum likelihood estimation, *Technometrics*, *21*, 71–78.

Anscombe, F. J. (1956). On estimating binomial response relations, *Biometrika*, *43*, 461–464.

Antle, C. E. and Bain, L. J. (1969). A property of maximum likelihood estimators of location and scale parameters, *SIAM Review*, *11*, 251–253.

Antle, C. E., Klimko, L., and Harkness, W. (1970). Confidence intervals for the parameters of the logistic distribution, *Biometrika*, *57*, 397–402.

Ashford, J. R. and Sowden, R. D. (1970). Multivariate probit analysis, *Biometrics*, *26*, 535–546.

Bain, L. J., Eastman, J., and Engelhardt, M. (1973). A study of life-testing models and statistical analyses for the logistic distribution, *Technical Reports of Aerospace Research Laboratories*, *73-0009*, v + 39 pp.

Bartlett, M. S. (1935). Contingency table interactions, *Journal of the Royal Statistical Society*, Supplement 2, 248–252.

Begg, C. B. and Gray, R. (1984). Calculation of polychotomous logistic regression parameters using individualized regressions, *Biometrika*, *71*, 11–18.

Bennett, S. (1983). Log-logistic regression models for survival data, *Applied Statistics*, *32*, 165–171.

Berkson, J. (1944). Application of the logistic function to bio-assay, *Journal of the American Statistical Association*, *37*, 357–365.

Berkson, J. (1951). Why I prefer logits to probits, *Biometrics*, 7, 327–339.

Berkson, J. (1953). A statistically precise and relatively simple method of estimating the bio-assay with quantal response, based on the logistic function, *Journal of the American Statistical Association*, *48*, 565–599.

Berkson, J. (1955). Maximum likelihood and minimum chi-square estimates of the logistic function, *Journal of the American Statistical Association*, *50*, 130–162.

Berkson, J. (1956). Estimation by least squares and by maximum likelihood, in *Proceeding of the Third Berkeley Symposium on Mathematical Statistics and Probability*, *1*, 1–11.

Berkson, J. (1957). Tables for the maximum likelihood estimate of the logistic function, *Biometrics*, *13*, 28–34.

Berkson, J. (1960). Nomograms for fitting the logistic function by maximum likelihood, *Biometrika*, *47*, 121–141.

Berkson, J. (1968). Application of minimum logit chi-squared to a problem of Grizzle with a notation on the problem of no interaction, *Biometrics*, *24*, 75–95.

Berkson, J. (1972). Minimum discrimination information, the 'no interaction' problem, and the logistic function, *Biometrics*, *28*, 443–468.

Birnbaum, A. and Dudman, J. (1963). Logistic order statistics, *The Annals of Mathematical Statistics*, *34*, 658–672.

Blom, G. (1958). *Statistical Estimates and Transformed Beta Variables*, Wiley, New York.

Bonney, G. E. (1986). Regressive logistic models for familial disease and other binary traits, *Biometrics*, *42*, 611–625.

Breslow, N. and Powers, W. (1978). Are there two logistic regressions for retrospective studies? *Biometrics*, *34*, 100–105.

Chen, E. H. and Dixon, W. J. (1972). Estimates of parameters of a censored regression sample, *Journal of the American Statistical Association*, *67*, 664–671.

Clayton, D. G. (1974). Some odds ratio statistics for the analysis of ordered categorical data, *Biometrika*, *61*, 525–531.

Cook, R. D. and Weisberg, S. (1982). *Residuals and Influence in Regression*, Chapman and Hall, New York.

Copas, J. B. (1983). Plotting p against x, *Applied Statistics*, *32*, 25–31.

Cornfield, J. and Mantel, N. (1950). Some new aspects of the application of maximum likelihood to the calculation of the dosage response curve, *Journal of the American Statistical Association*, *45*, 181–210.

Cox, D. R. (1966). Some procedures connected with the logistic qualitative response curve. *Research Papers in Statistics*, F. N. David (ed.), 55–71, Wiley, New York.

Cox, D. R. (1970). *The Analysis of Binary Data*, Methuen, London.

Cox, D. R. and Snell, E. J. (1968). A general definition of residuals (with discussion), *Journal of the Royal Statistical Society, Series B*, *30*, 248–275.

D'Agostino, R. B. and Lee, A. F. S. (1976). Linear estimation of the logistic parameters for complete or tail censored samples, *Journal of the American Statistical Association*, *71*, 462–464.

Davis, H. T. (1941). *The Analysis of Economic Time Series*, Principia Press, Bloomington.

Davis, L. J. (1985). Modification of the empirical logit to reduce bias in simple linear logistic regression, *Biometrika*, *72*, 199–202.

Day, N. E. and Kerridge, D. F. (1967). A general maximum likelihood discriminant, *Biometrics*, *23*, 313–323.

Durbin, J. (1973). Weak convergence of the sample distribution function when parameters are estimated, *Annals of Statistics*, *1*, 279–290.

Dyke, G. V. and Patterson, H. D. (1952). Analysis of factorial arrangements when the data are proportions, *Biometrics*, *8*, 1–12.

Emmens, C. W. (1941). The dose-response relation for certain principles of the pituitary gland, and of the serum and urine of pregnancy, *Journal of Endocrinology*, *2*, 194–225.

Engelhardt, M. (1975). Simple linear estimation of the parameters of the logistic distribution from a complete or censored sample, *Journal of the American Statistical Association*, *70*, 899–902.

Fisher, R. A. (1925). Theory of statistical estimation, *Proceedings of the Cambridge Philosophical Society*, *22*, 700–725.

Formann, A. K. (1982). Linear logistic latent class analysis, *Biometrical Journal*, *24*, 171–190.

Forsythe, G. E., Malcolm, M. A., and Moler, C. B. (1977). *Computer Methods for Mathematical Computations*, Prentice-Hall, Engelwood Cliffs.

Gart, J. J., Pettigrew, H. M., and Thomas, D. G. (1985). The effect of bias, variance estimation, skewness and kurtosis of the empirical logit on weighted least squares analyses, *Biometrika*, *72*, 179–190.

Glasbey, C. A. (1979). Correlated residuals in non-linear regression applied to growth data, *Applied Statistics*, *28*, 251–259.

Greenland, S. (1985). An application of logistic models to the analysis of ordinal responses, *Biometrical Journal*, *27*, 189–197.

Grizzle, J. E. (1961). A new method of testing hypotheses and estimating parameters for the logistic model, *Biometrics*, *17*, 372–385.

Grizzle, J. E. (1971). Multivariate logistic analysis, *Biometrics*, *27*, 1057–1062.

Gupta, S. S. and Gnanadesikan, M. (1966). Estimation of the parameters of the logistic distribution, *Biometrika*, *53*, 565–570.

Gupta, S. S., Qureishi, A. S., and Shah, B. K. (1967). Best linear unbiased estimators of the parameters of the logistic distribution using order statistics, *Technometrics*, *9*, 43–59.

Hall, I. J. (1975). One-sided tolerance limits for a logistic distribution based on censored samples, *Biometrics*, *31*, 873–880.

Hamilton, M. A. (1977). Estimating the logistic curve from grouped data, *Journal of Statistical Computation and Simulation*, *5*, 279–301.

Harter, H. L. and Moore, A. H. (1967). Maximum likelihood estimation, from censored samples, of the parameters of a logistic distribution, *Journal of the American Statistical Association*, *62*, 675–684.

Hilden, J., Habbema, J. D. F., and Bjerregaard, B. (1978). The measurement of performance in probabilistic diagosis. II: Trustworthiness of the exact values of the diagnostic probabilities, *Meth Inform Med*, *17*, 227–237.

Hitchcock, S. E. (1962). A note on the estimation of the parameters of the logistic function, using the minimum logit chi-square method, *Biometrika*, *49*, 250–252.

Hodges, J. L., Jr. (1958). Fitting the logistic by maximum likelihood, *Biometrics*, *14*, 453–461.

Hotelling, H. (1927). Differential equations subject to error and population estimates, *Journal of the American Statistical Association*, *22*, 283–314.

Imrey, P. B., Koch, G. G., and Stokes, M. E. (1981). Categorical data analysis: some reflections on the log linear model and logistic regression. I: Historical and methodological overview, *International Statistical Review*, *49*, 265–283.

Imrey, P. B., Koch, G. G., and Stokes, M. E. (1982). Categorical data analysis: some reflections on the log linear model and logistic regression. II: Data analysis, *International Statistical Review*, *50*, 35–63.

Jennings, D. E. (1986). Outliers and residual distributions in logistic regression. *Journal of the American Statistical Association*, *81*, 987–990.

Johnson, W. (1985). Influence measures for logistic regression: another point of view, *Biometrika*, *72*, 59–65.

Jung, J. (1955). On linear estimates defined by a continuous weight function, *Arkiv Matematik*, *3*, 199–209.

Kastenbaum, M. A. and Lamphiear, D. E. (1959). Calculation of chi-square to test the three-factor no interaction hypothesis, *Biometrics*, *15*, 107–115.

Kay, R. and Little, S. (1986). Assessing the fit of the logistic model: A case study of children with the haemolytic uraemic syndrome, *Applied Statistics*, *35*, 16–30.

Kneale, G. W. (1971). Problems arising in estimating from retrospective study data the latent period of juvenile cancers initiated by obstetric radiography, *Biometrics*, *27*, 563–590.

Ku, H. H. and Kullback, S. (1968). An application of minimum discrimination information estimation to a problem of Grizzle and Berkson on the test of 'no interaction' hypothesis, unpublished paper.

Lachenbruch, P. A. (1980). Note on combining risks using a logistic discriminant function approach, *Biometrical Journal*, *22*, 759–762.

Landwehr, J. M., Pregibon, D., and Shoemaker, A. C. (1984). Graphical methods for assessing logistic regression models, *Journal of the American Statistical Association*, *79*, 61–71.

Leach, D. (1981). Re-evaluation of the logistic curve for human populations, *Journal of the Royal Statistical Society, Series A*, *144*, 94–103.

Lehmann, E. L. (1959). *Testing Statistical Hypotheses*, Wiley, New York.

Lord, F. M. (1965). A note on the normal ogive or logistic curve in item analysis, *Psychometrika*, *30*, 371–372.

Mantel, N. (1966). Models for complex contingency tables and polychotomous dosage response curves, *Biometrics*, *22*, 83–95.

Mantel, N. and Brown, C. C. (1973). A logistic reanalysis of Ashford and Sowden's data on respiratory symptoms in British coal miners, *Biometrics*, *29*, 649–665.

Mantel, N. and Hankey, B. F. (1978). A logistic regression analysis of response time data where the hazard function is time dependent, *Communications in Statistics—Theory and Methods*, *A7*, 333–347.

McCullagh, P. (1977). A logistic model for paired comparisons with ordered categorical data, *Biometrika*, *64*, 449–453.

Morgan, B. S. T. (1985). The cubic logistic model for quantal assay data, *Applied Statistics*, *34*, 105–113.

Myers, M. H., Hankey, B. F., and Mantel, N. (1973). A logistic exponential model for use with response-time data involving regressor variables, *Biometrics*, *29*, 257–269.

Neyman, J. (1935). Su un teorema concernente le cosiddette statistiche sufficienti, *C. Ist. Ital. Attuari, Anno VI*, *13*, 3–17.

Norton, H. W. (1945). Calculation of chi-square from complex contingency tables, *Journal of the American Statistical Association*, *40*, 251–258.

Office of Population Censuses and Surveys (1980). *Population Projections, No. 11, Series PP2*, HMSO, London.

Ogawa, J. (1951). Contributions to the theory of systematic statistics. I, *Osaka Mathematical Journal*, *3*, 175–213.

Oliver, F. R. (1964). Methods of estimating the logistic growth function, *Applied Statistics*, *13*, 57–66.

Oliver, F. R. (1966). Aspects of maximum likelihood estimation of the logistic growth function, *Journal of the American Statistical Association*, *61*, 697–705.

Oliver, F. R. (1969). Another generalisation of the logistic growth function, *Econometrica*, *37*, 144–147.

Oliver, F. R. (1982). Notes on the logistic curve for human populations, *Journal of the Royal Statistical Society*, *Series A*, *145*, 359–363.

Pagano, M. and Tritchler, D. L. (1983). On obtaining permutation distributions in polynomial time, *Journal of the American Statistical Association*, *78*, 435–440.

Paiva-Franco, M. A. (1984). A log logistic model for survival time with covariates, *Biometrika*, *71*, 621–623.

Pearl, R. and Reed, L. J. (1920). On the rate of growth of the population of the United States since 1790 and its mathematical representation, in *Proceedings of the National Academy of Sciences*, *6*, 275–288.

Pearl, R. and Reed, L. J. (1924). *Studies in Human Biology*, Williams and Wilkins, Baltimore.

Pearl, R., Reed, L. J., and Kish, J. F. (1940). The logistic curve and the census count of 1940, *Science*, *92*, 486–488.

Plackett, R. L. (1959). The analysis of life test data, *Technometrics*, *1*, 9–19.

Pregibon, D. (1979). Data analytic methods for generalized linear models, unpublished Ph.D. Thesis, University of Toronto, Canada.

Pregibon, D. (1981). Logistic regression diagostics, *Annals of Statistics*, *9*, 705–724.

Prentice, R. (1976). Use of the logistic model in retrospective studies, *Biometrics*, *32*, 599–606.

Reed, L. J. and Berkson, J. (1929). The application of the logistic function to experimental data, *Journal of Physical Chemistry*, *33*, 760–779.

Sanathanan, L. (1974). Some properties of the logistic model for dichotomous response, *Journal of the American Statistical Association*, *69*, 744–749.

Santner, T. J. and Duffy, D. E. (1986). A note on A. Albert and J. A. Anderson's conditions for the existence of maximum likelihood estimates in logistic regression models, *Biometrika*, *73*, 755–758.

Schafer, R. E. and Sheffield, T. S. (1973). Inferences on the parameters of the logistic distribution, *Biometrics*, *29*, 449–455.

Schultz, H. (1930). The standard error of a forecast from a curve, *Journal of the American Statistical Association*, *25*, 139–185.

Scott, A. J. and Wild, C. J. (1985). Fitting logistic models in case-control studies, *Bulletin of the International Statistical Institute*, *51*, 12.3, 15pp.

Scott, A. J. and Wild, C. J. (1986). Fitting logistic models under case-control or choice based sampling, *Journal of the Royal Statistical Society*, *Series B*, *48*, 170–182.

Silverstone, H. (1957). Estimating the logistic curve, *Journal of the American Statistical Association*, *52*, 567–577.

Singleton, R. C. (1969). An algorithm for computing the mixed radix fast Fourier transform, *IEEE Transactions on Audio and Electroacoustics*, *17*, 93–103.

Stephens, M. A. (1979). Tests of fit for the logistic distribution based on the empirical distribution function, *Biometrika*, *66*, 591–595.

Tintner, G. (1952). *Econometrics*, Wiley, New York.

Tritchler, D. L. (1984). An algorithm for exact logistic regression, *Journal of the American Statistical Association*, *79*, 709–711.

Tsiatis, A. A. (1980). A note on the goodness-of-fit for the logistic regression model, *Biometrika*, *67*, 250–251.

Tsutakawa, R. K. (1972). Design of experiments for bio-assay, *Journal of the American Statistical Association*, *67*, 584–590.

Tsutakawa, R. K. (1980). Selection of dose levels for estimating a percent point of a logistic quantal response curve, *Applied Statistics*, *29*, 25–33.

Van Montfort, M. A. J. and Otten, A. (1976). Quantal response analysis: enlargement of the logistic model with a kurtosis parameter, *Biometrische Zeitschrift*, *18*, 371–380.

Verhulst, P. F. (1845). Recherches Mathématiques sur la loi d'accroissement de la population, *Nouveaux Mémoires de l'Académie Royale des Sciences et Belles Lettres de Bruxelles*, *18*, 1–38.

Vieira, S. and Hoffmann, R. (1977). Comparison of the logistic and the Gompertz growth function considering additive and multiplicative error terms, *Applied Statistics*, *26*, 143–148.

Wedderburn, R. W. M. (1976). On the existence and uniqueness of the maximum likelihood estimates for certain generalized linear models, *Biometrika*, *63*, 27–32.

Welch, B. L. (1939). Note on discriminant functions, *Biometrika*, *31*, 218–220.

Wijesinha, A., Begg, C. B., Funkenstein, H. H., and McNeil, B. J. (1983). Methodology for the differential diagnosis of a complex data set: a case study using data from routine CT-scan examinations, *Medical Decision Making*, *3*, 133–154.

Williams, D. A. (1976). Improved likelihood ratio tests for complete contingency tables, *Biometrika*, *63*, 33–37.

Wilson, E. B. and Worcester, J. (1943). The determination of L.D.50 and its sampling error in bio-assay, in *Proceedings of the National Academy of Sciences*, *29*, 79–85.

Woolf, B. (1955). On estimating the relation between blood group and disease, *Annals of Human Genetics*, *19*, 251–253.

18

Some Other Applications

18.1 APPLICATIONS TO ORDERED CATEGORICAL DATA

Kenneth J. Koehler
Iowa State University, Ames, Iowa

James T. Symanowski
Lilly Research Laboratories, Greenfield, Indiana

18.1.1 Introduction

A bivariate logistic regression model is presented in this section which is suitable for the analysis of bivariate ordered categorical data with potentially nonzero correlation between the bivariate responses. These type of data, often referred to as ordered two-way contingency tables, arise in many situations. In clinical trials, for example, two different categorical traits may be observed on each subject or measurements of a single trait may be taken on each subject at two different time points. In other situations correlated responses may arise from social or genetic relationships between subjects or from biological relationships between traits measured on individual subjects.

The model presented in this section is based upon the partitioning of a bivariate logistic distribution (discussed in Section 18.1.2) which is associated with a correlation coefficient that can range from $-3\pi^{-2}$ to 1. This distribution has a distribution function possessing an everywhere differentiable closed form expression. Consequently, maximum likelihood estimation can be easily and efficiently accomplished with a direct application of a Fisher scoring algorithm.

Two applications of this model are considered. The first (discussed in

Section 18.1.3) pertains to the analysis of two-way tables when no concomitant information is available. In this situation the inferences of primary interest include comparisons of the means and variances of the underlying marginal logistic distributions and an assessment of the strength of association between the bivariate responses. The latter is made possible through the estimation of the correlation coefficient for the underlying bivariate logistic distribution.

The second application (discussed in Section 18.1.4) addresses the analysis of bivariate categorical data when concomitant information is available in the form of either discrete or continuous explanatory variables. In this situation the importance of the covariates is assessed by the estimation of logistic regression parameters. These parameters are easily interpretable since the fitting of the bivariate logistic regression model is tantamount to the simultaneous fitting of a univariate logistic regression model to each margin. Again, estimation of the underlying correlation coefficient enables one to assess the strength of association between the bivariate responses.

18.1.2 A Bivariate Logistic Distribution

Cook and Johnson (1986) introduce the following family of bivariate uniform distributions

$$
\begin{aligned}
H(u_1, u_2) = {} & (1 + \omega)(u_1^{-\alpha^{-1}} + u_2^{-\alpha^{-1}} - 1)^{-\alpha} + \omega(2u_1^{-\alpha^{-1}} + 2u_2^{-\alpha^{-1}} - 3)^{-\alpha} \\
& - \omega(2u_1^{-\alpha^{-1}} + u_2^{-\alpha^{-1}} - 2)^{-\alpha} - \omega(u_1^{-\alpha^{-1}} + 2u_2^{-\alpha^{-1}} - 2)^{-\alpha}, \quad (18.1.2.1)
\end{aligned}
$$

where $\alpha > 0$ and $-1 \leq \omega \leq 1$ jointly determine the dependency between u_1 and u_2. This distribution function can be used to construct bivariate distributions with arbitrary marginal distribution functions $\{F_i(x_i): i = 1, 2\}$ by replacing u_i with $F_i(x_i)$ in (18.1.2.1). A bivariate distribution function with logistic marginal distributions is obtained by replacing u_i with $F_i(x_i) = (1 + \exp\{-(x_i - \mu_i)/\sigma_i\})^{-1}$, $i = 1, 2$. The ith marginal distribution has mean μ_i and variance $\sigma_i^2\pi^2/3$. The basic properties of this bivariate distribution are investigated in Symanowski and Koehler (1989). For the standardized case where $\mu_i = 0$ and $\sigma_i = 1$, for $i = 1, 2$, the bivariate logistic distribution function is

$$
\begin{aligned}
F(x_1, x_2) = {} & (1 + \omega)[(1 + \exp\{-x_1\})^{\alpha^{-1}} + (1 + \exp\{-x_2\})^{\alpha^{-1}} - 1]^{-\alpha} \\
& + \omega[2(1 + \exp\{-x_1\})^{\alpha^{-1}} + 2(1 + \exp\{-x_2\})^{\alpha^{-1}} - 3]^{-\alpha} \\
& - \omega[2(1 + \exp\{-x_1\})^{\alpha^{-1}} + (1 + \exp\{-x_2\})^{\alpha^{-1}} - 2]^{-\alpha} \\
& - \omega[(1 + \exp\{-x_1\})^{\alpha^{-1}} + 2(1 + \exp\{-x_2\})^{\alpha^{-1}} - 2]^{-\alpha}, \\
& \alpha > 0, \quad -1 \leq \omega \leq 1, \quad -\infty < x_i < \infty, \quad i = 1, 2. \quad (18.1.2.2)
\end{aligned}
$$

The corresponding joint density function is

$$f(x_1, x_2) = \frac{\alpha + 1}{\alpha} \exp\{-x_1 - x_2\}[(1 + \exp\{-x_1\})(1 + \exp\{-x_2\})]^{(1-\alpha)/\alpha}$$
$$\times [(1 + \omega)((1 + \exp\{-x_1\})^{\alpha^{-1}} + (1 + \exp\{-x_2\})^{\alpha^{-1}} - 1)^{-\alpha-2}$$
$$+ 4\omega(2(1 + \exp\{-x_1\})^{\alpha^{-1}} + 2(1 + \exp\{-x_2\})^{\alpha^{-1}} - 3)^{-\alpha-2}$$
$$- 2\omega(2(1 + \exp\{-x_1\})^{\alpha^{-1}} + (1 + \exp\{-x_2\})^{\alpha^{-1}} - 2)^{-\alpha-2}$$
$$-2\omega((1 + \exp\{-x_1\})^{\alpha^{-1}} + 2(1 + \exp\{-x_2\})^{\alpha^{-1}} - 2)^{-\alpha-2}] \tag{18.1.2.3}$$

The correlation coefficient (ρ), approaches a value of one as α tends to zero (regardless of the value of ω) and approaches a minimum value of $-3\pi^{-2}$ for $\omega = -1$ as α tends to ∞. Independence between x_1 and x_2 is approached as α approaches ∞ when $\omega = 0$. A closed-form expression for ρ as a function of α and ω is not generally available. One exception is the special case for $\alpha = 1$, where $\rho(\omega) = 0.5 + (\log 2)^2 3\pi^{-2}\omega$.

The number of parameters in the bivariate logistic distribution can be reduced by one with essentially no reduction of the range of possible correlations. This is done by expressing α and ω as functions of a single unrestricted dependency parameter λ. The expressions

$$\omega(\lambda) = (1 - \exp\{-\lambda\})(1 + \exp\{-\lambda\})^{-1} \tag{18.1.2.4}$$

and

$$\alpha(\lambda) = 40((\exp\{-2.5\lambda\} - 1)(\exp\{2.5\lambda\} + 1)^{-1} + 1), \tag{18.1.2.5}$$

where $-\infty < \lambda < \infty$, allow the correlation to be accurately approximated by

$$\rho(\lambda) = \frac{-c_1 + (c_1 + a_1\lambda + a_2\lambda^2)\exp\{a_3\lambda\}}{1 + (c_1 + a_1\lambda + a_2\lambda^2)\exp\{a_3\lambda\}} \tag{18.1.2.6}$$

($c_1 = 3\pi^{-2}, a_1 = -0.2131, a_2 = 0.0930, a_3 = 1.3739$). This approximation covers the entire range of possible correlations; i.e., $\lim_{\lambda\to\infty} \rho(\lambda) = 1$ and $\lim_{\lambda\to-\infty} \rho(\lambda) = -3\pi^{-2}$. The reduction from ($\alpha$, ω) to λ also stabilizes the numerical evaluation of maximum likelihood estimates.

Figures 18.1.2.1(a) through 18.1.2.1(d) show contours of constant density for various values of λ. The distributions have been standardized so that marginal means and variances are 0 and 1, respectively. Clearly these are in general nonelliptical contours. Figure 18.1.2.1(a) corresponds to a large value of λ (which coincides with a correlation of 0.9) and a peaked density surface while Figure 18.1.2.1(d) corresponds to a large negative

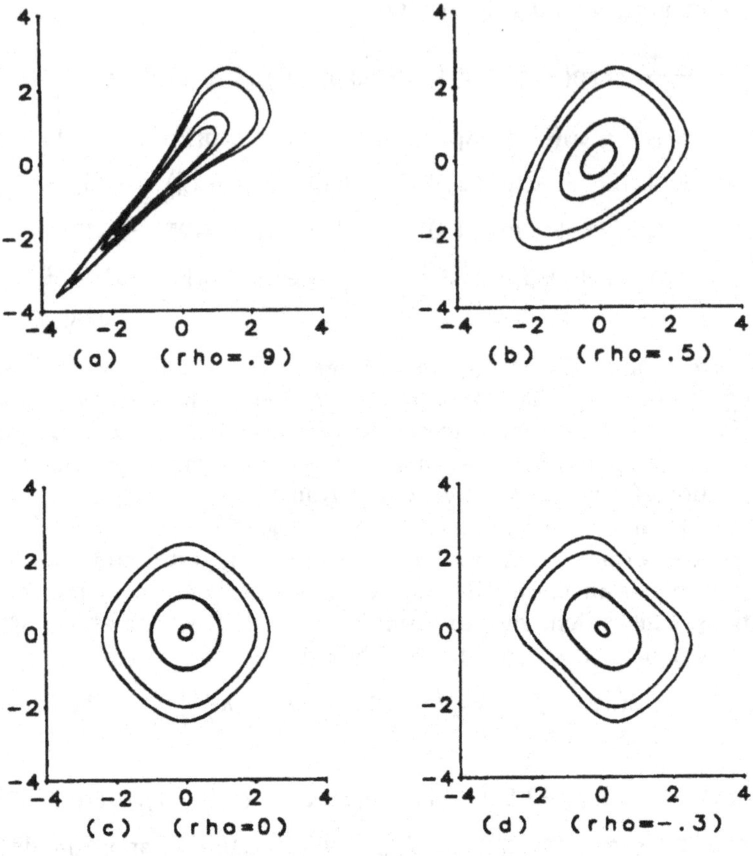

Fig. 18.1.2.1 Contours of constant density for selected values of λ.

value of λ (which coincides with a correlation of −0.3) and a less peaked density surface. There is a striking similarity between the contour plot in Figure 18.1.2.1(b) for $\lambda = 1.5326$ and $\rho = 0.5$ and the contour plot associated with the first of Gumbel's bivariate logistic distributions shown in Figure 11.2.1 (which also has a correlation equal to 0.5). Figure 18.1.2.1(c) provides contours for $\lambda = 0$ and corresponds to the case where X_1 and X_2 are nearly independent.

18.1.3 Application to Two-Way Contingency Tables with Ordered Categories

In this application the actual value of the bivariate logistic random variable (X_1, X_2) is not observed. Instead, X_1 is recorded as being in one of r distinct ordered categories and X_2 is recorded as being in one of another set of s distinct ordered categories. Then the responses for a sample of N units can be displayed as counts in an $r \times s$ contingency table where n_{ij} denotes the observed number of units in the category corresponding to the ith row and jth column of the table.

Denote the boundaries of the categories for X_1 by $c_0 < c_1 < c_2 < \cdots < c_{r-1} < c_r$ and the boundaries of the categories for X_2 by $d_0 < d_1 < d_2 < \cdots < d_{s-1} < d_s$. Here, the definitions $c_0 = d_0 = -\infty$ and $c_r = d_s = \infty$ are used for convenience; the other boundary values may not be known. Using (18.1.2.2) and (18.1.2.4)–(18.1.2.6), the probability that a random response falls into the category corresponding to the ith row and jth column of the contingency table is

$$P_{ij} = F(v_i, w_j) - F(v_{i-1}, w_j) - F(v_i, w_{j-1}) + F(v_{i-1}, w_{j-1}), \qquad (18.1.3.1)$$

where $v_i = (c_i - \mu_1)\sigma_1^{-1}$ and $w_j = (d_j - \mu_2)\sigma_2^{-1}$, and μ_i and $\sigma_i^2\pi^2/3$ are the mean and variance of X_i, respectively. If the N bivariate responses are independent (e.g., respondents are selected using simple random sampling with replacement) the kernel of the log-likelihood function for the corresponding multinomial distribution for the counts is

$$l(\delta) = \sum_{i=1}^{r} \sum_{j=1}^{s} n_{ij} \log(P_{ij}), \qquad (18.1.3.2)$$

where $\delta' = (\mu_1, \mu_2, \sigma_1, \sigma_2, \lambda, c_1, \ldots, c_{r-1}, d_1, \ldots, d_{s-1})$. Differences in the two sets of boundary values, however, cannot be completely distinguished from differences in the means and variances and δ is an unidentified parameter as defined by Kendall and Stuart (1967). This problem can be remedied by including additional restrictions in the model. As an illustration, consider the situation where the boundaries are unknown, but the same boundaries are used for each logistic variable; i.e., $r = s$ and $c_i = d_i$ for $i = 0(1)\,s$. This often arises when a single trait is monitored at two time points for each sampled unit. Full identification is achieved with the following $s + 2$ parameters:

$$\lambda, \ \eta = (\mu_2 - \mu_1)\sigma_1^{-1}, \quad \phi = \sigma_2\sigma_1^{-1}, \quad \Theta_i = (c_i - \mu_1)\sigma_1^{-1}, \quad i = 1(1)(s-1). \qquad (18.1.3.3)$$

The interpretations of the parameters are clear; Θ_i, $i = 1(1)(s - 1)$, are categorical boundaries for standardized values of the first logistic variable, η is a standardized mean shift parameter, and ϕ is the ratio of standard deviations. This parameterization is referred to as a mean shift model. In order to estimate all $s + 2$ parameters in (18.1.3.3), there must be at least three categories for each categorical variable.

Using (18.1.3.3) the log-likelihood given by (18.1.3.2) is fully identified with

$$P_{ij} = F(\Theta_i, (\Theta_j - \eta)\phi^{-1}) - F(\Theta_{i-1}, (\Theta_j - \eta)\phi^{-1}) - F(\Theta_i, (\Theta_{j-1} - \eta)\phi^{-1}) + F(\Theta_{i-1}, (\Theta_{j-1} - \eta)\phi^{-1}), \quad (18.1.3.4)$$

and parameter vector $\delta' = (\lambda, \eta, \phi, \Theta_1, \Theta_2, \ldots, \Theta_{s-1})$.

Another way to achieve full identification is with a parameterization that allows the boundaries for the two categorizations to be unknown and different. This is achieved with parameters $\lambda, \Theta_{11}, \ldots, \Theta_{(r-1)1}, \Theta_{12}, \ldots, \Theta_{(s-1)2}$, where

$$\Theta_{i1} = (c_i - \mu_1)\sigma_1^{-1}, \quad i = 1(1)(r-1), \qquad \Theta_{j2} = (d_j - \mu_2)\sigma_2^{-1}, \quad j = 1(1)(s-1), \quad (18.1.3.5)$$

are category boundaries for standardized values of the first and second logistic variables, respectively. Under this parameterization, the smallest tables that will allow unrestricted estimation are 2×3 and 3×2, respectively.

Once a parameterization is chosen so that full identification is achieved, then (18.1.3.2) can be maximized with respect to δ using an iterative computational procedure. This is easily done by using the Fisher scoring method which only requires the evaluation of expected values of second partial derivatives of the log-likelihood function. See Kennedy and Gentle (1980) and Rao (1973) for further details.

The bivariate logistic model with parameterization (18.1.3.3) provides an effective way to analyze the results of a clinical study of 524 patients suffering from allergy symptoms caused by ragweed. Thisted (1991) analyzed these data using a similarly parameterized bivariate probit model. The patients were separated into a treatment group and a placebo group and during the trial patients in both groups were allowed to use nonprescription decongestants. To determine the effectiveness of the treatment, it was necessary to compare the degree of nonprescription decongestant usage at the beginning and end of the trial for both the treatment group and the placebo group. For each patient, decongestant usage was classified by clinical technicians into four distinct ordered categories (none, intermittent, regular, full dosage) at the beginning and at the end of the trial. The bivariate logistic model provides a means of summarizing changes in decongestant usage during the trial.

The data for the treatment group are shown in Table 18.1.3.1. Maximum likelihood estimates for the parameters Θ_j, $j = 1, 2, 3, \eta$, and ϕ in the mean shift model (18.1.3.3) are presented in Table 18.1.3.2. The estimate for ρ is obtained by evaluating (18.1.2.6) at the maximum likelihood estimator for λ and the corresponding standard error is obtained via the delta method.

Table 18.1.3.1 Treatment Group Decongestant Usage

Pretrial usage	Posttrial usage				
	None	Interm.	Reg.	Full	
None	237	46	6	2	291
Interm.	26	9	2	0	37
Reg.	13	4	2	1	20
Full	20	6	2	3	31
	296	65	12	6	379

The estimated value of ρ suggests a moderate positive correlation between decongestant usage at the beginning and the end of the trial. The estimates for η and ϕ indicate that the average decongestant usage increased slightly during the trial while there was a substantial decrease in the variation in decongestant usage among subjects by the end of the trial. This reflects a reduction in high levels of decongestant usage and an increase in intermediate usage. It is interesting to note that when the mean shift model was fit to these data with ϕ restricted to 1, a large negative estimate of η was obtained corresponding to the decline in high levels of decongestant usage. However, the likelihood ratio test rejected this model in favor of the unrestricted model.

Comparison of the observed counts with the estimates of the expected counts, shown in Table 18.1.3.3, supports the adequacy of the unrestricted bivariate logistic model. The value of the Pearson chi-square test statistic is $X^2 = 9.8$ and the value of the log-likelihood ratio test statistic is $G^2 = 8.6$, both with 9 degrees of freedom. Neither statistic indicates that the model is inadequate.

The data for the placebo group are shown in Table 18.1.3.4. The parameter estimates for these data, shown in Table 18.1.3.5, are similar to those for the treatment group. Comparison of the expected counts in Table 18.1.3.6 to the observed counts in Table 18.1.3.4 supports the adequacy of the bivariate logistic model for the placebo group. It appears the average decongestant usage increased during the trial while the variation in usage decreased. Since these changes are not significantly different from those

Table 18.1.3.2 Maximum Likelihood Estimates of the Bivariate Logistic Model Parameters for the Treatment Group Decongestant Data

	θ_1	θ_2	θ_3	η	ϕ	ρ
MLE	1.19	1.88	2.40	0.67	0.41	0.37
std. err.	0.12	0.15	0.18	0.16	0.07	0.13

Table 18.1.3.3 Estimates of Expected Counts for the Treatment Group Decongestant Data

Pretrial usage	Posttrial usage None	Interm.	Reg.	Full	
None	237.7	41.7	8.1	3.3	290.7
Interm.	26.0	9.3	2.1	0.9	38.1
Reg.	12.2	4.9	1.1	0.5	18.7
Full	19.8	8.7	2.1	0.9	31.5
	295.6	64.7	13.4	5.5	379

estimated for the treatment group, it appears that the treatment had no real effect on changes in decongestant usage. The only significant deviation from the results for the treatment group is the small estimated value of ρ that suggests that the level of decongestant usage at the end of the trial is essentially unrelated to the level of decongestant usage at the beginning of the trial for the placebo group.

18.1.4 A Bivariate Logistic Regression Model

Univariate logistic regression has been extensively used to model effects of explanatory variables (covariates) on the probabilities of the possible outcomes a categorical response. Early applications of logistic regression models to bioassay were introduced by Berkson (1944, 1953). More extensive discussions that include applications to many other fields are given by Cox and Snell (1989), Maddala (1983), Agresti (1984), and Hosmer and Lemeshow (1989). The bivariate logistic regression model developed in this section provides a means of simultaneously modeling the effects of a set of covariates on a pair of correlated categorical responses.

Table 18.1.3.4 Placebo Group Decongestant Usage

Pretrial usage	Posttrial usage None	Interm.	Reg.	Full	
None	91	20	2	3	116
Interm.	8	5	1	0	14
Reg.	4	0	0	0	4
Full	9	2	0	0	11
	112	27	3	3	145

Table 18.1.3.5 Maximum Likelihood Estimates of the Bivariate Logistic Model Parameters for the Placebo Group Decongestant Data

	Θ_1	Θ_2	Θ_3	η	ϕ	ρ
MLE	1.39	2.17	2.49	0.88	0.41	0.06
std. err.	0.21	0.27	0.31	0.28	0.12	0.17

The logistic regression model can be viewed as a member of the class of quantal response models. For a binary response, let P_{1i} denote the probability of observing a response in the first category and let $P_{2i} = 1 - P_{1i}$ denote the probability of observing a response in the second category from an individual characterized by the values of the p covariates in the vector $x_i = (x_{1i}, \ldots, x_{pi})'$. In a quantal response model the binary response is a realization of an unobservable latent variable Z_i with some distribution function F and conditional mean $-X_i'\beta$, where β is a vector of p regression parameters. A response in the first category is observed if Z_i is smaller than some boundary value, say Θ_1, and it follows that

$$P_{1i} = \Pr(Z_i \leq \Theta_1) = F(\Theta_1 + X_i'\beta). \qquad (18.1.4.1)$$

Using the logistic distribution function $F(Y) = [1 + \exp(-Y)]^{-1}$ for the latent random variable provides a symmetric sigmoidal curve for the relationship between P_{1i} and $X_i'\beta$. This relationship is very similar to what is obtained by using a normal distribution for the latent variable, but it approaches the extremes, 0 and 1, more slowly. Usually there is no discernible difference in how well the logistic and normal (probit) models fit a set of data, but an important practical advantage of the logistic distribution is that it has a closed-form expression which facilitates the numerical evaluation of maximum likelihood estimators. Furthermore, the log-odds of observing a response in the first category under the conditions charac-

Table 18.1.3.6 Estimates of Expected Counts for the Placebo Group Decongestant Data

Pretrial usage	Posttrial usage None	Interm.	Reg.	Full	
None	90.4	20.8	2.5	2.2	116.0
Interm.	10.5	2.9	0.4	0.3	14.2
Reg.	2.8	0.8	0.1	0.1	3.7
Full	8.1	2.4	0.3	0.3	11.1
	112.0	26.9	3.2	2.9	145

terized by X_i is conveniently expressed as

$$\log\left[\frac{P_{1i}}{P_{2i}}\right] = \Theta_1 + X_i'\beta, \qquad (18.1.4.2)$$

which allows the elements of β to be interpreted with respect to linear associations between the covariates and the log-odds.

There are several ways to extend logistic regression to a single multicategory response. The model proposed by Walker and Duncan (1967) was found to be adequate for the data analyzed in this section. For this model, a response is observed in the jth category when $\Theta_{j-1} < Z_i \le \Theta_j$, where $-\infty = \Theta_0 < \Theta_1 < \cdots < \Theta_{c-1} < \Theta_c = \infty$ are the category boundaries and Z_i is a latent logistic random variable. Under the conditions characterized by the vector of covariate values, X_i, the conditional log-odds of observing a response in the jth category or lower is modeled as

$$\log\left[\sum_{l=1}^{j} P_{li} \Big/ \sum_{l=j+1}^{c} P_{li}\right] = \Theta_j + X_i'\beta,$$

and the conditional probability of observing a response in a particular category follows from

$$\sum_{l=1}^{j} P_{li} = [1 + e^{-\Theta_j - X_i'\beta}]^{-1}, \qquad j = 1(1)c - 1. \qquad (18.1.4.3)$$

This model reduces to (18.1.4.2) when there are only two categories.

The need for a bivariate logistic regression model arises in situations where a pair of correlated categorical responses are taken from each respondent. The bivariate logistic distribution (18.1.2.2) provides a means of constructing such a model. Let (Z_{i1}, Z_{i2}) denote the pair of random latent logistic random variables associated with a pair of responses from an individual characterized by the values of the p covariates in the vector X_i. The conditional mean of the univariate logistic distribution for Z_{ik} is modeled by $-X_i'\beta_k$, where β_k is a p-dimensional vector of regression parameters. Then, the probability of simultaneously observing the first response in the rth category and the second response in the sth category is the joint probability that $\Theta_{(r-1)1} < Z_{i1} < \Theta_{r1}$ and $\Theta_{(s-1)2} < Z_{i1} < \Theta_{s2}$, where $-\infty = \Theta_{0k} < \Theta_{1k} < \cdots < \Theta_{c_k k} = \infty$ are the boundary points defining the c_k categories for the kth response ($k = 1, 2$). Using (18.1.2.2), this probability can be written as

$$P_{irs} = F(\Theta_{r1} + X_i'\beta_1, \Theta_{s2} + X_i'\beta_2) - F(\Theta_{(r-1)1} + X_i'\beta_1, \Theta_{s2} + X_i'\beta_2)$$
$$- F(\Theta_{r1} + X_i'\beta_1, \Theta_{(s-1)2} + X_i'\beta_2) + F(\Theta_{(r-1)1} + X_i'\beta_1, \Theta_{(s-1)2} + X_i'\beta_2).$$
$$(18.1.4.4)$$

This model uses a separate Walker and Duncan model (18.1.4.3) for the marginal distribution of each of the categorical responses, but the models are fit simultaneously so that the correlation between the paired responses can be estimated. This can also greatly improve the efficiency of comparisons between corresponding elements of β_1 and β_2 when there is a substantial positive correlation between the paired responses. The boundary parameters are generally unknown and need to be estimated when this model is applied to a particular set of data.

Given the pairs of categorical responses from a random sample of $N = \Sigma_{i=1}^{I} \Sigma_{s=1}^{c_2} \Sigma_{r=1}^{c_1} n_{irs}$ individuals, the log-likelihood function for the counts in the resulting $c_2 \times c_1$ contingency table has kernel

$$l(\delta) = \sum_{i=1}^{I} \sum_{s=1}^{c_2} \sum_{r=1}^{c_1} n_{irs} \log(P_{irs}), \tag{18.1.4.5}$$

where δ' is a $(c_1 + c_2 + 2p - 1)$-vector of parameters $(\lambda, \Theta_1', \Theta_2', \beta_1', \beta_2')$ with $\Theta_k' = (\Theta_{1k}, \ldots, \Theta_{(c_k-1)k})$, $k = 1, 2$. Maximization of (18.1.4.5) for (18.1.4.4) is possible by direct application of the Fisher scoring procedure.

Denote the maximum likelihood estimator of δ by $\hat{\delta}$ and the true value by δ_0. Since the usual regularity conditions (Cox and Hinkley, 1974) are satisfied, it follows that $\sqrt{N}\,(\hat{\delta} - \delta_0) \xrightarrow{L} N(0, V)$, where $V = \lim_{N\to\infty} N(I(\delta_0)^{-1})$ and $I(\delta_0)$ is the Fisher information matrix associated with (18.1.4.4) and (18.1.4.5). δ_0 can be replaced by $\hat{\delta}$ in $I(\delta_0)$ in order to obtain consistent estimates of the asymptotic variances and covariances of the maximum likelihood estimates. An approximate maximum likelihood estimate for the underlying correlation, denoted by $\hat{\rho}$, is obtained by replacing λ_0 with $\hat{\lambda}$ in (18.1.2.6). The asymptotic variance of $\hat{\rho}$ can be approximated via the delta method.

A test for independence within the bivariate pairs can be done with the likelihood ratio test on one degree of freedom that restricts λ to be zero under the null hypothesis. If a large number of subjects are sampled for each combination of covariate levels, which is plausible if all of the concomitant variables are classificatory, then a separate estimate of the dependency parameter can be obtained for each combination. This would permit a host of testable hypotheses concerning the correlation structure by restricting some or all of the dependency parameters to be equal to each other and, perhaps, equal to some constant value. A test for model adequacy is easily done by comparing observed cell counts to expected cell counts where the expected counts are calculated from the estimated probabilities, $\hat{P}_{irs}$. The latter are obtained from (18.1.4.4) with the model parameters replaced by the maximum likelihood estimates.

Two applications of the bivariate logistic regression model are presented in this section. The first data to be analyzed are field ornithological data that were previously analyzed by Anderson and Pemberton (1985). These data consist of a color rating of the upper (UM) and lower (LM) mandibles recorded for each of 90 first-year blackbirds. The rating was an ordered scale ranging from mostly black (category 1) toward mostly yellow (category 3). Different birds were observed at different time points, and a time covariate with three levels (3.5, 9.5, 15.5) was formed where lower values correspond to earlier points in the year. It is known that as the birds mature, their mandible coloring changes from black to yellow. Since birds observed later in the year tend to be older than those observed earlier, time was suspected to be an important covariate.

In addition to investigating the importance of time, the question of whether or not there are independent biological mechanisms determining the color change of the two mandibles is also of interest. The presence of a nonzero underlying correlation would suggest the mechanisms determining the coloration of the two mandibles are not functioning independently. The data for this field study are given in Table 18.1.4.1.

Using the notation in (18.1.4.4), the boundary points for the lower mandible coloring are Θ_{11} and Θ_{21} and those for the upper mandible are Θ_{12} and Θ_{22}. Time is the single covariate in this analysis, and the regression coefficients are β_1 and β_2 for the lower and upper mandibles, respectively. The maximum likelihood estimates for these parameters and the correlation

Table 18.1.4.1 Blackbird Coloring Data

Time	Lower mandible	Upper mandible: One	Two	Three
3.5	1	36	0	0
	2	2	1	1
	3	1	0	0
9.5	1	19	0	0
	2	4	2	0
	3	1	1	4
15.5	1	4	0	1
	2	1	1	0
	3	1	1	9

Table 18.1.4.2 Maximum Likelihood Estimates of the Bivariate Logistic Model Parameters for the Ornithological Data

	LM			UM			
	θ_{11}	θ_{21}	β_1	θ_{12}	θ_{22}	β_2	ρ
MLE	2.79	3.66	−0.249	3.88	4.46	−0.288	0.836
std. err.	0.56	0.62	0.054	0.73	0.77	0.062	0.051

coefficient are given in Table 18.1.4.2. This model fit the data quite well. The value of the log-likelihood chi square test for the fit of this model is $G^2 = 18.6$ on 19 degrees of freedom.

The importance of the joint effect of time for each mandible can be assessed by testing the hypothesis H_0: $\beta_1 = \beta_2 = 0$. This is easily done by the likelihood ratio test. If l_0 is the value of the log-likelihood function maximized under H_0 and l_a is the value of the unrestricted maximized log-likelihood function, then $X^2 = 2(l_a - l_0)$ is asymptotically distributed as a chi-square random variable on 2 degrees of freedom. The observed value of the test statistic for the blackbird data is 26.7, which gives a strong indication that the time of year is a significant covariate for at least one of the traits. The importance of the marginal time effects on the individual mandibles is indicated by the standardized regression coefficients, −4.7 for the lower mandible and −4.6 for the upper mandible. The negative values of the coefficients indicate that as time increases the conditional probability of observing yellow coloring increases for each mandible, which is consistent with a priori expectations. The bivariate approach also provides an estimate of the covariance between the estimates of β_1 and β_2, which is needed to determine if the rates at which yellowing occurs are the same for the upper and lower mandibles. The maximum likelihood estimate of $\beta_1 - \beta_2$ is 0.039, and the corresponding estimate of the asymptotic standard error is 0.053. Therefore, it appears that the upper and lower mandibles change from black to yellow at about the same rate.

The large value of the correlation, $\hat{\rho} = 0.836$, relative to its standard error suggests that there are not independent biological mechanisms determining the color changes in the two mandibles. It should be noted that the apparent strong correlation is present conditional on the time of year the birds are observed. In this case, ignoring the time of year inflates the estimate of the correlation between the color change of the upper and lower mandibles by only a small amount. In other situations, however, ignoring important covariates can severely distort inferences made about correlations.

The second illustration considers a cross-classification of alleged crimes committed in conjunction with homicides in Florida between 1973 and 1977. Each case was cross-classified according to determinations made by the police department and the prosecuting attorney of record as to what degree each believes a felony was committed concurrently with the homicide (no felony, possible felony, felony). The cross-classification are partitioned into four groups according to the race combination of the defendant and victim (black/white, white/white, black/black, and white/black).

These data, initially reported in Radelet and Pierce (1983), were subsequently analyzed by Agresti (1984) with log-linear models and are reproduced here in Table 18.1.4.3. Agresti made use of the ordered responses with a row-effects model for the conditional association between race group and the prosecution classification (hereafter referred to as the court classification) given the police classification. In Agresti's analysis integer-scores were assigned to the crime classification categories which implicitly assumes that the distance between no felony and felony is exactly twice the distance between no felony and possible felony. Several issues can be addressed more easily and efficiently with a bivariate logistic regression approach. These include the quantification of the agreement between the police and

Table 18.1.4.3 Police and Court Classifications of Homicides

		Court classification		
Race defendant/victim	Police classification	No felony	Possible felony	Felony
Black/white	No felony	7	1	3
	Possible felony	0	2	6
	Felony	5	5	109
White/white	No felony	236	11	26
	Possible felony	7	2	21
	Felony	25	4	101
Black/black	No felony	328	6	13
	Possible felony	7	2	3
	Felony	21	1	36
White/black	No felony	14	1	0
	Possible felony	6	1	1
	Felony	1	0	5

court classifications after controlling for estimated race effects, direct quantifications of race effects on the log-odds of police and court classifications of possible felonies, and a powerful way to test if race effects are the same for the police and court classifications. Furthermore, the bivariate logistic regression approach avoids the use of arbitrary scores.

Using (18.1.2.2), the proportion of cases classified by the police into the rth category and by the court into the sth category is modeled as

$$P_{irs} = F(\Theta_{r1} + \beta_{i1}, \Theta_{s2} + \beta_{i2}) - F(\Theta_{(r-1)1} + \beta_{i1}, \Theta_{s2} + \beta_{i2}) - F(\Theta_{r1} + \beta_{i1}, \Theta_{(s-1)2} + \beta_{i2}) - F(\Theta_{(r-1)1} + \beta_{i1}, \Theta_{(s-1)2} + \beta_{i2}), \tag{18.1.4.6}$$

where $\Sigma_{i=1}^{4} \beta_{i1} = \Sigma_{i=1}^{4} \beta_{i2} = 0$ and $\beta_{1k}, \beta_{2k}, \beta_{3k}, \beta_{4k}$ correspond to the effects of the black/white, white/white, black/black, white/black levels, respectively, of the defendant/victim race variable on the police ($k = 1$) and court ($k = 2$) classifications. The boundary points $(\Theta_{11}, \Theta_{21})$, for the police classification, and $(\Theta_{12}, \Theta_{22})$, for the court classification, are unknown and must be estimated. Using P_{ir+}, $r = 1, 2, 3$, to denote the marginal probabilities for the police classifications, this model uses the logistic regression model

$$\log\left(\frac{P_{i1+}}{P_{i2+} + P_{i3+}}\right) = \Theta_{11} + \beta_{i1}, \qquad \log\left(\frac{P_{i1+} + P_{i2+}}{P_{i3+}}\right) = \Theta_{21} + \beta_{i1}, \qquad i = 1, 2, 3, 4, \tag{18.1.4.7}$$

for the police classifications. Similarly, it uses the logistic regression model

$$\log\left(\frac{P_{i+1}}{P_{i+2} + P_{i+3}}\right) = \Theta_{12} + \beta_{i2} \qquad \log\left(\frac{P_{i+1} + P_{i+2}}{P_{i+3}}\right) = \Theta_{22} + \beta_{i2}, \qquad i = 1, 2, 3, 4 \tag{18.1.4.8}$$

for the log-odds of court classifications of no felony and the log-odds of no or possible felony, respectively. The model was modified to include separate correlations between police and court classifications (ρ_1, ρ_2, ρ_3, ρ_4) for the four defendant/victim categories. Maximum likelihood estimates for the model parameters are given in Table 18.1.4.4 along with the associated standard errors.

Since $\hat{\Theta}_{21} - \hat{\Theta}_{22} = -0.02$ with a standard error of 0.09, it appears that the overall boundary for a felony classification is essentially the same for the police and the court. The estimated overall boundary between the

Table 18.1.4.4 Maximum Likelihood Estimate of the Bivariate Logistic Model Parameters for the Homicide Classification Data

PC	θ_{11}	θ_{21}	β_{11}	β_{21}	β_{31}	β_{41}
MLE	0.09	0.43	−2.18	0.46	1.49	0.22
std. err.	0.12	0.12	0.21	0.14	0.15	0.28
CC	θ_{12}	θ_{22}	β_{12}	β_{22}	β_{32}	β_{42}
MLE	0.24	0.45	−2.20	0.23	1.52	0.46
std. err.	0.12	0.12	0.21	0.14	0.15	0.30
	ρ_1	ρ_2	ρ_3	ρ_4		
MLE	0.64	0.82	0.89	0.91		
std. err.	0.07	0.02	0.02	0.04		

no felony and possible felony classes was somewhat lower for the police ($\hat{\theta}_{11} - \hat{\theta}_{12} = -0.15$ with a standard error of 0.09), as might be expected, but the difference is not significant. There appears to be no significant overall tendency for prosecutors to either upgrade or downgrade police classifications.

The relatively large negative values of $\hat{\beta}_{11}$ and $\hat{\beta}_{12}$ clearly indicate that both the police and the prosecutors were more likely to make a felony classification for a case with black defendants and white victims. In particular, $\exp(\hat{\beta}_{11}) = 0.11$ and $\exp(\hat{\beta}_{12}) = 0.11$ suggest that both the odds of classification as a felony versus no felony or a possible felony and the odds of classification of at least a possible felony versus no felony are increased by about a factor of 9 for both the police and the court. Conversely, the large values of $\hat{\beta}_{31}$ and $\hat{\beta}_{32}$ suggest that both the police and the prosecutors were less likely to make a felony classification for cases with a black defendant and a black victim (the odds are reduced by approximately a factor of 4 from the overall odds). Since the standard errors for $\hat{\beta}_{11} - \hat{\beta}_{12} = 0.02$ and $\hat{\beta}_{31} - \hat{\beta}_{32} = -0.03$ are 0.42 and 0.12, respectively, there is no indication of a special tendency for the prosecutors to either upgrade or downgrade police classifications in these cases. However, $\hat{\rho}_1 = 0.64$ is significantly smaller than $\hat{\rho}_3 = 0.89$, suggesting that there is less agreement between police and prosecutors for individual cases with black defendants and white victims than for individual cases with black defendants and black victims.

For the police classifications $\hat{\beta}_{i1}$ is closest to zero for $i = 4$, so the classifications of cases with white defendants and black victims are closest to the overall police classifications for the four defendant/victim race groups. For the court classifications, the results for the cases with white defendants and white victims are closest to the overall results for the four groups. In fact, $\hat{\beta}_{21} - \hat{\beta}_{22} = 0.23$ with standard error 0.10 indicates that prosecutors tend to upgrade the police classifications for cases with white defendants and white victims. Conversely, $\hat{\beta}_{41} - \hat{\beta}_{42} = -0.24$ with standard error 0.18 is not quite as suggestive of a tendency for prosecutors to downgrade police classifications for cases with white defendants and black victims.

REFERENCES

Agresti, A. (1984). *Analysis of Ordinal Categorical Data*, Wiley, New York.

Anderson, J. A. and Pemberton, J. D. (1985). The grouped continuous model for multivariate ordered categorical variables and covariate adjustment, *Biometrics*, *41*, 875–885.

Berkson, J. (1944). Applications of the logistic function to bioassay, *J. Am. Statist. Assoc.*, *39*, 357–365.

Berkson, J. (1953). A statistically precise and relatively simple method of estimating the bioassay with quantal response based on the logistic function, *J. Am. Statist. Assoc.*, *48*, 565–599.

Cook, R. D. and Johnson, M. E. (1986). Generalized Burr-pareto-logistic distributions with applications to a uranium exploration data set, *Technometrics*, *28*, 123–131.

Cox, D. R. and Hinkley, D. V. (1974). *Theoretical Statistics*, Chapman and Hall, London.

Cox, D. R. and Snell, E. J. (1989). *Analysis of Binary Data*, 2nd ed., Chapman and Hall, London.

Hosmer, D. W. and Lemeshow, S. (1989). *Applied Logistic Regression*, Wiley, New York.

Kendall, M. G. and Stuart, A. (1967). *Advanced Theory of Statistics*, Hafner, New York.

Kennedy, W. J. and Gentle, J. E. (1980). *Statistical Computing*, Marcel Dekker, New York.

Maddala, G. S. (1983). *Limited-Dependent and Qualitative Variables in Econometrics*, Cambridge University Press, London.

Radelet, M. and Pierce, G. L. (1983). *Race and prosecutorial discretion in homicide cases*, presented at annual meeting of Am. Sociol. Assoc., Detroit, Michigan.

Rao, C. R. (1973). *Linear Statistical Inference and Its Applications*, Wiley, New York.

Symanowski, J. T. and Koehler, K. J. (1989). A bivariate logistic distribution with applications to categorical responses, *Technical Report Number 89-29*, Department of Statistics, Iowa State University, Ames, Iowa.

Walker, S. H. and Duncan, D. B. (1967). Estimation of the probability of an event as a function of several independent variables, *Biometrika*, *54*, 167–179.

18.2 *k*TH NEAREST NEIGHBORS AND THE GENERALIZED LOGISTIC DISTRIBUTION

Colleen D. Cutler
University of Waterloo, Waterloo, Ontario, Canada

Let $W_{1,1}, \ldots, W_{1,n}$ and $W_{2,1}, \ldots, W_{2,mn}$ be independent samples of i.i.d. observations from a distribution m in $\mathbf{R}^N$. We show, under regularity conditions on m, that a certain "density-free" function of the kth nearest-neighbor distances to a point x in space asymptotically follows a generalized logistic distribution whose variance decreases as a function of k. This illustrates the greater stability of higher-order nearest neighbors in estimating the local scaling behavior of m.

18.2.1 Introduction

Let m be a probability measure on the Borel sets of $\mathbf{R}^N$, and suppose the limit

$$\sigma(x) = \lim_{r \to 0} \frac{\log m(B(x, r))}{\log r}$$

exists at the point x, where $B(x, r)$ denotes the closed ball of radius r centered at x. Then $\sigma(x)$ describes the scaling behavior of m at x; specifically, we can write

$$m(B(x, r)) = r^{\sigma(x) + \phi(x,r)},$$

where the error or correction term $\phi(x, r)$ approaches 0 as $r \to 0$. If m has a probability density function $p(x)$ (equivalently, m is absolutely continuous with respect to Lebesgue measure in $\mathbf{R}^N$) then we will observe $\sigma(x) = N$ for m—almost all x; moreover,

$$\lim_{r \to 0} r^{\phi(x,r)} = \kappa_N p(x) \quad m\text{—a.s.,} \qquad \text{where } \kappa_N = \frac{\pi^{N/2}}{\Gamma(1 + N/2)}$$

is a constant. Interest in continuous probability distributions which lack densities (such distributions potentially scale at values of $\sigma(x) < N$) appears to arise in two ways. One is that certain components in a vector of variables may in fact be functions of one another; consequently it may be possible to reduce the number of distinct variables under consideration (this includes the possibility that a density function $p(x)$ may exist on a lower-dimensional subspace; we will expand on this shortly). The study of the long-term behavior of chaotic dynamical systems (a topic of considerable recent interest among physicists and applied mathematicians) is a second, less mundane, source of curiosity about the scaling behavior of singular distributions. Many chaotic systems exhibit existence of a low-dimensional attracting set (or attractor) which is equipped with a probability distribution m describing the asymptotic behavior of trajectories of the system. The scaling behavior of m is a fundamental feature of the system asymptotics. (Many attractors seem to possess some fractal properties, and the scale values $\sigma(x)$ of the associated distributions are often noninteger. A very simple example of this phenomenon is provided by the uniform distribution across the usual Cantor set in [0, 1]; we observe $\sigma(x) = \log 2/\log 3$ m—a.s.) In this section we concentrate on the asymptotic theory for the case where $\sigma(x)$ is nonfractal or integer.

In order to estimate $\sigma(x)$ from a random sample $W_1, \ldots, W_n$ of observations from m, Cutler and Dawson (1989) considered using the nearest-neighbor distance

$$d_n(x) = \min_{1 \le j \le n} \|W_j - x\|$$

as the basis for an estimation procedure. It was shown that the statistic

$$l_n(x) = \frac{\log d_n(x)}{\log 1/n}$$

provides a consistent estimator of $1/\sigma(x)$, and that for a certain class of probability measures on $\mathbf{R}^N$, the appropriate transformation and formulation of a test statistic, based on nearest-neighbor distances from two samples, leads asymptotically to a logistic distribution. The purpose of this section is to extend these results to statistics based on the kth nearest-neighbor distance. We will see that in this case we asymptotically obtain *generalized* logistic distributions and that the advantage of using higher-order nearest neighbors is that the resulting test statistic has a smaller asymptotic variance.

A detailed discussion of the family of generalized logistic distributions and corresponding notation is provided in part B of the appendix. Part A of the appendix discusses the family of log gamma distributions; the log gamma arises naturally in what follows. We suggest that the reader be familiar with the contents and notation of the appendix before proceeding.

18.2.2 Results

Cutler and Dawson (1989) proved that, for m—almost all x, $\lim_{n\to\infty} l_n(x) = 1/\sigma(x)$ w.p.1 (with probability 1). Here we extend this result to the statistic

$$l_{k,n}(x) = \frac{\log d_{k,n}(x)}{\log k/n},$$

where $d_{k,n}(x)$ is the distance from x to its kth nearest neighbor among $W_1, \ldots, W_n$. To be notationally consistent, we will write $d_{1,n}(x)$ for $d_n(x)$ and $l_{1,n}(x)$ for $l_n(x)$. We will need the following elementary lemma.

Lemma 18.2.2.1 $\sum_{n=1}^{\infty} n^{\alpha}(1 - n^{-\beta})^n < \infty$ for all $\alpha \in \mathbf{R}$ and $0 \leq \beta < 1$.

Theorem 18.2.2.1 Suppose the pointwise limit $\sigma(x)$ exists for m—almost all x. Let $W_1, \ldots, W_n$ be i.i.d. observations from m. Then $\lim_{n\to\infty} l_{k,n}(x) = 1/\sigma(x)$ w.p.1 m—a.s.

Proof. From Theorem 3.2 of Cutler and Dawson (1989) it follows that, for any constants $a > 0$, $b > 0$,

$$\lim_{n\to\infty} \frac{\log ad_{1,n}(x)}{\log b/n} = \frac{1}{\sigma(x)} \quad \text{w.p.1 } m\text{—a.s.}$$

Since $d_{k,n}(x) \geq d_{1,n}(x)$, we immediately obtain, for all x in the topological support of m,

$$\limsup_{n\to\infty} \frac{\log d_{k,n}(x)}{\log k/n} \leq \frac{1}{\sigma(x)} \quad \text{w.p.1 } m\text{—a.s.}$$

We need only verify

$$\liminf_{n\to\infty} \frac{\log d_{k,n}(x)}{\log k/n} \geq \frac{1}{\sigma(x)}.$$

We will consider only the case $0 < \sigma(x) < \infty$. (The case $\sigma(x) = \infty$ is immediate; the case $\sigma(x) = 0$ can be proved using minor modifications of the method for $\sigma(x) > 0$.) Let

$$A_\epsilon = \bigcap_{j=1}^{\infty} \bigcup_{n=j}^{\infty} [l_{k,n}(x) < (1 - \epsilon)/\sigma(x)].$$

We will show that $P(A_\epsilon) = 0$ for every $\epsilon > 0$. Note that for any $r > 0$ we can write

$$P([d_{k,n}(x) > r]) = \sum_{j=0}^{k-1} \binom{n}{j} m(B(x, r))^j(1 - m(B(x, r)))^{n-j}. \quad (18.2.2.1)$$

Let

$$r_n = (k/n)^{(1-\epsilon)/\sigma(x)}.$$

Then we have

$$\begin{aligned} P\left(\left[l_{k,n}(x) < \frac{1-\epsilon}{\sigma(x)}\right]\right) &= P([d_{k,n}(x) > r_n]) \\ &= \sum_{j=0}^{k-1} \binom{n}{j} m(B(x, r_n))^j(1 - m(B(x, r_n)))^{n-j} \\ &= \sum_{j=0}^{k-1} \binom{n}{j} m(B(x, r_n))^j(1 - r_n^{\sigma(x)+\phi(x,r_n)})^{n-j} \\ &\leq \sum_{j=0}^{k-1} \binom{n}{j}\left(1 - \left(\frac{k}{n}\right)^{(1-\epsilon)(1+\phi(x,r_n)/\sigma(x))}\right)^{n-j}. \end{aligned}$$

(18.2.2.2)

Now since $\lim_{n\to\infty} \phi(x, r_n) = 0$ and $\binom{n}{j} \leq n^j \leq n^{k-1}$, it follows that for sufficiently large n, the last line of (18.2.2.2) is dominated by $Cn^{k-1}(1 - n^{-\beta})^n$ for some $0 < \beta < 1$ and $C > 0$. Consequently, there exists $C' > 0$ such that

$$\sum_{n=1}^{\infty} P\left(\left[l_{k,n}(x) < \frac{1-\epsilon}{\sigma(x)}\right]\right) \leq C' + C\sum_{n=1}^{\infty} n^{k-1}(1 - n^{-\beta})^n < \infty$$

by Lemma 18.2.2.1. Hence $P(A_\epsilon) = 0$. This proves the theorem. ■

Cutler and Dawson (1989) introduced a class of probability measures on $\mathbf{R}^N$ called *D-regular* distributions; this class is discussed in detail in that paper so we will provide only the basic definitions here. Let $1 \leq D \leq N$ be a positive integer. A subset $S \subseteq \mathbf{R}^N$ is called *absolutely D-regular* if, for every $x \in S$,

$$\lim_{r\to 0} \frac{H^D(S \cap B(x, r))}{\kappa_D r^D} = 1,$$

where $\kappa_D = \pi^{D/2}/\Gamma(1 + D/2)$ and H^D is the Hausdorff D-measure on the Borel sets of $\mathbf{R}^N$. (See, for example, Falconer, 1985. A good discussion of Hausdorff measure as generalized volume can also be found in Billingsley, 1986.) A probability measure m is called *D-regular* if there exists an absolutely D-regular set S such that $m(S) = 1$ and m is absolutely continuous with respect to the restriction of H^D to S. The Radon-Nikodym derivative $p = dm/dH^D$ we call the *D-density* of m. In case $D = N$ the notion of N-density coincides with the usual notion of density (as H^N agrees with Lebesgue measure on $\mathbf{R}^N$). Basically the class of D-regular distributions can be described as those probability measures which have densities with respect to the natural Lebesgue measure over some "nice" subset of $\mathbf{R}^N$; for example, the uniform distribution over the surface of a sphere in $\mathbf{R}^N$ is $(N - 1)$-regular. D-regular distributions can be characterized by the fact that the distribution of an appropriate normalization of $l_{1,n}(x)$ approaches an *extreme value* distribution as $n \to \infty$ (this is a special case of the log gamma with $\alpha = 1$). We now extend this result to the kth nearest neighbor.

Theorem 18.2.2.2 Suppose m is a D-regular distribution. Then $\sigma(x) = D$ m—a.s. and, for m—almost all x, $L_{k.n}(x) = D(\log n/k)(l_{k.n}(x) - 1/D)$ converges in distribution to the log gamma $\mathrm{LG}(\mu_k + \log k\kappa_D p(x), \theta_k^2; k)$, where $p(x)$ is the D-density of m at x.

Proof. Cutler and Dawson (1989) proved that, for D-regular distributions, $\sigma(x) = D$ m—a.s. Let $r_n(s) = n^{-(1/D+(s-\log k)/D\log n)}$. Then

$$
\begin{aligned}
P([L_{k.n}(x) < s]) &= P([d_{k.n}(x) > r_n(s)]) \\
&= \sum_{j=0}^{k-1} \binom{n}{j} m(B(x, r_n(s)))^j (1 - m(B(x, r_n(s))))^{n-j} \\
&= \sum_{j=0}^{k-1} \binom{n}{j} r_n(s)^{j(D+\Phi(x, r_n(s)))} (1 - r_n(s)^{D+\Phi(x, r_n(s))})^{n-j} \\
&= \sum_{j=0}^{k-1} \binom{n}{j} \left(\frac{e^{-(s-\log k)}}{n}\right)^j (r_n(s)^{\Phi(x, r_n(s))})^j \\
&\quad \times \left(1 - \left(\frac{e^{-(s-\log k)}}{n}\right) r_n(s)^{\Phi(x, r_n(s))}\right)^{n-j}.
\end{aligned}
$$

Now for D-regular distributions we have (as noted earlier in the paper for the absolutely continuous case)

$$
\lim_{n \to \infty} r_n(s)^{\Phi(x, r_n(s))} = \kappa_D p(x) \qquad \text{for } m\text{—almost all } x.
$$

Since $\binom{n}{j} n^{-j} \to 1/j!$, we obtain

$$
\lim_{n \to \infty} P([L_{k.n}(x) < s]) = \sum_{j=0}^{k-1} \frac{e^{-j(s-\log k\kappa_D p(x))}}{j!} \exp(-e^{-(s-\log k\kappa_D p(x))}).
$$

Comparing with (18.2.A.4) of the appendix, we see that this is the distribution function of a log gamma with parameter k and location shift given by $a = \log k\kappa_D p(x)$. If we let $\bar{\mu}_k$ denote the mean of this distribution, we see from (18.2.A.5) that

$$
\bar{\mu}_k = \gamma - h_1(k - 1) + \log k + \log \kappa_D p(x) = \gamma - c_k + \log \kappa_D p(x)
$$

where $|c_k - \gamma| = O(1/k)$. (This last follows from the identity $h_1(k - 1) = \log k + c_k$.) Thus for large k the asymptotic distribution is centered essentially at $\log \kappa_D p(x)$. From (18.2.A.6) we see that the asymptotic variance $\theta_k^2 = \pi^2/6 - h_2(k - 1) \to 0$ as $k \to \infty$. This illustrates (under the assumption of moment convergence, which we establish below) the greater stability of higher-order nearest neighbors.

We now show, subject to a mild local smoothness constraint on the supporting set S and D-density p, that all moments of $L_{k,n}(x)$ converge to the corresponding moments of the log gamma distribution. The key step is the following lemma.

Lemma 18.2.2.2 Let m have compact support and be absolutely continuous with respect to Lebesgue measure in $\mathbf{R}^N$ with bounded density function $p(z)$. Suppose p is continuous and strictly positive at the point x in $\mathbf{R}^N$. Then the density function $p_n(x, s)$ of $L_{k,n}(x)$ is uniformly bounded (in n and s) by a multiple of a log gamma density. That is, there exist $Q > 0$ and $C > 0$ such that, for sufficiently large n and all s, $p_n(x, s) \le Ce^{-ks}\exp(-Qe^{-s})$.

Proof. Let $h(r) = m(B(x, r))$. Note that for all r we can write

$$h(r) = \kappa_N p(x) r^N + b(r),$$

where

$$b(r) = \int_{B(x,r)} (p(z) - p(x))\, dz$$

is absolutely continuous and satisfies $|b'(r)| \le N\kappa_N \|p\| r^{N-1}$ with $\|p\| = \sup_z p(z)$. This will be used shortly. Since $D = N$ set

$$r_n(s) = n^{-(1/N + (s - \log k)/N \log n)} = n^{-1/N} e^{-(s - \log k)/N}.$$

We have

$$\begin{aligned} p_n(x, s) &= \frac{d}{ds} \sum_{j=0}^{k-1} \binom{n}{j} h(r_n(s))^j (1 - h(r_n(s)))^{n-j} \\ &= \left(\frac{d}{ds} h(r_n(s))\right)\left[\sum_{j=1}^{k-1} \binom{n}{j} j h(r_n(s))^{j-1}(1 - h(r_n(s)))^{n-j}\right. \\ &\quad \left. - \sum_{j=1}^{k} \binom{n}{j} j h(r_n(s))^{j-1}(1 - h(r_n(s)))^{n-j}\right] \\ &= -\left(\frac{d}{ds} h(r_n(s))\right)\left[\binom{n}{k} k h(r_n(s))^{k-1}(1 - h(r_n(s)))^{n-k}\right]. \end{aligned} \tag{18.2.2.3}$$

Now

$$\left|\frac{d}{ds}h(r_n(s))\right| = |h'(r_n(s))|\,|r_n'(s)|$$
$$\leq (2N\kappa_N\|p\|r_n(s)^{N-1})\left(\frac{r_n(s)}{N}\right) = 2\kappa_N\|p\|r_n(s)^N$$
$$= \frac{2\kappa_N\|p\|e^{-(s-\log k)}}{n} \leq \frac{Ae^{-s}}{n}$$

for some constant $A > 0$. Letting $a(n, k) = (n - 1)(n - 2) \cdots (n - k + 1)/n^{k-1}$ and substituting into the last line of (18.2.2.3), we obtain

$$p_n(x, s) \leq \frac{Aa(n, k)}{(k - 1)!e^{-s}(nh(r_n(s)))^{k-1}(1 - h(r_n(s)))^{n-k}}$$
$$\leq Ae^{-s}(nh(r_n(s)))^{k-1}(1 - h(r_n(s)))^{n-k}. \tag{18.2.2.4}$$

At this point we use the compactness of the support of m. Let r_0 be such that $h(r_0) = 1$. Note it follows then from the assumptions on the density p at x that the ratio $h(r)/r^N$ is bounded above and bounded away from 0 over the interval $(0, r_0]$. That is, there exist Q_1 and Q_2 such that $0 < Q_1 < h(r)/r^N < Q_2 < \infty$ for all $0 < r \leq r_0$. Using this and the inequality $(1 - u)^y \leq \exp(-yu)$ (for $y > 0$ and $0 \leq u \leq 1$) in (18.2.2.4), we obtain for all s:

$$p_n(x, s) \leq Ae^{-s}(nr_n(s)^NQ_2)^{k-1}\exp(-(n - k)r_n(s)^NQ_1)$$
$$= Ae^{-s}(Q_2e^{-(s-\log k)})^{k-1}\exp\left(-\frac{n - k}{n}Q_1e^{-(s-\log k)}\right)$$
$$\leq Ce^{-ks}\exp(-Qe^{-s}) \qquad \text{for } n \geq k + 1 \quad \text{and some} \quad C, Q > 0.$$

(Note that when $r_n(s) > r_0$ we have $p_n(x, s) = 0$ which proves the above inequality for that case.) This proves the lemma. ■

Since all moments of the log gamma exist, this shows that under the hypotheses of Lemma 18.2.2.2, for each $j \in \mathbf{Z}^+$, $|L_{k,n}(x)|^j$, $n \geq k + 1$, is uniformly integrable. Hence all moments of $L_{k,n}(x)$ converge to the corresponding moments of the limiting log gamma distribution in this case. To extend this result to a much more general situation we can use the same techniques used in Theorems 5.2 and 5.3 of Cutler and Dawson (1989). We state the resulting theorem without proof.

Theorem 18.2.2.3 Let the probability distribution m on $\mathbf{R}^N$ have finite ϵ-moment for some $\epsilon > 0$. (That is, if X has distribution m then $E(\|X\|^\epsilon) <$

∞.) Let $x \in \mathbf{R}^N$ and suppose one of the following conditions holds:

(a) m is absolutely continuous and the density function p is continuous at x with $p(x) > 0$.

(b) m is D-regular ($D < N$), x belongs to the absolutely D-regular support S, and the restriction $p|S$ is continuous at x with $p(x) > 0$. Further suppose that S satisfies the following smoothness constraint: there exist open sets $G \subseteq \mathbf{R}^D$, $U \subseteq \mathbf{R}^N$ with $x \in U$, and a C^1-mapping F: $G \to \mathbf{R}^{N-D}$ such that $S \cap U = \{(z, F(z)) | z \in G\}$. Then all moments of $L_{k,n}(x)$ converge to the corresponding moments of LG(μ_k + log $k\kappa_D p(x)$, θ_k^2; k). ■

The smoothness constraint on S in (b) of the above theorem enables us to locally (in a neighborhood of x) identify $W_1, \ldots, W_n$ with absolutely continuous observations $W_1^*, \ldots, W_n^*$ in $\mathbf{R}^D$, thereby reducing (b) to (a). The results of Pickands (1968) (for the case $k = 1$) appear to make this constraint on S unnecessary.

The preceding asymptotic theory shows that a nearest-neighbor approach to estimating D naturally involves the unknown D-density in the role of a nuisance parameter, since it determines the location of the limiting distribution. To eliminate this local density effect, Cutler and Dawson (1989) constructed a statistic based on two independent random samples $W_{1,1}, \ldots, W_{1,n}$ and $W_{2,1}, \ldots, W_{2,mn}$ (where $m > 1$) which converges to a logistic distribution centered at $1/D$. The extension of this technique to the kth nearest neighbor produces a generalized logistic distribution (with parameter k) centered at $1/D$. Specifically, let $d_{k,n}^{(1)}(x)$ and $d_{k,mn}^{(2)}(x)$ denote the kth nearest-neighbor distance to x from each of the two samples. Set

$$R_{k,n}(x) = \frac{1}{\log m} \log\left(\frac{d_{k,n}^{(1)}(x)}{d_{k,mn}^{(2)}(x)}\right).$$

Then $R_{k,n}(x)$ is a density-free statistic asymptotically centered at $1/D$. These results are made precise in the following theorem.

Theorem 18.2.2.4 Let m and x satisfy the hypotheses of Theorem 18.2.2.3. Then $R_{k,n}(x)$ converges in distribution to the generalized logistic $L(1/D, \sigma_k^2/D^2(\log m)^2; k)$ as $n \to \infty$; furthermore all moments of $R_{k,n}(x)$ converge correspondingly.

Proof. Note that we can write

$$R_{k,n}(x) = \frac{L_{k,mn}^{(2)}(x) - L_{k,n}^{(1)}(x)}{D \log m} + \frac{1}{D},$$

where $L^{(1)}_{k,n}(x)$, $L^{(2)}_{k,mn}(x)$ are computed from the first and second samples, respectively. The result now follows from Theorems 18.2.2.2 and 18.2.2.3 and the fact that (see Appendix B) the difference of two i.i.d. log gamma distributions produces a generalized logistic distribution centered at the origin. ∎

The above result can be used as the basis for constructing tests of hypothesis and/or confidence intervals for D. Since the variance $\sigma_k^2 \rightarrow 0$ as $k \rightarrow \infty$ this suggests that inference based on higher-order nearest neighbors will be more efficient; this is of course balanced by the fact that larger sample sizes are required in order for the asymptotic theory to be valid for larger k. However, since the decrease in variance is marginal past the third or fourth nearest neighbor (which can easily be seen from the form of σ_k^2; see (18.2.B.3) and (18.2.A.7)) it seems that procedures and sample sizes can be effectively based on a maximum value of $k \cong 4$.

In the case of D-regular distributions (in which case m—almost all points x share the same scaling behavior D) one approach is to randomly sample c "basepoints" $X_1, \ldots, X_c$ from m, and then compute the quantities $R_{k,n}(X_j)$, $j = 1, \ldots, c$. Statistical procedures can then be based on the mean $\bar{R}_{k,n}$. It is not difficult to show that, for m—almost all points $x_1, \ldots, x_c$ in $\mathbf{R}^N$, the variables $R_{k,n}(x_j)$, $j = 1, \ldots, c$, are asymptotically independent. For random basepoints $X_1, \ldots, X_c$ we therefore observe that $R_{k,n}(X_j)$, $j = 1, \ldots, c$ are asymptotically i.i.d. A procedure based on these ideas (for the case $k = 1$) is more fully developed in Cutler (1991) where the case of nonregular or fractal scaling exponents is also considered.

REFERENCES

Abramowitz, M. and Stegun, I. A., (eds.), (1970). *Handbook of Mathematical Functions with Formulas, Graphs, and Mathematical Tables*, Dover, New York.

Billingsley, P. (1986). *Probability and Measure*, 2nd ed., Wiley, New York.

Cutler, C. D. (1991). Some results on the behavior and estimation of the fractal dimensions of distributions on attractors. *J. Statist. Phys.*, *62*, 651–708.

Cutler, C. D. and Dawson, D. A. (1989). Estimation of dimension for spatially-distributed data and related limit theorems, *J. Multivariate Anal.*, *28*, 115–148.

Falconer, K. J. (1985). *The Geometry of Fractal Sets*, Cambridge University Press, New York.

Pickands III, J. (1968). Moment convergence of sample extremes, *Ann. Math. Statist.*, *39*, 881–889.

APPENDIX

18.2.A The Log Gamma Distribution

Let $\Gamma(\alpha) = \int_0^\infty u^{\alpha-1}e^{-u}\,du$ denote the usual gamma function. A random variable X_α is said to have a *standard log gamma* distribution with parameter $\alpha > 0$ if the probability density function of X_α is given by:

$$g(s;\alpha) = \frac{e^{-\alpha s}\exp(-e^{-s})}{\Gamma(\alpha)}, \qquad -\infty < s < \infty. \tag{18.2.A.1}$$

The corresponding moment generating function is

$$M(t;\alpha) = \frac{\Gamma(\alpha - t)}{\Gamma(\alpha)}, \qquad -\infty < t < \alpha, \tag{18.2.A.2}$$

so that the general form of the jth moment is

$$M^{(j)}(0;\alpha) = \frac{(-1)^j\Gamma^{(j)}(\alpha)}{\Gamma(\alpha)}. \tag{18.2.A.3}$$

The moments can be obtained by successively differentiating the *psi function* $\psi(t) = \Gamma'(t)/\Gamma(t)$. We will let μ_α and θ_α^2 denote, respectively, the mean and variance of the log gamma distribution with parameter α. The complete class of log gamma distributions comprises all affine transformations $W = a + bX_\alpha$. Notationally we express this as $W \sim \text{LG}(a + b\mu_\alpha, b^2\theta_\alpha^2; \alpha)$.

In the special case that $\alpha = k$, a positive integer, the distribution function of X_k can be obtained in closed form:

$$G(s;k) = \sum_{j=0}^{k-1} \frac{e^{-js}}{j!}\exp(-e^{-s}), \tag{18.2.A.4}$$

and the mean μ_k and variance θ_k^2 of X_k $(k \geq 2)$ are

$$\mu_k = -\psi(k) = \gamma - h_1(k-1), \tag{18.2.A.5}$$

$$\theta_k^2 = \psi'(k) = \frac{\pi^2}{6} - h_2(k-1), \tag{18.2.A.6}$$

where γ = Euler's constant $\cong 0.577$ and

$$h_1(k) = \sum_{j=1}^{k}\frac{1}{j}, \qquad h_2(k) = \sum_{j=1}^{k}\frac{1}{j^2}. \tag{18.2.A.7}$$

These follow from the recursive relations developed for $\psi(k)$ and $\psi'(k)$ in Abramowitz and Stegun (1970). In the case $k = 1$ we obtain a standard extreme value distribution $G_1(s) = \exp(-e^{-s})$ with mean $\mu_1 = \gamma$ and variance $\theta_1^2 = \pi^2/6$.

18.2.B The Generalized Logistic Distribution

The *standard generalized logistic* distribution with parameter $\alpha > 0$ has probability density function

$$f(s; \alpha) = \frac{\Gamma(2\alpha)}{\Gamma(\alpha)^2}\left(\frac{e^{-s}}{(1 + e^{-s})^2}\right)^{\alpha}, \qquad -\infty < s < \infty, \qquad (18.2.B.1)$$

which is symmetric about the origin. The usual logistic distribution corresponds to the case $\alpha = 1$. If $Y_\alpha = X_\alpha - X'_\alpha$, where X_α and X'_α are independent log gamma variables with parameter $\alpha > 0$, then a simple transformation of variable shows that Y_α has a generalized logistic distribution with the same parameter α. The complete class of generalized logistic distributions comprises all affine transformations $Z = a + bY_\alpha$, and we denote this by $Z \sim L(a, b^2\sigma_\alpha^2; \alpha)$, where σ_α^2 is the variance of Y_α.

When $\alpha = k$, a positive integer, the distribution function of Y_k can be found in closed form:

$$F(s; k) = \sum_{j=0}^{k-1} \binom{j + k - 1}{j} q(s)^j(1 - q(s))^k, \qquad (18.2.B.2)$$

where $q(s) = e^{-s}/(1 + e^{-s})$ and the variance of Y_k is

$$\sigma_k^2 = \frac{\pi^2}{3} - 2h_2(k - 1). \qquad (18.2.B.3)$$

For more details on this distribution, one may refer to Chapter 9.

18.3 ANALYSIS OF BIOAVAILABILITY DATA WHEN SUCCESSIVE SAMPLES ARE FROM LOGISTIC DISTRIBUTION

B. K. Shah
Pearl River, New York

18.3.1 Introduction

The purpose of bioavailability trials of an investigational compound is to ascertain how much of a substance is being absorbed in a given patient (or a group of patients for statistical purposes) over time. The underlying concept in such studies is to understand the relationship, if any, between the amount of substance absorbed in the patient under treatment, clinical variables, and the dosage regimen of the investigational compound. In order to measure the content of substance in a patient, a clinical investigator administers a substance orally or intravenously to a randomly selected patient and draws serum samples from the patient at a specified interval of time. The amount of substance present in a given serum sample is ascertained by a chemical assaying method. Let X_i be the concentration of substance in serum of a given patient at the ith point in time, $i = 1, 2, \ldots, k$. Without loss of generality assume that the interval of drawing serum samples is unity. An area under the concentration (AUC) is formed by joining the k concentration points of a given patient. Assume that the concentration of substance at the $(k + 1)$th time point is inmeasurable or zero. Then ΣX_i and $\Sigma \mu_i$, respectively, represents a sample and population area under the concentration for a given patient. The quantity ΣX_i, known as the measure of bioavailability for a given patient, is of interest to the investigator, since it helps him or her to make decisions concerning a treatment plan (increase or decrease or maintain the same regimen) for a patient.

Several authors (Westlake, 1976; Hauck and Anderson, 1984; Clayton and Leslie, 1981; and others) have developed a confidence interval for bioavailability (or comparative bioavailability of two substances) under the assumption that serum samples are normally and identically distributed with the same mean and the same variance σ^2. However, it has been experimentally and theoretically (Aiache et al., 1989; Wagner, 1971; Shah, 1976) found that most bioavailability data are drawn from distributions with varying population means such that they have a nonlinear trend. Shah (1988) has developed confidence limits for bioavailability under the assumption that the serum concentration X_i is normally distributed with mean μ_i and variance σ^2, such that $\mu_i > 0$ and has a trend.

The motivation of this section is to draw inferences or develop confidence limits for a population AUC for a single patient (or for a group of patients for statistical purposes), when serum samples are drawn from a logistic distribution with varying population means such that they constitute a trend.

Section 18.3.2 provides general formulas for calculating expected value and variance of $s^2 = \Sigma (X_i - \overline{X})^2/(k - 1)$ and $\delta^2 = \Sigma (X_{i+1} - X_i)^2/2(k - 1)$, which are the estimators of population variance σ^2. In Section 18.3.3, biases, variances of s^2 and δ^2, and relative efficiencies of δ^2 with respect to s^2 are obtained for logistic variables with varying population means μ_i such that μ_i forms a trend.

Section 18.3.4 provides simulated critical points of $t = \overline{X}\sqrt{k}/s$ and $t' = \overline{X}\sqrt{k}/\delta$ under the null hypothesis that bioavailability is zero (i.e., $H_0\colon \Sigma \mu_i = 0 \Leftrightarrow H_0\colon \mu_i = 0$ for all i). These critical values help to set up confidence limits for bioavailability for a patient (or a group of patients for statistical purposes). An example is provided to demonstrate the impact of different statistical assumptions on the outcome of an experiment.

18.3.2 Estimator of σ^2

If it is known or if one validates the statistical assumption through a statistical test procedure that successive serum concentrations are drawn from the same homogeneous population with the same mean μ and variance σ^2, no new statistical problem would arise. However, in bioavailability studies, it is theoretically (Wagner, 1976; Shah, 1976) and experimentally (Aiache et al., 1989) known that serum concentration data can be represented by two-term or multiterm exponential models, which indicates that serum concentration data are not drawn from the same homogeneous population with the same mean μ and variance σ^2. Further, if the investigational substance is a new formulation (generic products) of a standard formulation (i.e., brand name products), then prior studies of a standard formulation would also indicate whether population means of a standard formulation constitute a trend. Thus, having validated the statistical assumption that sampling of serum concentration is from populations with varying population means such that they form a linear or nonlinear trend, the question of interest is, what is the best estimator of σ^2? In this area of research von Neumann et al. (1941) have stated:

> ". . . There are cases, however, where the standard deviation may be held constant, but the mean varies from one observation to the next. If no correction is made for such variation of the mean, and the standard deviation is computed from the data in the conventional way, then the estimated standard deviation will tend to be larger than the true population variance.* When the variation in the mean is gradual, so that a trend (which need not be linear) is shifting the mean of the population, a rather simple method of minimizing the effect of trend on dispersion is to estimate standard deviation from difference."

However, in Section 3 of their paper, the authors proved the opposite result, namely, the sample variance $s^2 = \Sigma (X_i - \overline{X})^2/(k - 1)$ is a better estimator of σ^2 than half the mean square successive difference. Before proceeding to prove von Neumann's result for logistic samples, it is important to provide general formulas for obtaining mean and variance of s^2 and δ^2 for nonidentically distributed variables.

18.3.3 Moments of s^2 and δ^2 From Nonidentically Distributed Variables

Let the rth raw moment of a random variable X_i, whose mean is μ_i and variance is σ^2, be denoted by $E(X_i^r) = \mu_r'(i)$, $i = 1, 2, \ldots, k$. It is simple

*It should be "standard deviation" and not "variance" as quoted by von Neumann et al. on page 153.

to verify that

$$E(s^2) = \sigma^2 + \frac{\Sigma_1^k (\mu_1'(i) - \overline{\mu}_1')^2}{k - 1},$$

$$E(\delta^2) = \sigma^2 + \frac{\Sigma_1^{k-1} \{\mu_1'(i + 1) - \mu_1'(i)\}^2}{2(k - 1)},$$

$$\begin{aligned}
\operatorname{Var}(s^2) = {} & \frac{1}{k^2}\left[\sum_1^k \mu_4'(i) + 2\sum_{i<j} \mu_2'(i)\mu_2'(j)\right] + \frac{4}{k^2(k-1)^2} \\
& \times \left[\sum_{i<j} \mu_2'(i)\mu_2'(j) + 2\sum_{i<j<r} \mu_2'(i)\mu_1'(j)\mu_1'(r)\right. \\
& + 2\sum_{i<j<r} \mu_1'(i)\mu_2'(j)\mu_1'(k) + 2\sum \mu_1'(i)\mu_1'(j)\mu_2'(r) \\
& \left. + 6\sum_{i<j<r<s} \mu_1'(i)\mu_1'(j)\mu_1'(r)\mu_1'(s)\right] \\
& - \frac{4}{k^2(k-1)}\left[\sum_{i<j} \mu_3'(i)\mu_1'(j)\right. \\
& + \sum_{i<j} \mu_1'(i)\mu_3'(j) + \sum_{i<j<r} \mu_2'(i)\mu_1'(j)\mu_1'(r) \\
& + \sum_{i<j<r} \mu_1'(i)\mu_2'(j)\mu_1'(r) \\
& \left. + \sum_{i<j<r} \mu_1'(i)\mu_1'(j)\mu_2'(r)\right] - \{E(s^2)\}^2,
\end{aligned}$$

$$\begin{aligned}
\operatorname{Var}(\delta^2) = {} & \frac{1}{4(k-1)^2}\left[4\sum_1^k \mu_4'(i) - 3\{\mu_4'(i) + \mu_4'(k)\}\right. \\
& + 12\sum_{i<j} \mu_2'(i)\mu_2'(j) + 6\sum_{i<j<r<s} \mu_1'(i)\mu_1'(j)\mu_1'(r)\mu_1'(s) \\
& - 4\mu_2'(1)\sum_2^k \mu_2'(j) - 4\mu_2'(k)\sum_1^{k-1} \mu_2'(j) \\
& - 8\left\{\sum_{i<j}^k \mu_3'(i)\mu_1'(j) + \sum_{i<j}^k \mu_1'(i)\mu_3'(j)\right\} \\
& + 2\mu_2'(1)\mu_2'(k) + 4\mu_3'(1)\sum_2^k \mu_1'(j) \\
& + 4\mu_3'(k)\sum_1^{k-1} \mu_1'(j) + 4\mu_2'(1)\sum_{i<j=2}^k \mu_1'(i)\mu_1'(j)
\end{aligned}$$

$$+ 4\mu_2'(k) \sum_{i<j=1}^{k-1} \mu_1'(i)\mu_1'(j)\Bigg]$$

$$- \{E(\delta^2)\}^2,$$

where $\bar{\mu}_1' = \Sigma_1^k \mu_1'(i)/k$.

The above formulas are general and true for any nonidentically distributed variables. If $\mu_i = \mu$ for all i, these formulas reduce to those given in textbooks.

18.3.4 Biases and Variances of s^2 and δ^2 when $X_i \sim L(\mu_i, \sigma^2)$

Let X_i, $i = 1, 2, \ldots, k$, be successive measurements following a logistic distribution (denoted by $L(\mu_i, \sigma^2)$), if its cumulative distribution function (cdf) is

$$F(X; \mu_i, \sigma^2) = [1 + \exp\{-c(X - \mu_i)/\sigma\}]^{-1}, \quad \text{where } c = \pi/\sqrt{3},$$
$$-\infty < X < \infty, \quad \mu_i > 0, \quad \text{and } \sigma > 0.$$

Below we give a proof that δ^2 is a better estimator of σ^2 when $X_i \sim \text{IL}(\mu_i, \sigma^2)$ such that μ_i has a linear or nonlinear trend (where IL stands for independently and logistically distributed variables). We prove this result for $k = 3$.

Let

$$E(X(i))^k = \mu_k'(i) \qquad \text{and} \qquad E(X(i))^2 = \sigma^2 + \mu_1'^2(i),$$

$$\text{Var}(s^2) = \frac{1}{9}\Bigg[\frac{63}{5}\sigma^2 + 6\sigma^2 \sum_1^3 \mu_1'^2(i) + \sum_1^3 \mu_1'^4(i)$$

$$+ 3\sum_{i<j} (\sigma^2 + \mu_1'^2(i))(\sigma^2 + \mu_1'^2(j))$$

$$- \sum_{i\neq j\neq r} (3\sigma^2\mu_1'(i) + \mu_1'^3(i))(\mu_1'(j) + \mu_1'(k))\Bigg]$$

$$- \{E(s^2)\}^2.$$

$$\text{Var}(\delta^2) = \frac{1}{16}\Bigg[\frac{126}{5}\sigma^4 + \mu_1'^2(1)\{6\sigma^2 + \mu_1'^2(1)\}$$

$$+ 4\mu_1'^2(2)\{(\sigma^2 + \mu_1'^2(2)\}$$

$$+ \mu_1'^2(3)\{6\sigma^2 + \mu_1'^2(3)\}$$

$$+ 8(\sigma^2 + \mu_1'^2(1))(\sigma^2 + \mu_1'^2(2))$$

$$+ 8(\sigma^2 + \mu_1'^2(2))(\sigma^2 + \mu_1'^2(3))$$

$$+ 2(\sigma^2 + \mu_1'^2(1))(\sigma^2 + \mu_1'^2(3))$$

Table 18.3.1 Bias and Efficiency of δ^2 When Sampling Is from $L(\mu_i, \sigma^2)$ Such That $\mu_i = \alpha + \beta i$

α	β	σ^2	$E(s^2)$	$E(\delta^2)$	Var(s^2)	Var(δ^2)	Eff = 100 Var(s^2)/Var(δ^2)
1	1	1.0	2.0	1.5	3.40	2.20	154.55
1	1	1.5	2.5	2.0	6.15	4.57	134.43
1	1	2.0	3.0	2.5	9.60	7.80	123.08
1	1	3.0	4.0	3.5	18.60	16.80	110.71
2	1	2.0	3.0	2.5	9.60	7.80	123.08
2	1	2.5	3.5	3.0	13.75	11.88	115.79
2	1	3.0	4.0	3.5	18.60	16.80	110.71
2	1	4.0	5.0	4.5	30.40	29.20	104.11
1	2	5.0	9.0	7.0	75.00	52.50	142.86
1	2	5.5	9.5	7.5	86.35	62.42	138.33
1	2	6.0	10.0	8.0	98.40	73.20	134.43
1	2	7.0	11.0	9.0	124.60	97.30	128.06
1	3	10.0	19.0	14.5	320.00	215.00	148.84
1	3	10.5	19.5	15.0	343.35	234.68	146.31
1	3	11.0	20.0	15.5	367.40	255.20	143.97
1	3	12.0	21.0	16.5	417.60	298.80	139.78

$$- 4\mu_1'(2)\{\mu_1'(1) + \mu_1'(3)\}\{10\sigma^2 + \mu_1'^2(1) + 2\mu_1'^2(2) + \mu_1'^2(3)\}$$

$$+ 8\mu_1'(1)\mu_1'(3)(\sigma^2 + \mu_1'^2(2))\Big] - \{E(\delta^2)\}^2.$$

Table 18.3.1 provides the efficiencies of δ^2 when $\mu_i = \alpha + \beta i$ ($i = 1, 2, 3$). The efficiency has been defined as the ratio of two variances, since biases are independent of σ^2 and since both s^2 and δ^2 can be considered as a measure of residual mean sum of squares in a regression of x on i. The table shows that half the mean square successive difference is a better estimator of population variance than the sample variance when sampling is from logistically distributed variables such that the population means μ_i have a trend. Note that the Cramer-Rao lower bound for the estimator σ^2 does not hold in this case since the sampling process is not from the same homogeneous population.

18.3.5 Simulated Critical Points of $t_L = \overline{X}\sqrt{k}/s$ and $t_L' = \overline{X}\sqrt{k}/\delta$

Let $X_i \sim L(\mu_i, \sigma^2)$, and let the null hypothesis of interest be that their area under the concentration, which is a measure of bioavailability of a substance in a given patient, is zero. In statistical terminology, the null hypothesis can be written as H_0: $\Sigma\ \mu_i = 0 \Leftrightarrow H_0$: $\mu_i = 0$ for all i. Thus X_i becomes $L(0, \sigma^2)$ when H_0 is true. Let $t_L = \overline{X}\sqrt{k}/s$ and $t'_L = \overline{X}\sqrt{k}/\delta$ be respectively Student's t and t' statistics (obtained by Shah, 1988) for a logistic distribution. It is well known that $\overline{X}$ and s^2 are independently distributed when $X_i \sim N(\mu, \sigma^2)$. Shah (1988) has shown that $\overline{X}$ and δ^2 are statistically independent when $X_i \sim \text{IN}(\mu_i, \sigma^2)$. Since it is extremely difficult to obtain the distribution of t_L or t'_L, a simulation study was undertaken to tabulate the critical values of t_L and t'_L by taking 20,000 random samples of size k from a logistic distribution. The result of this study is provided in Table 18.3.2.

When reading the critical values of t_L and t'_L for a given sample size, note that the column headed by k refers to the actual value of the sample size and should not be interpreted as degrees of freedom.

A confidence interval for a single AUC is $\Sigma\ X_i \pm t'_{L,\alpha}\sqrt{k\delta^2}$, where $t'_{L,\alpha}$ is the critical value of t'_L at the α level of significance.

Example Let 0.05, 1.50, 1.10, 0.80, 0.30, 0.01 be the concentration of an investigational drug circulating in the blood of a patient. Then $\overline{X} = 0.6267$, $s^2 = 0.3673$, $\delta^2 = 0.26867$, $t = \overline{X}\sqrt{k}/s = 2.53292$ and $t' = \overline{X}\sqrt{k}/\delta = 2.9614$. If one assumes that the data are drawn from the same normal distribution, then one finds that the observed $t = 2.53292$ is less than the theoretical t (=2.571) at the 0.05 level, indicating that the mean AUC is not significantly different from zero. However, if one validates the assumption that the sampling is from nonidentically distributed variables, then one should use t' to test the null hypothesis. The observed t' is greater than 2.82, which is the critical value obtained by Shah (1988) under the normality assumption, while it is less than the simulated value of 2.72 (see

Table 18.3.2 Critical Values of t_L and t'_L

	t_L		t'_L	
k	5%	1%	5%	1%
5	2.631948	3.50775	2.895168	3.873752
6	2.479518	3.134041	2.722032	3.479162
7	2.364815	2.982332	2.539887	3.277855
8	2.351886	2.916975	2.486978	3.190600
9	2.269975	2.760224	2.398816	2.989305
10	2.267455	2.769197	2.365604	2.922003
12	2.200667	2.701699	2.314446	2.867922
15	2.112572	2.567523	2.191934	2.712371

Table 18.3.2) under the logistic distribution assumption. Thus a contradictory conclusion is reached under various statistical assumptions, even assuming that the serum data are from nonidentically distributed random variables.

REFERENCES

Aiache, J., Pierre, V., Beyssac, E., Prasad, V., and Skelly, J. (1989). New results on an vitro model for the study of the influence of fatty meals on the bioavailability of theophylline controlled-release formulations, *J. Pharm. Sc.*, *78*, 261–263.

Clayton, A. and Leslie, A. (1981). The bioavailability of erythromycin stearate versus enteric-coated erythromycin base when taken immediately before and after food, *Journal of Internal Med. Res.*, *9*, 470–477.

Hauck, W. W. and Anderson S. (1984). A new statistical procedure for testing equivalence in two-group comparative bioavailability trials, *Journal of Pharmacokinetics and Biopharmaceutics*, *12*, 83–91.

Shah, B. K. (1976). Data analysis problem in the area of pharmacokinetics research, *Biometrics*, *32*, 145–157.

Shah, B. K. (1988). On the generalization of *t*-distribution for nonidentically distributed variables and applications, *Gujarat Statistical Review*, *15*, 45–50.

Von Neumann, J., Kent, R. H., Bellinson, H. R., and Hart B. I. (1941). The mean square successive difference, *Ann. Math. Statist.*, *12*, 153–162.

Wagner, J. (1971). *Biopharmaceutics and Relevant Pharmacokinetics*, Drug Intelligence Publications, Hamilton, IL.

Westlake, W. J. (1976). Symmetrical confidence intervals for bioequivalence trials, *Biometrics*, *32*, 741–744.

18.4 NORMAL SCALE MIXTURE APPROXIMATIONS TO $F^*(z)$ AND COMPUTATION OF THE LOGISTIC-NORMAL INTEGRAL

John F. Monahan and Leonard A. Stefanski
North Carolina State University, Raleigh, North Carolina

In this section we exploit a normal scale mixture representation of the logistic distribution to obtain a convenient method of computing the logistic-normal integral

$$G(\eta, \tau) = \int_{-\infty}^{\infty} F^*(z)\tau^{-1}\phi\left(\frac{z - \eta}{\tau}\right) dz, \tag{18.4.1}$$

where $\phi(\cdot)$ is the standard normal density function.

The latter integral arises frequently in statistical modelling although our interest in the problem is due primarily to its appearance in certain logistic regression models with random effects. Two such models motivating our work are discussed in detail later in this section.

The integral in (18.4.1) cannot be evaluated in closed form and approximations to $G(\eta, \tau)$ are necessary. Among the standard numerical methods employed to evaluate $G(\eta, \tau)$, Gauss-Hermite quadrature is the most widely used and best known among statisticians. In a recent paper Crouch and Spiegelman (1990) show that 20-point Gauss-Hermite quadrature is not sufficiently accurate for certain applications and they propose an alternative form of quadrature that provides an approximation to any prescribed degree of accuracy. However, significantly greater accuracy is obtained only at the expense of greatly increased computational time relative to Gauss-Hermite quadrature.

We exploit a normal-scale mixture representation of $F^*(z)$ to obtain a new method of approximating $G(\eta, \tau)$. Our new method represents a compromise between Gauss-Hermite quadrature and Crouch and Spiegelman's (1990) method with respect to both accuracy and computational time. The primary and significant advantage of our method over other numerical methods is the ease with which it can be implemented in standard statistical computing software packages. We approximate $G(\eta, \tau)$ with a finite location-scale mixture of normal distribution functions, $\Phi(\cdot)$, and thus the method is easily programmed in any software package in which $\Phi(\cdot)$ is an intrinsically defined function.

Central to our method are approximations to $F^*(z)$ having the form

$$F_k^*(z) = \sum_{i=1}^{k} p_{k,i}\Phi(zs_{k,i}) \qquad (k = 1, 2, \ldots), \tag{18.4.2}$$

where $\{p_{k,i}, s_{k,i}\}_{i=1}^{k}$ are chosen to minimize

$$\Delta_k^* = \sup_z |F^*(z) - F_k^*(z)|, \tag{18.4.3}$$

for $k = 1, 2, \ldots$. We replace $F^*(z)$ in (18.4.1) with $F_k^*(z)$, obtaining the approximation

$$G_k(\eta, \tau) = \int_{-\infty}^{\infty} F_k^*(z)\tau^{-1}\phi\left(\frac{z - \eta}{\tau}\right) dz$$

$$= \sum_{i=1}^{k} p_{k,i}\Phi\left(\frac{\eta s_{k,i}}{\sqrt{1 + \tau^2 s_{k,i}^2}}\right). \tag{18.4.4}$$

It follows easily from (18.4.3) that

$$\sup_{\eta,\tau} |G_k(\eta, \tau) - G(\eta, \tau)| \le \Delta_k^*.$$

The acceptability of $G_k(\eta, \tau)$ as an approximant to $G(\eta, \tau)$ follows from the fact that $F_k^*(z)$ approximates $F^*(z)$ remarkably well for k as small as 3 and improves significantly with successive k. Table 18.4.1 contains values of $\{p_{k,i}, s_{k,i}\}$ and Δ_k^* ($k = 1, \ldots, 8$).

The following theorem provides explanation for the quality of the approximations and also justification for the class of approximants in (18.4.2) and (18.4.4).

Theorem Let

$$L(\sigma) = 1 - 2\sum_{j=1}^{\infty} (-1)^{j+1} \exp(-2j^2\sigma^2), \tag{18.4.5}$$

denote the Kolmogorov-Smirnov distribution and define $q(\sigma) = (d/d\sigma)L(\sigma/2)$. Then

$$F^*(z) = \int_0^\infty \Phi\left(\frac{z}{\sigma}\right) q(\sigma)\, d\sigma. \tag{18.4.6}$$

A detailed derivation of the theorem has been given elsewhere (Stefanski, 1990). A short proof of the key identity (18.4.6) follows.

Proof. Consider the integral equation

$$f^*(z) = \int_0^\infty \sigma^{-1}\phi\left(\frac{z}{\sigma}\right) q(\sigma)\, d\sigma. \tag{18.4.7}$$

Note that for $z > 0$, $F^*(z) = (1 + e^{-z})^{-1} = \sum_{n=0}^{\infty} (-1)^n \exp(-nz)$. Upon differentiation and appeal to symmetry it follows that

$$f^*(z) = \sum_{n=1}^{\infty} (-1)^{(n+1)} n \exp(-n|z|). \tag{18.4.8}$$

Thus for $q(\sigma)$ as defined by (18.4.5)

$$\int_0^\infty \sigma^{-1}\phi\left(\frac{z}{\sigma}\right) q(\sigma)\, d\sigma = \sqrt{\frac{2}{\pi}} \sum_{n=1}^{\infty} (-1)^{(n+1)} n^2 \int_0^\infty \exp\left(\frac{-z^2}{2\sigma^2} - \frac{n^2\sigma^2}{2}\right) d\sigma$$

Table 18.4.1 Least Maximum Approximants

k	Δ_k^*	$p_{k,i}$	$s_{k,i}$
1	9.5(−3)	1.00000 00000 00000	0.58763 21328 24711
2	5.1(−4)	0.56442 48434 56014	0.76862 11512 01617
		0.43557 51565 43986	0.43525 51800 69097
3	4.4(−5)	0.25220 15780 98282	0.90793 08374 49693
		0.58522 50592 35736	0.57778 72761 40136
		0.16257 33626 65982	0.36403 77294 79770
4	4.7(−6)	0.10649 89926 56952	1.02313 55805 00914
		0.45836 12270 14536	0.69877 43559 46609
		0.37418 90669 14829	0.47512 75246 40229
		0.06095 07134 13683	0.32106 46558 34542
5	6.0(−7)	0.04433 31519 39163	1.12271 60446 01626
		0.29497 33769 77114	0.80390 11123 64718
		0.42981 24819 00555	0.57619 76474 16307
		0.20758 95057 57111	0.41091 37884 75565
		0.02329 14834 26056	0.29147 98950 13730
6	8.4(−8)	0.01844 61051 35654	1.21126 19038 17665
		0.17268 13809 23308	0.89734 91017 45990
		0.37393 07960 25243	0.66833 39140 55091
		0.31696 99558 13251	0.49553 64080 73851
		0.10889 73000 53481	0.36668 89730 48750
		0.00907 44620 49063	0.26946 19401 81722
7	1.3(−8)	0.00771 18334 44756	1.29153 69623 80838
		0.09586 53530 40290	0.98192 29519 23815
		0.28194 83083 10964	0.75277 49130 79201
		0.34546 28488 09089	0.57547 48297 39348
		0.20984 86860 83383	0.43861 91442 39000
		0.05556 46179 00566	0.33410 42759 71241
		0.00359 83524 10953	0.25221 62211 45020
8	2.1(−9)	0.00324 63432 72134	1.36534 08062 96348
		0.05151 74770 33972	1.05952 39710 16916
		0.19507 79126 73858	0.83079 13137 65644
		0.31556 98236 32818	0.65073 21666 39391
		0.27414 95761 58423	0.50813 54253 66489
		0.13107 68806 95470	0.39631 33451 66341
		0.02791 24187 27972	0.30890 42522 67995
		0.00144 95678 05354	0.23821 26164 09306

$$= \sqrt{\frac{2}{\pi}} \sum_{n=1}^{\infty} (-1)^{(n+1)} n^2 \left\{ \frac{\sqrt{2\pi}\, \exp(-n|z|)}{2n} \right\}$$

$$= \sum_{n=1}^{\infty} (-1)^{(n+1)} n \exp(-n|z|), \tag{18.4.9}$$

proving the theorem upon appeal to (18.4.8). The integral evaluation in

(18.4.9) employs a standard change of variables; alternatively, see Gradshteyn and Ryzhik (1980), Formula 3.325, p. 307.

Our interest in the logistic-normal integral is due to its appearance in two important statistical models of current interest to statisticians. We discuss logistic regression measurement error models first and then random effects logistic regression models.

In one form of a logistic regression measurement error model, the binary outcome D is related to a p-dimensional predictor X, via $\mathrm{pr}(D = 1|X = x) = F^*(\beta^T x)$. Observable data consist of observations on (D, Z) where $X|(Z = z)$ is normally distributed with mean $\mu(z)$ and covariance matrix $\Omega(z)$. The conditional distribution of $D|(Z = z)$ is then given by

$$\mathrm{pr}(D = 1|Z = z) = G(\eta, \tau), \tag{18.4.10}$$

where $\eta = \beta^T\mu(z)$ and $\tau = \sqrt{\beta^T\Omega(z)\beta}$. This model and closely related models have been the topic of several recent papers: Carroll et al. (1984), Armstrong (1985), Stefanski and Carroll (1985), Prentice (1986), Schafer (1987), and Spiegelman (1988). We propose approximating $G(\cdot, \cdot)$ with $G_k(\cdot, \cdot)$, yielding a computationally convenient method for analyzing data with the logistic regression measurement error model (18.4.10).

Now consider the random-effects logistic regression model

$$\mathrm{pr}(D_{i,j} = 1|X_i = x_i, \epsilon_i) = F^*(\eta_i + \tau\epsilon_i), \qquad j = 1, \ldots, n_i, \quad i = 1, \ldots, N, \tag{18.4.11}$$

where $\eta_i = \beta^T x_i$ and $\{\epsilon_i\}_{i=1}^N$ is a sequence of independent standard normal variates. Models for correlated binary responses have been studied by several authors: Pierce and Sands (1975), Williams (1982), Ochi and Prentice (1984), Prentice (1986), Zeger, Liang, and Albert (1988), Prentice (1988), Stiratelli, Laird, and Ware (1984), and McCullagh (1989). They arise frequently in animal experiments wherein ϵ_i in (18.4.11) represents a random interlitter effect.

Quasilikelihood/variance-function estimation for this model requires the first two moments of $D_{i,\cdot}|(X_i = x_i)$:

$$E(D_{i,\cdot}|X_i = x_i) = n_i\mu_i = n_iG(\eta_i, \tau); \tag{18.4.12}$$

$$\mathrm{Var}(D_{i,\cdot}|X_i = x_i) = \Sigma_i = n_i^2\mu_i(1 - \mu_i) - n_i(n_i - 1) \times \int_{-\infty}^{\infty} F^{*(1)}(z)\tau^{-1}\phi\left(\frac{z - \eta_i}{\tau}\right) dz. \tag{18.4.13}$$

The identity $F^{*(1)} = F^*(1 - F^*)$ is used to derive (18.4.13).

Again we propose approximating $G(\cdot, \cdot)$ with $G_k(\cdot, \cdot)$ in (18.4.12), yielding an approximate mean $\tilde{\mu}_i$. Substitution of $F_k^{*(1)}$ for $F^{*(1)}$ in (18.4.13) yields an approximate variance

$$\tilde{\Sigma}_i = n_i^2\tilde{\mu}_i(1 - \tilde{\mu}_i) - n_i(n_i - 1)\int_{-\infty}^{\infty} F_k^{*(1)}(z)\tau^{-1}\phi\left(\frac{z - \eta_i}{\tau}\right) dz$$

$$= n_i^2\tilde{\mu}_i(1 - \tilde{\mu}_i) - n_i(n_i - 1)\sum_{j=1}^{k} \frac{p_{k.j}s_{k.j}}{\sqrt{1 + \tau^2 s_{k.j}^2}}\phi\left(\frac{\eta_i s_{k.j}}{\sqrt{1 + \tau^2 s_{k.j}^2}}\right). \tag{18.4.14}$$

The fact that neither the covariate nor the random effect varies with j in (18.4.11) is responsible for the simplicity of (18.4.14). More complex random effects logistic regression models have

$$\text{pr}(D_i = 1|X_i = x_i, \epsilon_i) = F(\eta_i + \epsilon_i), \qquad i = 1, \ldots, n, \tag{18.4.15}$$

where $\eta_i = \beta^T x_i$ and $\epsilon^T = (\epsilon_1, \ldots, \epsilon_n)$ has a multivariate normal distribution with mean zero and covariance matrix $\Omega = \{\omega_{r,m}\}$ containing unknown parameters. The marginal expectation $E(D_i|X_i = x_i)$ is given by μ_i in (18.4.12) with τ replaced by $\sqrt{\omega_{ii}}$ and can be approximated as before. Computation of the marginal covariance of D_r and D_m requires

$$\mu_{r.m} = E(D_rD_m|X_r = x_r, X_m = x_m)$$
$$= E\{F^*(\eta_r + \epsilon_r)F^*(\eta_m + \epsilon_m)\}.$$

Substituting F_k^* for F^* above yields the approximation

$$\tilde{\mu}_{r.m} = \sum_{i=1}^{k}\sum_{j=1}^{k} p_{k.i}p_{k.j}E\{\Phi(\eta_r s_{k.i} + \epsilon_r s_{k.i})\Phi(\eta_m s_{k.j} + \epsilon_m s_{k.j})\}$$

$$= \sum_{i=1}^{k}\sum_{j=1}^{k} p_{k.i}p_{k.j} \times$$

$$\Phi_2\left\{\frac{s_{k.i}\eta_r}{\sqrt{1 + s_{k.i}^2\omega_{r.r}}}, \frac{s_{k.j}\eta_m}{\sqrt{1 + s_{k.j}^2\omega_{m.m}}}, \frac{\omega_{r.m}s_{k.i}s_{k.j}}{\sqrt{(1 + s_{k.i}^2\omega_{r.r})(1 + s_{k.j}^2\omega_{m.m})}}\right\},$$

where $\Phi_2(\cdot, \cdot, \rho)$ is the bivariate standard normal distribution function with correlation ρ.

In some statistical software packages $\Phi_2(\cdot, \cdot, \cdot)$ is an intrinsically defined function and with these programs quasilikelihood/variance-function estimation for (18.4.15) using $\tilde{\mu}_{r.m}$ may be feasible when not too many of the $\omega_{r.m}$ are nonzero.

We now discuss the basic approximation in (18.4.2), describe the methods used to find $\{p_{k.i}, s_{k.i}\}_{i=1}^{k}$ minimizing Δ_k^*, and discuss the error in the approximation of $G(\cdot, \cdot)$ with $G_k(\cdot, \cdot)$.

The *least-maximum approximant* F_k^* minimizes $\Delta_k^* = \sup_z |\Delta_k(z)|$, where $\Delta_k(z) = F^*(z) - F_k^*(z)$ subject to the parametric constraints on F_k^*. Since $\Delta_k(z)$ is skew-symmetric, Δ_k^* is determined by the restriction of Δ_k to $[0, \infty)$.

Least-maximum approximations form the basis for computing all non-arithmetic functions wherein the approximant is usually a polynomial or rational function. In the case of a polynomial of degree m, the solution to the optimization problem is characterized by Chebyshev's alternation theorem (Hart et al., 1968), p. 45, which states that the difference function must have $m + 1$ roots and $m + 2$ local extrema, equal in magnitude and alternating in sign.

For our problem the approximant F_k^* is not polynomial nor is it linear in the $m = 2k - 1$ parameters $\{(p_{k,i}, s_{k,i})_{i=1}^k\colon 0 \le s_{k,i}; 0 \le p_{k,i} \le 1, \Sigma_{i=1}^k p_{k,i} = 1\}$. Chebyshev's theorem does not apply, but we conjecture that the least-maximum approximant has local extrema equal in magnitude and alternating in sign as in the polynomial case. Thus we solved for approximants F_k^* having difference functions Δ_k possessing m roots in addition to the constrained roots at 0 and ∞, and $m + 1$ local sign-alternating extrema.

We adapted Remes' second algorithm (Hart et al., 1968), p. 45 for finding polynomial least-maximum approximants in our problem. Given initial values $\{p_{k,i}^{(0)}, s_{k,i}^{(0)}\}_{i=1}^k$ such that the initial difference function $\Delta_k^{(0)}$ has $m + 1$ sign-alternating extrema not necessarily of equal magnitude, m roots of $\Delta_k^{(0)}$, $z_i^{(0)}$ $(i = 1, \ldots, m)$ were found. Then, letting $z_0^{(0)} = 0$ and $z_{m+1}^{(0)} = z^*$ for some large z^*, the $m + 1$ extrema $x_i^{(0)}$ of $\Delta_k^{(0)}$ in the intervals (z_{i-1}, z_i) $(i = 1, \ldots, m + 1)$ were found. Updated parameters $\{p_{k,i}^{(1)}, s_{k,i}^{(1)}\}_{i=1}^k$ were obtained by solving the $m + 1$ nonlinear equations

$$\Delta_k^{(0)}(x_i^0) = (-1)^{i+1}\Delta \qquad (i =, \ldots, m + 1), \qquad (18.4.16)$$

as functions of the $m + 1$ arguments $\{p_{k,i}, s_{k,i}\}_{i=1}^k$ and Δ. This three-step procedure was iterated until convergence. At the stationary point, Δ_k had the requisite number of sign-alternating, equal-magnitude extrema, giving us reason to believe that the least-maximum approximant on $[0, z^*)$ was found. Since for z^* sufficiently large, Δ_k is negative and monotonically increasing to zero, the restriction to $[0, z^*)$ is not binding and our error estimates apply to $[0, \infty)$.

The parameters of the approximations, and the maximum absolute differences Δ_k^* for $k = 1, \ldots, 8$ are given in Table 18.4.1. Note that $s_{1,1} = 0.58763$ is slightly smaller than the logistic-normal scale correction given by Cox (1970), p. 28 and agrees closely with the factor $16\sqrt{3}/15\pi = 0.58808$ given by Johnson and Kotz (1970), p. 6.

Figure 18.4.1 displays a graph of the functions

$$J_k(t) = \frac{\Delta_k(t)}{\Delta_k^*} + 2k - 1 \qquad (k = 1, \ldots, 8), \qquad (18.4.17)$$

illustrating the oscillations in $\Delta_k(\cdot)$ $(k = 1, \ldots, 8)$.

Two aspects of our problem complicated the implementation of Remes' second algorithm: (1) finding initial parameter values; and (2) solving (18.4.16). For $k < 3$, suitable starting parameters were found by trial and error. For larger k starting parameters were found using a nonlinear least-squares routine to minimize a discrete approximation to the L_{2r} norm of Δ_k. Acceptable starting values were obtained with $r = 1$; however, better starting values were found for $r = 2, 3, 4$, a consequence of the approximate equivalence of the sup norm and L_{2r} norm for r large.

In the usual applications of the algorithm, the approximant is either a polynomial or a rational function in which cases the equations in (18.4.16) are linear or can be replaced with a linear system in all of the parameters except Δ (Cody, Fraser, and Hart, 1968). In our application the equations are nonlinear in all the parameters, thereby complicating this step of the algorithm.

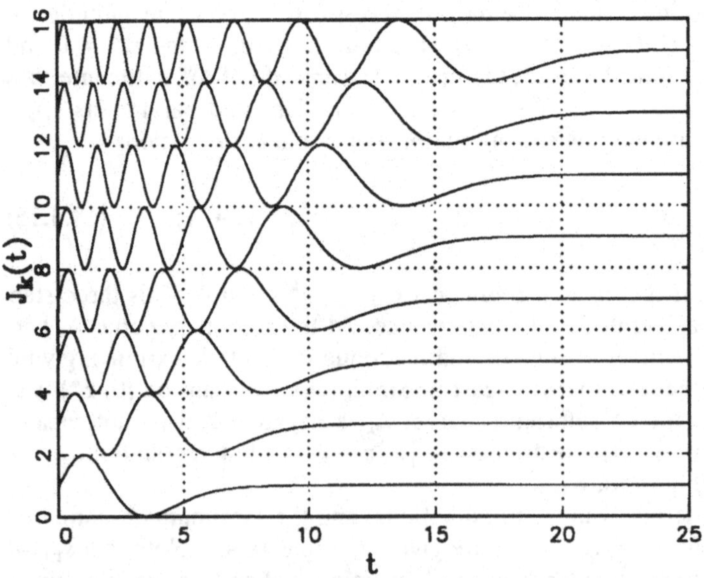

Fig. 18.4.1 Graph of the scaled and shifted difference functions in (18.4.17) showing the equal-magnitude sign-alternating behavior of the local extrema.

We now compare the accuracy of our approximation $G_k(\cdot, \cdot)$ to 20-point Gauss-Hermite quadrature and to two additional approximations that have been proposed in the measurement-error modeling literature for cases in which τ is small.

When applied to the logistic regression measurement error model, the general approximations given by Fuller (1987), p. 262 yield the approximant

$$G_{TS}(\eta, \tau) = F^*(\eta) + \frac{\tau^2}{2} F^{*(2)}(\eta). \tag{18.4.18}$$

Carroll and Stefanski (1990) proposed a range-preserving modification to the above approximation:

$$G_{RPTS}(\eta, \tau) = F^*\left\{\eta + \frac{\tau^2 F^{*(2)}(\eta)}{2F^{*(1)}(\eta)}\right\}. \tag{18.4.19}$$

The approximations in (18.4.18) and (18.4.19) are compared to those in (18.4.2) in Fig. 18.4.2. Here we have graphed the functions

$$\Delta_A^{**}(\tau) = \log_{10}\left\{\sup_\eta |G(\eta, \tau) - G_A(\eta, \tau)|\right\}$$

for the various approximations over the range $0.01 < \tau < 5$. The eight nearly linear graphs correspond to our approximations in (18.4.2), $k = 1, \ldots, 8$, top to bottom; the approximations in (18.4.18) and (18.4.19) are drawn with dashed lines; the remaining curve (dot-dash) corresponds to the approximation obtained using 20-point Gauss-Hermite quadrature. The "exact" evaluation of $G(\cdot, \cdot)$ was obtained using the method proposed by Crouch and Spiegelman (1990) with accuracy set at 10^{-15} and also by an adaptive Simpson's rule with convergence criterion set at 10^{-15}. The latter two differed by less than 10^{-14} across the range of τ values. The two Taylor series approximations are exact at $\tau = 0$ and are thus more accurate for very small τ; however, they quickly breakdown at τ increases from zero. It is noteworthy that the approximations $G_k(\eta, \tau)$ *improve* as τ increases.

The logistic normal integral (18.4.1) arises frequently in regression models for binary response data. Among the common methods of computing $G(\eta, \tau)$ none are easily programmed in standard statistical software packages and this limits their usefulness. This criticism applies to Gauss-Hermite quadrature, Crouch and Simpson's (1990) method, as well as to most other standard numerical methods.

Our approximations, obtained by exploiting the standard normal distribution function, are tailor-made for programming in statistical software

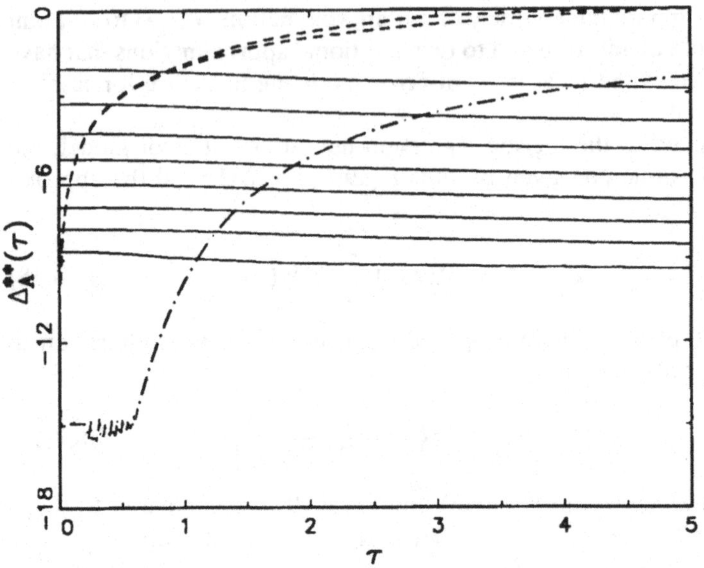

Fig. 18.4.2 Graph of the $\log_{10}$ error functions $\Delta_A^{**}(\tau)$. Solid lines, G_k, $k = 1, \ldots, 8$; dashed lines, Taylor series approximations; dot-dashed line, 20-point Gauss-Hermite quadrature.

packages. Note that numerical estimation and large-sample inference in the logistic regression models described above involves evaluation of not only the integral in (18.4.1) but also the first- and possibly second-order partial derivatives of (18.4.1) with respect to η and τ. The approximations in (18.4.4) have the advantage that all the required derivatives have closed-form expressions.

We have used the approximant $G_k(\eta, \tau)$ to fit logistic regression measurement error models to data using iteratively reweighted nonlinear least squares. This method is easily implemented in most major statistical software packages. For many applications a third- or fourth-order approximation ($k = 3$ or 4) will provide adequate accuracy as well as reasonable computational speed.

REFERENCES

Armstrong, B. (1985). Measurement error in the generalized linear model, *Communications in Statistics B*, *14*, 529–544.

Carroll, R. J., Spiegelman, C., Bailey, K., Lan, K. K. G., and Abbott, R. D. (1984). On errors-in-variables for binary regression models, *Biometrika*, *71*, 19–26.

Carroll, R. J. and Stefanski, L. A. (1990). Approximate quasilikelihood estimation in measurement error models, *Journal of the American Statistical Association*, *85*, 652–663.

Cody, W. J., Fraser, W., and Hart, J. F. (1968). Rational Chebyshev approximations using linear equations, *Numerische Mathematik*, *12*, 242–251.

Cox, D. R. (1970). *Binary Regression*, Chapman and Hall, London.

Crouch, A. and Spiegelman, D. (1990). The evaluation of integrals of the form $\int f(t)e^{-t^2}\,dt$. Application to logistic-normal models, *Journal of the American Statistical Association*, *85*, 464–469.

Fuller, W. A. (1987). *Measurement Error Models*, Wiley, New York.

Gradshteyn, I. S. and Ryzhik, I. W. (1965). *Tables of Integrals, Series and Products*, Academic Press, New York.

Hart, J. F., Cheney, E. W., Lawson, C. L., Maehley, H. J., Mesztenyi, C. K., Rice, J. R., Thatcher, H. G., and Witzgall, C. (1968), *Computer Approximations*, Wiley, New York.

Johnson, N. L. and Kotz, S. (1970). *Distributions in Statistics, Continuous Univariate Distributions*, Vol. 2, Boston, Houghton-Mifflin.

McCullagh, P. (1989). Binary data, dispersion effects and the sexual behaviour of Appalachian salamanders, *Technical Report No. 227*, Department of Statistics, University of Chicago.

Ochi, Y. and Prentice, R. L. (1984). Likelihood inference in a correlated probit regression model, *Biometrika*, *71*, 531–543.

Pierce, D. A. and Sands, B. R. (1975). Extra-binomial variation in binary data. *Technical Report No. 46*, Department of Statistics, Oregon State University.

Prentice, R. L. (1986). Binary regression using an extended beta-binomial distribution, with discussion of correlation induced by covariate measurement errors, *Journal of the American Statistical Association*, *394*, 321–327.

Prentice, R. L. (1988). Correlated binary regression with covariates specific to each binary observation, *Biometrics*, *44*, 1033–1048.

Schafer, D. (1987). Covariate measurement error in the generalized linear model, *Biometrika*, *74*, 385–391.

Spiegelman, D. (1988). Design and analysis strategies for epidemiologic research when the exposure variable is measured with error, unpublished Ph.D. thesis, Department of Biostatistics, Harvard University.

Stefanski, L. A. and Carroll, R. J. (1985). Covariate measurement error in logistic regression, *Annals of Statistics*, *13*, 1335–1351.

Stefanski, L. A. (1990). A normal scale mixture representation of the logistic distribution, *Statistics and Probability Letters*, *11*, 69–70.

Stiratelli, R., Laird, N. M., and Ware, J. H. (1984). Random-effects models for serial observations with binary response, *Biometrics*, *40*, 961–971.

Williams, D. A. (1982). Extra-binomial variation in logistic linear models, *Applied Statistics*, *31*, 144–148.

Zeger, S. L., Liang, K. Y., and Albert, P. S. (1988). Models for longitudinal data: A generalized estimating equation approach, *Biometrics*, *44*, 1049–1060.

18.5 SOME LOGISTIC GROWTH MODELS IN ECOLOGY

P. R. Parthasarathy and B. Krishna Kumar
Indian Institute of Technology, Madras, India

18.5.1 Introduction

Many phenomena in biology and medicine display regularities which can be described mathematically. Mathematical models are constructed not only to describe and explain quantitative relationships, but also to predict aspects of the phenomena which are not yet observed or even observable.

The density-dependent population dynamics describing population growth in a limited environment has been the subject of much theoretical and experimental investigation. In many ecological problems such as animal and cell populations, epidemics, plant tissues and learning processes, chemical reactions, and competition between species growth patterns are influenced by population size. Such populations do not increase indefinitely, but are limited by, for example, lack of food, cannibalism, and overcrowding. There are many deterministic models which describe density-dependent logistic population growth. A comprehensive treatment can be found in Hallam (1986) and in Murray (1989). Some interesting applications are developed in this volume by Tan; see Chapter 15.

In the following section, we outline the developments in the logistic equation and its generalizations. Its usefulness is illustrated in the field of tumor growth. Stochastic analogues are discussed. These are compared with mean-population-size-dependent models. The effects of random influences on the growth of populations are indicated.

18.5.2 Logistic Equation

In situations where there is complete overlap between generations the population changes in a continuous manner. A continuous-time model for a single species is of the form

$$\frac{dN(t)}{dt} = r(N(t))N(t), \tag{18.5.2.1}$$

where $N(t)$ is the number of individuals at time t and r, the intrinsic growth rate, is the excess of births over deaths as a function of the number of individuals. When the intrinsic growth rate is a constant, the population grows (or shrinks) exponentially. This simple model is associated with Malthus, who was interested in the ability of humans to reproduce faster than evolving technology could support.

A population model which represents the intrinsic rate of growth as a linearly decreasing function of the population size is

$$\frac{dN(t)}{dt} = rN(t)\left(1 - \frac{N(t)}{K}\right). \tag{18.5.2.2}$$

This is the Verhulst logistic equation, which has a global equilibrium point at $N(t) = K$, the carrying capacity of the environment. This process characterizes to a close approximation, the growth of many populations of organisms (bacterial, animal, human, etc.).

The experimental growth data for many embryos, organs and tumours fit more accurately the growth described by the Gompertz equation

$$\frac{dN(t)}{dt} = rN(t)\log\left(\frac{K}{N(t)}\right). \tag{18.5.2.3}$$

Here K is the saturation level. The differential equation

$$\frac{dN(t)}{dt} = rN^b, \qquad 0 < b < 1, \tag{18.5.2.4}$$

also fits well with certain tumor systems.

18.5.3 Generalizations

Allowing r and/or K in the logistic equation (18.5.2.2) to be time-dependent is a way of modeling environmental fluctuations, changes in the growth rates and other effects of perturbation of the system. Hallam and Clarke (1981) have examined the biological consequences of the application of this equation in a completely deterministic environment. Rosen (1984) has given a set of necessary and sufficient conditions for a biological population to be governed by this equation.

Another generalized form of the logistic equation is

$$\frac{dN(t)}{dt} = rN(t)\left\{1 - \left(\frac{N(t)}{K}\right)^{\theta}\right\}, \tag{18.5.3.1}$$

where θ represents the asymmetry of the growth curve. When $\theta = 1$, the equation is logistic and the growth curve is symmetrical about $N = K/2$. For $\theta < 1$, the maximum growth occurs for N less than $K/2$ but greater than K/e (see Gilpin et al., 1976). Rosen (1985) demonstrates that the evolutionary tendency for θ is to decrease if associated changes in r and K are also featured.

Turner (1969) has described the logistic function in which the maximum population size is allowed to vary. Getz (1984) has assumed that the per capita population growth rate is dependent on the per capita availability of the resources. Webb (1986) considers a general model of age-structured population dynamics with logistic type nonlinearity.

There has been considerable interest attached to population growth equations with built-in developmental time delays as exemplified by the logistic equation

$$\frac{dN(t)}{dt} = rN(t)\left\{1 - \frac{N(t - T)}{K(t)}\right\} \tag{18.5.3.2}$$

(May, 1980). In a real ecosystem with self-renewing resources, the actual level of the resources available to population $N(t)$ at time t will depend in the Verhulst fashion $N(t - T)$, where T is the "developmental time" associated with species maturation to full consumption. If T is large compared to r^{-1}, this equation leads to divergent oscillations and species extinction. It leads to a limit cycle if $\lambda T > \pi/2$. Rosen (1987) brings out the connection between this model and the θ-model described earlier. Gurney et al. (1980) have given a simple time delay model which explains the appearance of narrow "discrete" generations in cycling populations. Cohen and Rosenblat (1982) consider a logistic equation with distributed delay where the growth rate oscillates sinusoidally about a positive mean value.

In harvesting models, an exogenous force removes members of a population at some specified rate. Such a population is governed by the equation

$$\frac{dN(t)}{dt} = N(t)(a - bN(t)) - h. \tag{18.5.3.3}$$

These models have played an important role in the management of renewable resources. Some interesting applications are developed by Conrad (1986).

The main limiting features of the budworm population are food supply and the effect of parasites and predators. Ludwig et al. (1978) have chosen the logistic form to describe the former. The effect of predation is included by subtracting a term from the right-hand side of the logistic equation. This results in an S-shaped functional response.

A number of discrete single species population models appearing in the literature are collected by Bellows (1981) and May (1983). The simplest nonlinear difference equation arising in population dynamics is given by

$$N(t+1) = N(t)\left\{1 + r\frac{1 - N(t)}{K}\right\}. \tag{18.5.3.4}$$

The behavior of $N(t)$ is determined by the magnitude of the parameter r and undergoes abrupt changes at so-called critical values of r. May describes the result of the dynamical behavior of this model as chaos where $r > 2.828$ . . . . A detailed discussion is found in Nussey (1984).

Cooke and Nussey (1987) study in a qualitative way discrete models with harvesting. Cooke et al. (1988) consider several harvest management policies.

18.5.4 Tumor Growth

The literature on molecular biology of cancer is growing explosively. Bühler (1970) has given a survey of the work that has been done in tumor research.

Moolgavkar (1986) discusses briefly some models of carcinogenesis that relate fundamental cellular processes to the epidemics of cancer in human populations. Most human tissues show a steady increase in size during childhood. Once adult size is reached, the epithelia of these tissues continue to shed and replenish themselves. The growth curve of such tissues can be reasonably well represented by a Gompertz curve. The growth of some tissues, such as sex organs, may be represented by a logistic curve. Certain tissues show a sudden burst of growth in early life followed by a greatly decreased rate of cell division in later life, as in lymphoid tissue or by virtually no cell division, as in neural tissue.

Hansen and Tier (1982) have derived a stochastic model with a Gompertz growth law as the deterministic part. Kendal (1985) proposes a model relating the growth of tumors to the degree of their cellular heterogeneity. Frenzen and Murray (1986) develop a biologically plausible maturity time cell kinetics model which evolves as a Gompertz form.

Kolker (1987) has distinguished two regions of growth, free growth and exerted growth. The former growth rate is constrained by some internal reaction, while in the latter external substrata level controls the growth rate. He has derived a piecewise linear model which is shown to be a good approximation, useful for some research and practical needs. Gyllenberg and Webb (1989) have employed quiescence as a mechanism to explain characteristic Gompertz-type growth curves. Tan and Singh (1990) have presented a mixed model of carcinogenesis and have derived the incidence function of the number of tumors.

18.5.5 Stochastic Analogues

In the last section several deterministic models were described. The underlying concept in these models is essentially the same. Namely, that given certain initial conditions and rates, it is possible to predict the process level at time t. This may be valid for large stable biological populations over relatively short periods of time. It is an open question whether this can have much descriptive or predictive value when related to small populations or to long periods of time. The view that stochastic models are better suited to describe the development of biological populations rather than their deterministic counterparts is gaining support.

A stochastic analogue of the logistic process was introduced by Prendville (1948) as an example of a regulated birth and death process in which the state space is confined to a preassigned strip $0 \le n_1 \le n \le n_2$, by a suitable dependence of the birth and death rates on the population size. Specifically,

$$b_n = \alpha\left(\frac{n_2}{n} - 1\right), \quad d_n = \beta\left(1 - \frac{n_1}{n}\right), \qquad n_1 \le n \le n_2 \tag{18.5.5.1}$$

where α and β are positive constants and $n_1 < n_2$ are positive integers. Standard procedure yields explicit expression for the probability generating function, $F(s, t)$ of the number of particles at time t. Its mean $m(t)$ is given by

$$m(t) = \frac{\alpha n_2 + \beta n_1}{\alpha + \beta} - \{(\alpha n_2 + \beta n_1) - j(\alpha + \beta)\}e^{-(\alpha+\beta)t}, \tag{18.5.5.2}$$

where $X(0) = j$. As $t \to \infty$, an equilibrium region asymptotically is reached by the population and

$$\lim_{t\to\infty} F(s, t) = \frac{s^{n_1}(\alpha s + \beta)^{n_2-n_1}}{(\alpha + \beta)^{n_2-n_1}}, \tag{18.5.5.3}$$

$$\lim_{t\to\infty} m(t) = \frac{\alpha n_2 + \beta n_1}{\alpha + \beta}. \tag{18.5.5.4}$$

Holgate (1967) and Morgan (1976) have used this model to study clustering-splitting processes. McQuarrie (1967) considers this process for reversible unimolecular chemical reactions.

Norden (1982) has taken the birth and death rates to be

$$b_n = \lambda n\left(1 - \frac{n}{N}\right), \qquad d_n = \mu n\left(1 + \beta\frac{n}{N}\right) \qquad (18.5.5.5)$$

when the population size is n. He shows that this process is the stochastic analogue of the deterministic logistic model

$$\frac{dX(t)}{dt} = \lambda X(t)\left(1 - \frac{X(t)}{N}\right) - \mu X(t)\left(1 + \frac{\beta X(t)}{N}\right). \qquad (18.5.5.6)$$

He investigated the distribution of the extinction times from both the numerical and theoretical standpoint. Extending these results, Kryscio and Lefevre (1989) obtain an approximation to extinction for the standard S-I-S epidemic model.

As we have seen in Section 18.5.4, the growth of many cancer cells is best described by Gompertz growth curves. Since the Gompertz function is bounded above, stochastic Gompertz growth can explain the unbounded growth. Tan (1986) shows that the stochastic Gompertz birth and death process is a special case of nonhomogeneous birth and death processes. Hsu and Wang (1986) propose a stochastic model for the adhesion of bacteria to a surface based on the linear rates of adsorption and desorption, namely,

$$b_n = K_1(N - n), \quad d_n = K_2 n, \qquad n = 0, 1, 2, 3, \ldots, N. \qquad (18.5.5.7)$$

The mean number $m(t)$ of the number of cells absorbed to the surface is given by

$$m(t) = N\frac{K_1}{K_1 + K_2}(1 - e^{-(K_1+K_2)t}). \qquad (18.5.5.8)$$

Brendel and Perelson (1987) have found the probability that n cells are absorbed to the surface.

Smith and Tuckwell (1974) have modeled the growth of nonsaturating and saturating populations using stochastic differential equations. An explicit solution of the population size in terms of the modified Bessel functions has been given by Parthasarathy (1988) assuming that the birth and death rates are inversely proportional to the population size. Depending on whether the birthrate is greater than, equal to, or less than the death rate, the mean population size grows linearly, tends to a constant or zero. A random number of immigrants has been introduced into this process whenever the population size reaches zero by Parthasarathy and Krishna Kumar (1990a).

18.5.6 Dependence on the Mean Population

A stochastic model usually regarded as the analogue of the logistic relationship is described in Pielou (1977). Pielou uses the birth and death rates

$$b_n = c - dn, \qquad d_n = e + fn \tag{18.5.6.1}$$

when the population size is n. Its mean behavior satisfies

$$\frac{dm(t)}{dt} < rm(t)\left[1 - \frac{m(t)}{K}\right], \tag{18.5.6.2}$$

which is not logistic. One way to overcome this is to choose the birth and death rates suitably to achieve a certain stipulated mean growth rate. It is natural to assume that the birth and death rates depend on the mean population size and *not* on the exact population size.

Tognetti and Winley (1980) have considered the birth and death rates

$$b\left(1 - \frac{r\alpha_1 m(t)}{bK}\right) \quad \text{and} \quad \left(1 + \frac{r\alpha_2 m(t)}{mK}\right),$$

with $r = b - m$. This model has mean logistic behavior provided $\alpha_1 + \alpha_2 = 1$. Such approximations have been used by Renshaw (1973) for interconnected populations, by Lewis (1975) for the parasitic disease bilharziasis and by Parthasarathy and Mayilsamy (1982) for a compartmental model.

The dependence of the birth and death rates on the mean population size $m(t)$ has been so chosen by Parthasarathy and Krishna Kumar (1990b) that $m(t)$ satisfies the logistic equations of Verhulst and Gompertz. Exact expressions for the probability generating functions and the nontrivial stationary distributions are obtained. For the deterministic case, Jensen (1975) has studied the logistic equation in which the instantaneous rate of increase is identified with birthrate rather on the difference in birth and death rates. An application to fish population is given. A stochastic version of the Gompertz model with a constant death rate and mean-population-size-dependent birthrate is considered by Parthasarathy and Krishna Kumar (1990c). Extinction probabilities are discussed.

18.5.7 Random Fluctuations

Stochastic models of population growth form a basis for environmental impact estimates. There has been much interest in the problems connected with the effects of random influences on the growth of populations, whose size $N(t)$ satisfies a stochastic differential equation of the form

$$dN(t) = KN(t)G\left(\frac{N(t)}{K}\right) dt + N(t)\, dW(t), \qquad (18.5.7.1)$$

where $G(\cdot)$ is a saturation function and $W(t)$ is a Wiener process with zero mean. Several stochastic treatments of the logistic process have appeared, some treating the intrinsic growth rate as a random process, others regarding the carrying capacity K as a fluctuating parameter or adding a noise term. Details and references can be found in Goel et al. (1971).

Feldman and Roughgarden (1975) have obtained the stationary solution of

$$\frac{dN(t)}{dt} = N(t)[K_0 + dW(t) - N(t)]. \qquad (18.5.7.2)$$

The time-dependent solution has been obtained by Prajneshu (1980). Ginzburg (1982) assumes that the growth rate $r(t)$ has the form

$$r(t) = r + \zeta\epsilon(t), \qquad (18.5.7.3)$$

where $\epsilon(t)$ is the standard white noise and ζ characterizes the amplitude of the fluctuations. Hanson and Ryan (1989) consider a nonlinear stochastic population model that has a logistic drift instead of exponential.

Various diffusion models to population growth have been proposed by suitably parameterizing first-order differential equations (see Kurtz, 1981). Tuckwell and Koziol (1984) have summarized different diffusion processes employed modeling logistic growth. Tier and Hanson (1981) have studied persistence using diffusion theory with logistic population models which include small random fluctuations due to demographic and environmental stochasticity.

This is a bird's-eye view of the various developments of the logistic process in the growth patterns that occur in ecology. Obviously, this is *not* exhaustive.

ACKNOWLEDGMENT

One of the authors (B.K.K.) thanks the Council of Scientific and Industrial Research, India for their financial support during this research.

REFERENCES

Bellows, T. S. (1981). The descriptive properties of some models for density dependence, *J. Animal Ecol.*, *50*, 139–156.

Brendel, V. and Perelson, A. S. (1987). A note on stochastic models for bacterial adhesion, *J. Theor. Biol.*, *126*, 247–249.

Bühler, W. J. (1970). Genesis, growth and therapy of turmours: Mathematical models, *Methods of Inf. Med.*, *9*, 53–57.

Cohen, D. S. and Rosenblat, S. (1982). A delay logistic equation with variable growth rate, *SIAM J. Appl. Math.*, *42*, 608–624.

Conrad, J. M. (1956). Bioeconomics and management of renewable resources, in *Mathematical Ecology, an Introduction*, T. G. Hallam and S. A. Levin, (eds.), *Biomathematics*, Vol. 17, 381–403, Springer-Verlag, Heidelberg.

Cooke, K. L. and Nussey, H. E. (1987). Analysis of the complicated dynamics of some harvesting models, *J. Math. Biol.*, *25*, 521–542.

Cooke, K. L., Elderkin, R., and Witten, M. (1988). Harvesting procedures with management policy in iterative density dependent population models, *Natural Resource Modelling*, *2*, 383–420.

Feldman, M. W. and Roughgarden, J. (1975). A populations stationary distribution and chance of extinction in a stochastic environment with remarks on the theory of species packing, *Theor Pop. Biol.*, *7*, 197–207.

Frenzen, C. L. and Murray, J. D. (1986). A cell kinetic justification for Gompertz equation, *SIAM J. Appl. Math.*, *46*, 614–629.

Getz, W. M. (1984). Population dynamics: a per capita resource approach, *J. Theor. Biol.*, *108*, 623–643.

Gilpin, M. E., Case, T. J. and Ayala, F. J. (1976). θ-selection, *Math. Biosci.*, *32*, 131–139.

Ginzburg, L. R., Slobokin, L. B., Johnson, K., and Bindman, A. G. (1982). Quasiextinction probability as a measure of impact on population growth, *Risk Analysis*, *2*, 171–181.

Goel, N. S., Maitra, S. C., and Montroll, E. W. (1971). *Nonlinear Models of Interacting Populations*, Academic Press, New York.

Gurney, W. S. C., Blytha, S. P., and Nisbet, R. M. (1980). Nicholson's blowflies revisited, *Nature*, *287*, 17–21.

Gyllenberg, M. and Webb, G. F. (1989). Quiescence as an explanation of Gompertzian tumour growth, *Helsinki Univ. Tech., Inst. Math. Res. Reports*, A 264.

Hallam, T. G. (1986). Population dynamics in a homogeneous environment, in *Mathematical Ecology, An Introduction*, T. G. Hallam and S. A. Levin, (eds.), *Biomathematics*, Vol. 17, 61–94, Springer-Verlag, Heidelberg.

Hallam, T. G. and Clarke, C. E. (1981). Non-autonomous logistic equations as models of populations in a deteriorating environment, *J. Theor. Biol.*, *93*, 303–311.

Hanson, F. B. and Tier, C. (1982). A stochastic model of tumour growth, *Math. Biosci.*, *61*, 73–100.

Hanson, F. B. and Ryan, D. (1989). Mean and quasideterministic equivalence for linear stochastic dynamics, *Math. Biosci.*, *93*, 1–14.

Holgate, P. (1967). The size of elephant herds, *Math. Gaz.*, *51*, 302–304.

Hsu, J. P. and Wang, H. H. (1986). A stochastic analysis of bacterial adhesion, *J. Theor. Biol.*, *119*, 435–444.

Jensen, A. L. (1975). Comparison of logistic equations for population growth, *Biometrics*, *31*, 853–862.

Kendal, W. S. (1985). Gompertzian growth as a consequence of tumour heterogeneity, *Math. Biosci.*, *73*, 103–107.

Kolker, Y. (1987). A piece-wise linear growth model: Comparison with competing forms in batch culture, *J. Math. Biol.*, *25*, 543–551.

Kryscio, R. J. and Lefevre, (1989). On the extinction of the S-I-S stochastic epidemic, *J. Appl. Prob.*, *27*, 685–694.

Kurtz, T. G. (1981). *Approximation of Population Processes*, SIAM, Philadelphia.

Lewis, T. (1975). A model for the parasitic disease bilharziasis, *Adv. Appl. Prob.*, *7*, 705–766.

Ludwig, D., Jones, D. D., and Holling, C. S. (1978). Qualitative analysis of insect outbreak systems: The spruce budworm and forest, *J. Animal Ecol.*, *47*, 315–332.

May, R. M. (1980). Mathematical models in whaling and fisheries management, in *Lectures on Math. in the Life Sci.*, *13*, 1–63, Amer. Math. Soc.

May, R. M. (1983). Nonlinear problems in ecology and resource management, in *Chaotic behaviour of deterministic systems*, G. Iooss, R. H. G. Hellman, R. Stora, (eds.), 513–563, North-Holland, Amsterdam.

McQuarrie, D. A. (1967). Stochastic approach to chemical kinetics, *J. Appl. Prob.*, *4*, 478.

Moolgavkar, S. H. (1986). Carcinogenesis modeling: from molecular biology to epidemiology, *Ann. Rev. Public Health*, *7*, 151–169.

Morgan, B. S. T. (1976). Stochastic models of grouping changes, *Adv. Appl. Prob.*, *8*, 30–57.

Murray, J. D. (1989). *Mathematical Biology*, Biomathematics, Vol. 19, Springer-Verlag, Heidelberg.

Norden, R. H. (1982). On the distribution of the time to extinction in the stochastic logistic population model, *Adv. Appl. Prob.*, *14*, 687–708.

Nussey, H. E. (1984). Complicated dynamical behaviour in discrete population models, *Nieuw Arch. Wisk.*, *2*, 43–81.

Parthasarathy, P. R. (1988). Density-dependent Markov branching processes, *Mathematical Ecology*, T. G. Hallam, L. J. Gross, and S. A. Levin, (eds.), 569–579, World Sci., Singapore.

Parthasarathy, P. R. and Krishna Kumar, B. (1990a). Density-dependent birth and death processes with state dependent immigration, To appear in *Int. J. Math. Comp. Mod.*

Parthasarathy, P. R. and Krishna Kumar, B. (1990b). Two stochastic analogues of the logistic process, To appear in *Ind. J. Pure Appl. Math.*

Parthasarathy, P. R. and Krishna Kumar, B. (1990c). A birth and death process with logistic mean population, To appear in *Comm. Stat., Theory and Methods*.

Parthasarathy, P. R. and Mayilswami, P. (1982). Stochastic compartmental models with birth, death, and immigration of particles, *Comm. Stat., Theory and Methods*, *11*, 1625–1642.

Pielou, E. C. (1977). *Mathematical Ecology*, Wiley, New York.

Prendville, B. J. (1949). Discussion of symposium on stochastic processes, *J. Roy. Stat. Soc.*, *B 11*, 273.

Prajneshu (1980). Time-dependent solution of the logistic model of population growth in random environment, *J. Appl. Prob.*, *17*, 1083–1086.

Renshaw, E. C. (1973). Interconnected population processes, *J. Appl. Prob.*, *10*, 1–14.

Rosen, G. (1984). Characterizing conditions for generalized Verhulst logistic growth of a biological population, *Bull. Math. Biol.*, *46*, 963–965.

Rosen, G. (1985). Parameter evolution in the θ-model, *Evolution*, *39*, 707–708.

Rosen, G. (1987). Time delays produced by essential nonlinearity in population growth models, *Bull. Math. Biol.*, *49*, 253–255.

Smith, C. E. and Tuckwell, H. C. (1974). Some stochastic growth processes, in *Mathematical Problems in Biology*, Lecture Notes in Biomathematics, Vol. 2, 211–224.

Tan, W. Y. (1986). A stochastic Gompertz birth and death process, *Stat. Prob. Lett.*, *4*, 25–28.

Tan, W. Y. and Singh, K. P. (1990). A mixed model of carcinogenesis with applications to retinblastoma, *Math. Biosci.*, *98*, 211–225.

Thisted, R. A. (1991). Assessing the effect of allergy medications: modeled for paired comparisons on ordered categories. Technical Report #307, Department of Statistics, The University of Chicago, Chicago, Illinois.

Tier, C. and Hanson, F. B. (1981). Persistence in density dependent stochastic populations, *Math. Biosci.*, *53*, 89–117.

Tognetti, K. and Winley, G. (1980). Stochastic growth models with logistic mean population. *J. Theor. Biol.*, *82*, 167–169.

Tuckwell, H. C. and Koziol, J. A. (1987). Logistic population growth under random dispersal, *Bull. Math. Biol.*, *49*, 495–506.

Turner, M. E. (1969). A generalization of the logistic law of growth, *Biometrics*, *25*, 577–579.

Webb, G. F. (1986). Logistic models of structured population growth, *Comp. Math. Appl.*, *A*, *12*, 527–539.

Bibliography

Abramowitz, M. and Stegun, I. A. (Eds.) (1965). *Handbook of Mathematical Functions with Formulas, Graphs, and Mathematical Tables.* Dover Publications, New York.

Agresti, A. (1984). *Analysis of Ordinal Categorical Data.* John Wiley & Sons, New York.

Ahuja, J. C. and Nash, S. W. (1967). The generalized Gompertz-Verhulst family of distributions. *Sankhyā, Ser. A, 29*, 141–156.

Aiache, Pierre, Beyssac, Prasad, and Skelly (1989). New results on an vitro model for the study of the influence of fatty meals on the bioavailability of Theophylline controlled-release formulations, *J. Pharm. Sci., 78*, 261–263.

Aitchison, J. and Shen, S. M. (1980). Logistic-normal distributions: Some properties and uses. *Biometrika, 67*, 261–272.

Alam, K. (1970). A two-sample procedure for selecting the population with the largest mean from *k* normal populations. *Ann. Inst. Statist. Math., 22*, 127–136.

Albert, A. and Anderson, J. A. (1984). On the existence of maximum likelihood estimates in logistic regression models. *Biometrika, 71*, 1–10.

Ali, M. Masoom and Umbach, D. (1989). Estimation of quantiles of symmetrically truncated logistic distribution using a few optimally selected order statistics. *J. Info. Opt. Sci., 10*(2), 303–307.

Ali, M. Masoom and Umbach, D. (1988). Tables of BLUE's for quantiles of symmetrically truncated logistic distribution. *Technical Report No. 81*, Department of Mathematical Sciences, Ball State University.

Ali, M. Masoom, Umbach, D., and Hassanein, K. M. (1981a). Estimation of quantiles of exponential and double exponential distributions based on two order statistics. *Commun. Statist.—Theor. Meth.*, *10*(*19*), 1921–1932.

Ali, M. Masoom, Umbach, D., and Hassanein, K. M. (1981b). Small sample quantile estimation of Pareto populations using two order statistics. *Aligarh J. Statist.*, *1*(*2*), 139–164.

Ali, M. Masoom, Umbach, D., and Saleh, A. K. Md. E. (1982). Small sample quantile estimation of the exponential distribution using optimal spacings. *Sankhyā*, *B 44*(*2*), 135–142.

Ali, M. Masoom, Umbach, D., Saleh, A. K. Md. E., and Hassanein, K. M. (1983). Estimating quantiles using optimally selected order statistics. *Commun. Statist.—Theor. Meth.*, *12*(*19*), 2261–2271.

Ali, M. M. and Khan, A. H. (1987). On order statistics from the log-logistic distribution. *J. Statist. Plan. Inf.*, *17*, 103–108.

Ali, M. M., Mikhail, N. N., and Haq, M. S. (1978). A class of bivariate distributions including the bivariate logistic. *J. Multivar. Anal.*, *8*, 405–412.

Anderson, J. A. (1972). Separate sample logistic discrimination. *Biometrika*, *59*, 19–35.

Anderson, J. A. (1973). Logistic discrimination with medical applications, In *Discriminant Analysis and Applications* (Ed., T. Cacoullos), 1–15, Academic Press, Boston.

Anderson, J. A. (1974). Diagnosis by logistic discriminant function: Further practical problems and results. *Appl. Stat.*, *23*, 397–404.

Anderson, J. A. and Richardson, S. C. (1979). Logistic discrimination and bias correction in maximum likelihood estimation. *Technometrics*, *21*, 71–78.

Anderson, T. W. and Darling, D. A. (1954). A test of goodness-of-fit. *J. Amer. Statist. Assoc.*, *49*, 765–769.

Anderson, J. A. and Pemberton, J. D. (1985). The grouped continuous model for multivariate ordered categorical variables and covariate adjustment. *Biometrics*, *41*, 875–885.

Andrews, D. F., Bickel, P. J., Hampel, F. R., Huber, P. J., Rogers, W. H., and Tukey, J. W. (1972). *Robust Estimators of Location*, Princeton University Press, Princeton, New Jersey.

Anscombe, F. J. (1956). On estimating binomial response relations, *Biometrika*, *43*, 461–464.

Antle, C. E. and Bain, L. J. (1969). A property of maximum likelihood estimators of location and scale parameters, *SIAM Review*, *11*, 251–253.

Antle, C., Klimko, L., and Harkness, W. (1970). Confidence intervals for the parameters of the logistic distribution. *Biometrika*, *57*, 397–402.

Armstrong, B. (1985). Measurement error in the generalized linear model. *Commun. Statist.—Simul. Comp.*, *B14*, 529–544.

Arnold, B. C. (1975). Multivariate exponential distributions based on hierarchical successive damage. *J. App. Probab.*, *12*, 142–147.

Arnold, B. C. (1983). *Pareto Distributions*, Int. Co-op. Pub. House, Baltimore.

Arnold, B. C. (1987). Bivariate distributions with Pareto conditionals. *Statist. Probab. Lett.*, *5*, 263–266.

Arnold, B. C. (1989). A logistic process constructed using geometric minimization, *Statist. Probab. Lett.*, *7*, 253–257.

Arnold, B. C. (1990). A flexible family of multivariate Pareto distributions, To appear in *J. Statist. Plan. Inf.*

Arnold, B. C. and Balakrishnan, N. (1989). *Relations, Bounds and Approximations for Order Statistics*, Lecture notes in Statistics No. 53, Springer-Verlag, New York.

Arnold, B. C. and Meeden, G. (1975). Characterization of distributions by sets of order statistics, *Ann. Statist.*, *3*, 754–758.

Arnold, B. C. and Robertson, C. A. (1989a). Autoregressive logistic processes, *J. App. Probab.*, *26*, 524–531.

Arnold, B. C. and Robertson, C. A. (1989b). Elliptically contoured distributions with logistic marginals. *Technical Report No. 180*, Department of Statistics, University of California, Riverside, California.

Ashford, J. R. and Sowden, R. D. (1970). Multivariate probit analysis. *Biometrics*, *26*, 535–546.

Azlarov, T. A. and Volodin, N. A. (1982). *Characterization Problems Associated with the Exponential Distribution* (in Russian), Fan, Tashkent. Translated into English in 1986, and published by Springer-Verlag, Berlin.

Bain, L. J. (1967). Joint tolerance regions for the exponential distribution. *IEEE Trans. in Reliab.*, *R-16*, 111–113.

Bain, L. J. (1978). *Statistical Analysis of Reliability and Life-testing Models—Theory and Practice*, Marcel Dekker, New York.

Bain, L. J., Eastman, J., and Engelhardt, M. E. (1973). A study of life-testing models and statistical analysis for the logistic distribution. *Aerospace Research Laboratories Report ARL 73-0009*, Air Force Systems Command, USAF, Wright Patterson AFB, Ohio.

Balakrishnan, N. (1985). Order statistics from the half logistic distribution, *J. Statist. Comp. Simul.*, *20*, 287–309.

Balakrishnan, N. (1988a). Relations and identities for the moments of order statistics from a sample containing a single outlier. *Commun. Statist.—Theor. Meth.*, *17*(7), 2173–2190.

Balakrishnan, N. (1988b). Recurrence relations among moments of order statistics from two related outlier models. *Biomet. J.*, *30*, 741–746.

Balakrishnan, N. (1988c). Recurrence relations for order statistics from n independent and non-identically distributed random variables. *Ann. Inst. Statist. Math.*, *40*, 273–277.

Balakrishnan, N. (1989a). Recurrence relations among moments of order statistics from two related sets of independent and non-identically distributed random variables. *Ann. Inst. Statist. Math.*, *41*, 323–329.

Balakrishnan, N. (1989b). A relation for the covariances of order statistics from n independent and non-identically distributed random variables. *Statistische Hefte*, *30*, 141–146.

Balakrishnan, N. (1990). Approximate maximum likelihood estimation for a generalized logistic distribution. *J. Statist. Plan. Inf.*, *26*, 221–236.

Balakrishnan, N. (1991). Best linear unbiased estimates of the location and scale parameters of logistic distribution for complete and censored samples of sizes 2(1) 25(5)40. *Submitted for publication*.

Balakrishnan, N. and Ambagaspitiya, R. S. (1988). Relationships among moments of order statistics in samples from two related outlier models and some applications. *Commun. Statist.—Theor. Meth.*, *17*(7), 2327–2341.

Balakrishnan, N. and Chan, P. S. (1991). Estimation for the scaled half logistic distribution under Type II censoring. To appear in *Comp. Stat. Data. Analy.*

Balakrishnan, N., Chan, P. S., Ho, K. L., and Lo, K. K. (1991). Means, variances, and covariances of logistic order statistics in the presence of an outlier. To appear in *Selected Tables in Mathematical Statistics*.

Balakrishnan, N. and Cohen, A. C. (1990). *Order Statistics and Inference: Estimation Methods*, Academic Press, Boston.

Balakrishnan, N. and Joshi, P. C. (1983a). Single and product moments of order statistics from symmetrically truncated logistic distribution, *Demon. Math.*, *16*, 833–841.

Balakrishnan, N. and Joshi, P. C. (1983b). Means, variances and covariances of order statistics from symmetrically truncated logistic distribution. *J. Statist. Res.*, *17*, 51–61.

Balakrishnan, N. and Kocherlakota, S. (1985). On the double Weibull distribution: Order statistics and estimation. *Sankhyā*, *Ser. B*, *47*, 161–178.

Balakrishnan, N. and Kocherlakota, S. (1986). On the moments of order statistics from the doubly truncated logistic distribution. *J. Statist. Plan. Inf.*, *13*, 117–129.

Balakrishnan, N. and Leung, M. Y. (1988a). Order statistics from the Type I generalized logistic distribution. *Commun. Statist.—Simul. Comp.*, *17*(*1*), 25–50.

Balakrishnan, N. and Leung, M. Y (1988b). Means, variances and covariances of order statistics, BLUE's for the Type I generalized logistic distribution, and some applications. *Commun. Statist.—Simul. Comp.*, *17*(*1*), 51–84.

Balakrishnan, N. and Malik, H. J. (1986). A note on moments of order statistics. *Amer. Statist.*, *40*, 147–148.

Balakrishnan, N. and Malik, H. J. (1987). Moments of order statistics from truncated log-logistic distribution. *J. Statist. Plann. Inf.*, *17*, 251–267.

Balakrishnan, N. and Malik, H. J. (1991). Means, variances and covariances of logistic order statistics for sample sizes up to fifty. To appear in *Selected Tables in Mathematical Statistics.*

Balakrishnan, N., Malik, H. J., and Puthenpura, S. (1987). Best linear unbiased estimation of location and scale parameters of the log-logistic distribution. *Commun. Statist.—Theor. Meth.*, *16*, 3477–3495.

Balakrishnan, N. and Puthenpura, S. (1986). Best linear unbiased estimators of location and scale parameters of the half logistic distribution. *J. Statist. Comput. Simul.*, *25*, 193–204.

Barāth and Rasch, D. (1989). A CADEMO "szakértöi rendszer" alkalmazasa növényinövekedésvizsgálatok modellezésére. *Növénytermelés*, *38*, 281–288.

Baringhaus, L. (1980). Eine simultane Charakterisierung ger geometrischen Verteilung und der logistchen Verteilung. *Metrika*, *27*, 237–242.

Barlow, R. E. and Gupta, S. S. (1969). Selection procedures for restricted families of distributions. *Ann. Math. Statist.*, *40*, 905–917.

Barlow, R. E., Gupta, S. S., and Panchapakesan, S. (1969). On the distribution of the maximum and minimum of ratios of order statistics. *Ann. Math. Statist.*, *40*, 918–934.

Barndorff-Nielsen, O. (1964). On the limit distribution of the maximum of a random number of independent random variables. *Acta Math. Acad. Sci. Hung.*, *15*, 399–403.

Barnett, V. (1975). Probability plotting methods and order statistics. *J. Roy. Statist. Soc.*, *Ser. C*, *24*, 95–108.

Barr, D. R. and Davidson, T. (1973). A Kolmogorov-Smirnov test for censored samples. *Technometrics*, *15*, 739–757.

Bartlett, M. S. (1935). Contingency table interactions. *J. Roy. Statist. Soc.*, *Suppl.*, *2*, 248–252.

Bechhofer, R. E. (1954). A single-sample multiple decision procedure for ranking means of normal populations with known variances. *Ann. Math. Statist.*, *25*, 16–39.

Bechhofer, R. E., Kiefer, J., and Sobel, M. (1968). *Sequential Identification and Ranking Procedures (with special reference to Koopman-Darmois populations)*, The University of Chicago Press, Chicago and London.

Bechhofer, R. E. and Kulkarni, R. (1982). Closed adaptive sequential procedures for selecting the best of $k \geq 2$ Bernoulli populations, In *Statistical Decision Theory and Related Topics—III*, *1*, (Eds., S. S. Gupta and J. O. Berger) 61–108, Academic Press, New York.

Begg, C. B. and Gray, R. (1984). Calculation of polychotomous logistic regression parameters using individualized regressions. *Biometrika*, *71*, 11–18.

Bellows, T. S. (1981). The descriptive properties of some models for density dependence. *J. Animal Ecol.*, *50*, 139–156.

Bennett, S. (1983). Log-logistic regression models for survival data. *Appl. Statist.*, *32*, 165–171.

Berkson, J. (1944). Application of the logistic function to bio-assay. *J. Amer. Statist. Assoc.*, *37*, 357–365.

Berkson, J. (1951). Why I prefer logits to probits. *Biometrics*, 7, 327–339.

Berkson, J. (1953). A statistically precise and relatively simple method of estimating the bio-assay with quantal response, based on the logistic function. *J. Amer. Statist. Assoc.*, *48*, 565–599.

Berkson, J. (1955). Maximum likelihood and minimum chi-square estimates of the logistic function. *J. Amer. Statist. Assoc.*, *50*, 130–162.

Berkson, J. (1956). Estimation by least squares and by maximum likelihood. *Proc. of the Third Berkeley Symp. on Math. Statist. and Probab.*, *1*, 1–11.

Berkson, J. (1957). Tables for the maximum likelihood estimate of the logistic function. *Biometrics*, *13*, 28–34.

Berkson, J. (1960). Nomograms for fitting the logistic function by maximum likelihood. *Biometrika*, *47*, 121–141.

Berkson, J. (1968). Application of minimum logit chi-squared to a problem of Grizzle with a notation on the problem of no interaction. *Biometrics*, *24*, 75–95.

Berkson, J. (1972). Minimum discrimination information, the 'no interaction' problem, and the logistic function. *Biometrics*, *28*, 443–468.

Bernardo, J. M. (1976). Psi (digamma) function, Algorithm AS 103. *Appl. Statist.*, *25*, 315–317.

Bhandari, S. K. and Chaudhuri, A. R. (1988). On two conjectures about two-stage selection problem. *Sankhyā*, *Series B*.

Billingsley, P. (1986). *Probability and Measure*, Second edition, John Wiley & Sons, New York.

Birnbaum, A. and Dudman, J. (1963). Logistic order statistics. *Ann. Math. Statist.*, *34*, 658–663.

Bliss, C. I. (1967). *Statistics in Biology, Vol. 1*, McGraw-Hill, New York.

Block, H. W. (1975). Physical models relating to multivariate exponential and negative binomial distributions. *Modeling and Simulation*, *6*, 445–450.

Blom, G. (1958). *Statistical Estimates and Transformed Beta Variables*, John Wiley & Sons, New York.

Bonney, G. E. (1986). Regressive logistic models for familial disease and other binary traits. *Biometrics*, *42*, 611–625.

Bouver, H. and Bargmann, R. E. (1974). Tables of the standardized percentage points of the Pearson systems in terms of β_1 and β_2. *THEMIS Technical Report, No. 32*, Department of Statistics, University of Georgia, Athens, Georgia.

Bowman, K. O. and Shenton, L. R. (1981). Explicit accurate approximations for fitting the parameters of L_U. In *Statistical Dist. in Scientific Work, Vol. 5*, 231–240, D. Reidel, Dordrecht.

Bowman, K. O. and Shenton, L. R. (1988). Solutions to Johnson's S_B and S_U. *Commun. Statist.—Simul. Comput.*, *17*, 343–348.

Bowman, K. O. and Shenton, L. R. (1989). S_B and S_U distributions fitted by percentiles. A general criterion. *Commun. Statist.—Simul. Comput.*, *18*, 1–13.

Box, G. E. P. and Lucas, H. L. (1959). Design of experiments in nonlinear situations. *Biometrika*, *46*, 77–90.

Brendel, V. and Perelson, A. S. (1987). A note on stochastic models for bacterial adhesion. *J. Theor. Biol.*, *126*, 247–249.

Breslow, N. and Powers, W. (1978). Are there two logistic regressions for retrospective studies? *Biometrics*, *34*, 100–105.

Buhler, W. J. (1970). Genesis, growth and therapy of tumours: Mathematical models. *Methods of Inf. Med.*, *9*, 53–57.

Bukač, J. (1972). Fitting S_B curves using symmetrical percentile points. *Biometrika*, *59*, 688–690.

Büringer, H., Martin, H., and Schriever, K.-H. (1980). *Nonparametric Sequential Procedures*, Birkhauser, Boston.

Burr, I. W. (1942). Cumulative frequency functions. *Ann. Math. Statist.*, *13*, 215–232.

Carroll, R. J., Spiegelman, C., Bailey, K., Lan, K. K. G., and Abbott, R. D. (1984). On errors-in-variables for binary regression models. *Biometrika*, *71*, 19–26.

Carroll, R. J. and Stefanski, L. A. (1990). Approximate quasilikelihood estimation in measurement error models. To appear in *J. Amer. Statist. Assoc.*

Chan, L. K. (1969). Linear quantile estimates of the location and scale parameters of the logistic distribution. *Statistische Hefte*, *10*, 277–282.

Chan, L. K., Chan, N. N., and Mead, E. R. (1971). Best linear unbiased estimates of the parameters of the logistic distribution based on selected order statistics. *J. Amer. Statist. Assoc.*, *66*, 889–892.

Chan, L. K., Chan, N. N., and Mead, E. R. (1973). Tables for the best linear unbiased estimate based on selected order statistics from the normal, logistic, Cauchy and double exponential distribution. *Math. Comp.*, *27*, 445–446.

Chan, L. K. and Cheng, S. W. (1972). Optimum spacing for the asymptotically best linear estimate of the location parameter of the logistic distribution when samples are complete or censored. *Statistische Hefte*, *13*, 41–57.

Chan, L. K. and Cheng, S. W. (1974). An algorithm for determining the asymptotically best linear estimate of the mean from multiply censored logistic data. *J. Amer. Statist. Assoc.*, *69*, 1027–1030.

Chan, L. K. and Cheng, S. W. (1982). The best linear unbiased estimates of parameters using order statistics. *Soochow J. Math. & Natural Sci.*, *8*, 1–13.

Chen, E. H. and Dixon, W. J. (1972). Estimates of parameters of a censored regression sample. *J. Amer. Statist. Assoc.*, *67*, 664–671.

Cheng, S. W. (1975). A unified approach to choosing optimum quantiles for the ABLE's. *J. Amer. Statist. Assoc.*, *70*, 155–159.

Chernoff, H. (1971). A note on optimal spacings for systematic statistics, *Technical Report, No. 70*, Department of Statistics, Stanford University.

Chernoff, H. and Lehmann, E. L. (1954). The use of maximum-likelihood estimates in χ^2 test for goodness of fit. *Ann. Math. Statist.*, *25*, 579–586.

Chew, V. (1968). Some useful alternatives to the normal distribution. *Amer. Statist.*, *22*, 22–24.

Clark, C. E. and Williams, G. T. (1958). Distribution of members of an ordered sample. *Ann. Math. Statist.*, *29*, 862–870.

Clayton, A. and Leslie, A. (1981). The bioavailability of Erythromycin stearate versus Enteric-coated Erythromycin base when taken immediately before and after food. *J. of Internal Med. Res.*, *9*, 470–477.

Clayton, D. G. (1974). Some odds ratio statistics for the analysis of ordered categorical data. *Biometrika*, *61*, 525–531.

Cody, W. J., Fraser, W., and Hart, J. F. (1968). Rational Chebyshev approximations using linear equations. *Numerische Mathematik*, *12*, 242–251.

Cohen, A. and Sackrowitz, H. B. (1975). Unbiasedness of the chi-square, likelihood ratio, and other goodness of fit tests for the equal cell case. *Ann. Statist.*, *4*, 959–964.

Cohen, D. S. (1959). *A Two-sample Decision Procedure for Ranking Means of Normal Populations with a Common Known Variance*. M.S. Thesis, Department of Operations Research, Cornell University, Ithaca, New York.

Cohen, D. S. and Rosenblat, S. (1982). A delay logistic equation with variable growth rate. *Siam J. Appl. Math.*, *42*, 608–624.

Conrad, J. M. (1956). Bioeconomics and management of renewable resources, In *Mathematical Ecology, an Introduction* (Eds., T. G. Hallam and S. A. Levin), 381–403, Lecture Notes in Biomathematics No. 17, Springer-Verlag, Heidelberg.

Cook, R. D. and Johnson, M. E. (1986). Generalized Burr-Pareto-logistic distributions with applications to a uranium exploration data set. *Technometrics*, *28*, 123–131.

Cook, R. D. and Nachtsheim, C. J. (1980). A comparison of algorithms for constructing exact D-optimal designs. *Technometrics*, *22*, 315–323.

Cook, R. D. and Weisberg, S. (1982). *Residuals and Influence in Regression*. Chapman and Hall, New York.

Cooke, K. L., Elderkin, R., and Witten, M. (1988). Harvesting procedures with management policy in iterative density dependent population models. *Natural Resource Modelling*, *2*, 383–420.

Cooke, K. L. and Nussey, H. E. (1987). Analysis of the complicated dynamics of some harvesting models. *J. Math. Biol.*, *25*, 521–542.

Copas, J. B. (1983). Plotting *p* against *x*. *Appl. Statist.*, *32*, 25–31.

Cornfield, J. and Mantel, N. (1950). Some new aspects of the application of maximum likelihood to the calculation of the dosage response curve. *J. Amer. Statist. Assoc.*, *45*, 181–210.

Count, E. W. (1943). Growth pattern of human physique: an approach to kinetic anthropometry. *Human Biol.*, *15*, 1–32.

Cox, D. R. (1966). Some procedures connected with the logistic qualitative response curve. In *Research Papers in Statistics* (Ed., F. N. David), 55–71, John Wiley & Sons, New York.

Cox, D. R. (1970). *The Analysis of Binary Data*. Methuen, London.

Cox, D. R. (1970). *Binary Regression*. Chapman and Hall, London.

Cox, D. R. and Hinkley, D. V. (1974). *Theoretical Statistics*. Chapman and Hall, London.

Cox, D. R. and Snell, E. J. (1968). A general definition of residuals (with discussion). *J. Roy. Statist. Soc., Ser. B, 30*, 248–275.

Cox, D. R. and Snell, E. J. (1989). *Analysis of Binary Data, Second edition.* Chapman and Hall, London.

Cran, G. W., Martin, K. J., and Thomas, G. E. (1977). A remark on algorithms, Algorithm AS 109. *Appl. Statist., 26*, 111–114.

Crouch, A. and Spiegelman, D. (1990). The evaluation of integrals of the form $\int f(t)e^{-t^2}\,dt$. Application to logistic-normal models, To appear in *J. Amer. Statist. Assoc.*

Crow, E. L. and Siddiqui, M. M. (1967). Robust estimation of location. *J. Amer. Statist. Assoc., 62*, 353–389.

Csorgo, S. and Horvath, L. (1981). On the Koziol-Green model for random censorship. *Biometrika, 68*, 391–401.

Cutler, C. D. (1991). Some results on the behavior and estimation of the fractal dimensions of distributions on attractors, *J. Statist. Phys., 62*, 651–708.

Cutler, C. D. and Dawson, D. A. (1989). Estimation of dimension for spatially-distributed data and related limit theorems. *J. Multivar. Analy., 28*, 115–148.

D'Agostino, R. B. and Lee, A. F. S. (1976). Linear estimation of the logistic parameters for complete or tail censored samples. *J. Amer. Statist. Assoc., 71*, 462–464.

D'Agostino, R. B. and Stephens, M. A. (1986). *Goodness-of-fit Techniques*. Marcel Dekker, New York.

Dahiya, R. C. and Gurland, J. (1972). Pearson chi-square test of fit with random intervals. *Biometrika, 59*, 147–153.

David, F. N. and Johnson, N. L. (1954). Statistical treatment of censored data. I. Fundamental formulae, *Biometrika, 41*, 228–240.

David, H. A. (1981). *Order Statistics, Second edition.* John Wiley & Sons, New York.

David, H. A., Kennedy, W. J., and Knight, R. D. (1977). Means, variances, and covariances of normal order statistics in the presence of an outlier. *Selected Tables in Mathematical Statistics, 5*, 75–204.

David, H. A. and Shu, V. S. (1978). Robustness of location estimators in the presence of an outlier. In *Contributions to Survey Sampling and Applied Statistics: Papers in Honor of H. O. Hartley* (Ed., H. A. David), 235–250. Academic Press, Boston.

Davidson, R. R. (1980). Some properties of a family of generalized logistic distributions. In *Statistical Climatology, Developments in Atmospheric Science, 13* (Ed., S. Ikeda et al.). Elsevier Science Publishing Company, Amsterdam.

Davis, D. J. (1952). An analysis of some failure data. *J. Amer. Statist. Assoc.*, *47*, 113–150.

Davis, H. T. (1935). *Tables of the Higher Mathematical Functions*, Vols. 1 and 2, Principia Press, Bloomington.

Davis, H. T. (1941). *The Analysis of Economic Time Series*. Principia Press, Bloomington.

Davis, L. J. (1985). Modification of the empirical logit to reduce bias in simple linear logistic regression. *Biometrika*, *72*, 199–202.

Day, N. E. and Kerridge, D. F. (1967). A general maximum likelihood discriminant. *Biometrics*, *23*, 313–323.

De Haan, L. (1975). *On Regular Variation and Its Application to Weak Convergence of Sample Extremes*, Mathematical Center Tracts, 32, (Third edition). Amsterdam.

Devroye, L. (1986). *Non-uniform Random Variate Generation*. Springer-Verlag, New York.

Dixon, W. J. (1957). Estimates of the mean and standard deviation of a normal population. *Ann. Math. Statist.*, *28*, 806–809.

Dixon, W. J. (1960). Simplified estimation from censored normal samples. *Ann. Math. Statist.*, *31*, 385–391.

Dixon, W. J. and Tukey, J. W. (1968). Approximate behavior of the distribution of Winsorized *t* (trimming/Winsorization 2). *Technometrics*, *10*, 83–98.

Downton, F. (1966a). Linear estimates with polynomial coefficients. *Biometrika*, *53*, 129–141.

Downton, F. (1966b). Linear estimates of parameters in the extreme value distribution. *Technometrics*, *8*, 3–17.

Dubey, S. D. (1966). Transformations for estimation of parameters. *J. Indian Statist. Assoc.*, *4*, 109–124.

Dubey, S. D. (1969). A new derivation of the logistic distribution. *Naval Res. Logist. Quart.*, *16*, 37–40.

Duchrau, P., Frischmuth, K., Rasch, D., and Schimke, E. (1986). Optimum experimental design in growth curve analysis. *Rostock Math. Kolloq.*, *30*, 93–104.

Duchrau, P. and Frischmuth, K. (1990). A numerical search algorithm for experimental design. To appear in *Rostock Math. Kolloq.*

Dudewicz, E. J. and Koo, J. O. (1982). *The Complete Categorized Guide to Statistical Selection and Ranking Procedures*, Series in Mathematical and Management Sciences, Vol. 6. American Sciences Press, Columbus, Ohio.

Dufour, R. and Maag, U. R. (1978). Distribution results for modified Kolmogorov-Smirnov statistics for truncated or censored data. *Technometrics*, *20*, 29–32.

Dumonceaux, R. H. (1969). *Statistical inferences for location and scale parameter distributions*, Doctoral Thesis, University of Missouri-Rolla, Rolla, Missouri, U.S.A.

Durbin, J. (1973). Weak convergence of the sample distribution function when parameters are estimated. *Ann. Statist.*, *1*, 279–290.

Dyke, G. V. and Patterson, H. D. (1952). Analysis of factorial arrangements when the data are proportions. *Biometrics*, *8*, 1–12.

Eastman, J. A. (1972). *Statistical studies of various time-to-fail distributions*. Doctoral Thesis, University of Missouri-Rolla, Rolla, Missouri, U.S.A.

Eisen, M. (1979). *Mathematical Models in Cell Biology and Cancer Chemotherapy*. Lecture Notes in Biomathematics No. 3. Springer-Verlag, Berlin and New York.

Emmens, C. W. (1941). The dose-response relation for certain principles of the pituitary gland, and of the serum and urine of pregnancy. *J. of Endocrinology*, *2*, 194–225.

Engelhardt, M. (1975). Simple linear estimation of the parameters of the logistic distribution from a complete or censored sample. *J. Amer. Statist. Assoc.*, *70*, 899–902.

Engelhardt, M. E., Bain, L. J., and Smith, R. M. (1974). A further study of life-testing models and simple estimates for the logistic distribution. *Aerospace Research Laboratories Report*, *ARL 74-008*. Air Force Systems Command, USAF, Wright Patterson AFB, Ohio.

Falconer, K. J. (1985). *The Geometry of Fractal Sets*. Cambridge University Press, London and New York.

Farewell, V. T. and Prentice, R. L. (1977). A study of distributional shape in life testing. *Technometrics*, *19*, 69–76.

Feldman, M. W. and Roughgarden, J. (1975). A population's stationary distribution and chance of extinction in a stochastic environment with remarks on the theory of species packing. *Theor. Popul. Biol.*, *7*, 197–207.

Feller, W. (1940). On the logistic law of growth and its empirical verification in biology. *Acta Biomathematica*, *5*, 51–66.

Finney, D. J. (1947). The principles of biological assay. *J. Roy. Statist. Soc.*, *Ser. B*, *9*, 46–91.

Finney, D. J. (1952). *Statistical Methods in Biological Assay*, Hafner Publications, New York.

Fisher, R. A. (1924). The conditions under which χ^2 measures the discrepancy between observation and hypothesis. *J. Roy. Statist. Soc.*, *87*, 442–450.

Fisher, R. A. (1925). Theory of statistical estimation. *Proc. of Camb. Philos. Soc.*, *22*, 700–725.

Fisk, P. R. (1961). The graduation of income distributions. *Econometrica*, *29*, 171–185.

Fleishman, A. I. (1978). A method for simulating non-normal distributions. *Psychometrika*, *43*, 521–532.

Formann, A. K. (1982). Linear logistic latent class analysis. *Biomet. J.*, *24*, 171–190.

Forsythe, G. E., Malcolm, M. A., and Moler, C. B. (1977). *Computer Methods for Mathematical Computations*. Prentice-Hall, Englewood Cliffs.

Frenzen, C. L. and Murray, J. D. (1986) A cell kinetic justification for Gompertz equation. *Siam J. Appli. Math.*, *46*, 614–629.

Fuller, W. A. (1987). *Measurement Error Models*. John Wiley & Sons, New York.

Galambos, J. (1975). Characterizations of probability distributions by properties of order statistics I. *Statistical Distr. in Scientific Work*, Vol. 3, 71–88, D. Reidel, Dordrecht.

Galambos, J. (1976). A remark on the asymptotic theory of sums with random size. *Math. Proc. Cambridge Philos. Soc.*, *79*, 531–532.

Galambos, J. (1987). *The Asymptotic Theory of Extreme Order Statistics, Second edition*. Krieger, Melbourne, Florida.

Galambos, J. (1988). *Advanced Probability Theory*. Marcel Dekker, New York.

Galambos, J. and Kotz, S. (1978). *Characterizations of Probability Distributions*, Lecture Notes in Mathematics, No. 675. Springer-Verlag, Berlin.

Gart, J. J., Pettigrew, H. M., and Thomas, D. G. (1985). The effect of bias, variance estimation, skewness and kurtosis of the empirical logit on weighted least squares analyses. *Biometrika*, *72*, 179–190.

Gastwirth, J. L. and Cohen, M. L. (1970). Small sample behavior of robust linear estimators of location. *J. Amer. Statist. Assoc.*, *65*, 946–973.

Genest, C. and Mackay, J. (1986). The joy of copulas: bivariate distributions with uniform marginals. *Amer. Statist.*, *40*, 280–283.

George, E. O., El-Saidi, M., and Singh, K. (1986). A generalized logistic approximation of the Student *t* distribution. *Commun. Statist.—Simul. Comput.*, *15*, 1199–1208.

George, E. O. and Mudholkar, G. S. (1981). A characterization of the logistic distribution by a sample median. *Ann. Inst. Statist. Math.*, *33*, 125–129.

George, E. O. and Mudholkar, G. S. (1981). Some relationships between the logistic and the exponential distributions. *Statistical Distr. in Scientific Work*, *Vol. 4*, 401–409. D. Reidel, Dordrecht.

George, E. O. and Mudholkar, G. S. (1982). On the logistic and exponential laws. *Sankhyā*, *Ser. A*, *44*, 291–293.

George, E. O. and Ojo, M. O. (1980). On a generalization of the logistic distribution. *Ann. Inst. Statist. Math.*, *32*, 161–169.

George, E. O. and Rousseau, C. C. (1987). On the logistic midrange. *Ann. Inst. Statist. Math.*, *39*, 627–635.

George, E. O. and Rousseau, C. C. (1990). On the asymptotics of range and midrange. *Submitted for publication*.

George, E. O. and Singh, K. (1987). An approximation of *F* distribution by binomial probabilities. *Statist. Probab. Lett.*, *5*, 169–173.

Getz, W. M. (1984). Population dynamics: a per capita resource approach. *J. Theor. Biol.*, *108*, 623–643.

Gibbons, J. D., Olkin, I., and Sobel, M. (1977). *Selecting and Ordering Populations: A New Statistical Methodology*. John Wiley & Sons, New York.

Gilpin, M. E., Case, T. J., and Ayala, F. J. (1976). θ-selection. *Math. Biosci.*, *32*, 131–139.

Ginzburg, L. R., Slobokin, L. B., Johnson, K., and Bindman, A. G. (1982). Quasiextinction probability as a measure of impact on population growth. *Risk Analysis 2*, 171–181.

Glasbey, C. A. (1979). Correlated residuals in non-linear regression applied to growth data. *Appl. Statist.*, *28*, 251–259.

Glaz, J. (1979). Probabilities and moments for absorption in finite homogeneous birth-death processes. *Biometrics*, *35*, 813–816.

Gnedenko, B. V. and Gnedenko, D. V. (1982). On Laplace and logistical distributions as limits in the theory of probability (in Russian). *Serdika Bolgarska Math.*, *8*, 229–234.

Goel, N. S., Maitra, S. C., and Montroll, E. W. (1971). *Nonlinear Models of Interacting Populations*. Academic Press, New York.

Goel, P. K. (1975). On the distribution of standardized mean of samples from the logistic population. *Sankhyā B*, *37*, 165–172.

Govindarajulu, Z. (1963). On moments of order statistics and quasi-ranges from normal populations. *Ann. Math. Statist.*, *34*, 633–651.

Govindarajulu, Z. (1963). Relationships among moments of order statistics in samples from two related populations. *Technometrics*, *5*, 514–518.

Govindarajulu, Z., Huang, J. S., and Saleh, A. K. M. E. (1975). Expected value of the spacings between order statistics. *Statistical Distr. in Scientific Work*, *Vol. 3*, 143–147, D. Reidel, Dordrecht.

Gradshteyn, I. S. and Ryzhik, I. W. (1965). *Tables of Integrals, Series and Products*. Academic Press, New York.

Greenland, S. (1985). An application of logistic models to the analysis of ordinal responses. *Biomet. J.*, *27*, 189–197.

Grizzle, J. E. (1961). A new method of testing hypotheses and estimating parameters for the logistic model. *Biometrics*, *17*, 372–385.

Grizzle, J. E. (1971). Multivariate logistic analysis. *Biometrics*, *27*, 1057–1062.

Grubbs, R. E. (1971). Approximate fiducial bounds for the reliability of a series system for which each component has an exponential time-to-fail distribution. *Technometrics*, *13*, 865–871.

Gumbel, E. J. (1944). Ranges and midranges. *Ann. Math. Statist.*, *15*, 414–422.

Gumbel, E. J. (1960). *Statistics of Extremes*. Columbia University Press, New York.

Gumbel, E. J. (1961). Bivariate logistic distributions. *J. Amer. Statist. Assoc.*, *56*, 335–349.

Gupta, S. S. (1956). On a decision rule for a problem in ranking means. *Mimeograph Series*, *No. 150*, Institute of Statistics, University of North Carolina, Chapel Hill, North Carolina.

Gupta, S. S. (1963). Probability integrals of the multivariate normal and multivariate *t*. *Ann. Math. Statist.*, *34*, 792–828.

Gupta, S. S. (1965). On some multiple decision (selection and ranking) rules. *Technometrics*, *7*, 225–245.

Gupta, S. S. and Gnanadesikan, M. (1966). Estimation of the parameters of the logistic distribution. *Biometrika*, *53*, 565–570.

Gupta, S. S. and Han, S. (1990). An elimination type two-stage procedure for selecting the population with the largest mean from *k* logistic populations. To appear in *Amer. J. Math. Management Sci.*

Gupta, S. S. and Huang, D.-Y. (1981). *Multiple Decision Theory: Recent Developments*, Lecture Notes in Statistics, No. 6. Springer-Verlag, New York.

Gupta, S. S. and Liang, T. (1987). On some Bayes and empirical Bayes selection procedures, In *Probability and Bayesian Statistics* (Ed., R. Viertl). Plenum Publishing Corporation, New York.

Gupta, S. S. and Liang, T. (1990). On a lower confidence bound for the probability of a correct selection: analytical and simulational studies. To appear in *Proc. of the First Intl. Conf. on Statistical Computing* held in Turkey, Mar. 30–Apr. 2, 1987.

Gupta, S. S. and McDonald, G. C. (1970). On some classes of selection procedures based on ranks. In *Nonparametric Techniques in Statistical Inference* (Ed., M. L. Puri), 491–514. Cambridge University Press, London.

Gupta, S. S., Nagel, K., and Panchapakesan, S. (1973). On the order statistics from equally correlated normal random variables. *Biometrika*, *60*, 403–413.

Gupta, S. S. and Panchapakesan, S. (1972). On a class of subset selection procedures. *Ann. Math. Statist.*, *43*, 814–822.

Gupta, S. S. and Panchapakesan, S. (1974). Inference for restricted families: (a) multiple decision procedures; (b) order statistics inequalities. In *Reliability and Biometry: Statistical Analysis of Lifelength* (Eds., F. Proschan and R. J. Serfling), 503–596. SIAM, Philadelphia.

Gupta, S. S. and Panchapakesan, S. (1975). On a quantile selection procedure and associated distribution of ratios of order statistics from a restricted family of probability distributions. In *Reliability and Fault Tree Analysis: Theoretical and Applied Aspects of System Reliability and Safety Assessment* (Eds., R. E. Barlow, J. B. Fussell and N. D. Singpurwalla), 558–576. SIAM, Philadelphia.

Gupta, S. S. and Panchapakesan, S. (1979). *Multiple Decision Procedures: Theory and Methodology of Selecting and Ranking Populations*. John Wiley & Sons, New York.

Gupta, S. S. and Panchapakesan, S. (1985). Subset selection procedures, review and assessment. *Amer. J. Math. Management Sci.*, *5*, 235–311.

Gupta, S. S. and Panchapakesan, S. (1988). Selection and ranking procedures in reliability models. Chapter 9 in *Handbook of Statistics—7: Quality Control and Reliability* (Eds., P. R. Krishnaiah and C. R. Rao), 131–156. North-Holland, Amsterdam.

Gupta, S. S. and Panchapakesan, S. (1990). On sequential ranking and selection procedures. To appear in *Handbook of Sequential Methods* (Eds., B. K. Ghosh and P. K. Sen). Marcel Dekker, New York.

Gupta, S. S., Qureishi, A. S., and Shah, B. K. (1967). Best linear unbiased estimators of the parameters of the logistic distribution using order statistics. *Technometrics*, *9*, 43–56.

Gupta, S. S. and Shah, B. K. (1965). Exact moments and percentage points of the order statistics and the distribution of the range from the logistic distribution. *Ann. Math. Statist.*, *36*, 907–920.

Gupta, S. S. and Sohn, J. K. (1990). Selection and ranking procedures for Tukey's generalized lambda distributions. To appear in *Frontiers of Modern Statistical Inference Procedures—II* (Eds., E. J. Dudewicz and E. Boffinger). Proc. of Second IPASRAS Conference held in Sydney, Australia, August 9–14, 1987.

Gurney, W. S. C., Blytha, S. P., and Nisbet, R. M. (1980). Nicholson's blowflies revisited. *Nature*, *287*, 17–21.

Gyllenberg, M. and Webb, G. F. (1989). Quiescence as an explanation of Gompertzian tumour growth. *Inst. Math. Res. Reports*, *A-264*, Helsinki Univ. Tech.

Hahn, G. J. and Shapiro, S. (1967). *Statistical Models in Engineering*. John Wiley & Sons, New York.

Hailey, D. C. (1952). Estimation of the dosage mortality relationship when the cost is subject to error. *Technical Report, No. 15*. Applied Mathematics Laboratory, Stanford University.

Hall, I. J. (1975). One-sided tolerance limits for a logistic distribution based on censored samples. *Biometrics*, *31*, 873–879.

Hall, I. J. and Sampson, C. B. (1974). One-sided tolerance limits for a normal population based on censored samples. *J. Statist. Comput. Simul.*, *2*, 317–324.

Hallam, T. G. (1986). Population dynamics in a homogeneous environment, In *Mathematical Ecology, an Introduction* (Eds., T. G. Hallam and S. A. Levin), 61–94, Lecture Notes in Biomathematics, No. 17. Springer-Verlag, Heidelberg.

Hallam, T. G. and Clarke, C. E. (1981). Non-autonomous logistic equations as models of populations in a deteriorating environment. *J. Theor. Biol.*, *93*, 303–311.

Hamilton, M. A. (1977). Estimating the logistic curve from grouped data. *J. Statist. Comput. Simul.*, *5*, 279–301.

Hampel, F. R. (1974). The influence curve and its role in robust estimation. *J. Amer. Statist. Assoc.*, *69*, 383–393.

Han, S. (1987). *Contributions to Selection and Ranking Theory with Special Reference to Logistic Populations*, Ph.D. Thesis (also Technical Report No. 87-38), Department of Statistics. Purdue University, West Lafayette, Indiana.

Hanson, F. B. and Tier, C. (1982). A stochastic model of tumour growth. *Math. Biosci.*, *61*, 73–100.

Hanson, F. B. and Ryan, D. (1989). Mean and quasideterministic equivalence for linear stochastic dynamics. *Math. Biosci.*, *93*, 1–14.

Hart, J. F., Cheney, E. W., Lawson, C. L., Maehley, H. J., Mesztenyi, C. K., Rice, J. R., Thatcher, H. G., and Witzgall, C. (1968). *Computer Approximations*. John Wiley & Sons, New York.

Harter, H. L. (1970). *Order Statistics and their Uses in Testing and Estimation*, Vol. 2. U.S. Government Printing Office, Washington, D.C.

Harter, H. L. and Moore, A. H. (1966). Iterative maximum-likelihood estimation of the parameters of normal populations from singly and doubly censored samples. *Biometrika*, *53*, 205–213.

Harter, H. L. and Moore, A. H. (1967). Maximum-likelihood estimation, from censored samples, of the parameters of a logistic distribution. *J. Amer. Statist. Assoc.*, *62*, 675–684.

Hassanein, K. M. (1969). Estimation of the parameters of the logistic distribution by sample quantiles. *Biometrika*, *56*, 684–687.

Hassanein, K. M. (1974). Linear estimation of the parameters of the logistic distribution by selected order statistics for very large samples. *Statistische Hefte*, *15*, 65–70.

Hassanein, K. M. and Sebaugh, J. L. (1973). Estimation of the parameters of the logistic distribution from grouped samples. *Skand. Aktuarietidskr.*, *56*, 1–10.

Hauck, W. W. and Anderson, S. (1984). A new statistical procedure for testing equivalence in two-group comparative bioavailability trials. *Journal of Pharmacokinetics and Biopharmaceutics*, *12*, 83–91.

Herd, G. R. (1960). Estimation of reliability from incomplete data. *Proc. of the Sixth National Symp. on Reliability and Quality Control*, 202–217.

Hilden, J., Habbema, J. D. F., and Bjerregaard, B. (1978). The measurement of performance in probabilistic diagnosis: II. Trustworthiness of the exact values of the diagnostic probabilities. *Meth. Inform. Med.*, *17*, 227–237.

Hitchcock, S. E. (1962). A note on the estimation of the parameters of the logistic function, using the minimum logit chi-square method. *Biometrika*, *49*, 250–252.

Hodges, J. L., Jr. (1958). Fitting the logistic by maximum likelihood. *Biometrics*, *14*, 453–461.

Hoeffding, W. (1953). On the distribution of the expected values of the order statistics. *Ann. Math. Statist.*, *24*, 93–100.

Holgate, P. (1967). The size of elephant herds. *Math. Gaz.*, *51*, 302–304.

Hosmer, D. W. and Lemeshow, S. (1989). *Applied Logistic Regression*. John Wiley & Sons, New York.

Hotelling, H. (1927). Differential equations subject to error and population estimates. *J. Amer. Statist. Assoc.*, *22*, 283–314.

Howlader, H. A. and Weiss, G. (1989). Bayes estimators of the reliability of the logistic distribution. *Commun. Statist.—Theor. Meth.*, *18(1)*, 245–259.

Hsu, J. P. and Wang, H. H. (1986). A stochastic analysis of bacterial adhesion. *J. Theor. Biol.*, *119*, 435–444.

Huang, J. S. (1974). On a characterization of the exponential distribution by order statistics. *J. Appl. Prob.*, *11*, 605–608.

Huang, J. S. (1975). Characterizations of distributions by the expected values of the order statistics. *Ann. Inst. Statist. Math* , *27*, 87–93.

Huang, J. S. (1989). Moment problem of order statistics: a review. *Intern. Statist. Review*, *57*, 59–66.

Huber, P. J. (1970). Studentizing robust estimates. In *Nonparametric Techniques in Statistical Inference* (Ed., M. L. Puri). Cambridge University Press, Cambridge, England.

Hwang, J. S. (1978). A note on Bernstein and Müntz-Szasz theorems with applications to the order statistics. *Ann. Inst. Statist Math.*, *30*, 167–176.

Hwang, S. and Zelterman, D. (1986). On the distribution of EM estimates. *Proc. of the Joint Statistical Meetings—Statistical Computing Section*, 144–146.

Imrey, P. B., Koch, G. G., and Stokes, M. E. (1981). Categorical data analysis: some reflections on the log linear model and logistic regression. I: historical and methodological overview. *Internat. Statist. Rev.*, *49*, 265–283.

Imrey, P. B., Koch, G. G., and Stokes, M. E. (1982). Categorical data analysis: some reflections on the log linear model and logistic regression. II: data analysis. *Internat. Statist. Rev.*, *50*, 35–63.

Jennings, D. E. (1986). Outliers and residual distributions in logistic regression. *J. Amer. Statist. Assoc.*, *81*, 987–990.

Jennrich, R. J. (1969). Asymptotic properties of nonlinear least squares estimators. *Ann. Math. Statist.*, *40*, 633–643.

Jensen, A. L. (1975). Comparison of logistic equations for population growth. *Biometrics*, *31*, 853–862.

Johnson, N. L. (1949). Systems of frequency curves generated by methods of translation. *Biometrika*, *36*, 149–176.

Johnson, N. L. (1954). Systems of frequency curves derived from the first law of Laplace. *Trabajos de Estadistica*, *5*, 285–291.

Johnson, N. L. and Kotz, S. (1970). *Distributions in Statistics: Continuous Univariate Distributions*, *Vol. 2*. John Wiley & Sons, New York.

Johnson, N. L. and Kotz, S. (1975). On some generalized Farlie-Gumbel-Morgenstern distributions. *Commun. Statist.*, *4*, 415–427.

Johnson, N. L. and Kotz, S. (1977). On some generalized Farlie-Gumbel-Morgenstern distributions, II; Regression, correlation and further generalizations. *Commun. Statist.*, *6*, 485–496.

Johnson, N. L. and Kotz, S. (1989). Characterization based on conditional distributions. *J. Indian Statist. Assoc.*, *27*.

Johnson, W. (1985). Influence measures for logistic regression: another point of view. *Biometrika*, *72*, 59–65.

Joiner, B. L. and Rosenblatt, J. R. (1971). Some properties of the range in samples from Tukey's symmetric lambda distributions. *J. Amer. Statist. Assoc.*, *66*, 394–399.

Joshi, P. C. (1971). Recurrence relations for the mixed moments of order statistics. *Ann. Math. Statist.*, *42*, 1096–1098.

Joshi, P. C. (1972). Efficient estimation of the mean of an exponential distribution when an outlier is present. *Technometrics*, *14*, 137–144.

Jung, J. (1955). On linear estimates defined by a continuous weight function. *Arkivfoer Matematik*, *3*, 199–209.

Kadane, J. B. (1974). A characterization of triangular arrays which are expectations of order statistics. *J. Appl. Prob.*, *11*, 413–416.

Kakosyan, A. V., Klebanov, L. B., and Melamed, J. A. (1984). *Characterization of Distributions by the Method of Intensively Monotone Op-*

erators, Lecture Notes in Mathematics, No. 1088. Springer-Verlag, Berlin.

Kalbfleisch, J. D. and Prentice, R. L. (1980). *The Statistical Analysis of Failure Time Data*. John Wiley & Sons, New York.

Kale, B. K. and Sinha, S. K. (1971). Estimation of expected life in the presence of an outlier observation. *Technometrics*, *13*, 755–759.

Kaplan, E. L. and Meier, P. (1958). Non-parametric estimation from incomplete observations. *J. Amer. Statist. Assoc.*, *53*, 457–481.

Kastenbaum, M. A. and Lamphiear, D. E. (1959). Calculation of chi-square to test the three-factor no interaction hypothesis. *Biometrics*, *15*, 107–115.

Kay, R. and Little, S. (1986). Assessing the fit of the logistic model: A case study of children with the Haemolytic Uraemic Syndrome. *Appl. Statist.*, *35*, 16–30.

Kendal, W. S. (1985). Gompertzian growth as a consequence of tumour heterogeneity. *Math. Biosci.*, *73*, 103–107.

Kendall, M. G. and Stuart, A. (1961). *The Advanced Theory of Statistics*, Vol. 2. Charles Griffin and Co., London.

Kendall, M. G. and Stuart, A. (1967). *Advanced Theory of Statistics*. Hafner Publishing Co., New York.

Kennedy, W. J. and Gentle, J. E. (1980). *Statistical Computing*. Marcel Dekker, New York.

Kimura, M. (1964). Diffusion models in population genetics. *J. Appl. Probab.*, *1*, 177–232.

Kjelsberg, M. O. (1962). *Estimation of the Parameters of the Logistic Distribution Under Truncation and Censoring*. Doctoral Thesis, University of Minnesota, U.S.A.

Kneale, G. W. (1971). Problems arising in estimating from retrospective study data the latent period of juvenile cancers initiated by obstetric radiography. *Biometrics*, *27*, 563–590.

Kolker, Y. (1987). A piece-wise linear growth model: Comparison with competing forms in batch culture. *J. Math. Biol.*, *25*, 543–551.

Kolmogorov, A. N. (1933). Sulla determinazione empirica di una legge di distribuziane. *Giorna. Inst. Attuari.*, *4*, 83–91.

Koutrovelis, I. A. (1981). Large sample quantile estimation in Pareto laws. *Commun. Statist.—Theor. Meth.*, *10(2)*, 189–201.

Koziol, J. A. (1980). Goodness-of-fit tests for randomly censored data. *Biometrika*, *67*, 693–696.

Koziol, J. A. and Byar, D. P. (1975). Percentage points of the asymptotic distributions of one and two sample K-S statistics for truncated or censored data. *Technometrics*, *17*, 507–510.

Koziol, J. A. and Green, S. B. (1976). A Cramer-von Mises statistic for randomly censored data. *Biometrika*, *63*, 465–474.

Kryscio, R. J. and Lefevre, J. (1989). On the extinction of the S-I-S stochastic epidemic. *J. Appl. Probab.*, *27*, 685–694.

Ku, H. H. and Kullback, S. (1968). An application of minimum discrimination information estimation to a problem of Grizzle and Berkson on the test of "no interaction" hypothesis. *Unpublished Paper*.

Kubat, P. and Epstein, B. (1980). Estimation of quantiles of location-scale distribution based on few ordered observations. *Technometrics*, *12*, 345–361.

Kuiper, N. H. (1960). Tests concerning random points on a circle. *Proc. Koninkl. Neder. Akad. van. Wetenschappen.*, *A 63*, 38–47.

Kulldorff, G. (1964). Optimum spacing of sample quantiles from a logistic distribution and best linear unbiased estimates of its parameters. *Technical Report*. University of Lund, Sweden.

Kurtz, T. G. (1981). *Approximation of Population Processes*. SIAM, Philadelphia.

Lachenbruch, P. A. (1980). Note on combining risks using a logistic discriminant function approach. *Biomet. J.*, *22*, 759–762.

Landwehr, J. M., Pregibon, D., and Shoemaker, A. C. (1984). Graphical methods for assessing logistic regression models. *J. Amer. Statist. Assoc.*, *79*, 61–71.

Larntz, K. (1978). Small-sample comparisons of exact levels for chi-squared goodness-of-fit statistics. *J. Amer. Statist. Assoc.*, *73*, 253–263.

Leach, D. (1981). Re-evaluation of the logistic curve for human populations. *J. Roy. Statist. Soc.*, *Ser. A*, *144*, 94–103.

Lehmann, E. L. (1959). *Testing Statistical Hypotheses*. John Wiley & Sons, New York.

Lewis, T. (1975). A model for the parasitic disease bilharziasis. *Adv. Appl. Probab.*, *7*, 705–766.

Liang, T. and Panchapakesan, S. (1990). Isotonic selection with respect to a control: a Bayesian approach, To appear in *Frontiers of Modern Statistical Inference Procedures—II* (Eds., E. J. Dudewicz and E. Boffinger). Proc. of the Second IPASRAS Conference held in Sydney, Australia, August 9–14, 1987.

Lin, G. D. (1984). A note on equal distributions. *Ann. Inst. Statist. Math.*, *36*, 451–453.

Lin, G. D. (1988). Characterizations of distributions via relationships between two moments of order statistics. *J. Statist. Plann. Inf.*, *19*, 73–80.

Lindley, D. V. (1980). Approximate Bayesian methods (with discussants). *Trabajos de Estadistica y de Investigacion Operativa*, *31*, 232–245.

Lindley, D. V. and Singpurwalla, N. D. (1986). Multivariate distributions for the life lengths of components of a system sharing a common environment. *J. Appl. Probab.*, *23*, 418–431.

Lippert, S. (1986). *Untersuchungen des t-Tests bei Nichtlinearität und Nichtnormalität am Beispiel der logistischen Funktion* Master Thesis, Department of Mathematics, University of Rostock.

Lloyd, D. K. and Lipow, M. (1962). *Reliability: Management, Methods, and Mathematics*. Prentice-Hall, New Jersey.

Lloyd, E. H. (1952). Least-squares estimation of location and scale parameters using order statistics. *Biometrika*, *39*, 88–95.

Lockhart, R. A., O'Reilly, F. J., and Stephens, M. A. (1986). Tests of fit based on normalized spacings. *J. Roy. Statist. Soc.*, *Ser. B*, *48*, 344–352.

Loeve, M. (1977). *Probability Theory I, Fourth edition*. Springer-Verlag, New York.

Lomax, K. S. (1954). Business failures: Another example of the analysis of failure data. *J. Amer. Statist. Assoc.*, *49*, 847–852.

Lord, F. M. (1965). A note on the normal ogive or logistic curve in item analysis. *Psychometrika*, *30*, 371–372.

Lorenzen, T. J. and McDonald, G. C. (1981). Selecting logistic populations using the sample medians. *Commun. Statist. A—Theor. Meth.*, *10*, 101–124.

Lotka, A. J. (1931). The structure of a growing population. *Human Biol.*, *3*, 459–493.

Ludwig, D., Jones, D. D., and Holling, C. S. (1978). Qualitative analysis of insect outbreak systems: The spruce budworm and forest. *J. Animal Ecol.*, *47*, 315–332.

Lurie, D. and Hartley, H. O. (1972). Machine-generation of order statistics for Monte Carlo computations. *Amer. Statist.*, *26*, 26–27.

Maddala, G. S. (1983). *Limited-dependent and Qualitative Variables in Econometrics*. Cambridge University Press, London.

Madreimov, I. and Petunin, Y. I. (1983). A characterization of the uniform distribution with the aid of order statistics. *Theor Prob. Math. Statist.*, *27*, 105–110.

Mage, D. T. (1980). An explicit solution for S_B parameters using four percentile points. *Technometrics*, *22*, 247–251.

Malik, H. J. (1980). Exact formula for the cumulative distribution function of the quasi-range from the logistic distribution. *Commun. Statist.—Theor. Meth.*, *A9(14)*, 1527–1534.

Malik, H. J. (1985). Logistic Distribution, In *Encyclopedia of Statistics* (Eds., S. Kotz and N. L. Johnson). John Wiley & Sons, New York.

Malik, H. J. and Abraham, B. (1973). Multivariate logistic distributions. *Ann. Statist.*, *1*, 588–590.

Mann, H. B. and Wald, A. (1942). On the choice of the number of class intervals in the application of the chi-square test. *Ann. Math. Statist.*, *13*, 306–317.

Mann, N. R. and Fertig, K. W. (1973). Tables for obtaining Weibull confidence bounds and tolerance bounds based on best linear invariant estimates of parameters of the extreme-value distribution. *Technometrics*, *15*, 87–102.

Mantel, N. (1966). Models for complex contingency tables and polychotomous dosage response curves. *Biometrics*, *22*, 83–95.

Mantel, N. and Brown, C. C. (1973). A logistic reanalysis of Ashford and Sowden's data on respiratory symptoms in British coal miners. *Biometrics*, *29*, 649–665.

Mantel, N. and Hankey, B. F. (1978). A logistic regression analysis of response time data where the hazard function is time dependent. *Commun. Statist.—Theor. Meth.*, *A7*, 333–347.

Massaro, J. M. and D'Agostino, R. B. (1992). To appear.

May, R. M. (1980). Mathematical models in whaling and fisheries management, In *Lectures on Math. in the Life Sci.*, *13*, 1–63, American Mathematical Society.

May, R. M. (1983). Non linear problems in ecology and resource management, In *Chaotic Behaviour of Deterministic Systems* (Eds., G. Iooss, R. H. G. Hellman, and R. Stora), 513–563. North-Holland, Amsterdam.

McCullagh, P. (1977). A logistic model for paired comparisons with ordered categorical data. *Biometrika*, *64*, 449–453.

McCullagh, P. (1989). Binary data, dispersion effects and the sexual behaviour of Appalachian salamanders. *Technical Report, No. 227*, Department of Statistics, University of Chicago.

McCullagh, P. and Nelder, J. A. (1989). *Generalized Linear Models*, Second edition, Chapman and Hall, London.

McQuarrie, D. A. (1967). Stochastic approach to chemical kinetics. *J. Appl. Probab.*, *4*, 478.

Meade, N. (1988). A modified logistic model applied to human populations. *J. Roy. Statist. Soc., Ser. A*, *151*, 491–498.

Miller, R. S. and Botkin, D. B. (1974). Endangered species: models and predictions, *Amer. Scient.*, *62*, 172–181.

Milton, R. C. (1963). Tables of equally correlated multivariate normal probability integral. *Technical Report, No. 27*, Department of Statistics, University of Minnesota, Minneapolis, Minnesota.

Mogyorodi, J. (1967). On the limit distribution of the largest term in the order statistics of a sample of random size (in Hungarian). *Magyar Tud. Akad. Mat. Fiz. Oszt. Kozl.*, *17*, 75–83.

Mood, A. M. (1950). *Introduction to the Theory of Statistics*. McGraw-Hill, New York.

Moolgavkar, S. H. (1986). Carcinogenesis modeling: from molecular biology to epidemiology. *Ann. Rev. Publ. Heal.*, *7*, 151–169.

Morgan, B. S. T. (1976). Stochastic models of grouping changes. *Adv. Appl. Probab.*, *8*, 30–57.

Morgan, B. S. T. (1985). The cubic logistic model for quantal assay data. *Appl. Statist.*, *34*, 105–113.

Mosteller, F. (1946). On some useful "inefficient" statistics. *Ann. Math. Statist.*, *17*, 377–407.

Mudholkar, G. S. and George, E. O. (1978). A remark on the shape of the logistic distribution. *Biometrika*, *65*, 667–668.

Murray, J. D. (1989). *Mathematical Biology*. Lecture Notes in Biomathematics, No. 19. Springer-Verlag, Heidelberg.

Myers, M. H., Hankey, B. F., and Mantel, N. (1973). A logistic exponential model for use with response-time data involving regressor variables. *Biometrics*, *29*, 257–269.

Nelson, W. (1972). Theory and applications of hazard plotting for censored failure data. *Technometrics*, *14*, 945–966.

Neyman, J. (1935). Su un teorema concernente le cosiddette statistiche sufficienti. *C. Ist. Ital. Attuari.*, *Anno. VI*, *13*, 3–17.

Norden, R. H. (1982). On the distribution of the time to extinction in the stochastic logistic population model. *Adv. Appl. Probab.*, *14*, 687–708.

Norden, R. H. (1984). On the numerical evaluation of the moments of the distribution of states at time *t* in the stochastic logistic process. *J. Statist. Comput. Simul.*, *20*, 1–20.

Norton, H. W. (1945). Calculation of chi-square from complex contingency tables. *J. Amer. Statist. Assoc.*, *40*, 251–258.

Nürnberg, G. (1985, 1986). Robustness of two-sample tests for variances. In *Robustness of Statistical Methods and Nonparametric Statistics* (Eds., D. Rasch and M. L. Tiku), 75–82. VEB Deutscher Verlag der Wissenschaften, Berlin.

Nussey, H. E. (1984). Complicated dynamical behaviour in discrete population models. *Nieuw Arch. Wisk.*, *2*, 43–81.

Oakes, D. (1989). Bivariate survival models induced by frailties. *J. Amer. Statist. Assoc.*, *84*, 487–493.

Ochi, Y. and Prentice, R. L. (1984). Likelihood inference in a correlated probit regression model. *Biometrika*, *71*, 531–543.

Office of Population Censuses and Surveys (1980). *Population Projections, No. 11, Series PP2*. HMSO, London.

Ogawa, J. (1951). Contributions to the theory of systematic statistics. *I, Osaka Math. J.*, *3*, 175–213.

Oliveira, J. T. De (1961). La representation des distributions extremales bivariees. *Bull. Intl. Statist. Inst.*, *33*, 477–480.

Oliver, F. R. (1964). Methods of estimating the logistic growth function. *Appl. Statist.*, *13*, 57–66.

Oliver, F. R. (1966). Aspects of maximum likelihood estimation of the logistic growth function. *J. Amer. Statist. Assoc.*, *61*, 697–705.

Oliver, F. R. (1969). Another generalisation of the logistic growth function. *Econometrica*, *37*, 144–147.

Oliver, F. R. (1982). Notes on the logistic curve for human populations. *J. Roy. Statist. Soc., Ser. A*, *145*, 359–363.

Owen, D. B. (1967). Variables sampling plans based on the normal distribution. *Technometrics*, *9*, 417–423.

Owen, D. B. (1988). The starship. *Commun. Statist.—Simul. Comput.*, *17*, 315–323.

Pagano, M. and Tritchler, D. L. (1983). On obtaining permutation distributions in polynomial time. *J. Amer. Statist. Assoc.*, *78*, 435–440.

Paiva-Franco, M. A. (1984). A log logistic model for survival time with covariates. *Biometrika*, *71*, 621–623.

Parthasarathy, P. R. (1988). Density-dependent Markov branching processes, In *Mathematical Ecology* (Eds., T. G. Hallam, L. J. Gross, and S. A. Levin), 569–579. World Sci., Singapore.

Parthasarathy, P. R. and Krishna Kumar, B. (1990a). Density-dependent birth and death processes with state dependent immigration, To appear in *Intl. J. Math. Comp. Mod.*

Parthasarathy, P. R. and Krishna Kumar, B. (1990b). Two stochastic analogues of the logistic process, To appear in *Ind. J. Pure and Appl. Math.*

Parthasarathy, P. R. and Krishna Kumar, B. (1990c). A birth and death process with logistic mean population, To appear in *Commun. Statist.—Theor. Meth.*

Parthasarathy, P. R. and Mayilswami, P. (1982). Stochastic compartmental models with birth, death and immigration of particles. *Commun. Statist.—Theor. Meth.*, *11*, 1625–1642.

Pearl, R. (1925). *The Biology of Population Growth*. Knopf, New York.

Pearl, R. (1940). *Medical Biometry and Statistics*. W. M. Sanders Co., Philadelphia.

Pearl, R. and Reed, L. J. (1920). On the rate of growth of the population of the United States since 1790 and its mathematical representation. *Proc. of National Acad. Sci.*, *6*, 275–288.

Pearl, R. and Reed, L. J. (1924). *Studies in Human Biology*. Williams and Wilkins, Baltimore.

Pearl, R., Reed, L. J., and Kish, J. F. (1940). The logistic curve and the census count of 1940. *Science*, *92*, 486–488.

Pearson, E. S. and Hartley, H. O. (1970). *Biometrika Tables for Statisticians, Vol. I, Third edition*. Cambridge University Press, England.

Pearson, E. S., Johnson, N. L., and Burr, I. W. (1979). Comparisons of the percentage points of distributions with the same first four moments, chosen from eight different systems of frequency curves. *Commun. Statist.—Simul. Comput.*, *8*, 191–229.

Pearson, K. (1934). *Tables of the Incomplete B-Function*. Cambridge University Press, England.

Perks, W. F. (1932). On some experiments in the graduation of mortality statistics. *J. of the Institute of Actuaries*, *58*, 12–57.

Pettitt, A. N. and Stephens, M. A. (1976). Modified Cramer-von Mises statistics for censored data. *Biometrika*, *63*, 291–298.

Pettitt, A. N. and Stephens, M. A. (1977). The Kolmogorov-Smirnov goodness-of-fit statistic with discrete and grouped data. *Technometrics*, *19*, 205–210.

Pickands III, J. (1968). Moment convergence of sample extremes. *Ann. Math. Statist.*, *39*, 881–889.

Pielou, E. C. (1977). *Mathematical Ecology*. John Wiley & Sons, New York.

Pierce, D. A. and Sands, B. R. (1975). Extra-binomial variation in binary data. *Technical Report No. 46*, Department of Statistics, Oregon State University.

Plackett, R. L. (1958). Linear estimation from censored data. *Ann. Math. Statist.*, *29*, 131–142.

Plackett, R. L. (1959). The analysis of life test data. *Technometrics*, *1*, 9–19.

Pohl, S. (1988). Lokal-optimale Versuchspläne für spezielle nichtlineare Regressions-funktionen Diplomarbeit. *Sekt. Mathematik*. WPU, Rostock.

Prajneshu, D. (1980). Time-dependent solution of the logistic model of population growth in random environment. *J. Appl. Probab.*, *27*, 1083–1086.

Pregibon, D. (1979). *Data Analytic Methods for Generalized Linear Models*. Ph.D. Thesis, University of Toronto, Toronto, Canada.

Pregibon, D. (1981). Logistic regression diagnostics. *Ann. Statist.*, *9*, 705–724.

Prendville, B. J. (1949). Discussion of symposium on stochastic processes. *J. Roy. Statist. Soc.*, *Ser. B*, *11*, 273.

Prentice, R. L. (1976). Use of the logistic model in retrospective studies. *Biometrics*, *32*, 599–606.

Prentice, R. L. (1976). A generalization of the probit and logit methods for dose response curves. *Biometrics*, *32*, 761–768.

Prentice, R. L. (1986). Binary regression using an extended beta-binomial distribution, with discussion of correlation induced by covariate measurement errors. *J. Amer. Statist. Assoc.*, *83*, 321–327.

Prentice, R. L. (1988). Correlated binary regression with covariates specific to each binary observation. *Biometrics*, *44*, 1033–1048.

Pyke, R. (1965). Spacings. *J. Roy. Statist. Soc.*, *Ser. B*, *27*, 395–449.

Radelet, M. and Pierce, G. L. (1988). Race and prosecutorial discretion in homicide cases, Presented at the *Annual Meeting of Amer. Sociol. Assoc.*, Detroit, Michigan.

Ragab, A. and Green, J. (1987). Estimation of the parameters of the log logistic distribution based on order statistics. *Amer. J. Math. Management Sci.*, 7, 307–323.

Raghunandanan, K. and Srinivasan, R. (1970). Simplified estimation of parameters in a logistic distribution. *Biometrika*, *57*, 677–678.

Raghunandanan, K. and Srinivasan, R. (1971). Simplified estimation of parameters in a double exponential distribution. *Technometrics*, *13*, 689–691.

Ramberg, J. S. and Schmeiser, B. W. (1972). An approximate method for generating symmetric random variables. *Commun. ACM*, *15*, 987–990.

Ramberg, J. S. and Schmeiser, B. W. (1974). An approximate method for generating asymmetric random variables. *Commun. ACM*, *17*, 78–82.

Rao, C. R. (1973). *Linear Statistical Inference and Its Applications*. John Wiley & Sons, New York.

Rasch, D. (1984). Robust confidence estimation and tests for parameters of growth functions, In *Szamitasttechnikai es kibernetikai modszerek alkalmazasa az orvostodomanyban es a biologiaban Szeged* (Ed., I. Gyoeri), 306–313.

Rasch, D. (1985). Finite sample behaviour of an asymptotic *t*-test in exponential regression. *Statistics*, *16*, 121–124.

Rasch, D. (1986). Optimum experimental design for fitting growth curves in cattle. *Arch. f. Tierzucht*, *29*, 85–91.

Rasch, D. (1988). Recent results in growth analysis. *Statistics*, *19*, 585–604.

Rasch, D. (1991). *Einführung in die Mathematische Statistik, Band II Varianzanalyse, Regressionsanalyse und weitere Anwendungen*. VEB Deutscher Verlag der Wissenschaften, Berlin.

Rasch, D., Nürnberg, G., and Busch, K. (1988). CADEMO—Ein Expertensystem zur Versuchsplanung, In *Fortschritte der Statistik—Software 1* (Eds., F. Faulbaum und H. M. Ühlinger), 193–201. Fischer Verl., Stuttgart and New York.

Rasch, D., Guiard, V., Nürnberg, G., Rudolph, P. E., and Teuscher, F. (1987). The expert system CADEMO. Computer Aided Design of Experiments and Modelling. *Statistical Software Newsletter*, *13*, 107–114.

Rasch, D., Rudolph, P. E., und Schimke, E. (1985). Optimale versuchspläne in der nichtlinearen regression. *Probleme der angewandten Statistik*, *13*, 93–121.

Rasch, D. and Schimke, E. (1985, 1986). A test for exponential regression, In *Robustness of Statistical Methods and Nonparametric Statistics* (Eds., D. Rasch and M. L. Tiku), 104–112. VEB Deutscher Verlag der Wissenschaften, Berlin.

Rasch, D. and Tiku, M. L. (Eds.) (1985, 1986). *Robustness of Statistical Methods and Nonparametric Statistics*. VEB Deutscher Verlag der Wissenschaften (1985) and Reidel Publ. Co., Dortrecht (1986).

Read, T. R. C. (1984). Small sample comparisons for the power divergence goodness-of-fit statistics. *J. Amer. Statist. Assoc.*, *79*, 929–935.

Reed, L. J. and Berkson, J. (1929). The application of the logistic function to experimental data. *J. Physical Chemistry*, *33*, 760–779.

Renshaw, E. C. (1973). Interconnected population processes. *J. Appl. Probab.*, *10*, 1–14.

Renyi, A. (1970). *Probability Theory*. North-Holland. Amsterdam.

Rosen, G. (1984). Characterizing conditions for generalized Verhulst logistic growth of a biological population. *Bull. Math. Biol.*, *46*, 963–965.

Rosen, G. (1985). Parameter evolution in the θ-model. *Evolution*, *39*, 707–708.

Rosen, G. (1987). Time delays produced by essential nonlinearity in population growth models. *Bull. Math. Biol.*, *49*, 253–255.

Saleh, A. K. Md. E. (1981). Estimating quantiles of exponential distribution, In *Statistics and Related Topics* (Eds., M. Csorgo, D. A. Dawson, J. N. K. Rao and A. K. Md. E. Saleh), 279–283. North Holland, Amsterdam.

Saleh, A. K. Md. E., Ali, M. Masoom, and Umbach, D. (1983). Estimating the quantile function of location-scale family of distributions based on a few selected order statistics. *J. Statist. Plann. Inf.*, *8*, 75–87.

Saleh, A. K. Md. E., Ali, M. Masoom, and Umbach, D. (1985). Large sample estimation of Pareto quantiles using selected order statistics. *Metrika*, *32*, 49–56.

Sanathanan, L. (1974). Some properties of the logistic model for dichotomous response. *J. Amer. Statist. Assoc.*, *69*, 744–749.
Santner, T. J. (1975). A restricted subset selection approach to ranking and selection problems. *Ann. Statist.*, *3*, 334–349.
Santner, T. J. and Duffy, D. E. (1986). A note on A. Albert and J. A. Anderson's conditions for the existence of maximum likelihood estimates in logistic regression models. *Biometrika*, *73*, 755–758.
Sarhan, A. E. and Greenberg, B. G. (Eds.) (1962). *Contributions to Order Statistics*. John Wiley & Sons, New York.
Sarndal, C. E. (1962). *Information From Censored Samples*. Almqvist and Wiksell, Stockholm, Sweden.
Satterthwaite, S. P. and Hutchinson, T. P. (1978). A generalization of Gumbel's bivariate logistic distribution. *Metrika*, *25*, 163–170.
Schafer, D. (1987). Covariate measurement error in the generalized linear model. *Biometrika*, *74*, 385–391.
Schafer, R. E. and Sheffield, T. S. (1973). Inferences on the parameters of the logistic distribution. *Biometrika*, *29*, 449–455.
Schneider, B. E. (1978). Trigamma function, Algorithm AS 121. *Appl. Statist.*, *27*, 97–99.
Schorr, B. (1974). On the choice of the class intervals in the application of the chi-square test. *Math. Operationsforsch. und Statist.*, *5*, 357–377.
Schultz, H. (1930). The standard error of a forecast from a curve. *J. Amer. Statist. Assoc.*, *25*, 139–185.
Scott, A. J. and Wild, C. J. (1985). Fitting logistic models in case-control studies. *Bull. of Intern. Statist. Inst.*, *51*, 12.3, 15 pp.
Scott, A. J. and Wild, C. J. (1986). Fitting logistic models under case-control or choice based sampling. *J. Roy. Statist. Soc.*, *Ser. B*, *48*, 170–182.
Shah, B. K. (1966). On the bivariate moments of order statistics from a logistic distribution. *Ann. Math. Statist.*, *37*, 1002–1010.
Shah, B. K. (1970). Note on moments of a logistic order statistics. *Ann. Math. Statist.*, *41*, 2151–2152.
Shah, B. K. (1976). Data analysis problem in the area of pharmacokinetics research. *Biometrics*, *32*, 145–147.
Shah, B. K. (1988). On the generalization of *t*-distribution for non-identically distributed variables and applications. *Gujarat Statist. Rev. XV*, *No. 2*, 45–50.
Shah, B. K. and Dave, P. H. (1963). A note on log-logistic distribution. *J. of M. S. University of Baroda* (*Sci. Number*), *12*, 21–22.
Shohoji, T. and Sasaki, H. (1985). An aspect of growth analysis of weight in savannah baboon. *Growth*, *49*, 300–309.
Shoukri, M. M., Mian, I. U. H., and Tracy, D. S. (1988). Sampling prop-

erties of estimates of the log-logistic distribution, with application to Canadian precipitation data. *Canad. J. Statist.*, *16*, 223–226.

Shu, V. S. (1978). *Robust estimation of a location parameter in the presence of outliers*. Ph.D. Thesis, Iowa State University, Ames, Iowa.

Silverstone, H. (1957). Estimating the logistic curve. *J. Amer. Statist. Assoc.*, *52*, 567–577.

Singleton, R. C. (1969). An algorithm for computing the mixed radix fast Fourier transform. *IEEE Trans. on Audio and Electroacoustics*, *17*, 93–103.

Slifker, J. F. and Shapiro, S. S. (1980). The Johnson system: Selection and parameter estimation. *Technometrics*, *22*, 239–246.

Smith, C. E. and Tuckwell, H. C. (1974). Some stochastic growth processes. In *Mathematical Problems in Biology*, 211–224, Lecture Notes in Biomathematics No. 2. Springer-Verlag, Heidelberg.

Sohn, J. (1985). *Multiple Decision Procedures for Tukey's Generalized Lambda Distributions*, Ph.D. Thesis (also Technical Report No. 85-20). Department of Statistics, Purdue University, West Lafayette, Indiana.

Spiegelman, D. (1988). *Design and analysis strategies for epidemiologic research when the exposure variable is measured with error*, Ph.D. Thesis. Department of Biostatistics, Harvard University.

Stefanski, L. A. and Carroll, R. J. (1985). Covariate measurement error in logistic regression. *Ann. Statist.*, *13*, 1335–1351.

Stefanski, L. A. (1990). A normal scale mixture representation of the logistic distribution. To appear in *Statist. Probab. Lett.*

Stephens, M. A. (1970). Use of the Kolmogorov-Smirnov, Cramer-von Mises and related statistics without extensive tables *J. Roy. Statist. Soc.*, *Ser. B*, *32*, 115–122.

Stephens, M. A. (1979). Tests of fit for the logistic distribution based on the empirical distribution function. *Biometrika*, *66*, 591–595.

Stiratelli, R., Laird, N. M., and Ware, J. H. (1984). Random-effects models for serial observations with binary response. *Biometrics*, *40*, 961–971.

Strauss, D. J. (1979). Some results on random utility. *J. Mathematical Psychology*, *20*, 35–52.

Stukel, T. (1988). Generalized logistic models. *J. Amer. Statist. Assoc.*, *83*, 426–431.

Symanowski, J. T. and Koehler, K. J. (1989). A bivariate logistic distribution with applications to categorical responses. *Technical Report No. 89–29*. Department of Statistics, Iowa State University, Ames, Iowa.

Tadikamalla, P. R. and Johnson, N. L. (1982). Systems of frequency curves generated by transformation of logistic variables. *Biometrika*, *69*, 461–465.

Tadikamalla, P. R. and Johnson, N. L. (1982). Tables to facilitate fitting L_U distribution. *Commun. Statist.—Simul. Comput.*, *11*, 249–271.

Tamhane, A. C. (1986). A survey of literature on quantal response curves with a view toward application to the problem of selecting the curve with the smallest q-quantile (ED100q). *Commun. Statist.—Theor. Meth.*, *15*, 2679–2718.

Tamhane, A. C. and Bechhofer, R. E. (1977). A two-stage minimax procedure with screening for selecting the largest normal mean. *Commun. Statist. A—Theor. Meth.*, *6*, 1003–1033.

Tamhane, A. C. and Bechhofer, R. E. (1979). A two-stage minimax procedure with screening for selecting the largest mean (II): an improved PCS lower bound and associated tables. *Commun. Statist. A—Theor. Meth.*, *8*, 337–358.

Tan, W. Y. (1975). On the absorption probability and absorption time to finite homogeneous birth-death processes by diffusion approximation. *Metron*, *33*, 389–401.

Tan, W. Y. (1976). On the absorption probability and absorption time of finite homogeneous birth-death processes. *Biometrics*, *32*, 745–752.

Tan, W. Y. (1977). On the distribution of second degree polynomial in normal random variables. *Canad. J. Statist.*, *5*, 241–250.

Tan, W. Y. (1983). On the distribution of number of mutants at the hypoxanthine-quanine phorsphoribosal transferase locus in Chinese hamster ovary cells. *Math. Biosci.*, *67*, 175–192.

Tan, W. Y. (1984). On the first passage probability distributions in continuous time Markov processes. *Util. Math.*, *26*, 89–102.

Tan, W. Y. (1986). A stochastic Gompertz birth and death process. *Statist. Probab. Lett.*, *4*, 25–28.

Tan, W. Y. and Piantadosi, S. (1990). On stochastic growth processes with application to stochastic logistic growth, To appear in *Statistica Sinica*.

Tan, W. Y. and Singh, K. P. (1990). A mixed model of carcinogenesis with applications to retinblastoma. *Math. Biosci.*, *98*, 211–225.

Tarter, M. E. (1966). Exact moments and product moments of the order statistics from the truncated logistic distribution. *J. Amer. Statist. Assoc.*, *61*, 514–525.

Tarter, M. E. and Clark, V. A. (1965). Properties of the median and other order statistics of logistic variates. *Ann. Math. Statist.*, *36*, 1779–1786. Correction *8*, 935.

Thoman, D. R., Bain, L. J., and Antle, C. E. (1970). Maximum likelihood estimation, exact confidence intervals for reliability, and tolerance limits in the Weibull distribution. *Technometrics*, *12*, 363–372.

Tier, C. and Hanson, F. B. (1981). Persistence in density dependent stochastic populations. *Math. Biosci.*, *53*, 89–117.

Tierney, L. and Kadane, J. (1986). Accurate approximations for posterior moments and marginals. *J. Amer. Statist. Assoc.*, *81*, 82–86.

Tiku, M. L. (1967). Estimating the mean and standard deviation from a censored normal sample. *Biometrika*, *54*, 155–165.

Tiku, M. L. (1968). Estimating the parameters of normal and logistic distributions from censored samples. *Austral. J. Statist.*, *10*, 64–74.

Tiku, M. L. (1980). Robustness of MML estimators based on censored samples and robust test statistics. *J. Statist. Plann. Inf.*, *4*, 123–143.

Tiku, M. L. (1980). Goodness of fit statistics based on the spacings of complete or censored samples. *Austral. J. Statist.*, *22*, 260–275.

Tiku, M. L., Tan, W. Y., and Balakrishnan, N. (1986). *Robust Inference*. Marcel Dekker, New York.

Tintner, G. (1952). *Econometrics*. John Wiley & Sons, New York.

Tognetti, K. and Winley, G. (1980). Stochastic growth models with logistic mean population. *J. Theor. Biol.*, *82*, 167–169.

Tritchler, D. L. (1984). An algorithm for exact logistic regression. *J. Amer. Statist. Assoc.*, *79*, 709–711.

Tsiatis, A. A. (1980). A note on the goodness-of-fit for the logistic regression model. *Biometrika*, *67*, 250–251.

Tsutakawa, R. K. (1972). Design of experiments for bio-assay. *J. Amer. Statist. Assoc.*, *67*, 584–590.

Tsutakawa, R. K. (1980). Selection of dose levels for estimating a percent point of a logistic quantal response curve. *Appl. Statist.*, *29*, 25–33.

Tuckwell, H. C. and Koziol, J. A. (1987). Logistic population growth under random dispersal. *Bull. Math. Biol.*, *49*, 495–506.

Tukey, J. W. (1960). The practical relationship between the common transformations of percentages or fractions and of amounts. *Technical Report, No. 36*, Statistical Research Group, Princeton.

Tukey, J. W. (1962). The future of data analysis. *Ann. Math. Statist.*, *33*, 1–67.

Tukey, J. W. and McLaughlin, D. H. (1963). Less vulnerable confidence and significance procedures for location based on a single sample: trimming/Winsorization 1. *Sankhyā, Ser. A*, *25*, 331–352.

Turner, M. E. (1969). A generalization of the logistic law of growth. *Biometrics*, *25*, 577–579.

Umbach, D., Ali, M. Masoom, and Hassanein, K. M. (1981a). Small sample estimation of exponential quantiles with two order statistics. *Aligarh J. Statist.*, *1*(2), 113–120.

Umbach, D., Ali, M. Masoom, and Hassanein, K. M. (1981b). Estimating Pareto quantiles using two order statistics. *Commun. Statist.—Theor. Meth.*, *10(19)*, 1933–1941.

van der Laan, P. (1989). Selection from logistic populations. *Statist. Neerlandica*, *43*, 169–174.

van Montfort, M. A. J. and Otten, A. (1976). Quantal response analysis: enlargement of the logistic model with a kurtosis parameter. *Biomet. Zeit.*, *18*, 371–380.

Vaughan, R. J. and Venables, W. N. (1972). Permanent expressions for order statistics densities. *J. Roy. Statist. Soc., Ser. B*, *34*, 308–310.

Verhulst, P. J. (1838). Notice sur la lois que la population suit dans sons accroissement. *Corr. Math. et Physique*, *10*, 113–121.

Verhulst, P. J. (1845). Recherches mathématiques sur la loi d'accroissement de la population. *Académie de Bruxelles*, *18*, 1–38.

Vieira, S. and Hoffmann, R. (1977). Comparison of the logistic and the Gompertz growth function considering additive and multiplicative error terms. *Appl. Statist.*, *26*, 143–148.

Von Mises, R. (1936). La distribution de la plus grande de *n* values, Reprinted in *Selected Papers II*, *Amer. Math. Soc.*, (1954), 271–294.

von Neumann, J., Kent, R. H., Bellinson, H. R., and Hart, B. I. (1941). The mean square successive difference. *Ann. Math. Statist.*, *12*, 153–162.

Voorn, W. J. (1987). Characterization of the logistic and log-logistic distributions by extreme value related stability with random sample size. *J. Appl. Prob.*, *24*, 838–851.

Wagner, J. (1971). *Biopharmaceutics & Relevant Pharmacokinetics*. Drug Intelligence Publications, Illinois.

Walker, S. H. and Duncan, D. B. (1967). Estimation of the probability of an event as a function of several independent variables. *Biometrika*, *54*, 167–179.

Watson, G. S. (1961). Goodness-of-fit tests on a circle. *Biometrika*, *48*, 109–114.

Webb, G. F. (1986). Logistic models of structured population growth. *Comp. and Math. with Applications*, *12A*, 527–539.

Wedderburn, R. W. M. (1976). On the existence and uniqueness of the maximum likelihood estimates for certain generalized linear models. *Biometrika*, *63*, 27–32.

Welch, B. L. (1939). Note on discriminant functions. *Biometrika*, *31*, 218–220.

Westlake, W. J. (1976). Symmetrical confidence intervals for bioequivalence trials. *Biometrics*, *32*, 741–744.

Wheeler, R. E. (1980). Quantile estimators of Johnson curve parameters. *Biometrika*, *67*, 725–728.

Wijesinha, A., Begg, C. B., Funkenstein, H. H., and McNeil, B. J. (1983). Methodology for the differential diagnosis of a complex data set: A case study using data from routine CT-scan examinations. *Medical Decision Making*, *3*, 133–154.

Williams, D. A. (1976). Improved likelihood ratio tests for complete contingency tables. *Biometrika*, *63*, 33–37.

Williams, D. A. (1982). Extra-binomial variation in logistic linear models. *Appl. Statist.*, *31*, 144–148.

Wilson, E. B. and Worcester, J. (1943). The determination of L. D. 50 and its sampling error in bio-assay. *Proc. of National Acad. Sci.*, *29*, 79–85.

Woolf, B. (1955). On estimating the relation between blood group and disease. *Ann. of Human Genetics*, *19*, 251–253.

Yule, G. U. (1925). The growth of population and the factor which controls it. *J. Roy. Statist. Soc., Ser. A*, *88*, 1–58.

Zeger, S. L., Liang, K. Y., and Albert, P. S. (1988). Models for longitudinal data: A generalized estimating equation approach. *Biometrics*, *44*, 1049–1060.

Zelterman, D. (1987a). Parameter estimation in the generalized logistic distribution. *Comput. Statist. Data Analy.*, *5*, 177–184.

Zelterman, D. (1987b). Estimation of percentage points by simulation. *J. Statist. Comput. Simul.*, *27*, 107–125.

Zelterman, D. (1989). Order statistics of the generalized logistic distribution. *Comput. Statist. Data Analy.*, 7, 69–77.

Author Index

Abbott, R. D., 539, 559
Abraham, B., 12, 16, 237, 238, 261, 575
Abramowitz, M., 6, 15, 22, 46, 520, 521, 553
Agresti, A., 502, 511, 553
Ahuja, J. C., 212, 220, 553
Aiache, 523, 524, 529, 553
Aitchison, J., 475, 487, 553
Alam, K., 154, 163, 553
Albert, A., 461, 462, 487, 553
Albert, P. S., 533, 540, 586
Ali, M. M., 248, 260, 554
Ali, M. M., 201, 207, 554
Ali, M. Masoom, 98, 112, 113, 553, 554, 580, 584
Ambagaspitiya, R. S. 263, 288, 556
Anderson, J. A., 461, 462, 481, 487, 506, 511, 553, 554
Anderson, S., 523, 529, 570
Anderson, T. W., 334, 359, 554
Andrews, D. F., 264, 273, 288, 554
Anscombe, F. J., 455, 456, 487, 554
Antle, C., 123, 129, 142, 143, 395, 487, 555, 583
Armstrong, B., 533, 538, 555
Arnold, B. C., 27, 29, 37, 46, 114, 116, 119, 170, 173, 175, 184, 237, 247, 248, 257–260, 263, 267, 288, 555
Ashford, J. R., 480, 487, 555
Ayala, F. J., 548, 566
Azlarov, T. A., 170, 184, 555

Bailey, K., 539, 559
Bain, L. J., 123, 129, 138, 142, 143, 395, 483, 487, 555, 564, 583
Balakrishnan, N., 1, 17–19, 22, 25, 27, 29, 33, 37, 44, 46, 47, 49, 59, 67, 78–80, 82–85, 87,

[Balakrishnan, N.]
96, 98, 112–114, 116, 119, 120, 123, 129, 143, 173, 178, 184, 201, 207, 209, 212, 216–220, 227, 233, 234, 263, 264, 267–269, 273, 275, 279, 280, 282, 283, 288, 290, 373, 374, 383, 387, 395, 396, 555–557, 584
Baráth, 431, 433, 445, 557
Bargmann, R. E., 200, 206, 207, 559
Baringhaus, L., 177, 184, 231, 234, 557
Barlow, R. E., 159, 161, 163, 557
Barndorff-Nielsen, O., 179, 184, 557
Barnett, V., 300, 359, 557
Barr, D. R., 341, 359, 557
Bartlett, M. S., 480, 487, 557
Bechhofer, R. E., 145, 147, 152–154, 157, 158, 163, 165, 166, 558, 583
Begg, C. B., 481, 482, 487, 494, 558, 586
Bellinson, H. R., 529, 585
Bellows, T. S., 543, 548, 558
Bennett, S., 483, 487, 558
Berkson, J., 1, 15, 16, 449, 453–455, 457, 479, 480, 487, 488, 492, 502, 511, 558, 580
Bernardo, J. M., 22, 47, 558
Beyssac, 529, 553
Bhandari, S. K., 155, 163, 558
Bickel, P. J., 288, 554
Billingsley, P., 515, 520, 559
Bindman, A. G., 548, 566
Birnbaum, A., 18, 44, 47, 87, 96, 475, 488, 559
Bjerregaard, B., 486, 490, 570
Bliss, C. I., 394, 396, 559
Block, H. W., 249, 261, 559
Blom, G., 46, 47, 95, 96, 488, 559
Blytha, S. P., 548, 568
Bonney, G. E., 485, 488, 559
Botkin, D. B., 398, 424, 575
Bouver, H., 200, 206, 207, 559
Bowman, K. O. 197, 205–207, 559
Box, G. E. P., 441, 445, 559
Brendel, V., 545, 548, 559
Breslow, N., 483, 488, 559
Brown, C. C., 480, 491, 575
Buhler, W. J., 543, 548, 559
Bukač, J., 202, 207, 559
Büringer, H., 147, 163, 559
Burr, I. W., 191, 207, 208, 559, 578
Busch, K., 446, 580
Byar, D. P., 339, 360, 572

Carroll, R. J., 533, 537, 539, 540, 559, 560, 582
Case, T. J., 548, 566
Chan, L. K., 85, 87, 90, 94, 96, 97, 560
Chan, N. N., 85, 87, 96, 560
Chan, P. S., 288, 556
Chaudhuri, A. R., 155, 163, 558
Chen, E. H., 488, 560
Cheney, E. W., 539, 569
Cheng, S. W., 84, 85, 87, 90, 94, 97, 560
Chernoff, H., 102, 113, 327, 359, 560
Chew, V., 3, 15, 141, 143, 560
Clark, C. E., 44, 45, 47, 560
Clark, V. A., 18, 48, 583
Clarke, C. E., 541, 549, 569
Clayton, A., 523, 529, 560
Clayton, D. G , 484, 488, 560
Cody, W. J., 536, 539, 561
Cohen, A., 327, 359, 561
Cohen, A. C., 49, 59, 78, 80, 84, 85, 96, 98, 112, 116, 119, 120, 233, 234, 374, 396, 556

Cohen, D. S., 154, 163, 542, 548, 561
Cohen, M. L., 273, 275, 289, 565
Conrad, J. M., 543, 548, 561
Cook, R. D., 258, 261, 445, 466, 471, 488, 496, 511, 561
Cooke, K. L., 543, 548, 561
Copas, J. B., 485, 488, 561
Cornfield, J., 454, 488, 561
Count, E. W., 428, 445, 561
Cox, D. R., 456, 461, 463, 464, 466, 467, 481, 488, 489, 502, 505, 511, 535, 539, 561, 562
Cran, G. W., 20, 47, 562
Crouch, A., 530, 537, 539, 562
Crow, E. L., 273, 289, 562
Csörgo, S., 339, 359, 562
Cutler, C. D., 512–516, 518–520,562

D'Agostino, R. B., 291, 300, 309, 315, 317, 320, 324, 326–328, 337–339, 345, 356, 358–360, 489, 562, 575
Dahiya, R. C., 328, 360, 562
Darling, D. A., 334, 359, 554
Dave, P. H., 191, 208, 581
David, F. N., 44, 45, 47, 562
David, H. A., 27, 29, 30, 37, 47, 80, 84, 202, 207, 263, 264, 269, 271, 273, 278, 289, 562
Davidson, R. R., 212, 213, 220, 562
Davidson, T., 341, 359, 557
Davis, D. J., 10, 15, 58, 78, 83, 84, 124, 141, 143, 298, 306, 360, 395, 396, 563
Davis, H. T., 6, 15, 22, 47, 459, 460, 476, 489, 563
Davis, L. J., 458, 459, 563
Dawson, D. A., 513–516, 518–520, 562
Day, N. E., 461, 481, 489, 563
de Haan, L., 226, 234, 563
de Oliveira, J. T., 244, 261, 577
Devidas, M., 223
Devroye, L., 255, 256, 261, 563
DiCroce, P., 449
Dixon, W. J., 79, 84, 120, 122, 274, 289, 488, 560, 563
Downton, F., 79, 84, 113, 114, 116, 119, 120, 563
Dubey, S. D., 181, 184, 191, 207, 211, 212, 218, 220, 563
Duchrau, P., 445, 563
Dudewicz, E. J., 147, 163, 563
Dudman, J., 18, 44, 47, 87, 96, 475, 488, 559
Duffy, D. E., 462, 492, 581
Dufour, R., 339, 341, 360, 563
Dumonceaux, R. H., 129, 139, 143, 564
Duncan, D. B., 504, 512, 585
Durbin, J., 475, 489, 564
Dyke, G. V., 2, 15, 484, 489, 564

Eastman, J., 123, 142, 143, 395, 483, 487, 555, 564
Eisen, M., 397, 398, 423, 564
Elderkin, R., 548, 561
El-Saidi, M., 15, 213, 220, 565
Emmens, C. W., 1, 15, 479, 489, 564
Engelhardt, M., 123, 142, 143, 395, 483, 487, 489, 555, 564
Epstein, B., 98, 113, 573

Falconer, K. J., 515, 520, 564
Farewell, V. T., 214, 220, 564
Feldman, M. W., 547, 548, 564
Feller, W., 428, 445, 564
Fertig, K. W., 373, 396, 575
Finney, D. J., 1, 15, 564

Fisher, R. A., 326, 360, 454, 489, 564
Fisk, P. R., 1, 15, 304, 307, 332, 360, 565
Fleishman, A. I., 440, 445, 565
Formann, A. K., 475, 489, 565
Forsythe, G. E., 466, 489, 565
Fraser, W., 536, 539, 561
Frenzen, C. L., 543, 548, 565
Frischmuth, K., 445, 563
Fuller, W. A., 537, 539, 565
Fung, K., 129, 143, 373
Funkenstein, H. H., 494, 586

Galambos, J., 167–169, 172–174, 177, 180–182, 184, 185, 226, 227, 235, 565
Gart, J. J., 458, 489, 565
Gastwirth, J. L., 273, 275, 289, 565
Genest, C., 251, 256, 261, 565
Gentle, J. E., 500, 511, 572
George, E. O., 8, 15, 16, 18, 47, 182, 183, 185, 212–214, 220, 221, 223, 226, 227, 229, 235, 245, 261, 565, 566, 576
Getz, W. M., 542, 548, 566
Gibbons, J. D., 147, 163, 566
Gilpin, M. E., 542, 548, 566
Ginzburg, L. R., 547, 548, 566
Glasbey, C. A., 476, 489, 566
Glaz, J., 406, 423, 566
Gnanadesikan, M., 85, 90, 95–97, 490, 567
Gnedenko, B. V., 181, 185, 566
Gnedenko, D. V., 181, 185, 566
Goel, N. S., 547, 548, 566
Goel, P. K., 162, 163, 566
Govindarajulu, Z., 29, 42, 47, 175, 185, 234, 235, 566
Gradshteyn, I. S., 533, 539, 566
Gray, R., 481, 482, 487, 558
Green, J., 201, 208, 579
Green, S. B., 339, 360, 573
Greenberg, B. G., 57, 78, 82, 84, 99, 113, 581
Greenland, S., 484, 489, 566
Grizzle, J. E., 2, 15, 480, 484, 489, 490, 566, 567
Grubbs, R. E., 138, 143, 567
Guiard, V., 446, 580
Gumbel, E. J., 12, 15, 23, 47, 213, 220, 237–239, 244, 251, 261, 567
Gupta, S. S., 17–22, 25, 30, 31, 44, 47, 57, 58, 67, 78, 82, 84, 85, 87, 90, 95–97, 119, 120, 145, 147, 150, 151, 154–156, 158–165, 269, 273, 282, 289, 374, 387, 394–396, 480, 490, 557, 567, 568
Gurland, J., 328, 360, 562
Gurney, W. S. C., 542, 548, 568
Gyllenberg, M., 544, 548, 568

Habbema, J. D. F., 486, 490, 570
Hahn, G. J., 203, 207, 568
Hailey, D. C., 57, 78, 568
Hall, I. J., 128, 143, 373–375, 381, 387, 396, 490, 569
Hallam, T. G., 540, 541, 549, 569
Hamilton, M. A., 456, 457, 490, 569
Hampel, F. R., 269, 288, 289, 554, 569
Han, S., 149, 152–156, 162, 164, 165, 567, 569
Hankey, B. F., 483, 484, 491, 575, 576
Hanson, F. B., 543, 547, 549, 551, 569, 583
Haq, M. S., 248, 260, 554
Harkness, W., 123, 142, 487, 555
Hart, B. I., 529, 585
Hart, J. F., 535 536, 539, 561, 569

Harter, H. L., 49, 50, 53, 54, 56, 57, 78, 490, 569
Hartley, H. O., 20, 48, 375, 396, 574, 578
Hassanein, K. M., 85, 97, 98, 112, 113, 554, 569, 584
Hauck, W. W., 523, 529, 570
Herd, G. R., 317, 360, 570
Hilden, J., 486, 490, 570
Hinkley, D. V., 505, 511, 561
Hitchkock, S. E., 455, 457, 490, 570
Ho, K. L., 288, 556
Hodges, J. L., Jr., 454, 490, 570
Hoeffding, W., 173, 185, 570
Hofmann, R., 476, 493, 585
Holgate, P., 545, 549, 570
Holling, C. S., 549, 574
Horvath, L., 339, 359, 562
Hosmer, D. W., 502, 511, 570
Hotelling, H., 459, 490, 570
Howlader, H. A., 140, 142, 143, 570
Hsu, J. P., 545, 549, 570
Huang, D.-Y., 147, 164, 567
Huang, J. S., 172, 173, 175, 185, 566, 570
Huber, P. J., 282, 288, 289, 554, 570
Hutchinson, T. P., 258, 261, 581
Hwang, J. S., 175, 185, 570
Hwang, S., 217, 218, 220, 570

Imrey, P. B., 484, 490, 571

Jennings, D. E., 466, 472, 490, 571
Jennrich, R. J., 432, 445, 571
Jensen, A. L., 398, 400, 424, 546, 549, 571
Johnson, K., 548, 566
Johnson, M. E., 258, 261, 496, 511, 561
Johnson, N. L., 8, 11, 15, 16, 44, 45, 47, 171, 183, 185, 189, 190, 196–198, 205, 207, 208, 211, 221, 252, 261, 535, 539, 562, 571, 578, 582, 583
Johnson, W., 483, 490, 571
Joiner, B. L., 162, 165, 571
Jones, D. D., 549, 574
Joshi, P. C., 19, 29, 33, 42, 44, 46, 47, 263, 289, 556, 571
Jung, J., 95–97, 490, 571

Kadane, J., 140, 143, 173, 185, 571, 584
Kakosyan, A. V., 176, 177, 185, 571
Kalbfleisch, J. D., 211, 214, 220, 572
Kale, B. K., 263, 289, 572
Kaplan, E. L., 317, 339, 360, 572
Kastenbaum, M. A., 480, 490, 572
Kay, R., 485, 486, 490, 572
Kendal, W. S., 543, 549, 572
Kendall, M. G., 65, 78, 499, 511, 572
Kennedy, W. J., 289, 500, 511, 562, 572
Kent, R. H., 529, 585
Kerridge, D. F., 461, 481, 489, 563
Khan, A. H., 201, 207, 554
Kiefer, J., 147, 163, 558
Kimura, M., 417, 424, 572
Kish, J. F., 16, 492, 578
Kjelsberg, M. O., 18, 19, 48, 572
Klebanov, L. B., 185, 571
Klimko, L., 123, 142, 487, 555
Kneale, G. W., 483, 491, 572
Knight, R. D., 289, 562
Koch, G. G., 484, 490, 571
Kocherlakota, S., 19, 33, 46, 227, 234, 556, 557

Koehler, K. J., 258, 261, 495, 496, 512, 582
Kolker, Y., 543, 549, 572
Kolmogorov, A. N., 334, 360, 572
Koo, J. O., 147, 163, 563
Kotz, S., 8, 16, 168–172, 181, 183, 185, 190, 198, 207, 211, 221, 227, 235, 252, 261, 535, 539, 565, 571
Koutrovelis, I. A., 98, 113, 572
Koziol, J. A., 339, 360, 547, 551, 572, 573, 584
Krishna Kumar, B., 540, 546, 550, 577
Kryscio, R. J., 545, 549, 573
Ku, H. H., 480, 491, 573
Kubat, P., 98, 113, 573
Kuiper, N. H., 334, 360, 573
Kulkarni, R., 158, 163, 558
Kullback, S., 480, 491, 573
Kulldorff, G., 85, 97, 573
Kurtz, T. G., 547, 549, 573

Lachenbruch, P. A., 484, 491, 573
Laird, N. M., 533, 540, 582
Lamphiear, D. E., 480, 490, 572
Lan, K. K. G., 539, 559
Landwehr, J. M., 466, 471, 472, 491, 573
Larntz, K., 327, 360, 573
Lawson, C. L., 539, 569
Leach, D., 427, 445, 476, 477, 491, 573
Lee, A. F. S., 489, 562
Lefevre, J., 545, 549, 573
Lehmann, E. L., 327, 359, 464, 491, 560, 573
Lemeshow, S., 502, 511, 570
Leslie, A., 523, 529, 560
Leung, M. Y., 25, 47, 178, 184, 209, 212, 216, 218–220, 557
Lewis, T., 546, 549, 573
Liang, K. Y., 533, 540, 586
Liang, T., 162, 164, 165, 567, 573
Lin, G. D., 175, 185, 573
Lindley, D. V., 140, 143, 244, 261, 574
Lipow, M., 136, 143, 574
Lippert, S., 438, 445, 574
Little, S., 485, 486, 490, 572
Lloyd, D. K., 136, 143, 574
Lloyd, E. H., 80, 84, 85, 97, 574
Lo, K. K., 288, 556
Lockhart, R. A., 350, 351, 360, 574
Loeve, M., 270, 289, 574
Lomax, K. S., 191, 208, 574
Lord, F. M., 475, 491, 574
Lorenzen, T. J., 148, 149, 150, 153, 160, 162, 165, 574
Lotka, A. J., 428, 446, 574
Lucas, H. L., 441, 445, 559
Ludwig, D., 543, 549, 574
Lurie, D., 375, 396, 574

Maag, U. R., 339, 341, 360, 563
MacKay, J., 251, 256, 261, 565
Maddala, G. S., 502, 511, 574
Madreimov, I. 175, 186, 574
Maehley, H. J., 539, 569
Mage, D. T., 202, 208, 574
Maitra, S. C., 548, 566
Malcolm, M. A., 466, 489, 565
Malik, H. J., 12, 16, 18, 19, 22, 25, 27, 29, 30, 32, 44, 47, 48, 87, 96, 119, 120, 201, 207, 237, 238, 261, 269, 288, 557, 574, 575
Mann, H. B., 327, 360, 575
Mann, N. R., 373, 396, 575
Mantel, N., 454, 480, 483, 484, 488, 491, 561, 575, 576
Martin, H., 147, 163, 559
Martin, K. J., 20, 47, 562

Massaro, J., 291, 328, 360, 575
May, R. M., 542, 543, 549, 575
Mayilswami, P., 546, 550, 577
McCullagh, P., 428, 446, 484, 491, 533, 539, 575
McDonald, G. C., 148–150, 153, 160, 162, 164, 165, 567, 574
McLaughlin, D. H., 274, 290, 584
McNeil, B. J., 494, 586
McQuarrie, D. A., 545, 549, 575
Mead, E. R., 85, 87, 96, 560
Meade, N., 428, 446, 575
Meeden, G., 175, 184, 555
Meier, P., 317, 339, 360, 572
Melamed, J. A., 185, 571
Mesztenyi, C. K., 539, 569
Mian, I. U. H., 200, 208, 581
Mikhail, N. N., 248, 260, 554
Miller, R. S., 398, 424, 575
Milton, R. C., 158, 165, 575
Mogyoródi, J., 179, 186, 576
Moler, C. B., 466, 489, 565
Monahan, J. F., 529
Montroll, E. W., 548, 566
Mood, A. M., 139, 143, 576
Moolgavkar, S. H., 398, 424, 543, 549, 576
Moore, A. H., 49, 50, 53, 54, 56, 57, 78, 490, 569
Morgan, B. S. T., 480, 491, 545, 550, 576
Mosteller, F., 87, 97, 98, 113, 576
Muldholkar, G. S., 8, 16, 182, 183, 185, 212, 221, 227, 229, 235, 245, 261, 565, 576
Murray, J. D., 540, 543, 548, 550, 565, 576
Myers, M. H., 483, 491, 576

Nachtsheim, C. J., 445, 561
Nagel, K., 158, 160, 164, 567
Nash, S. W., 212, 220, 553
Nelder, J. A., 428, 446, 461, 575
Nelson, W., 317, 360, 576
Neyman, J., 454, 491, 576
Nisbet, R. M., 548, 568
Norden, R. H., 399, 400, 424, 545, 550, 576
Norton, H. W., 480, 491, 576
Nürnberg, G., 440, 446, 576, 580
Nussey, H. E., 543, 548, 550, 561, 576

Oakes, D., 237, 250, 261, 576
Ochi, Y., 533, 539, 577
Ogawa, J., 84, 87, 97, 492, 577
Ojo, M. O., 8, 15, 213, 214, 220, 566
Oliver, F. R., 1, 16, 460, 476–478, 492, 577
Olkin, I., 147, 163, 566
O'Reilly, F. J., 350, 351, 360, 574
Otten, A., 493, 585
Owen, D. B., 199, 208, 373, 392, 393, 396, 577

Pagano, M., 464, 492, 577
Paiva-Franco, M. A., 483, 492, 577
Panchapakesan, S., 144, 147, 151, 158–165, 557, 567, 568, 573
Parthasarathy, P. R., 540, 545, 546, 550, 577
Patterson, H. D., 2, 15, 484, 489, 564
Pearl, R., 1, 16, 398, 424, 427, 446, 450, 458, 475, 476, 492, 577, 578
Pearson, E. S., 20, 48, 208, 578
Pearson, K., 20, 48, 578
Pemberton, J. D., 506, 511, 554
Perelson, A. S., 545, 548, 559
Perks, W. F., 1, 16, 211, 221, 578
Pettigrew, H. M., 489, 565

Pettitt, A. N., 339, 341, 361, 578
Petunin, Y. I., 175, 186, 574
Piantadosi, S., 400, 410, 424, 583
Pickands III, J., 519, 520, 578
Pielou, E. C., 397, 400, 424, 546, 550, 578
Pierce, D. A., 533, 539, 578
Pierce, G. L., 508, 511, 579
Pierre, 529, 553
Plackett, R. L., 1, 16, 18, 19, 23, 44, 48, 123, 143, 483, 492, 578
Pohl, S., 446, 578
Powers, W., 483, 488, 559
Prajneshu, D., 547, 550, 578
Prasad, 529, 553
Pregibon, D., 466–468, 471, 486, 491, 492, 573, 578, 579
Prendville, B. J., 544, 550, 579
Prentice, R. L., 209, 211, 213, 214, 220, 221, 483, 492, 533, 539, 564, 572, 577, 579
Puthenpura, S., 201, 207, 557
Pyke, R., 350, 361, 579

Oureishi, A. S., 18, 47, 57, 78, 82, 84, 85, 87, 97, 119, 120, 289, 374, 387, 394–396, 480, 490, 568

Radelet, M., 508, 511, 579
Ragab, A., 201, 208, 579
Raghunandanan, K., 79, 80, 84, 120–122, 274, 282, 288, 289, 579
Ramberg, J. S., 162, 165, 579
Rao, C. R., 404, 424, 500, 511, 579
Rasch, D., 427, 429, 431–433, 435, 441, 443–446, 557, 563, 579, 580
Read, T. R. C., 327, 361, 580
Reed, L. J., 1, 16, 398, 424, 427, 446, 449, 450, 458, 475, 476, 492, 578, 580
Renshaw, E. C., 546, 550, 580
Renyi, A., 294, 361, 580
Rice, J. R., 539, 569
Richardson, S. C., 481, 487, 554
Robertson, C. A., 259, 260, 555
Rogers, W. H., 288, 554
Rosen, G., 541, 542, 550, 580
Rosenblat, S., 542, 548, 561
Rosenblatt, J. R., 162, 165, 571
Roughgarden, J., 547, 548, 564
Rousseau, C. C., 18, 47, 226, 229, 235, 566
Rudolph, P. E., 446, 580
Ryan, D., 547, 549, 569
Ryzhik, I. W., 533, 539, 566

Sackrowitz, H. B., 327, 359, 561
Saleh, A. K. Md. E., 98, 112, 113, 185, 554, 566, 580
Sampson, C. B., 373, 396, 569
Sanathanan, L., 475, 492, 581
Sands, B. R., 533, 539, 578
Santner, T. J., 147, 153, 165, 462, 492, 581
Sarhan, A. E., 57, 78, 82, 84, 99, 113, 581
Sarndal, C. E., 102, 113, 581
Sasaki, H., 428, 447, 581
Satterthwaite, S. P., 258, 261, 581
Schafer, D., 533, 539, 581
Schafer, R. E., 123, 143, 493, 581
Schimke, E., 435, 445, 446, 563, 580
Schmeiser, B. W., 162, 165, 579
Schneider, B. E., 22, 48, 581
Schorr, B., 361, 581
Schriever, K.-H., 147, 163, 559

Schultz, H., 1, 16, 459, 460, 476, 493, 581
Scott, A. J., 484, 493, 581
Sebaugh, J. L., 85, 97, 569
Shah, B. K., 18–22, 25, 30, 31, 44, 47, 48, 57, 78, 82, 84, 85, 87, 97, 119, 120, 191, 208, 269, 289, 374, 387, 394–396, 480, 490, 522–524, 528, 529, 568, 581
Shapiro, S. S., 202, 203, 206–208, 568, 582
Sheffield, T. S., 123, 143, 493, 581
Shen, S. M., 475, 487, 553
Shenton, L. R., 197, 205–207, 559

Shoemaker, A. C., 466, 471, 491, 573
Shohoji, T., 428, 447, 581
Shoukri, M. M., 200, 208, 581
Shu, V. S., 263, 269, 271, 273, 278, 289, 562, 582
Siddiqui, M. M., 273, 289, 562
Silverstone, H., 456, 493, 582
Singh, K., 15, 213, 214, 220, 544, 551, 565, 566, 583
Singleton, R. C., 465, 493, 582
Singpurwalla, N. D., 244, 261, 574

Sinha, S. K., 263, 289, 572
Skelly, 529, 553
Slifker, J. F., 202, 206, 208, 582
Slobokin, L. B., 548, 566
Smith, C. E., 545, 550, 582
Smith, R. M., 123, 143, 564
Snell, E. J., 467, 489, 502, 511, 562
Sobel, M., 147, 163, 558, 566
Sohn, J. K., 162, 165, 568, 582
Sowden, R. D., 480, 487, 555
Spiegelman, C., 539, 559
Spiegelman, D., 530, 533, 537, 539, 562, 582
Srinivasan, R., 79, 80, 84, 120–122, 274, 282, 288, 289, 579
Stefanski, L. A., 529, 531, 533, 537, 539, 540, 560, 582
Stegun, I. A., 6, 15, 22, 46, 520, 521, 553
Stephens, M. A., 300, 309, 315, 317, 320, 324, 326, 327, 337–339, 341, 345, 347, 350, 351, 356, 358–361, 473–475, 493, 562, 574, 578, 582
Stiratelli, R., 533, 540, 582
Stokes, M. E., 484, 490, 571
Strauss, D. J., 244, 261, 582
Stuart, A., 65, 78, 499, 511, 572
Stukel, T., 209, 221, 582
Symanowski, J. T., 258, 261, 495, 496, 512, 582

Tadikamalla, P. R., 189, 190, 197, 205, 208, 582, 583
Tamhane, A. C., 147, 154, 156–158, 165, 166, 583
Tan, W. Y., 290, 397, 398, 400, 404, 405, 410, 424, 425, 544, 545, 551, 583, 584
Tarter, M. E., 18, 19, 48, 583
Teuscher, F., 446, 580
Thatcher, H. G., 539, 569
Thoman, D. R., 129, 143, 583
Thomas, D. G., 489, 565
Thomas, G. E., 20, 47, 562
Tier, C., 543, 547, 549, 551, 569, 583
Tierney, L., 140, 143, 584
Tiku, M. L., 67, 78, 264, 273, 274, 283, 290, 351, 352, 361, 446, 580, 584
Tintner, G., 459, 460, 476, 493, 584
Tognetti, K., 546, 551, 584

Tracy, D. S., 200, 208, 581
Tritchler, D. L., 464–466, 492, 493, 577, 584
Tsiatis, A. A., 466, 469, 470, 493, 584
Tsokos, C. P., 449
Tsutakawa, R. K., 480, 481, 493, 584
Tuckwell, H. C., 545, 547, 550, 551, 582, 584
Tukey, J. W., 162, 166, 274, 288–290, 554, 563, 584
Turner, M. E., 542, 551, 584

Umbach, D., 98, 112, 113, 553, 554, 580, 584

van der Laan, P., 152, 166, 585
Van Montfort, M. A. J., 493, 585
Vaughan, R. J., 264, 267, 290, 585
Venables, W. N., 264, 267, 290, 585
Verhulst, P. J., 1, 14, 16, 397, 424, 427, 447, 449, 475, 476, 493, 585
Vieira, S., 476, 493, 585
Volodin, N. A., 170, 184, 555
von Mises, R., 226, 235, 585
von Neumann, J., 524, 529, 585
Voorn, W. J., 177, 186, 231, 235, 585

Wagner, J., 523, 524, 529, 585
Wald, A., 327, 360, 575
Walker, S. H., 504, 512, 585
Wang, H. H., 545, 549, 570
Ware, J. H., 533, 540, 582
Watson, G. S., 335, 361, 585
Webb, G. F., 542, 544, 548, 551, 568, 585
Wedderburn, R. W. M., 461, 493, 585
Weisberg, S., 466, 471, 488, 561
Weiss, G., 140, 142, 143, 570
Welch, B. L., 461, 494, 585
Westlake, W. J , 523, 529, 585
Wheeler, R. E., 206, 208, 585
Wijesinha, A., 481, 494, 586
Wild, C. J., 484, 493, 581
Williams, D. A. 472, 494, 533, 540, 586
Williams, G. T , 44, 45, 47, 560
Wilson, E. B., 1, 16, 454, 479, 494, 586
Winley, G., 546, 551, 584
Witten, M., 548, 561
Witzgall, C., 539, 569
Woolf, B., 480, 494, 586
Worcester, J., 1, 16, 454, 479, 494, 586

Yule, G. U., 427, 447, 586

Zeger, S. L., 533, 540, 586
Zelterman, D., 209, 212, 216–221, 570, 586

Subject Index

Acceptance sampling plans, 392, 393
Approximate maximum likelihood estimates, 60–77, 273–288
  asymptotic variances and covariance, 65–76
  bias, variances and covariance, 63–66
  comparison with other estimates, 68, 69
  derivation of estimates, 60–65
  examples, 74–77
  robustness, 273–288

Bioavailability data, 522–529
  estimation of parameters, 524–527
    biases and variances, 526, 527
    moments, 524–526
[Bioavailability data]
  introduction, 522, 523
  simulated critical points, 527–529
Bivariate logistic distribution, 496–502
  applications to two-way contingency tables, 498–502
Bivariate logistic regression model, 502–511

Characterizations, 167–184
  introduction, 167–170
  miscellaneous, 181–184
  model for growth, 170–172
  properties of order statistics, 172–175
  random sample size, 175–181

Double Weibull distribution, 227

Extended logistic (Type I generalized) distribution, 232

Generalized hypergeometric function, 230
Goodness-of-fit tests, 291–359, 466–475
  empirical distribution function, 293–299
    graphical, 297–299
  formal, 324–359
    chi-squared tests, 325–333
    empirical distribution function statistics, 334–349
    tests based on normalized spacings, 349–353
    tests based on regression and correlation, 353–359
  introduction, 291, 292
  probability plots, 299–324
    censored samples, 313–317
    deviations from linearity, 308–313
    estimation of parameters, 300–305
    examples, 306, 307, 317–324
    informal test, 300–305
    summary, 324
  recommendation, 359

Half logistic distribution, 232–234
Hausdorff dimension, 512–520

Johnson's system, 189–191

Linear estimation, 80–122
  best unbiased estimation, 80–112
[Linear estimation]
    censored samples, 80–82
    examples, 82, 83
    quantiles, 98–112
    selected order statistics, 84–96
  polynomial coefficients, 113–119
  simplified, 120–122
Log-gamma distribution, 516–521
Logistic function, 449–486
  applications, 475–486
    bioassay, 479–481
    medical diagnosis, 481, 482
    population growth, 476–479
    public health, 483–486
  estimation of parameters, 452–458
  goodness-of-fit techniques, 466–469
  introduction, 449–452
Logistic growth models, 427–445, 458–460, 540–547
  applications to tumor growth, 543, 544
  experimental design problems, 440–445
    D-optimal designs, 443, 444
    example, 444, 445
    nonlinear regression functions, 440–443
  generalizations, 541–543
  hypothesis testing and interval estimation, 435–440
    confidence estimation, 436, 437
    simulational results, 437–440
    test procedures, 435, 436
  introduction, 427–429, 540
  logistic equation, 541
  parameter estimation, 429–435, 458–460

[Logistic growth models]
asymptotic covariance matrix, 433–435
examples, 430–433
stochastic analogues, 544–547
dependence on the mean population, 546
random fluctuations, 546, 547
Logistic-normal integral, 529–538
error comparison, 538
Gauss-Hermite quadrature, 537, 538
least maximum approximants, 530–538
Taylor series approximation, 538
Logistic regression model, 461–473
estimation and discrimination procedures, 461–466
goodness-of-fit techniques, 469–473

Maximum likelihood estimates, 50–59, 124–142
location and scale, 50–59, 124–128
asymptotic variances and covariance, 54–56
bias, variances and covariance, 52, 53, 124, 125
confidence intervals, 125–128
derivation of estimates, 50–52, 124, 125
examples, 57–59, 142
reliability, 136–140
example, 142
mean and variance, 140
tolerance limits, 128–136
example, 141
Midrange, 226, 229–231
characteristic function, 230, 231
Multivariate extreme distribution, 244
Multivariate generalizations, 237–260
classical, 238–242
conditional distributions, 257, 258
differences of extremes, 244–246
extremes, 253–255
Farlie-Gumbel-Morgenstern form, 252
flexible, 252, 253
frailty and archimedean, 250, 251
geometric minima and maxima, 246–250
introduction, 237, 238
miscellaneous notes, 258–260
mixture representation, 242, 243
parameter estimation, 256, 257
simulation, 255, 256

Nearest neighbors, 512–520

Order statistics, 17–30, 227–230
characteristic function, 227–230
cumulants, 22, 23
density function, 17
joint density function, 17
joint moment generating function, 23–25
modes, 21
moment generating function, 21
moments, 21
percentage points, 19,20
product moments, 25
recurrence relations, 25–30
Outliers, 263–288
distributions of order statistics, 264–267

[Outliers]
functional behavior of order statistics, 269–272
introduction, 263, 264
moments of order statistics, 267–269
robustness of
approximate maximum likelihood estimators, 273–288
best linear unbiased estimators, 273–288
estimators of location, 273–281
estimators of scale, 282–288
Gastwirth mean, 273–281
linearly weighted means, 273–281
median, 273–281
modified maximum likelihood estimators, 273–288
sample mean, 273–281
sample standard deviation, 282–288
simplified linear estimators, 273–288
trimmed estimators, 273–288
Winsorized estimators, 273–281

Plackett's approximation, 44–46
Power distribution, 227

Quasi-range, 31, 32
distribution, 32

Ranking and selection, 145–162
comments, 161, 162
formulations, 145–147
from a family portially ordered, 158–161
[Ranking and selection]
indifference zone approach, 151–156
elimination-type two-stage, 154–156
single-stage, 151–153
subset approach, 148–154
restricted single-stage, 153, 154
unrestricted single-stage, 148–151
with smallest q-quantile, 156–158
Related distributions, 223–234
double exponential, 224–230
exponential, 224–230
extreme value, 224–227
geometric, 231, 232
half logistic, 232–234
density function, 233
distribution function, 233
order statistics, 233
recurrence relations, 233, 234
introduction, 223, 224

Sample range, 30–32
density function, 31, 32
distribution function, 30, 31
Stochastic growth models, 397–423
conclusions, 423
diffusion approximation, 410–423
absorption distribution and moments, 418–423
stationary distribution, 415
introduction, 397, 398
logistic, 399–410
moments, 407–410
probability distribution of first absorption time, 404–407
transition probabilities and absolute probabilities, 400–404

Tolerance limits, 373–395
  illustrative examples, 393–395
  introduction, 373
  one-sided, 374–387
    large-sample approximation, 382–387
  two-sided, 387–391
Translated families, 189–207
  fitting, 196–199
    maximum likelihood, 198, 199
    moments, 197
    percentiles, 197, 198
  introduction, 189–191
  L_B system, 193–195
    fitting, 201–205
  L_L (log-logistic) system, 191–193
    example, 200, 201
    fitting, 199
  L_U system, 195, 196
    fitting, 205, 206
[L_U system]
    selection, 206, 207
Truncated logistic distribution, 33–43
  density function, 33
  distribution function, 33
  order statistics, 33–43
    recurrence relations, 34–43

Univariate generalizations, 209–214, 520–522
  introduction, 209
  Type I, 209–212, 520–522
    characterizations, 211, 212
    order statistics, 218–220
    parameter estimates, 214–218
  Type II, 210–212
  Type III, 210–213
  Type IV, 210–214